Bacterial Ge[illegible]s and Genomics

Understanding of bacterial genetics and genomics is fundamental to understanding bacteria and higher organisms, as well. Novel insights in the fields of genetics and genomics are challenging the once clear borders between the characteristics of bacteria and other life. Biological knowledge of the bacterial world is being viewed under a new light with input from genetics and genomics. The replication of bacterial circular and linear chromosomes, coupled (and uncoupled) transcription and translation, multiprotein systems that enhance survival, wide varieties of ways to control gene and protein expression, and a range of other features all influence the diversity of the microbial world. This text acknowledges that readers have varied knowledge of genetics and microbiology. Therefore, information is presented progressively, to enable all readers to understand the more advanced material in the book.

This second edition of *Bacterial Genetics and Genomics* updates the information from the first edition with advances made over the past five years. This includes descriptions for 10 types of secretion systems, bacteria that can be seen with the naked eye, and differences between coupled transcription–translation and the uncoupled runaway transcription in bacteria. Topic updates include advances in bacteriophage therapy, biotechnology, and understanding bacterial evolution.

Key Features

- Genetics, genomics, and bioinformatics integrated in one place
- Over 400 full-colour illustrations explain concepts and mechanisms throughout and are available to instructors for download
- A section dedicated to the application of genetics and genomics techniques, including a chapter devoted to laboratory techniques, which includes useful tips and recommendations for protocols, in addition to troubleshooting and alternative strategies
- Bulleted key points summarize each chapter
- Extensive self-study questions related to the chapter text and several discussion topics for study groups to explore further

This book is extended and enhanced through a range of digital resources that include:

- Interactive online quizzes for each chapter
- Flashcards that allow the reader to test their understanding of key terms from the book
- Useful links for online resources associated with Chapters 16 and 17

Bacterial Genetics and Genomics

Second Edition

Lori A.S. Snyder

Second edition published 2024
by CRC Press
2385 NW Executive Center Drive, Suite 320, Boca Raton FL 33431

and by CRC Press
4 Park Square, Milton Park, Abingdon, Oxon, OX14 4RN

CRC Press is an imprint of Taylor & Francis Group, LLC

Library of Congress Cataloging-in-Publication Data

Names: Snyder, Lori, author.
Title: Bacterial genetics and genomics / Lori Snyder.
Description: Second edition. | Boca Raton : CRC Press, 2024. | Includes
bibliographical references and index.
Identifiers: LCCN 2023040585 (print) | LCCN 2023040586 (ebook) | ISBN
9781032461786 (hardback) | ISBN 9781032461779 (paperback) | ISBN
9781003380436 (ebook)
Subjects: LCSH: Bacterial genetics. | Bacteria--Physiology.
Classification: LCC QH434 .S595 2024 (print) | LCC QH434 (ebook) | DDC
579.3/135--dc23/eng/20231222
LC record available at https://lccn.loc.gov/2023040585
LC ebook record available at https://lccn.loc.gov/2023040586

ISBN: 978-1-032-46178-6 (hbk)
ISBN: 978-1-032-46177-9 (pbk)
ISBN: 978-1-003-38043-6 (ebk)

DOI: 10.1201/9781003380436

Typeset in Utopia
by Evolution Design & Digital Ltd (Kent)

Contents in Brief

Contents

Chapter 6 Transcriptomes 105

Part III Proteins, Structures, and Proteomes 121

Chapter 7 Proteins 123

Chapter 17 Genome Analysis Techniques 291

Chapter 18 Laboratory Techniques 309

Author's Thoughts on the Second Edition

I am delighted to be revisiting *Bacterial Genetics and Genomics.* In approaching revisions for the Second Edition, I want to thank everyone who has given myself and my editors feedback on your experiences with the First Edition of the book. Good educators know that all feedback is valuable. Thank you. My primary motivation with the First Edition was that I would be able to effectively communicate to students and from what I have been hearing, I have achieved that.

Understanding the needs of students is why this book starts without any (or many) assumptions about prior knowledge. Students and educators have told me that they appreciate this feature of the book. Whether this is the first time a student is taught the information in Chapter 1 or the early chapters are used by students to refresh their memories, these fundamentals are an essential learning resource.

Feedback has also been that readers appreciate my writing style. I am told that readers find even the most complex topic are clearly described. Throughout I have made an effort to repeat complex concepts, explaining them in more than one way because from my experience this enhances the learning potential for the reader. Different people learn in different ways. This is what I kept in mind in writing the First Edition and working on the Second Edition, to make it for all of you. For this reason, topics are in logically formed Parts and Chapters, divided by subheadings with new glossary terms in bold, and with concepts backed-up by phenomenal figures, further enhancing what is described in the text in the figure legend.

As academics, lecturers, and professors with many years of teaching experience behind us, it is easy to forget how much we didn't know when we were students. We can also come to believe that everything we learned as students and since is set in stone, yet the field of bacterial genetics and genomics is dynamic and changeable, just like the bacteria themselves. Indeed, in this Second Edition I describe in Chapter 2 research that breaks the coupled transcription–translation paradigm of bacteria, adding another to the many ways bacteria break the rules. The book includes other new insights, discoveries, concepts, and terminologies that were not around when many of us were doing our PhDs. These have all been thoroughly researched and updated for the Second Edition with new findings and new journal articles to explore under Further Reading at the end of each chapter.

Please remember as well, no matter what your age or experience level, it can be easy for any of us to miss a concept, be unaware of a new finding or term, or get turned around on what something means, particularly if it is new or if new information means that it has changed. This book will support student learners, educators, and researchers with an interest in bacterial genetics and genomics, providing them with the information and tools needed to be able to confidently explore the subject in greater depth on their own, as suggested in the Discussion Topics at the end of each chapter.

Note to Reader

This book is organized into chapters that are grouped into parts, made up of three chapters that go together. This organization is designed to help group key ideas together and to build on concepts presented in earlier chapters as more advanced topics are discussed in later chapters.

Whenever a new term is encountered, this will be shown in bold. The definition of the term will be in the text in context where it is used, but it will also be in the Glossary at the back of the book. The term will only be in bold the first time, so if there is an unfamiliar term, it is worth checking the Glossary for a definition.

In addition to a standard Glossary, there is also a Glossary of Bacterial Species at the back of this book. This has been included as a quick reference guide so that readers can easily find out about the bacterial species that are mentioned. There are 97 species referred to in this book and it is not expected that readers will have heard of all of these. Bacterial genetics and genomics have developed far beyond *Escherichia coli*, which is reflected in the writing here.

Where possible, interesting and useful information is also explained in figures and sometimes in tables or other illustrations. These will be accompanied by explanatory text that is slightly different from the main text. By explaining the concepts in two slightly different ways and including an illustration, this should clarify these points.

Each chapter ends with a list of the Key Points covered in the chapter. This list can be used to determine if the topics have all been clearly understood, before moving on. It also summarizes what has been covered in the chapter.

At the end of each chapter there are also some questions about the topics and terms covered in the chapter to aid in study and revision. There are three types of questions. First, terms introduced in the chapter are reviewed to ensure understanding. Next, questions are asked about the material in the chapter to test understanding of what has been read. Finally, some topics are presented for discussion, which extend the subject beyond what has been covered in the chapter itself.

Lastly, each chapter ends with a list of Further Reading. These are provided for students who would like to do some further reading in the subject area. The references provided may be useful in completing the discussion questions at the end of the chapter and in satisfying curiosity.

As a useful reference and guide, other information that might come in handy when studying and applying your knowledge of bacterial genetics and genomics is printed on the final pages of this book after the Index.

Instructor and Student Resources

The Instructor and Student Resources which accompany this book can be accessed on the book's companion website: www.routledge.com/cw/snyder. These are the resources available:

- Interactive quiz questions
- Links to external sites which support Chapters 16 and 17
- Answers to the end-of-chapter questions
- Hints to the discussion questions
- Flashcards from the online glossary
- Figures from the book available for download (as PDF and PPT)
- Link to affiliated blog, which provides information on advances in bacterial genetics and genomics

Acknowledgments

Throughout the writing of the first edition of this book I received valuable assistance from my editors, Liz Owen and Jordan Wearing, who diligently read through my chapters, provided feedback, formatting advice, and guidance and for this they have my thanks. Jordan has continued as editor for this second edition, providing feedback from the first edition to help me improve upon it. Thanks as well to the other staff at CRC Press/Taylor & Francis who made this publication possible, including Chuck Crumly, and the returning illustrator, Patrick Lane, who has once again brought my pencil sketches to life. I would also like to thank the reviewers of the first edition—those who provided helpful comments on the original proposal for *Bacterial Genetics and Genomics*, those who provided detailed feedback on the written chapters, and those who wrote reviews and provided feedback after its publication.

Finally, special thanks go to my husband Alex and our child Wil for their continued patience as I tackled the revisions for this edition. Your support and love are very much appreciated. I am also thankful to my parents who inspired me with science and supported me throughout their lives.

About the Author

I grew up in Clinton, Connecticut, USA, and it is there that my interest in biology started, due in large part to my parents. My mother and father met at Yale University Hospital in New Haven, Connecticut, where he was a respiratory therapist, and she was a nurse. Our conversations over family dinner made it clear to me that the science lessons I was learning in school had direct application in the real world. In middle school, my father and I made a spirometer for measuring the vital capacity of the lungs in our basement workshop. When it came time for me to decide on universities, my mother and I took a road trip to visit the top places on my list, which solidified my decision to go to the College of William and Mary in Virginia, USA. There I studied both Biology and Music. When I started, I was determined to be a veterinarian, and also to continue to study guitar, which I had been playing since I was 10 years old. As I started learning more in my course, I developed a passion for genetics and decided to do a PhD. It was recommended that I first do a master's degree, due to the competitive nature of human genetics PhDs. So, I decided to stay at William and Mary for a two-year program, which came with a stipend and paid my tuition fees. A condition of the stipend was that I taught laboratory classes; this was an excellent opportunity to gain some teaching experience. I wasn't convinced I would enjoy teaching before I started, but I found it very rewarding and ever since then I have sought to find a balance in my career between teaching and research, both aspects of academic life that I enjoy. It was during my master's degree that I started my research in bacteriology in a class with Professor Carl Vermeulen. The summer before my class with Dr. V, one of his students had made a discovery and he offered us the chance to pursue this discovery as a research project. Because of this opportunity, I embarked on my microbiology career, working on *Escherichia coli* lipopolysaccharide. I was able to present this research at the American Society for Microbiology General Meeting, an awe-inspiring experience due to the size and scope of the conference. When I finished my master's, I applied for PhDs that had strong microbiology and molecular genetics aspects. I found this at Emory University in Atlanta, Georgia, USA. As before, this came with a stipend and my fees paid, in exchange for a bit of teaching. I found an ideal home in which to do my PhD with my supervisor, Professor Bill Shafer. His group was working on antibiotic resistance, which was a subject important to me, and his research involved molecular genetics. The subject of the research was perfect. My short rotation in the first year went really well. I got some good results and got along well with the other people in the lab and with Bill. My decision to complete my PhD in his group has shaped my career going forward. His research focused on the bacterial species Neisseria gonorrhoeae, which has remained my focus. A year or so after I started my PhD, the first few fragments of the genome sequence data of this species arrived in Bill's lab and I started doing the analysis. This began my love of bacterial genetics and genomics, which has been a common thread in my research ever since, ultimately leading to the writing of this book. As I was writing my PhD thesis for Bill, I remember him remarking on the quality of my writing and that I might have a future in academic writing. I have always carried that supportive comment from him with me and it has been an inspiration throughout the writing of this book and other works. Since my PhD at Emory, I have been involved in several genomics projects at the University of Oxford, UK, and University of Birmingham, UK, during my post-doctoral training and now at Kingston University, where I have been a Senior Lecturer, Reader, Associate Professor, and now Professor.

Lori A.S. Snyder

Welcome to the World of Bacterial Genetics and Genomics

Bacteria are small, single-celled organisms that play various important roles on the planet and in the lives of human beings. Although they are microscopic, bacteria have had a tremendous influence upon the other organisms of the world, which are often more complex in their cellular make up and in their genetics.

The study of genetics is the study of the code of life that is carried within the DNA of living things. Bacteria have been the subject of extensive research into their genetics, both because of the roles for some of these in human disease and also because these organisms are easy to grow, easy to manipulate, and therefore a great deal can be learned about the principles of genetics.

Our understanding of bacterial genetics and the ways in which we investigate bacterial genetics has changed as the genomics field has advanced. Genomics is wider in scope than genetics, being concerned with all of the DNA in an organism, rather than a single gene or small portion. These two research fields, bacterial genetics and bacterial genomics, complement each other and influence each other. Where once we did not know that DNA was the genetic material, now we can lay bare the whole genome of any bacteria using sequencing methods. Genetics and genomes can no longer be separated, and under the novel insights from genetics and genomics, once-believed borders in biology start to fade: most biological knowledge can be viewed under a new light when applied to the bacterial world.

Once it was believed that genome size correlated with the complexity of the organism. Now we know that some viral genomes are not only larger than some bacterial genomes, but also more complex. Some genomes of higher organisms (eukaryotes) are smaller than some bacterial genomes, challenging our old assumptions. Bacterial plasmids can be as big as chromosomes and can carry essential genes, required by the bacteria, just like chromosomes. The difference between a bacterium and an organelle within a eukaryotic cell are blurred, as further evidence supports organelle evolution from bacteria. Genomics has even revealed the transfer of genes between bacteria and higher organisms. Bacteria can even be larger than eukaryotic cells, visible to the naked eye, or can form multicellular life forms, once believed to be the realm of higher organisms. There are exceptions to the standard rules that are thought of as differentiating bacteria and eukaryotes, including the identification of bacteria that contain linear chromosomes and those that do not use coupled transcription–translation.

The species concept, generally relying on sexual reproduction to classify different species, cannot accurately describe the differences and relations within and between groups of bacteria. Bacterial cells do not undergo sexual reproduction and their genomes can be influenced by horizontal gene transfer from unrelated bacteria. Although we define bacteria into species, largely identified by characteristics observed in the laboratory, in the field, or outcomes of human infection, the whole concept of what constitutes a species is hotly debated when dealing with bacteria.

The bacterial world is incredibly diverse. Assumptions made about the lines between bacteria, viruses, and eukaryotes have been challenged. Similarly, assumptions that all bacteria behave the same must be cast aside. Even within a bacterial species, there can be significant variation in the genes within the genome, the expression pattern of those genes, and their behavior. These concepts will be introduced and explored, both through comparing different bacterial species and exploring the diversity that can be seen within species and within populations.

Further reading

Ku C, Nelson-Sathi S, Roettger M, Sousa FL, Lockhart PJ, Bryant D, Hazkani-Covo E, McInerney JO, Landan G, Martin WF. Endosymbiotic origin and differential loss of eukaryotic genes. *Nature*. 2015; *524(7566)*: 427–432.

Raoult D, La Scola B, Birtles R. The discovery and characterization of mimivirus, the largest known virus and putative pneumonia agent. *Clin Infect Dis*. 2007; *45(1)*: 95–102.

Vivarès CP, Méténier G. Towards the minimal eukaryotic parasitic genome. *Curr Opin Microbiol*. 2000; *3(5)*: 463–467.

Part I

DNA, Genes, and Genomes

The three chapters that start off this book in Part I focus on the fundamentals of genetics and genomics, starting with a detailed discussion of DNA. This is built upon in the following chapter where genes are introduced, followed by a chapter that addresses genomics. Some readers may find that some of the material presented in Part I is review; this is important so that all readers have the same basis for understanding the material in the rest of the book. Each of these first three chapters includes historic information on their topics and also new insights that have been gained. What is presented here will be built on throughout the book.

DOI: 10.1201/9781003380436-1

Chapter 1

DNA

This chapter will introduce the biologically important molecule **DNA**, **deoxyribonucleic acid**. DNA is the genetic material, carrying within its structure the code of life; thus, it is important to understand how this molecule arose on Earth, how scientists determined that DNA held the code of life, what the structure of the molecule is, and how this holds the genetic information. Terms and concepts that will be explored in more detail in this book are introduced here, forming the fundamental background and concepts of bacterial genetics and genomics.

Life originated from RNA with DNA evolving later

There once was a time when there was no DNA. Billions of years in the past, before life emerged on the planet, a primordial soup of chemicals and compounds held the promise of what we have on the Earth today. In our study of genetics and genomics today, we analyze the sequence of DNA, but when investigating the distant past, it is believed that it was not DNA that was at the origin, but **RNA** (**ribonucleic acid**).

According to the **RNA World hypothesis**, the first nucleic acids on our planet were RNAs, which were first established as catalyst and genetic material. Early life on Earth is believed to have been RNA-based, rather than DNA-based. Today, there exist viruses with RNA genomes that serve as examples and evidence that RNA can be the carrier of the genetic code. The discovery of **ribozymes**, RNA enzymes (covered in Chapter 4), has provided further supporting evidence for the RNA World hypothesis by showing that RNA can be a biological catalyst. Therefore, the necessary functions for early life on Earth can all have been accomplished by RNA in the absence of DNA. DNA-based storage of genetic information, which is seen in living cells today, came later in the evolution of life. The RNA World model, with DNA arising on Earth later, is supported by chemistry and the complexity of enzymes needed for DNA to be synthesized and therefore used as the code of life.

It is believed that DNA arose after templated protein synthesis, whereby proteins are made based on the sequence of RNA. There are some suggestions that DNA may have emerged before this process arose, although these theories still start with RNA. It is also unclear whether DNA came about before or after the **Last Universal Common Ancestor** (**LUCA**), that is, the lifeform that came before the lineage split that resulted in today's **bacteria**, **eukaryotes**, and **archaea**. If DNA arose after the LUCA, this would explain fundamental key differences in the way in which errors are corrected in the genetic code between the different lineages. If, however, DNA came about before the bacteria, eukaryotes, and archaea lineages split, then the differences between the genes responsible for accurately copying DNA in bacteria versus the other lineages would have to be due to the complete replacement of these genes in the other lineages later in evolution. If DNA containing these genes and other features was present before the split, then it would follow that these genes should continue to carry similarity with one another across the three lineages. Because they do not, either DNA arose after the LUCA, or it arose before, and the copying accuracy genes were later swapped out for those with similarity to what is seen in the lineages today.

DOI: 10.1201/9781003380436-2

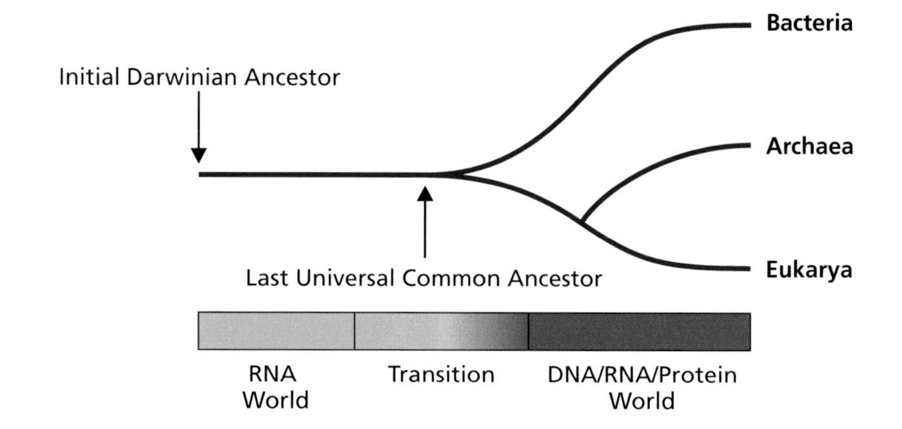

Figure 1.1 Evolution of life as we know it according to the RNA World hypothesis. The world we understand now evolved from an earlier form. One hypothesis, strongly supported, is that life began as an RNA World. It is believed that 4 billion years ago, the Initial Darwinian Ancestor arose, the first life form that started evolution on Earth. From this later arose the LUCA, which diverged into the lineages we describe today as bacteria, archaea, and eukaryotes. There was a transition to the world of today, a DNA/RNA/Protein World, either before or after the LUCA.

Over 4 billion years ago, the first life on Earth appeared. This first life form was the initial replicator, capable of making more of itself, and it was the beginning of evolution as the **Initial Darwinian Ancestor** (**Figure 1.1**). The LUCA came much later. The RNA World hypothesis would mean that Darwinian evolution occurred in two phases: first as the RNA World of ribozymes catalyzing metabolism and RNA replication starting with the Initial Darwinian Ancestor and then as the DNA/RNA/Protein World of DNA, ribosomal RNA (rRNA), messenger RNA (mRNA), transfer RNA (tRNA), ribosomes, and protein enzymes for metabolism, RNA synthesis, protein synthesis, and DNA replication, which continues today. Evolution, based on RNA, would have occurred before DNA arose and before the split into the prokaryote, eukaryote, and archaea lineages, in the first form(s) of life on Earth. As these progressed, evolution led to the LUCA.

Nucleic acids are made of nucleoside bases attached to a phosphate sugar backbone

The complexity of life is determined by **nucleic acids**. These biological molecules are what make up DNA and RNA, the blueprints for life. All nucleic acids have two aspects: the **backbone** and the **nucleoside bases**. The nucleic acid backbone itself is made of two parts, alternating between five-carbon sugars and phosphates that form a chain. Each sugar in the backbone chain is attached to a nucleoside base, also referred to as a **nitrogenous base** due to the inclusion of nitrogens in the structure.

There are four different nucleoside bases that can be included in each nucleic acid molecule, which in turn can have hundreds, thousands, or millions of bases depending on its length. The order, pattern, and combination of the nucleoside bases determine the genetic information that is carried by the nucleic acid.

RNA contains ribose sugars and is single-stranded

The first nucleic acid on the planet, according to the RNA World hypothesis, was RNA. The sugar within the RNA backbone is **ribose**, a monosaccharide that is present here in a five-member ring form (**Figure 1.2**). The nucleotide bases attached to the ribose are **adenine** (A), **cytosine** (C), **guanine** (G), and **uracil** (U). These bases are either **purines** (A and G), which have two rings in their structure, or **pyrimidines** (U and C), which have one ring. RNA is single-stranded, meaning that it has one backbone, although in some cases the RNA molecule may bend back upon itself, forming structures that are necessary for the function of some RNAs. RNA is the genetic material of some viruses and **bacteriophages** (viruses that infect bacteria), encoding the whole of their genome without the need for DNA. Even in cells with DNA genomes, RNA is important for protein synthesis. It is the molecule that is the intermediate between DNA and protein in the biosynthetic process of expressing the code of life in DNA as functional and structural components of the cell in proteins. RNA is the template by which the order of amino acids in a protein is directly interpreted.

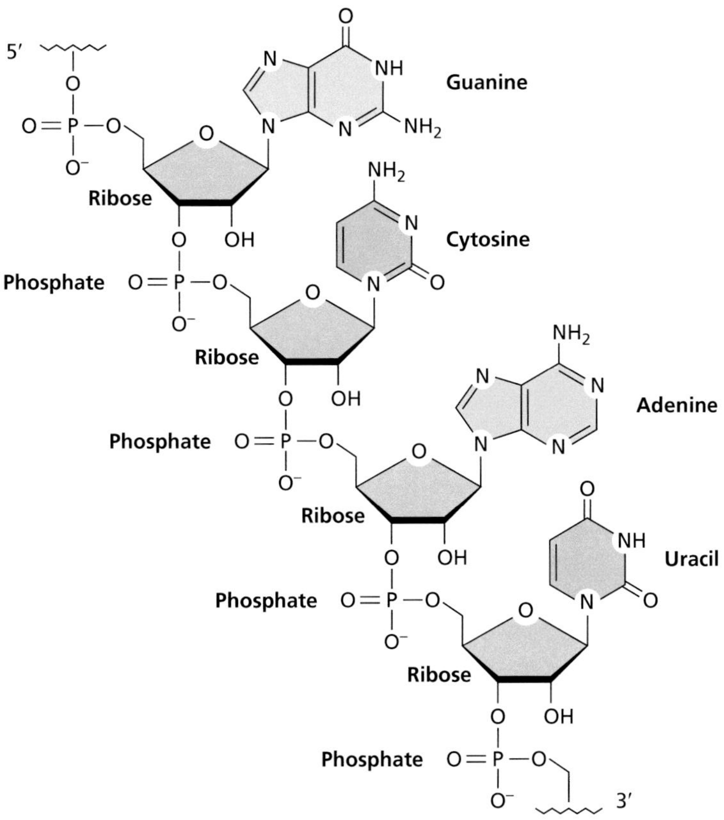

Figure 1.2 The chemical structure of RNA. The phosphates are joined to the ribose sugars by phosphodiester bonds to make up the backbone of the RNA. The nucleotide bases guanine, cytosine, adenine, and uracil attach to the ribose sugar at carbon 1.

DNA contains deoxyribose sugars and is double-stranded

DNA stores the genetic information for everything, aside from some viruses and bacteriophages. In DNA, the sugar is **deoxyribose**, another monosaccharide with a five-member ring that is similar to ribose, although lacking one OH group. The nucleotide bases in DNA are adenine (A), cytosine (C), guanine (G), and thymine (T). In DNA, the pyrimidine thymine is present instead of uracil. These differences, that is, the presence of the deoxyribose sugar and of thymine, are characteristic of DNA and set it apart from RNA. DNA is double-stranded, meaning that there are two backbones going in opposite directions in a structure referred to as **antiparallel**. The directionality of each strand is defined by the bonds in the backbone on the 3rd (3′) and 5th (5′) carbons on the sugars. The direction of each strand impacts the processes of **replication** (the copying of DNA), and **transcription** (the generation of RNA based on the DNA sequence). The two strands of DNA are another key characteristic of this molecule. Although single-stranded DNA can be found in nature and generated in the laboratory, DNA's native form is double-stranded. In this form, the nucleotide bases covalently attached to one backbone are joined by **hydrogen bonds** to the nucleotide bases covalently attached to the other backbone (**Figure 1.3**). These hydrogen bonds between the nucleotide bases form the double-stranded structure. Adenine forms two hydrogen bonds with thymine, and guanine forms three hydrogen bonds with cytosine on the opposite strand of the DNA structure. The double-stranded structure of DNA twists into a double helix (**Figure 1.4**).

The double helix of DNA is by now famous, included today in works of art as well as scientific textbooks. It has been the inspiration for sculptures (**Figure 1.5**), paintings, fashion, jewelry, tattoos, and other creative works. The DNA double helix has even been featured on the back of the UK £2 coin (**Figure 1.6**). The structural model of the DNA double helix created by James Watson and Francis Crick when they first unraveled the structure of the molecule from its chemical composition is displayed in the Science Museum in London. It is interesting to note that, at the time of its construction, it was believed that there were two

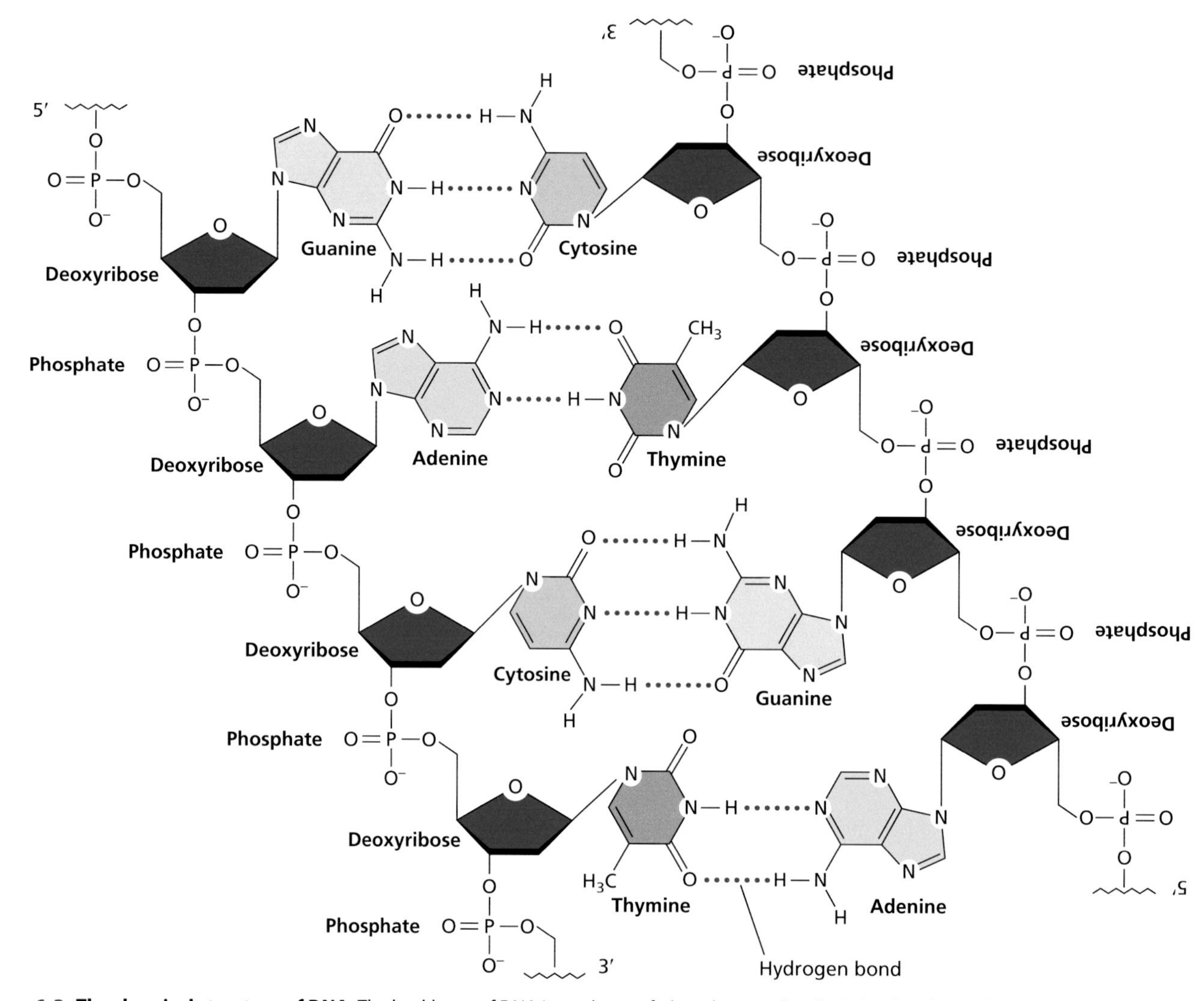

Figure 1.3 The chemical structure of DNA. The backbone of DNA is made up of phosphate molecules joined to deoxyribose sugars by phosphodiester bonds. Note that the deoxyribose sugars lack an OH group on carbon 2 that is present on the ribose sugars of RNA. The nucleotide bases guanine, cytosine, adenine, and thymine attach to the deoxyribose sugar at carbon 1. Hydrogen bonds between complementary bases connect the two strands of the DNA structure.

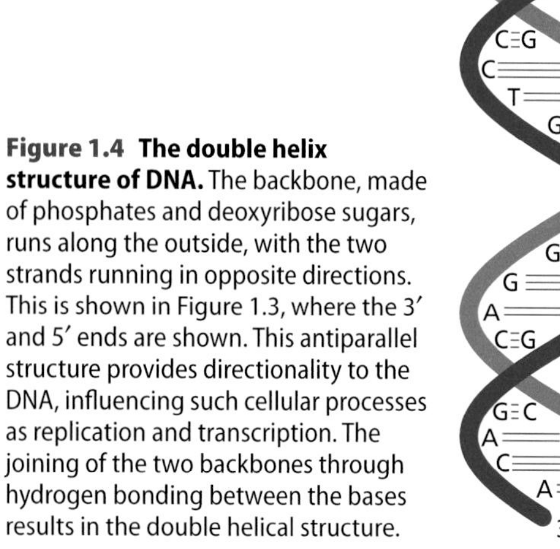

Figure 1.4 The double helix structure of DNA. The backbone, made of phosphates and deoxyribose sugars, runs along the outside, with the two strands running in opposite directions. This is shown in Figure 1.3, where the 3′ and 5′ ends are shown. This antiparallel structure provides directionality to the DNA, influencing such cellular processes as replication and transcription. The joining of the two backbones through hydrogen bonding between the bases results in the double helical structure.

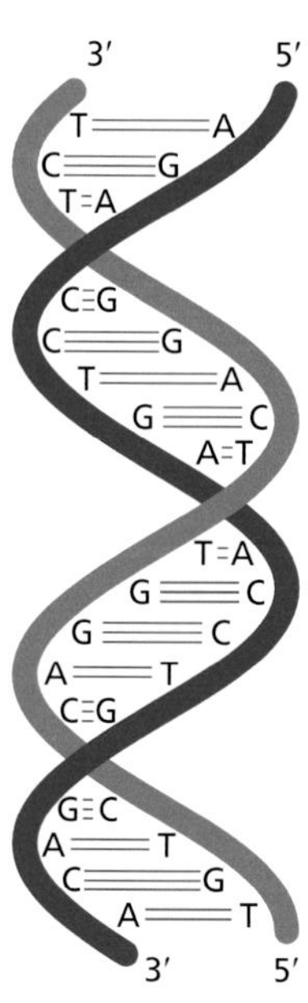

Figure 1.5 The DNA double helix as an inspiration for art. This is "The Double Helix—Mutation of Increased Compassion" by Franco Castelluccio. The structure of the double helix is here inspiration for a sculpture balancing male and female forms as the backbone of DNA. Artistic works such as these inspire discussions about evolution, mutation, and metamorphosis.

hydrogen bonds between cytosine (C) and guanine (G), rather than three. The DNA double helix model in the Science Museum is therefore inaccurate.

Figure 1.6 The DNA double helix on the £2 coin. In 2003, £2 coins were minted in the United Kingdom to commemorate the description of the structure of the DNA double helix.

Nucleoside bases have additional roles within the cell

Nucleoside bases can have roles within the cell in addition to their inclusion in DNA and RNA. For instance, adenines make up **ATP, adenosine triphosphate**, which stores and carries energy for cells (**Figure 1.7**). ATP stores and releases energy by means of a string of three phosphates attached to the sugar, hence the name triphosphate. Loss of the last of the three phosphates releases energy for the cell from the breaking of the high-energy bond between the phosphates. The resulting molecule after the loss of one phosphate is **ADP, adenosine diphosphate**, because there are now two phosphates attached to the sugar. Loss of the second phosphate to release more energy produces **AMP, adenosine monophosphate**, with just a single phosphate. The energy generated in the cell, from metabolism, is stored in part by adding phosphates back to produce ATP again.

The structures of the purine nucleosides guanine and adenine include a double ring, whereas cytosine, thymine, and uracil have a single ring as pyrimidines. Cell signaling can involve purine-based molecules cGMP, cyclic AMP (cAMP), and guanosine tetraphosphate (ppGpp). Some nucleosides are also enzyme cofactors, including coenzyme A, FAD, FMN, NAD, and NADP+.

Artificial nucleic acids have been created in the laboratory

Artificial nucleic acids are known as **xeno nucleic acids** or **XNA**. In these artificial forms, the sugar is neither ribose nor deoxyribose and is instead an alternative sugar. The XNAs that have been produced include FANA (containing 2′-fluoroarabinose), ANA (containing arabinose), TNA (containing threose), CeNA (containing cyclohexene), GNA (containing glycol), and HNA (containing hexitol). A locked nucleic acid (LNA) has also been generated in the laboratory and shows promise for use in biotechnology applications. In LNA there is a ribose, as in RNA; however, the oxygens in the sugar are connected by an additional covalent bond bridge. This effectively locks the sugar into this form, rather than other forms the ribose can take in nature. PNA, peptide nucleic acids, have been made, which have a pseudo-peptide containing backbone. There are also artificial nucleotides or unnatural base pair (UBP) combinations, including d5SICS and dNaM.

DNA was discovered in 1869 and identified as the genetic material 75 years later

Contrary to popular culture, DNA was not discovered by James Watson and Francis Crick in 1953. Swiss chemist Friedrich Miescher (**Figure 1.8**) discovered DNA in 1869, 84 years before Watson and Crick.

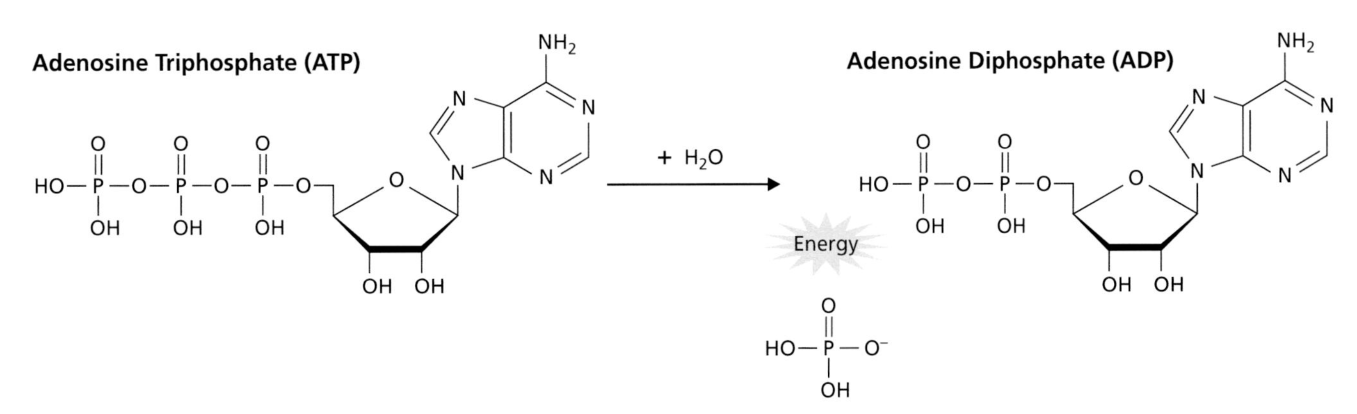

Figure 1.7 The chemical structure of ATP and ADP. Notice that, in the name, triphosphate is the clue to the structure having three phosphates. The addition of water (H_2O), releases one of these phosphates, generating the structure adenosine diphosphate with two phosphates. This process produces a free phosphate and releases energy.

Figure 1.8 The man who discovered DNA, Friedrich Miescher (1844–1895). Working with bandages containing human pus, Friedrich Miescher was the first person to isolate and describe DNA. (From Wellcome Collection. Published under CC BY 4.0.)

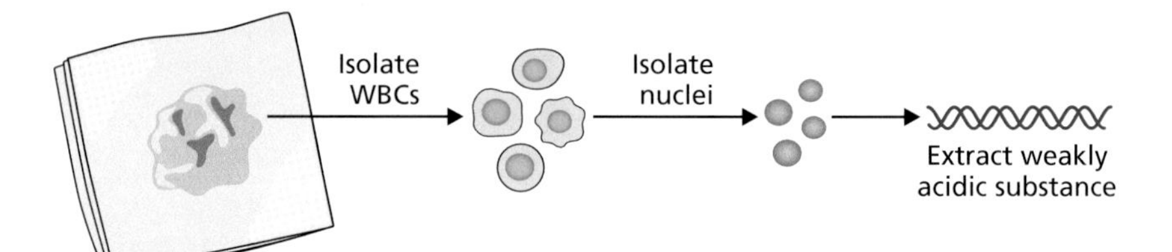

Figure 1.9 Miescher's process that first isolated DNA. Friedrich Miescher, a chemist from the mid-1800s, interested in the components within white blood cells (WBCs), collected bandages covered in pus. He was able to isolate from the bandages intact white blood cells and from these isolate the nuclei from within the cells. Miescher extracted from the nuclei a weakly acidic substance, which later became known as DNA.

Working on the components within white blood cells, leukocytes, Miescher discovered a substance that he called nuclein within the nuclei of these cells. Miescher was attempting to isolate white blood cell proteins. From a local hospital, he obtained bandages containing fresh pus from wounds (**Figure 1.9**). This pus, in the era before antibiotics, would have been an abundant source of fairly pure leukocytes.

Miescher found a weakly acidic substance in the nuclei that had a high phosphorous content and that did not dissolve in any solution that would have dissolved proteins. He therefore knew that this was not a protein and something newly discovered. Nuclein became known as nucleic acid and is now called deoxyribonucleic acid (DNA).

The first person to visualize DNA in a living cell was Walther Flemming. In 1879, through microscopy of nuclei, Flemming revealed threadlike structures that are the chromosomes. Flemming investigated how these chromatids, as he called them, separate during cell division. In 1881, Albrecht Kossel showed that Miescher's nuclein was a nucleic acid and named it deoxyribonucleic acid. Kossel isolated and identified the five bases of nucleic acids, A, T, G, C, and U. At the time, isolation of nucleic acids would not have generated pure DNA; therefore, it is not unexpected that Kossel found uracil in his samples, as they would have also included RNA. In the early 1900s there was a re-discovery of Gregor Mendel's experiments in crossbreeding plants and observations of the inheritance of traits from the mid-1800s. In the late 1900s, two scientists working independently on different model systems, Walter Sutton and Theodor Boveri, suggested that Mendel's units of inheritance were contained within chromosomes.

Phoebus Levene (**Figure 1.10**) was the first to discover some of the key properties of Miescher's nuclein. Publishing in 1919, Levene described the three components of nucleic acids (phosphate, sugar, and base) providing further definition and detail on the three parts that make up DNA, beyond Kossel's identifications of the bases. Levene was also the first to discover the sugar component of both RNA (ribose) and DNA (deoxyribose). By breaking down yeast nucleic acids, Levene discovered that nucleic acids are made of a series of nucleotides made up of phosphates, sugars, and nitrogenous bases. At the time, DNA was not yet recognized as the vehicle for genetics and inheritance, yet Levene studied the molecule in depth, identifying how the three components of nucleic acids form together into nucleotides, which in turn join into chains.

Figure 1.10 Phoebus Levene (1869–1940) uncovered the molecular make-up of nucleic acids. Levene discovered that nucleic acids are made up of phosphates, sugars, and bases and that DNA contains deoxyribose as its sugar and RNA contains ribose as its sugar.

In 1928, Frederick Griffith's investigations revealed important insight into the transfer of genetic traits in bacteria. Using two different morphologies of *Streptococcus pneumoniae*, one that was virulent and killed mice and one that was not, Griffith explored the transfer of genetic traits in bacteria. The virulent version of the bacteria had a smooth morphology due to its bacterial capsule. The non-virulent version of *S. pneumoniae* had a rough appearance when grown on media plates and had no capsule. When the virulent bacteria were heat killed, they did not kill mice. However, when the heat-killed virulent bacteria were injected into mice along with the nonvirulent bacteria, the mice died. Bacteria could be isolated from the dead mice and these showed that the isolated bacteria were now smooth in appearance, although the only living bacteria injected into the mice

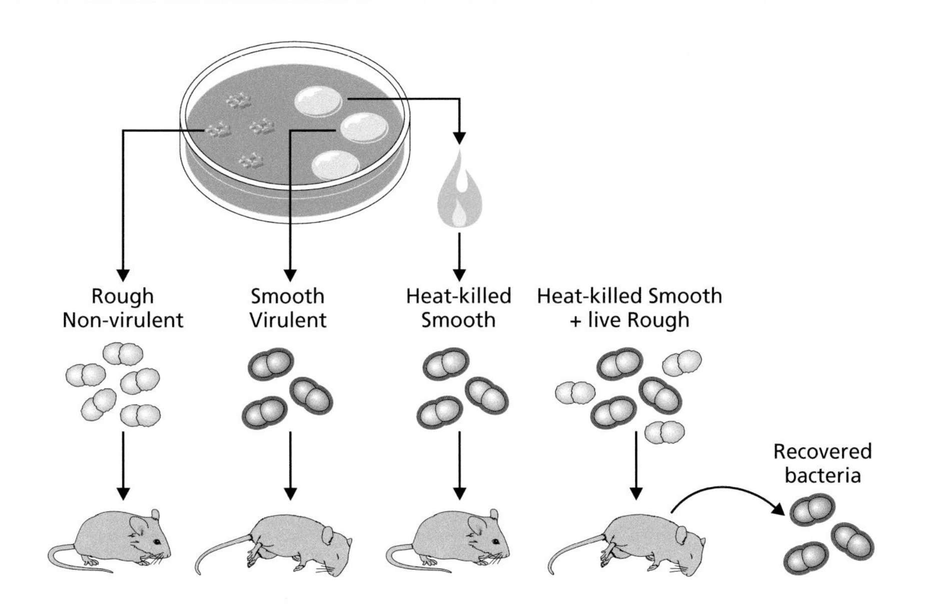

Figure 1.11 The isolation of the transforming principle by Frederick Griffith. In 1928, Frederick Griffith conducted experiments with *S. pneumoniae*. Injecting rough, non-virulent bacteria into mice did not kill the mice. Injecting smooth, virulent bacteria into mice killed the mice. If, however, the smooth, virulent bacteria were first heat killed and then injected into mice, they lived. This showed that the virulent bacteria could be rendered safe by heat killing them. In the pivotal last part of the experiment, the rough, nonvirulent bacteria were injected into mice along with the smooth, virulent bacteria, which had been heat killed. These mice died and bacteria were recovered that were smooth in appearance. The trait of being smooth and virulent was transferred from the dead bacteria to the living rough, nonvirulent bacteria.

in the experiment were the rough nonvirulent *S. pneumoniae*. The living rough, nonvirulent bacteria had been transformed by something within the heat-killed bacteria into smooth, virulent *S. pneumoniae* that killed the mice (**Figure 1.11**). Griffith had identified what he called the transforming principle, a substance capable of transferring the traits of one cell to another. At the time, the prevalent theory was that proteins were the genetic material, because DNA was believed to be too simplistic a molecule to be the blueprint for life. The earliest experiments on the genetic material and the transfer of genetic traits were done in bacteria, paving the way for advances in understanding bacterial genetics and genomics.

In 1944, Oswald Avery, Colin MacLeod, and Maclyn McCarty showed that the hereditary units, such as those described by Mendel, are DNA, not protein. Using a similar experimental design as had been employed by Griffith, the transfer of traits between *S. pneumoniae* was investigated in an attempt to answer whether DNA or protein or something else within the cell was responsible for the virulence characteristic of the bacteria. In Avery, MacLeod, and McCarty's experiments, the smooth, virulent bacteria were heat killed and then subjected to enzymatic digestion. The enzymes used would selectively degrade different components of the heat-killed cells: carbohydrase to degrade carbohydrates, lipase to degrade lipids, protease to degrade proteins, RNase to degrade RNA, and DNase to degrade DNA. The end products of these enzymatic digestions were combined with rough, nonvirulent *S. pneumoniae* and the resulting transformants were plated onto agar media. In all cases except one, both rough, nonvirulent bacteria and smooth, virulent bacteria were recovered, the latter having been generated through the acquisition of the genetic traits for the capsule, and thus virulence, from the cellular material of the dead, heat-killed, enzymatically digested bacteria.

However, when the enzyme involved was DNase, no smooth, virulent bacteria were recovered. This demonstrated that DNA is required for the transfer of genetic traits; therefore, DNA is the genetic material (**Figure 1.12**).

The first X-ray images of DNA were taken in 1937 with the structure finally solved in 1953

In 1937, William Astbury, who had trained with the scientists who invented the methodology, was the first to use X-ray crystallography to show that DNA had a regular structure. The attribution of a double helix structure uncovered by X-ray crystallography was later claimed by Rosalind Franklin in 1953, who was able to produce high-resolution images showing the phosphates on the outside. However, the first X-rays of DNA by Astbury were previous to this and already suggested a regular structure to the nucleic acid.

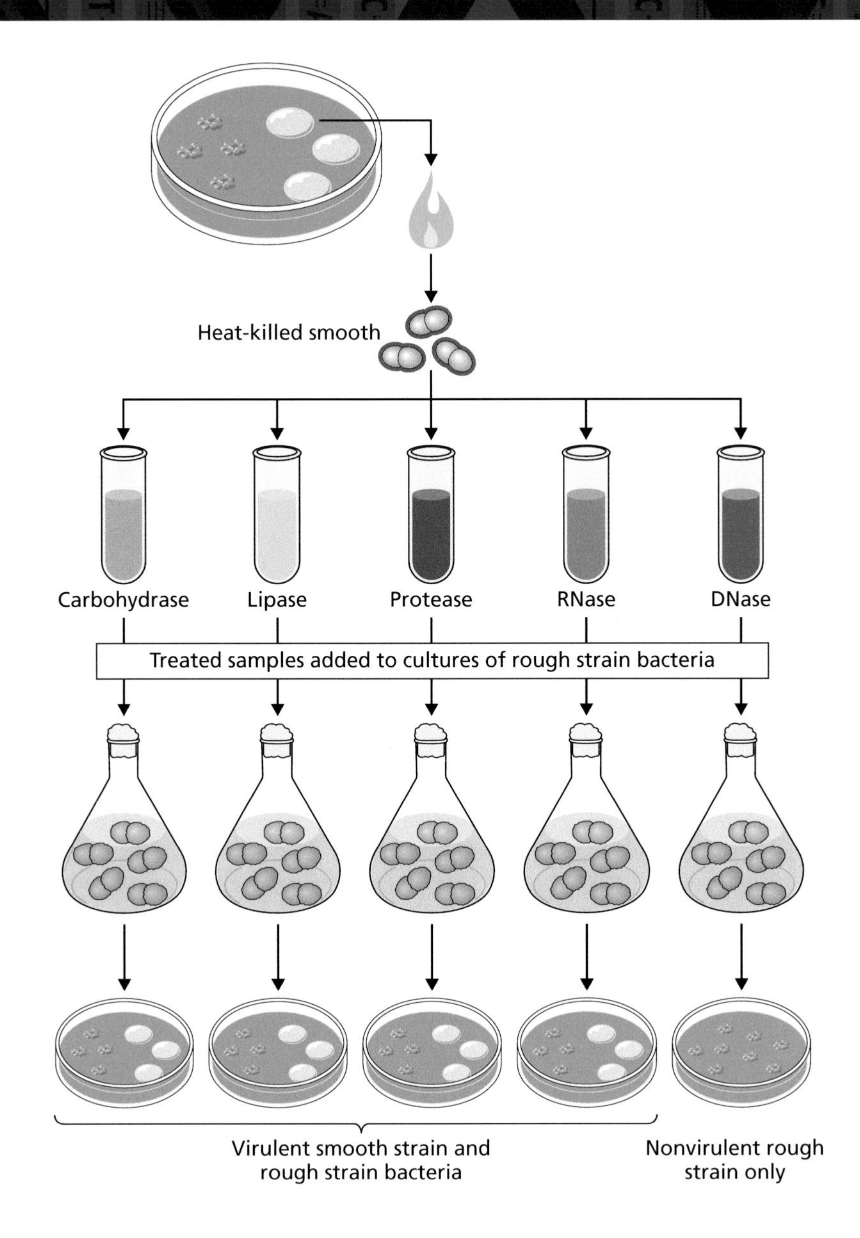

Figure 1.12 The experiments of Avery, MacLeod, and McCarty to identify the transforming principle in *S. pneumoniae*. Following on from the experiments done by Griffith (see Figure 1.11), Oswald Avery, Colin MacLeod, and Maclyn McCarty took the heat-killed smooth, virulent bacteria and treated this material with carbohydrase, lipase, protease, RNase, and DNase. When these treatments were combined with the rough, nonvirulent bacteria, the transforming principle was retained after all treatments except DNase. Both smooth and rough bacteria were isolated from all other treatments; however, only rough bacteria could be recovered from the combination of the DNase-treated, heat-killed, virulent bacterial mixture and the rough, nonvirulent bacteria. These experiments demonstrated that DNA is the genetic material.

After the work of Avery and colleagues, it became known that the transforming principle, the component genetically conferring traits, is encoded in the nucleic acids of cells. This work interested Erwin Chargaff, who researched the DNA composition within cells. In 1950, he showed that DNA of different species were different, containing different concentrations of the four nitrogenous bases. Chargaff also showed that although the composition of DNA was different between species, some characteristics were universal. The quantity of As within the cell was roughly equal to the Ts, and the Gs were nearly equal to the Cs, so the amount of purines (A+G) and pyrimidines (C+T) were nearly the same. This became known as "Chargaff's Rule."

The findings of Avery, MacLeod, and McCarty were confirmed in 1952 by the work of Alfred Hershey and Martha Chase, who demonstrated that T4 bacteriophage DNA enters bacterial cells, not proteins. When a bacteriophage infects a bacterial cell, it can subvert the cellular systems to make more bacteriophage. Therefore, the genetic information that orchestrates this and copies the bacteriophage genome for packaging into new bacteriophage particles is within the DNA, as this is all that enters the bacterial cell. This body of work, and others, contributed to the knowledge available at the time that Watson and Crick, and others around the world, were attempting to uncover the structure of DNA.

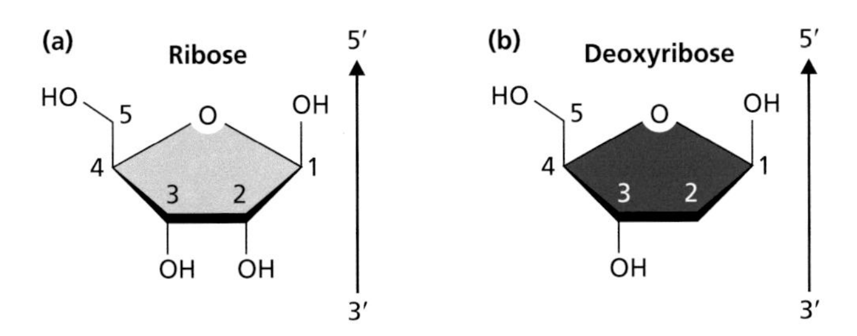

Figure 1.13 The structure of the sugars in (a) RNA and (b) DNA. The five carbons within these structures are numbered as shown. The 5th carbon is used to indicate directionality along the attached phosphate group onward, the 5′ end, and the 3rd carbon is used to indicate the opposite directionality of the strand, the 3′ end. The nitrogenous bases, A, U, G, and C for RNA and A, T, G, and C for DNA, attach to the 1st carbon.

DNA consists of two bidirectional strands joined by deoxyribose sugars and nucleotide bases

In RNA the sugar is ribose and in DNA the sugar is deoxyribose. The five carbons on these sugars are numbered such that the carbon outside the ring is the 5th carbon, the next carbon back on the ring is the 4th carbon, the carbon after that is the 3rd carbon, and so on (**Figure 1.13**).

The numbering of the carbons on the sugars is how we get our description of directionality of a DNA strand. The backbone ribose or deoxyribose can be drawn with the 3rd carbon either on the top or on the bottom. It does not matter which way it is drawn, and it should be remembered that molecules exist in three-dimensional space, not the flat two-dimensional plane that we use to illustrate them. To differentiate the directionality of the DNA strand, it is important to know the orientation of the molecule. Thus, we use 5′ → 3′ or 3′ → 5′ to be clear on the molecular direction of the DNA or RNA strand. The phosphate bonds to the sugar (ribose for RNA, deoxyribose for DNA). The creation of a bond between the phosphate and the sugar, at either the 5′ or the 3′ carbon, releases one water molecule (H_2O) (**Figure 1.14**).

The backbone is composed of an alternating series of phosphate, sugar, phosphate, sugar, with each set joined by a **phosphodiester bond** formed when the phosphate of one group bonds with the 3rd carbon of the sugar of another group. The nitrogenous bases connect to the 1′ carbon, where there is an OH group on the sugar. This displaces the OH from the sugar and an H from the base, forming H_2O (**Figure 1.15**).

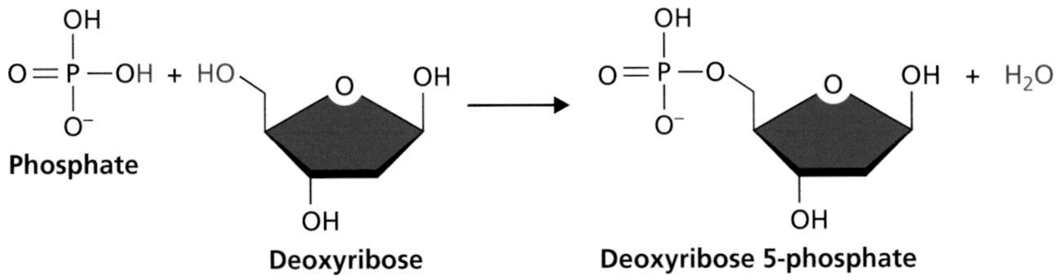

Figure 1.14 Formation of the bond between a deoxyribose sugar and phosphate. An OH group from the phosphate and the OH group on either the 5th carbon or the 3rd carbon of the deoxyribose (DNA) are the site of the bond between the phosphate and deoxyribose (or ribose) to create the backbone of DNA (or RNA). The formation of the bond with the 5′ carbon is shown here. The creation of this bond liberates two hydrogens and one oxygen (red), which together form a single water molecule, H_2O (red).

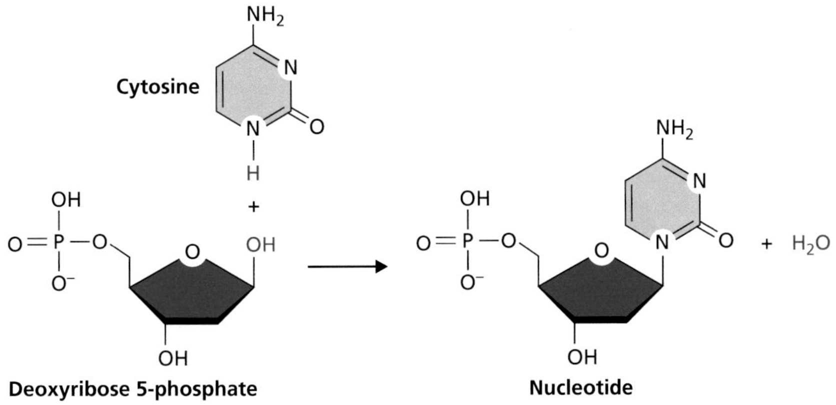

Figure 1.15 Formation of the bond between a nitrogenous base and deoxyribose phosphate to form a nucleotide. The OH group on the 1st carbon of the deoxyribose 5-phosphate and the H on the nitrogenous base (cytosine shown here) are the site of the bond that creates a nucleotide. The formation of the bond liberates two hydrogens and one oxygen (red), which together form a single water molecule, H_2O (red).

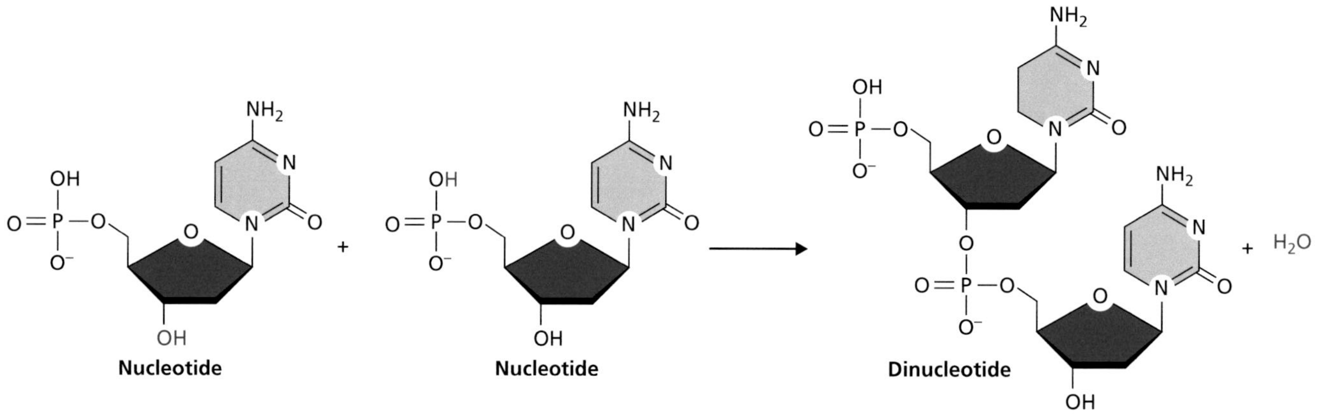

Nucleotide

Nucleotide

Dinucleotide

Figure 1.16 Formation of a dinucleotide. The OH of a phosphate group on one nucleotide and the OH group on the 3′ carbon of the other nucleotide are the site of the bond that creates a dinucleotide. The formation of the bond liberates two hydrogens and one oxygen (red), which together form a single water molecule, H_2O (red).

One phosphate, one sugar, and one nitrogenous base make a nucleotide (see Figure 1.15). A **dinucleotide** is two nucleotides bonded together. The phosphate of one joins to the 3′ carbon of the other in a phosphodiester bond and releases H_2O (**Figure 1.16**). An **oligonucleotide** is a few nucleotides joined together in a strand. In molecular biology, the term oligonucleotide is used to describe short synthetic pieces of DNA.

The bases across from one another in the DNA double helix are joined by hydrogen bonds. There are two hydrogen bonds between adenine (A) and thymine (T) and three hydrogen bonds between guanine (G) and cytosine (C), making the bond stronger between guanine and cytosine.

When DNA is heated, these hydrogen bonds break, first at the adenine and thymine bonds and then at the guanine and cytosine bonds. Therefore, the sequence influences the melting temperature that separates the two strands. Sequences with more guanines and cytosines have higher melting temperatures than those with many adenines and thymines. We will see later in this book how this has been exploited in molecular biology, and how some bacterial species have a higher level of guanines and cytosines or a higher content of adenines and thymines.

Production of nucleotides in the cell is essential so that processes such as replication of DNA and transcription of RNA, described later in this book, can happen. Nucleotides are also required for maintaining and repairing DNA in the cell. But if there is an excess of nucleotides, this wastes energy for the cell. The production of nucleotides is therefore tightly regulated to make the right amount of each that is needed.

DNA is copied semiconservatively every time a cell divides

The copying of DNA by a cell is called replication. This process is essential. If the DNA within the bacterial cell is not copied, then during cell division, where a single bacterial cell is split into two daughter cells, there will not be enough genetic material for both daughter cells.

The division of individual bacterial cells into daughter cells enables a population of bacteria to expand. The population, if it has grown from a single bacterial cell, is clonal, meaning that all of the cells are copies of the original cell. Any variations in the DNA between cells are mutations, which are likely to have been generated during replication, but can occur by other mechanisms, which will be discussed in later chapters. The concept of DNA replication, the process by which DNA is copied, is simple; however, the various components and mechanisms involved require complex coordination of nucleic acids and proteins.

Matthew Meselson and Franklin Stahl conducted experiments that showed in 1958 that DNA is copied in a semiconservative manner, meaning that one strand of the double helix comes from the original DNA and one strand is newly synthesized. *Escherichia coli* was grown in media containing ^{15}N, rather than the normally found ^{14}N (**Figure 1.17**). This allowed the incorporation of this heavy nitrogen into the molecules within the cell, including DNA. Once the ^{15}N-containing DNA had been established in the culture, the bacteria were put into media containing normal ^{14}N. When the *E. coli* DNA was replicated in the ^{14}N-containing media, any new nucleotides created would no longer contain the heavy ^{15}N. In this way, it was shown that replication generates a double helix with one original strand, in this experiment labeled with ^{15}N, and one newly synthesized strand, in this experiment containing ^{14}N.

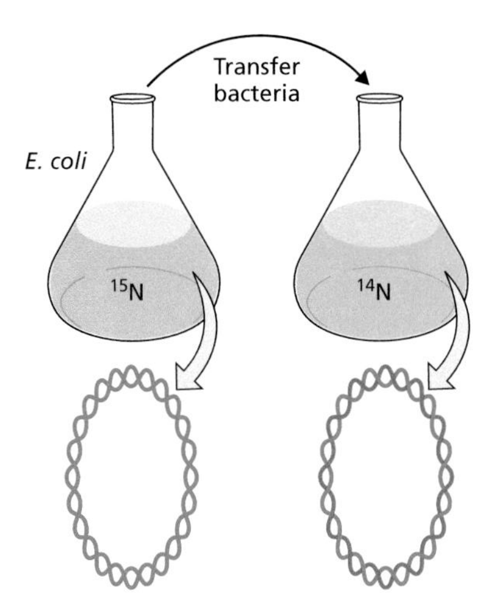

Figure 1.17 The Meselson–Stahl experiment using *E. coli* in media containing heavy nitrogen (^{15}N). *E. coli* were grown in media containing ^{15}N (blue flask). Once the bacteria had fully integrated ^{15}N into the DNA (circular double helix with two blue strands), the bacteria were collected and transferred into fresh media with ^{14}N (orange flask). The results showed that the copied DNA was half from the original DNA, labeled with ^{15}N (blue strand), and half newly synthesized DNA, which used the normal ^{14}N (orange strand). Therefore, we know that when DNA is copied in replication, the two new double helices are made. In each, one strand originates from the starting DNA and one strand is newly synthesized.

In the process of DNA replication, the hydrogen bonds in a small section of the DNA are broken and the strands come apart. It should be remembered that the rest of the double helix is still together; the entire double helix of the chromosome does not come apart. Free nucleotides in the cell, as **deoxyribonucleoside triphosphates**, are able to hydrogen bond to the now available nucleotides in the DNA (**Figure 1.18**). The three phosphates on the bases to be used for replication are essential. The 3′ OH on the sugar attacks the alpha phosphate, the one farthest in, and displaces the other two phosphates (beta and gamma). This is called nucleophilic attack and produces a phosphodiester bond.

Replication starts at the origin of replication and requires primers

Replication starts at a place in the bacterial chromosome called the **origin of replication**. Helicase recognizes the origin of replication and breaks the hydrogen bonds between the DNA bases at this location. The split in the DNA double helix caused by helicase forms what is referred to as the replication fork, a split that progresses along the DNA as it is replicated. To prevent the reformation of the hydrogen bonds, which would bring the DNA strands back together, single-strand binding proteins attach to the strands that were separated by helicase (**Figure 1.19**). Nucleotides are needed for replication to occur, both deoxyribonucleoside triphosphates and also ribonucleoside triphosphates, plus extra ATP to provide energy to fuel the action of the replication proteins.

The major protein involved in DNA replication is an enzyme called **DNA polymerase**. Unfortunately, this enzyme can only extend an existing nucleic acid strand. When the origin separates due to the breaking of hydrogen bonds between bases, there is nothing DNA polymerase can do. DNA polymerase needs there to be some nucleotides on the new strand to get started.

Therefore, to start the process of DNA replication, an enzyme called **primase**, which is able to freshly incorporate nucleotides on a new strand, forms a short oligonucleotide of ribonucleic acids. This is an RNA primer, which has been synthesized complementary to the original strand of DNA. With this RNA primer in place, DNA polymerase can add deoxyribonucleotide bases, progressing the synthesis of the new DNA strand (**Figure 1.20**).

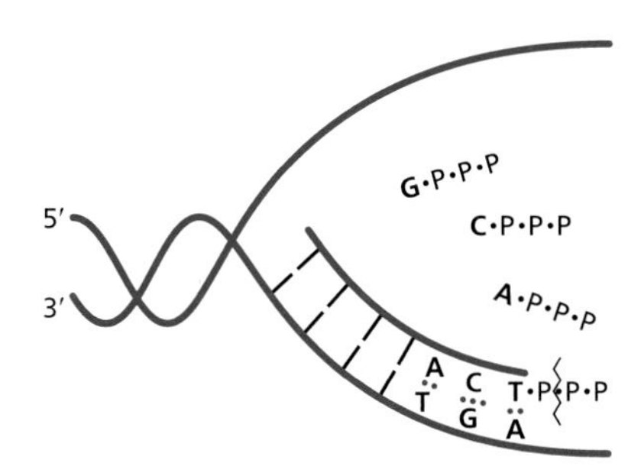

Figure 1.18 Addition of free nucleotides to a DNA strand. Free nucleotides in the cell are available as GTP, CTP, ATP, and TTP, each with three phosphates, deoxyribose, and the base. When added to a DNA strand, two of the phosphates are removed, providing the energy for the formation of the phosphodiester bond between the nucleotide and the DNA backbone.

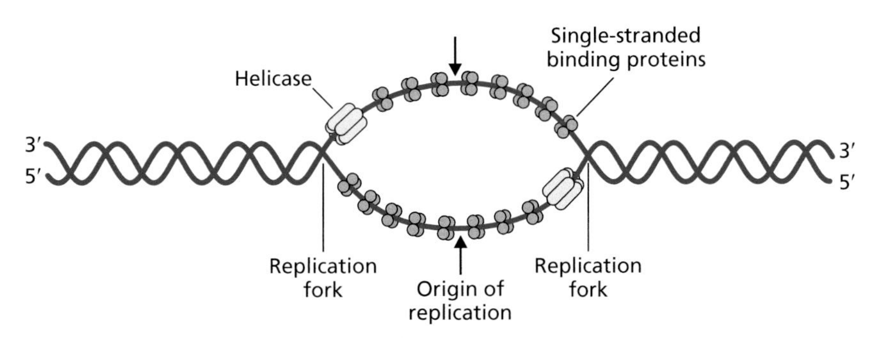

Figure 1.19 Helicase opens the DNA double helix to initiate replication at the origin of replication. The protein helicase opens the DNA double helix at the origin of replication (arrows) by breaking the hydrogen bonds between the nucleotide bases. It progresses toward the replication fork, where the DNA is still together, breaking hydrogen bonds as it goes. In its wake, single-stranded binding proteins coat the single-stranded DNA to prevent the hydrogen bonds reforming.

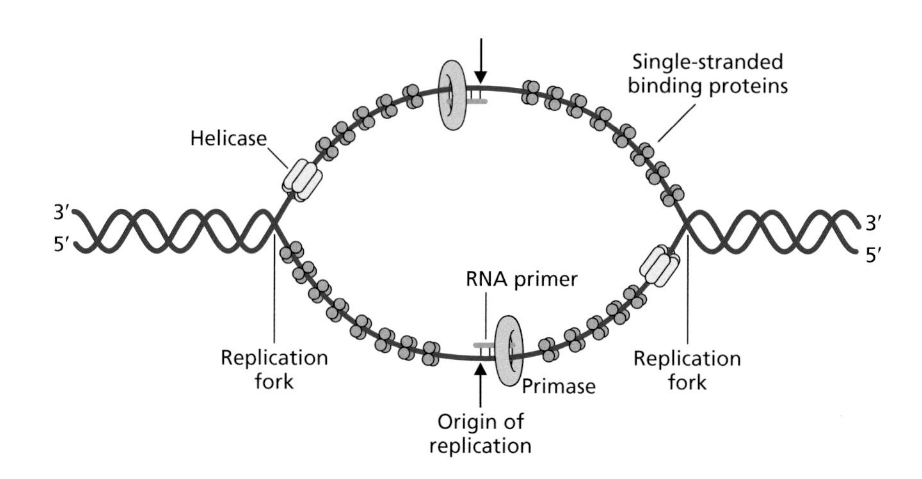

Figure 1.20 Primase incorporates ribonucleic acids to generate a short RNA primer. The enzyme primase initiates the synthesis of the new DNA strands by incorporating ribonucleotides, making a short RNA-based primer. These are required because DNA polymerase III can only add on to an established strand, not create a new one, like primase.

DNA polymerase can only add bases in the 5′ to 3′ direction

The new DNA strand is built 5′ to 3′ through the action of DNA polymerase III, a specialist polymerase for generation of DNA copies. It cannot function in the reverse direction, 3′ to 5′. This process works fine for one strand, but not for the other, which is in the opposite orientation. On one strand, replication can progress smoothly toward the **replication fork**, 5′ to 3′, progressing from the point where the original two DNA strands come apart to be copied to where the original strands are still hydrogen bonded to one another. However, on the opposite DNA strand, replication must progress in short bursts that move away from the replication fork, in the 5′ to 3′ direction (**Figure 1.21**).

On the **leading strand**, going toward the replication fork, primase needs to initiate the copying of the DNA strand only once through the creation of a single RNA oligonucleotide stretch. On the **lagging strand**, however, RNA primers are made repeatedly, as are the short sections of DNA that connect to them. This enables DNA polymerase to add new nucleotides in a 5′ to 3′ direction, even though the replication fork is moving in the opposite direction. Lagging strand fragments are about 1,000 bases of DNA until it comes to the next primer. These fragments are called **Okazaki fragments**. The RNA primer is removed by a specialized DNA polymerase called DNA polymerase I, replacing the RNA bases with DNA bases. The gap left behind is filled by DNA polymerase I, which now uses the Okazaki fragment as a primer.

The final bond between these extended fragments, now completely made of DNA, is joined by an enzyme called **ligase**. Ligase covalently joins the 3′ OH on a 3′ nucleotide to the 5′ phosphate adjacent. Although DNA polymerase can add new nucleotides to the growing DNA strand, it cannot join two backbones of nucleotides that have been incorporated at either end of a fragment. This final bond must be made by the separate ligase enzyme.

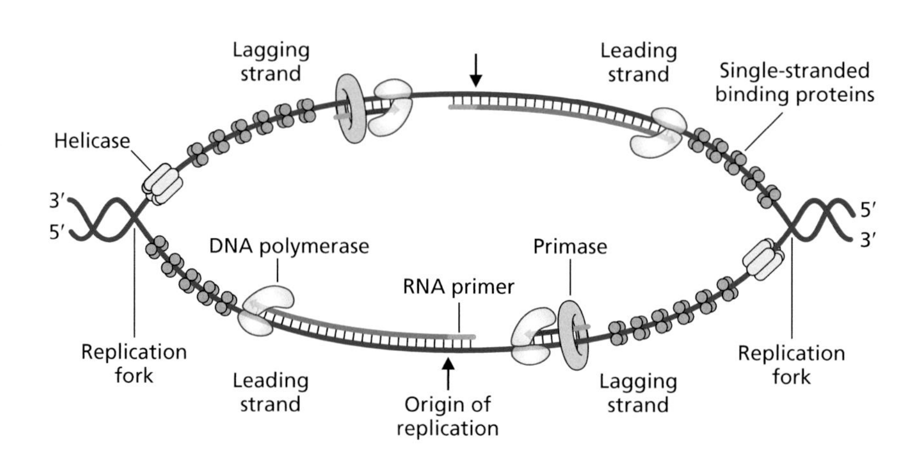

Figure 1.21 Deoxyribonucleotides are added in replication by DNA polymerase III. The RNA primers are extended by DNA polymerase III, incorporating complementary deoxyribonucleotides in the 5′ to 3′ direction along the DNA. On the leading strand, which heads toward the replication fork, DNA is replicated continuously. On the lagging strand, DNA is replicated in short fragments that head away from the replication fork.

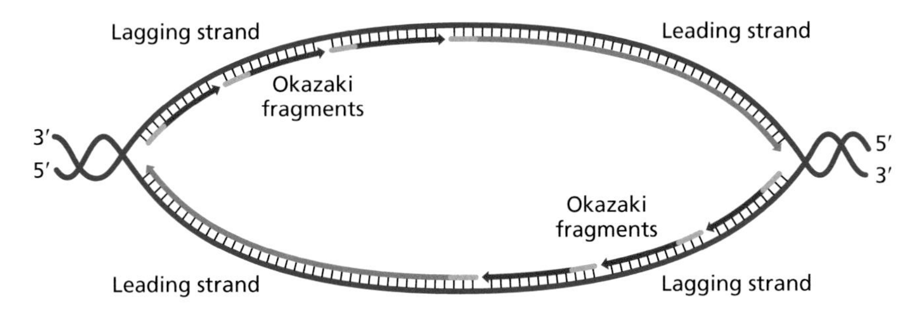

Figure 1.22 The lagging strand in DNA replication is made of Okazaki fragments. While the DNA polymerase progressing toward the replication fork makes a continuous leading strand, the lagging strand is made in short pieces. These Okazaki fragments start as part RNA primer and part DNA. Primase creates new RNA primers as the lagging strand progresses, providing the means for DNA polymerase III to add complementary DNA bases. The RNA primers are removed and replaced with DNA by DNA polymerase I. The gaps in the backbone between the Okazaki fragments are joined by the ligase enzyme.

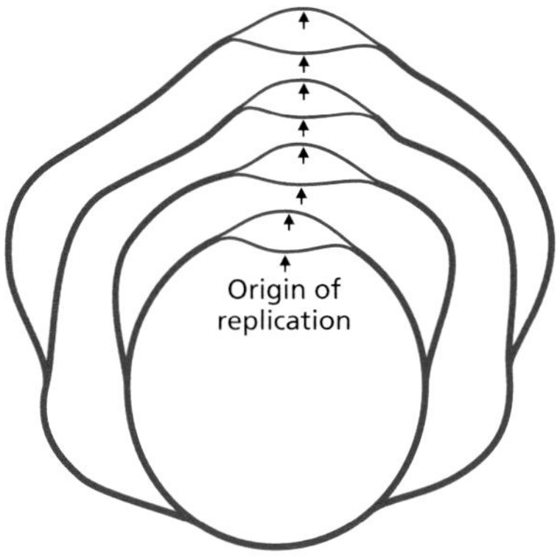

In bacteria, two replication forks are made at the origin of replication, one traveling in each direction around the circular chromosome (**Figure 1.22**). Therefore, one newly formed strand is the leading strand going toward one replication fork and is also simultaneously the lagging strand going toward the other replication fork. As the new strands are synthesized, the replication forks move around the circular chromosome. This process creates stress in the structure of the chromosome. The double helix structure of the DNA must be untwisted to allow the fork to continue to progress. This is accomplished by a topoisomerase II, known as **DNA gyrase**.

Once the replication forks have moved away from the origin of replication, another round of replication can begin. This means that it is not necessary to wait until the complete chromosome is replicated before another replication cycle is begun. Therefore, the DNA at the origin of replication is often present in more than one copy, indeed multiple copies, which may influence the expression and function of any associated genes (**Figure 1.23**).

Figure 1.23 Progression of the two replication forks around the bacterial chromosome. At the start of replication of most circular bacterial chromosomes, two replication forks are created. As replication progresses, these move away from each other and around the circle of DNA. Another round of replication can begin once the first has moved away from the origin of replication, meaning that there can be several copies of the origin of replication and the nearby genes in the cell at any one time.

Bacterial DNA can occur in several different forms

In general, bacteria have one circular chromosome that replicates bidirectionally. The **chromosome** therefore contains all of the genetic material for the bacteria to survive and reproduce. During DNA replication it is this chromosome that is copied, so that the genetic traits can be passed to both daughter cells during bacterial cellular division. For the majority of bacteria, the majority of essential genetic features are contained in this one circular chromosome (**Figure 1.24**). There are exceptions to this description, of course, and it should not be assumed that all bacteria are the same. The discovery of bacteria with multiple chromosomes, linear chromosomes, and essential genes outside of what is believed to be the chromosome has caused some review and reassessment of the genetic material within some species of bacteria.

All of the genetic material within the bacterial cell contained on the chromosomes is considered to be the **genome**. Regardless of the number of chromosomes involved, the genetic code contained on them is collectively known as the genome. Any accessory DNA that is extra-chromosomal, not on the chromosome, is considered by many as not part of the genome because extra-chromosomal elements can be easily lost or gained by the bacteria. This genetic material is generally on **plasmids**.

Plasmids are small circular extra-chromosomal DNA molecules found in some bacterial cells. In general, plasmids do not contain essential genes, but may carry genes that could be beneficial to the bacterial cell. These extra-chromosomal pieces of DNA replicate independently and segregate themselves into the daughter cells during division independently of the chromosome. Plasmids

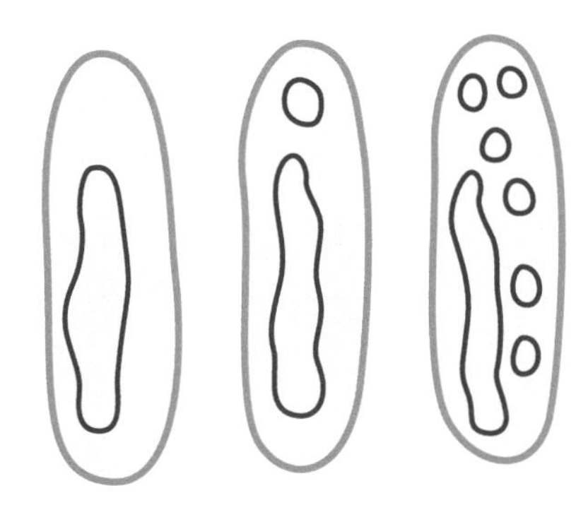

Figure 1.24 The DNA within a bacterial cell. Most bacterial cells contain a circular chromosome. This is illustrated here as a large circle within the cell (left). Actually, the chromosome is much larger than the cell itself and is coiled up tightly within the cell, so what is shown here is representative of a chromosome and not to scale with the rest of the cell. In addition, the bacterial cell may contain one (center) or more plasmids (right) that are present in single or multiple copies.

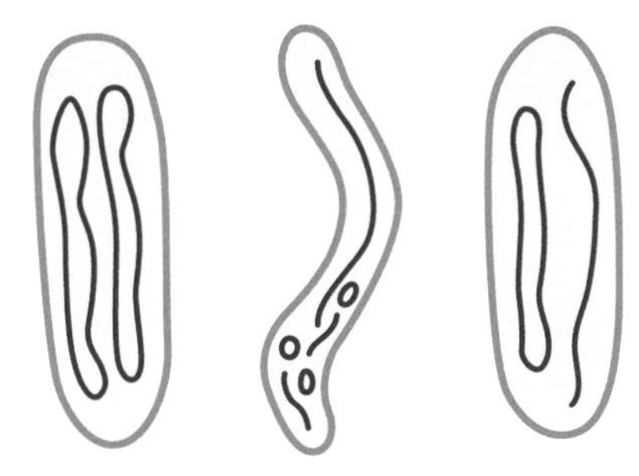

Figure 1.25 Different chromosomal arrangements in bacterial cells. In *Rhodobacter sphaeroides* and other bacteria, two circular chromosomes are present in the cell (left). These are here illustrated as large circles within the bacterial cell; however, the chromosomes are actually much larger than the cell and are tightly wound to fit within the cell. *Borrelia burgdorferi* contains a linear chromosome, as well as several linear and circular plasmids (center). In *Agrobacterium tumefaciens*, both a linear chromosome and a circular chromosome are present (right).

can be of varying sizes, but are generally much smaller than the chromosome. Plasmids can contribute genes to the bacterial repertoire that are in addition to those found in the genome.

As with all "rules" in biology, there are, of course, exceptions, where it has been found that essential genes are present on plasmids, where bacteria have more than one chromosome, and where bacteria contain linear chromosomes (**Figure 1.25**). Two circular chromosomes have been found in *Rhodobacter sphaeroides*, *Brucella* spp., and *Vibrio* spp. *Paracoccus denitrificans* has three circular chromosomes. *Borrelia burgdorferi* was the first bacteria discovered to contain a linear chromosome, and *Streptomyces* spp. also have a linear chromosome. There is one linear chromosome and one circular chromosome in *Agrobacterium tumefaciens*. Therefore, caution should be taken when assessing the DNA within a bacterial cell and making assumptions about the chromosomes, plasmids, and significance of the genetic material.

Additional genetic material that can be found within bacterial cells can come from bacteriophages. Bacteriophages are small, virus-like particles that infect bacterial cells. These can either kill the bacteria outright by causing lysis in the **lytic phase** or incorporate their DNA into the genetic material of the bacteria, entering a **lysogenic phase.** When bacteriophage genomes, or fragments thereof, are integrated into bacterial genomes, the bacteriophage genome is then referred to as a **prophage**.

Prophage genomes can remain dormant in bacterial genomes for many generations. It is possible that these prophage genomes will accumulate mutations that prevent the bacteriophage from activating. Inactive prophage sequences mean that no new bacteriophage particles are made and lytic phase is never achieved. The genetic material from the bacteriophage remains in the bacterial genome as a prophage, where the genes may contribute to the biology of the bacteria, or they may not.

After replication, the two chromosomes produced need to be segregated into what will become the new daughter cells. Each daughter cell must have one complete genome to be viable. It is therefore necessary for the bacterial cell to ensure that the chromosomes are moved into the two halves of the cell before division is complete. To accomplish this, partitioning proteins, which have different names in different bacterial species, bind to the chromosome near the origin of replication. These proteins then aid in ensuring that the chromosomes are segregated into the daughter cells. In some species this is accomplished through association with the poles of the dividing cells and in others at locations near the division.

The replication of plasmids is independent of the replication of the chromosome(s)

The processes involved in copying the DNA are similar; however, the replication of plasmids is a separate, independent event. An origin of replication, just as is seen in the chromosome, is present in the plasmid and from here replication around the circle is initiated. At any one time, one or more complete copies of a plasmid may be present in a bacterial cell. The copy number of a plasmid is dictated by the plasmid itself. Due to their independent replication outside the chromosome, their potential to vary in copy number, and their relative ease of isolation and manipulation in the laboratory due to their size, plasmids have been exploited by researchers who use molecular biology techniques to better understand biology and to make it work to achieve desired outcomes.

It should be remembered when studying the genetics and genomics of bacteria that the genome is often in one chromosome, but not necessarily. Likewise, it should be remembered that the chromosome is often circular, although not always. Therefore, when assessing a DNA code, the circle has been broken and the first base of the code was actually preceded by the last base of the genome sequence.

Key points

- DNA and RNA are nucleic acids, made up of sugars, phosphates, and nitrogenous bases.
- RNA has ribose sugars, phosphates, and adenine, cytosine, guanine, and uracil bases.
- DNA has deoxyribose sugars, phosphates, and adenine, cytosine, guanine, and thymine bases.
- DNA forms a double helix with a backbone joined by phosphodiester bonds and a core joined by hydrogen bonds between bases on opposite strands.
- There are two hydrogen bonds between adenine and thymine and three hydrogen bonds between guanine and cytosine.
- Free nucleotides, not part of DNA or RNA, have roles in bacterial cells for energy storage, for signaling, and as enzyme cofactors.
- DNA was discovered in 1869 by Friedrich Miescher and 75 years later, in 1944, DNA was shown to be the genetic material by Avery, McLeod, and McCarty.
- Copying of DNA in replication is semi-conservative, meaning one strand is from the previous DNA and one strand is new.
- Generally, bacteria have one circular chromosome, although some may have more, and some may have linear chromosomes.
- In some bacteria, there is extra-chromosomal DNA (plasmids), which can carry additional genetic information.

Key terms

Define the following terms introduced in this chapter. Check your answers using the definitions in the Glossary. These terms are also available as Flashcards online.

Adenine	Deoxyribonucleic acid	Leading strand
Adenosine triphosphate	Deoxyribose	Ligase
Bacteriophage	Dinucleotide	Lysogenic phase
Chromosome	Genome	Lytic phase
Cytosine	Guanine	Nitrogenous base
DNA gyrase	Hydrogen bond	Nucleic acid
DNA polymerase	Lagging strand	Nucleoside base
Nucleotide base	Primase	Replication fork
Okazaki fragment	Prophage	Ribonucleic acid
Oligonucleotide	Purine	Ribose
Origin of replication	Pyrimidine	Thymine
Plasmid	Replication	Uracil

Questions and discussion topics

Self-study questions

Answer each question using 50–100 words or a table or labeled diagram. Advice on where to find answers to these questions is available online.

1. In the RNA World hypothesis, RNA came first, later followed by DNA. Explain the main structural differences between RNA and DNA.
2. Draw the chemical structure of a segment of RNA, including the linkages of the backbone, the structure of the ribose, and at least one base.
3. What are the names of the four nucleotides found in RNA?
4. What is the name of the force that holds together the two strands of the DNA double helix and how does it vary along the length of the DNA?
5. When and by whom was DNA first discovered?
6. What did Griffith's experiment with *Streptococcus pneumoniae* show?
7. How did Avery, McCarty, and MacLeod build on Griffith's findings?
8. The antiparallel strands of DNA are labeled 5′ or 3′. Explain this designation.
9. When a dinucleotide forms, water is released. Draw the chemical structures involved in this process.
10. Draw the key features involved at the initiation of replication of DNA.
11. Replication requires the introduction of short segments of RNA, so that DNA polymerase can add deoxyribonucleotide bases. What protein is involved, what term is used to describe the RNA, and how frequently is the RNA made in replication?
12. Draw the progression of replication from the origin of replication toward the replication forks, showing the leading and lagging strands.

Discussion topics

These topics are presented for discussion in study groups, as part of class discussions, or on your own. These questions go beyond what is directly covered in this part of the book. Use the research literature and other reading to explore these topics in more depth. Tips to help prepare for topic discussions are available online.

1. The Last Universal Common Ancestor, known also as LUCA, was a lifeform present before the split in evolutionary lineages into bacteria, eukaryotes, and archaea. What common features still exist in the RNA and DNA processes between these three distinct lineages that support a common ancestry?
2. There is a long period of history between the isolation of DNA from pus on bandages to the demonstration that it is the genetic material in *S. pneumoniae* experiments. During that time, it was believed that proteins were the genetic material; DNA was too simplistic. Discuss how overcoming long-standing assumptions such as this have contributed to experimental design and theuse of positive and negative controls today.
3. When thinking of replication, it is often easy to think of it as happening once around a circular chromosome. Discuss the ways in which this is not the way replication occurs in a typical bacterial cell, keeping in mind all of the various kinds of DNA that can be present in a bacterial cell and the variety of DNA in different bacterial species.

Online quiz questions

To further self-assess your understanding of the chapter material, please visit the following link, where you can participate in a range of interactive quiz questions:

www.routledge.com/cw/snyder

Further reading

Life originated from RNA with DNA evolving later

Bowman JC, Hud NV, Williams LD. The ribosome challenge to the RNA world. *J Mol Evol.* 2015; *80*: 143–161.

Orgel LE, Crick FHC. Anticipating an RNA world some past speculations on the origin of life: Where are they today? *FASEB J.* 1993; *7(1)*: 238–239.

Poole AM. Getting from an RNA world to modern cells just got a little easier. *BioEssays.* 2006; *28*: 105–108.

Poole AM, Horinouchi N, Catchpole RJ, Si D, Hibi M, Tanaka K, Ogawa J. The case for an early biological origin of DNA. *J Mol Evol.* 2014; *79*: 204–212.

Saito H. The RNA world "hypothesis." *Nat Rev Mol Cell Biol.* 2022; *23(9)*: 582.

Yarus M. Getting past the RNAWorld: The initial Darwinian ancestor. *Cold Spring Harb Perspect Biol.* 2011; *3*: a003590.

Nucleic acids are made of nucleoside bases attached to a phosphate sugar backbone

Pinheiro VB, Taylor AI, Cozens C, Abramov M, Renders M, Zhang S. Synthetic genetic polymers capable of heredity and evolution. *Science.* 2012; *336(6079)*: 341–344.

Watson JD, Crick FH. Genetical implications of the structure of deoxyribonucleic acid. *Nature.* 1953; *171(4361)*: 964–967.

DNA was discovered in 1869 and identified as the genetic material 75 years later

Avery OT, MacLeod CM, McCarty M. Studies of the chemical nature of the substance inducing transformation of pneumococcal types. Induction of transformation by a desoxyribonucleic acid fraction isolated from Pneumococcus Type III. *J Exp Med.* 1944; *79*: 137–158.

Dahm R. Discovering DNA: Friedrich Miescher and the early years of nucleic acid research. *Hum Genet.* 2008; *122*: 565–581.

Levene PA. The structure of yeast nucleic acid. IV. Ammonia hydrolysis. *J Biol Chem.* 1919; *40*: 415–424.

Miescher F. Ueber die chemische Zusammensetzung der Eiterzellen. Med.-Chem. *Unters.* 1871; *4*: 441–460.

Miko I. Gregor Mendel and the principles of inheritance. *Nat Educ.* 2008; *1(1)*: 134.

O'Connor C. Isolating hereditary material: Frederick Griffith, Oswald Avery, Alfred Hershey, and Martha Chase. *Nat Educ.* 2008; *1(1)*: 105.

Wolf G. Friedrich Miescher: The man who discovered DNA. *Chem Herit.* 2003; *21*: 10–11, 37–41.

The first X-ray images of DNA were taken in 1937 with the structure finally solved in 1953

Chargaff E. Chemical specificity of nucleic acids and mechanism of their enzymatic degradation. *Experientia.* 1950; *6*: 201–209.

Hershey AD, Chase M. Independent functions of viral protein and nucleic acid in growth of bacteriophage. *J Gen Physiol.* 1952; *36(1)*: 39–56.

Pray LA. Discovery of DNA structure and function: Watson and Crick. *Nat Educ.* 2008; *1(1)*: 10.

Watson JD, Crick FHC. A structure for deoxyribose nucleic acid. *Nature.* 1953; *171*: 737–738.

DNA is copied semiconservatively every time a cell divides

Hajduk IV, Rodrigues CD, Harry EJ. Connecting the dots of the bacterial cell cycle: Coordinating chromosome replication and segregation with cell division. *Semin Cell Dev Biol.* 2016; *53*: 2–9.

O'Donnell M, Langston L, Stillman B. Principles and concepts of DNA replication in bacteria, archaea, and eukarya. *Cold Spring Harb Perspect Biol.* 2013; *5(7)*. pii: a010108. doi: 10.1101/cshperspect.a010108.

Bacterial DNA can occur in several different forms

Algora-Gallardo L, Schniete JK, Mark DR, Hunter IS, Herron PR. Bilateral symmetry of linear streptomycete chromosomes. *Microb Genom.* 2021; *7(11)*: 000692.

Baril C, Richaud C, Baranton G, Saint-Girons I. Linear chromosome of *Borrelia burgdorferi*. *Res Microbiol.* 1989; *140*: 507–516.

Hinnebusch J, Tilly K. Linear plasmids and chromosomes in bacteria. *Mol Microbiol.* 1993; *10(5)*: 917–922.

Suwanto A, Kaplan S. Physical and genetic mapping of the Rhodobacter sphaeroides 2.4.1 Genome: Presence of two unique circular chromosomes. *J Bacteriol.* 1989; *171(11)*: 5850–5859.

Toro E, Shapiro L. Bacterial chromosome organization and segregation. *Cold Spring Harb Perspect Biol.* 2010; *2(2)*: a000349.

Chapter

2

Genes

This chapter will describe the genetic features within DNA that are called genes. When most people think of DNA, they tend to think of the genes that it carries. These are the features that in human beings determine genetic traits like eye color, hair color, and blood type. In bacteria, DNA also contains genes: portions of the double helix that determine bacterial genetic traits like shape and size. Not every nucleotide within DNA is part of a gene, but genes are abundant and major features, responsible for the proteins within the cell. In this chapter, we will explore terms and concepts that relate to genes, including their purpose and how they are expressed. It is important to understand what a gene is, how it is defined, how it can be recognized in a DNA sequence, how it can be interpreted, and the ways in which its expression can be modulated, because genes are often a major focus in the study of bacterial genetics and genomics.

Genes are features in the DNA that encode proteins

DNA, as we have learned in Chapter 1, is a string of nucleotides in a double helix structure, where the two strands are complementary to one another. The nucleotides are present in the DNA in a specific order, which is referred to as the **sequence**. The sequence of one strand of the DNA double helix is complementary to the sequence on the opposite strand. These sequences of DNA nucleotides, As, Ts, Gs, and Cs, are interpreted by the cell for various functions. The most widely studied function of the DNA sequence within living cells is in its role as **genes**.

Genes are the units of heredity. They are what constituted the transforming principle in the early studies by Griffith, Avery, McCarty, and MacLeod. Genes carry the features and traits that we see in humans, plants, and other organisms on the planet. Genes accomplish their function as units of heredity by carrying instructions on how to make **proteins**. Through the specific sequence of nucleotides in the DNA, a gene encodes the information necessary to make one or more proteins. As we will see later, proteins are important for the structure, maintenance, metabolism, and other processes of the bacterial cell.

The word gene comes from the Ancient Greek *genos*, and it was first used by Wilhelm Johannsen at the beginning of the twentieth century. The related term genetics was used a few years earlier by William Bateson, taking the term from the ancient Greek *genetikos*. Johannsen also applied specific terms to the study of the outward appearance of the organism and the genetics of the organism. We use the terms **genotype** and **phenotype** to differentiate these concepts.

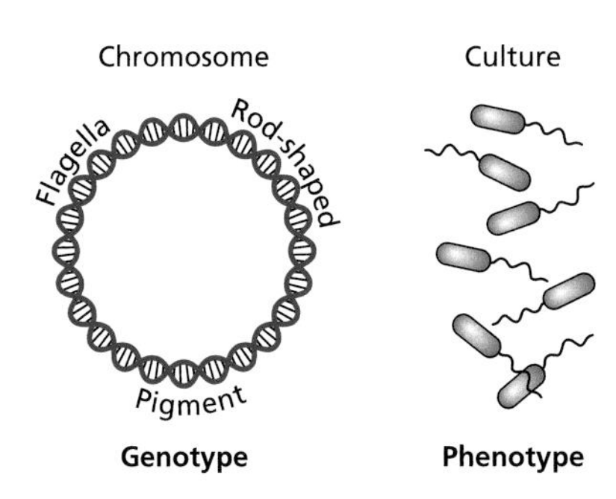

Figure 2.1 The genotype and phenotype of bacteria. The genotype is carried in the bacterial DNA. It is all of the traits that are possible for the organism. For example, the DNA may code for flagella, rod-shaped cells, and production of pigment. The features that are currently displayed by the bacterial cell are the phenotype. For example, these bacteria in culture have flagella, are rod-shaped, and are pigmented.

An individual, specific DNA sequence is the **genotype**, encoding the potential to express a trait (**Figure 2.1**). The genotype of a bacterial cell is the potential to display various traits encoded in the DNA. The genetic make-up of the bacteria determines the different traits and characteristics that the cell is potentially able to display and express. In general, the genotype largely remains the same, even as the displayed features of the bacteria may change as needed for optimal growth and survival. Changes to the DNA can arise, often through mutation, which can alter the genotype and therefore impact upon the potential features that can be expressed.

DOI: 10.1201/9781003380436-3

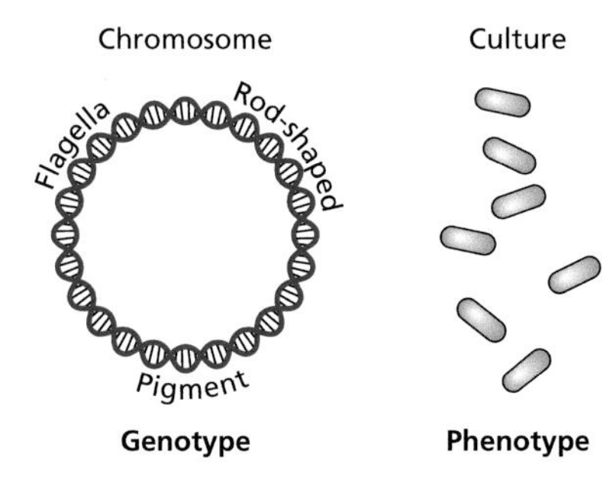

Figure 2.2 The genotype does not always match the phenotype. It may be that a bacterial cell possesses the genetics for a trait, but it is not expressing it. The genotype may not be the same as the phenotype. For example, the chromosome may have the genes for flagella and the production of pigment, but these may not actually be present in the culture of bacteria.

The **phenotype** is the expression of the genetic traits in the genotype (see Figure 2.1). It is different from the genotype in that it describes the features that are actually expressed. It is not the case that all genes are being used to make the proteins they encode at all times; therefore, phenotype reflects only those parts of the DNA sequence that are displaying the features it encodes. Phenotypes may change in response to the environment, nutrient availability, presence of antimicrobials, and other factors. The phenotype is dictated by the potential of the genetic traits encoded in the genotype, but it does not necessarily reveal the precise nature of the genes responsible for the genotype.

As an example of the difference between genotype and phenotype, consider the expression of bacterial flagella. Some bacterial cells possess one or more appendages that enable them to be motile; these are called flagella. Through the spinning action of the flagella, bacteria are able to move through their environment. To be able to express flagella and use them for motility, the bacteria must have the genes for flagella. There are many genes involved and these encode a variety of proteins that must all be displayed in order for the bacterial cell to successfully express flagella. But, having the genes is not enough. The flagella proteins, flagella assembly proteins, and accessory proteins to create functional flagella must all be present. Having the genes does not necessarily ensure the flagella are made. In some instances, bacteria may find it beneficial to not express flagella, while in other instances motility using flagella is useful. Therefore, the bacteria could contain the genetics for making flagella, but also have mechanisms that could prevent expression of the flagella genes (**Figure 2.2**).

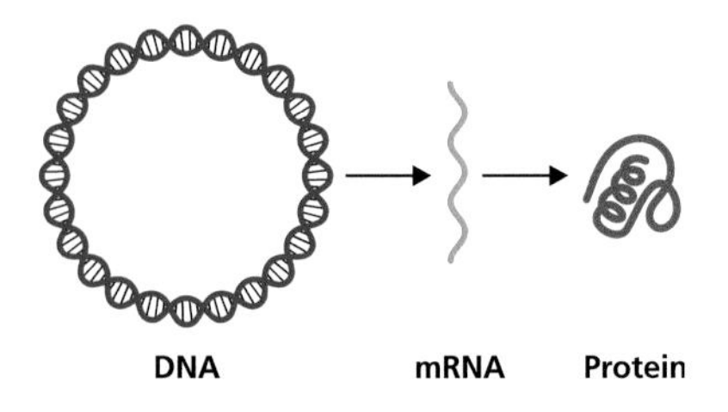

Figure 2.3 DNA is the template for the production of mRNA, which is the template for the production of protein. DNA is used to make messenger RNA during transcription. This mRNA is then used to make proteins. Thus, the genetic information from the DNA encodes the information to make proteins.

Bacterial transcription generates RNA based on the DNA sequence

Although DNA is the code of life, the genotype is not expressed as a phenotype directly from the DNA sequence. The genetic information contained in the DNA must pass through RNA before it can be decoded into protein. To accomplish this, the DNA must be used as a template to produce the needed segment as RNA.

To make a protein, a specific type of RNA, **messenger RNA** or **mRNA**, is needed. A process called **transcription** makes mRNA based on the nucleotide sequence within DNA. The sequence is copied from the DNA to the mRNA intermediate, which is then used to direct protein synthesis (**Figure 2.3**). Transcription occurs in three phases: **initiation**, **elongation**, and **termination**.

Initiation of transcription

Transcription of the DNA sequence of a gene into mRNA is initiated at the **start point of transcription**, a point at a specific base that sits before a gene, namely the 5′ end of the coding sequence of the gene, also referred to as **upstream**, meaning before the gene. This specific base is designated +1. Based on the +1 start point of transcription, the **downstream** bases, going toward the gene in the 3′ direction, are positively numbered. The bases upstream of the start point of transcription are negatively numbered (**Figure 2.4**). This numbering scheme becomes important when discussing the features that come before the start point of transcription and which drive the initiation of the transcription process. The terms upstream and downstream are relative to the gene. One gene's upstream can be another gene's downstream and can be a source of confusion; therefore, generally this book will use 5′ and 3′ indications of directionality, rather than upstream and downstream.

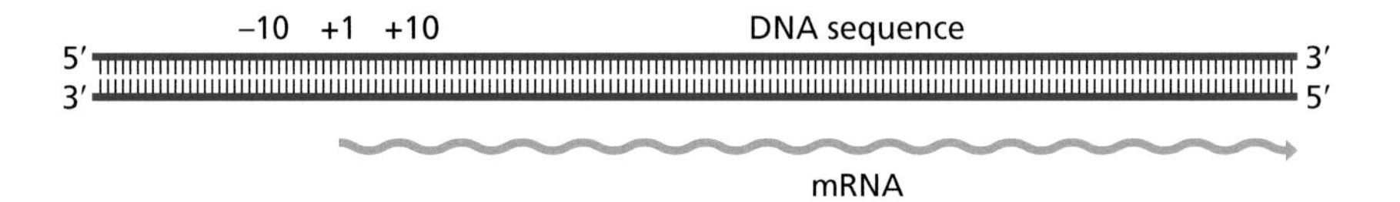

Figure 2.4 The start point of transcription is before the start of the gene. When identifying features in the DNA, the point at which transcription starts, generating mRNA, is labeled as +1. Those bases before it are negatively numbered and those after are positively numbered. This allows for the recognition of other features in the DNA.

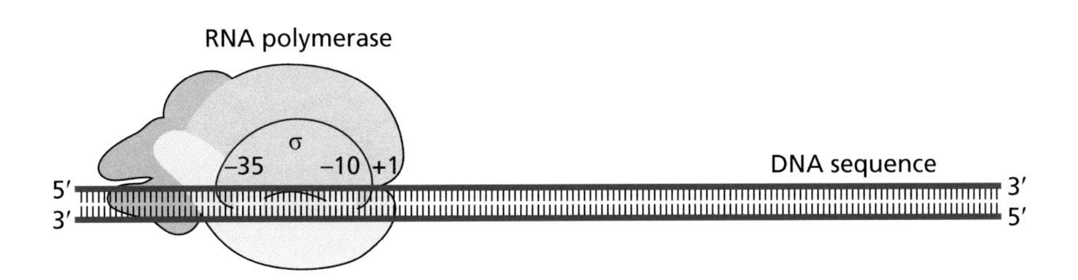

Figure 2.5 RNA polymerase binds to DNA at the −10 and −35 regions. To initiate transcription, RNA polymerase binds to the DNA before a gene at two sites where the sequence is recognized by the RNA polymerase protein. These are the −10 and −35 regions, identified as being before the +1 start point of transcription. It is the σ subunit of RNA polymerase that recognizes and binds to the −10 and −35 sequences.

The main enzyme involved in the process of transcription is **RNA polymerase**. Bacteria have one type of RNA polymerase, which produces mRNA and most other types of RNA in the cell. The whole of RNA polymerase is referred to as a **holoenzyme**, which is the RNA polymerase including the **sigma subunit** (σ). Although variations in the subunits of the RNA polymerase can alter the recognition of the sequences required for transcriptional initiation, most genes are transcribed by RNA polymerase with a σ^{70} (sigma 70) subunit.

The sigma subunit of RNA polymerase recognizes the sequences at the −10 and −35 positions that are 5′ of the gene. This is the **promoter region** (**Figure 2.5**). The σ^{70} subunit is the type of σ subunit associated with most genes. It directs RNA polymerase to the specific sequence 5′ of genes, so that those gene sequences will be transcribed and then expressed via translation. RNA polymerase holoenzyme binds to the −10 and −35 sequences in a **closed complex**, so named because the double helix is still joined by hydrogen bonds between the complementary bases.

The RNA polymerase enzyme then opens about 12–15 bases of the DNA, breaking the hydrogen bonds. This makes an **open complex**, allowing for transcription to begin (**Figure 2.6**). RNA polymerase works in the same direction and functions in the same directionality as DNA polymerase, working from the 5′ to the 3′ end of the DNA strand. Like the leading strand in replication by DNA polymerase, RNA polymerase produces a continuous new strand of nucleotides.

Transcriptional elongation

Transcription begins at the open complex with the incorporation of the first **ribonucleotide triphosphate** by RNA polymerase. This ribonucleotide triphosphate retains all three of the phosphates at the 5′ carbon. RNA polymerase progresses on to incorporate the next ribonucleotide, which this time loses its two phosphates in the process of integration (**Figure 2.7**). Thus, the growing strand of mRNA is made based on the template DNA, through hydrogen bonding of the ribonucleotides to the open complex DNA bases (**Figure 2.8**).

After about 10 bases are transcribed, the σ^{70} subunit dissociates from the rest of the RNA polymerase enzyme. The σ^{70} subunit is no longer needed for mRNA extension, to build the complete mRNA sequence; it is only required for the initiation of transcription. The core RNA polymerase enzyme remains associated with the DNA until the completion of transcription (**Figure 2.9**).

As transcription continues in the elongation phase, the RNA polymerase core enzyme progresses along the DNA template, unwinding the double helix as it goes. In the wake of transcription, the complementary base pairs in the DNA once again hydrogen bond to one another and the double helix structure reforms.

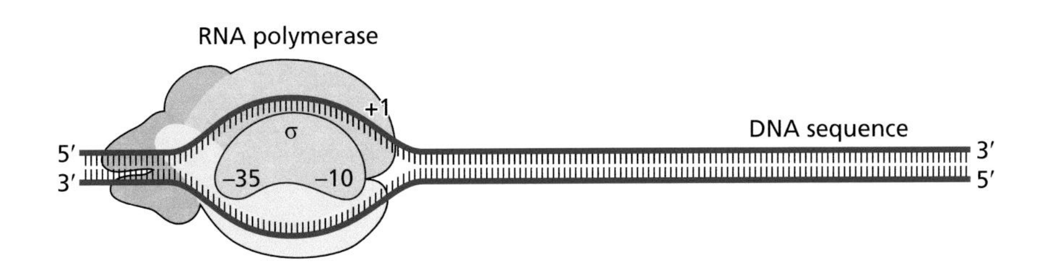

Figure 2.6 The RNA polymerase forms an open complex with the DNA. At the −10 and −35 regions, RNA polymerase binds to the DNA. The RNA polymerase then breaks the hydrogen bonds between the DNA strands and opens the DNA double helix. This open complex gives the RNA polymerase access to the nucleotide bases of the DNA and allows the initiation of transcription.

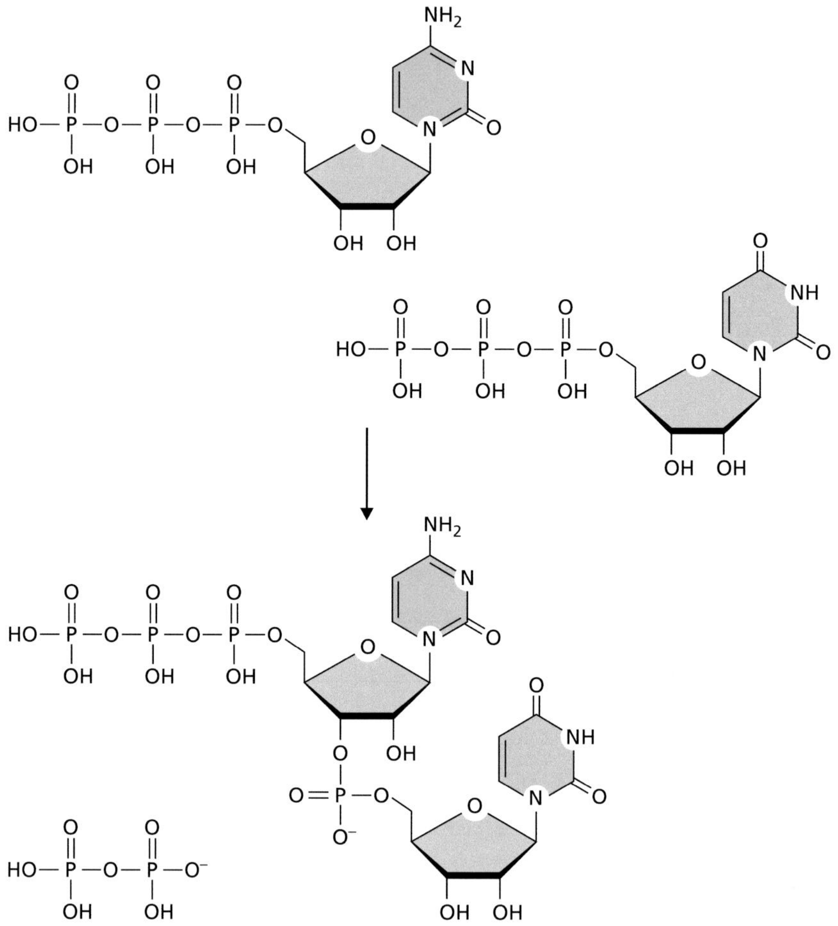

Figure 2.7 The structure of the growing mRNA strand. As transcription occurs, mRNA is generated. The first ribonucleotide base of the mRNA strand retains the three phosphates attached to the ribose. When the next base is added, two of the phosphates on the second ribonucleotide are lost as the phosphodiester bond is created.

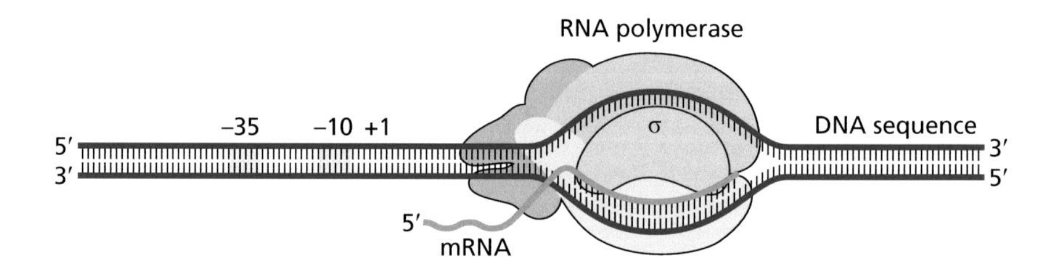

Figure 2.8 RNA polymerase progresses along the DNA, incorporating bases into the mRNA. Once the DNA double helix is in an open complex formation, the RNA polymerase can start to generate mRNA based on the DNA sequence. From here, RNA polymerase progresses along the DNA, adding ribonucleotide bases to the mRNA.

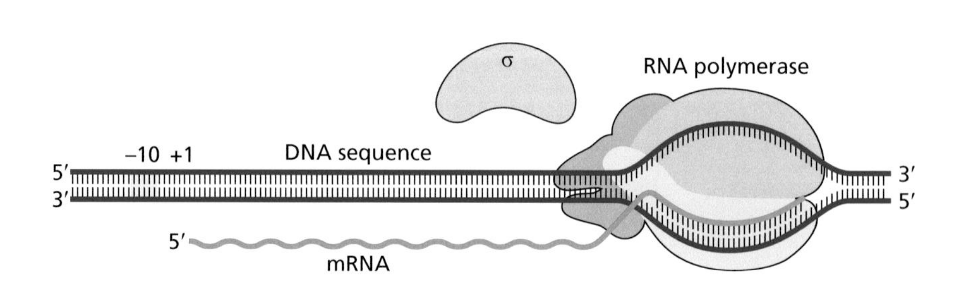

Figure 2.9 As elongation progresses, the sigma subunit of RNA polymerase disassociates. The sigma subunit of RNA polymerase, the σ factor, is only needed for the initiation of transcription. The sigma subunit is released from the RNA polymerase holoenzyme once 10 or so bases have been transcribed into mRNA. The RNA polymerase core enzyme then continues the process of elongation, while the σ factor can associate with another RNA polymerase and initiate a new round of transcription.

RNA polymerase continues transcription, through one or more genes, until it encounters a transcriptional termination signal. At the end of transcription, the mRNA is released and the RNA polymerase enzyme dissociates from the DNA.

Transcriptional termination

Termination of transcription comes in two types: **intrinsic transcriptional termination** and **Rho-dependent transcriptional termination**. Intrinsic transcriptional termination is also known as Rho-independent transcriptional termination because it does not involve the protein Rho. In intrinsic transcriptional termination, there are features within the mRNA that signal RNA polymerase to stop transcription. Usually, the features in the mRNA are **palindromic** sequences, which can be read the same forward and backward such as the words kayak, level, and refer. These palindromic mRNAs form stem-loop structures followed by poly-U regions causing dissociation of the RNA polymerase from DNA (**Figure 2.10**). On the other hand, Rho-dependent transcriptional termination uses the Rho factor, a protein, to stop the progression of RNA polymerase and production of the **transcript** at specific sites. The transcript is the product of transcription: the RNA made by RNA polymerase from the DNA template. Rho binds to the mRNA strand at a Rho utilization site and then slides along the mRNA toward the RNA polymerase enzyme. If there is a stem-loop structure formed in the mRNA, this will cause the RNA polymerase to pause, giving Rho time to catch up to the RNA polymerase. The Rho protein catches up and causes dissociation of RNA polymerase and DNA (see Figure 2.10).

Bacterial translation produces proteins based on the mRNA sequence

mRNA is made through the process of transcription, as are other forms of RNA, including **transfer RNA (tRNA)** and **ribosomal RNA (rRNA)**. There is more on RNAs, including their structures, in Chapter 4. DNA is the template for these RNAs, the sequences of which are encoded on the chromosome. mRNA is the message that carries the code for making a protein. tRNA transfers the message from a DNA nucleic acid sequence to an amino acid sequence for a protein. rRNA is contained within the **ribosomes**, the core of translation activity. All three of these forms of RNA participate in bacterial **translation**, the generation of protein based on the sequence of the mRNA that has been transcribed from a gene (or genes).

Proteins are made up of sequences of **amino acids**. The mRNA carries the instructions for which amino acids are in the protein and in which order they appear in the protein chain. Initially the code for a protein is in the DNA

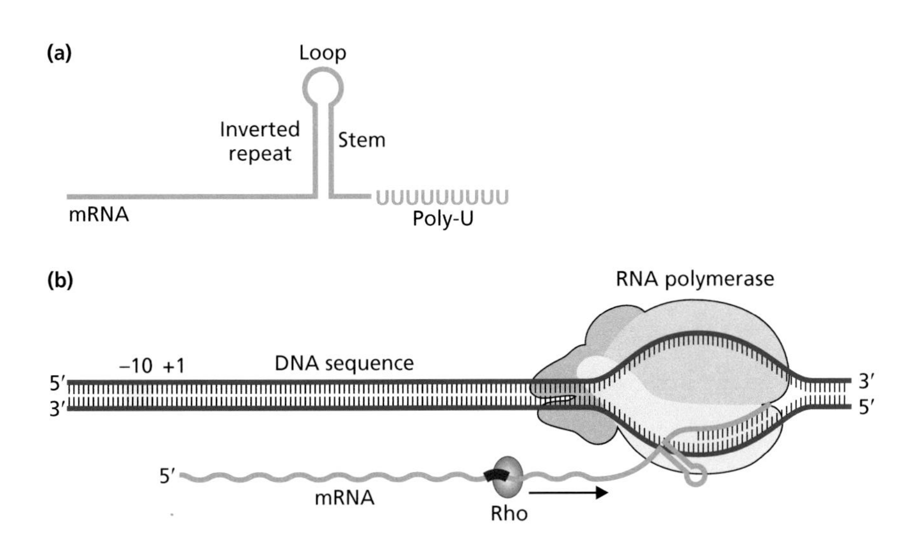

Figure 2.10 Transcriptional terminators. (a) Rho-independent transcriptional terminators generate a stem-loop structure in the mRNA, which is followed by a region of several Us. The stem-loop of mRNA forms due to complementary base pairing within an inverted repeat sequence. (b) Rho-dependent transcriptional terminators rely on the Rho protein, which is able to bind to mRNA at a specific sequence and then slide along the mRNA toward RNA polymerase. If the Rho catches up to the RNA polymerase, this causes transcription to terminate. In some cases a stem-loop in the mRNA can cause the RNA polymerase to pause giving the Rho protein time to catch up.

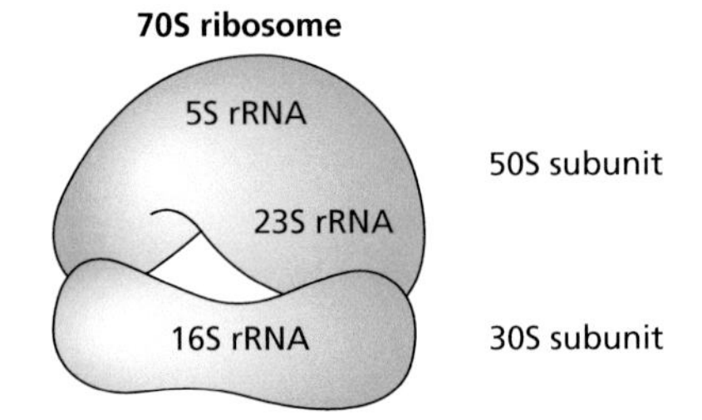

Figure 2.11 The 70S bacterial ribosome is made up of subunits of proteins and ribosomal RNAs. The larger subunit is 50S. It contains proteins and the 5S rRNA and the 23S rRNA. The smaller subunit is 30S and contains proteins and the 16S rRNA. These subunits come together during bacterial translation.

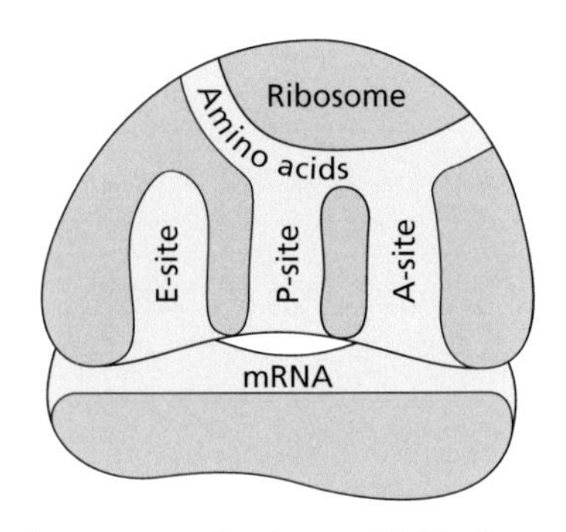

Figure 2.12 The three tRNA binding sites of a bacterial ribosome. A cross section shows that there are three locations within the 70S ribosome of bacteria where the tRNAs involved in translation can be found. The first is the A-site, where a tRNA carrying an amino acid enters. This tRNA also carries an anticodon, which matches the mRNA sequence at the A-site. The amino acid on this tRNA will then form a peptide bond to the amino acid(s) attached to the tRNA in the P-site, growing the polypeptide chain by one more amino acid. The tRNA in the E-site has released its amino acid and is ready to exit the process when the ribosome moves along the mRNA to the next codon.

as a gene. Then, it is in the mRNA as a transcript. Then, it is translated into the protein sequence. The translation of the mRNA sequence into protein involves the decoding of the nucleotides into the amino acids that they represent. Sequences of three bases, called **codons**, within the mRNA correspond to the amino acids of the protein, as well as other signals important for translation.

Ribosomes

Bacterial translation requires specialized complexes of protein and ribonucleotides, called ribosomes. These complexes are the locations of the synthesis of proteins, based on the genetic sequence carried by mRNA. The enzymatic activity of ribosomes catalyzes the formation of **peptide bonds**, which link one amino acid to another in the growing **peptide chain** as the protein is formed.

There are two parts of the ribosome, a large subunit and a small subunit. In bacteria, the large subunit is referred to as the **50S subunit** and contains proteins as well as the **5S rRNA** and the **23S rRNA**. The 50S subunit is responsible for the peptide chain formation. The small subunit in bacteria is referred to as the **30S subunit** and contains proteins as well as the **16S rRNA**. The 30S subunit is responsible for the decoding of the mRNA. Together, they are referred to as the **70S ribosome**. Two-thirds of the bacterial ribosome is made of RNA and one-third is made of protein (**Figure 2.11**).

Within the ribosome there are three binding sites for tRNAs: the **aminoacyl site (A-site)**, the **peptidyl site (P-site)**, and the **exit site (E-site)** (**Figure 2.12**).

Translation initiation

There are four phases of translation: **initiation**, **elongation**, **termination**, and **ribosome recycling**. To initiate translation, the ribosomal subunits come together on the mRNA at a location on the transcript called the **translation initiation region**. This region contains the ribosome binding site, a span of 30 bases that can occur more than once on a bacterial transcript, particularly when that transcript encodes more than one protein (**Figure 2.13**).

Initiation of translation is the most tightly regulated of the phases of translation, limited by the rate of association of the ribosome with the mRNA, a complex that can be prevented through several regulatory mechanisms. It involves the mRNA strand that will be translated, the 50S ribosomal subunit, the 30S ribosomal subunit, three **initiation factor proteins** (IF1, IF2, IF3), and the first tRNA that will be involved in the process. The initiation factor protein IF1 binds to and blocks the A-site of the 30S subunit of the ribosome. Initiation factor protein IF2, the mRNA, and fMet-$tRNA_f^{Met}$ join the 30S subunit. An fMet-$tRNA_f^{Met}$ is a specific type of tRNA that carries a methionine that can be the first amino acid of a peptide chain. A sequence on the mRNA of 5–9 bases, called the **Shine–Dalgarno sequence or ribosome binding site**, is 5′ of the initiation codon of the gene, the very first set of three nucleotide bases that signals the start of the polypeptide chain encoded by the gene. In *Escherichia coli* the ribosome binding site has a **consensus sequence** of GGAGG; generally, a sequence similar to this makes up the ribosome binding site. The Shine–Dalgarno sequence on the mRNA hybridizes to the anti-Shine–Dalgarno sequence on the 16S rRNA within the 30S ribosomal subunit. The alignment of the Shine–Dalgarno sequence, aided by the initiation factor protein IF3, puts the initiation codon in the P-site, where the first tRNA, (fMet-$tRNA_f^{Met}$), can interact with it (**Figure 2.14**).

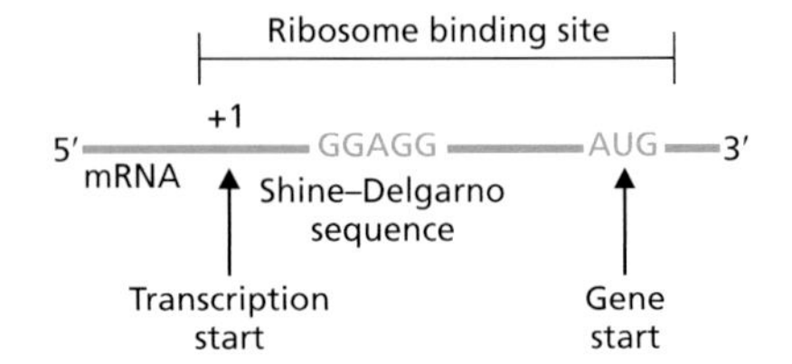

Figure 2.13 The ribosome binds to the mRNA between the start of the transcript and the start of the gene. The ribosome recognizes a sequence between the start of the mRNA transcript and the start of the encoded gene and binds to this region. Specifically, there is a consensus sequence, the Shine–Dalgarno sequence, that recruits the ribosome to begin translation.

The first tRNA used in translation is fMet-tRNA$_f^{Met}$, referred to as the initiator tRNA$_f^{Met}$. It is bound to a special methionine amino acid that has been *N*-formylated. In this way it is different from the tRNAs carrying methionine that may be used later in the process of translation. Elongator tRNA^{Met} also binds to methionine, but is involved in adding this amino acid within the growing peptide chain during the elongation phase of translation, rather than at the start during translation initiation. Each tRNA contains within its RNA sequence a specific three nucleotide sequence that is the reverse complement of codons found in the mRNA. These are **anticodons** and each corresponds to a specific amino acid that is loaded onto the tRNA (see Figure 2.14). The anticodon on the fMet-tRNA$_f^{Met}$ hybridizes to the initiation codon of the mRNA.

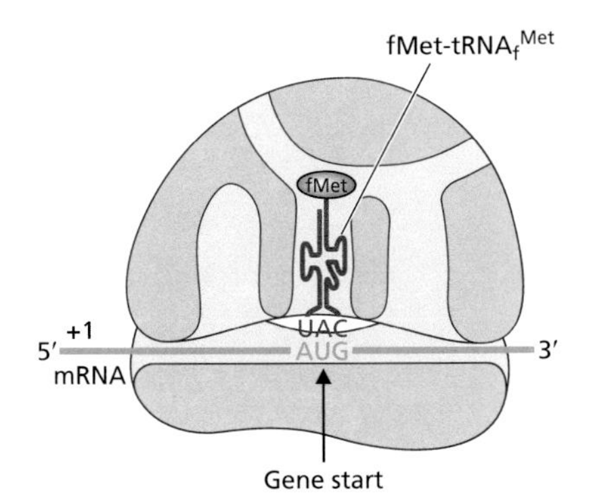

Figure 2.14 The first tRNA recruited during translation is fMet-tRNA$_f^{Met}$. A specific tRNA starts the process of translation. This is fMet-tRNA$_f^{Met}$ and it is recruited straight into the P-site of the ribosome. It contains the anticodon for the initiation codon that is at the start of the gene sequence, which is carried by the mRNA.

Translational elongation

Based on the sequence of the next codon, a tRNA carrying the corresponding anticodon and loaded with the corresponding amino acid enters the A-site to begin the elongation phase of translation. During translation elongation, the mRNA sequence serves as the template for protein production, with the ribosome sliding along the mRNA. With both the P-site and A-site loaded with tRNAs carrying amino acids, the α-amino group of the A-site tRNA attacks the carbonyl group of the P-site peptidyl group (**Figure 2.15**). This removes the amino acid from the tRNA in the P-site. Depending on how long translation has been happening, this may result in one amino acid bonding to the initial methionine or it may result in one amino acid forming a peptide bond with the growing amino acid chain. The growing peptide chain has now moved to the tRNA in the A-site. The ribosome then slides along the mRNA, resulting in the tRNA from the P-site, with no amino acids attached, moving to the E-site and the tRNA at the A-site, with the associated peptide chain, moving to the P-site. This movement of the ribosome along the mRNA results in the A-site becoming free for the next tRNA to enter. Elongation continues at a rate of approximately 20 amino acids per second, with the ribosome sliding along the codons of the mRNA and the tRNAs moving from A-site to P-site to E-site as the peptide chain grows longer. This continues until the ribosome encounters a signal to terminate translation (**Figure 2.16**).

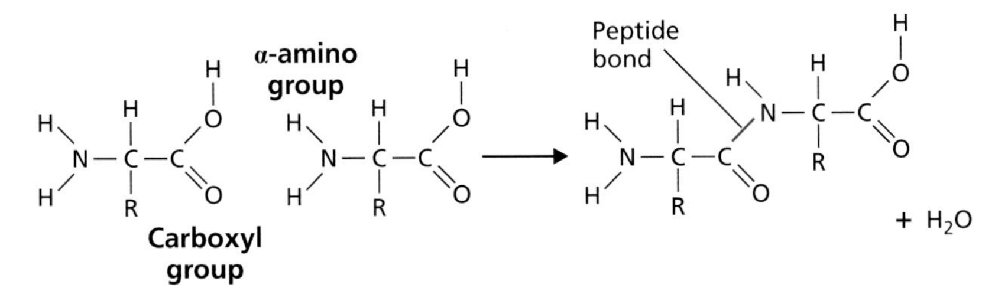

Figure 2.15 Peptide bond formation. Peptide bonds are the bonds that join two amino acids together. Peptide bonds form inside the ribosome as a new amino acid is transferred from the tRNA to the growing peptide chain. The peptide bond (red) is formed between the C of the carbonyl group of one amino acid and the N of the α-amino group of the other amino acid. The OH from the carbonyl group of one amino acid and an H from the α-amino group of the other amino acid are released as H_2O during peptide bond formation.

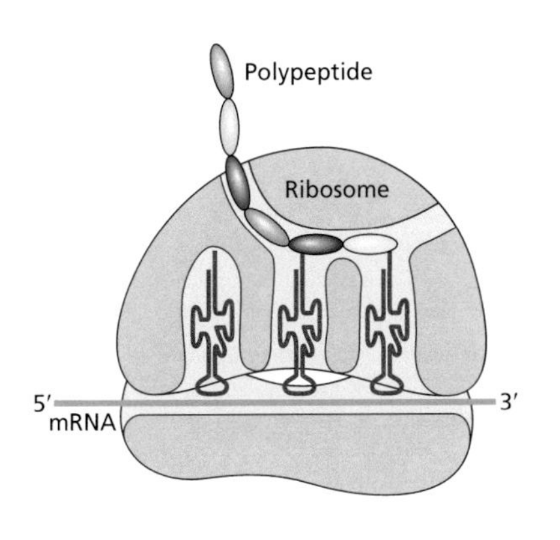

Figure 2.16 The amino acid chain grows during elongation. As the ribosome progresses during translational elongation, new codons are accessible at the A-site. A tRNA with the corresponding anticodon will enter the A-site, bringing with it the associated amino acid. Present in the P-site is the previously used tRNA, now carrying the whole of the amino acid chain that has been formed. At the E-site is the tRNA previous to the one in the A-site, which has released the amino acids to the tRNA at the P-site and is ready to exit the ribosome.

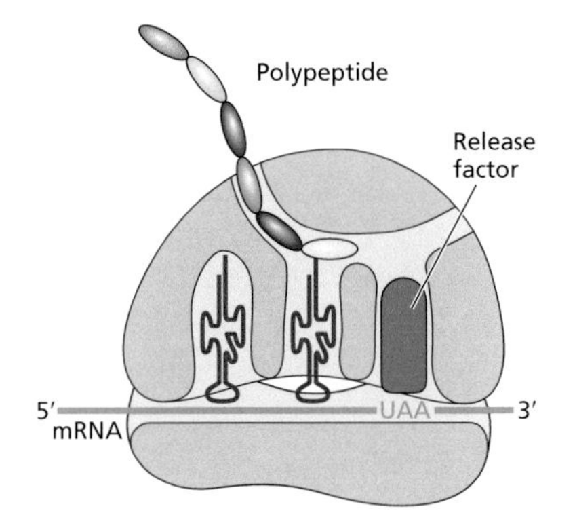

Figure 2.17 Translation ends when a termination codon is reached by the ribosome. Rather than a tRNA bringing another amino acid to add, the presence of a termination codon at the A-site signals the ribosome to stop making the polypeptide. Release factor enters the A-site and triggers the end of translation, where the polypeptide is released from the tRNA in the P-site and both the tRNAs in the P-site and E-site are released. In the final phase, the ribosome dissociates and prepares to start a new round of translation.

Translational termination and ribosome recycling

The termination phase of translation is brought about when the ribosome encounters a **termination codon**, also called a **stop codon**, which is encoded within the mRNA. Because there is no tRNA with an anticodon that corresponds to the stop codons, instead of a tRNA entering the A-site, a **release factor** binds to the A-site. The presence of the release factor in the A-site triggers the release of the newly synthesized polypeptide chain of amino acids from the ribosome (**Figure 2.17**).

Ribosome recycling is the final phase of translation and is the stage that allows the process to begin again at the next ribosome binding site. Once the peptide chain is released, the ribosomal subunits dissociate and release the mRNA strand. IF3 binds to the 30S subunit, promoting the separation of the subunits and preparing them for the new round of initiation.

Once a ribosome has moved through the initiation phase and the ribosome has progressed sufficiently along the mRNA to make the ribosome binding site available, a second ribosome can join the process of translation from the same mRNA strand. In this way, several peptide chains can be made from one mRNA at the same time. Recall that several mRNAs can be made from the same gene at the same time. In this way, multiple copies of a protein can be quickly and efficiently generated through coupled transcription–translation.

Coupled transcription–translation in bacteria has mRNA being made and used to produce proteins in tandem

Translation takes place in concert with transcription in many bacteria. This is referred to as **coupled transcription–translation** and, until recently, was believed to be a defining feature of bacterial cells. It is now known that not all bacterial species engage in coupled transcription–translation. For those that do, like *E. coli*, this process is important. It is likewise important that we understand the interactions between transcription and translation that occur in this process.

In coupled transcription–translation, before the mRNA has even finished being made by transcription, it is being interpreted by translation into a protein (**Figure 2.18**). Indeed, more than one protein chain can be made from each mRNA strand. It has been observed that several proteins can be simultaneously generated by translation in the wake of RNA polymerase's action in transcription (**Figure 2.19**). Although coupled transcription–translation is often cited as a classic differentiator of bacterial cells and eukaryotic cells, it has been shown to also occur within the nuclei of mammalian cells at sites of transcription "factories," where polypeptides are also made. Coupled transcription–translation in bacteria was clearly demonstrated in the 1970s from electron microscopy images of *E. coli*, showing transcription and translation, including several events of translation occurring off several transcripts originating from the DNA (see Figure 2.19). However, more

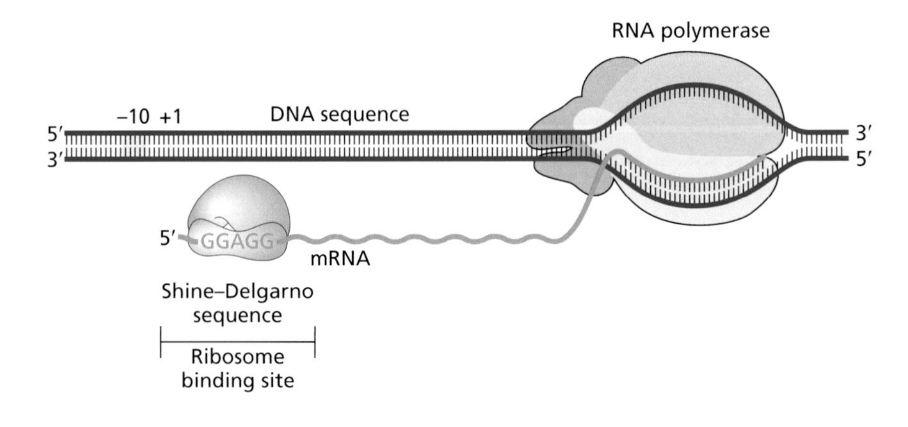

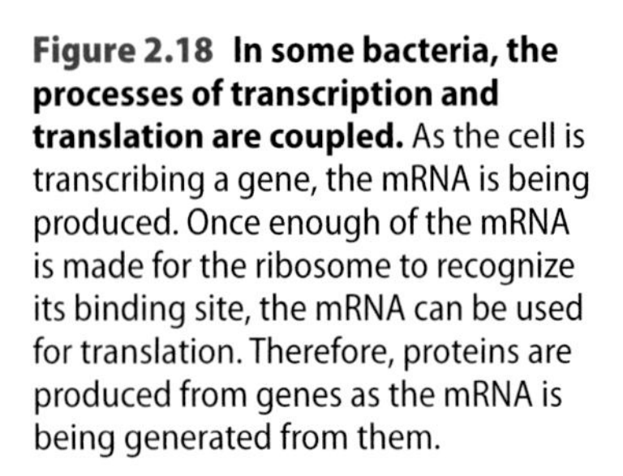

Figure 2.18 In some bacteria, the processes of transcription and translation are coupled. As the cell is transcribing a gene, the mRNA is being produced. Once enough of the mRNA is made for the ribosome to recognize its binding site, the mRNA can be used for translation. Therefore, proteins are produced from genes as the mRNA is being generated from them.

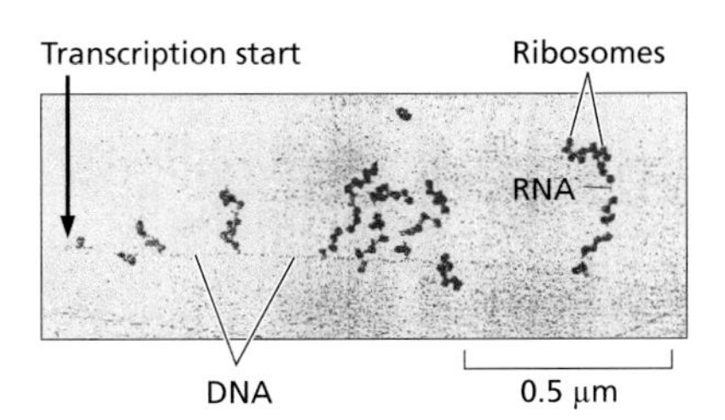

Figure 2.19 Electron micrograph of coupled transcription–translation in *E. coli*. Using high-resolution electron microscopes, it is possible to visualize the process of transcription and translation, demonstrating that in bacteria these processes happen together. Here the DNA strand can be seen, as well as the mRNA that is being transcribed from it. Note that several mRNAs are being made simultaneously. The mRNA is then coated with ribosomes, which translate the mRNA into protein. Note how many ribosomes are attached to the mRNA, simultaneously making several peptide chains from a single mRNA.

recently it has been appreciated that this is one of many assumptions about bacteria in general that is an extrapolation of findings in *E. coli.*

Coupled transcription–translation is not necessarily the rule for all bacteria

In ground-breaking research published by Johnson, Lalanne, Peters, and Li in 2020, it would appear that coupled transcription–translation is not the universal truth for all bacterial species. As will be encountered repeatedly in this book, much of what we know about bacterial genetic and genomics comes from studies done with *E. coli* and it has largely been assumed that these observations will be the same, or similar, in other bacterial species. As more explorations are being done with species beyond *E. coli, Staphylococcus aureus, Bacillus subtilis,* and other heavily studied organisms, we are finding that not all bacteria do things the same way.

This is the case for transcription and translation. In *B. subtilis,* **runaway transcription** occurs where the RNA polymerase conducting transcription runs more quickly than the translating ribosomes, generating a large space along the mRNA between RNA polymerase and the first ribosome (**Figure 2.20**). This uncouples the processes of transcription and translation in *B. subtilis* and some other bacterial species. The relatively more rapid rate of transcription in runaway transcription means that the elongating mRNA has a much longer section that is ribosome-free than in bacteria that have coupled transcription–translation. On a practical level, these differences between coupled transcription–translation and runaway transcription mean that there are also differences in mechanisms of gene regulation (see Chapter 5). Previously one of the defining features of bacterial cells as opposed to other cells was coupled transcription–translation. As we learn more about the diverse array of bacterial species and the differences between them, we find more and more evidence that is breaking down our assumptions about these microbes.

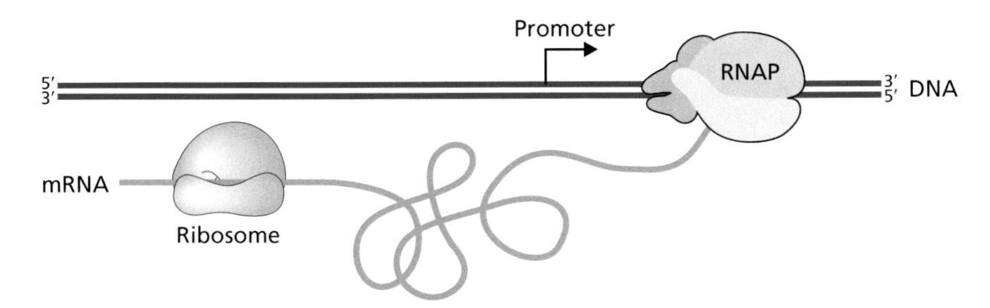

Figure 2.20 Until fairly recently it was believed that all bacteria engaged in coupled transcription–translation. New research has shown that in species such as *B. subtilis*, runaway transcription happens instead. In coupled transcription–translation, the transcribing RNA polymerase (RNAP) generating the mRNA and the translating ribosome generating the protein work at a similar pace. However, in runaway transcription the RNA polymerase is transcribing the DNA into mRNA much faster than the ribosomes are working. This means that there is a large gap on the mRNA between these two cellular machines. Proteins are still being produced from genes as the mRNA is being generated from them, as in coupled transcription–translation, but there is a greater distance along the mRNA between the ribosome and RNA polymerase in runaway transcription than was previously appreciated to happen in bacteria.

Bacteria can have more than one gene on an mRNA strand and form operons

Thus far, we have considered that a transcript encodes a single gene. This is not always the case. In bacteria, more than one gene can be carried on a single mRNA transcript. Therefore, two or more different proteins can be translated from the same transcript. Each protein-encoding portion of the mRNA will have its own ribosome binding site, with initiation, elongation, termination, and ribosome recycling occurring separately for each of the protein-encoding genes carried by the message. These are referred to as **polycistronic units**, that is, sets of genes that are transcribed together (**Figure 2.21**).

In some cases, differences in the promoters used or other signals in the sequence may mean that at times a gene is part of a polycistronic unit and at other times it is transcribed into mRNA on its own. Bacterial mRNA can be polycistronic, but does not have to be. Genes that are together in a piece of polycistronic mRNA may be an **operon**.

An operon is a region of DNA where a set of genes with related functions are controlled by a common regulatory element and as such generate an mRNA transcript containing these genes. Operons have distinct elements that control the transcription of the genes. This can ensure that a set of genes of related function are expressed in equivalent numbers and/or at the same time, being derived from the same mRNA strand. In cases where a gene can be part of a polycistronic unit or can be transcribed separately from a second promoter, this gene is at times part of the controlled operon and at other times not.

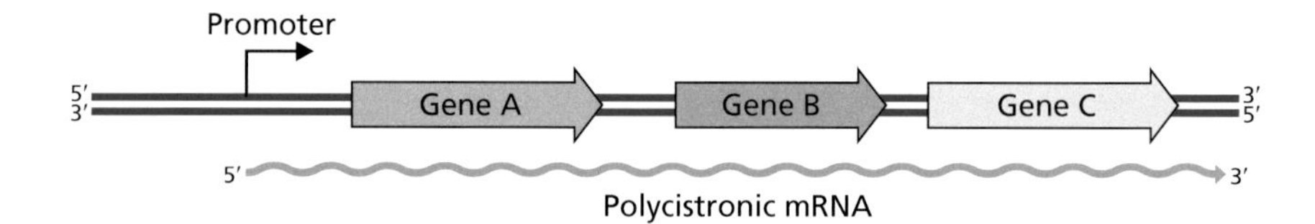

Figure 2.21 Genes as polycistronic units. Groups of genes that are adjacent to one another can be transcribed together when they are part of a polycistronic unit. These are two or more genes that are transcribed into mRNA from the same promoter, generating one mRNA that encodes more than one gene.

Open reading frames are regions of the DNA between termination codons

DNA contains nucleotide bases arranged to encode proteins, RNAs, and to provide genetic signals to the cell. The most common function of the DNA in bacteria is to encode proteins. To do so, a transcribed strand of mRNA must be translated into a polypeptide chain of amino acids that make a protein. The three nucleotide codes within the mRNA corresponds to the amino acids of the protein it encodes, making the codon.

There are 61 tRNAs with different anticodons, sequences within the tRNA that will hybridize to the codon on the mRNA. Three codons have no tRNA with complementary anticodons. In translation, the ribosome will stop adding amino acids to the growing peptide chain when it reaches a codon for which there is no corresponding amino acid. These triplet codes of nucleotides are called termination codons. Each termination codon has a nucleotide sequence in the DNA, which is transcribed into the mRNA, and is known by a semi-precious gem stone name: TAG stop codon in the DNA, UAG in the mRNA, amber; TGA stop codon in the DNA, UGA in the mRNA, opal; and TAA stop codon in the DNA, UAA in the mRNA, ocher. These termination codons signal the release factor to enter the ribosome at the A-site, terminating translation.

There is no tRNA with an anticodon for UAG, UGA, or UAA; however, most rules have exceptions, including this one. The exceptions will be explored in Chapter 7. Except in rare circumstances, the presence of a TAG, TGA, or TAA in the DNA will signal the termination of translation. Therefore, it is possible to identify regions containing genes, which will be translated into long polypeptide chains by finding

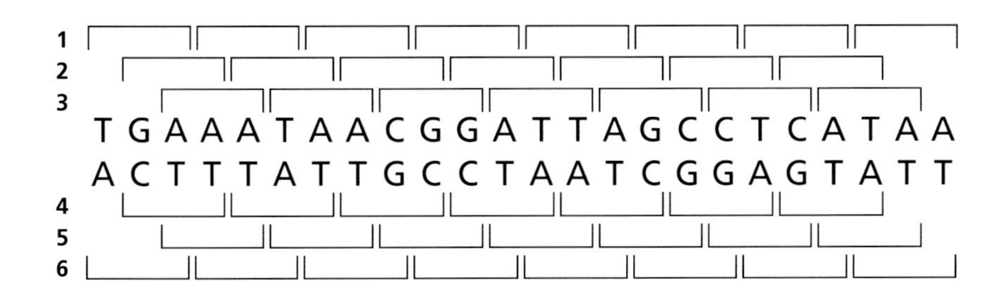

Figure 2.22 DNA has six reading frames. There are six reading frames in DNA, with three on the forward strand and three on the reverse. Reading frames are the sets of three nucleotides that make up a codon. There are therefore three different combinations of bases on the forward strand and three on the reverse strand.

areas of the DNA sequence without stop codons. DNA is double stranded, with a **positive strand** running in one direction and the **negative strand** in the opposite direction along the double helix (**Figure 2.22**). The designation positive strand refers to the strand of DNA encoding a gene, also known as the **sense strand**, while the negative strand is the **antisense strand** containing the complementary bases. These designations are appropriate when the gene is known, but it may be that when trying to identify regions with potential genes, the genes are on either strand. At the genomic level (Chapter 3) these designations do not apply, because genes are found on both strands.

Each codon is made of three bases. To read a codon code, three bases are read together, then the next three, then the next. A different codon code can be read by starting one base farther along and a third different codon code using the base after that. Therefore, there are three different codon codes that can be read on each of the two strands of DNA. These are referred to as **reading frames**. In order for a protein to be made, the reading frame must not contain any stop codons, otherwise these would prematurely signal the end of translation of the protein. The area between two stop codons in the same frame is called the **open reading frame** or **ORF**. Most ORFs are very short and do not encode proteins. Long ORFs, that is, stretches of sequence where there are many amino acid codons but no termination codons, are likely locations for genes (see Figure 2.19).

To be a gene, however, an ORF must have a start codon as well as having a stretch of amino acids before reaching a stop codon. An ORF that has a start codon at the beginning of the frame is called a **coding sequence (CDS)** (**Figure 2.23**). There are three possible initiation codons; however, they are not all used in equal proportions and not all codons that carry the same sequence can be taken as the start of translation. The most abundant initiation codon is the ATG start codon, which is decoded to *N*-formylated methionine at the start of the amino acid string. This is present in the mRNA transcript as AUG and is seen in 90% of the coding regions in *E. coli*. The second most abundant initiation codon is the GTG start codon. This is present in the mRNA transcript GUG and decodes to a *N*-formylated methionine. It is the start codon for 8% of the coding regions in *E. coli*. The most rare start codon is the TTG initiation codon, present in the mRNA transcript as the sequence UUG. Like GTG, this also decodes to *N*-formylated methionine; therefore, all polypeptide chains start with *N*-formylated methionine as they are generated on the ribosome.

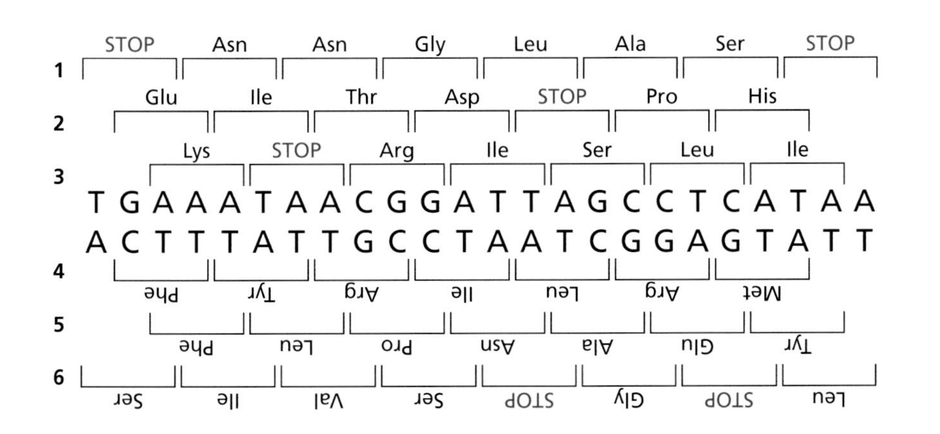

Figure 2.23 An ORF that is also a CDS. An ORF is a reading frame in a section of DNA where there are no stop codons. When there is a start codon early in the ORF, then this feature is a CDS. Note the ATG start codon present immediately after the TGA stop codon at the start of the sequence.

There are genes that contain what could be interpreted as more than one initiation codon. They may have a TTG codon sequence that is then followed later in the ORF by an ATG codon. Which codon is used as the initiation codon by the translation machinery is determined by the location of the ribosome binding site, responsible for initiation of translation. It is therefore important to remember the relative abundance of each initiation codon when determining the location of a CDS. If there is a GTG or TTG present, this may be the initiation codon, but this choice over an ATG should be supported by the location of a ribosome binding site or, ideally, experimental evidence.

When a CDS is a gene

When identified in the DNA sequence, or in the mRNA sequence, a region with both an initiation and termination codon is a CDS. Only experimental evidence showing the expression of the CDS's encoded protein can classify it as a gene. Quite often, evidence of expression of the protein encoded by the CDS is obtained in one strain of one bacterial species and then this information is extrapolated to other strains and even other species of bacteria, provided there is sufficient similarity between the sequences. This similarity may be at the DNA level, where two CDSs share the same nucleotides in many of the same locations. This is **DNA identity** and is a measure of the **homology** of the sequences, or the degree to which they are the same. In some cases, the similarity between two features is most apparent at the protein level of the sequence. In this case, we are comparing the amino acid similarity. There are two measures of amino acid homology, the identity and the similarity. **Amino acid identity**, similarly to DNA identity, is a comparison of the exact sequence. The **amino acid similarity** takes into account the similar nature of some amino acids, due to their charge or interaction with other molecules.

To determine the amino acid identity and similarity from the information in a DNA sequence, the nucleotide sequence must be translated into an amino acid sequence. In this case, it is not bacterial translation, using ribosomes and tRNAs that translate the DNA, via mRNA, into the amino acid strand. Instead, a **codon table** can be used to determine what would happen in bacterial translation, and therefore artificially translate the nucleotide sequence into an amino acid sequence. There are six reading frames in a strand of DNA that can be translated, three on the positive strand and three on the negative strand (**Figure 2.24**). The amino acid sequence amongst these will be the one that is framed by the start and stop codons, the initiation and termination codons.

The codons between the start and stop codons within the CDS, the three-letter codes, one following another, encode the amino acids of the corresponding protein. To determine the amino acid sequence, a codon table is used. This is a table containing all of the possible combinations of three nucleotide bases and indicates the amino acid to which they correspond or highlights that the code is a termination codon. Often, there is more than one codon that encodes a particular amino acid. This is shown in the codon table, where several codons correspond to the same amino acid. For some of these, there is what is referred to as a **third base wobble**. This means that the first two bases are the key to deciphering the amino acid and the third base in a codon can be different bases, yet still encoding the same amino acid (**Table 2.1**).

Figure 2.24 The area between two stop codons in a reading frame is an ORF. The six reading frames of DNA are composed of codons. Some of these codons encode amino acids and some signal the termination of translation, a stop codon. The region between two stop codons in a reading frame is referred to as an open reading frame. A short ORF is shown here in reading frame 1, where a short segment of DNA has no stop codons in this frame.

TABLE 2.1 Codon table

First base	Codon	Amino acid	Codon	Amino acid	Codon	Amino acid	Codon	Amino acid	Third base
T	TTT	Phe (F)	TCT	Ser (S)	TAT	Tyr (Y)	TGT	Cys (C)	T
	TTC	Phe (F)	TCC	Ser (S)	TAC	Tyr (Y)	TGC	Cys (C)	C
	TTA	Leu (L)	TCA	Ser (S)	TAA	Stop	TGA	Stop	A
	TTG	Leu (L)	TCG	Ser (S)	TAG	Stop	TGG	Trp (W)	G
C	CTT	Leu (L)	CCT	Pro (P)	CAT	His (H)	CGT	Arg (R)	T
	CTC	Leu (L)	CCC	Pro (P)	CAC	His (H)	CGC	Arg (R)	C
	CTA	Leu (L)	CCA	Pro (P)	CAA	Gln (Q)	CGA	Arg (R)	A
	CTG	Leu (L)	CCG	Pro (P)	CAG	Gln (Q)	CGG	Arg (R)	G
A	ATT	Ile (I)	ACT	Thr (T)	AAT	Asn (N)	AGT	Ser (S)	T
	ATC	Ile (I)	ACC	Thr (T)	AAC	Asn (N)	AGC	Ser (S)	C
	ATA	Ile (I)	ACA	Thr (T)	AAA	Lys (K)	AGA	Arg (R)	A
	ATG	Met (M)	ACG	Thr (T)	AAG	Lys (K)	AGG	Arg (R)	G
G	GTT	Val (V)	GCT	Ala (A)	GAT	Asp (D)	GGT	Gly (G)	T
	GTC	Val (V)	GCC	Ala (A)	GAC	Asp (D)	GGC	Gly (G)	C
	GTA	Val (V)	GCA	Ala (A)	GAA	Glu (E)	GGA	Gly (G)	A
	GTG	Val (V)	GCG	Ala (A)	GAG	Glu (E)	GGG	Gly (G)	G
Second base →	T		C		A		G		

Note: This table shows the amino acids that are encoded by the various three base codon combinations, as well as those for translational termination. The DNA version of the code is shown there, including Ts. During translation mRNA is used; therefore, the Ts would have been transcribed into Us. Both the three-letter amino acid code and the single-letter amino acid code are shown.

Mutations that change the DNA sequence may not change the encoded amino acid. As can be seen in the codon table, there is scope for bases to change and to still encode the same amino acid. When describing mutations, such as **single-nucleotide polymorphisms (SNPs)**, the terms **synonymous** and **non-synonymous** are often used. The SNP has changed just one base in a region of DNA. If that one base change is positioned within a codon such that it still encodes the same amino acid, then this is a synonymous change. Although the DNA code has altered, the codon code has not, and the resulting protein will carry the same amino acid sequence. If, however, the base that has been changed does influence the amino acid sequence, causing a change in the codon that means a different amino acid will be placed in the protein during translation, then this is called non-synonymous. The SNP change will have altered the DNA sequence in such a way that the amino acid sequence is also altered (**Figure 2.25**). This may have consequences for the function of the protein. Later in this book, events that give rise to SNPs are described, as well as methods that can be used to introduce them.

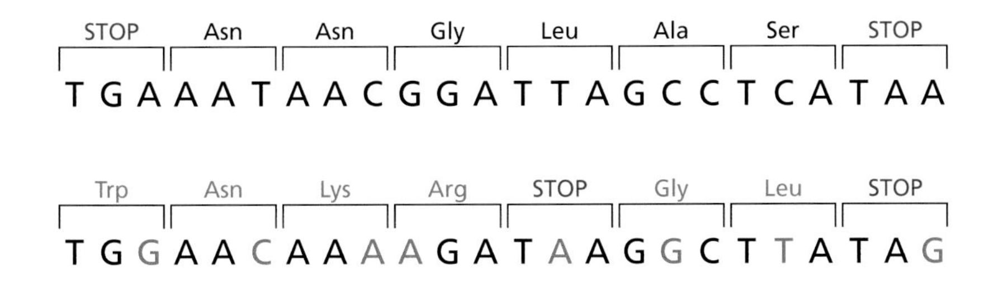

Figure 2.25 Single base changes may change the amino acid sequence. SNPs are changes in a single base of DNA. Depending on the position of the change in the codon, this may change the encoded amino acid, or it may not. As in the codon table (Table 2.1), the same amino acid can be encoded by more than one codon. These differ by one or two bases. As shown, when the bases change, shown in orange on the bottom DNA sequence, the amino acids stay the same in some cases and change in others. SNPs can also affect stop codons.

Although a synonymous SNP will not alter the sequence of the encoded protein, it may alter the rate at which the protein is made. Not all tRNAs are generated within the cell at the same level. Some tRNAs are very rare within the bacterial cell. The use of these codons is selected against evolutionarily; generally, the bacterial genome sequence will favor some codons over others based on the availability of the corresponding tRNAs. If a nonstandard codon that is not commonly found in the species is present within a gene, it may cause a pause in the translation process, as the ribosome awaits the arrival of the correct, rare tRNA. This can result in fewer proteins being made than could be made with a standard codon in this place.

Expression of genes is controlled: not all genes are on all of the time

Promoters are regions of the DNA sequence where transcription is initiated to generate mRNA. These regions are near the start point of transcription in bacteria, generally located just 5′ of the transcriptional start site. Promoters are also on the same strand of the DNA double helix as the start point of transcription. The directionality of the –10 and –35 sequences in the promoter dictate the direction of RNA polymerase transcription. The strand containing the promoter and the start point of transcription is referred to as the sense strand. The promoter is where RNA polymerase begins its work in transcription.

The promoter region is where all transcription initiation factors and RNA polymerase bind to specific DNA sequences to initiate transcription. In addition, other transcription factors may bind to the promoter region and modify transcript generation.

All bacterial promoters must have an RNA polymerase binding site, which is generally the –10 and –35 sequences that are present 5′ of the gene's transcriptional start site. The **–10 promoter sequence** is also called the **Pribnow box**. The sequence of the –10 is usually similar to the consensus sequence TATAAT. The **–35 promoter sequence** is usually similar to the consensus sequence TTGACA, although this sequence is less conserved than that of the –10 sequence.

There are, however, some promoters that completely lack the –35 portion of the promoter region and instead have an **extended –10 region**. This extension to the –10 sequence compensates for the absence of the –35 region and allows RNA polymerase to bind and initiate transcription.

Other promoters may contain what is referred to as an **UP element**. When it is present, the UP element is 5′ of –35, and therefore farther away from the gene than the rest of the promoter (**Figure 2.26**). This additional sequence enhances transcription of the associated gene(s) due to it imparting better binding of the alpha subunit of RNA polymerase. This enhancement efficiently recruits RNA polymerase to the UP element containing promoter and increases the level of transcription.

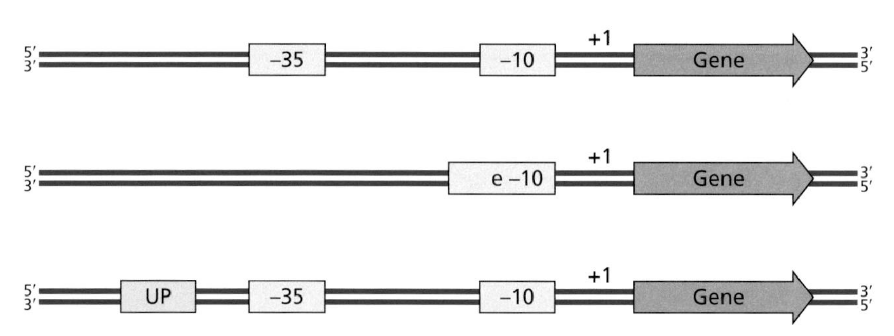

Figure 2.26 The features of a promoter region. Promoter regions frequently include the –10 and –35 sequences that are recognized by the σ^{70} subunit of RNA polymerase. In some instances, an extended –10 region (e – 10) may be present, which compensates for the absence of a –35. An UP element can also be present upstream of the –35, which enhances RNA polymerase binding.

Sigma factors are responsible for RNA polymerase promoter recognition

The specificity of RNA polymerase for the promoter region sequence is determined by the enzyme's **sigma subunit** (σ). This is one part of RNA polymerase; the sigma factor and the other subunit components of RNA polymerase make up the **holoenzyme** (**Figure 2.27**). Not all RNA polymerase within the bacterial cell will contain identical σ subunits. By changing the σ subunit, the specificity of RNA polymerase for the promoter region sequences is changed.

The σ^{70} subunit is the primary sigma factor in bacterial cells, responsible for the majority of the genes and those housekeeping genes essential for life. Other sigma factors are responsible for other functions in the cell. Unlike the primary sigma factor, which is in abundance in the cell, the production of these alternative sigma factors is tightly regulated, thus ensuring controlled regulation of the genes associated with the promoters they recognize. Sigma factors enable the simultaneous expression and regulation of many genes across the bacterial genome.

The σ^{54} subunit, for example, is involved in the nitrogen response and therefore controls the relevant genes that possess a different promoter sequence, recognized by σ^{54}. Alternative sigma factors are responsible for several important processes in the bacterial cell and can include response to the environment, response to and survival during conditions of stress, or the development of the cell. Bacterial cells are not static. They must change in accordance with their environment and the situations they encounter. These changes can result in the differentiation of the cell. The process of sporulation, whereby some bacteria make spores containing their genetic material that are resistant to stress, includes control of the transcription of sporulation-specific genes by alternative sigma factors.

Different species have different numbers of sigma factors. Some species have tens of different sigma factors, all recognizing a different promoter region sequence. Other species have only a few sigma factors, relying for the most part on σ^{70} and regulating gene expression through other means. It therefore cannot be assumed that a particular bacterial species possesses a particular sigma factor, even when that sigma factor is present in other species and has been demonstrated to serve a vital role there.

The alternative sigma factors have different binding sites from the consensus −10 and −35 used by σ^{70}. Therefore, when analyzing DNA sequence data, if no −10 and −35 sequences are found 5′ of a gene, it could be that an alternative sigma factor is responsible for initiation of transcription for this gene. Where the sequences of the alternative sigma factor binding sites are known, it can be beneficial to investigate their presence as well. In some cases, the same gene can be transcribed by RNA polymerase holoenzyme carrying a σ^{70} or by one with an alternative sigma factor.

The sigma factor component of the RNA polymerase holoenzyme is only required for the initiation of transcription. Once the transcription process enters elongation, the sigma subunit is shed and is able to initiate transcription again at the same site or at another promoter through association with another RNA polymerase.

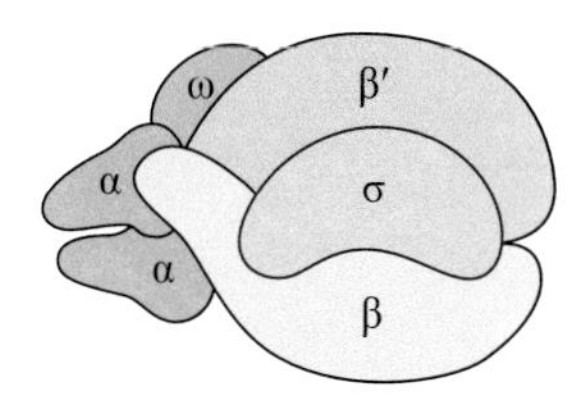

Figure 2.27 The subunits of RNA polymerase. RNA polymerase is composed of six subunits as the RNA polymerase holoenzyme, with two α subunits, a β subunit, a β′ subunit, a ω subunit, and the σ subunit. The presence of the σ subunit completes the holoenzyme. Without the σ subunit, the RNA polymerase core enzyme remains.

Regulatory proteins change the level of transcription

In addition to RNA polymerase, there are proteins that can bind to the promoter regions in bacteria. These DNA-binding proteins are called **regulators**. These proteins can either bind in association with RNA polymerase and enhance its ability to initiate transcription, as activators, or bind to a region of the promoter that excludes the binding of RNA polymerase, repressing the initiation of transcription.

Repressor proteins prevent or reduce the transcription of genes

Repressors are regulatory proteins that bind to specific DNA-binding sites in promoter regions and prevent transcription by RNA polymerase, or in some cases significantly reduce transcription of the 3′ gene(s). Some repressors bind to a region that overlaps the –10 and/or –35 regions of the promoter. This prevents RNA polymerase from binding to the promoter (**Figure 2.28**).

Other repressors bind to sites on the DNA that overlap an activator protein binding site. In these situations, both proteins, repressor and activator, are trying to bind to the same location on DNA and only one can do so. When it is the repressor that binds, this prevents RNA polymerase binding to the promoter region or in some cases prevents the open complex formation necessary for initiation of transcription.

A repressor may also bind to more than one site in association with the gene for which it represses transcription. Once bound, the multiple copies of the repressor protein interact. This can cause bends, loops, or hairpins in the DNA. These topology changes stop RNA polymerase binding, open complex formation, activator binding, or elongation.

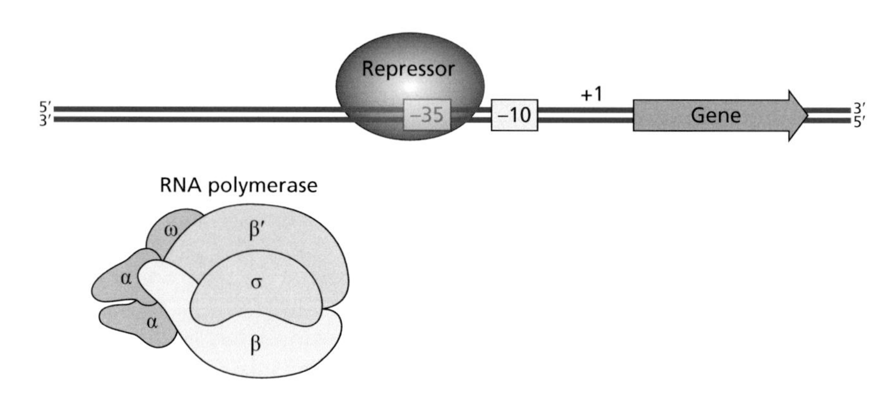

Figure 2.28 The binding of a repressor protein to a promoter region. Repressor proteins are DNA-binding proteins that are capable of binding to a promoter region and preventing the binding of RNA polymerase, thereby blocking the initiation of transcription.

Activator proteins contribute to the expression or increased expression of genes

Regulatory proteins that are **activators** are capable of inducing transcription of genes. Activators, as DNA-binding proteins, can bind near or at promoters and thus assist in the binding of RNA polymerase to the promoter for the initiation of transcription. In some cases, repressors also bind to the promoter and the two opposing proteins compete. If the activator binds, the gene(s) are transcribed (**Figure 2.29**).

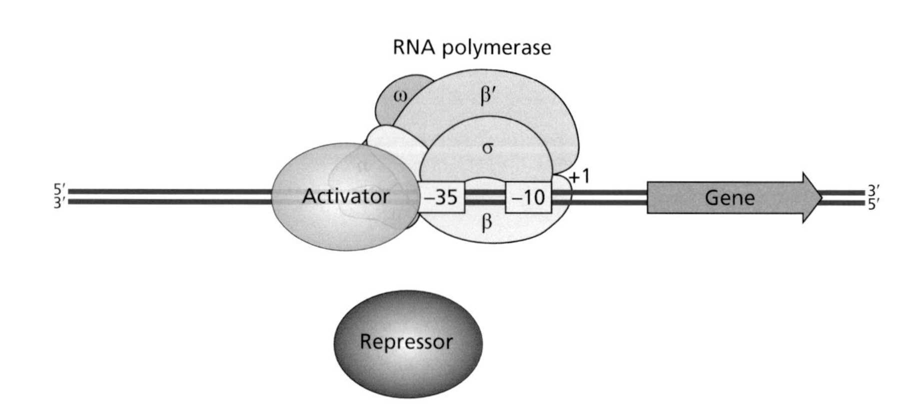

Figure 2.29 The binding of an activator can compete with the binding of a repressor. Activator proteins are DNA-binding proteins that assist in the binding of RNA polymerase to a promoter. When the binding sites for activators and repressors are close to one another or overlapping, only one of the proteins can bind. When the repressor binds, it prevents RNA polymerase from binding. When the activator binds, it enhances the binding of RNA polymerase and the initiation of transcription.

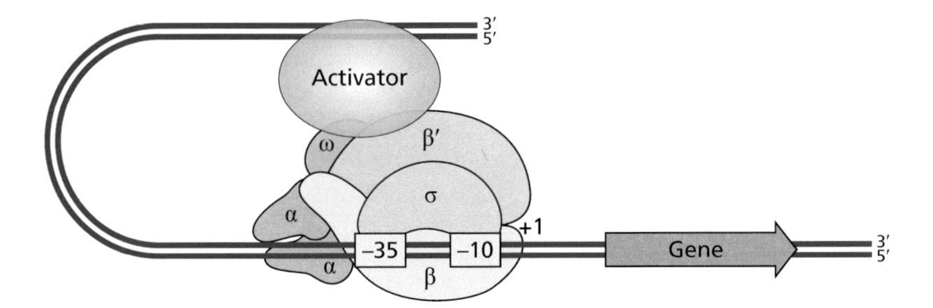

Figure 2.30 **Some activators bind distantly from the promoter.** In some cases, the activator protein binds to a region that is upstream of the −35 region by some distance. It contributes to RNA polymerase binding, enhancing transcription, through the bending of DNA, which brings the activator together with the RNA polymerase.

Activators can also bind to **enhancers**, which are sequence regions that enhance transcription. Enhancers are regions of DNA that can be 5′ of a gene, within a gene, 3′ of a gene, or distantly. These regions in the DNA need not be in the promoter. The binding of an activator to an enhancer often results in the DNA bending, enabling the activator protein to interact with the transcriptional machinery at the promoter (**Figure 2.30**).

Regulatory RNAs may have an impact upon transcription

There are **small RNAs** made within bacterial cells that are not mRNAs, tRNAs, or rRNAs. These other RNAs are regulatory RNAs and although many act on the level of translation there are some that regulate transcription. Most of these act in ***trans***, which means that they are acting on elements that are remote. In some examples, an antisense regulatory RNA, which has the complementary sequence to part of the gene, binds to a complementary region on an mRNA leading to premature transcriptional termination, often between two genes in an operon. Thus, the more 3′ encoded genes of the operon have a lower level of transcription than do those genes at the 5′ start of the operon.

Regulatory RNAs can also act through **transcription interference**. This occurs when divergent or tandemly transcribed promoters interfere with each other. Since genes can be on either strand of the DNA, in any of the six reading frames, some regions between genes contain **divergent promoters**, with one transcribing on one strand and the other on the other strand and in the opposite direction (**Figure 2.31**). In divergent promoters, a process called collision results in one or both RNA polymerases dissociating and stopping transcription. There can also be more than one promoter orientated in the same direction along a stretch of DNA, driving transcription on the same strand. A process called **promoter occlusion** can occur at tandem promoters. Strong transcription from the first promoter prevents formation of an initiation complex at the second promoter, thereby occluding the function of the second promoter. Transcription interference can also occur where the second promoter has formed an open complex and is "knocked off" by the progression of the stronger transcript and progression of RNA polymerase from the first promoter. These processes can be influenced by the binding of regulatory RNAs and their impact upon RNA polymerase binding and progression.

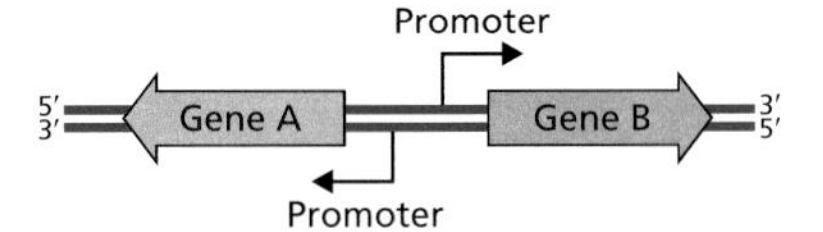

Figure 2.31 **Divergent promoters drive transcription of genes on opposite strands.** Some promoters overlap one another or are near to one another when the genes for which they drive transcription are on opposite strands. The placing of divergent promoters and their associated protein binding sites can mean that RNA polymerase for one promoter interferes with the binding of RNA polymerase for the other. Likewise, the binding of activators and repressors may influence transcription from both promoters.

Riboswitches alter the transcript that includes their sequence

Regulation of the expression of genes can also be achieved through the action of **riboswitches** that are encoded in the DNA that is transcribed into mRNA. These regulatory elements are contained within the mRNA that they regulate. Their presence alters the transcription of the mRNA. Usually, riboswitches are within the 5′ **untranslated region**, or **UTR**, of the mRNA, a region between the start site of transcription and the start codon of translation. As such, they act in ***cis***, meaning they act on the gene(s) nearby.

Riboswitches act in response to changes in conditions for the bacterial cell. This can include the presence within the cell of uncharged tRNAs (those without amino acids loaded), the temperature of the cell, and the occurrence of ribosome stalling. Ribosomes generally stall due to a time lag in recruiting a tRNA during the translation process. Typically, the tRNA that causes a stall in the ribosome is one that is carrying a nonstandard codon for the species, thus being rare in the cell. Stalled ribosomes change the mRNA structure and signal premature transcriptional termination. Temperature changes can activate RNA thermometers, which are sequences within mRNA that fold in a temperature-sensitive manner, altering transcription. Riboswitches, including RNA thermometers, are discussed in more detail in Chapter 4.

Global regulators control the transcription of multiple genes

Regulator proteins can influence the ability of RNA polymerase to transcribe a gene. Some of these act on just one transcriptional unit, but there are several regulators of transcription that work upon many genes together for a coordinated response. These are called **global regulators** because their action influences transcription at several locations in the genome. Unlike regulators that act on a single operon, global regulators are capable of influencing the transcription of many genes in the cell, regardless of their location. Global regulators can be repressors or activators and can also act as either, depending upon the promoter regions to which they bind. Generally, global regulators are factors in addition to any alternative sigma factors, although sigma factors do have a role across the genome.

Although not present in all bacteria, global regulator H-NS has an important impact on transcription regulation. H-NS is histone-like nucleotide structuring protein and controls expression of numerous genes due to its presence binding to regions of the DNA. Regulation via H-NS has been demonstrated for bacterial cellular metabolism, expression of fimbriae on the surface of the bacteria, flagella expression, and maintaining homeostasis. There is more on H-NS in Chapter 5.

Global regulators can also act on transcription via signaling molecules like cyclic AMP (cAMP) and cyclic-di-GMP. cAMP is made from ATP when the glucose concentrations within the bacterial cell are low. The generated cAMP binds to the protein Crp (cAMP receptor protein), together making an active transcription factor. Regulation involving cAMP has been shown for cellular processes such as catabolite regulation, biofilm formation, and bacterial growth. There is more on global regulators and these signaling molecules in Chapter 5.

Global regulators can be responsible for changes to the cell that are necessary to survive in the changing conditions of the environment in which the bacteria live. In *E. coli*, the FNR regulator (fumarate and nitrate reductase) is responsible for the regulation of genes that are necessary during oxygen starvation conditions. When oxygen is low, bacteria that are able to live anaerobically can use FNR to switch from genes that are used in aerobic metabolism to those that can conduct anaerobic metabolism.

Other global regulators include **two-component regulatory systems**. These enable the bacteria to regulate genes in response to environmental changes. One component of these systems senses the change, while the other regulates the associated genes. The two components are two proteins that each function together to turn on the necessary genes in response to the environment. There is a part of the two-component regulatory system that is the sensor, which is embedded within the bacterial cell membrane. When the sensor receives a signal from the environment, it activates the other component of the system, the response regulator. The regulator can then bind to promoter regions of genes across the genome, changing their expression in response to the detected environmental signal. Two-component regulatory systems are explored in more detail in Chapters 5 and 13, where the roles of specific systems in example bacterial species are discussed.

Some global regulatory proteins exert their influence over the expression of genes by binding to and changing the topology of DNA. One such protein is the integration host factor, known as IHF. This protein can bind along DNA, causing the strand to bend and perhaps bring together distant parts of the DNA, which may influence the binding of RNA polymerase or other proteins.

Not all global regulators include proteins. There is a class of RNA referred to as **noncoding RNAs (ncRNAs)**. Like tRNAs and rRNAs, these are functional, rather than encoding the message for protein production, like mRNA. The ncRNAs are often involved in post-transcriptional regulation, meaning that the transcript is made, but its expression as a protein is somehow altered due to influences on translation or other processes that happen after the generation of the mRNA. There are, however, some ncRNAs that regulate transcription and do so for several transcripts, therefore making them global regulators of transcription. Regulation by ncRNAs is described in greater depth in Chapter 4.

Essential genes and accessory genes

Essential genes are those required for life, while **accessory genes** provide additional function. Because different bacteria live in different niches, encounter different environments, and experience different changes in conditions, different bacteria contain different genes. What is essential for life in one species may be an accessory gene in another.

The number of genes in a bacterial genome varies widely, dependent on species and lifestyle. Both the environment that the bacteria lives in and any transitions between environments that it experiences will influence the genes that have evolved to be within the genome of the species. The majority of differences in gene numbers between species are due to the environment(s) the bacteria live in and the need to respond to changes. For intracellular bacteria, which live inside eukaryotic cells, there tend to be fewer genes due to the consistency of the environment and the ability to use some processes from the host cell.

During the course of evolution, genes have been lost from some species (**Figure 2.32**), most evident in intracellular bacteria that survive within eukaryotic cells. Thus, endosymbiosis tends to reduce the number of genes needed. In other bacteria, gene numbers have increased in the course of evolution through

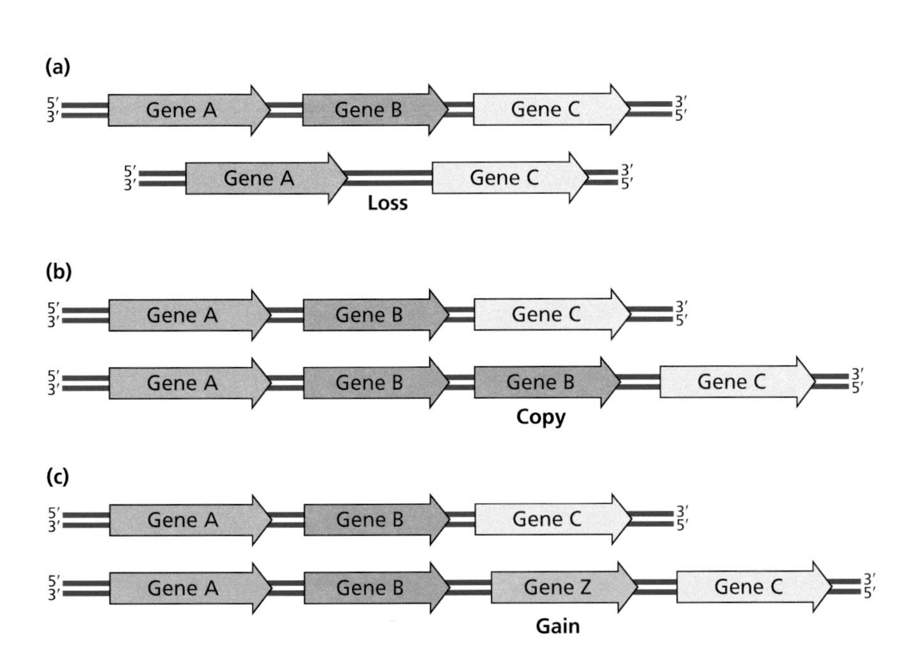

Figure 2.32 Deletion of genes, duplication of genes, and addition of horizontally transferred genes. (a) During evolution, regions of DNA can be deleted from the genome, resulting in the loss of genetic material, and potentially whole genes. (b) During evolution, regions of DNA can be duplicated, making a second copy of that segment of DNA. This can result in the gain of a second copy of some genes. (c) New genetic material can be added into the genome when DNA enters the bacteria from another cell. Once incorporated, this horizontally transferred material may result in a gain in function(s) for the bacteria through acquisition of horizontally transferred genes.

either gene duplication and modification of duplicates to new functions or the acquisition of genes from other bacteria (see Figure 2.32), called **horizontal gene transfer**.

Of course, the gene repertoire within a bacterial cell would be influenced by the environment in which the bacteria grow, availability of nutrients and other resources, and other factors. These conditions can be artificially controlled in the laboratory. It has therefore been hypothesized that there is a minimal bacterial genome, containing only those genes that are absolutely required. Researchers in synthetic biology are working to understand the minimal bacterial genomes through the creation of bacterial chromosomes engineered to contain only specific genes.

Key points

- The cellular process of bacterial transcription takes the genetic information from DNA and generates mRNA based in this sequence. There are three phases to bacterial transcription: initiation, elongation, and termination.
- Multiple rounds of transcription can occur simultaneously for the same gene.
- Bacterial translation uses the genetic code carried by mRNA to generate proteins with the assistance of tRNAs carrying the amino acids that will make the proteins and rRNA contained within the ribosomes that orchestrate protein production.
- Whilst DNA and RNA are made up of nucleic acids, proteins are made up of amino acids, which come together in polypeptide chains based on the gene sequence.
- There are four phases of translation that share similar names with the phases of transcription, being initiation, elongation, termination, and ribosome recycling; therefore, it is important when discussing initiation to know which process is being described.
- The ribosome slides along the mRNA during translation elongation, providing three base codons for the corresponding tRNA anticodons and recruiting amino acids to the growing polypeptide chain.
- In bacterial cells, transcription and translation happen together, with the mRNA being made and bound by ribosomes to make proteins. In some bacteria there is coupled transcription–translation, where the RNA polymerase and ribosome are very close on the mRNA strand. In others, where runaway transcription occurs, they are farther apart.
- Genes that are adjacent to one another may be transcribed into the same mRNA, forming a polycistronic unit. A set of genes in a polycistronic unit that share the same regulatory controls can be considered to be an operon, where the products of the genes are expressed together in an organized manner.
- There are six potential reading frames of three base codons encoded in the DNA, three on the positive strand and three on the negative strand. Included are three termination codons (TAG, TGA, and TAA) that define ORFs and three initiation codons (ATG, GTG, and TTG) that define a CDS.
- Repressor regulatory proteins can prevent or reduce the transcription from a promoter and activator regulatory proteins can cause or enhance the transcription from a particular promoter.

Key terms

Define the following terms introduced in this chapter. Check your answers using the definitions in the Glossary. These terms are also available as Flashcards online.

Accessory gene	Peptide bond	Shine–Dalgarno sequence
Activator	Phenotype	Sigma subunit
Amino acid	Polycistronic units	Stop codon
Anticodon	Promoter region	Termination codon
Coding sequence	Protein	Transcription
Codon	RNA polymerase	Transcription elongation
Coupled transcription–translation	Regulator	Transcription initiation
Essential gene	Repressor	Transcription termination
Gene	Ribosomal RNA	Transfer RNA
Genotype	Ribosome	Translation
Homology	Ribosome binding site	Translation elongation
Messenger RNA	Runaway transcription	Translation initiation
Open reading frame	Sequence	Translation termination

Questions and discussion topics

Self-study questions

Answer each question using 50–100 words or a table or labeled diagram. Advice on where to find answers to these questions is available online.

1. What is the specific term for a sequence of DNA that encodes a protein and what features distinguish protein-encoding sequences of DNA from other sequences?
2. For proteins to be made from DNA, first an RNA template has to be made. What kind of RNA is this, what is the name of the process that generates it, and what are the three stages in the process of it being made?
3. Ribosomes are made up of protein and RNA. What kind of RNA is in a ribosome and what are the designations of the different RNAs in a bacterial ribosome?
4. Draw the ribosome in cross section, showing the binding sites, RNA channel, and channel for the growing peptide chain.
5. What is the name of the process where a sequence of RNA becomes a protein and what are the three stages in the process of the protein being made?
6. What are the specific key elements that are involved at the start of translation?
7. Draw the formation of a peptide bond.
8. Explain how polycistronic units coordinate expression of multiple genes.
9. Differentiate between an ORF and a CDS.
10. Using the codon table (Table 2.1) translate this DNA sequence into its corresponding amino acid sequence: 5′ - GGTGAAAATGAGACGATTTGTTCT - 3′. Using the single letter amino acid abbreviations, this should spell a word.
11. If the 14th base in the DNA sequence in question 10 became a T, how would this SNP alter the encoded protein sequence?
12. Compare designations of essential genes and accessory genes.

Discussion topics

These topics are presented for discussion in study groups, as part of class discussions, or on your own. These questions go beyond what is directly covered in this part of the book. Use the research literature and other reading to explore these topics in more depth. Tips to help prepare for topic discussions are available online.

1 There are many instances where bacterial genotypes do not match their phenotypes, where the physical features of the bacteria are different from what is encoded in its genome because one or more proteins are not being expressed. Discuss the reasons why a bacterial cell may not express particular proteins and what advantages this may have. Why would the genome continue to carry genes that are not being expressed?

2 Bacteria engage in coupled transcription–translation and runaway transcription. Explore the linking of transcription and translation in each of these forms of transcription and how they are both similar and different from one another. Compare these bacterial processes also to what occurs in eukaryotic cells where these processes are physically separated.

3 Gene expression using alternative sigma factors can play a key role in virulence for many bacterial species. Discuss one species of interest where this is the case, considering the sigma factor, the genes it regulates, and the phenotypes this generates in the bacterial population.

Online quiz questions

To further self-assess your understanding of the chapter material, please visit the following link, where you can participate in a range of interactive quiz questions:

www.routledge.com/cw/snyder

Further reading

Genes are features in the DNA that encode proteins

Letunic I, Bork P. Interactive tree of life (iTOL) v3: An online tool for the display and annotation of phylogenetic and other trees. *Nucleic Acids Res.* 2016; *44*(*W1*):W242–W245.

Snyder LA, Loman NJ, Fütterer K, Pallen MJ. Bacterial flagellar diversity and evolution: Seek simplicity and distrust it? *Trends Microbiol.* 2009; *17*(*1*):1–5.

Waite RD, Struthers JK, Dowson CG. Spontaneous sequence duplication within an open reading frame of the pneumococcal type 3 capsule locus causes high-frequency phase variation. *Mol Microbiol.* 2001; *42*(*5*): 1223–1232.

Bacterial transcription generates RNA based on the DNA sequence

Iborra FJ, Jackson DA, Cook PR. Coupled transcription translation within nuclei of mammalian cells. *Science.* 2001; *293*: 1139–1142.

Laursen BS, Sørensen HP, Mortensen KK, Sperling-Petersen HU. Initiation of protein synthesis in bacteria. *Microbiol Mol Biol Rev.* 2005; *69*(*1*): 101–123.

Bacterial translation produces proteins based on the mRNA sequence

Miller Jr OL, Hamkalo BA, Thomas Jr CA. Visualization of bacterial genes in action. *Science.* 1970; *169*: 392–395.

Coupled transcription–translation is not necessarily the rule for all bacteria

Johnson GE, Lalanne J-B, Peters ML, Li GW. Functionally uncoupled transcription–translation in *Bacillus subtilis*. *Nature.* 2020; *585*: 124–128.

Bacterial mRNA can contain more than one gene and form operons

Lewis M. Allostery and the lac operon. *J Mol Biol.* 2013; *425*(*13*): 2309–2316.

Sigma factors are responsible for RNA polymerase promoter recognition

Feklístov A, Sharon BD, Darst SA, Gross CA. Bacterial sigma factors: A historical, structural, and genomic perspective. *Ann Rev Microbiol.* 2014; *68*: 357–376.

Kazmierczak MJ, Wiedmann M, Boor KJ. Alternative sigma factors and their roles in bacterial virulence. *Microbiol Mol Biol Rev.* 2005; *69*(*4*): 527–543.

Global regulators control the transcription of multiple genes

Perrenoud A, Sauer U. Impact of global transcriptional regulation by ArcA, ArcB, Cra, Crp, Cya, Fnr, and Mlc on glucose catabolism in *Escherichia coli. J Bacteriol.* 2005; *187*(*9*): 3171–3179.

Essential genes and accessory genes

Hutchison CA 3rd, Chuang RY, Noskov VN, Assad-Garcia N, Deerinck TJ, Ellisman MH. Design and synthesis of a minimal bacterial genome. *Science.* 2016; *351*(*6280*): aad6253.

Preska Steinberg A, Lin M, Kussell E. Core genes can have higher recombination rates than accessory genes within global microbial populations. *Elife.* 2022; *11*: e78533.

Chapter

Genomes

3

This chapter will discuss the DNA within a bacterial cell that makes up the genome, including the contribution to the genome of genes that are essential to life. Genetic features within the bacterial genome, such as the rRNAs and tRNAs that are noncoding yet essential to life processes will also be discussed. This chapter will explore the concept of the genome in bacteria, where there is generally one circular chromosome that contains the whole genome, and how this is not true of all bacterial species. Some bacteria have multiple chromosomes that together make up the whole genome. Some bacteria have linear chromosomes as their genome or linear chromosomes that together with circular chromosomes make up the genome. Extra-chromosomal DNA can also be present within the bacterial cell, generally present as plasmids, which can contribute additional traits to the bacteria. The structure and composition of bacterial genomes will be explored in this chapter, including how these structures can contribute to the expression of genes and the consequences of transcription, translation, and replication happening simultaneously within the bacterial cell. Changes in the genome are also discussed, including how these can arise and therefore drive evolution of bacteria.

Bacterial chromosomes carry the genetic material of the organism

All of the DNA within a bacterial cell has the ability to contribute to the traits expressed by the organism. All genes within a cell have the potential to be expressed as proteins. This is the genetic material of the organism, the vast majority of which is within the bacterial chromosome, or chromosomes. The DNA present in the cell has to be replicated to continue on to the next generation of bacteria. The genetic features present in the DNA have to be transcribed to contribute to the traits expressed by the bacteria and the transcribed DNA has to be translated for the encoded proteins to influence the bacterial characteristics. Through replication, transcription, and translation the DNA is sustained in the species and can contribute to its characteristics.

When the genetic material within a cell is discussed, the term **genome** is used. A genome is generally considered to be all of the genetic material within a cell. In a bacterial cell, the genome may be contained within one molecule of DNA or several separate and discrete molecules of DNA. The genome is what makes up the **genotype** of the cell, being all of the genetic material in the cell encoding everything that the bacterial cell has the potential to express as its **phenotype**.

A **chromosome** is a discrete molecule of DNA, which can be in either a circular form or a linear form (**Figure 3.1**). In bacteria, chromosomes are most commonly found as circular DNA molecules, where the DNA double helix has no ends and forms a continuous circle. However, in some bacterial species linear chromosomes are present, where there are ends to the DNA double helix molecule in structures that are largely similar to what is seen in eukaryotes with their linear chromosomes.

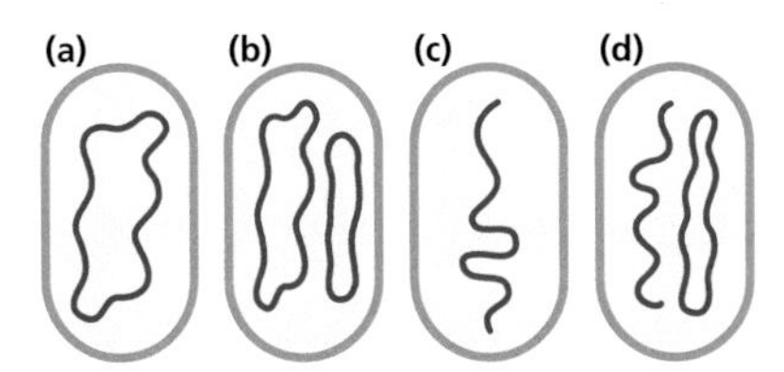

Figure 3.1 Bacterial chromosomes. Most bacterial species have a single circular chromosome (a). Some species have two or more circular chromosomes that make up the genome (b). A few bacterial species, however, have linear chromosomes that may constitute the whole of the genome (c) or be combined with circular chromosomes as the genome of these species (d).

DOI: 10.1201/9781003380436-4

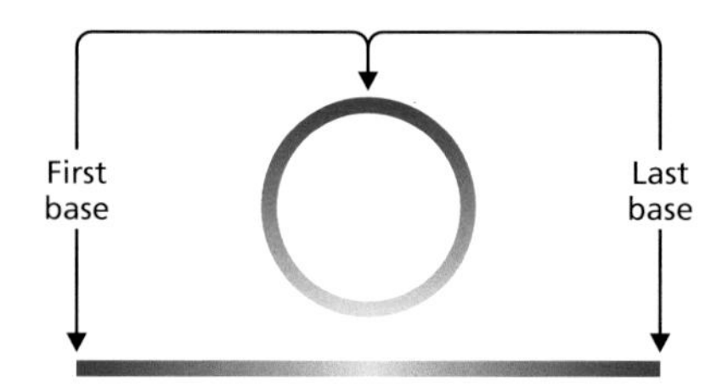

Figure 3.2 A circular bacterial chromosome represented in different formats. Bacterial genome sequence data are represented in circular and linear formats. A circular display format reflects the circular nature of the bacterial chromosomal DNA (top). A linear display can often be easier to use for analysis, comparisons, and visualizations (bottom), where the circular DNA data have been linearized by selecting a first and last base. It must be remembered that the DNA is actually present as a closed circle in the cell; therefore, features identified in analysis that are at the beginning (left end) are actually near to those at the end (right end) in the cell, where the chromosome is circular.

Circular bacterial chromosomes may be depicted as circular maps, which show the relative location of genetic features around the circle of the DNA molecule (**Figure 3.2**). However, it is also common to represent a circular chromosome as a linear molecule (see Figure 3.2). This is commonly done to make comparisons and alignments between circular chromosomes easier. It must be remembered in such cases that in actuality, the chromosome forms a circle, so the first base of the linearized form is actually next to the last base. In the past, it was common when reporting the sequence data to split a circular chromosome at the first base of the conserved gene *dnaA*, but this convention is not always observed. In cases where otherwise similar chromosomes are linearized starting at different places, the resulting alignments can look as if there has been a change in the genome, when this is really an artificial introduction of a break in the DNA sequence data and the presentation of a circular chromosome in an artificial linear form. Genome sequence data in the public databases will be linear in their format, regardless of the type of chromosome, with the first base chosen by the researchers designated as base 1.

Some bacteria have multiple chromosomes

The genome can be made up of more than one circular or linear chromosome or a combination of circular and linear chromosomes, depending on the bacterial species. For example, as mentioned in Chapter 1, *Brucella* and *Vibrio* species have two circular chromosomes and *Agrobacterium tumefaciens* has one linear chromosome and one circular chromosome. Regardless of the form of the chromosomes, in order to persist as the genetic material of the species, all of the chromosomes must be carried on to the next generation of bacterial cells. Each chromosome within the cell, being discrete and distinct portions of the genome, is replicated separately. As a result, each chromosome has its own origin of replication (discussed in Chapter 1), where the process of replication is initiated, and is copied by its own set of replication machinery.

Having multiple chromosomes that replicate independently, each with their own origin of replication, can mean that part of the genome is present in a higher copy number than the rest of the genome. One of the chromosomes may replicate more rapidly than the other, so that instead of there being one large circular chromosome and one small circular chromosome, the small chromosome is actually present in two or more copies, often due to the relative speed of copying each chromosome. In addition, with multiple origins of replication, more of the genome will be under the dosage effect of proximity to the origin. The concept of dosage effects and the origin of replication will be explored in more detail later in this chapter.

When multiple chromosomes are in a bacterial cell, whether circular or linear, the genome is divided between them. Part of the genetic material of the species is carried on one chromosome and part on another chromosome. This raises the question of what defines the DNA within a bacterial cell as a chromosome and therefore part of the genome. In order for the DNA molecule to be called a chromosome, it must carry essential genes. In some bacteria, we see all of the genes for certain essential functions, such as replication or transcription, on one chromosome while other essential genes are on another. The key factor in the chromosome definition being that there is at least one essential gene carried on that particular molecule of DNA.

The largest bacteria with many copies of its genome

In 2022, a group of researchers discovered what is believed to be the largest bacteria. *Thiomargarita magnifica* was found on the underside of leaves in a mangrove forest. Although these bacteria are single cells, they are nearly 1 cm long and can be seen with the naked eye, being roughly the size and shape of an eyelash. The *T. magnifica* cell has multiple copies of its genome, up to 700,000 copies, enabling the cell to be as large as it is; these chromosomes are distributed along the cell inside structures that are **pseudoorganelles**. Unlike the organelles seen in eukaryotic cells, these bacterial pseudoorganelles are formed by folds of the cytoplasmic membrane. With the ability to see these bacteria without the need for 100× magnification and the possession of organelle-like structures containing their DNA, this bacterial species challenges some long held beliefs about features that differentiate bacteria and eukaryotes.

Plasmids contribute additional genetic features

In the case of bacteria, it could be argued that **plasmids** are part of the genome because plasmids contain genetic material that is inside the cell. It would also be argued that plasmids are not part of the genome because plasmids are separate from chromosomes, can be acquired from other bacteria as a whole, and can be transient and therefore lost from a cell. Indeed, some plasmids can be gained or lost from a bacterial cell without impacting upon the organism. Other plasmids have evolved within the bacteria and carry genes that contribute to the characteristic traits that are attributed to the organism. Whether a plasmid should therefore be considered part of the genome or not is likely to be dependent on the bacteria in question and even then may be the subject of debate.

Plasmids contribute to evolution and diversity in bacteria through their presence. These circular DNA molecules bring to the bacterial cell alternative genes that are not present in the genome, which may open up new niches and enhance survival of the bacteria. In turn, the enhanced survival of the bacteria provides for the persistence of the plasmid, providing an environment in which the plasmid DNA is replicated and maintained. By carrying a trait that can increase survival of the bacteria, such as antibiotic resistance, the plasmid can ensure not only its continuation, but also its spread throughout the bacterial population. Such spread is then only limited by compatibility of the plasmid and its genes with the bacterial cell and its systems.

Additionally, some plasmids ensure their persistence through mechanisms such as the toxin/antitoxin systems, whereby a bacterial cell is destroyed if the plasmid is lost. These systems produce a long-lasting toxin within the cell that is neutralized by a short-lived antitoxin. So long as the genes for both are within the bacterial cell, it can survive. But, if the plasmid is lost, the action of the antitoxin stops before the action of the toxin and there is no genetic material from which to make more antitoxin. Therefore, the bacteria in the population that lose the plasmid are killed by the toxin in an effective mechanism that ensures that plasmid is kept in the bacterial population once it has been acquired. This is referred to as **post-segregational killing** (**PSK**) because it happens after the plasmid DNA has been segregated into the dividing bacterial cells.

When DNA that looks like a plasmid may actually be a chromosome

Some plasmids have been found that contain essential genes. Although plasmids that carry essential genes are likely to be much smaller in size than the chromosome(s) within the bacterial cell, it may be that the best designation for these sorts of plasmids is chromosome. The presence of essential genes differentiates a circular secondary chromosome from a **megaplasmid**, which is a very large plasmid. Like plasmids, chromosomes may have various numbers within cells. Chromosomes carry essential genes and genetic features, regardless of size. The designation of chromosome is therefore dependent on our ability to analyze the DNA sequence and identify essential genes. As the features of plasmids are explored in greater depth and the functions of the genes encoded are better defined, more essential genes may be identified. It is therefore possible that some DNA that was initially designated as a plasmid may be better described as a small chromosome due to the presence of genes found to be essential.

Like a chromosome, each plasmid has its own origin of replication and goes through the process of replication independently of other DNA in the cell. Plasmid replication results in variation in **plasmid copy number**, or the number of copies of the plasmid that are present in the cell at any one time. Some plasmids may have just one copy in the bacterial cell, similar to the chromosome, while others may be present in 100–200 copies.

Prophages add bacteriophage genomes to a bacterial genome

Within the bacterial genome there may be one or more **prophages**, which are the genomes of bacteriophages, the small virus-like particles that can infect bacterial cells. During the bacteriophage lysogenic phase, its genome is integrated into the bacterial genome, where it is then carried, replicated, and passed on to new bacterial cells. The bacteriophage genome is thus able to make use of the biosynthetic systems within the bacterial cell to copy itself by becoming a prophage within the bacterial genome.

In addition to the bacteriophage genes, which are responsible for making new bacteriophage particles, the prophage genome may contribute genes to the bacteria that can influence the bacterial phenotype. These additional genes in the prophage genome contribute to the genetic material of the bacterial cell in a manner similar to the contribution of genes on a plasmid. Some bacteriophages are known to contain virulence factors and toxin genes that may alter the characteristics of the bacterial species once it acquires the prophage. Some of the prophages in bacterial genomes, with their additional genes, create strain-specific genotypes that are a differentiating factor between different strains of the same species that display different virulence characteristics.

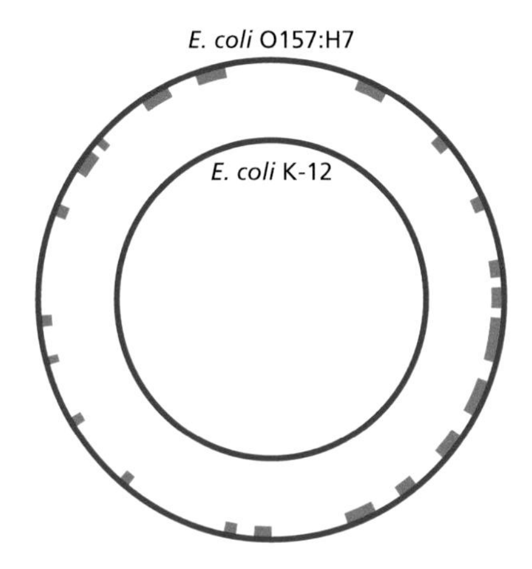

Figure 3.3 Differences in genome size between different strains of bacteria. Shown here is a schematic of the *E. coli* strain K-12 genome (inner circle) compared with the *E. coli* O157:H7 genome (outer circle). The regions in red are present only in the *E. coli* O157:H7 genome sequence and contribute to the difference in genome size between the bacteria of over 1 Mb.

Prophages can be major contributors to the content of a bacterial genome. In *Escherichia coli*, for example, the prophage genomes within the bacterial genome can account for up to **1 Mb** (**megabase**) of the DNA in the cell, differentiating some strains of *E. coli* from other strains of *E. coli* with smaller genomes (**Figure 3.3**). There will be more on the contributions of bacteriophages to bacterial genomes later in the text, including an entire chapter dedicated to bacteriophages (see Chapter 21).

The sizes of bacterial genomes are characteristic of bacterial species

Each species of bacteria has a typical genome size and a typical number of genes, although not all genomes of the same species have the same genes. Within a bacterial species there are different **strains**, that is, bacteria which vary from one another slightly on the genetic level, although they are all within the same species. The presence of prophage and other factors that may be strain-specific can alter the size of a genome beyond the scope of the typical genome size for the species. There will, however, be a typical size or size range for the genome of each bacterial species and comparisons against that size can reveal whether large changes in genome size, either insertions or deletions of genetic material, have occurred.

Regardless of the size of the genome itself, there is a general rule of thumb to approximate the number of genes within a bacterial genome. Due to the size of a typical bacterial gene and the length of DNA typically found between genes and associated with promoter regions, generally there is one bacterial gene per each **kilobase** (**kb**) of DNA. By this rule, it can be estimated that there are approximately 2,000 genes in a 2 Mb genome or 3,000 genes in a 3 Mb genome, which will be annotated as coding sequences (discussed in Chapter 2). Therefore, there is a correlation between the size of a bacterial genome and the number of genes that can be found within the genome (**Table 3.1**).

TABLE 3.1 Identified coding sequences and genome sequence size

Bacteria	Coding sequences	Genome size (base pairs)
Sorangium cellulosum	9,375	13,033,779
E. coli O157:H7	5,449	5,620,522
Anaeromyxobacter dehalogenans	4,346	5,013,479
E. coli K-12	4,140	4,641,652
Bacillus subtilis	4,175	4,215,606
Streptococcus pneumoniae	2,125	2,160,842
Neisseria gonorrhoeae	2,157	2,153,922
Haemophilus influenzae	1,610	1,830,138
Helicobacter pylori	1,469	1,667,867
Candidatus Zinderia insecticola	206	208,564
Candidatus Nasuia deltocephalinicola	137	112,091

Contributions of the core genome to defining the species and the accessory genome to defining the strain

Although genome sizes can vary within a species, due to strain-specific differences in the presence of genes, prophages, and other elements, the core content of the genome will remain the same. This core content is a set of genes and other genetic material that is conserved among all bacteria within a species. This is the **core genome** (Figure 3.4). Specifically, the genes that are conserved within a bacterial species are the **core genes** whereas the core genome includes other conserved genomic content, such as rRNAs and tRNAs.

It is the genes within the core genome that characterize and define the bacterial species, as modified and regulated by the other noncoding genetic features. This core genetic material will encode the traits that are typically expressed by all of the strains within a species and determine how and when these traits are typically expressed. The core genes tend to be fundamental to the bacterial species, including those that are essential for life and those that are important for survival in the typical niche occupied by the bacterial species.

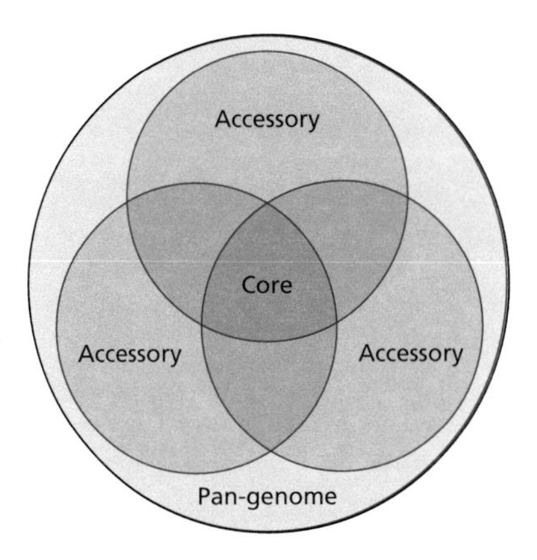

Figure 3.4 The core genome, accessory genome, and pan-genome. The core genome includes all of the genetic features that are included in all of the examples of a particular species. Each genome will, in addition, have other genetic features that contribute to the accessory genome. When all of the potential genes in a species are considered, everything from the core genome and all accessory genomes, even when not seen together in the same cell, this is the pan-genome.

Different strains within the same species may also possess other genes and other genetic material that may not be shared across the whole of the species. This is referred to as the **accessory genome** because it contains other genes and features that are not present in all representatives of the species (see Figure 3.4). These genes, considered accessory genes, can include virulence factors that vary between strains, producing particular virulence phenotypes such as the production of strain-specific toxins. The accessory genome can also include genes or regulatory systems that enable some strains within a species to survive in different environments from the standard niche for the species or features that enable some strains to withstand environmental changes in a niche that are not characteristics of the whole of the species.

All of the possible genes within a single bacterial species is the **pan-genome** (see Figure 3.4). It includes all of the core genome genes and all of the accessory genome genes that have been identified. Some accessory genome genes may never actually be found all together in the same bacterial cell and may indeed be mutually exclusive, where their presence in the same genome is not compatible. However, the species as a whole has the potential to carry any of the accessory genome genes along with the core genome genes; therefore, they are all possible genes of the species and included in the pan-genome.

Bacterial genomes are densely packed

Bacterial chromosomes are fairly densely packed with genes and other genetic features, including rRNA loci, small noncoding RNAs, and prophage. Compared with eukaryotic genomes, there is often little space, that is, very few base pairs (bp), between each of these features. Regions previously believed to be longer stretches of intergenic DNA, being between two genes, are being revealed to contain small noncoding RNAs and/or important regulatory sequences. As such, the bacterial genome is very densely packed, especially when all potential genetic features are considered. The distribution of features in a bacterial chromosome tends to be fairly even across the genome. This means that the architecture of the genome, including its content, homology, and **synteny** in comparison with other sequences, and the architecture of the individual chromosomes can be analyzed easily, as there will be very few gaps, and can contribute to understanding the nature of the genome and its contents.

DNA base composition differs between species

Most of the chromosome, as a whole, will share the same composition with regard to the proportion of each of the nucleotide bases that it contains: the number of adenines and thymines and of guanines and cytosines. Erwin Chargaff noted in 1950 that DNA was made of the bases adenine, cytosine, guanine, and thymine, but that these bases were not equally represented in all bacterial species. The variations in the proportions of adenines and thymines and of guanines and cytosines can be illuminating.

The number of guanines and cytosines (**G+C**) in bacterial genomes ranges from 14% in *Candidatus Zinderia insecticola* to 75% in *Anaeromyxobacter dehalogenans* of the total genomic composition. This means that most of the bases in *Z. insecticola* are adenine and thymine base pairs and that most of the bases in *A. dehalogenans* are guanine and cytosine base pairs. The characteristic G+C can give us insight into the bacterial species and its behaviors. In general, the G+C within the genomes of free-living bacteria is higher than that found in bacteria that cannot survive outside their hosts. Therefore, the G+C calculation can reveal something about the typical growth environment of the organism being studied.

The cost of synthesizing the raw materials required for guanines and cytosines, the GTP and CTP within the cell, is higher than the cost of synthesizing the ATP and TTP for adenines and thymines. This cost difference works against genomes having a high G+C content from an energy point of view. As a result, bacterial species that inhabit nutrient-limited niches where energy may be a limiting factor tend toward lower G+C content.

In addition, larger genomes, such as those greater than 3 Mb, tend to show higher G+C than smaller genomes. Therefore, the genome size should be taken into account when assessing the G+C and its likely implications, as this could be a reflection of size rather than just habitat.

Base composition differs between coding region

The G+C is not universal across the whole of the genome. For example, the conserved function of rRNA means that DNA sequences of these loci tend to have slightly different G+C from the rest of the genome. Coding sequences and intergenic regions, however, will tend to adhere to the overall G+C of the genome. Alteration of a gene sequence to suit the genomic G+C average is far less likely to have a detrimental impact upon the organism than would changing the sequence of the highly conserved rRNA. The availability of different codons corresponding to the same amino acid means that codons within gene sequences can be changed to synonymous codons without changing the amino acid sequence encoded by the gene, therefore providing scope for changing the gene to suit the G+C of the genome.

The origin of replication impacts the base composition

The **origin of replication** is the location along a chromosome where replication begins. It is a region that is recognized by the replication machinery and where replication will consistently start. This impacts upon the nature of the chromosome, including what genes are present and what is encoded in this region, the orientation of genes in this region, and the base composition of the DNA in this region.

Near the bacterial chromosome's origin of replication, the DNA sequence tends to be higher in guanines and cytosines, while there tend to be more adenines and thymines near the **terminus**, where replication ends. In addition, there is a bias for bases depending upon if they are on the **leading strand** or the **lagging strand** of replication and it is this bias that can help identify the origin of replication.

There are differences in the mutations that can accumulate on the leading and lagging strands of replication, introduced due to the differences in replication mechanisms on each strand. As a result, the leading strand tends to have more guanines, while the lagging strand tends to have more cytosines. Through recognition of these leading and lagging strand biases, it is possible to identify the origin of replication on the bacterial chromosome and the **termination of replication**: the location where a round of chromosomal replication ends.

The calculation of G–C/G+C for the leading and the lagging strand is the **GC skew**. Trends in the GC skew can identify the origin of replication. As seen in Chapter 1, the leading and lagging strands at the two replication forks are on opposite strands, resulting in the GC skew. Assessing the GC skew around a circular chromosome can create a recognizable pattern that assists in the definition of the origin of replication for the chromosome (**Figure 3.5**).

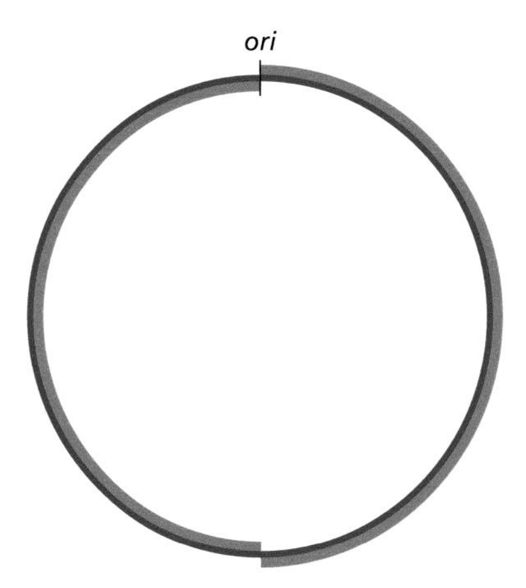

Figure 3.5 The GC skew of a circular bacterial chromosome can identify the origin of replication. The calculated G-C/G+C (red) is here mapped to the circular chromosome (blue). Mapping of the G-C/G+C data onto the circular chromosome reveals the bias in base composition for the replication leading (right) and lagging (left) strands. Therefore, the origin of replication can be inferred from these data (ori).

The influence of the leading strand and the lagging strand on gene orientation

There tend to be more genes on the leading strand of the bacterial chromosome versus the lagging strand, due to the continuous nature of replication on the leading strand. Although genes can be encoded on either strand, there is a bias toward genes being present on the leading strand, which is replicated continuously. The presence of a greater number of genes on the leading strand creates a strand bias for the orientation of genes in the chromosome, which can be observed when analyzing the genome sequence (**Figure 3.6**).

This abundance of genes on the leading strand is the result of evolutionary pressures. The observed strand bias against genes on the lagging strand is likely to be due to the progression of both DNA replication proteins and RNA transcription proteins on the same DNA strand. Inside the bacterial cell the processes of transcription and replication are happening at the same time, using the same chromosome as their templates for copying the DNA and for generating mRNA from the DNA. The processes progress at different rates along the DNA and as a result they are likely to collide as they are conducting replication and transcription.

When replication and transcription are both going in the same direction on the leading strand, there is a chance of collision between the proteins involved when they catch up to each other. However, collisions are perhaps either more frequent or more disruptive on the lagging strand because of the discontinuous nature of replication on the lagging strand, regardless of the speed of the processes. All collisions between replication and transcription on either strand will slow the progress of the replication fork.

Highly transcribed genes are likely to be being transcribed during replication. The transcription of highly expressed genes will therefore cause fewer collisions if the genes are located on the leading strand, rather than the lagging strand. Indeed, the most heavily transcribed regions of the genome are the rRNA loci, which are

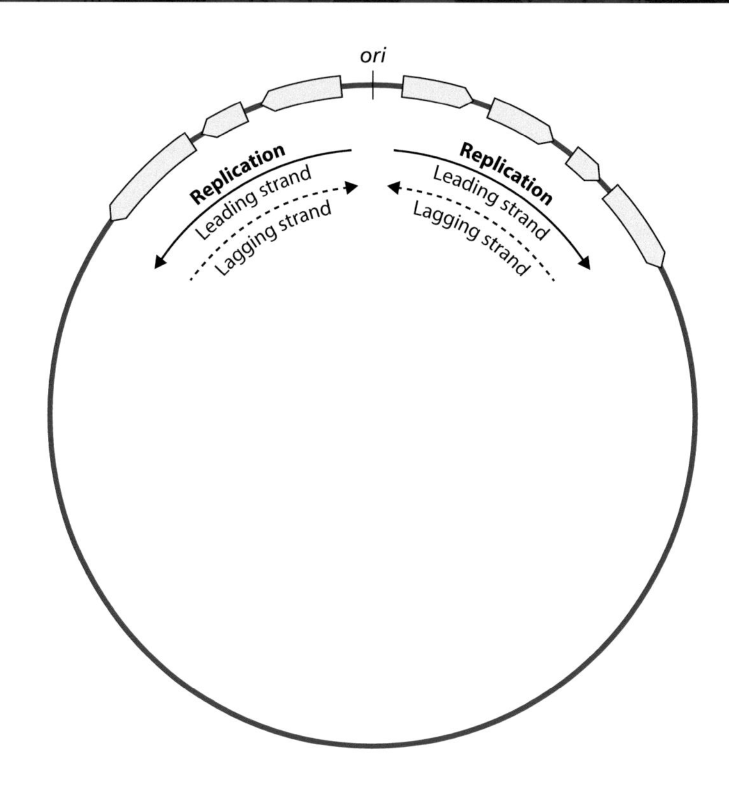

Figure 3.6 Presence of genes on the leading strand and the lagging strand. The origin of replication (*ori*) for this circular genome is shown and the direction for the leading strand of replication is indicated. The genes and their orientation on the chromosome are illustrated, showing that most of the genes are on the leading strand, being encoded divergently from the origin of replication. These leading strand genes are transcribed in an orientation that goes away from the origin of replication.

present on the leading strand. To avoid or minimize the effects of clashes between transcription mechanisms and replication mechanisms, most genes are also encoded on the leading strand.

Essential genes tend to be on the leading strand, regardless of their expression level. Mutations can occur as a result of replication stalls following a polymerase collision. Perhaps the positioning of essential genes on the leading strand is beneficial because this placement is more likely to avoid the introduction of mutations into essential genes during such replication stalls.

Genomic architecture can impact gene expression

The location of a gene within the chromosome can impact upon its expression. For example, during replication, genes that are near to the origin of replication are copied first. This creates two copies of these genes in the cell, increasing the gene dose. Replication does not necessarily wait for one round to end before another begins, so as further rounds of replication initiate, more copies of the genes near the origin are generated. The genes near the origin therefore experience a **dosage effect** and tend to be the most highly expressed genes in the genome (**Figure 3.7**).

Although the process of replication is about copying the genetic material within the cell, it can also impact upon the process of transcription, which generates mRNA. On the lagging strand, the progression of DNA polymerase during replication displaces any RNA polymerase that is in its way. This stops all transcription by the RNA polymerase from the genomic region until replication moves on. Therefore, the expression of genes on the lagging strand is prone to disruption. The impact on the leading strand is less disruptive.

Conservation of the order of genetic features between bacterial species

Homology is a measure of the similarity between sequences at the DNA or protein level. Another term, **synteny**, is used to describe similarity in the presence and order of features within a chromosome. Synteny refers more to the structure and content of regions than it does to their actual sequence.

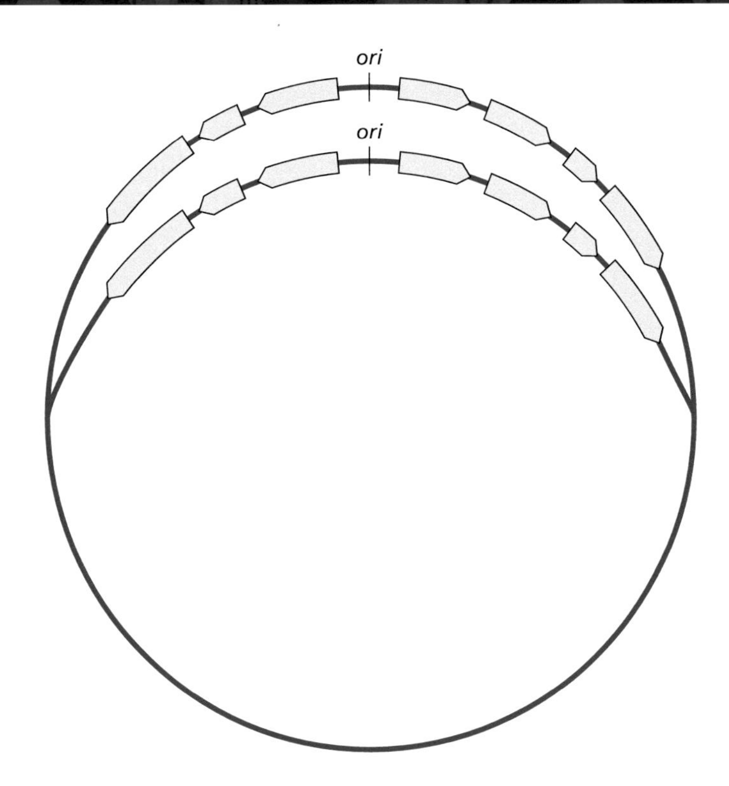

Figure 3.7 The dosage effect upon genes of their presence near the origin of replication. When genes are present close to the origin of replication (*ori*), this can impact the number of copies of the gene present in the bacterial cell. As replication progresses, the part of the bacterial chromosome near the origin of replication is present in more than one copy. Thus, the genes in this region are present in the cell in more than one copy, each capable of being a template for transcription. When there is more than one chromosome, gene dosage can affect more genes due to the availability of origins of replication on each chromosome.

Early in genomic investigations, it became apparent that genes encoding proteins involved in associated functions tended to cluster together on the chromosome. The close proximity within the chromosome of genes that contribute to related functions appears to be beneficial for the co-regulation of these genes. Adjacent genes may be co-transcribed into the same mRNA. These co-transcriptional units can form operons when they are co-regulated. There will be more on operons in Chapter 5. When analyzing a genome, it is important to consider whether adjacent genes may be co-transcribed, as this can reveal how the genes are regulated. Evolutionary pressures will tend to keep these blocks of co-transcribed and co-regulated genes together. For example, if a block of genes that function together are separated by a rearrangement of the order of genetic features in the chromosome, there may be a loss of fitness for the affected bacteria. The cells affected by the rearrangement will be selected against in the bacterial population and those that retained the original order of the genes will be selected for in the population. Therefore, evolutionarily, regions that contain clusters of genes with related functions are preserved and this synteny can be seen across even very diverse bacteria.

One striking example of synteny that is retained in many bacteria, even those that are evolutionarily very distant, comes from common features of bacterial cells. All bacterial cells must divide into daughter cells to grow and survive. Most bacterial cells have peptidoglycan cell walls. Construction of more of the peptidoglycan cell wall is a coordinated event that coincides with the process of cellular division; as the cell grows to enable it to then divide, a greater quantity of cell wall material is required. Therefore, in most bacterial species the genes involved in cellular division and those involved in cell wall synthesis are clustered together. Chromosomal synteny can be seen at the division and cell wall synthesis (*dcw*) locus. There is similarity in the genes and their organization across species barriers at the *dcw* cluster, even when very diverse bacteria that are evolutionarily distant are considered (**Figure 3.8**). The *dcw* clusters therefore have synteny, because the genes and their configuration are conserved between bacterial species, even if the sequence homology calculations show that the sequences are not very similar. Other gene clusters that remain syntenic across evolutionarily diverse species include the NADH dehydrogenase operon (*nuo* genes) and the ribosomal protein operons.

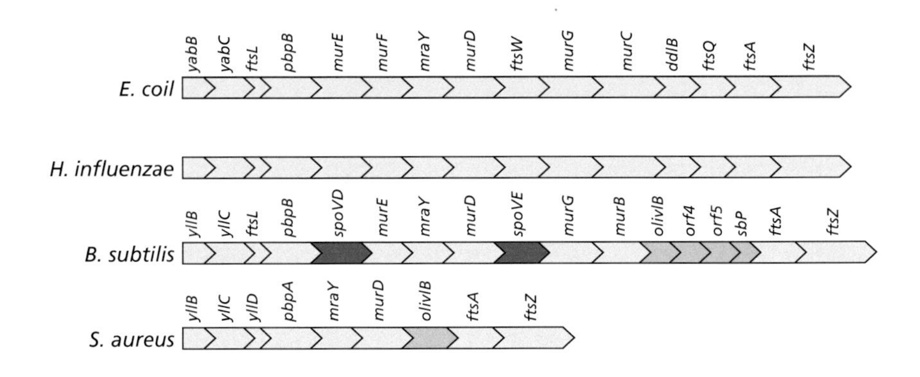

Figure 3.8 The chromosomal synteny of the *dcw* cluster between diverse species. Shown here are the division and cell wall synthesis clusters of genes for *E. coli*, *H. influenzae*, *B. subtilis*, and *Staphylococcus aureus*. These species are quite diverse; therefore, the homology between the genes of the different species can be quite low. However, the overall organization and composition of the *dcw* clusters are similar, making them a region of synteny between these species.

Distantly related species may only have a few key gene clusters that are conserved. Comparisons of closely related bacterial species will reveal greater synteny between the chromosomes. When assessing a region of a chromosome and comparing two or more species, synteny is one of the features investigated. If two species share synteny over a region, then they may share the function of the genes encoded in the syntenic regions. Comparing genomes, there may also be a gene that is present in just one strain or species in a region that otherwise has synteny; this may reveal something about differences between the bacteria. Changes in synteny, in the order in which genes are found in a chromosome, may indicate differences in the regulation and expression of genes between two strains or species.

Supercoiling can also influence gene expression

The DNA that makes up the genome and any additional genetic material, such as plasmids, must all be contained within the bacterial cell. Each 10 bases in a DNA double helix is 3.4 nm long. Bacterial genomes can range in size from 112,091 bases (in *Candidatus Nasuia deltocephalinicola*) to 14,782,125 bases (in *Sorangium cellulosum*); therefore, the genomes can range in size from approximately 38,111 to 5,025,923 nm. Bacterial cells range from 200 nm (in *Mycoplasma gallicepticum*) to 750,000 nm (in *Thiomargarita namibiensis*). The DNA that must be inside the cell is often far larger than the cell itself. In order to fit all of the genetic material inside the bacterial cell, the DNA must be compacted. To achieve this, bacterial cells will **supercoil** their DNA to make a very long molecule into a mass of DNA that is far more compact. There are some proteins that aid in the process of supercoiling, help retain supercoiling, or otherwise aid in bending and compacting the genome, which will be introduced in Chapter 5. To achieve a compact state and fit within the bacterial cell, most of the DNA is negatively supercoiled, that is, twisted against the normal turn of the helix such that it becomes much smaller and able to fit within the cell (**Figure 3.9**).

Unlike eukaryotic cells, bacterial cellular DNA is not contained within the membranes of a nucleus. Bacteria do not have membrane-bound organelles such as nuclei. However, due to the supercoiling of the DNA into a compact mass of genetic material, a **nucleoid** can be observed. Although the nucleoid is not a discrete organelle, the nucleoid is a location within the cell that contains the DNA (**Figure 3.10**).

The compact structure of the bacterial chromosomes can limit the availability of the sequences of some genes for transcription, depending upon where the genes are located within the mass of DNA. It has been observed in experiments with *E. coli* that when the cell is treated with a chemical to halt transcription and translation, the nucleoid becomes a very compact sphere. However, the processes of transcription and translation need access to the DNA strands, which become more loosely associated with the nucleoid, associating instead with transcription and translation machinery. This gives the nucleoid its irregular appearance in the cell. Supercoiling of the DNA can make it difficult for RNA polymerase to access promoter sites and initiate transcription for regions of the chromosome buried

Figure 3.9 Supercoiling of DNA compacts its size. Shown is the space taken up by a DNA molecule of 1 Mb if it is stretched out straight. Following supercoiling, the same 1 Mb of DNA takes up far less space and is able to fit inside the bacterial cell.

deep within the compacted nucleoid structure. Likewise, genes that are highly expressed can impose limits on the compaction of a genomic region, due to the continual presence of the coupled transcription–translation machinery.

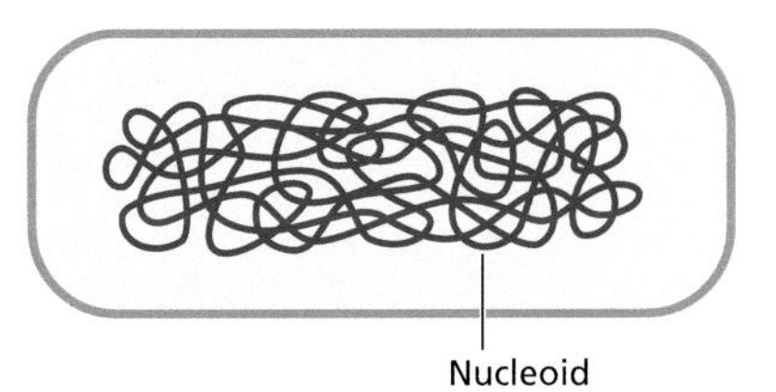

Figure 3.10 The mass of supercoiled DNA that is the bacterial nucleoid. Supercoiling of DNA compacts the genome so that it fits inside the cell, producing a mass of condensed genetic material in the cell that can be seen and is identified as the nucleoid.

Distribution of noncoding genetic features in the bacterial chromosome

In addition to genes, there are noncoding genetic features present within the chromosomes of bacteria, including the locations that encode the tRNAs. These features are carried within the DNA and transcribed to the tRNAs that carry amino acids to the ribosome. There can be multiple tRNA loci in different locations in the genome for each amino acid, with each tRNA carrying a unique anticodon that corresponds to the different codon possibilities for each amino acid.

The rRNAs that are present within the ribosomes are also carried in the DNA genome. These loci contain the sequences for all of the ribosomal RNA. The rRNA locus is transcribed and the rRNAs are incorporated into the bacterial 70S ribosome along with ribosomal proteins. Together these components make up the 50S ribosome, with its 23S and 5S rRNA, and the 30S ribosome, with its 16S rRNA. In bacteria, there can be several copies of the rRNA locus present around the genome (**Figure 3.11**), each carrying the sequences for 23S, 5S, and 16S rRNAs. These can be scattered around the single circular chromosome, creating regions within the chromosome of DNA sequence identity (or near identity).

Further, there are small RNAs present as genetic material within bacterial cells that are transcribed but not translated. These **noncoding RNAs (ncRNAs)** tend to be present in intergenic regions or as regions overlapping coding regions, but on the opposite strand. In some cases, ncRNAs have been identified as genes in error, being labeled in the genome sequence data as coding regions. Functional studies and increased understanding and recognition of ncRNAs can help clarify the assumptions made about the nature of features in genome sequence data. Regions previously believed to be "junk DNA," or devoid of genes, have been found to contain ncRNAs that serve vital regulatory functions within the cell. Further study of these and other yet to be discovered features of bacterial genomes may reveal even more within the sequence of DNA.

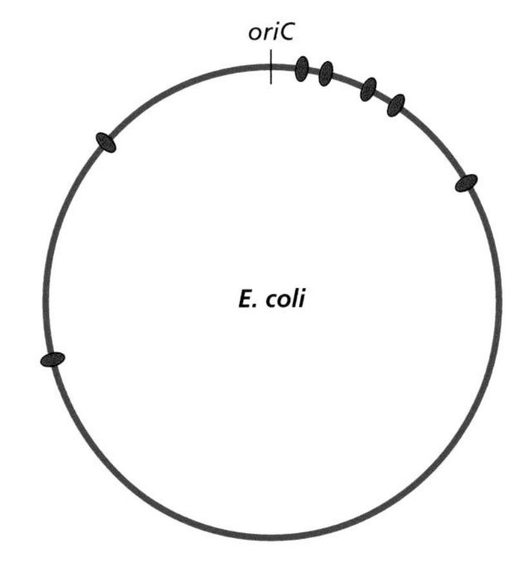

Figure 3.11 The presence of rRNA loci in bacterial genomes. Some bacterial genomes, like the *E. coli* shown here, have more than one rRNA locus in the genome. Here there are seven loci in this single circular chromosome, with each locus encoding the 5S rRNA, 16S rRNA, and 23S rRNA (indicated in red).

Mutations in the bacterial genome

The bacterial genome is subject to mutations: changes in the genetic code that may, or may not, alter the encoded features. **Mutation** drives bacterial evolution. Through the changes to the organism's genetics, bacteria are able to expand into new niches and display new characteristics. A change to the genome as a result of mutation can increase, decrease, or have no effect on the fitness of the organism.

Transposons, IS elements, and other mobile elements contribute to genomic changes

There are other genetic elements in the DNA that contribute to changes within a bacterial genome, many that differ in presence or location between strains of the same bacterial species. **Transposons**, for example, are mobile genetic elements capable of self-excision and reintegration into the chromosome. Transposons carry within their sequences all of the genes necessary for mobilization to new parts of the bacterial genome. Transposons can also carry other genes, such as antibiotic resistance determinants.

Insertion sequence (IS) elements are related to transposons in structure and mobilization; however, **IS elements** do not carry additional genes. Like transposons, IS elements can move from one part of a bacterial genome to another. Introduction of new genetic material at a site in the genome through the insertion of an IS element can disrupt the genes of the region or their regulation. Both transposons and IS elements can contribute to genetic mutation by disrupting genes in this way.

Other mobile elements include **prophages**, which are capable of self-insertion during the bacteriophage lysogenic phase and which excise from the chromosome during the lytic phase of bacteriophage growth. Prophage insertion into the genome may disrupt genes and regulatory networks in a manner similar to the insertion of transposons and IS elements. Like transposons, the introduction of a prophage genome may also bring with it new genes.

Small changes in the genome created by point mutations and indels

The process of replication is not perfect. As mentioned, there can be stalls in replication. These events may cause errors to be introduced into the DNA sequence. In addition, the replication machinery itself can sometimes introduce errors. While there are error correction systems within the bacteria to mitigate against the introduction of replication errors, not all errors are corrected. Therefore, changes to the sequence of the genome can be introduced during the process of replication. In addition, some bases can change spontaneously to another base. Unless it is corrected, the spontaneous change in a base can become fixed into the DNA sequence. These events create point mutations, that is, changes to single base pairs along the double helix.

The most common **point mutation** is cytosine to thymine. This change occurs as a result of the spontaneous **deamination** of the cytosine, forming uracil (**Figure 3.12**). Uracil is not a base that is normally found in DNA, so during

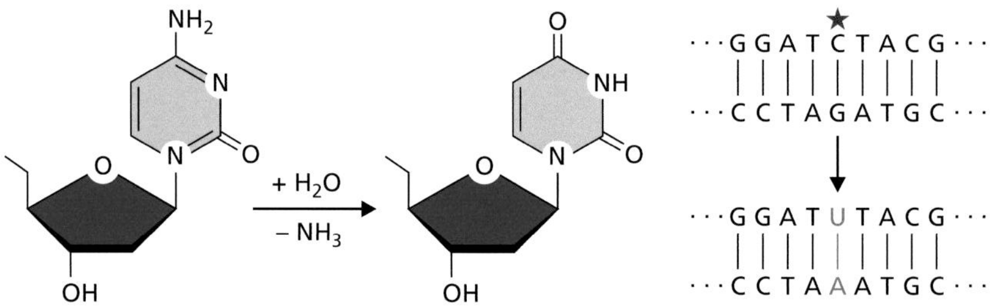

Figure 3.12 The deamination of cytosine to uracil generates a point mutation. Cytosine, by the addition of H_2O, loses NH_3 and becomes uracil. Uracil hydrogen bonds with adenine, creating a point mutation.

replication the uracil is copied as a thymine and thus there is a change from a cytosine to a thymine. These are not the only point mutations that occur. Any base can erroneously be changed, some as uncorrected errors in replication, some as chemical changes to bases, and some as errors introduced by repair mechanisms.

By the same mechanisms that introduce point mutations, another type of mutation called indels can occur. An **indel** is a short insertion or deletion of one or a few bases. These small changes can alter the reading frame of a coding sequence, frame-shifting a gene so that it prematurely reaches a termination codon. They can also disrupt regulatory regions, altering the binding sites for RNA polymerase and/or regulatory proteins.

Often when comparing genome sequences these changes can be observed, both single base changes in sequences and small insertions or deletions. Since it can be difficult when analyzing genome sequence data to correctly identify the original sequence, to determine if an insertion or a deletion has occurred, the term indel covers both possibilities. Likewise, a single base change is called a **single-nucleotide polymorphism** (**SNP**). The designation SNP is not necessarily a judgment on which base is mutated when comparing two genomes; it is simply an indication that the bases are different.

Translocations can change the order of genetic features in a genome

Translocation is movement of DNA from one place to another along or between DNA molecules. The transfer of DNA from one location on a DNA molecule to another location on the same molecule, from one chromosome to another, from chromosome to plasmid, or from plasmid to chromosome are all examples of translocation.

The key feature of translocations is the transposition within a chromosome or plasmid, where a segment of DNA is moved from one location to another. Translocation does not alter the genetic content of the cell, although genes may be disrupted in the process of the DNA moving if the borders of the translocation are within genes.

A change in the position of a gene within the chromosome can alter the expression of the gene through a variety of mechanisms. Altering the proximity of the gene from the origin of replication may either increase or decrease the gene dose, therefore providing more or fewer templates for transcription. A change in the strand upon which the gene is located can also impact upon its expression, with genes on the lagging strand being more prone to interference during transcription by the procession of the replication machinery. Translocation may also modify the regulation of the expression of a gene through causing a change in the relative location of genes from their regulatory elements. It may be that a promoter was disrupted by the translocation or that there has been an increase in the relative distance a regulatory protein would need to diffuse in the cell from where it was made by transcription and translation to its binding site.

Inversions flip the DNA strand upon which genetic features are located

Inversion is a form of translocation, where a DNA region is inverted, resulting in the DNA carried in the chromosome or the plasmid being in the opposite orientation of how it was originally. In this type of translocation, the DNA remains on the same DNA molecule, but is reintegrated into it so that the strands of the DNA are switched. What was on one of the strands will then be on the opposite strand (see Figure 3.12). Inversions can occur anywhere in the chromosome; however, due to the consequences of the inversion, not all remain fixed. Some inversions may reduce the fitness for survival of the bacteria and therefore are selected against evolutionarily.

Inversions that occur around the origin of replication can preserve both the synteny of the regions involved in the inversion and any gene dose effect produced

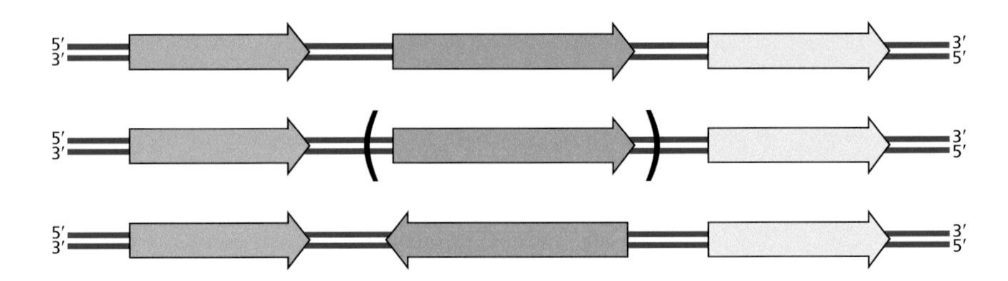

Figure 3.13 Inversion of DNA changes the strand of DNA upon which a sequence is located. Three genes are shown here, all encoded on the top strand. An inversion occurs, represented by the parentheses. The middle gene is now in the opposite orientation and on the bottom strand.

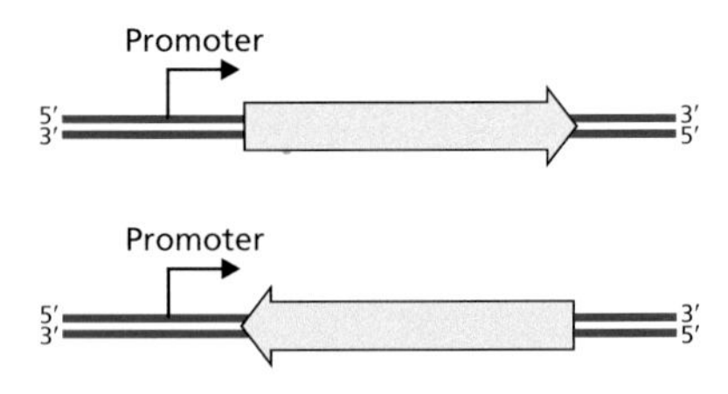

Figure 3.14 Programmed inversions can influence gene expression. In some bacteria, inversion events may place a gene, such as shown here, under the control of a promoter (bent arrow) in one orientation (top) or not in the inverted orientation (bottom). Therefore, the gene is only expressed in one configuration, subject to the inversion event.

by the proximity of the genes to the origin. In addition, the location of genes on either the leading or the lagging strands is also preserved when an inversion is centered on the origin of replication (**Figure 3.13**).

Inversions can be programmed events involved in gene-expression changes. In these cases, the bacterial cell possesses inversion mechanisms that are capable of recognizing the invertible region and completing its inversion. Generally, inversions can result in genes being placed and removed from the influence of a promoter region (**Figure 3.14**).

Recombination changes the genome

Recombination is a term to describe the exchange of genetic material between or within DNA molecules. This process makes possible rearrangements, deletions, inversions, and other changes to the DNA. Crucial to the process is the joining together of DNA strands that were not previously joined. In bacteria, this can occur within a chromosome, between chromosomes when the bacterial cell has more than one, between copies of the chromosome during replication, between chromosomes and plasmids, and between the bacterial DNA and fragments of DNA that have entered the cell.

Generally, **homology**, the similarity between sequences, can facilitate recombination. The homologous regions, where the DNA is the same or similar, will come together and be acted upon by the recombination machinery, resulting in a new DNA sequence. Non homologous recombination can also occur, bringing in the DNA sequence to a region with which it does not share sequence similarity.

Through recombination, the DNA within bacterial cells can change. Although these organisms do not reproduce sexually, they are capable of mixing, shuffling, and rearranging genetic information in their chromosomes (**Figure 3.15a**), which gives them scope for evolution and adaptation. In addition to the genetic material within the bacterial cell, recombination can act on DNA that has entered the cell from elsewhere, further enhancing the potential for change to the genome (**Figure 3.15b**).

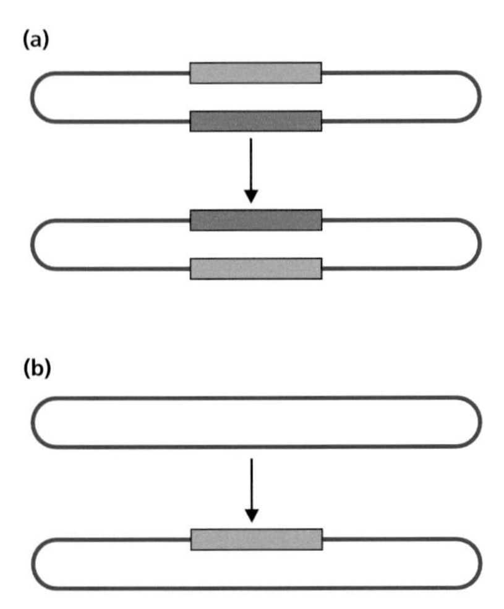

Figure 3.15 Recombination involves the exchange of DNA between two molecules. This can rearrange the sequence in the chromosome (a) or introduce new genetic material (b).

Horizontal gene transfer introduces new genetic material

Most of the changes to a bacterial genome can be attributed to **horizontal gene transfer** (**HGT**). This is a process whereby DNA from another organism is taken up into the bacterial cell and incorporated into the genome. In this way, very large changes can occur to the bacteria. DNA can be transferred between cells. Although bacteria do not go through sexual reproduction, HGT provides the scope for bacterial species to acquire new DNA or use DNA from other cells similar to themselves to repair their DNA.

HGT can result in sections of DNA that are different from the rest of the chromosome. These regions may only be present in some strains of a bacterial species, creating regions of difference between them. **Regions of difference** can also occur between strains and species when there is a duplication or deletion of genetic material; therefore, not all regions of difference can be assumed to be the result of HGT.

The clustering together in regions of the chromosome of genes that have related functions means that when these segments of DNA are horizontally transferred, the process is likely to transfer complete systems. This is advantageous when these genes work together for a shared function and are controlled by a shared regulatory system. Novel genetic traits can be obtained by the bacterial cell via DNA from other bacteria, which is then incorporated into its genome. It is also within the scope of HGT to provide a mechanism to repair a deleterious mutation by replacing it with DNA from an external source that carries the sequence without mutation.

HGT can result in incorporation of DNA into the genome that has a different composition from the rest of the genome. Markers of horizontally transferred DNA that originated from another bacterial species can include differences in the G+C content, changes in the GC skew relative to the trends of the chromosome, and differences in codon usage within the horizontally acquired region. Over time, the sequence of the horizontally acquired DNA may mutate to match the chromosome in which it is now incorporated, making it more difficult to identify. The DNA sequence may change to better accommodate the codon usage of the rest of the genome and the sequence may increase or decrease in its number of guanines and cytosines to better match the G+C of the genome.

There are limitations to high-frequency HGT in nature, both in terms of which bacteria engage in HGT and in terms of the impact of HGT. Not all instances of HGT will result in a change in the genome. DNA taken into the cell may not be incorporated and could instead be broken down, with its constituent nucleotides used to fuel replication. If HGT does result in the incorporation of a new gene or genes, each transfer event affects only one cell. That one cell may not survive. If, however, the horizontally acquired genetic traits increase the fitness of that bacterial cell, selective pressures will work to change the bacterial population.

In some cases, the regions of difference between strains of the same species are regions that are believed to have once been horizontally transferred. DNA can enter bacterial cells in three ways: transformation, conjugation, and transduction.

Transformation involves bacterial uptake of DNA from its surroundings

DNA from the environment can be taken up by some bacterial cells in a process called **transformation** (**Figure 3.16**). Some species do so naturally and are said to be naturally competent for transformation, while other bacterial species can be manipulated to perform transformation in the lab.

Species that are naturally competent for transformation include species such as *B. subtilis, H. influenzae, H. pylori, N. gonorrhoeae,* and *S. pneumoniae,* the latter being the bacterial species involved in the first transformation experiments that demonstrated that DNA was the genetic material (Chapter 1). *N. gonorrhoeae,* for example, takes up linear DNA from the environment, actively binding to this DNA and passing it through its membranes and into the cytoplasm where the bacterial genome resides. Some species do not take up linear DNA; however, they are competent to take circular plasmid DNA into their cells. Once inside, a plasmid can then become established within the strain, self-replicating and being

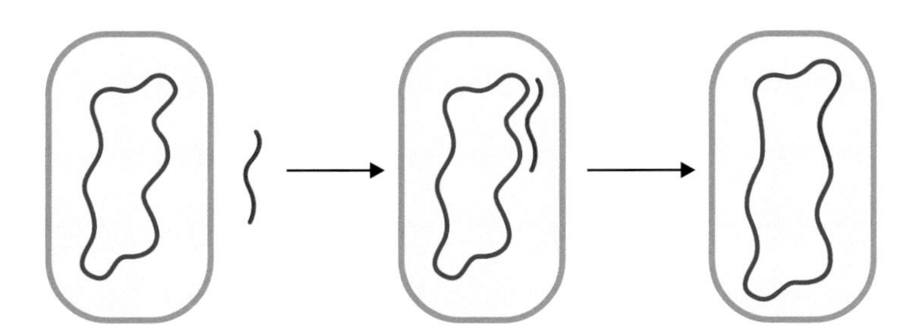

Figure 3.16 The process of bacterial transformation results in DNA from the environment entering into the bacterial cell. Once inside the cell, recombination can integrate the DNA into the chromosome.

passed on to the next generation of bacteria. Some species that can be naturally competent for transformation are not competent at all times. *S. pneumoniae*, for example, is only competent once the bacteria reach a phase of growth in which there are sufficient bacteria growing in the local environment. In other species, such as *B. subtilis*, competence is at the onset of stationary phase or at high cell density, but even then, only about 10% of the bacterial population is naturally competent.

E. coli is the work horse of molecular biology, being used as a tool for genetic manipulation of itself and other organisms. Plasmids can be constructed and then introduced into competent *E. coli*, where these bacterial cells will make more of the plasmid. Indeed, some plasmids are designed so that the *E. coli* cells will express the genes on the plasmid as proteins. All of this hinges on getting genetically engineered DNA into the *E. coli* cells, yet this is not a species that is naturally competent for transformation. Instead, we further manipulate the bacteria in the lab so that they will take up the desired DNA.

E. coli can be made competent for transformation using a number of established protocols that influence the surface of the bacteria and their membranes, allowing DNA to pass into the cells during a brief shock of higher temperature. Once created, competent cells can be cryostored in an ultra-cold freezer until they are needed. Such cells can also be purchased, ready-made, from laboratory supply companies. Alternatively, *E. coli* can be made electrocompetent using established protocols to make the membrane receptive to DNA that will enter the cell when transient pores are made in the cell membrane through the application of an electric shock. The processes of making competent and electrocompetent *E. coli* cells will be explored in Chapter 18.

Conjugation is an encoded mechanism for DNA transfer from one bacterial cell to another

Some bacteria participate in a process called **conjugation**. Bacteria possessing **conjugation pili** can directly inject DNA into other bacterial cells. Those with the pili and associated conjugation machinery are donors and those into which DNA is injected are recipients (**Figure 3.17**).

To be a donor for conjugation, the bacteria must possess certain genetic traits, referred to as the **fertility factor** or **F-factor**. Generally, this is contained on an **F plasmid**, which through the conjugation machinery is able to transfer a copy of itself from the donor bacterial cell to the recipient cell, therefore establishing the F plasmid in the recipient. The recipient then becomes a donor, a cell capable of conjugation.

Recombination can, on occasion, bring the F plasmid into the chromosome of the bacterial cell. These are referred to as **Hfr** strains for **high frequency of**

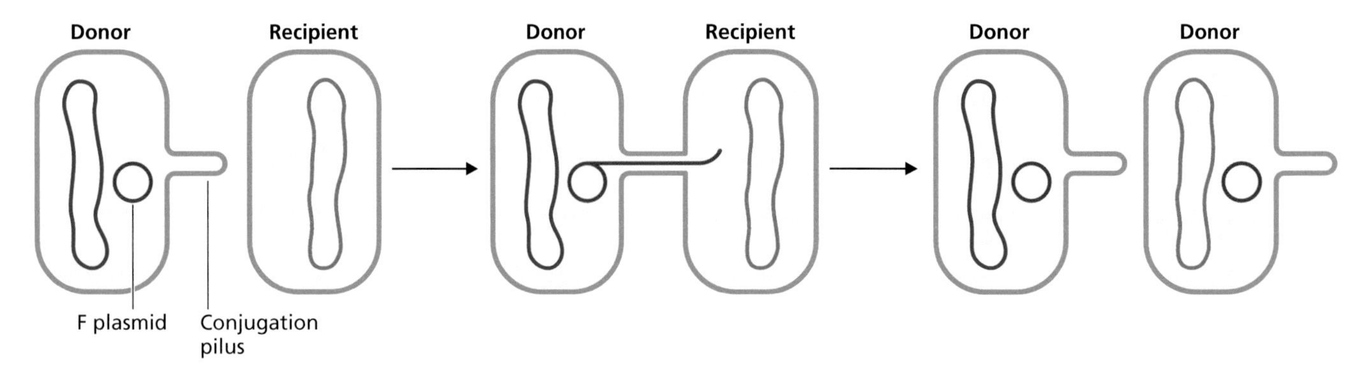

Figure 3.17 Conjugation is able to transfer DNA directly from one cell to another. A cell with the F plasmid, a donor cell, is able to make a conjugation pilus (left). With the pilus, it can attach to a recipient cell and transfer a copy of the F plasmid through the conjugation pilus into the recipient cell. Once complete, the recipient has a copy of the F plasmid and becomes a donor cell.

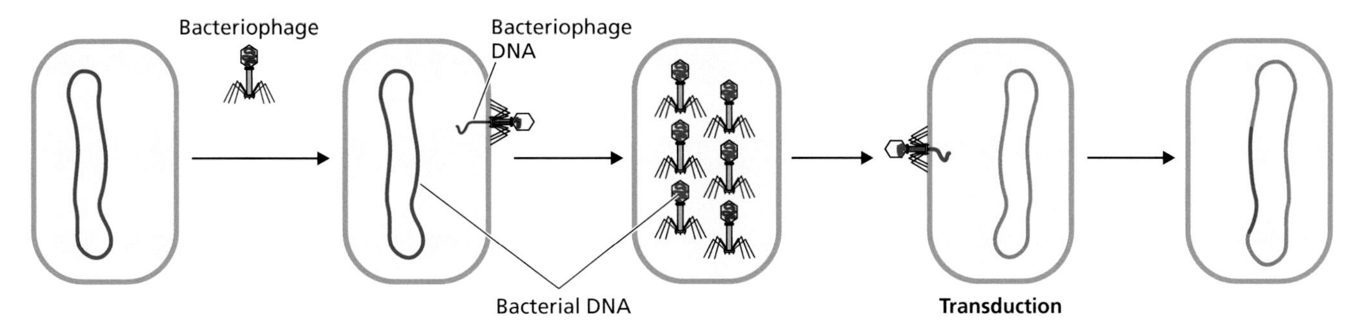

Figure 3.18 A bacteriophage can transfer DNA from one bacterial cell to another in a process called transduction. A bacteriophage attaches to a bacterial cell and transfers into the cell its bacteriophage genome (left images). The bacteriophage genome directs the bacterial cell to make more bacteriophage particles (middle image). These particles are packaged with bacteriophage DNA but may also package bacterial DNA. When these bacteriophages infect a new bacterial cell, they carry the bacterial genetic material to the new cell (right images).

recombination. When an Hfr strain engages in conjugation, it takes with it the DNA into which it has integrated, therefore transferring chromosomal DNA from one cell to another, rather than plasmid DNA. The transferred chromosomal DNA, with its homology to the chromosome within the cell, has a high frequency of recombination with the recipient chromosome.

The process of conjugation requires direct contact between the bacterial cells and therefore can be disrupted if the connection between the cells through the conjugation pilus is lost. This phenomenon has been exploited in mapping experiments to identify the location of different genetic features on a chromosome. When located in this way, the feature is said to be at a certain number of minutes on the chromosome, corresponding to the time it takes for a feature on an Hfr chromosome to transfer to a recipient cell.

The process of transduction can introduce bacteriophage DNA into a bacterial cell

Transduction can transfer DNA from one bacterial cell to another via bacteriophages. Often the transferred DNA is bacteriophage DNA; however, it is possible for bacterial genomic DNA to be transferred in this way (**Figure 3.18**).

During the process of making new bacteriophage particles within the bacterial cell, the bacteriophages are assembled and filled with DNA. This process may result in some bacterial DNA being packaged into a bacteriophage particle. In this way, the bacterial DNA can be moved from one bacterial cell to another. This process, unlike conjugation, does not require the bacteria to be in contact with one another, as the bacteriophage acts as the intermediary in the process of moving the DNA from cell to cell.

Bacteriophages dock with bacteria through interaction with the bacterial surface; therefore, these transfers are limited by bacteriophage specificity. Transfer of DNA from one bacterial cell of a species to another of the same species is more likely to occur in transduction than transfer to unrelated species.

Key points

- Bacteria can have one or more chromosomes that make up the genome, containing the genetic material of the cell.
- The size of a bacterial genome can be influenced by its environment, yet the size is generally characteristic of the species.
- Bacteria can have extra-chromosomal DNA, such as plasmids, that carry accessory genes.
- There is roughly one gene for each 1 kb of DNA sequence.

- The genes conserved in a species and found in all examples of the species make up the core genome, with other genes making up the accessory genome.
- The size of the genome and the environment in which the bacteria are generally found will influence the number of guanines and cytosines found in the DNA sequence.
- The origin of replication will influence the expression of genes through gene dose and through the orientation of genes on the leading and lagging strand.
- The introduction of transposons, IS elements, and prophages into regions of the genome may incorporate new genes into the bacterial cell and may disrupt existing genetic features.
- Small changes in the DNA sequences between closely related bacteria, such as SNPs and indels, are likely to have occurred due to replication errors and other mutagenic events.
- HGT can result in the incorporation of new DNA into the genome, resulting in regions of difference between strains of the same species.
- Transformation, conjugation, and transduction are naturally occurring mechanisms for bacterial uptake of DNA from outside sources.

Key terms

Define the following terms introduced in this chapter. Check your answers using the definitions in the Glossary. These terms are also available as Flashcards online.

Accessory genes	Horizontal gene transfer	Recombination
Accessory genome	Indel	Regions of difference
Conjugation	Inversion	Strain
Conjugation pili	Insertion sequence elements	Supercoil
Core genes	Kilobase	Synteny
Core genome	Megaplasmid	Termination of replication
Deamination	Mutation	Transduction
Dosage effect	Nucleoid	Transformation
F plasmid	Pan-genome	Translocation
G+C	Plasmid copy number	Transposon
GC skew	Point mutation	
High frequency of recombination	Pseudoorganelle	

Questions and discussion topics

Self-study questions

Answer each question using 50–100 words or a table or labeled diagram. Advice on where to find answers to these questions is available online.

1. What is the term for a bacteriophage genome that has been incorporated into a bacterial genome and what types of genes might be found there?
2. What are the typical size ranges of bacterial genomes and what features can alter genome size within related bacterial genomes?
3. How is it possible that the entire length of the bacterial genome is able to fit inside the bacterial cell?
4. In general, what is the most likely orientation of genes near the origin of replication?
5. What is the term used to describe the similarity between the presence and order of features in the genome and how does this compare to homology?

6 What do we now know is present in regions once believed to be "junk" DNA?

7 Two genetic elements are able to mobilize in bacterial genomes. What are these elements and how does their movement introduce mutations?

8 Draw the structures involved in the deamination of cysteine to uracil.

9 Compare and contrast the difference between an indel and an SNP.

10 What are the consequences of translocation of DNA in a bacterial genome?

11 Draw a diagram of how an inversion event can influence the expression of a bacterial gene.

12 Bacterial cells can acquire new DNA via three mechanisms. What are they, and explain briefly how do they work? Which one requires direct contact between bacterial cells?

Discussion topics

These topics are presented for discussion in study groups, as part of class discussions, or on your own. These questions go beyond what is directly covered in this part of the book. Use the research literature and other reading to explore these topics in more depth. Tips to help prepare for topic discussions are available online.

1 In Chapter 2, the concepts of essential genes and accessory genes were introduced. In this chapter, the concepts of core genes and the core genome were added to these terms. Compare the differences in meaning between the terms essential gene and core gene. Discuss the types of genes that might fall into both categories or only one.

2 The term pan-genome is used to describe all of the known genes of a particular species of bacteria, or sometimes more than one related species. As we sequence more genomes, are pan-genomes getting larger? Do you think we will achieve a definitive pan-genome for species that participate in HGT?

3 Mutation drives evolution. In bacterial species, there are many different mechanisms that can introduce mutations in the genome. Evaluate the various types of mutations and how they have contributed to bacterial evolution of key phenotypes, such as virulence or antibiotic resistance.

Online quiz questions

To further self-assess your understanding of the chapter material, please visit the following link, where you can participate in a range of interactive quiz questions:

www.routledge.com/cw/snyder

Further reading

Bacterial chromosomes carry the genetic material of the organism

Hayashi T, Makino K, Ohnishi M, Kurokawa K, Ishii K, Yokoyama K. Complete genome sequence of enterohemorrhagic *Escherichia coli* O157:H7 and genomic comparison with a laboratory strain K-12. *DNA Res.* 2001; *8*(*1*): 11–22.

Land M, Hauser L, Jun S-R, Nookaew I, Leuze MR, Ahn T-H. Insights from 20 years of bacterial genome sequencing. *Funct Integr Genomics.* 2015; *15*(*2*): 141–161.

DNA within the bacterial cell

Canchaya C, Fournous G, Brüssow H. The impact of prophages on bacterial chromosomes. *Mol Microbiol.* 2004; *53*(*1*): 9–18.

The largest bacteria with many copies of its genome

Voland J-M, Gonzalez-Rizzo S, Gros U, Tyml T, Ivanova N, Schulz F, Goudeau D, Elisabeth NH, *et al.* A centimeter-long bacterium with DNA contained in metabolically active, membrane-bound organelles. *Science.* 2022; *376*(*6600*): 1453–1458.

The sizes of bacterial genomes are characteristic of bacterial species

Database of Genome Size. www.cbs.dtu.dk/databases/DOGS.

Wilbanks EG, Doré H, Ashby MH, Heiner C, Roberts RJ, Eisen JA. Metagenomic methylation patterns resolve bacterial genomes of unusual size and structural complexity. *ISME J.* 2022; *16*(*8*): 1921–1931.

The origin of replication impacts the base composition

Pomerantz RT, O'Donnell M. What happens when replication and transcription complexes collide? *Cell Cycle*. 2010; *9*(*13*): 2537–2543.

Genomic architecture can impact gene expression

Rocha EPC. The organization of the bacterial genome. *Annu Rev Genet*. 2008; *42*: 211–233.

Supercoiling can also influence gene expression

Griswold A. Genome packaging in prokaryotes: The circular chromosome of *E. coli*. *Nat Educ*. 2008; *1*(*1*): 57.

Mutations in the bacterial genome

Block DH, Hussein R, Liang LW, Lim HN. Regulatory consequences of gene translocation in bacteria. *Nucleic Acids Res*. 2012; *40*(*18*): 8979–8992.

HGT introduces new genetic material

Barlow M. What antimicrobial resistance has taught us about horizontal gene transfer. *Methods Mol Biol*. 2009; *532*: 397–411.

Finkel SE, Kolter R. DNA as a nutrient: Novel role for bacterial competence gene homologs. *J Bacteriol*. 2001; *183*(*21*): 6288–6293.

Huddleston JR. Horizontal gene transfer in the human gastrointestinal tract: Potential spread of antibiotic resistance genes. *Infect Drug Resist*. 2014; *7*: 167–176.

Transformation involves bacterial uptake of DNA from its surroundings

Blokesch M. Natural competence for transformation. *Curr Biol*. 2016; *26*(*23*): 3255.

Part II

RNA, Transcriptional Regulation, and Transcriptomes

The following three chapters will explore in more detail the role of RNA in the bacterial cell, the regulation of transcription of various RNAs in the bacterial cell, and the study of the whole of the transcripts within the bacterial cell that make up the **transcriptome**. These chapters will build on material provided in Part I of this book, presenting the material introduced there in more depth and detail.

DOI: 10.1201/9781003380436-5

Chapter

4

RNA

RNA, ribonucleic acid, is a key biological molecule and in bacteria it is responsible for a variety of functions within the cell. The RNA from bacterial cells has been investigated using a variety of techniques, starting from the simple isolation of the nucleic acids away from the other components of the cell to understanding the sequence order of the ribonucleotides. Each of the types of RNA (mRNA, tRNA, rRNA, and ncRNA) has differences in their lengths, compositions, functions, and structures. These differences will be explored in this chapter, demonstrating how various RNAs contribute to the regulation of gene expression.

Bacterial mRNAs are translated into proteins as they are being transcribed

Bacterial mRNA, messenger RNA, is transcribed and then used immediately. Unlike eukaryotic mRNA, it is not packaged and it is not transported to another part of the cell. As a result, the opportunity for mRNA to interact with itself or with other molecules within the cell is reduced. Bacteria engage in either coupled transcription-translation or runaway transcription, which means that as the DNA is being transcribed to make mRNA, that mRNA is being translated to make the protein encoded by the gene on the DNA. There are very few changes to the mRNA transcript after it is made.

As a result, the mRNA transcript is ready to be translated into protein immediately (**Figure 4.1**). Transcription is initiated and the mRNA elongated. Before this process is complete and transcription can be terminated, translation can begin. The ribosome can initiate translation as soon as the ribosome binding site (RBS) is transcribed into mRNA. DNA is transcribed to mRNA and is translated to protein together, which is possible because bacterial cells do not process their mRNA or transport it away from the DNA before translation.

The size of mRNA is determined by the genes it encodes

Bacterial mRNAs carry the genetic code of the DNA to the ribosomes, where proteins are manufactured. The process of translation requires other RNAs; tRNAs bring amino acids to the ribosome and rRNAs are within the ribosome, while the process can be modulated by small noncoding RNAs. Bacterial mRNAs can carry one or more genes, depending on the proximity of genes to one another, the location of promoters that will drive transcription, and the presence of transcriptional terminators (see Chapter 2). Because bacterial mRNAs can have more than one gene, they can also have more than one initiation codon and more than one termination codon. These features influence the way in which the ribosomes will interact with the mRNA during translation.

The nucleotide length of a molecule of mRNA is approximately equal to the genes it encodes. For many mRNAs, which carry just one gene, this means that they are about the size of an average bacterial gene, 1,000–1,500 bp. There are some mRNA molecules that are polycistronic units, including two or more genes (Chapter 2). These are therefore longer, including the coding regions of the adjacent and co-transcribed genes, which are often separated by a few bases.

DOI: 10.1201/9781003380436-6

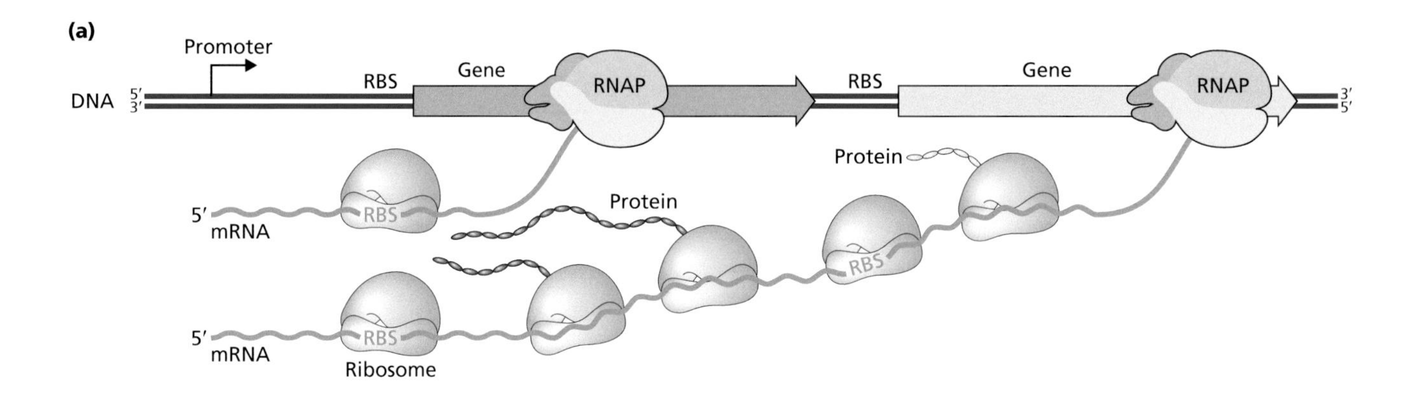

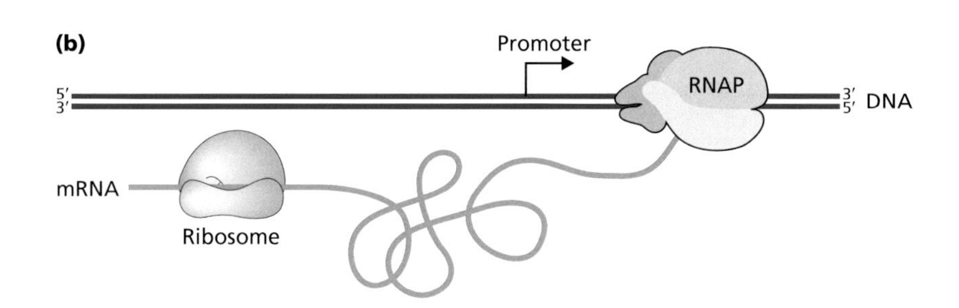

Figure 4.1 In (a) coupled transcription–translation and (b) runaway transcription in bacteria protein is translated as mRNA is transcribed. Transcription is initiated from a promoter (bent arrow) 5′ of co-transcribed genes. Once the ribosome binding site (RBS) for the first gene is represented in the mRNA, a ribosome can initiate translation into protein. RNA polymerase (RNAP) continues elongating the mRNA through the first gene and into the second. Once the RBS for the second gene is present on the mRNA, a ribosome can bind and protein can be made from the second gene as well. As one RNAP continues its elongation, another RNAP can transcribe more mRNA. Likewise, as one ribosome moves on from the RBS to translate the mRNA into protein, another ribosome can translate the same mRNA, generating more protein.

The mRNA is not just the gene, however. The regions flanking the gene are also transcribed into mRNA. These regions contain the signals needed for the translation of the mRNA into protein, including the ribosomal binding site before the start of the gene and the signals for translational termination at the end of the gene (**Figure 4.2**). An mRNA that starts with the initiation codon and ends with the termination codon will not be expressed as a protein due to the absence of the sequences needed to start and end translation.

The start of the 5′ end of mRNA is dictated by its promoter region

The start point of transcription is between the binding site for RNA polymerase at the promoter and the initiation codon of the gene (**Figure 4.3**). The beginning of the mRNA is dictated by the start point of transcription. At the initiation of transcription, RNA polymerase recruits the first nucleotide to the transcript, which is paired with the DNA base that is the start point of transcription. This first base retains all three of its phosphates at what will be the 5′ end of the mRNA. The second nucleotide is added by RNA polymerase, removing two of its three phosphates to form the phosphodiester bond between the ribonucleotides. The two free phosphates are called **pyrophosphate (PP_i)**. The mRNA elongates from

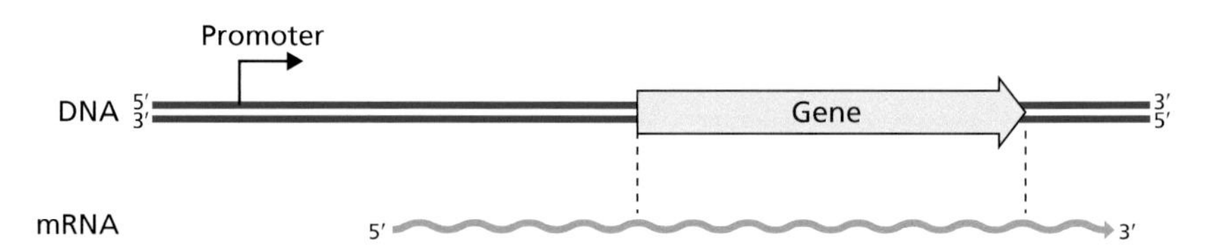

Figure 4.2 The size of mRNA is dictated by the gene it encodes. The size of an mRNA is based upon the bacterial gene that it encodes. Transcription of the gene sequence into mRNA begins at the promoter (bent arrow) located 5′ of the start of the gene. Transcription continues past the end of the gene to a transcriptional termination signal. The mRNA is therefore larger than the gene because it also contains sequences before and after the gene, based on the signals for initiation and termination of transcription.

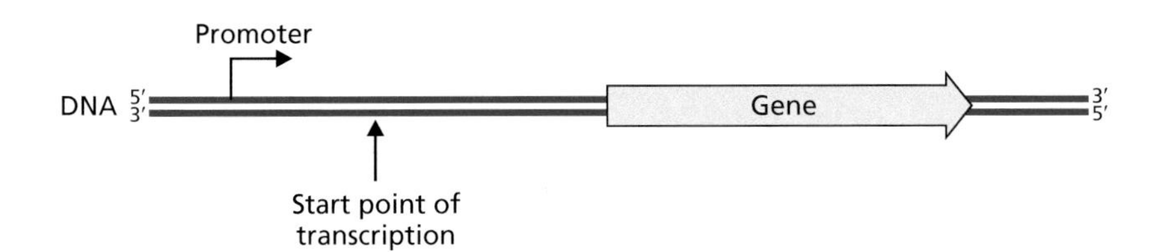

Figure 4.3 The start point of transcription is between the promoter and the gene. DNA is transcribed into mRNA by RNA polymerase, which recognizes and binds to the promoter region (represented here with a bent arrow). The mRNA does not start at the promoter, however. RNA polymerase starts transcription at a point between the promoter region and the gene. This is the start point of transcription.

5′ to 3′, like DNA polymerase, but without the need for the short primer sequence that DNA polymerase requires. Depending on the promoter used for the transcript, there can be different start points of transcription for the same gene.

Sigma factors are inherently involved in the 5′ end of the mRNA. The sigma factor protein recognizes and binds to the promoter region, recruiting RNA polymerase for transcription. If there is more than one promoter and the promoters overlap, the sigma factors will compete for occupation of the promoter region. The start point of transcription can therefore be different depending on the sigma factor driving transcription. The differences in the mRNA sequences at the 5′ end may be negligible, but could also be different enough that they impact the expression of the mRNA, in addition to the influence of the different sigma factors.

Not all transcripts are translated into proteins

Because the sequences that indicate the initiation and termination of transcription are different from those that initiate and terminate translation, a gene is transcribed regardless of whether it can be translated or not. For instance, a gene that contains a frameshift, placing it out of frame between the initiation codon and termination codon, will still be transcribed. Likewise, a gene containing a mutation that generates a premature termination codon will also be transcribed. As a result, when termination codons are arrived at prematurely there will be a large region of the mRNA that is not translated (**Figure 4.4**). Because the mRNA encoding a frameshift is not translated into a full-length protein, the mRNA is not covered with ribosomes and therefore not protected from degradation by RNases. This can perhaps impact upon the translation of any co-transcribed genes 3′ of the frameshifted gene. If the first gene in an mRNA is not translated, the mRNA devoid of ribosomes can be targeted by RNases and may be degraded before the second gene in the transcript can be translated.

Mutations outside the coding region can also prevent protein being made from an mRNA transcript. If there are, for instance, mutations in the RBS that alter the recognition of the Shine–Dalgarno sequence and prevent its recognition by the 16S rRNA within the ribosome, then ribosomes will not bind and initiate translation. If the mRNA is a polycistronic unit, it may be that other RBSs on the transcript are intact and function; however, if the first gene is not translated, the mRNA may be targeted by RNases for degradation, due to the lack of bound ribosomes. There are, however, some mRNAs that are capable of initiating translation even without an RBS. These transcripts rely on an unstructured region at the 5′ end of the mRNA to allow non specific binding that recruits the ribosome.

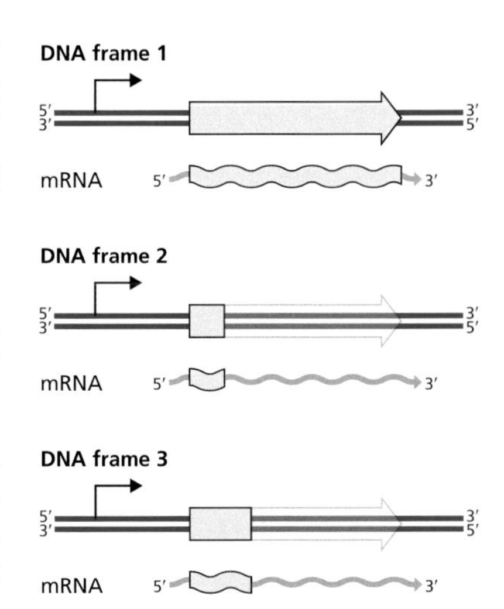

Figure 4.4 When a gene is frameshifted, it is still transcribed. The mRNA made from the gene is based on the signals for initiation and termination of transcription. If there is a frameshift in the gene, this influences translation, but it will not impact upon transcription. In reading frame 1, the initiation codon and termination codon are in the same frame and the whole gene will be translated (yellow region). In reading frames 2 and 3, a frameshift places a premature termination codon in-frame and only a portion of the gene is translated (yellow region). The transcript made from a frameshifted gene is the same whether it is in or out of frame.

There are untranslated regions at the 5′ end of the mRNA transcript

The mRNA transcript contains more than just the gene sequence. It has an **untranslated region** at the 5′ end of the mRNA, which is transcribed before the gene sequence and is not translated by the ribosomes. This untranslated region contains the RBS, to which the ribosome will bind to initiate the translation of the mRNA into protein. Following the RBS, the mRNA has the initiation codon and then the coding region of the gene. The gene sequence ends with the termination codon, which will signal the ribosome to stop translation (**Figure 4.5**). Within an mRNA, a termination codon may be followed by another coding region, starting with its own RBS and initiation codon (see Figure 4.1). Some mRNA in bacteria gets poly adenylated at the 3′ end, although this is a characteristic feature of eukaryotic mRNA and not often associated with bacteria.

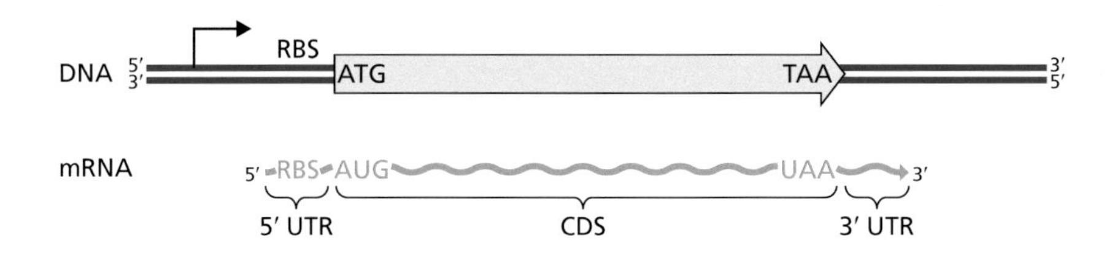

Figure 4.5 Structure of an mRNA. Transcription of mRNA is from 5′ to 3′ and begins before the coding sequence of the gene at the promoter (bent arrow). The mRNA includes the 5′ untranslated region (5′ UTR) containing the RBS, followed by the initiation codon of the gene (here ATG on the DNA and AUG on the mRNA), the coding sequence of the gene (CDS), and then the termination codon (here TAA on the DNA and UAA on the mRNA). There is then a 3′ untranslated region (3′ UTR).

The untranslated region at the 5′ end of the transcript may stabilize the mRNA and prevent its degradation. In *Escherichia coli*, for example, the untranslated region of the *ompA* transcript provides stability to this mRNA that is dependent upon the growth rate of the bacteria. It is believed that this stability to the mRNA is achieved by the formation under certain conditions of a secondary structure at the 5′ end of the transcript in the untranslated region. The mRNA structure that is believed to be formed by *E. coli ompA* mRNA is structurally similar to other secondary structures capable of stabilizing mRNA. Although the sequences that form mRNA stabilizing structures can be quite diverse, the resulting secondary structures are similar and offer similar protection from degradation for the mRNA.

Within the 5′ untranslated region of the mRNA is an RBS, which must be properly spaced from the initiation codon for effective translation to occur. The ribosome has to be able to bind to the RBS through recognition of the Shine–Dalgarno sequence. Once bound, the ribosome then has to align the initiation codon on the mRNA with its own A-site (see Chapter 2). Translation of the mRNA can be prevented by mutations to the initiation codon, such that it is no longer an ATG, GTG, or TTG codon that can be used to initiate translation.

Features at the 3′ end of the mRNA transcript can influence the expression of encoded genes

Although the termination codon is used by the bacterial cell's translation machinery, it can also impact upon the mRNA. The termination codon provides a signal for translation to stop, that is, for the ribosome to end its procession and dissociate from the mRNA. Premature termination codons, through mutation or frameshift (see Figure 4.4), can lead to stretches of mRNA that are not occupied by ribosomes, leading to their degradation by RNases (**Figure 4.6**). The RNases recognize stretches of mRNA that are not being translated by ribosomes and remove these transcripts, which do not encode full proteins.

If the ribosome gets to the end of the mRNA without encountering a termination codon signal to end translation, the ribosome stalls and is unable to dissociate from the transcript. This can occur when there is a mutation on the termination codon for the last (or only) coding region on the mRNA (**Figure 4.7**). The termination codon may be absent from the transcript if transcription has terminated prematurely, making a short mRNA that does not contain all of the coding region sequence.

When a ribosome stalls due to the absence of a termination codon, the ribosome remains on the mRNA with a peptidyl tRNA in the P-site. The tRNA, ribosome, and peptide are not released. The stalled ribosome on the mRNA

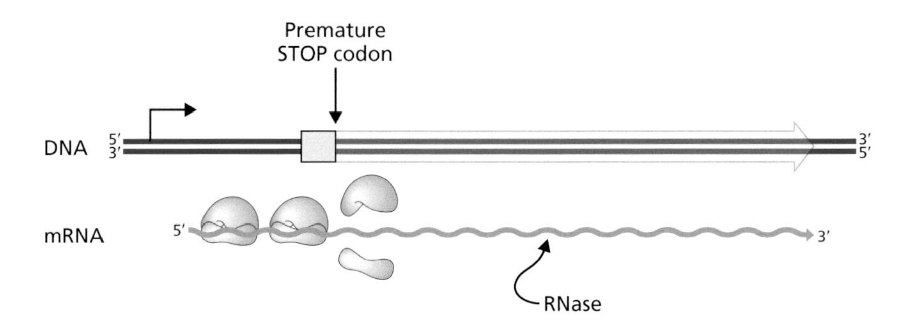

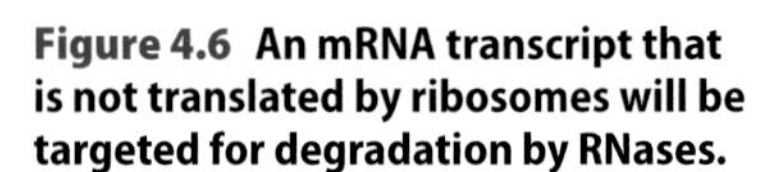
Figure 4.6 An mRNA transcript that is not translated by ribosomes will be targeted for degradation by RNases. When there is a premature stop codon in a coding region, the mRNA is still made; however, the full length of the mRNA is not translated into protein by the ribosomes. The ribosomes terminate translation, leaving a large portion of the mRNA unoccupied by ribosomes. When mRNA is not being translated, the lack of ribosomes along its length makes it a target for degradation by RNase enzymes. This removes transcripts from the cell that do not encode full proteins.

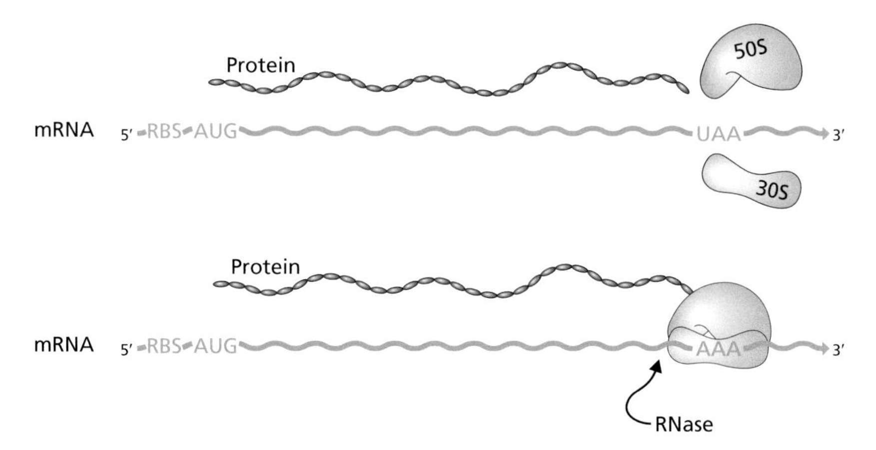

Figure 4.7 Non stop decay targets mRNA for degradation when the ribosome is stalled at the 3′ end. The ribosome binds to the mRNA at the RBS and initiates translation of the gene at the initiation codon (here AUG). The ribosome will continue with translational elongation until it reaches a termination codon (here UAA). It will then terminate translation and disassociate from the mRNA as the 50S and 30S ribosomal subunits (top). If, however, the ribosome reaches the 3′ end of the mRNA and has still not encountered a termination codon, the ribosome stalls at the end of the mRNA, unable to progress and unable to disassociate without a termination codon signal to stop translation (bottom). A stalled ribosome at the end of an mRNA can be recognized by RNases and triggers the degradation of the mRNA by RNases, which recognize the stalled ribosome mRNA complex.

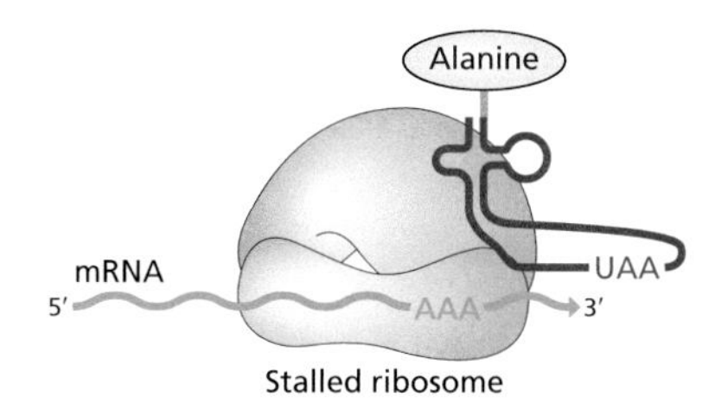

Figure 4.8 tmRNA resolves stalled ribosomes when there is no termination codon. Without a termination codon to signal translational termination, the translation machinery is unable to dissociate from the mRNA. A tmRNA is able to enter the A-site of the stalled ribosome, carrying an alanine, and complete translation, using a termination codon carried on the tmRNA.

triggers degradation of the mRNA by RNases. This process is called **non stop decay** and serves to protect the bacterial cell from truncated proteins that could be harmful to the cell (see Figure 4.7). However, triggering degradation of the mRNA by RNases does not help the other components of translation, that is, the ribosome, tRNA, and peptide. A specialized RNA present within the bacterial cell resolves the issues for the translation machinery. A **transfer messenger RNA (tmRNA)** carrying an alanine on its **tRNA-like domain (TLD)** enters the A-site of the stalled ribosome. The tmRNA then accepts the polypeptide chain from the tRNA that is stuck in the P-site of the ribosome. The tmRNA resumes translation, putting a C-terminal peptide tag onto the aberrant polypeptide. The sequence for this peptide tag is encoded on the tmRNA itself in its **mRNA-like domain (MLD)**, along with a termination codon to end the process of translation (**Figure 4.8**). The tmRNA encoded termination codon causes translation to end as it should have done, releasing the ribosome from the mRNA. This is a process called ***trans*-translation.**

The name tmRNA is in recognition of its dual function. It acts as both a tRNA, carrying the amino acid alanine to the A-site of the ribosome, and an mRNA, carrying sequence information that is translated into a peptide chain. The amino acid acceptor stem and T-stem-loop of the tmRNA are typical of tRNAs. However, the tmRNA D-stem-loop folds differently from that of tRNAs and there is no anticodon loop. There is instead an mRNA-like region that encodes the short peptide and stop codon (**Figure 4.9**). This short-peptide chain, added to the translated aberrant polypeptide, is a signal that targets the protein for degradation.

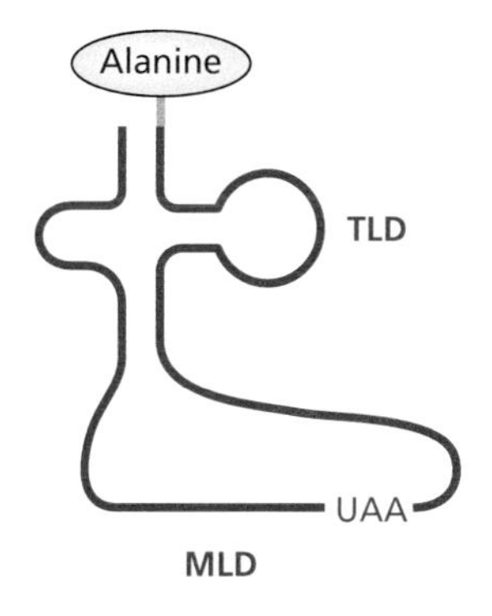

Figure 4.9 The structure of a tmRNA is similar to a tRNA, but also has an mRNA-like region. Like a tRNA (TLD), there is an acceptor stem and T-stem-loop on a tmRNA. Structurally, the D-stem-loop is different from a tRNA and there is no anticodon loop. Here there is a region that encodes a short peptide and a termination codon, like an mRNA (MLD).

Stability of RNA and its degradation by nucleases and hydrolysis

RNA is inherently unstable and liable to degradation. It can be degraded by **nucleases** acting within the mRNA strand, that is, **endonucleolytic** activity, or by nucleases that can remove one nucleotide at a time from either the 5′ or 3′ ends, through **exonucleolytic** action. For example, the major mRNA degradation enzyme RNase E acts by endonucleolytic cleavage of regions that are rich in adenines and uracils.

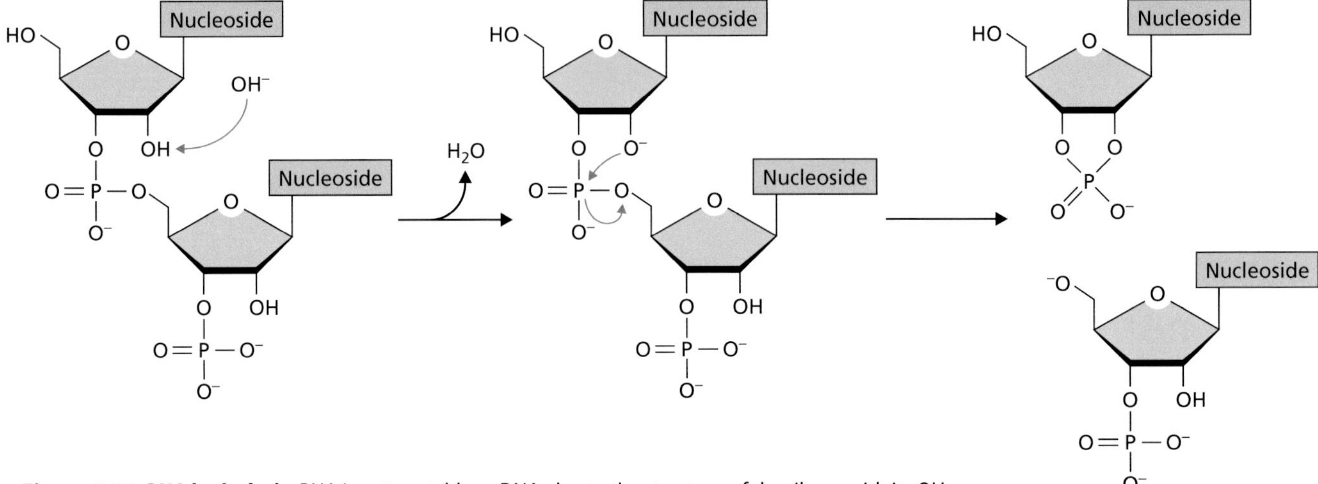

Figure 4.10 RNA hydrolysis. RNA is not as stable as DNA, due to the structure of the ribose with its OH group on the 2′ carbon. This OH group is liable to hydrolysis in the presence of free OH^- ions, resulting in the release of H_2O and the breaking of the ribose phosphate backbone.

In addition to the enzymatic degradation of RNA by **RNase** enzymes that target RNA and break it down, the structure of RNA means that it will pull itself apart. **Hydrolysis**, the spontaneous removal of a hydrogen from the 2′ OH results in a free H_2O and also results in the breaking of the phosphodiester bond between the phosphate and the next ribose in the backbone (**Figure 4.10**). This means that, even in the absence of nucleases, DNA lacking the 2′ OH that is targeted by hydrolysis is a much more stable structure than RNA. The increased stability of DNA makes it an ideal molecule for the storage of genetic material and the instability of RNA makes it ideal for the transmission of that information in the form of mRNA.

As a result of the action of RNases and its own instability, the half-life of bacterial mRNA is a matter of minutes. In *E. coli*, for example, the average mRNA half-life is 2 minutes. As a result of the short half-life of mRNA, the switch in the production of different proteins can be rapid, with protein translation stopping when the corresponding mRNA is degraded and new proteins are being made in concert with new transcription. The rapid turnover of mRNA enables the bacterial cell to respond to the need for different biological processes through changes in the types and levels of proteins being made. In this way, changes at the level of transcription can quickly impact the cell.

Secondary structures formed by mRNAs impact ribosome binding and translation initiation

Translation happens while transcription is occurring. The ribosomes immediately act upon the mRNA within a bacterial cell to initiate translation. There is, however, scope for the mRNA to form secondary structures. During translation, the mRNA strand becomes occupied by ribosomes, which keep the mRNA single stranded. Secondary structures can form in regions of mRNA that are not bound by ribosomes if the sequences are complementary to one another and capable of base pairing into regions of double-stranded mRNA. Therefore, the composition of the sequence of an mRNA can influence the progression of the ribosomes through the formation of secondary structures that impact upon translation.

The RBS Shine–Dalgarno sequence is often part of secondary structures that change from folded to unfolded to folded, fluctuating the availability of the Shine–Dalgarno sequence for ribosome binding (**Figure 4.11**). To gain access to bind the Shine-Dalgarno sequence at the 5′ end of the mRNA, the 30S subunit of the ribosome needs to be nearby when the mRNA unfolds. In *E. coli*, a detailed and in-depth analysis of mRNA in the cell revealed that there is a relatively unstructured

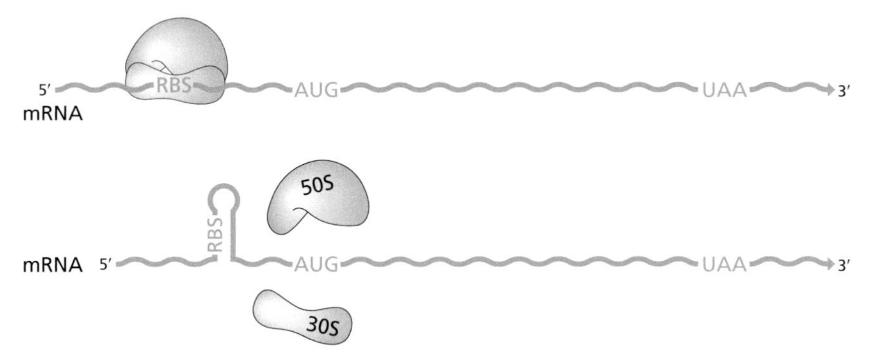

Figure 4.11 Secondary structures in mRNA can impact upon translation. Initiation of translation requires access to the RBS by the ribosome (top). If a secondary structure of the mRNA at the location of the RBS prevents 30S from gaining access to the site, translation cannot occur (bottom). Secondary structures in mRNA can interfere with the translation process; if the RBS is not available then translation of the mRNA does not occur.

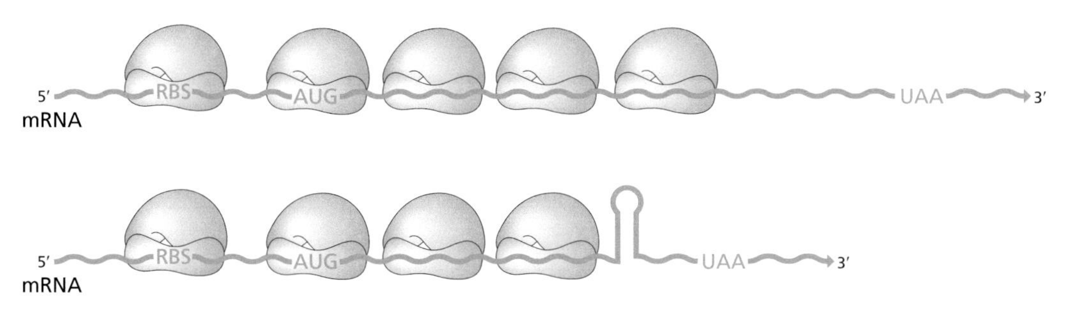

Figure 4.12 Secondary structures formed by mRNA can disrupt the progress of ribosomes along the mRNA during translation. These can influence the expression of the mRNA into protein. If the encoded amino acids correspond to abundant tRNAs, then the progression of the ribosome during translational elongation is rapid (top). If, however, the ribosome moves more slowly to await a less common tRNA, the secondary structure formed by the mRNA stalls the ribosome (bottom).

region at about 20 nucleotides 5′ of the initiation codon. It may be that the 30S subunit makes nonspecific contact here before the ribosome specifically binds at the Shine–Dalgarno site.

There are structural regions within some mRNAs that can stall ribosomes. However, the impact of the formation of these secondary structures can be canceled out by the inclusion of codons for common and abundant tRNAs. When tRNAs are abundant, the ribosome is able to progress quickly along the mRNA and the mRNA is completely occupied by ribosomes. For secondary structures to form, regions of mRNA without ribosomes are needed; therefore, secondary structures do not form when translation is happening efficiently. When the codons on the mRNA are common and their tRNAs are abundant, the ribosome is able to progress rapidly through the region. When there are plenty of tRNAs for the codon, because it is an abundant and commonly used codon, then translation is fast. The speed of translation and the resulting occupation of the mRNAs by ribosomes overcome the stall that may result due to the secondary structure of mRNA (**Figure 4.12**).

Secondary structures formed by mRNAs influence translational termination

There are secondary structures within the mRNA that can impact upon the termination of translation as well as its initiation and elongation. These secondary structures occur four to eight nucleotides 5′ of the UAA, UAG, or UGA termination codon, which is at the 3′ end of the mRNA encoded coding region.

Secondary structures present in mRNA can slow down the progress of the ribosome along the mRNA during elongation. When translation elongation is slow, it is more likely that translation will accurately terminate. The translation

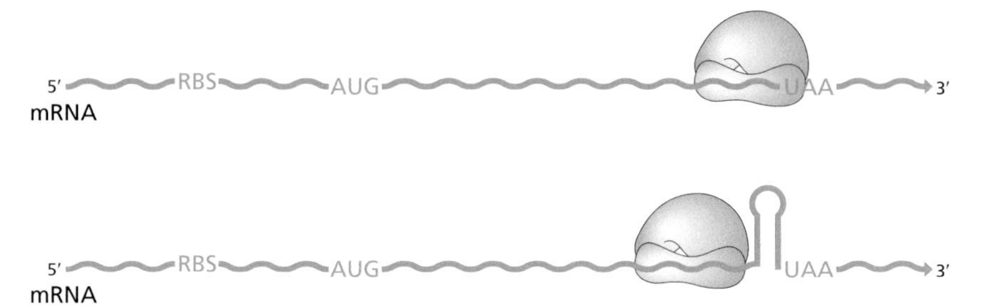

Figure 4.13 Translational termination and secondary structures. Translational termination occurs when the termination codon aligns within the A-site, triggering termination. If the ribosome progresses through translational elongation too rapidly, the correct alignment with the termination codon may not occur (top). In order for this process to happen efficiently, a secondary structure in the mRNA can slow the progression of the ribosome enough so that the termination codon aligns properly within the ribosomal A-site.

termination fidelity is therefore enhanced by the presence of secondary structures occurring just before the termination codon. When the termination codon enters the ribosome A-site at a slowed rate, it is more likely that the termination codon will accurately align within the ribosome and trigger translational termination (**Figure 4.13**).

Tertiary structures within mRNAs impact expression of the encoded gene

In addition to forming simple secondary structures in the form of folds and hairpins, mRNAs can form more complex tertiary structures. These mRNA tertiary structures are often involved in regulating the expression of the gene encoded by the mRNA.

Frequently present within the 5′ untranslated region of the mRNA are structures referred to as **riboswitches**. Riboswitches are able to regulate the expression of the genes that share their mRNA molecule. Unlike regulatory proteins, riboswitches are included in the mRNA itself. The structure of the riboswitch and the way in which this structure interacts with other components within the cell impacts upon gene expression. Although there are different forms of riboswitches, they all share the common feature of being present on the mRNA that they regulate.

Riboswitches include an **aptamer** and a regulatory region called the **expression platform**. The aptamer portion of the mRNA is between 70 and 170 nucleotides. The mRNA in this region forms a tertiary structure that makes up the riboswitch aptamer. The aptamer structure has the capacity to selectively and specifically bind to other molecules within the cell, such as metabolites. By binding to metabolites with its aptamer, the riboswitch is able to regulate the expression of metabolic genes in response to the needs of the cell. In this way, riboswitches can directly impact protein expression in the presence or absence of metabolites.

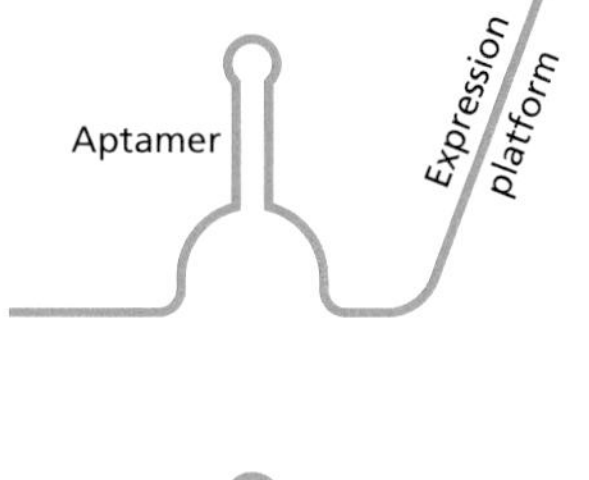

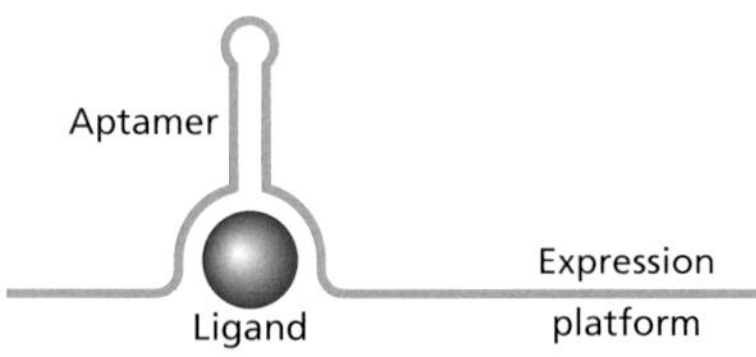

Figure 4.14 A riboswitch controls the expression of the mRNA that contains it. The aptamer region (top) binds to a specific metabolite (bottom), causing a structural change in the mRNA. The differences in the bound and unbound states influence the expression of the gene.

When the riboswitch aptamer binds to a specific metabolite within the cell, the structure of the aptamer region is stabilized into a tightly folded structure. This tight structure can interfere with either transcription or translation. These two structures, the aptamer and the aptamer bound to metabolite, are mutually exclusive, with one structure allowing the expression of the mRNA and the other preventing its translation into protein (**Figure 4.14**). In some mRNAs, the aptamer and metabolite structure forms a transcriptional terminator, ending transcription of the mRNA prematurely. This regulates expression of the protein at the level of transcription. In other mRNAs, the aptamer and metabolite structure itself makes the 5′ untranslated region unavailable for ribosome binding, which interferes with translation. The lack of translation acting on the mRNA means that it is targeted for degradation by RNases.

Some riboswitches have the ability to cleave themselves. This **ribozyme** activity, when an RNA is able to perform an enzyme-like activity such as cleavage, occurs when there is sufficient metabolite recognized by the aptamer. Cleavage of the mRNA prevents additional protein products from being made from the mRNA, stopping translation from the mRNA and targeting it for degradation by RNases.

RNA thermometers modify the expression of proteins from mRNA based on temperature

Bacterial cells are able to sense changes in temperature. One way that they can do this is using **RNA thermometers**. In response to changes in temperature, a region within the mRNA itself undergoes a conformational change that alters expression of the mRNA's encoded protein(s). Secondary structures form in the 5′ untranslated region of the mRNA. These secondary structures may make the RBS unavailable for binding by ribosomes. Higher temperatures cause the hydrogen bonds between the bases to break, changing or eliminating the secondary structure of the mRNA 5′ untranslated region (**Figure 4.15**). RNA thermometers can also be present in the untranslated regions between genes in a polycistronic unit. The structure of the folded RNA thermometer can be simple with one stem loop or more complex with several.

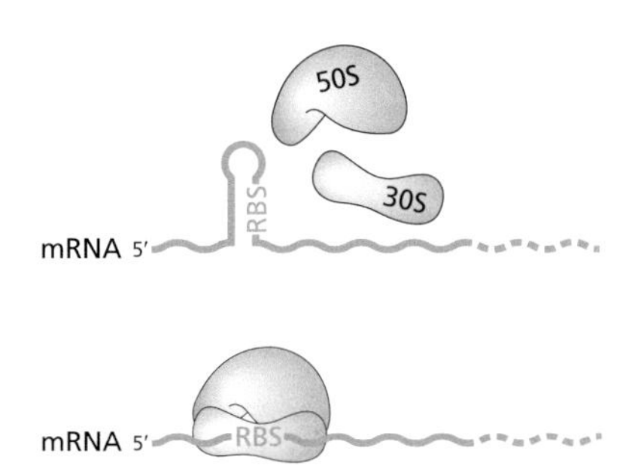

Figure 4.15 An RNA thermometer alters transcription based on temperature. The secondary structure of the mRNA in an RNA thermometer blocks the RBS from being accessible by the ribosome's 50S and 30S subunits (top). The mRNA is therefore not translated. At higher temperatures, the hydrogen bonds between the secondary structure forming nucleotides break, making the RBS accessible (bottom). The ribosome binds to the mRNA and translation is initiated.

Genes that are controlled by RNA thermometers include those that encode **heat-shock proteins**, which help protect the cell and cellular processes from increases in temperature, which can lead to protein misfolding and degeneration. The *rpoH* mRNA, encoding the heat-shock response orchestrating sigma factor σ^{32}, is an RNA thermometer. RNA thermometers can also be involved in the expression of cold-shock proteins when the cell experiences a decrease in temperature.

RNA thermometers can control the expression of virulence factors in pathogenic bacteria. In *Yersinia pestis*, for example, the *lcrF* mRNA encodes an activator of expression of virulence genes and is an RNA thermometer. Thus, the virulence factors are only expressed at high temperature, when the RNA thermometer changes structure and the RBS becomes available. In bacterial pathogens, an increase in temperature can be a signal to the cell that it has entered a mammalian host. This signals the cell to begin expression of virulence factors that in other circumstances would be a drain on the cell's energy and resources.

Polyadenylation of mRNA is not just for eukaryotes

Polyadenylation of mRNA, the post-transcriptional modification of mRNA to add adenines to the 3′ end, is often cited as being specific to eukaryotic cells. However, bacterial cells can polyadenylate mRNAs as well. The poly-A tails on bacterial mRNAs tend to be shorter than those on the mRNA of eukaryotic cells. There can be 14–60 adenines at the 3′ end in bacteria, compared with 60–200 in eukaryotic cells. The proportion of mRNA that is polyadenylated within a bacterial cell varies by species, with some having as little as 1% of mRNA transcripts with poly-A tails and others having up to 60%. Most mature mRNAs in eukaryotic cells are polyadenylated.

The poly-A tails are added to the 3′ OH end of the mRNA transcript or to the 3′ OH end created when mRNA is cleaved post-transcriptionally. There do not appear to be specific sequences within the mRNA itself that are targeted for polyadenylation. As a result, any RNA can become polyadenylated in a bacterial cell. The presence of the poly-A tail may stabilize the mRNA or it may target the mRNA for degradation, depending upon the context. For the most part, however, polyadenylation targets RNA for degradation. Therefore, incomplete transcripts of mRNA, regulatory small noncoding RNAs, and nonfunctional tRNAs are liable to undergo polyadenylation to promote their degradation. Active and functional RNAs may be protected from polyadenylation by virtue of such signals of active functionality as ribosome occupation of mRNAs and folding into tRNA structures.

Bacterial tRNAs are folded into tight structures

Following transcription, bacterial tRNA sequences are folded into specific, functional, tight structures that enable these molecules to perform their functions in translation within the bacterial cell. Each tRNA is made up of between 75 and 90 nucleotides, which form three **stem-loop** structures due to complementarity within the sequence of the tRNA. Stems are formed when self-complementary regions of RNA hydrogen bond and form a region of double-stranded RNA. Loops are formed by the region between the two complementary regions. The first stem of the tRNA, starting from the 5′ end of the strand, is about four nucleotides and is referred to as the **D-stem** and the corresponding loop is the **D-loop**. Next is the **anticodon stem**, made of about six nucleotides, and the **anticodon loop**, which includes the three-base anticodon. The final of the three stem-loop structures is the **T-stem**, of about five nucleotides, and the **T-loop**. In some tRNAs, there can be an additional **variable loop** between the anticodon stem and the T-stem (**Figure 4.16**).

The 5′ and 3′ ends of the tRNA strand pair together to form a fourth stem in the structure, the **amino acid acceptor stem**, made of about seven nucleotides. It is this fourth stem, made up of the ends of the tRNA, which accepts the amino acid, which the tRNA will carry to the ribosome during translation. There are more than 20,000 mature tRNA molecules within a bacterial cell and many of these are loaded with the corresponding amino acid. This abundance of charged tRNAs makes the molecules readily available to the ribosomes for translation. The 3′ end of the tRNA sequence contains a CCA motif, which is vital for the attachment of the amino acid. The adenine within this motif is where the amino acid attaches to the tRNA (**Figure 4.17**). **Aminoacyl-tRNA synthetases** add the amino acid to either the 2′ or 3′ hydroxyl group of the 3′ terminal adenine, which is at the end of the CCA motif. In some cases, the CCA is added to the tRNA post-transcriptionally; however, this varies by both the species and the tRNA.

The anticodon on the anticodon loop of the tRNA interacts with the codon on the mRNA within the ribosome. The anticodon also interacts with the aminoacyl-tRNA synthetase to ensure that the correct amino acid is loaded onto the end of the tRNA. This process ensures that, during translation, the correct amino acid is added to the growing peptide chain, based on the codons encoded in the gene.

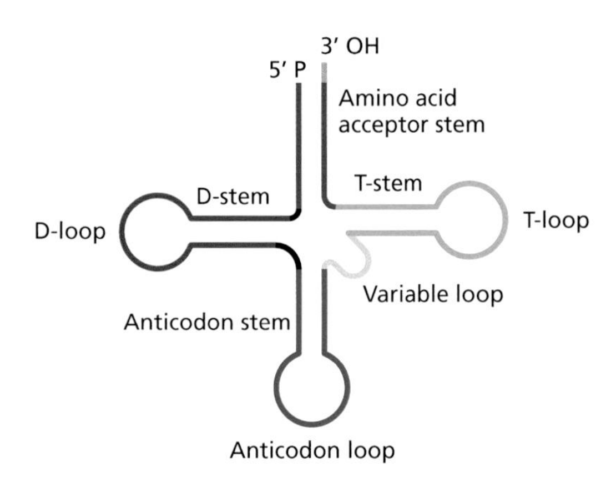

Figure 4.16 The structure of tRNAs. There are three stem-loop structures in a tRNA, and optionally an additional variable loop. Starting from the 5′ end of the tRNA, these are the D-stem, D-loop, anticodon stem, anticodon loop, optional variable loop, T-stem, and T-loop. The fourth stem, the amino acid acceptor stem, is formed by the beginning (5′) and end (3′) of the tRNA strand.

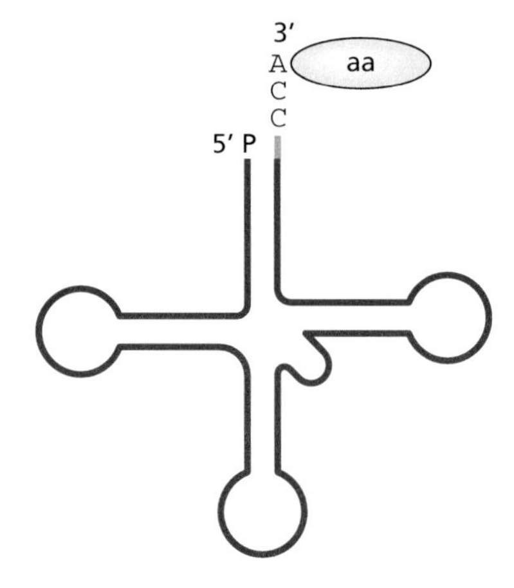

Figure 4.17 An amino acid is added to the adenine at the 3′ end of the tRNA by an aminoacyl-tRNA synthetase. The end of the tRNA contains a CCA motif, which may be part of the transcribed tRNA or it may be added later. The amino acid is added to either the 2′ or 3′ hydroxyl group of the adenine.

The features of tRNAs vary slightly, depending on the anticodon and the amino acid that they bind. For example, the extra variable loop is only present in tRNAs for leucine, serine, and tyrosine.

The tertiary structure of tRNA is L-shaped. This shape is achieved through the stacking of the stem-loop structures. The D-stem stacks on the anticodon stem. The amino acid acceptor stem stacks on the T-stem. These stacks of stems form the two arms of the L-shape. The bend in the L-shape is made of the D-loop and the T-loop (**Figure 4.18**).

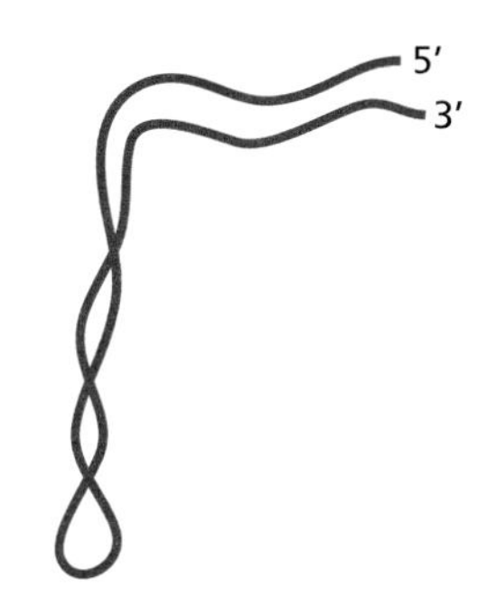

Figure 4.18 The tertiary structure of tRNA forms an L-shape. The two arms of the L-shape are made by the stacking of the stems, with the D-stem stacking on the anticodon stem and the amino acid acceptor stem stacking on the T-stem.

The tRNA has to be properly folded to functionally interact with the ribosome; therefore, the stems of the tRNA must have bases that are able to hydrogen bond to each other. The anticodon on the anticodon loop must remain conserved; mutations here can cause misincorporation of amino acids. There are also other nucleotides within the tRNA that are conserved based on the anticodon sequence. Additionally, some nucleotides within the anticodon loop are important for tRNA interactions with the ribosome.

tRNA transcripts undergo post-transcriptional processing

The transcription of genes into mRNA begins before the start of the gene and extends after its end. In the same way, tRNA transcription from the DNA template begins before and ends after the sequence of the final mature tRNAs in the cell that interact with ribosomes. Also, the loci encoding tRNA are often polycistronic with other tRNAs; the different tRNAs are transcribed together and then divided into separate tRNAs by RNases. Some of the regions of the bacterial genome that encode tRNAs are near the rRNA loci.

Most tRNAs are contiguous sequences, where the sequence that is encoded in the DNA is the same as the final sequence in the mature tRNA. Some bacteria, however, have **group I introns** within tRNA sequences, as well as within genes and rRNAs. Group I introns have ribozyme activity, which means that they are capable of self-splicing to remove the region of sequence that is not present in mature tRNA.

The initial transcript, the pre-tRNA, must be processed to become a mature tRNA. Both of the ends must be trimmed, removing the flanking regions that are part of the transcript. If it is not already present in the sequence of the tRNA, the CCA motif must be added by **nucleotidyl transferase** so that the amino acid can be added to the amino acid acceptor stem by aminoacyl-tRNA synthetases. If they are present, introns are removed.

RNase P cleaves pre-tRNA. This RNase is a site-specific endoribonuclease, meaning that, based on a recognition sequence within the pre-tRNA, the enzyme breaks the phosphodiester bond within the RNA cleaving the pre-tRNA. RNase P removes the 5′ leader sequence from the transcribed tRNA molecule. The RNase P enzyme contains a single RNA and a single protein that act together to recognize the RNA target and cleave it. The RNA–RNA interaction between the tRNA and the RNA within RNase P leads to cleavage.

Some of the nucleotides within the tRNA may be modified post-transcriptionally. These can include the conversion of adenine to **pseudouridine** (ψ), the conversion of adenine to **inosine** (I), and the conversion of uracil to **dihydrouridine** (D). Modifications enhance the structural stability of the tRNA, particularly in bacteria that encounter extremes or changes of temperature, and aid in correct folding. Pseudouridine is commonly found in tRNAs. It is chemically similar to uracil; however, the connection of the base to the ribose is a carbon–carbon bond, rather than the typical carbon–nitrogen bond seen in the standard nucleotides (**Figure 4.19**). Found only in tRNAs, inosine has a hypoxanthine attached to the 1′ carbon of its ribose ring (**Figure 4.20**). Dihydrouridine is formed when two additional hydrogens are added to uracil, which saturates the pyrimidine ring such that there are no longer any double bonds in the ring structure (**Figure 4.21**). Dihydrouridine can be found in tRNAs and rRNAs.

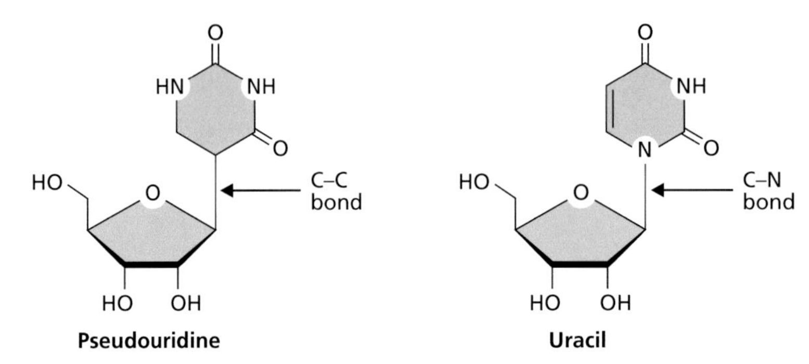

Figure 4.19 Some tRNAs contain the modified nucleotide pseudouridine. Pseudouridine (left) is similar to uracil (right); however, the bond between the ribose sugar and the base is a carbon–carbon (C–C) bond rather than a carbon–nitrogen (C–N) bond.

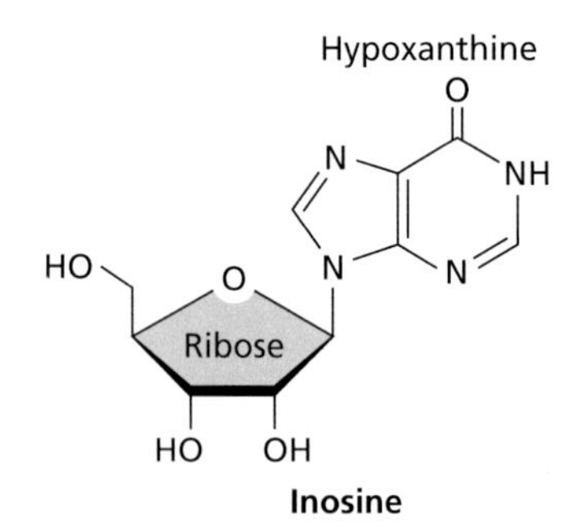

Figure 4.20 Some tRNAs contain the modified nucleotide inosine. Inosine is made when a hypoxanthine is attached to a ribose sugar.

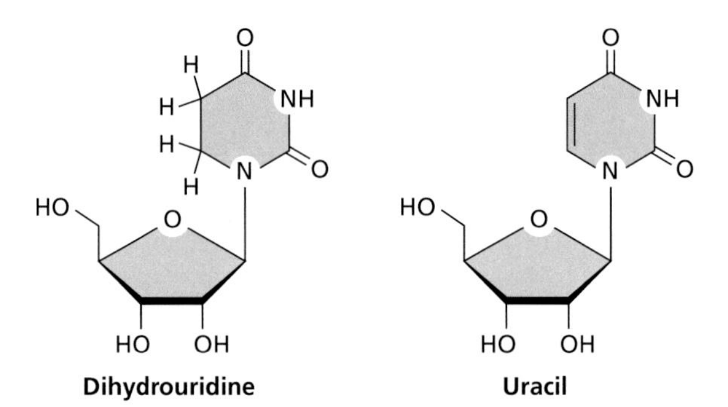

Figure 4.21 Some tRNAs contain the modified nucleoside dihydrouridine. Dihydrouridine is similar to uracil; however, the addition of two hydrogens means there are no longer any double bonds in the pyrimidine ring structure.

rRNAs are essential components of the ribosome

Ribosomes are made up of both proteins and rRNAs. The RNA components are essential for the function of the ribosome. There are three types of rRNA in bacterial cells: 5S rRNA, 16S rRNA, and 23S rRNA. All three fold into complex three-dimensional shapes and then combine with ribosomal proteins to form the ribosome. Typically, the 5S rRNA, 16S rRNA, and 23S rRNA are co-transcribed as one unit. There is often more than one locus in the genome for the rRNAs. For example, *E. coli* has seven rRNA loci.

The sequences for the rRNAs are about 120 nucleotides for 5S rRNA, about 1,500 nucleotides for 16S rRNA, and about 2,900 nucleotides for 23S rRNA. These sequences are transcribed into single-stranded RNAs, which fold into complex secondary structures through the pairing of nucleotides. The secondary structures formed differ between bacterial species, in accord with species-specific sequence differences in the rRNAs.

The rRNAs participate in the activities of the ribosome. The 16S rRNA is within the 30S subunit of the ribosome and its 3′ end binds to the 5′ end of the mRNA at the Shine-Dalgarno consensus sequence. The 5S rRNA is part of the 50S ribosomal subunit and links the functional aspects of the ribosomal P-site and A-site together, although the 5S rRNA does not make contact with the tRNAs in these sites. The 23S rRNA is also part of the 50S subunit of the ribosome and is a peptidyl transferase ribozyme.

The bacterial cell also contains noncoding RNAs that can regulate other RNAs

Noncoding RNAs (ncRNAs) do not encode proteins like mRNA, do not accept amino acids and recognize codons like tRNAs, and do not complex with ribosomal proteins like rRNAs. Typical bacterial ncRNAs range in size from 50 to 500 nucleotides. Although it was known from isolation that short RNAs existed, it was believed that short RNAs were incomplete transcripts of mRNA, tRNA, or rRNA. In 1981, Tomizawa and Itoh discovered that a short RNA of about 108 nucleotides was an **antisense RNA**, the first noncoding regulatory RNA identified. This ncRNA is complementary to a region within the ColEI plasmid DNA that is cleaved to produce the replication primer. When the antisense RNA binds to this region, replication of the ColEI plasmid is blocked. There is complementary base pairing between the antisense ncRNA and the RNA that is cleaved to produce the replication primer. Other ncRNAs have since been discovered that also act by base pairing with mRNA as antisense RNAs.

Either alone or in concert with other cellular components, ncRNAs can form secondary and tertiary structures. The structure of the ncRNA can impact upon its function as much as the sequence. By examining the ncRNA sequence for regions of complementarity, a structure can be predicted. The structures that can be formed due to hydrogen bonding between complementary sequences may be different at different temperatures and under different conditions, particularly in those ncRNAs involved in the regulation of environmental responses. Pairing of ncRNAs to mRNAs is structurally driven. The ncRNA, its mRNA target, or both will have a stem-loop region. This makes first contact with the other molecule. At the 3′ end, ncRNAs that pair with mRNAs tend to have a hairpin followed by a stretch of poly-U sequence. This structure generates an intrinsic transcriptional terminator. The ncRNA and the mRNA both contain regions where the nucleotides are complementary, which binds these two RNA molecules together. This type of ncRNA also tends to have an Hfq binding site.

The Hfq protein is a chaperone protein that stabilizes ncRNAs that carry an Hfq binding site and works in concert with the ncRNA to inactivate mRNAs. When the two RNAs come together to make a double-stranded RNA, the binding of Hfq stabilizes the structure and promotes RNase cleavage and degradation of the target mRNA (**Figure 4.22**). The Hfq binding protein is usually required for ncRNA and mRNA interactions in Gram-negative bacteria.

Although many ncRNAs regulate expression by binding to mRNA at the RBS and preventing the ribosome from initiating translation, some bind elsewhere. The inhibition of translation by other mechanisms includes the formation of secondary structures that inhibit translation elongation or that change the stability of the mRNA. There are ncRNAs that are encoded on the opposite strand to the gene they regulate. These are clear antisense RNAs, having the reverse complement sequence of the corresponding mRNA. When the antisense RNA base pairs with the mRNA, this forms double-stranded RNA that is targeted by RNases (**Figure 4.23**). It may also be that the transcription of an antisense RNA itself interferes with the ability to transcribe the gene.

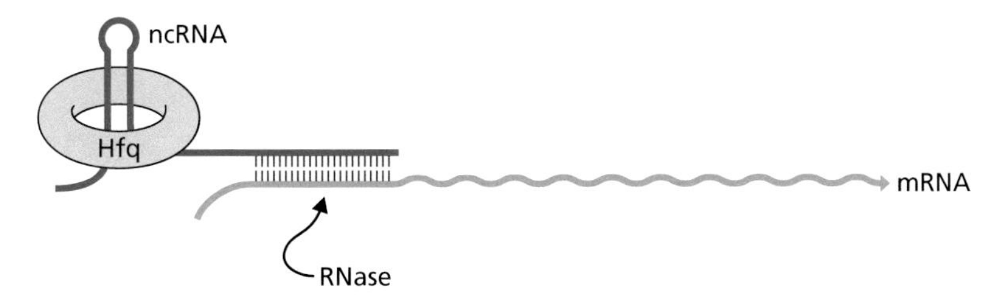

Figure 4.22 The Hfq protein assists in the binding of an ncRNA to its targeted mRNA. Base pairing between the ncRNA and mRNA forms double-stranded RNA. This is targeted by RNases within the bacterial cell, resulting in cleavage and degradation of the mRNA.

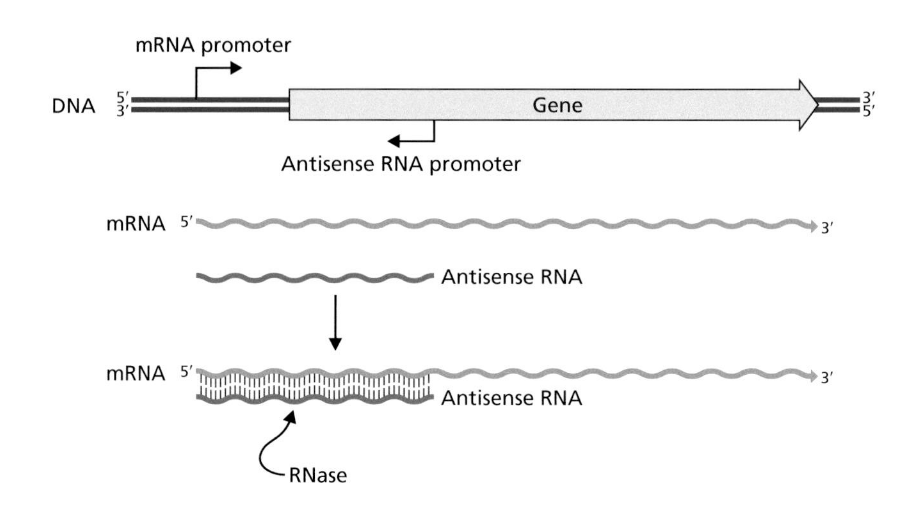

Figure 4.23 Antisense RNAs are encoded on the opposite strand of the gene they regulate. The sequence of the antisense RNA is perfectly complementary to the gene sequence that has been transcribed into the mRNA because it is generated from the same DNA template, just on the opposite strand. Base pairing between the antisense RNA and the mRNA from the gene forms double-stranded RNA, which is targeted for degradation by RNases.

Key points

- The length of bacterial mRNA is similar to the length of the gene in the DNA.
- There are regions at both the 5′ and 3′ ends of mRNA that are not translated into protein.
- Secondary structure formation in mRNA can prevent translation initiation by denying the 30S ribosomal subunit access to the Shine-Dalgarno sequence.
- Translation elongation can be slowed when there are secondary structures in the mRNA that interfere with the progression of the ribosome.
- At the 3′ of mRNA, secondary structures can slow the progress of the ribosome so that the termination codon aligns properly within the ribosomal A-site.
- Riboswitches and RNA thermometers enable an mRNA to sense changes in the cell and alter the expression of the gene(s) encoded on the same mRNA molecule.
- tRNAs are folded into tight structures, with the anticodon exposed at one end and an amino acid attached at the other.
- Three rRNAs are embedded within the ribosome and act together with ribosomal proteins during translation.
- Bacterial ncRNAs are able to influence the expression of genes through interaction with mRNA.

Key terms

Define the following terms introduced in this chapter. Check your answers using the definitions in the Glossary. These terms are also available as Flashcards online.

Anticodon loop	Hydrolysis	RNA thermometers
Anticodon stem	Introns	Stem-loop
Antisense RNA	mRNA-like domain	T-loop
Aptamer	Non stop decay	Transcriptome
D-loop	Nuclease	Transfer messenger RNA
D-stem	Nucleotidyl transferase	*trans*-translation
Endonucleolytic	Polyadenylation	tRNA-like domain
Exonucleolytic	Riboswitch	T-stem
Expression platform	Ribozyme	Untranslated region
Heat-shock proteins	RNase	Variable loop

Questions and discussion topics

Self-study questions

Answer each question using 50–100 words or a table or labeled diagram. Advice on where to find answers to these questions is available online.

1. A key feature of the production of proteins in bacteria is that genes are transcribed and then translated straight away. What is this process called and what features of the bacterial cell enable it?
2. What is the average size of bacterial mRNA and what dictates the size of bacteria mRNA?
3. What is the difference between the start point of transcription and the start point of translation?
4. Why might a gene be transcribed into mRNA, but not be translated into a protein?
5. What happens when there is no termination codon before the 3′ end of an mRNA and what is the name of the process that is used to recover from this?
6. Why is RNA less stable than DNA? Draw what happens.
7. Name at least three ways that secondary structures in the mRNA can interfere with the translation of the mRNA.
8. Compare the activity of a riboswitch with the activity of those that are ribozymes, considering when one might be more advantageous than the other.
9. Polyadenylation is generally thought of in the context of eukaryotic cells, but also occurs in prokaryotes. How does polyadenylation compare between eukaryotes and bacteria?
10. Draw the basic structure of a tRNA, including and labeling all of the loops and stems.
11. In Chapter 2, self-study question 3, you were asked about the designations of RNA in the ribosome. In which subunits are each of the rRNAs found?
12. How does the Hfq protein interact with ncRNA to lead to the degradation of mRNA?

Discussion topics

These topics are presented for discussion in study groups, as part of class discussions, or on your own. These questions go beyond what is directly covered in this part of the book. Use the research literature and other reading to explore these topics in more depth. Tips to help prepare for topic discussions are available online.

1. As in Figure 4.6, an mRNA transcript that is not translated by ribosomes is available to be targeted by RNases for degradation. Explore examples where this has been demonstrated to occur, as a result of a frameshift mutation, point mutation, or phase variation.
2. Explore the ways in which aptamers are being designed and used in either research or potential biotechnology applications. Discuss how their small size, specific structures, and specificity of binding make them powerful tools.
3. The 5′ and 3′ untranslated regions of a gene are included in the mRNA transcript and are involved in the process of translation. Discuss the ways in which the ribosome, as discussed here, and other proteins from your literature research interact with sequences in these untranslated regions. How do such interactions influence the expression of proteins?

Online quiz questions

To further self-assess your understanding of the chapter material, please visit the following link, where you can participate in a range of interactive quiz questions:

www.routledge.com/cw/snyder

Further reading

Bacterial mRNAs are translated into proteins as they are being transcribed

Del Campo C, Bartholomäus A, Fedyunin I, Ignatova Z. Secondary structure across the bacterial transcriptome reveals versatile roles in mRNA regulation and function. *PLoS Genet.* 2015; *11*(*10*): e1005613.

Irastortza-Olaziregi M, Amster-Choder O. Coupled transcription-translation in prokaryotes: an old couple with new surprises. *Front Microbiol.* 2021; *11*: 624830.

Bacterial tRNAs are folded into tight structures

Lorenz C, Lünse CE, Mörl M. tRNA modifications: impact on structure and thermal adaptation. *Biomolecules.* 2017; *7*(*2*): 35.

Shepherd J, Ibba M. Bacterial transfer RNAs. *FEMS Microbiol Rev.* 2015; *39*: 280–300.

rRNAs are essential components of the ribosome

Petrov AS, Bernier CR, Gulen B, Waterbury CC, Hershkovits E, Hsiao C. Secondary structures of rRNAs from all three domains of life. *PLOS ONE.* 2014; *9*(*2*): e88222.

The bacterial cell also contains noncoding RNAs that can regulate other RNAs

Jørgensen MG, Pettersen JS, Kallipolitis BH. sRNA-mediated control in bacteria: an increasing diversity of regulatory mechanisms. *Biochim Biophys Acta Gene Regul Mech.* 2020; *1863*(*5*): 194504.

Michaux C, Verneuil N, Hartke A, Giard JC. Physiological roles of small RNA molecules. *Microbiology.* 2014; *160*: 1007–1019.

Tomizawa J, Itoh T. Plasmid ColE1 incompatibility determined by interaction of RNA I with primer transcript. *Proc Natl Acad Sci USA.* 1981; *78*(*10*): 6096–6100.

Chapter

5 Transcriptional Regulation

All RNA is generated by transcription. The initiation of the process requires RNA polymerase to bind to the promoter region present in the DNA. Promoter regions are located 5′ of the region that will be transcribed into RNA. Transcriptional regulation modifies the frequency of transcription, regulating the amount of RNA made. For some RNAs, transcription occurs at a relatively consistent level. The coding regions encoded by these RNAs are said to have **constitutive expression**, which means that the genes are transcribed at roughly the same levels at all times. This does not mean that all constitutively expressed genes are expressed at the same levels relative to each other; the individual levels of expression vary depending on the gene. Constitutive expression refers only to the level of transcription of the gene being equal regardless of the state of the cell.

Genes encode different proteins, which may be required at different levels within the cell at different times. The amount of mRNA transcribed from some genes is therefore different at different times, depending on the state of the cell. Many genes are subject to **regulated expression**, where the levels of mRNA transcribed may be higher or lower at various times for the cell and under various conditions. Regulation of mRNA levels is often at the point of transcription initiation. If RNA polymerase cannot initiate transcription for a gene, then there will be little to no mRNA made and, without the mRNA carrying the message of the gene, it cannot be translated into protein. If binding of RNA polymerase to the promoter region is facilitated, then the initiation of transcription can occur more readily, producing an increase in the amount of the mRNA carrying the message encoded in the gene. Modulation in the amount of RNA synthesized is **transcriptional regulation**.

Regulation of gene expression at the level of transcription

Gene expression relies on a series of factors that must all be in place. The first is that the DNA encodes a functional copy of the gene. This gene sequence must then be transcribed into mRNA, so that it can be expressed as a protein through translation. The process of transcription can be controlled, resulting in genes being expressed when they are needed and being switched off when they are not needed. The result is that proteins are made when they are needed, rather than all of the time. By regulating mRNA production and protein synthesis, the energy of the cell is conserved.

Regulation of gene expression can be influenced by the condition of the bacterial cell. For example, the types and levels of nutrients available in the environment can change. Bacteria must respond to these changes in order to survive, making best use of the nutrients that are available. Response to a changing environment or conditions can result in alterations to the genes expressed at the level of transcription.

Transcriptional regulation of genes is often complex, requiring multiple components. The sigma subunit of RNA polymerase must be able to recognize the promoter region to initiate transcription. Different sigma factors will recognize different promoter region sequences. In addition, there may be multiple DNA-binding proteins with separate, overlapping, or the same binding sites within or proximal to the promoter. The promoter region can also have binding sites for

DOI: 10.1201/9781003380436-7

noncoding RNAs. Genes can be part of regulatory cascades, where activation or repression of transcription of one gene impacts upon the expression of another gene and so on down the chain. Proteins or ncRNAs may bind to the transcript itself, altering the progression of its transcription or the translation of the mRNA. The structure of the genome and the associated availability of promoter regions can impact the ability for RNA polymerase to access the DNA, influencing the initiation and completion of transcription.

The classic example of transcriptional regulation: the *lac* operon

The first example of transcriptional regulation was described by Jacques Lucien Monod and François Jacob and remains the classical example of the mechanisms involved in changes in transcription as a result of cellular conditions. In 1961, Monod and Jacob published a study describing their findings. They made a connection between the binding of proteins to DNA and the regulation of the amount of protein within the cell. It was only later that mRNA was shown to be the link between the DNA and its encoded protein.

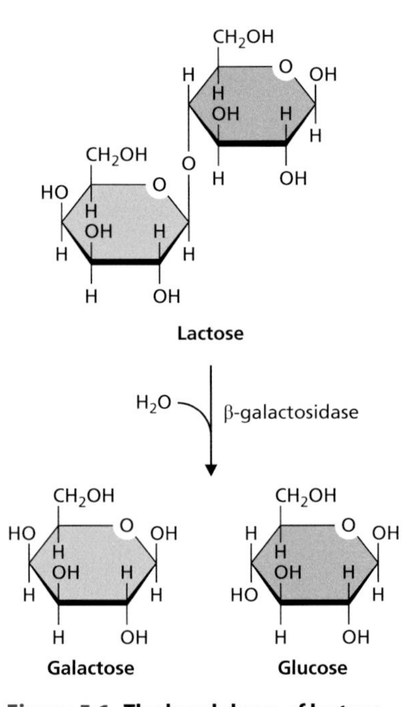

Figure 5.1 The breakdown of lactose. Within a bacterial cell such as *E. coli*, the enzyme β-galactosidase can break down the sugar lactose into its component sugars, galactose and glucose.

The regulation of gene expression in the classic *lac* operon system was investigated in *Escherichia coli*. Metabolically, *E. coli* and other bacteria require sugars to generate cellular energy stored as ATP. The disaccharide **lactose** is broken down within the bacterial cell by the enzyme **β-galactosidase** to its component sugars, **glucose** and **galactose** (**Figure 5.1**). The β-galactosidase enzyme also converts lactose into the metabolite allolactose. Lactose is transported into the bacterial cell by the transport protein **β-galactoside permease**. The genes encoding β-galactosidase and β-galactoside permease are adjacent to one another in the DNA of *E. coli*, encoded by *lacZ* and *lacY*, respectively. These genes are followed by *lacA* in a single transcriptional unit. Upstream of the three *lac* operon genes, collectively referred to as *lacZYA*, is the *lacI* gene, which encodes a **repressor protein**. Between the *lacZYA* and *lacI* genes is the promoter for the *lacZYA* genes and an **operator**, which is a binding site for proteins to bind to DNA and influence transcription (**Figure 5.2**). The *lacZYA* genes make up the *lac* **operon** and the associated features are involved in regulation of the *lac* operon. Operons are co-transcribed genes that are also co-regulated, such that these adjacent genes are transcribed together by RNA polymerase and are together regulated.

When there is no lactose in the *E. coli* cell, the repressor protein LacI binds to the operator, preventing the transcription of *lacZYA*. Without transcription of the genes, the proteins that they encode are not made. When the *E. coli* cell does contain lactose, the *lacZ* encoded β-galactosidase converts some of the lactose to allolactose, which binds to the LacI repressor protein. There is a conformational change in the LacI repressor protein when it binds allolactose, which causes the protein to no longer be able to bind to DNA. In this capacity, lactose as allolactose is an **effector molecule**, a cellular component that causes or prevents transcription. If mutations are made in the DNA such that there is no LacI repressor protein, then the *lacZYA* genes are always on, becoming constitutively expressed. Likewise, alteration or deletion of the operator sequence results in constitutive expression. If there is no promoter, then RNA polymerase has no binding sequence with which to initiate transcription and no mRNA carrying

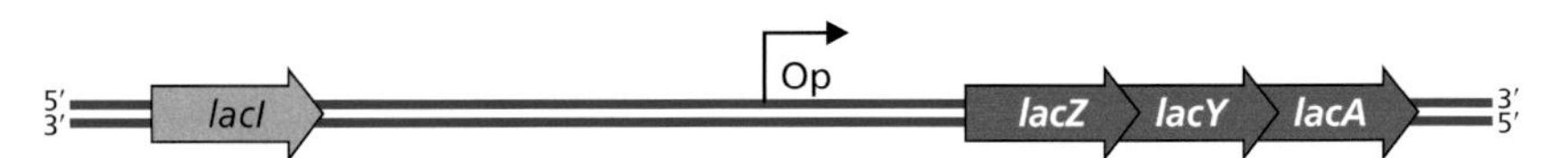

Figure 5.2 The structure of the *lac* operon. The *lacZ*, *lacY*, and *lacA* genes form one transcriptional unit. Upstream is the *lacI* gene encoding LacI, the repressor of transcription of *lacZYA*. The LacI repressor protein binds to a region between the *lacI* and *lacZYA* genes called the operator (Op). Binding of LacI to the operator prevents RNA polymerase from binding to the promoter (bent arrow) and initiating transcription of *lacZYA*.

the sequences of the *lacZYA* genes is made. Because the *lacZ* gene encodes the β-galactosidase, deletion of this gene means that lactose cannot be broken down into glucose and galactose and no allolactose is produced. A deletion of the *lacY* gene encoding the β-galactoside permease prevents the entry of lactose into the bacterial cell.

The *lac* operon is also subject to catabolite repression

The *E. coli* cell will use glucose by preference over other sugars. Capture of useable energy, mainly as ATP, is more efficient from glucose than from lactose or other sugars and thus the cell will use glucose first over other sugars available. The preference for certain molecules causes **catabolite repression**, where in this case the presence of glucose represses the expression of genes needed for the metabolism of lactose.

In addition to the operator binding site for LacI, the promoter region for *lacZYA* contains a binding site for CAP–cAMP. **CAP** is the **catabolite activator protein**, which is able to form a complex with **cyclic AMP (cAMP)**, a form of adenosine monophosphate where the phosphate is bound to both the 5′ and 3′ carbons (**Figure 5.3**). The binding of the combined CAP–cAMP to its *lac* operon binding site on the DNA is required for the transcription of *lacZYA*. While the LacI repressor has a negative influence on transcription, the binding of CAP–cAMP to the promoter region exerts positive control upon the *lac* operon (**Figure 5.4**). The binding of CAP–cAMP to DNA is associated with the presence of glucose within the bacterial cell. When the concentration of glucose is up, the concentration of cAMP is down. This is because glucose is efficiently processed to produce ATP. Because cAMP is scarce, it does not complex with the protein CAP and there is no binding of CAP–cAMP to activate transcription of the *lac* operon.

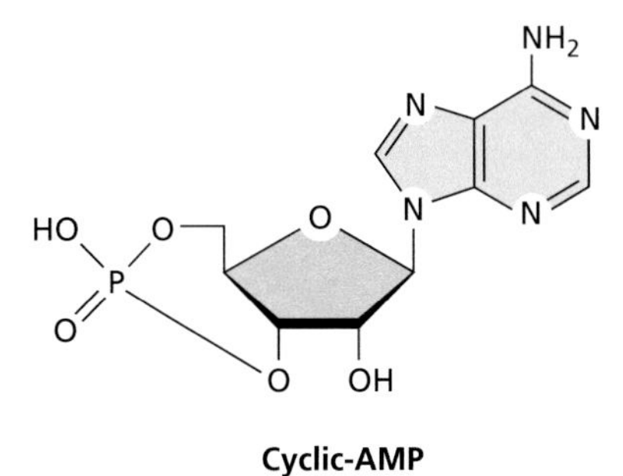

Figure 5.3 The structure of cAMP. AMP is adenosine monophosphate, which is generated when ATP, adenosine triphosphate, loses two phosphates and releases the energy for those bonds. cAMP is a form of AMP where the remaining phosphate is bonded to both the 5′ and the 3′ carbons on the ribose sugar.

If glucose is at low levels in the cell, but there is no lactose available, then the *lac* operon is off. It is not transcribed because the correct combination of control elements that would permit transcription of *lacZYA* is not present. When glucose is low, cAMP is high and is able to bind to CAP, which together are able to bind to the operator in the promoter region of the *lac* operon. This should allow the initiation of transcription; however, without lactose in the cell the LacI repressor protein is able to bind to its operator, preventing transcription. For there to be transcription of the *lacZYA* genes there must be lactose, as allolactose, present to bind to the LacI repressor protein and change its conformation, otherwise the repressor will bind to the operator and negate the positive influence of the binding of CAP–cAMP. When glucose is low or absent and lactose is present, the allolactose binds to the LacI repressor protein and the conformational change caused eliminates the DNA binding function of the repressor. Since low glucose levels result in abundant cAMP, there is plenty of it to bind to CAP. The CAP–cAMP complex is then able to bind to the operator in the *lac* operon and encourage the initiation of transcription of *lacZYA* by RNA polymerase. Therefore, when there is no glucose, but there is lactose, the enzymes that will enable the *E. coli* to use lactose are made. The two levels of control on the *lac* operon, which senses glucose and lactose availability, stops the bacterial cell wasting energy on making unnecessary enzymes for lactose metabolism when glucose is available and when there is no lactose (**Figure 5.5**).

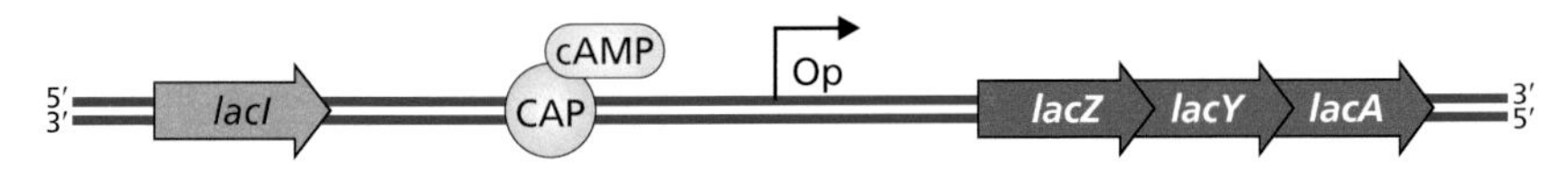

Figure 5.4 The binding of CAP–cAMP to the *lac* operon. When glucose is scarce, cAMP is abundant and binds to the CAP. This CAP–cAMP complex is able to bind to the intergenic region between *lacZYA* and *lacI*. This binding is required for the expression of *lacZYA* from its promoter (bent arrow).

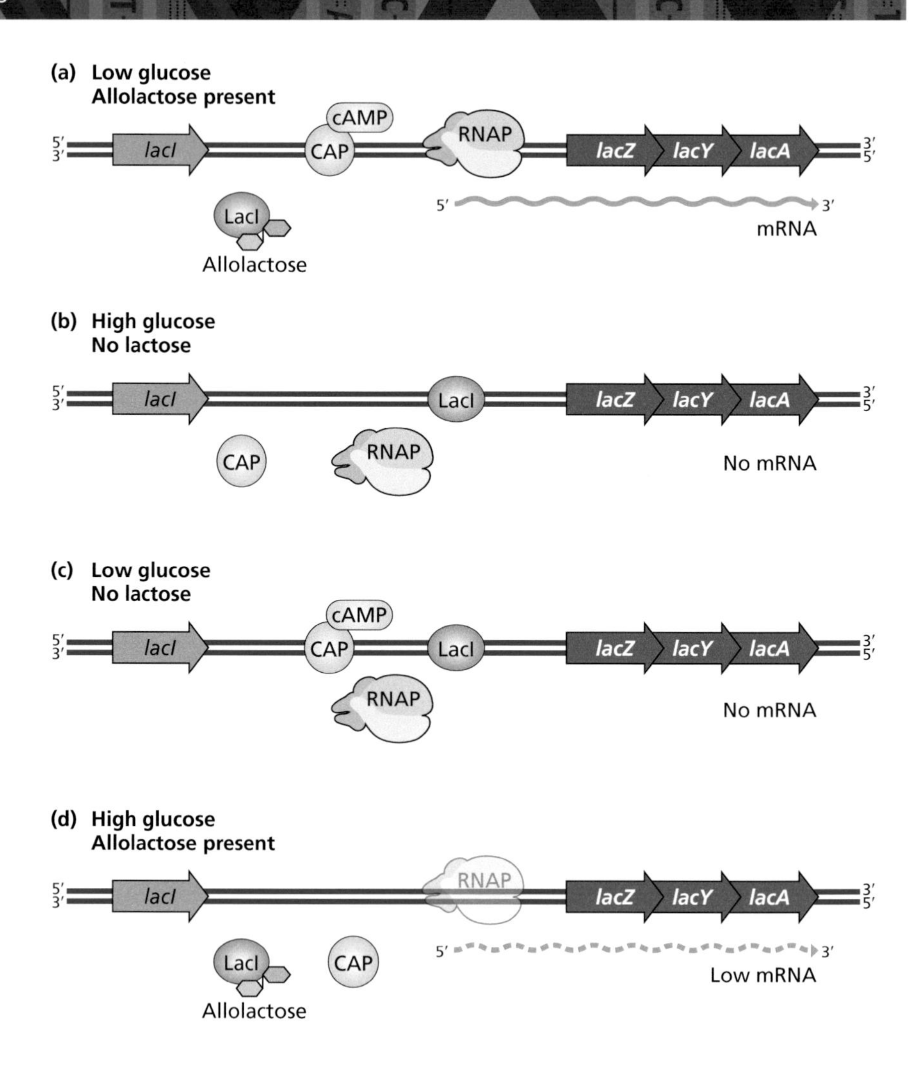

Figure 5.5 The regulation of the *lac* operon. When glucose is low and lactose is available (a), allolactose binds to the LacI repressor, preventing it from being able to bind to the operator. At the same time, cAMP is available to bind to CAP, resulting in the CAP–cAMP binding to the DNA and turning on transcription of the *lac* operon by RNA polymerase (RNAP). When glucose is high (b and d), cAMP is low and the CAP–cAMP complex is not made, so the positive signal for *lac* operon transcription is absent. If there is no lactose present (b and c), the LacI repressor protein will bind to the operator and prevent any transcription of the *lac* operon, even when glucose is low and CAP–cAMP binds to the DNA (c). In a cell high in glucose where lactose is also present (d), the *lac* operon is transcribed, but at a low level in the absence of CAP–cAMP binding.

The actions of the corepressor tryptophan on the *trp* operon

Tryptophan is an amino acid required for the translation of some proteins. Tryptophan can either be acquired from the environment or, if there is no tryptophan available, be synthesized by the bacterial cell. When the concentration of tryptophan within the bacterial cell is low, there is transcription of the *trp* operon. When the concentration of tryptophan is high, transcription of the *trp* operon is repressed. There are five co-transcribed genes in the *trp* operon, which encode enzymes that are able to produce tryptophan for the cell when it is scarce. The five genes, *trpEDCBA*, are preceded by a promoter and an operator (**Figure 5.6**). It is costly to the bacterial cell to make these enzymes and to produce tryptophan, so the *trp* operon is turned off when the cell can obtain sufficient tryptophan from its environment.

Regulation of the *trp* operon is via a repressor protein that is able to bind to tryptophan and block the transcription of the genes. In this system, tryptophan acts as a **corepressor**, which is a small molecule that is able to switch a repressor protein to its active DNA-binding conformation by binding to the protein. Transcription of the *trp* operon is not initiated in the presence of tryptophan, due to the action of the repressor protein and the corepressor, tryptophan.

An attenuation mechanism controls the expression of the *trp* operon

The *trp* operon is also subject to regulation by a system called **attenuation**. This process prevents the formation of a complete strand of mRNA from the *trp* operon.

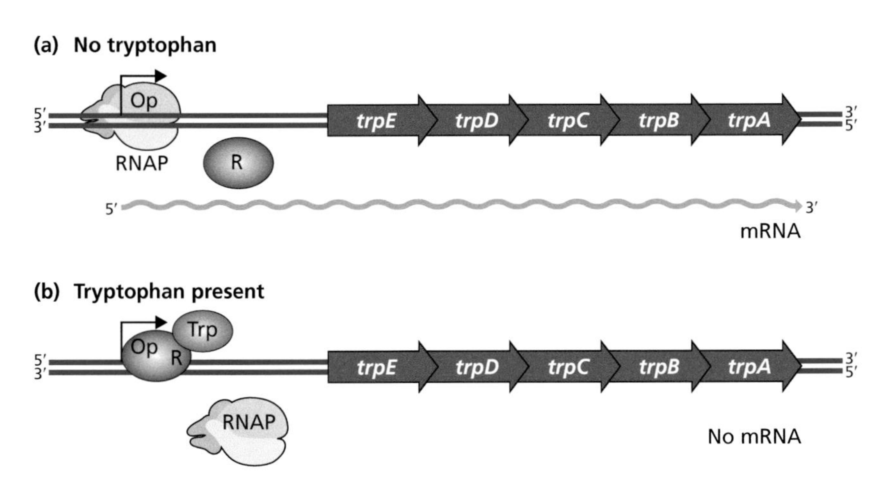

Figure 5.6 The structure of the *trp* operon and its transcription. The five genes of the *trp* operon, *trpEDCBA*, encode enzymes that are able to make tryptophan for the cell. These genes are transcribed from the promoter (bent arrow), when there is no tryptophan available to the bacterial cell (top). However, when tryptophan is present, a repressor protein (R) is able to bind to the operator (Op) in concert with the corepressor tryptophan (Trp) and prevent transcription by RNA polymerase (RNAP).

When tryptophan is abundant, RNA polymerase stops transcription prematurely. Between the operator and the *trpE* gene, there is a region called the **leader** followed by another region called the **attenuator**. The leader sequence encodes a short polypeptide, which is made from the leader sequence mRNA transcript. The attenuator sequence contains regions of self-complementary nucleotides and as such forms hairpins in the mRNA transcript. The types of hairpins that are formed by the attenuator depend on the translation of the leader. The influence of mRNA secondary structure on gene expression was discussed in more general terms in Chapter 4 (see Figure 4.7).

RNA polymerase, in the absence of the *trp* operon repressor protein, binds and initiates transcription of the *trpEDCBA* genes, as well as the leader and attenuator sequences that precede the genes (**Figure 5.7**). Shortly after the initiation of transcription, a ribosome binds to the mRNA and initiates translation during coupled transcription–translation. The first region to be translated is the leader sequence, which encodes the 14-amino acid leader peptide. The leader sequence includes two codons for tryptophan amino acids; therefore, during translation of the leader, the ribosome must incorporate two tryptophans into the leader peptide. If there is a high concentration of tryptophan in the bacterial cell, the ribosome is able to translate the leader peptide quickly, rapidly forming the polypeptide and dissociating from the mRNA. If the levels of tryptophan within the cell are low, the ribosome stalls while it waits for tRNAs carrying tryptophan to arrive at the A-site and continue translation. The leader polypeptide is therefore made slowly when there is little tryptophan present.

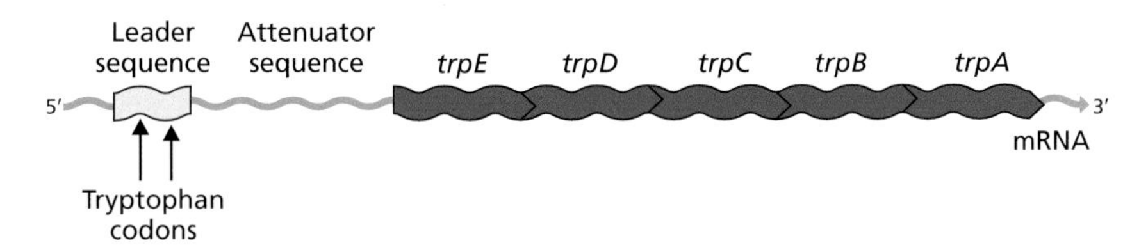

Figure 5.7 The features of the *trp* operon mRNA transcript. The mRNA transcribed from the *trp* operon includes, from 5′ to 3′, a leader sequence, an attenuator sequence, *trpE*, *trpD*, *trpC*, *trpB*, and *trpA*. The leader sequence is a short coding region, which encodes a peptide of 14 amino acids, including two tryptophans. The attenuator sequence contains regions of self-complementarity, where the mRNA forms hairpin secondary structures (**Figure 5.8**). The *trpEDCBA* genes encode proteins capable of synthesizing tryptophan.

(a)

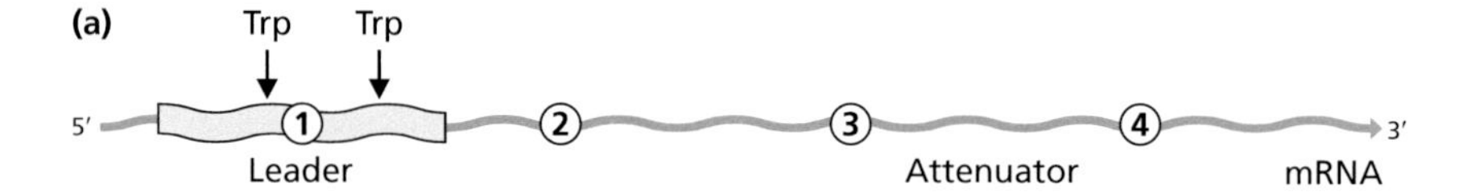

(b)

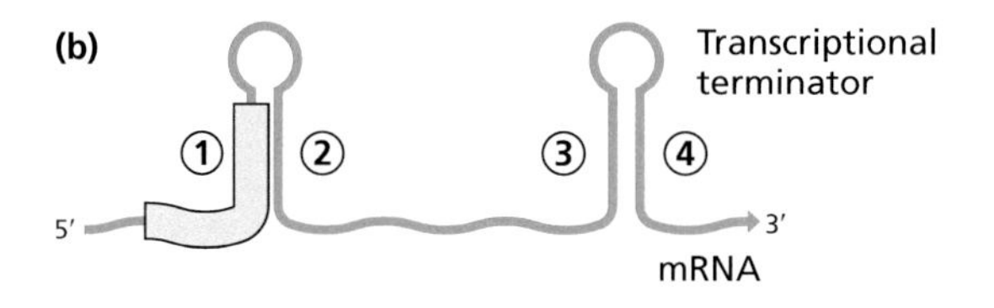

(c)

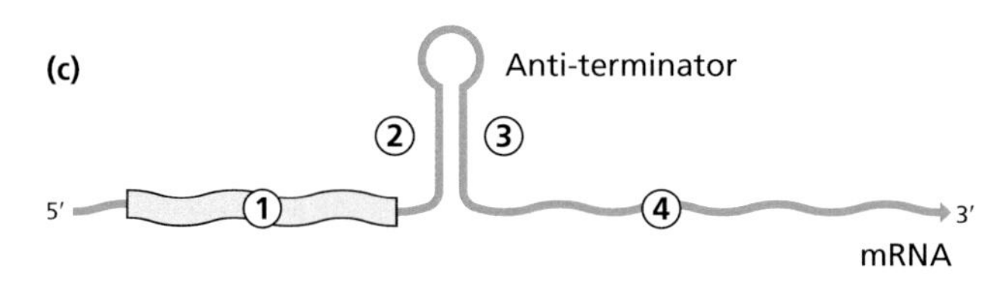

Figure 5.8 The secondary structures of the 5′ end of the *trp* operon mRNA. (a) The mRNA of the *trp* operon includes a leader sequence at the 5′ end that encodes a short peptide, followed by an attenuator sequence. There are four complementary regions that can base pair with one another forming secondary structures. (b) In one pairing, the leader sequence-containing region 1 pairs with region 2, and region 3 pairs with region 4. The pairing of 3 and 4 forms a transcriptional terminator. (c) Alternatively, regions 2 and 3 can pair together, forming an anti-terminator.

The attenuator sequence, at the 5′ end of the *trp* operon mRNA transcript, forms secondary structures in the mRNA. The attenuator region can form two different configurations of hairpins (see Figure 5.8). One conformation forms a transcriptional termination signal. The other conformation does not form a transcriptional termination signal; rather it stops the transcriptional terminator hairpin from forming. This mRNA hairpin is an **anti-terminator**. If the ribosome is slow to progress through the leader sequence codons due to the scarcity of the tryptophan amino acid, it pauses, and the pausing allows the formation of the anti-terminator hairpin and results in transcription of the complete *trp* operon mRNA (**Figure 5.9**). If tryptophan is abundant within the bacterial cell, the ribosome is able to quickly progress with translation. Upon the completion of translation of the leader sequence, the ribosome dissociates from the mRNA. This rapid progression of the ribosome allows the formation of the transcriptional termination hairpin and of a second hairpin in the attenuator sequence. These hairpins in the mRNA cause RNA polymerase to come off from the DNA, terminating transcription prematurely (**Figure 5.10**). In this configuration of hairpins, a complete transcript for the *trp* operon is not made, negatively regulating the expression of the tryptophan synthesis genes when tryptophan is abundant.

When tryptophan in the cell is high, the ribosome is quick through the leader, the transcriptional terminator forms, and transcription of the *trp* operon ends prematurely. When tryptophan is low, the ribosome is slow, the anti-terminator forms, and there is transcription of the complete *trp* operon, enabling tryptophan to be synthesized by the cell.

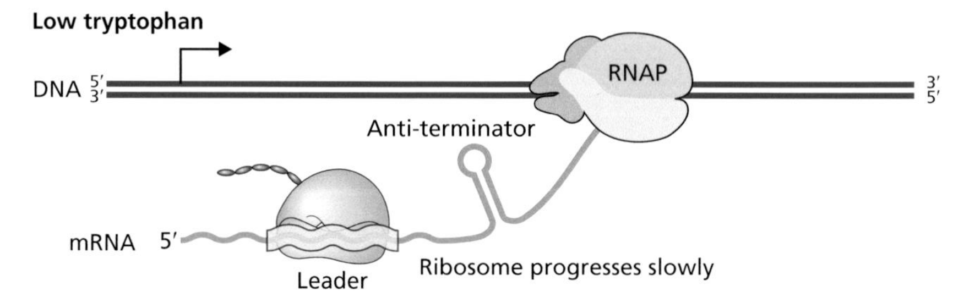

Figure 5.9 The translation of the *trp* operon leader polypeptide and anti-terminator formation. Within the 5′ end of the *trp* operon mRNA, there is a short leader sequence encoding a polypeptide containing two tryptophan amino acid codons. When there is little tryptophan in the cell, the ribosome translating the leader must pause to wait for $tRNA^{Trp}$. This pausing allows the adjacent attenuation sequence to form an anti-terminator hairpin. RNA polymerase (RNAP) continues to transcribe the *trp* operon.

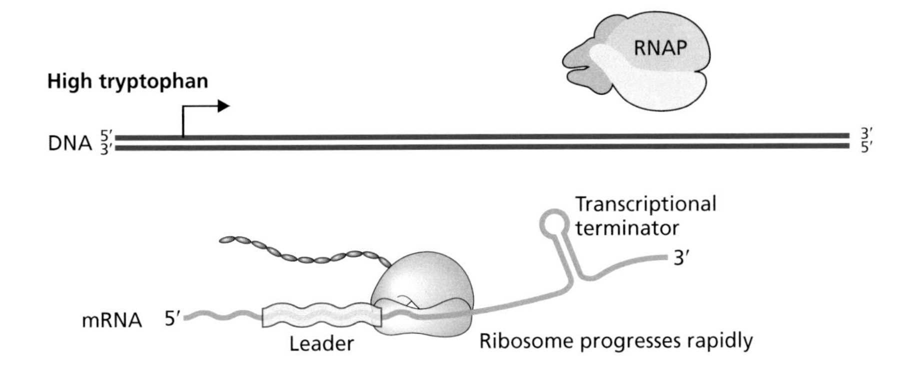

Figure 5.10 The translation of the *trp* operon leader peptide and transcriptional terminator formation. Within the 5′ end of the *trp* operon mRNA, there is a short leader sequence encoding a polypeptide containing two tryptophan amino acid codons. When tryptophan is abundant within the bacterial cell, the ribosome translating the leader will progress quickly. This rapid translation enables the attenuation sequence to form a transcriptional terminator hairpin. As a result, RNA polymerase (RNAP) ceases transcription of the *trp* operon and dissociates from the DNA.

Genes are regulated locally by *trans*-acting factors

The organization of genes into operons such as the *lac* operon and the *trp* operon facilitates the synchronous expression of these genes. However, distant genes can also be co-regulated through the use of the same regulatory proteins to orchestrate their expression. While the regulator protein for the *lac* operon is encoded near to its binding site for repression of *lacZYA*, regulatory proteins can be encoded anywhere in the genome and can bind anywhere in the genome. Although many regulatory proteins may act on transcription of genes near to where they are encoded on the chromosome, they can also act in *trans*, influencing the expression of distant genes. For example, in addition to the genes of the *lac* operon, transcriptome studies have shown that *E. coli* that are grown in the presence of lactose and the absence of glucose have altered the regulation of hundreds of other genes in the genome.

As seen in the *lac* and *trp* operons, there are DNA-binding proteins that are repressors of transcription and those that are activators of transcription. Some regulatory proteins bind to a metabolite or other small molecule within the bacterial cell and this binding, or lack thereof, changes the capacity of the regulatory protein to bind to DNA. Other regulatory proteins simply bind to their sequence-specific binding site on the DNA. In order to interrupt the influence of a regulatory protein that acts on its own to bind to DNA, the expression of the regulatory protein has to be stopped. This can be accomplished either by preventing the gene encoding the regulator from being transcribed or by preventing the transcribed mRNA from being translated into the protein. Influencing the transcription or translation of a regulatory protein creates a regulatory cascade, where the regulation of the level of expression of the regulatory protein impacts upon the level of expression of the genes that the regulatory protein regulates.

Repressors, activators, and inducers can influence the expression of many genes

Repressor proteins, which stop or reduce the expression of genes through binding to the DNA in or near the promoter region, can act on more than one promoter. Several genes can be co-regulated by a repressor protein. If, for example, the cell does not need to manufacture proteins for a particular biosynthetic process, the use of a single repressor can stop transcription of all of the genes involved in the process. Regardless of their location in the genome, the repressor protein can coordinate and turn off the expression of genes when their products are not needed, reducing the waste of energy involved in making proteins that are not needed.

Often working in opposition to repressors are activator proteins. These DNA-binding proteins assist in the binding and recruitment of RNA polymerase to the promoter. The actions of the activator protein result in an increased expression of the gene. In some cases, this may be through interaction between the activator protein and the RNA polymerase, where the activator bound to its

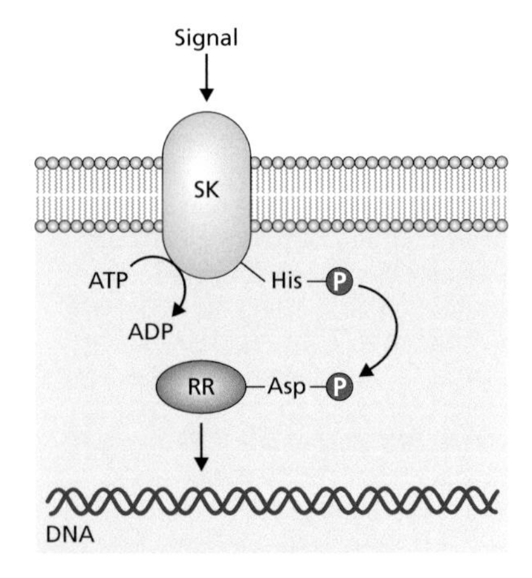

Figure 5.11 Two-component regulatory systems regulate transcription in response to external signals. The membrane-bound sensor kinase (SK) senses a signal from outside the cell and autophosphorylates, using a phosphate from ATP. This phosphoryl group is then passed on to a response regulator (RR). The phosphorylated response regulator in the two-component system undergoes a conformational change, enabling it to bind DNA and regulate gene expression in response to the external signal.

sequence-specific binding site may aid in the binding of the RNA polymerase to a weak promoter region. Like repressors, activators can regulate genes that are scattered throughout the genome. Synchronized activation of genes can be achieved when a set of genes are turned on by the same activator protein. This can ensure that all genes required for a biosynthetic process are turned on simultaneously.

Both repressors and activators can be subject to the effects of inducers. Inducers are molecules within bacterial cells that can bind to regulatory proteins and influence their activity. When an inducer binds to a repressor, the inducer prevents the repressor from binding to DNA and repressing expression. Allolactose is an example of an inducer that interacts with a repressor. In the *lac* operon, the allolactose binds to the LacI repressor, preventing it from repressing the transcription of the *lacZYA* genes. When an inducer binds to an activator it enhances the activation ability of the activator. It does this by aiding the activator in strongly binding to the DNA. Without the inducer, these activator proteins have very poor binding to the DNA. In the *lac* operon, cAMP binds to the regulatory protein CAP, enabling it to bind to DNA and activate transcription.

Two-component regulators sense change and alter transcription

Two-component regulatory systems are one of the main mechanisms used by bacterial cells to respond to stimuli. In response to stimulus, gene expression is changed. Cellular functions involving differential gene expression such as pathogenesis, biofilm formation, and motility can be regulated by two-component systems. There are two proteins involved in two-component regulatory systems; these two components give the system its name. The first protein component is a **sensor kinase**, which is a membrane-embedded receptor protein that recognizes the signal for the two-component regulatory system. When this sensor kinase binds to its ligand, this causes the sensor kinase to **autophosphorylate**, adding a phosphate from ATP to a specific histidine within the sensor kinase itself. The phosphorylated form of the sensor protein is able to transfer its phosphoryl group to the second protein of the two-component regulatory system. This second protein is the **response regulator** protein, which is only active when it is phosphorylated. The phosphate from the sensor kinase is transferred to an aspartic acid on the response regulator protein. This changes the conformation of the protein, so that the active response regulator is able to bind to DNA at the promoter of the genes it regulates and either increase or decrease transcription (**Figure 5.11**).

Phosphorelays are similar to two-component regulatory systems; however, the sensor kinase does not directly phosphorylate the protein that will bind the DNA and regulate expression. Instead, there is an expanded cascade involving four proteins, along which the phosphate is passed. After autophosphorylation in response to stimuli, the sensor kinase transfers its phosphoryl group to a response regulator that has no DNA binding ability, even when activated. This response regulator transfers the phosphoryl group to a phosphotransfer protein, which in turn transfers it to the terminal response regulator that binds DNA (**Figure 5.12**). In some cases, the sensor kinase and the first response regulator are fused into one protein, embedded in the membrane.

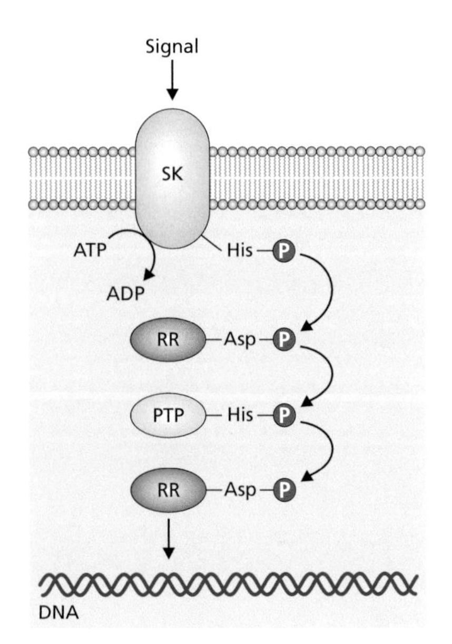

Figure 5.12 Phosphorelays regulate transcription in response to external signals. The membrane-bound sensor kinase (SK) in the phosphorelay system senses a signal from outside the cell and autophosphorylates, using a phosphate from ATP. This phosphoryl group is then passed on to a response regulator (RR) that lacks a DNA-binding domain. The phosphoryl group is passed on again, this time to a phosphotransfer protein (PTP), which transfers the phosphoryl group to a second RR. This last protein in the phosphorelay gains the ability to bind DNA following a phosphorylation-induced conformational change.

DNA changes in the promoter region locally regulate transcription in *cis*

Transcription relies upon DNA, not only to be the template for the transcription of mRNA, but also to contain all of the sequence signals needed to initiate transcription. The sequence information is able to therefore exert a *cis* effect upon the mRNA.

The clearest example of this is the promoter itself. Without a promoter, there is no sequence for the sigma factor to recognize and recruit RNA polymerase to initiate transcription. Therefore, any mutations in the promoter sequence DNA may impact the level of transcription by altering recognition of the promoter by the sigma factor (**Figure 5.13**). There may also be more than one promoter upstream of a gene and the location of these sequences in the DNA may influence which promoter is used to drive expression, when transcription is initiated in the cell, and at what level the region is transcribed.

Operators also exist within the DNA. These sequences are bound by regulatory proteins, and changes to their sequences may alter the binding of the DNA-binding proteins (**Figure 5.14**). The binding of regulatory proteins to their binding sites is sequence-specific. The protein has to be able to recognize its binding site to achieve effective binding and either prevent transcription by RNA polymerase or enhance it. Likewise, any changes to RNA binding sites, where regulatory RNAs will bind to the mRNA transcript, can change the expression level of the associated gene.

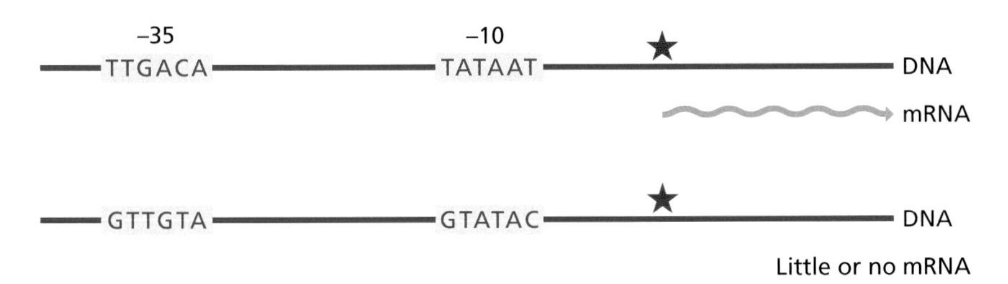

Figure 5.13 Promoter mutations alter the expression of genes. Mutations in the consensus promoter region recognized by the sigma factor can impact the binding of the sigma factor and thus RNA polymerase. When the –10 and –35 consensus promoter sequences are present, σ^{70} can bind to the promoter, recruit RNA polymerase, and initiate transcription of mRNA from the start point of transcription (★) (top). If a mutation reduces or eliminates binding of the sigma factor to the promoter, little to no transcript of the gene is made by RNA polymerase and the gene is not expressed. When the sequence is too different from the recognized consensus sequence (bottom), there will be no mRNA produced because transcription cannot be initiated.

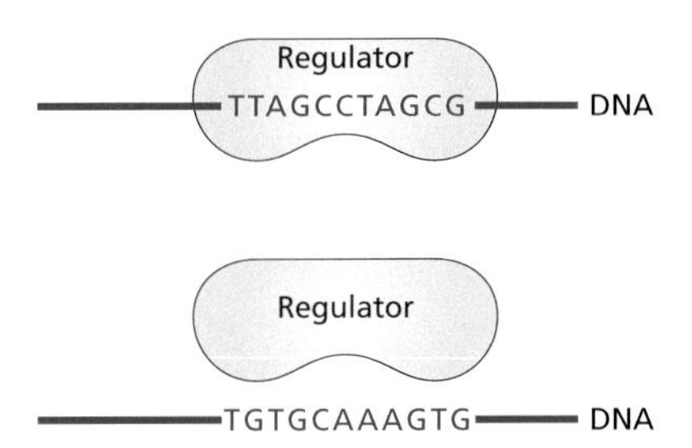

Figure 5.14 Binding of regulatory proteins to DNA is sequence-specific. The regulatory protein will bind to its binding site based on the DNA sequence (top). If there are changes in the DNA sequence, it may no longer be recognized by the regulatory protein, which will not bind (bottom). This can impact the expression of the associated gene through either lack of repression or lack of activation.

Programmed changes to DNA can alter transcription locally

Some mutations within promoter regions are programmed into the DNA itself. Due to the sequence features present, the level of expression from the promoter changes based on the promoter mutations. One mechanism for this is through changes in simple sequence repeats. These can be homopolymeric tracts of the same base or they can be repeated short sequences of bases. Changes in the lengths of these repeats are introduced during replication and impact upon the expression of the gene from the repeat associated promoter (**Figure 5.15**). This process is called **phase variation** and is used by some bacteria to generate diversity in gene expression within the population of bacteria.

Other mechanisms may alter the DNA upstream of a gene and influence the expression of that gene. There are some genes in some genomes that are driven by promoters that are affected by inversion events, such as *fimA* in *E. coli* involved in fimbriae expression. These are not random mutations, but mechanisms programmed into the DNA itself. Localized changes to the DNA, where a segment

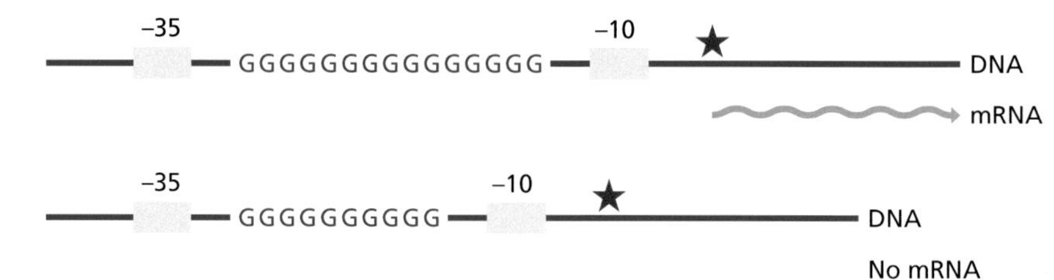

Figure 5.15 Changes in simple sequence repeats alters the promoter and causes phase variation. In bacterial species that use transcriptional phase variation, simple sequence repeats are associated with promoter regions. Changes in the length of these repeats alter the promoter, which may result in increased or decreased ability for the sigma factor to bind to its consensus, recruit RNA polymerase, and initiate transcription at the start point (★). Here the number of G bases between the −10 and −35 promoter sequences has changed. The number of G bases between the −10 and −35 promoter regions of the DNA creates ideal spacing of the promoter elements and RNA polymerase is able to initiate transcription and produce mRNA (top). A reduction in the number of G bases creates a promoter region where the −10 and −35 sequences are not ideally spaced, so RNA polymerase cannot recognize the promoter and no mRNA is made.

of DNA within the genome is excised, inverted, and reinserted at the same location, can change the orientation of a promoter so that it is no longer upstream of the gene. The inversion event may switch between a gene having a promoter and being expressed or not having a promoter and not being expressed (**Figure 5.16**). Promoter inversion may also alter the strength of the expression level, switching between a strong promoter in one orientation and a weak one in the other orientation, or switching between two promoters recognized by different sigma factors.

The DNA of bacterial cells can be methylated and these methyl groups may disrupt the ability of regulatory proteins to bind to their binding sites. **Methylation** in bacteria occurs at the C-5 or N-4 of cytosine or at the N-6 of adenine when a methyl group is added by a methyltransferase from *S*-adenosyl methionine. The methylation of an adenine or cytosine can alter the binding of regulatory proteins to the DNA at that site, either due to the physical presence of the methyl group or due to a change in the structure of the DNA that influences binding. The binding of proteins to DNA can also block the methylation of the DNA by methyltransferases.

In *E. coli*, expression of pili can switch on and off as a result of changes in methylation at the promoter for the pilus protein. There are two GATC methylation sites, $GATC^{prox}$ and $GATC^{dist}$, where the adenine can be methylated by DNA methylases. These two methylation sites overlap the binding sites for the Lrp regulatory protein and the PapI regulatory protein. When $GATC^{prox}$ is methylated but $GATC^{dist}$ is not, the gene expression of *papAB* encoding the pili is on. PapI and Lrp bind to the DNA at $GATC^{dist}$ and activate transcription. However, when $GATC^{dist}$ is methylated and $GATC^{prox}$ is not, expression of pili is off. Lrp binds to $GATC^{prox}$ and represses transcription by preventing RNA polymerase from accessing the promoter (**Figure 5.17**).

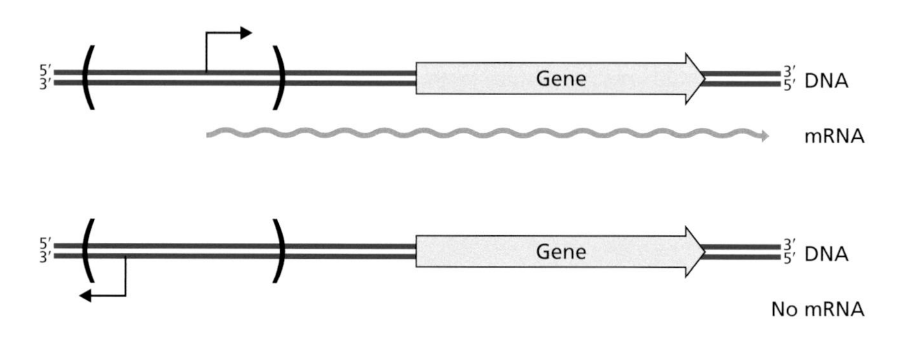

Figure 5.16 Gene expression may change due to inversion events. Inversion of a region of DNA, where the sequence is turned around in the opposite orientation, can impact upon expression. Promoters in the invertible region (marked with parentheses) are switched in their orientation relative to the gene. The inversion has switched the expression of the gene, with mRNA being produced when the promoter is in one orientation (top) and not when it is in the other (bottom).

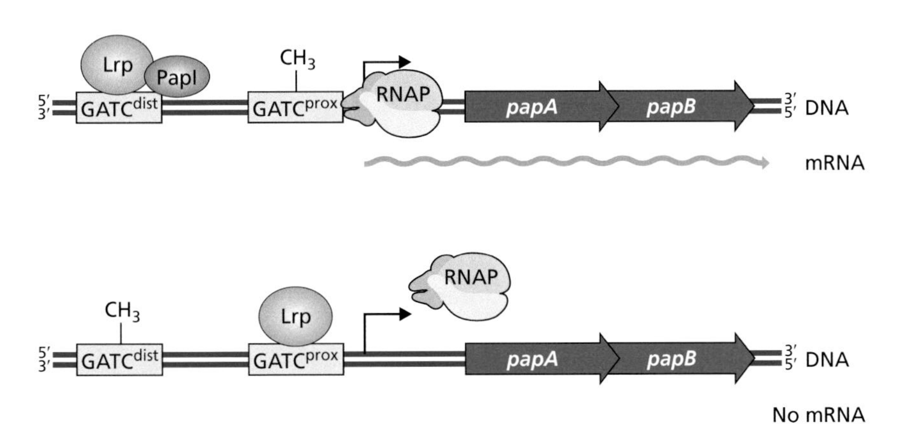

Figure 5.17 Methylation of DNA can influence the binding of regulatory proteins. The methylation state of DNA can change the binding of regulatory proteins, such as the binding of Lrp and PapI to the promoter region of the pilus genes *papAB* in *E. coli*. Some bases of DNA can become methylated, here adding a methyl group to an adenine. The presence or absence of the methylation can alter the capacity of regulatory proteins to recognize the binding site, bind, and modify the expression of the gene by RNA polymerase (RNAP). When GATC^{prox} is methylated, Lrp and PapI bind to the GATC^{dist} region, allowing RNAP to recognize the promoter (bent arrow) and initiate transcription (top). When GATC^{dist} is methylated, the Lrp regulatory protein binds instead to the GATC^{prox} region repressing transcription by preventing RNAP from accessing the promoter (bottom).

Some promoters overlap in divergent, bidirectional promoters

Researchers have demonstrated that 19% of transcriptional start sites in *E. coli* originate from divergent bidirectional promoters. Due to the symmetry of the sequences in the promoter and double-stranded nature of DNA, some regions form promoters on the opposite strands. The positions of the –10 and –35 can vary as can the transcriptional start site (**Figure 5.18**). In all cases, the promoter components on each strand overlap. Although identified and studied in *E. coli*, these types of promoters are present throughout the prokaryotes.

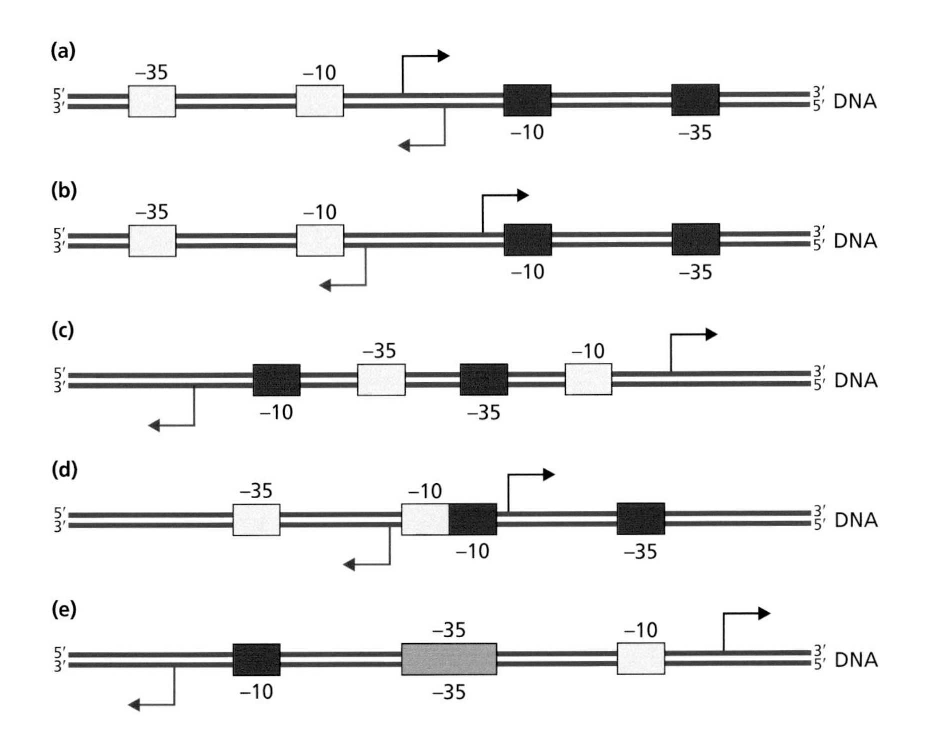

Figure 5.18 Bidirectional promoters with overlapping features. There are five different ways that the features organize in bidirectional promoters, which start transcription on opposite strands. First (a), the promoters may face one another, where the transcripts will overlap. If the features are a bit closer together, then the transcriptional start sites overlap (b). The –10 and –35 features may overlap as well, with the –35 from one strand present between the –10 and –35 regions of the promoter on the other strand (c). Features may also be shared between promoters on opposite strands, with a longer –10 region containing the sequence for the –10 on both strands (d) or the –35 being shared between the promoters on both strands (e).

Sigma factors are essential for the initiation of gene transcription

Sigma factors function by binding to promoters and combining with the RNA polymerase core enzyme to make the RNA polymerase holoenzyme. The RNA polymerase holoenzyme, made of the core enzyme and a sigma factor, is capable of initiating transcription. The availability and function of sigma factors may be altered by environmental conditions. The sigma factors, in addition to their role in initiating transcription, can therefore regulate transcription.

Genes can have two or more promoters, recognized by different sigma factors. There is a great deal of diversity within bacterial species, in terms of both the numbers of different sigma factors that the cell can make and the specificity of each of the sigma factors. Sigma factors each recognize a distinct, specific sequence within the promoter and bind to them. Different sigma factors therefore regulate the expression of genes through recognition of different promoters.

After the initiation of transcription, the sigma factor leaves the RNA polymerase holoenzyme and the RNA polymerase core enzyme continues with transcriptional elongation. This sigma factor is then able to go on to a promoter and initiate transcription by joining with another RNA polymerase core enzyme. There is a limit to the amount of RNA polymerase in the bacterial cell; therefore, the sigma factors compete with one another for a limited pool of RNA polymerase core enzyme with which to initiate transcription.

Sigma factors can orchestrate global regulation of gene transcription

Many genes are transcribed by RNA polymerase with the σ^{70} housekeeping gene sigma factor. Alternative sigma factors are able to initiate transcription of a variety of genes, often enabling the bacteria to survive stresses, respond to nutrient availability, and orchestrate morphological changes. Alternative sigma factors are often involved in survival in stress, which can be survival in hosts. Within the host, the expression of bacterial virulence factor genes and within host survival genes can be controlled by alternative sigma factors, such as σ^S involved in stress responses and σ^{32} involved in temperature responses. For example, the ECF sigma factor, for extracytoplasmic functions, is involved in the initiation of transcription of genes in response to cell envelope stress and the environment.

The promoters of heat-shock proteins are recognized by σ^{32}, the key regulator of the heat-shock response. The half-life of σ^{32} at the normal temperature for a bacterial species is about 1 minute, but at higher temperatures the half-life increases to about 5 minutes. Therefore, at higher temperatures, σ^{32} will stay in the cell longer and be able to participate in the **initiation** of transcription more often than at normal temperature. Once sufficient quantities of heat-shock proteins have been made, the activity and quantity of σ^{32} in the cell can be reduced. Heat-shock proteins DnaK, DnaJ, GrpE, and HflB are chaperones to misfolded proteins; however, as the number of misfolded proteins decreases, these heat-shock proteins instead interact with σ^{32} and target it for degradation. The half-life of the sigma factor and its interaction with heat-shock proteins that it up-regulates work together to optimize the amount of σ^{32} in the bacterial cell based on temperature and protein misfolding.

Control of sigma factor activity involves several components

Control of sigma factor expression can involve small noncoding RNAs. In some cases, these act at the level of translation, including translation from existing mRNA as well as newly synthesized mRNA. For example, generation of sigma factor protein σ^S (RpoS) is driven by stresses. The *E. coli* σ^S has about 500 genes for which it controls expression. The σ^S itself is regulated by at least four ncRNAs and requires the RNA chaperone Hfq. One ncRNA, DsrA, senses changes in temperature; another, RprA, senses stress at the cell surface; and a third, OxyS, responds to oxidative stress. Each increases translation of the σ^S encoding mRNA. The ncRNAs DsrA and RprA, together with Hfq, bind at the 5′ end of the *rpoS* mRNA, stopping the formation of a hairpin and making the ribosome binding site (RBS) accessible by ribosomes. The presence of σ^S within the cell is also controlled through regulation of its own transcription and differential degradation of the σ^S protein by **proteolysis**.

There are also **anti-sigma factors** that are able to prevent the action of sigma factors. Anti-sigma factors bind to sigma factors and block the portion of the sigma factor that binds to the RNA polymerase core enzyme. Often anti-sigma factors inhibit the activity of the alternative sigma factors, keeping them sequestered until they are needed. When the alternative sigma factors are released from anti-sigma factor control, anti-sigma factors may then inhibit the housekeeping sigma factor σ^{70} to prevent it from initiating transcription. The genes for anti-sigma factors tend to be co-transcribed with those for the sigma factor that they regulate.

Global regulation can be influenced by the binding of chromatin proteins to the DNA

Bacterial DNA exists in a compact mass, known as the nucleoid. One of the ways the compaction of DNA is achieved is through DNA supercoiling, using the turns of the double helix to compact the DNA into a small size. There are **chromatin proteins** associated with the DNA within the nucleoid. These proteins also contribute to the compactness of the DNA within the cell. Macromolecular crowding of all of the bacterial cellular components within the cell also plays a role in compacting the DNA. Bacterial DNA takes up only about a quarter of the interior of the bacterial cell, which is also full of enzymes, RNAs, and other small molecules that crowd the DNA into a small space.

The transcription in some regions, such as the loci for the rRNAs and tRNAs, is high. The DNA for these cellular features must be accessible to the transcription machinery locally, yet the DNA also has to be compact overall. Therefore, the structure of the DNA, its supercoiling, interaction with chromatin proteins, and macromolecular crowding are involved in the global regulation of the features on the chromosome. The main chromatin proteins that contribute to nucleoid compaction in most bacterial cells are **H-NS**, the **histone-like nucleoid structuring protein**; **HU**, the **heat unstable protein**; **Fis**, the **factor for inversion stimulation**; and **IHF**, the **integration host factor**.

The H-NS protein binds to DNA, making regions unavailable for transcription

The structure of the H-NS protein has an N-terminal dimerization domain that enables the protein to form **dimers** with another H-NS. The C-terminal domain is the DNA-binding domain of the protein, which is connected to the N-terminus by a flexible region within the H-NS protein. H-NS proteins preferentially bind to A+T-rich regions of the DNA. When it binds to these regions it silences them. Within the bacterial cell genome, if regions of horizontal transfer are A+T-rich, having originated from a species that has lower G+C, the H-NS binding can prevent expression of foreign genes that may be detrimental to the cell.

H-NS is present in approximately 20,000 copies of the protein per cell, making it abundant within the cell and a major protein component of the nucleoid. Regulation in gene expression due to H-NS is a result of the capacity of many copies of H-NS to together bind to contiguous regions of DNA and cause it to bend through interactions between H-NS proteins (**Figure 5.19**). The H-NS induced

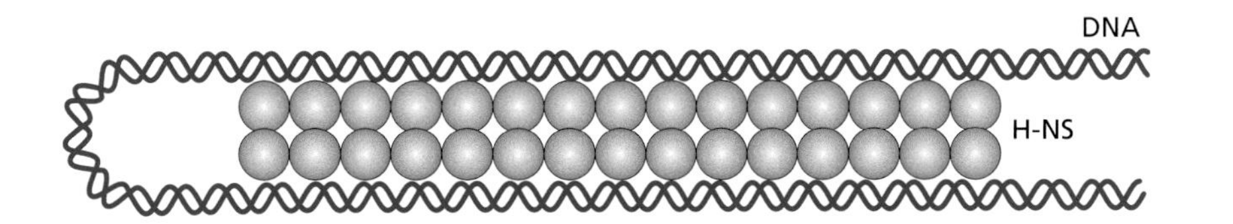

Figure 5.19 H-NS proteins bind to DNA, compacting it and bending it. Multiple copies of the H-NS protein are able to bind to the DNA and to each other, causing the DNA to form compact structures. The bends in the DNA created by H-NS binding brings together the nucleoid and condenses the DNA into a smaller space within the bacterial cell. The binding of DNA by H-NS and the resulting changes in DNA structure can impact the expression of genes by making regions of DNA inaccessible to RNA polymerase.

bending prevents RNA polymerase from accessing promoter regions to initiate transcription from the DNA. In addition, the binding of H-NS to regions of DNA can block the progression of RNA polymerase during transcriptional elongation.

HU and IHF are homologous proteins that act in a similar way upon DNA

The HU and IHF proteins are homologous to one another, being similar in sequence and activity, yet the IHF protein has sequence specificity in its DNA binding, while the HU protein does not. The binding of IHF to DNA generates a bend in the DNA of 140°–180° at the sites that are specifically recognized by IHF due to its sequence.

HU, with no sequence specificity, instead binds based on the DNA structure. Supercoiled DNA is bound by HU preferentially. HU also bends DNA, like its homologue IHF. HU also binds to structures in the DNA like nicks, gaps, and junctions. HU recognizes the bends that are caused by these structures and stabilizes them.

The Fis nucleoid protein is involved in the regulation of rRNA transcription

Fis binds to DNA and bends it, in bends of 50°–90°. Fis recognizes a binding sequence, but this sequence is poorly conserved, and the protein can also bind in a non specific way to DNA. In *E. coli*, Fis binds to the chromosome at roughly once every 230 bp.

In addition to its structural role within the nucleoid, Fis is involved in the activation of transcription of rRNAs. rRNAs are highly transcribed regions of the genome. Fis binds to three sites that are 5′ of the rRNAs and enhances transcription by RNA polymerase.

Quorum sensing causes transcriptional changes within the bacterial cell

For some bacteria, it is energetically expensive to express certain genes and their resulting proteins if there are too few bacteria in a local area. When the bacterial population is large, expression of the genes can be beneficial, often not just to the bacteria expressing these genes, but also to other cells within the population.

To achieve an assessment of the number of bacteria nearby, some bacterial species produce a chemical signal that nearby bacteria are able to sense. If the cell is alone, it will receive little to no signal; the only cell making the signal molecule is itself. If the cell is in a large population with other bacteria like itself, then it will receive a strong signal, indicating that the population size is large. This process is called **quorum sensing**.

Quorum sensing allows the bacterial population to adapt and survive in the environment in a different way when the population is large versus when it is small. The most common and well-studied signaling molecules are ***N*-acyl homoserine lactones**, but other signal molecules exist as well. The signaling molecule is exported from the cell into the extracellular environment. The bacteria are also able to take up the signaling molecule from the environment and recognize it as a quorum sensing molecule. If the concentration of signal is high, then this corresponds to a high population and the expression of genes is induced (**Figure 5.20**). More detail about how bacteria use quorum sensing is explored in Part V.

Biofilm formation is a specialized response to quorum sensing and other signals

Biofilms are formed by microorganisms that have come together as a community with each cell contributing to the whole. The biofilm structure is particularly

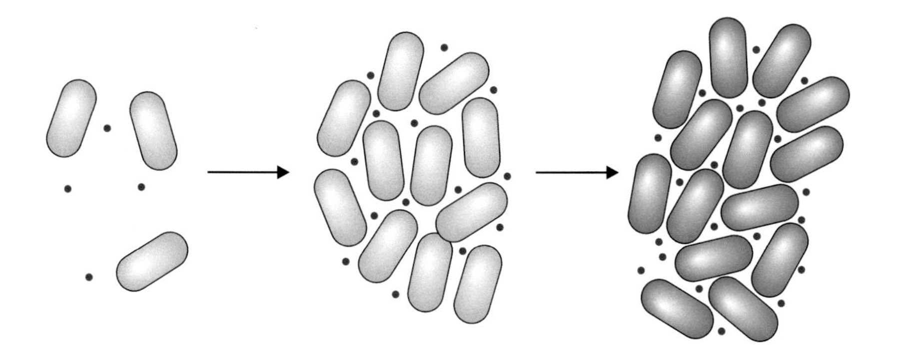

Figure 5.20 Quorum sensing enables bacteria to assess the size of its population. Diffusion of a small molecule (red) made by the bacteria can be sensed by the bacteria and used to account for the size of the population. As the bacterial population grows (left to right), the density of signaling molecules increases. Once the population density has reached a key point, the bacteria sense enough of the quorum sensing molecule nearby and alter their gene expression (denoted by change in color).

robust. It adheres to surfaces and due to its make-up is resilient to stresses, including antibiotics. Biofilm formation is complex, requiring the controlled regulation of several genes that must be expressed in the correct order.

Quorum sensing can trigger the formation of a biofilm. It is believed that the bacterial cells are better able to form a biofilm when there are enough of them in one place. The cell can assess this state using quorum sensing. Other signals can also be involved in the bacterial cells' switch from free-living organisms to a biofilm. These include the presence of antibiotics, the presence of a suitable surface on which to build the biofilm, and the availability of nutrients and host-derived cues.

The biofilm includes extracellular sugars as **exopolysaccharide**; therefore, the synthesis and export of polysaccharides must be up regulated for biofilm formation. There are also extracellular proteins and DNA within the biofilm, as well as the bacteria themselves. The specific biofilm matrix proteins are transcriptionally regulated, as is the DNA export system. DNA can be released from bacterial cells by lysis and in this way contribute to the extracellular DNA. Some bacteria are capable of autolysis upon receipt of certain cues. However, living bacterial cells can contribute DNA as well, through the release of membrane blebs containing DNA. The generation of DNA-containing blebs can also be up regulated. Bacterial biofilms are explored in more detail in Part V.

Cyclic di-GMP is involved in the regulation of a range of functions within the bacterial cell

The state of the environment that the bacteria cell is in and changes to that environment can be sensed in a number of ways. One is through the creation and degradation of a secondary molecule that acts as a signal for the state of the environment. Two molecules of GTP can come together to form **cyclic di-GMP** through the enzymatic action of **diguanylate cyclases**. Cyclic di-GMP is broken down by **phosphodiesterase** enzymes. Often the activity of these two enzymes is modulated by environmental conditions. The cyclic di-GMP is able to bind to a range of receptors that regulate gene expression, including riboswitches and transcription factors. In this way, cyclic di-GMP is able to regulate diverse cellular functions including biofilm formation, motility, and expression of virulence factors including those involved in adhesion.

The small molecule ppGpp is an indicator of the state of the bacterial cell

Within bacterial cells, the guanosine tetraphosphate (ppGpp) molecule varies in its quantity depending on the state of the cell. High levels of ppGpp accumulate within cells that are starved of amino acids. These cells enter a stringent response where synthesis of RNA is reduced including the inhibition of tRNA and ribosome synthesis, as well as the production of mRNAs. Some global regulators, including IHF, Lrp, and σ^S, rely on ppGpp for their expression.

Protein thermosensors regulate expression of proteins via transcriptional regulation

There are protein thermosensors, in addition to the RNA thermosensors from Chapter 4. Heat-shock proteins are needed as chaperones to prevent proteins from misfolding or degrading at increased temperatures. Many proteins in the bacterial cell are prone to misfolding at temperatures higher than normal, including regulatory proteins. The misfolding of a regulatory protein will change the structure of the regulator as the temperature increases. At high temperatures, the DNA binding activity of repressor proteins and activator proteins is poor, due to the temperature-induced structural changes in the protein. In the absence of stabilization of the regulator protein structure by heat-shock protein chaperones, the regulator fails to act. Temperature increases can either end the repression of a gene by the repressor protein or prevent the activation of a gene by an activator protein, influencing the transcription of the gene.

Stability and degradation of mRNA by ribonuclease III impacts upon whether a transcript is expressed as a protein

Several criteria can induce mRNA to be degraded by RNases, including the formation of bends in the mRNA and the presence of double-stranded RNA. The bacterial cell contains several different types of RNase enzymes that are able to recognize different features of RNA, including when it is double stranded, when it is folded into a hairpin, when it is occupied by ribosomes, and when a ribosome is stalled at the 3′ end of an mRNA inducing non-stop decay.

RNase III is a global regulator of gene expression. Through its action degrading mRNA, it stops the production of proteins. RNase III binds to double-stranded RNA and cleaves phosphodiester bonds on both strands of segments of double-stranded RNA. Antisense RNAs binding to mRNA form double-stranded RNA that is targeted by RNase III. The secondary structures formed by mRNA folding upon itself generates double-stranded RNA.

Regulation of gene expression can be through the action of RNA binding proteins

The regulation of expression from a transcript can occur through direct action that influences whether the complete transcript is made. Expression can be controlled both during and after transcription of the mRNA through the actions of RNA binding proteins. These proteins can change the susceptibility of an mRNA to degradation by RNases, change the availability of the RBS, aid in other cell components interacting with the mRNA, and influence the generation of the complete transcript through encouraging the formation of transcriptional terminators or anti-terminator secondary structures.

Some RNA binding proteins are RNases, which bind to RNA and cleave it, as previously discussed in this chapter. Other RNA binding proteins can block the action of RNases by binding to RNase recognition sequences, blocking them from being bound by RNase. RNA binding proteins can also change the secondary structure of an mRNA, protecting it from RNases that recognize such structures or making the mRNA more susceptible to degradation due to structural changes.

Unless the RBS is available for recognition by the complementary sequence in the 30S ribosomal subunit's 16S rRNA, translation cannot initiate. RNA binding proteins can influence translation by binding to the RBS region or causing the region to form secondary structures. If the mRNA is not covered by ribosomes engaged in translation, it is prone to degradation. RNA binding proteins can stabilize the formation of a transcriptional terminator, which prematurely stops transcriptional elongation, especially of mRNAs that are not being translated.

Alternatively, binding of an RNA binding protein can result in an anti-terminator structure forming, which stops the transcriptional terminator structure from forming.

Some RNA binding proteins bind to both mRNA and another molecule in the cell, a small noncoding RNA or protein. In the case of ncRNAs bound by RNA binding proteins, the protein aids in the complementary binding of the ncRNA to the mRNA RBS region.

Key points

- Transcriptional regulation alters the levels of mRNA in the cell for the genes encoded on the DNA.
- The regulation of expression of the co-transcribed *lac* operon genes *lacZYA* is influenced by the presence of glucose and lactose, mediated by the regulatory proteins LacI and CAP–cAMP.
- Transcription of the *trp* operon, encoding genes for the synthesis of tryptophan, is regulated by the presence of tryptophan within the cell, which alters the binding properties of the repressor protein and influences the formation of mRNA secondary structures.
- Regulatory proteins can orchestrate synchronized gene-expression changes across the genome through binding to DNA sites associated with distant genes.
- Two-component regulatory systems sense stimuli through the sensor kinase, which in response autophosphorylates and then passes the phosphoryl group to the response regulator, which regulates gene expression.
- Changes in the DNA sequence associated with the promoter region can influence the transcription initiated from the promoter.
- Alternative sigma factors can coordinate the expression of genes across the genome.
- DNA-binding proteins within the nucleoid can have a direct impact on gene expression due to the availability of the DNA for the initiation and elongation of transcription.
- Bacteria can sense the size of the population through quorum sensing and orchestrate the expression of population-dependent characteristics such as biofilm formation.

Key terms

Define the following terms introduced in this chapter. Check your answers using the definitions in the Glossary. These terms are also available as Flashcards online.

β-Galactosidase	Chromatin	Phosphorelay
β-Galactoside permease	Constitutive expression	Proteolysis
Anti-sigma factor	Corepressor	Quorum sensing
Anti-terminator	Cyclic AMP	Regulated expression
Attenuation	Effector molecule	Repressor protein
Attenuator	Exopolysaccharide	Response regulator
Autophosphorylate	Methylation	Sensor kinase
Biofilm	*N*-acyl homoserine lactones	Transcriptional regulation
Catabolite activator protein	Operator	Tryptophan
Catabolite repression	Operon	Two-component regulatory system

Questions and discussion topics

Self-study questions

Answer each question using 50–100 words or a table or labeled diagram. Advice on where to find answers to these questions is available online.

1. What cellular process must occur for a gene to be translated into a protein and how might the bacterial cell alter the levels of production of the protein encoded by a gene?
2. List three situations, not using examples from this chapter, where a bacterial cell may regulate its gene expression to ensure the correct or most efficient proteins are being produced.
3. Draw the genes involved in the *lac* operon locus, including the gene upstream of the *lac* operon itself. Label each of the genes.
4. Add to the diagram from question 3 labels for each of the features involved in the regulation of expression of the *lac* operon.
5. Explain what happens to the transcription of the *lac* operon in *E. coli* when glucose levels are low and lactose is present. Use the diagram you have drawn for reference.
6. Compare your answer to question 5 with what happens at the *lac* operon when glucose levels are high, first when there is no lactose and then when there is lactose present.
7. Draw the different ways in which the sequence 5′ of the *trp* operon can form hairpins when it is transcribed into mRNA. Label the features of this region.
8. Describe what happens in the regulation of the *trp* operon when levels of tryptophan are low. How does this influence the progression of both the ribosome and RNA polymerase and therefore the transcription of *trpEDCBA*?
9. Contrast your answer to question 8 with what would happen when levels of tryptophan are high. How does the progression of the ribosome and RNA polymerase differ and how therefore does transcription of the *trp* operon differ?
10. What are the key components of a two-component regulatory system and what are the steps that occur in such a system when a signal is detected?
11. How does a phosphorelay compare to a two-component regulatory system? Include features that are similar as well as those that are different.
12. How can sigma factors, a subunit of RNA polymerase, be involved in the control of gene expression?

Discussion topics

These topics are presented for discussion in study groups, as part of class discussions, or on your own. These questions go beyond what is directly covered in this part of the book. Use the research literature and other reading to explore these topics in more depth. Tips to help prepare for topic discussions are available online.

1. The *lac* operon and the *trp* operon are often taught as key examples of gene regulation in bacterial systems; however, there are many other regulatory systems in a variety of other bacterial species. Investigate one other bacterial operon and the way(s) in which it is regulated. Explore whether there are repressor proteins, activators proteins, protein binding sites, effector molecules, ncRNAs, riboswitches, and so on.
2. Two-component regulatory systems are often involved in important cellular processes related to host cell interaction or pathogenesis. Investigate a bacterial species of interest and identify a two-component regulatory system or phosphorelay that has been the topic of research. Discuss how the bacteria uses this system to sense its environment and orchestrate the expression of genes needed for the situation in which it finds itself.
3. Discuss the ways in which phase variation is different from gene-expression regulation, although both result in differential expression of genes through varying the transcription of DNA to mRNA.

Online quiz questions

To further self-assess your understanding of the chapter material, please visit the following link, where you can participate in a range of interactive quiz questions:

www.routledge.com/cw/snyder

Further reading

The classic example of transcriptional regulation: the *lac* operon

Jacob F, Monod J. Genetic regulatory mechanisms in the synthesis of proteins. *J Mol Biol.* 1961; *3*(*3*): 318–356.

Ralston A. Operons and prokaryotic gene regulation. *Nat Educ.* 2008; *1*(*1*): 216.

The actions of the corepressor tryptophan on the *trp* operon

Evguenieva-Hackenberg E. Riboregulation in bacteria: From general principles to novel mechanisms of the trp attenuator and its sRNA and peptide products. *Wiley Interdiscip Rev RNA.* 2022; *13*(*3*): e1696.

Yanofsky C. The different roles of tryptophan transfer RNA in regulating trp operon expression in *E. coli* versus *B. subtilis. Trends Genet.* 2004; *20*(*8*): 367–374.

Two-component regulators sense change and alter transcription

Tiwari S, Jamal SB, Hassan SS, Carvalho PVSD, Almeida S, Barh D, Ghosh P, Silva A, Castro TLP, Azevedo V. Two-component signal transduction systems of pathogenic bacteria as targets for antimicrobial therapy: an overview. *Front Microbiol.* 2017; *8*: 1878.

Programmed changes to DNA can alter transcription locally

Zelewska MA, Pulijala M, Spencer-Smith R, Mahmood HA, Norman B, Churchward CP, Calder A, Snyder LA. Phase variable DNA repeats in Neisseria gonorrhoeae influence transcription, translation, and protein sequence variation. *Microb Genom.* 2016; *2*(*8*): e000078.

Global regulation can be influenced by the binding of chromatin proteins to the DNA

Dame RT. The role of nucleoid-associated proteins in the organization and compaction of bacterial chromatin. *Mol Microbiol.* 2005; *56*(*4*): 858–870.

Wade JT, Grainger DC. Waking the neighbours: disruption of H-NS repression by overlapping transcription. *Mol Microbiol.* 2018; *108*(*3*): 221–225.

Protein thermosensors regulate expression of proteins via transcriptional regulation

Richards J, Belasco JG. Riboswitch control of bacterial RNA stability. *Mol Microbiol.* 2021; *116*(*2*): 361–365.

Tucker BJ, Breaker RR. Riboswitches as versatile gene control elements. *Curr Opin Struct Biol.* 2005; *15*(*3*): 342–348.

Chapter

6

Transcriptomes

The term transcriptome refers to all of the RNA transcripts that are present within the cell. In some instances, transcriptome may be used to refer to just the mRNAs within the cell, representing the genes that are being transcribed and thus likely to be expressed as proteins. Transcriptome can also include other RNAs within the bacterial cell, such as the ncRNAs that influence the expression of genes. Also transcribed in the cell, and therefore also encompassed in the term transcriptome, are the rRNAs and the tRNAs; however, the transcription of these tends to be more constitutive, while the transcription of mRNAs and ncRNAs can vary depending on the conditions impacting the cell. Regardless of the breadth of RNA types included, it is the variations in the amounts of the different distinct transcripts within the cell that is the focus of transcriptomics.

Only a portion of the genes in the genome are expressed at any one time. As discussed in Chapter 5, the regulation of genes means that the generation of some of the transcripts within the bacterial cell is controlled. As a consequence, the RNAs that constitute the transcriptome are not constant. The levels of rRNA and tRNA present within the bacterial cell are relatively constant; if they do change, these changes are not specific to a particular gene product. Changes in rRNA and tRNA may indicate an increase or decrease in general protein production, rather than being specific for any particular products. The mRNAs and ncRNAs are linked to specific gene products; therefore, increases or decreases in particular mRNA or ncRNA transcripts over time are expected within the transcriptome. Studying the transcriptome can provide insight into how bacterial cells use their genes and regulate the expression of their genes in changing environments.

The transcriptome changes over time

The transcriptome changes over time and in response to conditions impacting the bacterial cell. Changes to the transcriptome can occur as regulated genes are switched on and off, being up and down regulated in their expression. As seen in Chapter 5, the presence of lactose in the absence of glucose will cause the transcription of the *lac* operon. However, when glucose is present, the *lac* operon is not transcribed. Therefore, the presence in the transcriptome of the *lac* operon mRNA is not constant and is influenced by the presence of lactose and glucose within the bacterial cell. As the bacteria transition from stimulus to stimulus, they adapt to these changes and make best use of their resources by differentially expressing sets of genes (**Figure 6.1**). The transcriptome therefore changes over time as a result of the stimuli to which the bacterial cell is exposed and the corresponding activation and repression of transcription.

Changes to the transcriptome over time can also occur due to mutation, which changes the genomic template for transcription. Environmental pressures may select for mutations that alter the expression of transcripts encoding genes that are beneficial to bacterial growth in such an environment. These sorts of adaptations can occur as a result of the presence of antibiotics or during the process of host adaptation or adaptation to growth in a laboratory. A mutation that causes the loss of a regulatory protein or regulatory ncRNA can affect the presence of transcripts within the transcriptome that are created as a result of these regulators

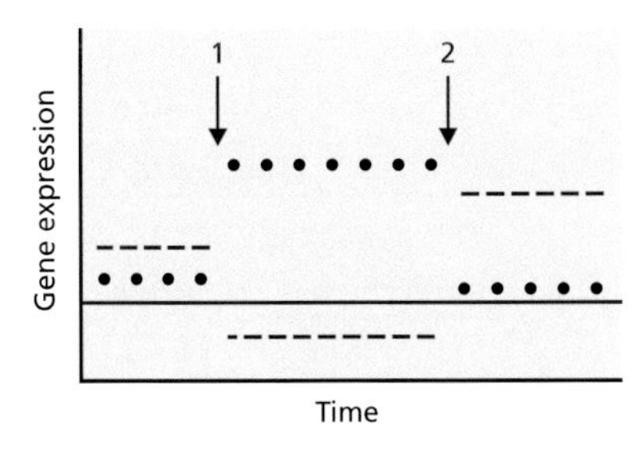

Figure 6.1 The transcriptome changes over time. As bacterial cells encounter various stimuli, the sets of genes that are expressed changes. In response to a stimulus (arrow 1), the expression of some genes will be activated (dotted line) and others repressed (dashed line). When the cell encounters a different stimulus (arrow 2), a different set of genes is turned on (dashed line) and others turned off (dotted line). Thus, the RNAs that make up the transcriptome change over time. Some RNAs that make up the transcriptome remain relatively unchanged, including the mRNAs for housekeeping genes, rRNAs, and tRNAs (solid line).

DOI: 10.1201/9781003380436-8

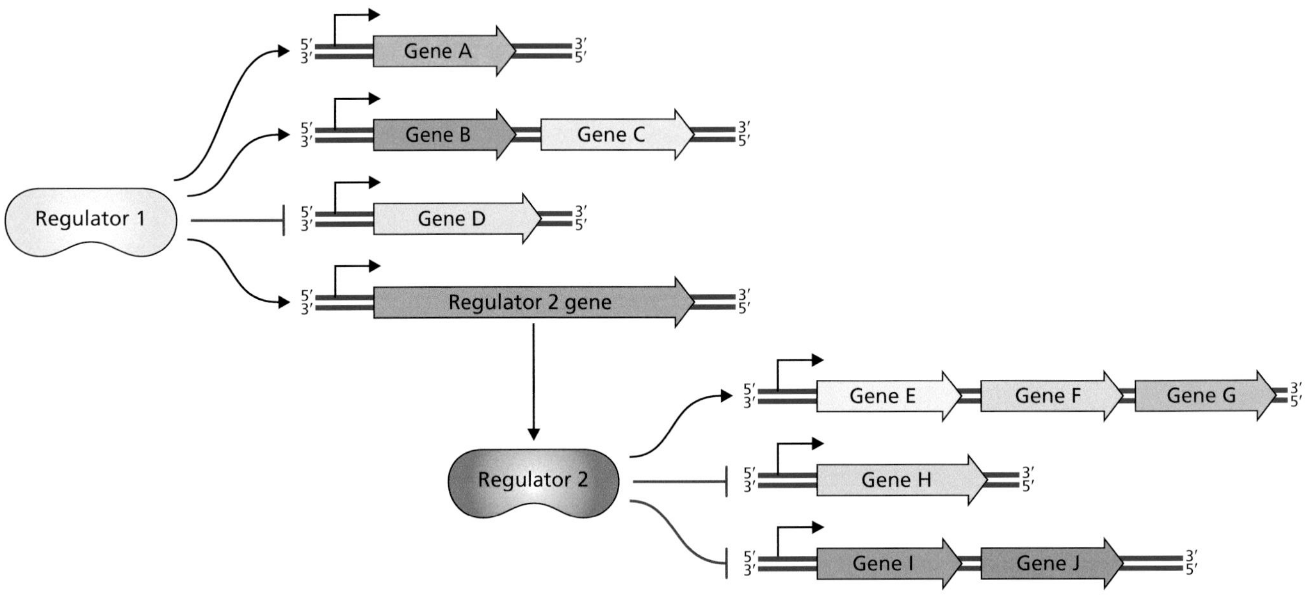

Figure 6.2 Regulatory networks alter the transcription of several genes. When a regulatory protein or regulatory ncRNA (regulator 1) is involved in the regulation of several genes, the activation (arrows) and repression (red lines) of the genes within this network are coordinated together. Regulators have the capacity to regulate other regulators (regulator 2), thus generating more complex regulatory networks.

(**Figure 6.2**). Changes to a regulator may influence the expression of just one gene or operon, or it may impact several genes if it is part of a **regulatory network**. Genomic changes such as sequence inversions, duplications, or deletions can alter the regulation and transcription of features associated with the affected region, changing the transcriptome. Mutations in the genome at the DNA level therefore have a knock-on effect to the levels of the various RNAs within the bacterial cell transcriptome.

Changes in the transcriptome can arise either through transcriptional control or through genetic mutation, with differing consequences for each. While changes over time due to changes in conditions encountered by the bacterial cell are transient, changes to the genome sequence that impact the transcriptome can be permanent. It is therefore important to put transcriptomic discoveries into the context of the genome. While the presence or absence of the *lac* operon mRNA in the transcriptome may be due to the presence of lactose and glucose, it may also be due to a mutation resulting in the loss of LacI or a change in the sequence of the operon promoter that changes the strength of that promoter for recruiting RNA polymerase. When assessing the transcriptome, it is important to consider the genome as well, to determine if changes are due to regulation or mutation.

The transcriptome changes due to changing conditions

The growth rate of bacterial cells is influenced by the synthesis of the ribosomes that are required for translation, enabling the cell to produce the proteins needed for growth. Ribosomes are made up of both rRNAs and also ribosomal proteins. The expression of ribosomal proteins requires the transcription of their genes into mRNA by RNA polymerase and subsequent translation by existing ribosomes. Transcription of the rRNAs is also required to form ribosomes (**Figure 6.3**). When the bacterial cell experiences stress, there is a decrease in the growth rate. The regulation of gene expression coordinates the growth rate of the bacterial population through orchestration of the levels of ribosomes, of the production of new cellular components needed for cell division, and of the cellular processes such as genome replication.

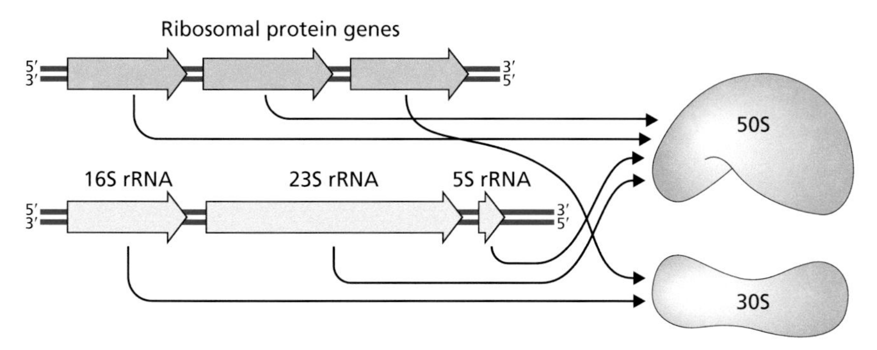

Figure 6.3 The formation of ribosomes from proteins and RNA. Ribosomes are required for translation of mRNAs into proteins. New ribosomes are synthesized within the bacterial cell through the combination of ribosomal proteins and ribosomal RNA (rRNA). The genes encoding ribosomal proteins are transcribed by RNA polymerase into mRNA and translated by ribosomes into ribosomal proteins. There are approximately 32 ribosomal proteins, depending on the species, in the 50S subunit, encoded by genes in various locations in the chromosome. There are approximately 22 ribosomal proteins in the 30S subunit of the ribosome. The genomic loci encoding the rRNAs are also transcribed by RNA polymerase. There is often more than one rRNA locus in a bacterial genome, each with 16S rRNA, 23S rRNA, and 5S rRNA loci. The rRNA and protein products form together into ribosomes.

Responses to a variety of environmental cues alter gene expression within the bacterial cell and these changes in the transcriptome dictate the rate at which bacteria can viably divide and the population expand. It was recognized by Torbjörn Caspersson, Jean Brachet, and Jack Schultz in 1939 that cells engaged in making an abundance of proteins, such as rapidly growing bacteria, were rich in RNA. By implication, this finding suggested that RNA is required for cells to make proteins, which was later shown to be due to the process of transcription. This early observation about the nature of RNA and differences in the levels of RNA within bacterial cells gave insight into how the transcriptome can change due to changes in cellular processes, just as studying transcriptomes can today.

In a growing population of bacterial cells, a variety of stresses may be encountered and must be overcome to ensure continued survival of the population. If a needed sugar or amino acid is scarce, the bacterial cell must compensate for this shortage of nutrients in order to survive. Each stress triggers the regulation of an associated cascade of gene expression, altering the overall transcriptome in response to the stress. For this reason, transcriptome studies can be complex. Slight changes in the environment can cause changes in the transcriptome; therefore, careful control of experimental conditions and replication of experiments are key to producing informative results.

Expression of genes outside the *lac* operon in response to glucose and lactose

The presence of glucose and lactose within the cell dictates the expression of the genes in the *lac* operon, as discussed in Chapter 5. At a transcriptomic level, genes outside the *lac* operon also vary in their expression level in response to the presence of these sugars. CAP–cAMP positively regulates the *lac* operon in the absence of glucose, but the *lac* operon is not the only feature regulated by CAP. The gene for CAP is not part of the *lac* operon; the gene is found elsewhere on the chromosome. In conjunction with cAMP, CAP activates several genes within the *Escherichia coli* genome, including those for sugar metabolism such as the *lac* operon, but also other genes that are expressed in response to low glucose (**Figure 6.4**). CAP–cAMP is also able to repress the expression of genes; therefore, it is also referred to as **CRP**, **cAMP receptor protein**.

Evolutionary theory suggests that cellular processes will be stopped if they are not needed, in order to conserve and best use cellular energy. One theory

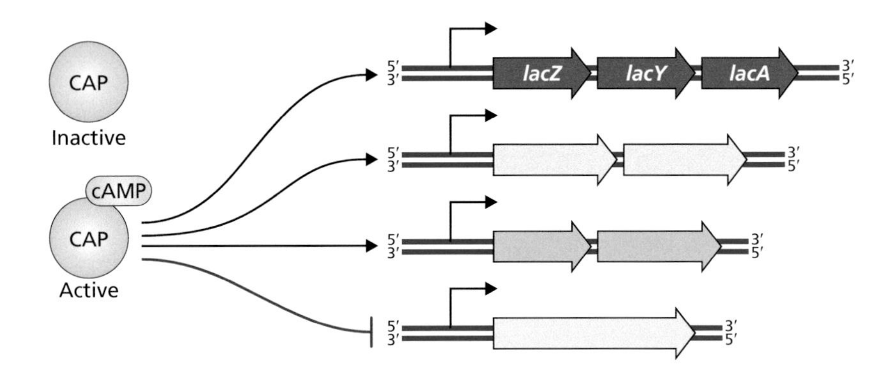

Figure 6.4 CAP–cAMP regulates the *lac* operon and also other genes in the *E. coli* genome. The regulatory protein CAP is inactive and does not regulate gene expression when it is alone. CAP becomes active when it is complexed with the metabolite cAMP, which is present in the bacterial cell when glucose is low. The CAP–cAMP complex activates the genes of the *lac* operon (arrow) and also regulates other genes in the *E. coli* genome (activation, arrow; repression, red line).

regarding the conservation of cellular energy is that the production of unnecessary proteins is a drain on the available pool of amino acids, which could be used for the translation of other proteins that could be beneficial for the cell. However, investigations into the *lac* operon suggest that it is the processes of transcription and particularly translation themselves that are costly. Presumably, this cost is due to the utilization of numerous ribosomes during the process of translation and of the RNA polymerase holoenzyme during transcription. There is therefore a general cost to transcription and translation of any genes, which has led to the evolution of complex regulatory networks. Repression of gene expression for genes that are not needed at all times and in all conditions, such as is observed on the *lac* operon, will conserve needed energy and apply it to cellular processes that are useful under the current growth conditions.

Tryptophan and its impact on the transcriptome

The *trp* operon, as described in Chapter 5, is regulated by a complex mechanism to ensure the proper expression of its genes during conditions in which gene expression is advantageous to the bacterial cell. There is both a regulatory protein, TrpR, and an attenuator system that control whether the *trp* operon is expressed. By its nature, the attenuator system is only able to impact upon the expression of the *trp* operon, whereas the repressor protein has the potential to bind to other promoter regions and regulate the expression of other genes. It has been demonstrated experimentally that there are a few genes elsewhere in the *E. coli* genome that are regulated by the tryptophan repressor protein (**Figure 6.5**). These include *trpR*, encoding the repressor protein itself, which is repressed by TrpR in the presence of tryptophan. This is **feedback regulation**, where the regulatory protein regulates the expression of its own gene. Also

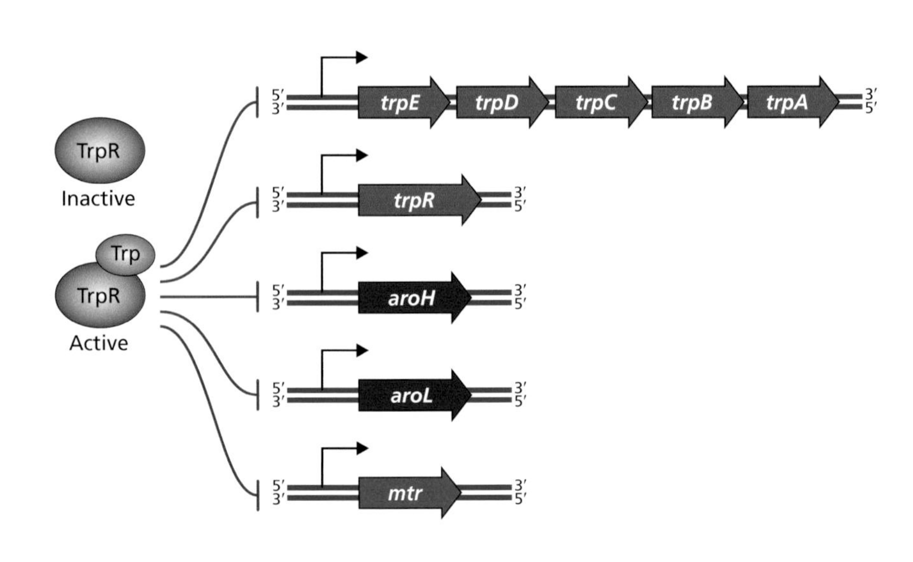

Figure 6.5 The TrpR repressor protein that regulates the *trp* operon also regulates other genes in the *E. coli* genome. When the repressor protein (TrpR) is bound by the amino acid tryptophan (Trp), it is able to bind to the operator on the DNA 5′ of the *trp* operon. It is also able to bind to a few other promoters in the *E. coli* genome, upstream of *trpR*, *aroH*, *aroL*, and *mtr*. This forms a small regulatory network and means that TrpR regulates its own expression through repression of *trpR* in the presence of tryptophan.

repressed are the genes *aroH*, *aroL*, and *mtr*, which are also involved in the uptake and biosynthesis of tryptophan.

While the *trp* operon attenuator system can only alter the expression of the mRNA in which it is contained, it is possible that other genes may be regulated by similar mechanisms, reliant on amino acid levels within the cell. The abundance of the amino acid tryptophan influences the transcriptome outside the *trp* operon. At a basic level, any gene that encodes a tryptophan codon can be attenuated in translation with the ribosomes pausing to await rare tRNAs charged with tryptophan. The various tRNAs are present within the cell at different relative amounts with some being quite common and others being rare. It may be that pauses to await the arrival of a rare tRNA influence the rate of translation and it may be that secondary structures are formed in the mRNA while the ribosome is paused, in a manner similar to the programmed response in the *trp* operon leader sequence. Transcriptomic investigations of gene expression suggest that there are several operons in the *E. coli* genome that are regulated in response to tryptophan. Comparisons between these tryptophan response results with those in *E. coli* that have lost the *trp* operon repressor protein, TrpR, show that changes in expression are not due to just the *trp* regulatory protein. When assessing transcriptomic data, it is important to consider mechanisms of gene regulation, such as attenuation, that do not rely on regulatory proteins.

Transcriptomic changes occur when bacterial cells contact host cells

For bacterial species that live in association with eukaryotic host cells, the contact with those cells can be an important signal to the bacterial cell to alter the expression of its genes. Protein expression that contributes to the attachment of the bacterial cell to the host cell surface, to internalization of the bacterial cell within the host cell for intracellular bacteria, and to the formation of **microcolonies** and biofilms on the host cell surface are important for the survival of the bacteria within the host.

For *Helicobacter pylori*, the bacterial cells express genes in response to association with the gastric epithelial cells that line the inside of the stomach. To make the most of living in this niche and allow the bacteria to establish what can be very long-term, often life-long, colonization of the host, *H. pylori* expresses a specific set of genes. Some of these are within a **pathogenicity island**, a region of the genome containing virulence or pathogenicity genes (**Figure 6.6**). The characteristics of some pathogenicity islands suggest that they originated from a different bacterial species and were acquired by horizontal gene transfer. Genes within this pathogenicity island and elsewhere in the *H. pylori* genome enable the bacteria to acquire needed nutrients and establish microcolonies on the host cell surface.

Biofilm formation can be induced upon contact with a number of surfaces, including host cells, but also environmental surfaces and the surfaces of medical devices. Regardless of the bacteria or the surface, the first stage in iofilm formation

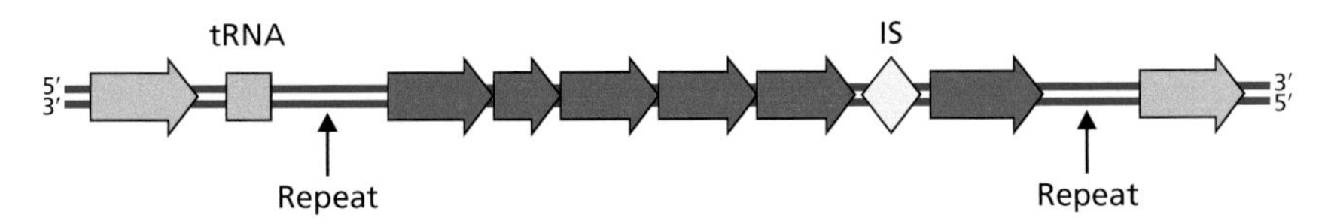

Figure 6.6 The structure and features of a pathogenicity island. Pathogenicity islands are regions within a bacterial genome that contain genes that are involved in the ability of the bacteria to cause disease. However, the characteristics of the DNA sequence within this region often suggest that the region has been acquired from another bacterial species, characterized by having a different %G+C (purple) than the core genome (orange). These genes may include virulence genes, toxin genes, genes for DNA integration and mobility, and IS elements (yellow diamond). Pathogenicity islands are often adjacent to a tRNA sequence in the genome and flanked by direct repeats.

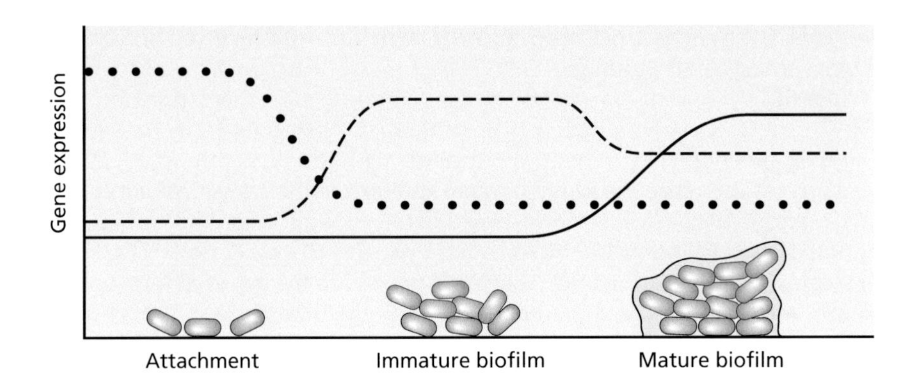

Figure 6.7 Some bacterial species form biofilms on the surfaces of host cells. The first stage of biofilm formation is the attachment of the bacteria to the surface of the host cells, requiring expression of surface adhesion proteins (dotted line). Later, other genes are expressed in response to the stimulus received when attachment is achieved, enabling the biofilm structure to begin to form (dashed line). Changes in the transcriptome lead to changes in the growing population of bacteria on the host cell surface and the formation of a mature biofilm, including the extracellular matrix (solid line).

is the attachment of the bacteria to this surface (**Figure 6.7**). Expression of surface proteins and surface structures allows the bacteria to attach to host cells and then make close contact with them. In response to contact with the host cell, gene expression changes, so that the bacteria can make best use of the available nutrients in the localized niche. Being able to take advantage of the resources available in the niche is important before biofilm formation. As the surface-attached bacteria grow and divide in this niche, increasing in numbers, the expression of genes involved in biofilm formation can be activated.

Changes in temperature can trigger changes in the transcriptome

Regulation of gene expression in response to temperature can be vital to the survival and transmission of bacteria. Studying the transcriptome can reveal how temperature is interpreted by the bacterial cell and can dictate the expression of genes. In particular, bacteria that are able to colonize different hosts that have different body temperatures may require switches in gene expression in response to the various temperatures to ensure optimal growth and gene expression in each of the hosts.

In *Yersinia pestis*, for example, the bacteria are able to grow in fleas, with a body temperature of 26 °C, and in humans, with a body temperature of 37 °C. *Y. pestis* causes plague in humans, yet does not kill the fleas it inhabits. By colonizing fleas, the bacteria are able to grow and expand their population. When the flea bites a human, some of the bacteria are transferred into the new host. For optimal growth in each host, different genes are required (**Figure 6.8**). The temperature of the host can be used by the bacteria as an indicator of the host environment in which it finds itself. There are over 400 genes that are differentially regulated at these two temperatures in ***Y. pestis***, which can be detected through analysis of the transcriptomes from each of the two conditions. The regulated genes include virulence factors and expression regulators that contribute to growth and survival of the bacterial cells within the different hosts.

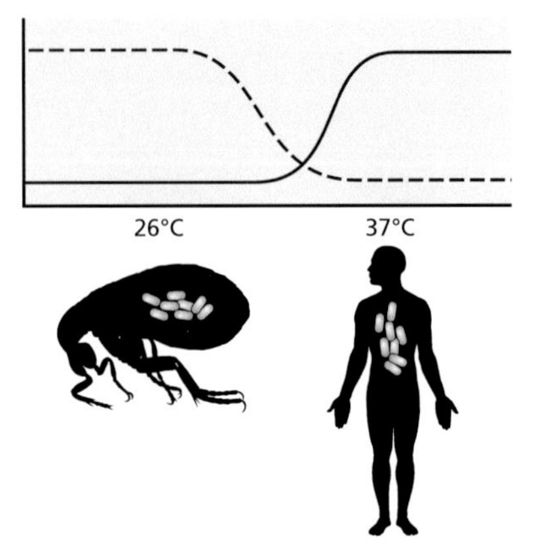

Figure 6.8 The bacterial pathogen *Y. pestis* regulates gene expression based on whether it is within a flea or a human. The regulation of genes differentially expressed in *Y. pestis* between fleas and humans is coordinated in response to temperature. The body temperature of a flea is 26 °C; when the bacteria are at this temperature, they express a set of genes optimal for growth in the flea (dashed line). Upon the shift to a human host, the temperature shifts to 37 °C and gene expression shifts accordingly, switching on the genes that cause the symptoms of plague in humans (solid line).

Host-associated bacteria are not the only ones that change their transcriptome in response to temperature. Free-living bacterial species can also regulate gene expression due to temperature differences. *Pseudomonas extremaustralis* is a bacterial species from the Antarctic and is therefore able to grow at low temperatures. At low temperatures, genes for metabolism of sugars and amino acids are repressed. At the same time, genes for ethanol oxidation are up regulated at low temperature, suggesting that this pathway is important for *P. extremaustralis* survival at temperatures such as 8 °C. When temperatures increase, the bacteria up regulate their metabolic systems to take advantage of the conditions.

Expression of key proteins can indicate a response to temperature change

The response of bacterial cells to temperature changes can include the expression of heat-shock proteins and **cold-shock proteins**, which are involved

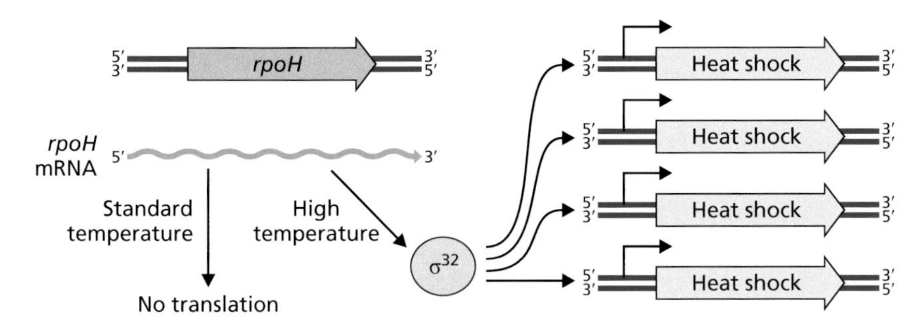

Figure 6.9 Transcription of heat-shock protein genes by σ^{32}, which is translationally regulated by temperature. The σ^{32} protein is encoded by *rpoH*. When this gene is transcribed, the resulting mRNA includes a temperature-sensitive region. At the standard temperature for the bacteria, the temperature-sensitive region prevents translation of the *rpoH* mRNA. At increased temperature, the *rpoH* mRNA is translated, generating σ^{32} proteins, which bind to the promoter regions of heat-shock protein genes, up regulating their expression in response to temperature.

in the adaptation of the cell to transient temperature changes. Expression of these proteins corresponds to changes in their transcription, reflected in the transcriptome. The genes for heat-shock proteins are turned on in response to an increase in temperature, rapidly transcribing the genes and translating them into proteins that enable the bacterial cell to withstand the temperature increase. Likewise, when shifted to a cold temperature, the transcription and translation of cold-shock protein genes are activated.

Heat-shock proteins include chaperones that are able to stabilize the structure and conformation of proteins to prevent or correct temperature-induced misfolding of proteins. Some heat-shock proteins act as proteases, cleaving and removing proteins from the cell that are misfolded and could be detrimental to the cell at the increased temperature. In *E. coli* and other bacterial species, the heat-shock response is mediated by σ^{32}, a sigma factor encoded by *rpoH*, which recruits RNA polymerase to the promoter regions of heat-shock protein genes. A temperature-sensitive region in the *rpoH* mRNA inhibits translation until the temperature increases and the secondary structure of the mRNA changes (**Figure 6.9**).

At low temperatures, membrane fluidity decreases, which can impact the function of membrane proteins. Enzymatic activity also decreases with decrease in temperature; enzymes operate optimally at the standard growth temperature for the bacterial species. DNA structures are also stabilized at lower temperatures, with hydrogen bonds being more robust and more liable to form at low temperatures, resulting in the formation of secondary structures. Each of these temperature-sensitive events can contribute to differences in the transcriptome at lower temperature, where genes are switched on or off to compensate for low-temperature changes to the cell. Cold-shock proteins are able to overcome the effects of the cold upon these processes. For example, the Csp family proteins interact with DNA and RNA secondary structures, enabling more efficient replication and transcription than would otherwise be possible at lower temperatures.

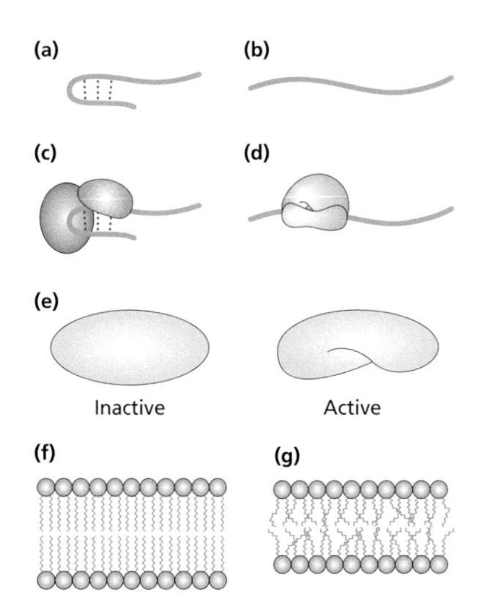

Figure 6.10 Thermosensors can occur in nucleic acids, proteins, and lipids. Different molecules within the bacterial cell are capable of sensing temperature changes. Hydrogen bonding between bases in DNA and RNA can form (a) or relax secondary structures (b), making regions available for binding by secondary structure recognizing proteins (c) or available for binding by RNA polymerase for transcription or ribosomes for translation (d). Proteins can undergo conformational changes due to temperature, altering their activity in response to temperature (e). Temperature changes can impact the bacterial membrane through influencing the lipids, being more compact and thicker at lower temperatures (f) and more dispersed and thinner at higher temperatures (g).

Different types of thermosensors can alter gene expression due to temperature

Regulation of transcription due to temperature can involve thermosensors (see Chapter 4), present as RNAs, proteins, and lipids that change secondary structures at different temperatures (**Figure 6.10**). These molecules are temperature sensitive and by this nature confer to the cell the signal that temperature has changed and are thus thermosensors. Hydrogen bonds are prone to breakage and formation as a result of temperature and form the basis of nucleic acid-based thermosensors. In DNA, the structures formed through hydrogen bonding can change in response to temperature and alter availability of promoters for transcription. In this way,

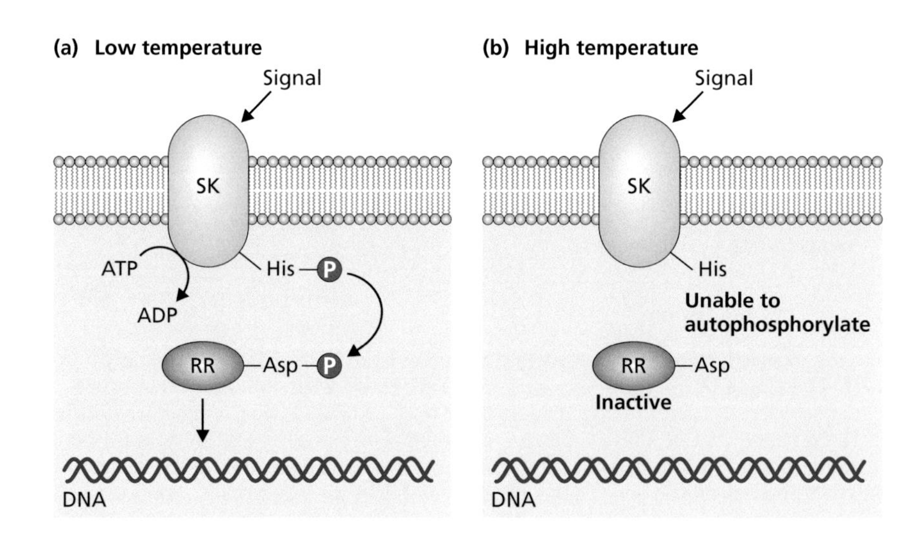

Figure 6.11 Temperature-sensitive autophosphorylation and sensor kinase activation. Some sensor kinases (SK) in two-component regulatory systems are only able to autophosphorylate in a temperature-sensitive manner. At low temperature (a), these sensor kinases autophosphorylate and are thus able to phosphorylate the response regulator (RR) of the two-component system. Phosphorylated response regulators are able to bind to DNA and regulate gene expression. At higher temperatures (b), autophosphorylation of the sensor kinase does not occur and the response regulator remains inactive.

the temperature sensitivity of DNA can impact upon the transcriptome. These temperature-associated structural changes in DNA may occur in concert with H-NS and the associated bending of DNA. At the RNA level, thermosensors may form hairpins and RNA structures that relax at increased temperatures allowing access to the ribosome binding site and initiation of transcription.

Thermosensors need not be nucleic acid-based. Protein dimer formation can be temperature sensitive, with some repressor proteins only being functional as dimers. One temperature permits dimer formation, and the regulatory protein is able to repress gene expression, while another temperature inhibits dimerization. Temperature can also change the ability of enzymes to function, due to conformational changes. It is these sorts of changes that heat-shock and cold-shock proteins can mediate against. However, for some enzymes these temperature-sensitive conformational changes can be thermosensors themselves, particularly when the enzyme involved is a sensor kinase of a two-component regulatory system or phosphorelay. Some sensor kinases will only autophosphorylate, and therefore activate, at lower temperature, providing a temperature sensor regulatory system where the response regulator is only active at lower temperature (**Figure 6.11**). Even if the sensor kinase is not itself temperature sensitive, the bacterial cell membrane is, and changes to the membrane due to temperature (see Figure 6.10) can alter the functions of embedded proteins, like sensor kinases.

At lower temperatures, membranes become thicker and are thinner at higher temperatures. These changes can cause conformational changes in the membrane proteins. In addition, the fluidity of the bacterial cell membrane is influenced by temperature. Membrane fluidity can alter the interactions that occur between membrane proteins and lipids. Regions of membrane proteins that influence a sensor kinase's autophosphorylation function may become exposed due to changes in membrane fluidity.

Global gene regulation can occur in response to iron

Coordination of gene expression to make best use of the nutrients available in the local niche is essential for survival of bacterial cells. Some environments can be nutrient restricted; therefore, genes involved in the acquisition of essential nutrients are up regulated. This enables the bacterial cell to scavenge rare nutrients or to make use of alternative forms of the nutrient that may need processing before the cell can make use of it. This is the case for iron-acquisition systems in bacterial cells, where iron may be limited or come in different forms. In Gram-negative bacteria, there is a conserved regulatory protein, the ferric uptake regulator Fur, which is involved in iron-acquisition gene regulation. The Fur protein also influences metabolism, growth of the cell, response to stress, and the pathogenicity of the bacteria.

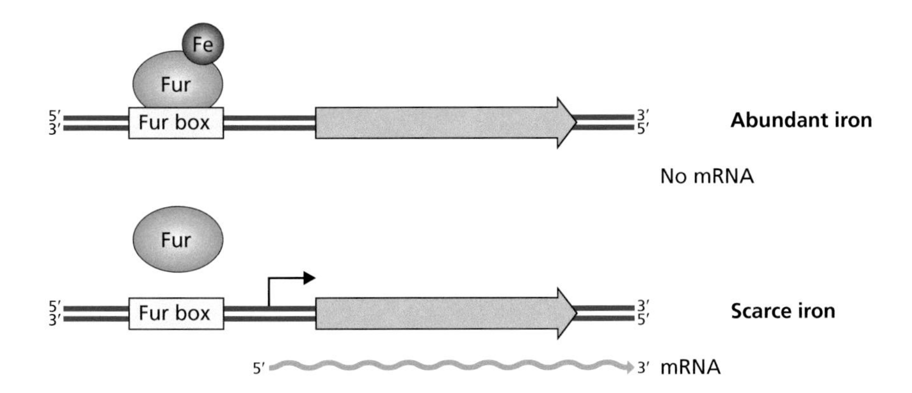

Figure 6.12 The Fur protein binds to iron and regulates iron-acquisition genes. When iron is abundant, it binds to the Fur protein within the bacterial cell. This Fe–Fur complex is able to bind to DNA, recognizing a sequence in the DNA called the Fur box. The binding of Fe–Fur to the Fur box represses the transcription of associated genes. When iron is scarce, Fur is not able to bind to DNA on its own. The genes associated with Fur boxes are then transcribed in response to there being low iron.

Within the host there is very little free iron. Iron in the host is bound by proteins or complexed in compounds, most notably heme. Almost 70% of the iron in the human body is within heme. In different niches within the host, different types of iron-containing compounds are present; therefore, different mechanisms for iron acquisition are required. Excess iron is also toxic to the bacterial cell. Regulation of iron-acquisition genes must be tightly regulated, so that they are expressed when appropriate, yet also repressed as needed.

When iron is bound to the Fur regulator, as Fe–Fur, the complex is able to bind to Fur-box recognition sequences within certain promoter regions in the bacterial genome. The binding of Fe–Fur to the promoter represses transcription by preventing RNA polymerase from having access to the promoter region. When iron is low, the Fur protein is not bound by iron and Fur on its own cannot repress transcription. This enables the transcription and then translation of genes that will orchestrate the acquisition of iron (**Figure 6.12**).

Bacterial cells require nutrients and regulate gene expression to get what they need

As previously discussed, bacterial cells require sugars, such as glucose or lactose, to grow. Genes such as those in the *lac* operon are regulated to achieve acquisition and metabolism of sugars for the cell. Bacterial cells also require amino acids, either from their environment or synthesized within the cell itself from other components available to it. Thus, expression of genes involved in amino acid acquisition or synthesis, like the *trp* operon, can be up regulated when needed. Other nutrients are also required and can be in limited supply at various times in the life of the bacterial cell. To optimize growth in a given condition, the expression of multiple gene systems across the bacterial genome has to be coordinated. Genes that are not required are repressed, so as not to be a drain on cellular resources such as the transcription and translation machinery. Genes that can allow the bacterial cell to take advantage of the available nutrients are activated.

Some nutrients, such as CO_2 and NH_3, are so small that they can readily diffuse across the bacterial cell membrane. When the concentration of such nutrients is higher in the environment than in the cell, passive diffusion is able to ensure that there is enough inside the cell (**Figure 6.13**). Diffusion does not require any cellular energy, which is beneficial for the cell when levels of such nutrients are high in the environment. However, when the nutrient levels are lower outside the bacterial cell than within it, diffusion cannot be relied upon to achieve sufficient intracellular concentrations. Passive diffusion will actually work against the bacterial cell, removing these small molecules from the cell into the environment, where the concentration is lower. Under these conditions, the bacterial cell must actively transport nutrients capable of diffusing through the membrane, in order to achieve higher intercellular concentrations.

To overcome passive diffusion of nutrients out of the cell and to take up nutrients that are too large to be able to diffuse across the membrane,

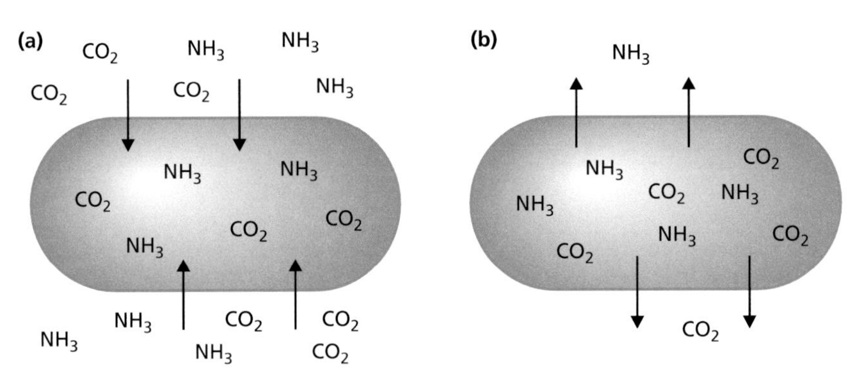

Figure 6.13 Passive diffusion can bring some nutrients through the cell membrane into the bacterial cell. Nutrients like CO_2 and NH_3 are small enough to pass through the bacterial cell membrane. These nutrients move into, and out of, the cell through diffusion. This means that when the extracellular concentration of the nutrient is higher than the concentration within the cell, the nutrient will move into the cell (a). However, when the extracellular concentration is lower than within the cell, the nutrient will leave and is then lost to the bacteria (b).

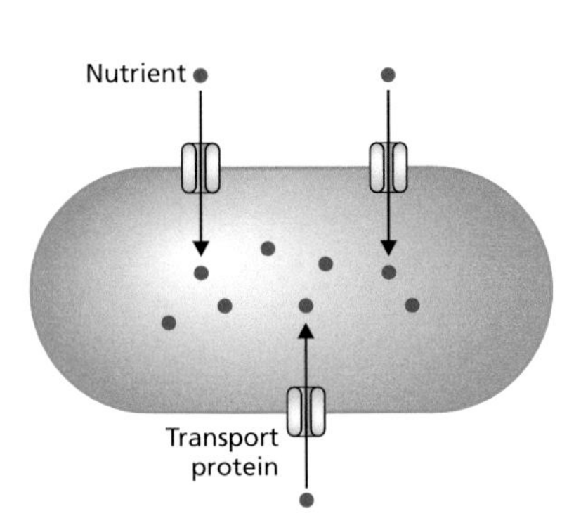

Figure 6.14 Bacterial cell transport proteins bring nutrients into the cell. Some nutrients are too large to diffuse across the bacterial cell membrane, and some are so small that they may diffuse out of the cell despite being needed. Transport proteins are able to move nutrients across the membrane (Gram positive) or membranes (Gram negative).

membrane-associated transport proteins are required. Recall that in Chapter 5 the *lac* operon includes *lacY* encoding β-galactoside permease for the transport of lactose into the bacterial cell. Membrane transport systems vary, depending on the bacterial species and the structure of its membranes. For example, in Gram-negative species, there is both an inner cytoplasmic membrane and an outer surface membrane that is exposed to the environment. Membrane transport systems bringing in nutrients and expelling unwanted material must be able to achieve transport across both membranes in Gram-negative bacteria. In some transport systems, the whole system is made of several proteins that span the two membranes. In other systems, transportation is across one membrane and into the periplasm and another transporter takes it through the second membrane (**Figure 6.14**).

Each bacterial cell in a culture is different

When bacteria are studied, they tend to be in artificial cultures, maintained in a laboratory. Assumptions are made that the entire population of bacteria within the culture are all behaving similarly and are expressing similar sets of genes. However, each cell is in a slightly different microenvironment that is all its own. On solid agar surfaces, the proximity of other bacterial colonies could deplete the nutrients in the agar in a localized region. This can be seen on the plate, where we have larger colonies growing when there are fewer other colonies nearby (**Figure 6.15**). In broth media cultures, bacteria toward the interface between the liquid media and the air are in a different microenvironment from those in other parts of the same liquid culture. Species that require oxygen to grow, **obligate aerobes**, will grow and divide close to the surface of the liquid culture, whereas

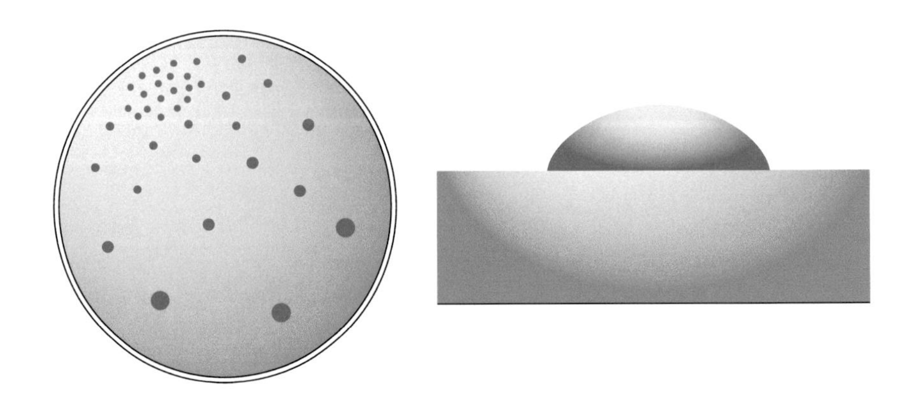

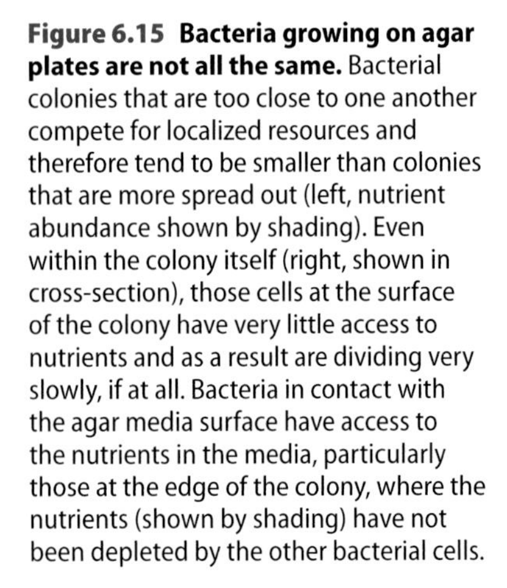

Figure 6.15 Bacteria growing on agar plates are not all the same. Bacterial colonies that are too close to one another compete for localized resources and therefore tend to be smaller than colonies that are more spread out (left, nutrient abundance shown by shading). Even within the colony itself (right, shown in cross-section), those cells at the surface of the colony have very little access to nutrients and as a result are dividing very slowly, if at all. Bacteria in contact with the agar media surface have access to the nutrients in the media, particularly those at the edge of the colony, where the nutrients (shown by shading) have not been depleted by the other bacterial cells.

those that can grow without oxygen, **facultative aerobes**, will be distributed throughout the broth (**Figure 6.16**). In addition to differences in nutrient availability across the whole of a culture, the individual bacterial cells within the culture may take different metabolic and behavioral pathways for growth and survival. Two cells may be achieving the same ends, yet they have got there through the use of different pathways requiring the expression of different sets of genes.

While free-living bacteria can be expressing different sets of genes within the same population in a common culture, the most striking example of differences in cells in a population occurs in biofilms. Biofilms are established and maintained by virtue of the attachments made by the bacterial cells to the surface and also through the attachments formed between bacterial cells. Depending on their place within the biofilm, some bacterial cells may not have direct contact with the surface, only with other bacterial cells. Those cells in contact with the surface may not be exposed to the environment, being in contact only with the surface and other bacterial cells. The regulation of genes needed for biofilm development includes those regulated by quorum sensing mechanisms, which determine that there are enough bacteria locally for a biofilm to be viable. The biofilm is maintained through the specific interaction of each cell with the environment immediately adjacent to it.

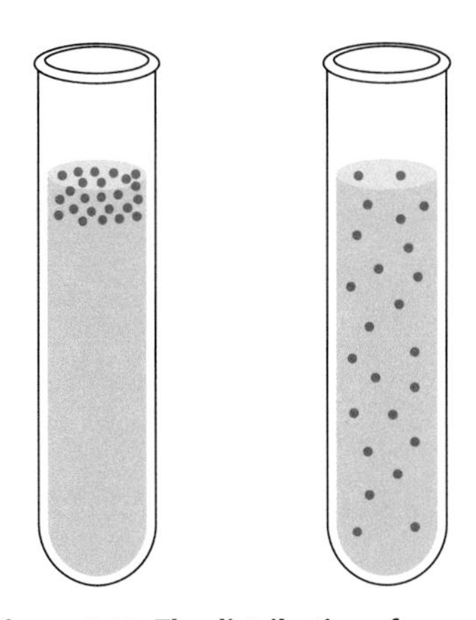

Figure 6.16 The distribution of bacteria within broth cultures can reveal whether they need oxygen or can thrive in its absence. Obligate aerobes must have oxygen to survive; therefore, they grow toward the surface of liquid cultures (left). Facultative aerobes can grow using oxygen or they can use alternative means for growth, so they are distributed throughout liquid culture media (right).

Expression profiling provides a population level understanding of regulation

Expression profiling is an assessment of the genes that are expressed, often determined by investigating the transcriptome. Expression profiling experiments begin with a population of bacterial cells that have been grown in culture or otherwise sampled. The extracted mRNA is therefore derived from a pool of bacterial cells, which may be expressing different sets of genes. It is therefore important when analyzing these sorts of results to remember that the data are an average transcriptome. The expression profile represents the general population and gives a population level understanding of regulation. Individual cells within the population may be expressing genes differently.

Single-cell analysis can get around the averaging of a population by looking specifically at the expression of genes in just one cell. Analysis of the transcriptome from a single cell reveals that the populations that appear to be homogeneous are actually quite diverse. To achieve this experimentally can be quite difficult. Single cells must be isolated and lyzed and the RNA captured for reverse transcription to complementary DNA (cDNA). This cDNA, generated from the RNA sequence by reverse transcriptase, is then amplified to make libraries for sequencing, so that there is sufficient sample for analysis. The amplification process may introduce bias. If too few individual cells are investigated, they may not give an accurate account of what is happening in the population; an outlier may have been captured for single-cell analysis, which would otherwise be part of a larger dataset in a population level transcriptome experiment.

The expression of multiple genes is coordinated together across the chromosome

To coordinate responses and ensure that genes encoding products that work together are expressed together, bacterial cells orchestrate gene regulation on a global scale. For example, strains of *Salmonella enterica* serovar *Enteritidis* can have similar genomes carrying similar genes, but they have very different phenotypes and pathogenicity. Those strains that are of higher pathogenicity have been shown to express higher levels of virulence genes. Those strains of lower pathogenicity have reduced expression of transcriptional regulators and reduced expression of genes regulated by the alternative σ^S. This illustrates that there is a network of genes within these strains that are up- and down regulated together, producing different outcomes from similar genetic material. The difference in expression of similar sets of genes has a direct influence on the pathogenicity of the two different strains.

The coordination of gene expression at several loci in response to change is very different to the control of 3–5 genes present together in an operon, as explored in Chapter 5. Regulatory networks occur when several genes at different locations are coordinated in their expression. Sigma factors can be the common factor involved in coordinated regulation. Genes that have promoters for the same alternative sigma factor will be transcribed by RNA polymerase in a coordinated manner, so that these genes are expressed together in response to events such as stress or temperature change. Promoters can also be acted upon by transcriptional regulators. These can be repressors or activators that control transcription from more than one promoter region. Regulatory RNAs also have a role in global gene expression, influencing the transcription of different genes in different locations across the genome.

The transcriptional network landscape can have topology

When thinking about genetic regulation, it is important to remember that the bacterial cell is a dynamic three-dimensional structure. Different parts of the DNA encoding different proteins are close to different parts of the cell, even if they are contained on the same chromosome. Due to the process of coupled transcription–translation, the product of a gene is made in close proximity to the DNA that encodes it. This is in striking contrast to eukaryotic cells, where mRNA is transported away from the DNA in the nucleus with ribosomes in the cytoplasm for expression.

Studies investigating the localization of global transcriptomes within bacterial cells show that bacterial mRNAs tend to be produced near to where the protein product is needed. For example, in *E. coli*, mRNAs for inner membrane proteins are present at the membrane (**Figure 6.17**). The DNA encoding these proteins is transcribed close to the final location of the protein, so that, as the mRNA is translated, the protein can be inserted into the membrane. As the N-terminus of the protein is produced by translation, the signal peptide present in this portion of the polypeptide is inserted into the membrane, anchoring the associated mRNA to the inner membrane.

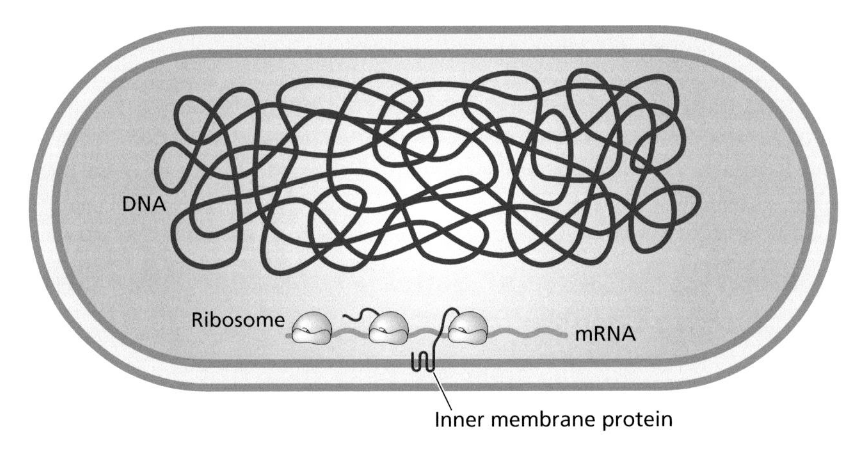

Figure 6.17 The bacterial cell is a three-dimensional structure. For some genes, the DNA is located near to where the protein that it encodes will be needed. This is particularly striking when inner membrane proteins are considered. The DNA containing the genes for inner membrane proteins is located near the inner surface of the membrane. When the gene is transcribed into mRNA, the mRNA is also near the inner surface of the bacterial membrane. Translation begins shortly after transcription, with the ribosome starting to make the inner membrane protein. Once enough of the protein has been synthesized by the ribosome, it is inserted into the membrane while translation is occurring.

For protein products that need to be exported through the inner membrane, such as outer membrane and periplasmic proteins, the mRNA transcripts are found dispersed in the cell. Localization of these proteins can happen after the conclusion of transcription and translation, with the protein then moving to where it is needed.

Key points

- The transcriptome is the RNA within the bacterial cell, just as the genome is the DNA.
- Unlike the genome, which is relatively stable, the transcriptome changes frequently as the bacterial cell switches on and off the expression of different genes.
- Regulatory networks coordinate the expression of sets of genes across the genome.
- Bacterial cells encounter a variety of stresses and alter the expression of their genes to overcome these stresses and enable the bacterial population to survive.
- The availability of nutrients such as glucose, lactose, tryptophan, and other nutrients are stimuli for the global regulation of sets of genes.
- Within the host, the regulation of the virulence genes within the bacterial genome influences the pathogenicity of the organism.
- Biofilm formation requires the orchestration of a range of genes, which must be up regulated and down regulated as the biofilm forms.
- Changes in the transcriptome can arise due to changes in temperature and can be mediated by the state of DNA, RNA, proteins, and the membrane at various temperatures.
- The transcriptome reveals when bacterial cells have up regulated the expression of nutrient transport protein genes.
- The expression profile within a transcriptome often represents the transcriptome from a population of bacterial cells.

Key terms

Define the following terms introduced in this chapter. Check your answers using the definitions in the Glossary. These terms are also available as Flashcards online.

cAMP receptor protein	Facultative aerobes	Obligate aerobes
Cold-shock proteins	Feedback regulation	Pathogenicity island
Expression profiling	Microcolonies	Regulatory network

Questions and discussion topics

Self-study questions

Answer each question using 50–100 words or a table or labeled diagram. Advice on where to find answers to these questions is available online.

1 What is the difference between gene regulation, such as that described in Chapter 5, and regulatory networks?

2 RNA is not all mRNA. What are the names designations of other types of RNA and where are they located in the bacterial cell?

3 How do the *lac* and *trp* regulators influence expression of genes outside the *lac* operon and the *trp* operon?

4 What are the key features of pathogenicity islands that define them and suggest they originated from a different bacterial species?

5 What are the three main stages of the formation of a bacterial biofilm and what are the main events that occur at each stage?

6 Give examples of how changes in temperature can modulate gene expression and how this is relevant to bacterial survival or virulence.

7 Describe at least one way in which nucleic acids, proteins, and lipids can each be involved in temperature sensing in bacteria.

8 Draw a diagram that shows how temperature signals can impact phosphorylation in a two-component regulatory system.

9 How does iron availability modify the level of transcription of iron-regulated genes?

10 Explain passive diffusion. How does a bacterial cell circumvent passive diffusion?

11 How do bacterial cells within a population growing on an agar plate differ from one another? How would this differ for a population growing in broth?

12 How does coupled transcription–translation relate to the localization of the proteins that are being produced?

Discussion topics

These topics are presented for discussion in study groups, as part of class discussions, or on your own. These questions go beyond what is directly covered in this part of the book. Use the research literature and other reading to explore these topics in more depth. Tips to help prepare for topic discussions are available online.

1 Explore a regulatory network of a bacterial species of interest to see how the regulatory proteins interact, interplay, and respond to the various changes experienced by the bacterial cells in laboratory experiments and how these equate to their natural environment.

2 Choose one bacterial species that forms biofilms. Discuss the advantages for the species in terms of survival and nutrient uptake when the cells are in biofilms compared with when they are not.

3 A variety of bacteria alter their transcriptomes in response to temperature changes. Discuss two related species that have been shown to differentially express genes upon change in temperature and compare between the two species the repertoire of genes where expression changes.

Online quiz questions

To further self-assess your understanding of the chapter material, please visit the following link, where you can participate in a range of interactive quiz questions:

www.routledge.com/cw/snyder

Further reading

The transcriptome changes over time

Pilla R, Suchodolski JS. The role of the canine gut microbiome and metabolome in health and gastrointestinal disease. *Front Vet Sci.* 2020; *6*: 498.

Shah DH. RNA sequencing reveals differences between the global transcriptomes of *Salmonella enterica* serovar *Enteritidis* with high and low pathogenicities. *Appl Environ Microbiol.* 2014; *80*(*3*): 896–906.

Expression of genes outside the *lac* operon in response to glucose and lactose

Stoebel DM, Dean AM, Dykhuizen DE. The cost of expression of *Escherichia coli lac* operon proteins is in the process, not in the products. *Genetics.* 2008; *178*(*3*): 1653–1660.

Tryptophan and its impact on the transcriptome

Khodursky AB, Peter BJ, Cozzarelli NR, Botstein D, Brown PO, Yanofsky C. DNA microarray analysis of gene expression in response to physiological and genetic changes that affect tryptophan metabolism in *Escherichia coli. Proc Natl Acad Sci USA.* 2000; *97*(*22*): 12170–12175.

Kumamoto AA, Miller WG, Gunsalus RP. *Escherichia coli* tryptophan repressor binds multiple sites within the *aroH* and *trp* operators. *Genes Dev.* 1987; *1*(*6*): 556–564.

Transcriptomic changes occur when bacterial cells contact host cells

La MV, Raoult D, Renesto P. Regulation of whole bacterial pathogen transcription within infected hosts. *FEMS Microbiol Rev.* 2008; *32*(*3*): 440–460.

Westermann AJ, Vogel J. Cross-species RNA-seq for deciphering host-microbe interactions. *Nat Rev Genet*. 2021; *22*(6): 361–378.

Changes in temperature can trigger changes in the transcriptome

Ben-Ari T, Neerinckx S, Gage KL, Kreppel K, Laudisoit A, Leirs H. Plague and climate: scales matter. *PLoS Pathog*. 2011; *7*(9): e1002160.

Lebreton A, Cossart P. RNA- and protein-mediated control of *Listeria monocytogenes* virulence gene expression. *RNA Biol*. 2017; *14*(5): 460–470.

Global gene regulation can occur in response to iron

Moreau MR, Massari P, Genco CA. The ironclad truth: how *in vivo* transcriptomics and in vitro mechanistic studies shape our understanding of *Neisseria gonorrhoeae* gene regulation during mucosal infection. *Pathog Dis*. 2017; *75*(5): ftx057.

Runyen-Janecky LJ. Role and regulation of heme iron acquisition in Gram-negative pathogens. *Front Cell Infect Microbiol*. 2013; *3*: 55.

Part III

Proteins, Structures, and Proteomes

The following three chapters will delve into the details of how proteins are made within the bacterial cell, the features of the amino acids that make up proteins, the structures of amino acids and proteins and how these influence function, and how protein levels within bacterial cells are regulated and maintained. These chapters will explore individual proteins as well as the whole **proteome**, made up of all the proteins within the cell. These chapters build upon topics that were introduced in earlier chapters when the DNA and RNA that encode and are involved in the creation of proteins were covered, presenting the material on proteins introduced there in more depth and detail.

DOI: 10.1201/9781003380436-9

Chapter

7

Proteins

Proteins are made from amino acids. The sequence of the amino acids within a protein is dictated by the DNA sequence that makes up the gene encoding the protein. This DNA sequence is transcribed into mRNA, which is translated into protein as an amino acid sequence. While genes are important to carry the code of life, it is the proteins that give life to the cells. Proteins form bacterial structures, either directly or through their actions as enzymes. To understand how proteins work in the bacterial cell, first the properties of the various amino acids that make up proteins must be discussed. This chapter will explore how the features of each amino acid contribute to the structure and function of the protein.

Amino acids contain an amine group, a carboxyl group, and a side chain

Just as DNA and RNA are made of individual nucleic acid units, proteins are made of individual amino acids. Whereas nucleic acids are joined together by phosphodiester bonds, the amino acids within proteins are joined to one another by peptide bonds. Proteins are formed during translation, where amino acids are added together, one by one, within the ribosome.

All amino acids have an **amine group** ($-NH_2$) and a **carboxyl group** ($-COOH$) that are characteristic of their makeup (**Figure 7.1**). Each amino acid has a specific side chain, generically referred to as **R**, which is found between the amine group and the carboxyl group (see Figure 7.1).

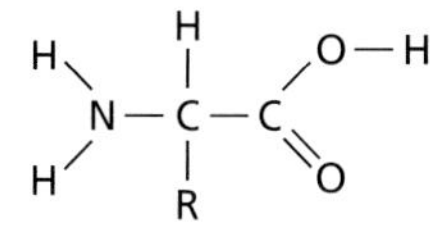

Amino acids are made of carbon, hydrogen, oxygen, and nitrogen, these elements being contained within the amine and carboxyl group. For some amino acids these might be their only elements. For other amino acids, there are other elements found in the side chain including sulfur in methionine and cysteine.

Figure 7.1 The general structure of an amino acid. All amino acids have an amine group on one end, the $-NH_2$, and a carboxyl group on the other end, the $-COOH$. Between these is the side chain, represented by R.

The production of functional proteins from amino acids

Amino acids join into peptides, which become longer through the addition of more amino acids, forming polypeptides, and then fold into structures that are proteins. The chains of amino acids created during translation are linear with the amine end of one amino acid linked by a peptide bond to the carboxyl group of the next and so forth. The peptide bond between amino acids is formed between the $-NH_2$ of one amino acid and the $-COOH$ group of another, releasing H_2O (**Figure 7.2**). Peptide bonds can be broken by hydrolysis, although this is a very slow process within a cell. Within the bacterial cell, proteins are broken down enzymatically by peptidases and proteases.

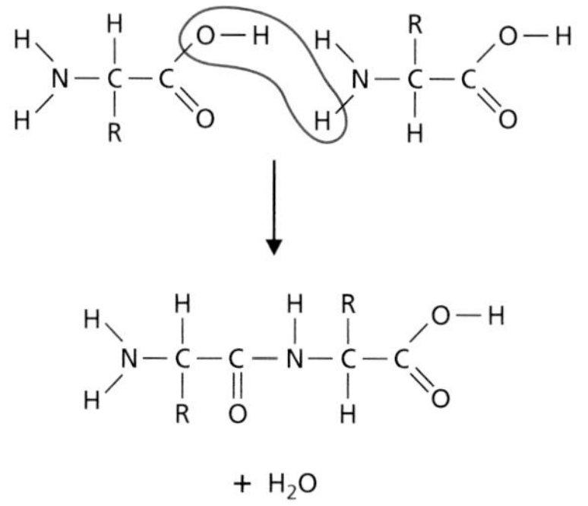

Peptide bonding, and thus peptide chain formation, starts at what will be the N-terminus of the protein. Subsequent amino acids are added onto the carboxyl group end during translation. The peptide bond is formed within the ribosome, as the amino acid attached to its tRNA in the A-site is added to the growing peptide chain that is attached to the tRNA in the P-site. The formation of the peptide bond accompanies the release of the peptide chain from the tRNA in the P-site. This tRNA then moves to the E-site while the tRNA now carrying the peptide chain moves from the A-site to the P-site ready for the next peptide bond formation to add another amino acid.

$+ H_2O$

Figure 7.2 Formation of a peptide bond between two amino acids. The carboxyl group of one amino acid (left) forms a peptide bond (red) with the amine group of another amino acid (right). This process releases a water molecule, formed from the oxygen and hydrogen on the carboxyl group and one of the hydrogens on the amine group (circled in red).

DOI: 10.1201/9781003380436-10

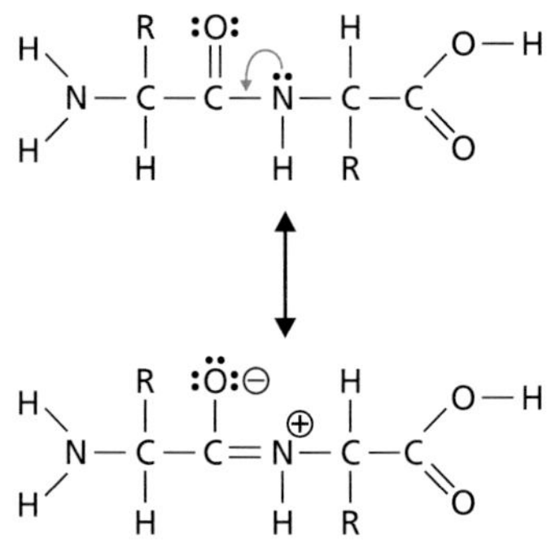

The inflexible nature of the peptide bond imposes limits on the amino acids

Peptide bonds are slightly stronger and slightly shorter than other covalent single bonds and therefore have more of a double bond character. This is due to the presence of a lone pair of electrons on the nitrogen that contribute to the peptide bond (**Figure 7.3**).

Figure 7.3 The partial double bond character of a peptide bond. Electron pairs (dots) from the amine end of one amino acid contribute to the peptide bond between the amine end of one amino acid and carboxyl end of another. This makes the nitrogen positively charged and the oxygen negatively charged, contributing to the strength of the covalent bond between the carbon and nitrogen.

Due to the double bond-like nature of the peptide bond, there is reduced rotation between two joined amino acids than there would be with a single bond. The peptide bond tends to be in a ***trans*** configuration, with side chains of adjacent amino acids on opposite sides from one another (**Figure 7.4**). This formation is favored due to **steric hindrance** where the amino acid side chains interfere with one another structurally and this interference cannot be overcome by rotation due to the inflexible peptide bond. The rigid nature of the peptide bond therefore tends to have amino acids in *trans*; however, proline tends to be in a ***cis*** configuration due to its odd structure (described later in this chapter), where the side chain connects back to the N of the amine group.

Rotation within the peptide chain is therefore associated with the α carbon, which retains its flexible single bond. The juxtaposition of the flexible bonds at the α carbon and the relatively inflexible peptide bond between the amino acids influences the protein structures that can form. Peptide chains tend to be drawn in *trans*, with the N-terminus on the left and C-terminus on the right (**Figure 7.5**).

Amino acids are generally present as zwitterions

The carboxyl group of an amino acid is a weak acid. It can lose its hydrogen to become $-COO^-$. The amino acid's amine group, in contrast, is a weak base. It can gain a hydrogen to become $-NH_3^+$. In pH range 2.2–9.4, which includes general physiological pH conditions for most bacteria, an amino acid is generally present with a $-COO^-$ end and an $-NH_3^+$ end. This is called a **zwitterion**, having a negatively charged end and a positively charged end, and a net charge of zero. The zwitterionic form is the usual state of an amino acid or peptide, although they are frequently shown with $-COOH$ and $-NH_2$ ends (**Figure 7.6**).

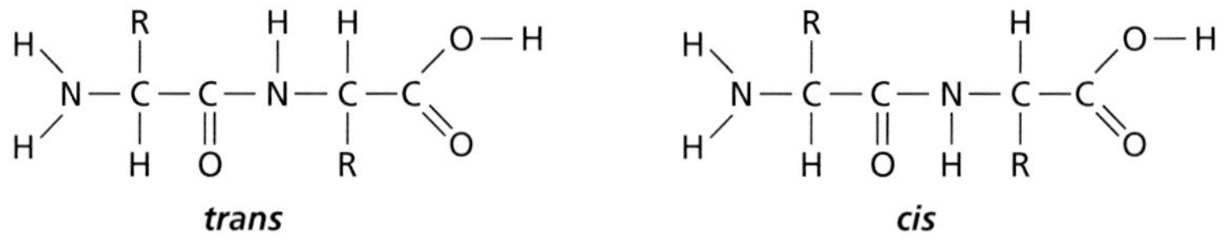

trans *cis*

Figure 7.4 *Trans* and *cis* conformations of amino acids. Due to the double bond character of the peptide bond between amino acids, the side chains are locked into one side or the other of the molecule. When two adjacent amino acids have side chains on opposite sides, this is *trans* (left). When both side chains are on the same side, this is *cis* (right).

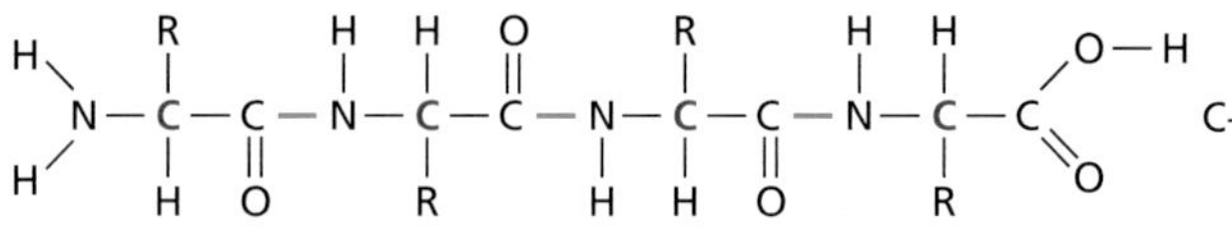

N-terminus C-terminus

Figure 7.5 Typical illustration of a peptide. The N-terminus is on the left and the C-terminus is on the right. The side chains are on opposite sides in the *trans* configuration, as is typical for most amino acids. Rotation around the α carbon (red) is more flexible than the peptide bond (blue).

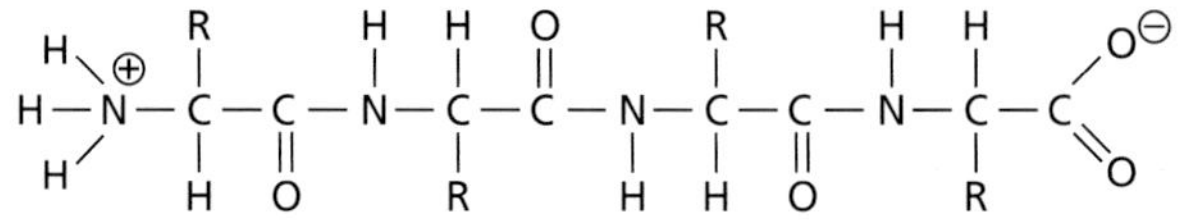

Figure 7.6 Zwitterionic form of a peptide. At normal pH, the carboxyl group will tend to lose a hydrogen and the amine group will tend to gain one. This gives the molecule a positive charge at the amine end and a negative charge at the carboxyl end, but a zero net charge overall. Generally, peptides are zwitterionic, with a positive charge on the N-terminus and a negative charge on the C-terminus.

There are 20 amino acids encoded in the standard genetic code of DNA

There are 20 standard amino acids that are encoded by codons in the DNA and carried by mRNA to be translated at the ribosome. Amino acids that can be found in proteins are called **proteinogenic amino acids**. Although hundreds of different amino acids exist, most are not encoded in DNA and therefore are not found in proteins. The specific amino acids within a protein are determined by the gene encoding it in its DNA. The codon sequences of three nucleotides each correspond to one of the 20 amino acids or are a signal to terminate translation.

The four nucleotides within DNA, adenine, thymine, guanine, and cytosine, can come together in 64 different three-base codon combinations in the mRNA as adenine, uracil, guanine, and cytosine. Three of these 64 combinations are the stop codons that terminate translation. The remaining 61 codons correspond to amino acids, but these 61 codons encode only 20 amino acids, which means that most amino acids have more than one codon. The exception is methionine, which has only the codon AUG, and tryptophan, which has only the codon UGG. Arginine, leucine, and serine have six codons, but most amino acids have two codons (see Table 2.1).

Most of the amino acids have more than one codon. This means that if a mutation in the nucleic acid sequence occurs, the new codon has the potential to still correspond to the same amino acid. For example, the codon GGU changing to the codon GGC due to a mutation that affects the last base of the codon does not change the encoded amino acid, as both of these are codons for glycine. If the nucleic acid sequence changes the codon to a different amino acid, that amino acid might still have similar characteristics to the original amino acid. For example, if the codon CUU encoding the amino acid leucine changes to the codon AUU encoding isoleucine, the resulting protein may be unaffected due to the similarity of these two amino acids. However, a hydrophilic amino acid could be changed into a hydrophobic amino acid based on the change in DNA and mRNA sequence, or it could be changed from an amino acid with an uncharged side chain to a charged one. Occasionally there is translational read through of stop codons, where mutated tRNAs may recognize a termination codon as an amino acid encoding codon. For example, UAA terminator is interpreted as CAA for Gln, UAU or UAC for Tyr, or AAA for Lys due to a mutation in the tRNA anticodon. A single base change in a tRNA could result in the stop codon UAG being recognized as CAG for Gln, UAC or UAU for Tyr, or AAG for Lys.

The codons present in the DNA sequence are species specific

It is possible to predict the amino acid sequence of a protein from the DNA sequence. The codons and their corresponding amino acids are known and therefore the ability to predict the protein sequence from a DNA sequence is robust. However, it is not possible to reverse such predictions and to accurately predict the DNA sequence from a protein amino acid sequence. Predicting a DNA sequence from a protein sequence is not straightforward because some amino acids have six codons, some have four codons, and most have at least two. Because there are options for different codons that will encode the same amino acid, variations in the use of codons are seen between the bacterial species.

Some codons are preferred over others, dependent upon the bacterial species. Therefore, although most amino acids can be encoded by more than one codon, there will be one that is used most often in the coding sequences of bacterial genes based on the species. If a codon is rare or used infrequently in a genome, there is likely to be less of the corresponding tRNA. The use of rare codons with few tRNAs within the cell results in decreased stability for mRNAs containing the rare codon. Slow or stalled translation as a result of the scarcity of the tRNA will mean that the mRNA is targeted for degradation. Because codon usage is species specific, analysis of the codon usage within regions of the genome can identify regions of

DNA where codon usage is different from the rest of the genome. Regions with different codon usage can indicate sections of the bacterial genome that have been acquired from another species.

The prevalence of the different termination codons, necessary for the proper ending of an amino acid chain, can be influenced by the species-specific G+C content of the genome. For instance, the frequency of TGA as a stop codon generally increases with the G+C content of the genome, whereas the stop codon TAA increases with lower genomic G+C. However, the G+C influence on stop codon usage generally does not apply to the termination codon at the end of a gene. The bias applies only to termination codons present in the other five reading frames, those that are not used by a coding sequence and therefore prevent erroneous proteins from being made from noncoding frames of the DNA. Stop codons are plentiful in the reading frames that are not used by the coding sequence and it is here that there will tend to be TGA in high G+C genomes and TAA in low G+C genomes.

The amino acids that can be made by bacteria are species specific

Some amino acids can be essential for the bacterial cell because the amino acid cannot be made by the cell from other components. The cell must therefore acquire the amino acid as it would other nutrients for growth. The amino acids that are essential, that is, those that cannot be made by the cell, differ depending on the species of bacteria. The requirement for essential amino acids in the growth media can help differentiate and identify bacterial species.

Most amino acids are L stereoisomer α-amino acids

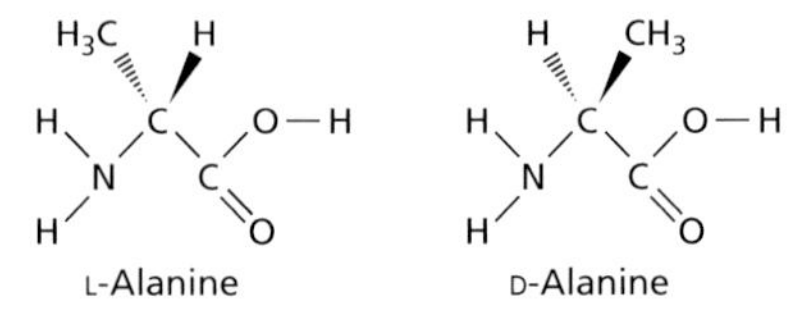

Figure 7.7 L and D stereoisomers of the amino acid alanine. In alanine the side chain attached to the α-carbon is $-CH_3$. Stereoisomers of alanine will have the $-CH_3$ side chains as mirror images in the structure, with the L stereoisomer having the side chain on the left (left) and the D stereoisomer having the side chain on the right (right).

The carbons in amino acids are numbered starting from the -COOH group. The second carbon along from the carboxyl group is called the α-carbon and this is where the side chain of the amino acid is found. Most amino acids found in nature have the $-NH_2$ group bonded to the α-carbon, making them **α-amino acids**.

Amino acids can occur in two different structures that are mirror images of each other (**Figure 7.7**). Glycine is the only amino acid that does not have stereoisomers. The structure of glycine has two hydrogens attached to its α-carbon; the R side chain is simply another hydrogen. Therefore, the mirror images are identical, so it does not have different stereoisomers. Most amino acids are **L stereoisomers**, referring to the three-dimensional structure of the amino acid and the position of the side chain. L amino acids are incorporated into proteins during translation. D amino acids, the mirror image of L amino acids, can be added to the protein during **post-translational modification**, a process where changes are made to a peptide after translation. D amino acids are also part of the bacterial peptidoglycan cell wall on the outside of Gram-positive bacteria and between the two membranes of Gram-negative bacteria.

The classification of an amino acid is determined by its side chain

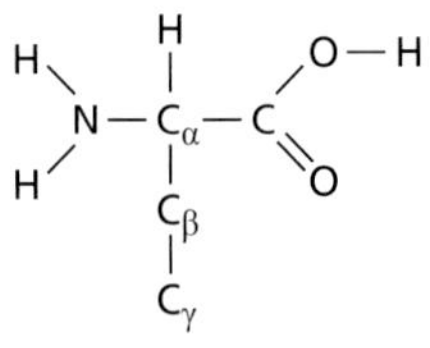

Figure 7.8 Designation of carbons in an amino acid. The carbon to which the amino acid side chain is attached is referred to as the α-carbon. Subsequent carbons attached to the α-carbon in the side chain are designated by Greek letters.

The side chains of amino acids are added onto the α-carbon, and any subsequent carbons in the structure are therefore designated by a Greek letter: β, γ, δ, and so on (**Figure 7.8**). Some amino acid side chains are linear but not all of them are. **Branched chain amino acids** that have side chains with branches in their chemical structure include leucine, isoleucine, and valine.

Amino acids are classified based on the properties possessed by the side chain, which include being weak acids or weak bases, being hydrophilic or hydrophobic, and being polar or nonpolar. These properties of the amino acids within the protein determine the function of the protein, the structure of the protein and how it folds, and the stability of the structure of the protein.

Hydrophobic side chains tend to be buried in the middle of proteins that are found in the bacterial cytoplasm. Hydrophobic amino acids are also found in proteins that are within the bacterial membrane, which is itself hydrophobic.

Proteins that act by binding negatively charged molecules, such as DNA, have positively charged amino acids on their surface. Therefore, the positively charged surface of a regulatory protein may aid in the binding of the protein to the operator, thereby influencing transcription from a promoter.

The features and classifications of amino acids contribute together to the overall features of proteins containing them. The structure of the protein is influenced by factors such as the inflexible nature of the peptide bond and whether that peptide bond is *trans* or *cis*. The potential for hydrogen bonding between the oxygen on the carboxyl group and the hydrogen on the amine groups of different amino acids within a peptide is influenced by the flexibility of the peptide chain and any steric crowding caused by the nature of the side chains, the charge of the side chains, and the hydrophobic or hydrophilic nature of the side chains.

Glycine is small and flexible

The most flexible of all of the amino acids structurally is glycine. It is able to easily rotate because the side chain is simply a hydrogen (**Figure 7.9**); therefore, it does not generate steric hindrance that might limit its ability to rotate. Structures including glycine are not restricted by interference from its side chain, unlike other amino acids, and therefore it can be found both inside and outside a protein. With such a small side chain, glycine is the smallest of the amino acids. Glycine is abbreviated as Gly or G and is encoded by four codons: GGU, GGC, GGA, and GGG.

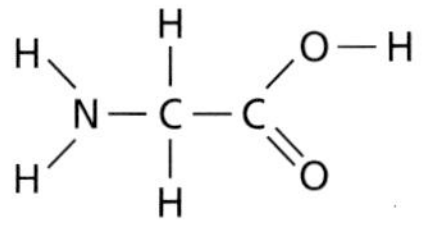

Figure 7.9 The structure of the amino acid glycine. Glycine is the smallest and most flexible of the amino acids, with a hydrogen for its side chain.

Alanine is abundant and versatile

Alanine is small like glycine, but is less flexible than glycine. It has -CH_3 as a side chain (**Figure 7.10**), which is hydrophobic and polar. Like glycine, the flexible nature of the alanine amino acid means that it is versatile. Alanine is also abundant in the bacterial cell; therefore, it can be found in many proteins. Alanine is abbreviated as Ala or A and is encoded by four codons: GCU, GCC, GCA, and GCG.

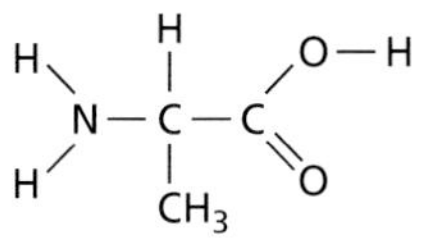

Figure 7.10 The structure of the amino acid alanine. Alanine has a hydrophobic side chain containing a carbon and three hydrogens.

Arginine has a positively charged side chain

The side chain of the amino acid arginine is positively charged, containing carbons, nitrogens, and hydrogens (**Figure 7.11**). Arginine is often found in proteins that bind to phosphorylated substrates. Due to similarities in the features of the side chain, arginine and lysine (see Figure 7.21) are similar. Arginine is abbreviated as Arg or R and is encoded by six codons: CGU, CGC, CGA, CGG, AGA, and AGG.

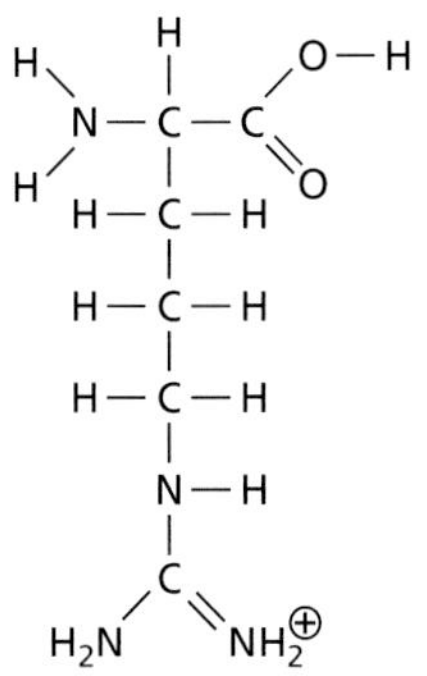

Figure 7.11 The structure of the amino acid arginine. Arginine has a positively charged side chain containing four carbons, three nitrogens, and 11 hydrogens.

Asparagine was the first amino acid identified

The first amino acid identified was asparagine, isolated from asparagus by Louis-Nicolas Vauquelin and Pierre Jean Robiquet in 1806. It is acidic, with an uncharged side chain (**Figure 7.12**), and is similar to aspartic acid. The amine within the asparagine side chain is easily hydrolyzed, which results in asparagine

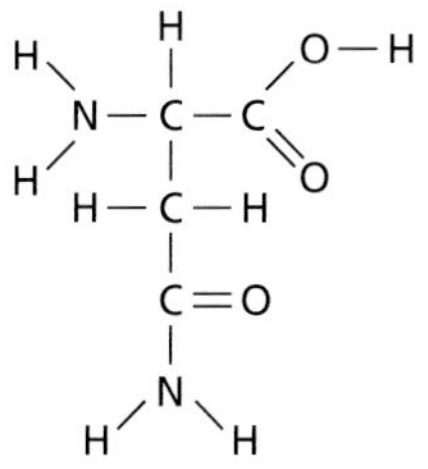

Figure 7.12 The structure of the amino acid asparagine. Asparagine has an uncharged side chain containing two carbons, one nitrogen, one oxygen, and four hydrogens.

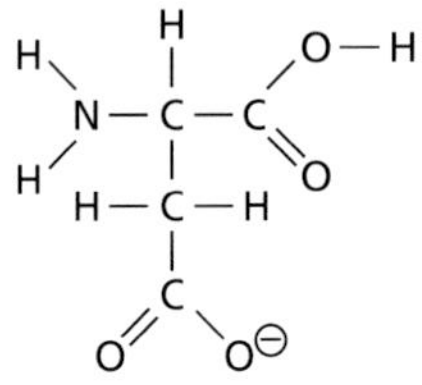

Figure 7.13 The structure of the amino acid aspartic acid. Aspartic acid has a negatively charged side chain containing two carbons, two oxygens, and two hydrogens.

changing to aspartic acid. Asparagine is involved in hydrogen bonding and is a common site to attach carbohydrates in glycoproteins. Asparagine is abbreviated as Asn or N and is encoded by two codons: AAU and AAC.

Aspartic acid is negatively charged and binds to positively charged molecules

The side chain of aspartic acid is negatively charged, containing carbons, oxygens, and hydrogens, and is acidic (**Figure 7.13**). This charge means that aspartic acid is able to bind to positively charged molecules and as such is one of two acidic amino acids involved in enzymatic active centers. It is both acidic and hydrophilic, often being found on the outer surface of proteins. Aspartic acid is abbreviated as Asp or D and is encoded by two codons: GAU and GAC.

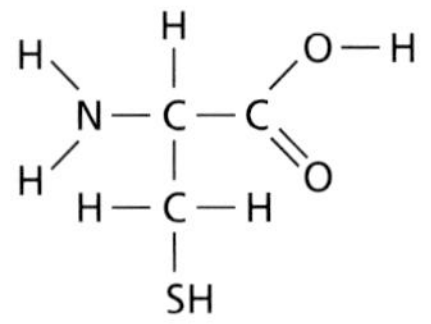

Figure 7.14 The structure of the amino acid cysteine. Cysteine has a side chain containing a carbon, a sulfide, and three hydrogens.

Cysteine forms disulfide bonds with other cysteines

The amino acid cysteine has sulfide in its side chain (**Figure 7.14**), which can form covalent **disulfide bonds** with other cysteines. The sulfides that are within the side chains of the cysteines come together and the resulting disulfide bonds stabilize the protein structure. The sulfur within the side chain of cysteine can also bond readily to heavy metal ions. Cysteine is abbreviated as Cys or C and is encoded by two codons: UGU and UGC.

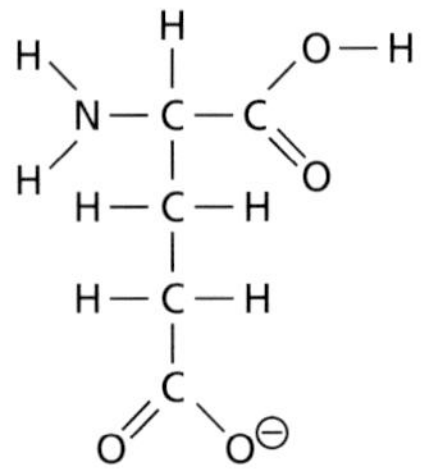

Figure 7.15 The structure of the amino acid glutamic acid. Glutamic acid has a negatively charged side chain containing three carbons, two oxygens, and four hydrogens.

Glutamic acid is a large, acidic amino acid

Glutamic acid is similar to aspartic acid in that it is an acidic amino acid with a negatively charged side chain (**Figure 7.15**). However, glutamic acid is large, with a longer side chain than aspartic acid and a side chain that is slightly more flexible. Glutamic acid is abbreviated as Glu or E and is encoded by two codons: GAA and GAG.

Glutamic acid can be easily converted to proline. This is accomplished in a three-step process in which glutamic acid becomes glutamate semialdehyde, which becomes Δ′-pyrroline-5-carboxylate, which then becomes proline (**Figure 7.16**).

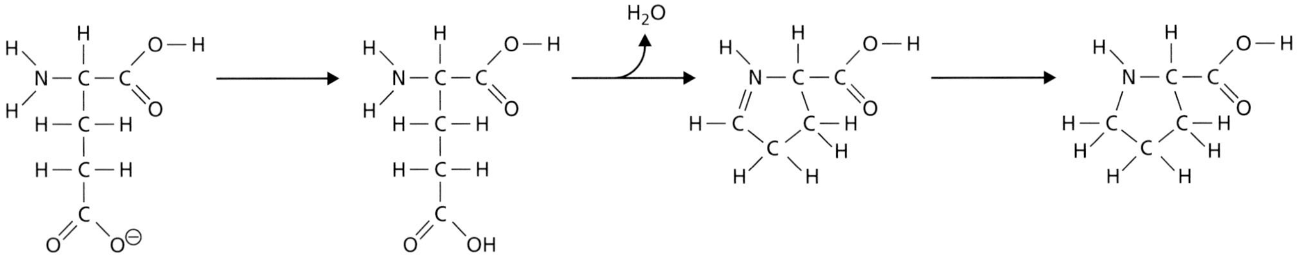

Glutamic acid → **Glutamate semialdehyde** → (H_2O) **Δ′-Pyrroline-5-carboxylate** → **Proline**

Figure 7.16 The conversion of glutamic acid to proline. First, glutamic acid is converted to glutamate semialdehyde. Loss of water from this intermediate produces Δ′-pyrroline-5-carboxylate. This then becomes the amino acid proline.

Glutamine has an uncharged side chain

Glutamine is similar to glutamic acid in its side chain, being acidic and large (**Figure 7.17**). However, glutamine has an uncharged side chain rather than the negative charge of glutamic acid. Glutamine is used to store ammonia in proteins. Glutamine is abbreviated as Gln or Q and is encoded by two amino acids: CAA and CAG.

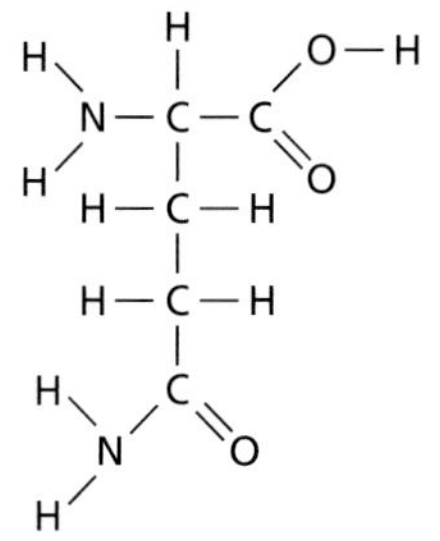

Figure 7.17 The structure of the amino acid glutamine. Glutamine has an uncharged side chain containing three carbons, a nitrogen, one oxygen, and six hydrogens.

Histidine has a large positively charged side chain containing a ring structure

The side chain of histidine includes a ring structure made of carbons and nitrogens (**Figure 7.18**). This gives it a large, positively charged side chain. Histidine is commonly found in enzymes and can stabilize the structure of a protein. In acidic conditions, protonation of the nitrogen within histidine occurs, which causes a conformational change in the protein. Histidine is abbreviated as His or H and is encoded by two codons: CAU and CAC.

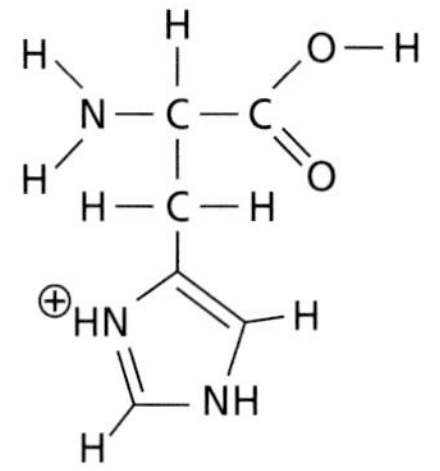

Figure 7.18 The structure of the amino acid histidine. Histidine has a positively charged side chain containing four carbons, two nitrogens, and six hydrogens.

Isoleucine has a branched side chain

Isoleucine has a branched side chain (**Figure 7.19**) and as such is usually interchangeable with leucine and sometimes valine. The branched side chain of isoleucine is large and hydrophobic. Due to its properties, isoleucine contributes to correct folding and protein stability and is important for ligand binding. Isoleucine is abbreviated as Ile or I and is encoded by three codons: AUU, AUC, and AUA.

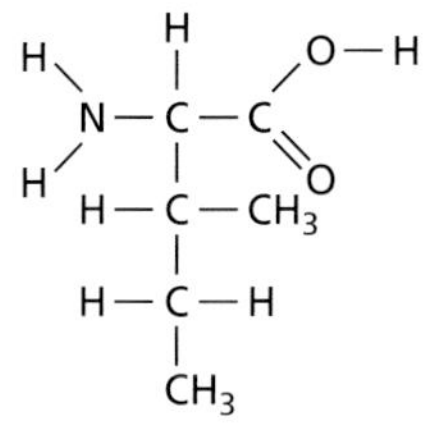

Figure 7.19 The structure of the amino acid isoleucine. Isoleucine has a hydrophobic side chain containing four carbons and nine hydrogens in a branched structure.

Leucine is similar to isoleucine, although its branched side chain is configured differently

Leucine has the same number of carbons and hydrogens in its side chain as isoleucine, although it is configured differently (**Figure 7.20**). With large-branched side chains, leucine is similar to isoleucine and valine. Generally, leucine is found buried within folded proteins due to the hydrophobic nature of its side chain. Leucine is abbreviated as Leu or L and is encoded by six codons: UUA, UUG, CUU, CUC, CUA, and CUG.

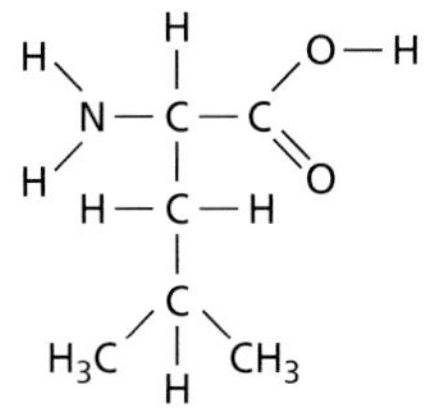

Figure 7.20 The structure of the amino acid leucine. Leucine has a hydrophobic side chain containing four carbons and nine hydrogens.

Lysine has a long, flexible side chain

Lysine has a long side chain made of four carbons, a nitrogen, and 11 hydrogens (**Figure 7.21**), which is flexible along its length. The side chain is also positively charged and as a result interacts with negatively charged molecules within the bacterial cell, such as DNA. Lysine is often found on the surface of hydrophilic proteins, such as those found in the cytoplasm. The features of lysine mean that it is also found to be active in enzymes. Lysine is abbreviated as Lys or K and is encoded by two codons: AAA and AAG.

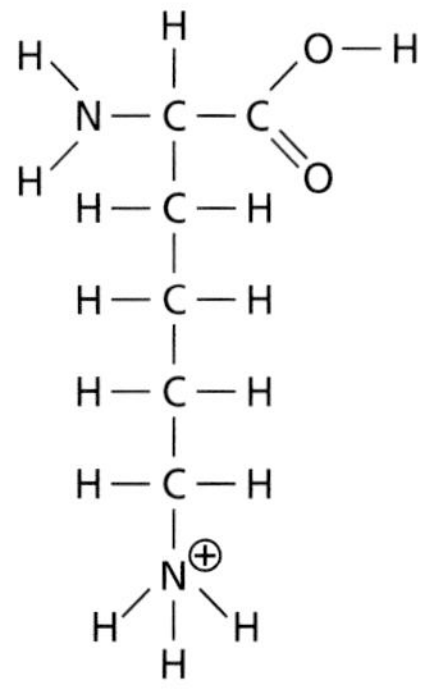

Figure 7.21 The structure of the amino acid lysine. Lysine has a positively charged side chain containing four carbons, a nitrogen, and 11 hydrogens.

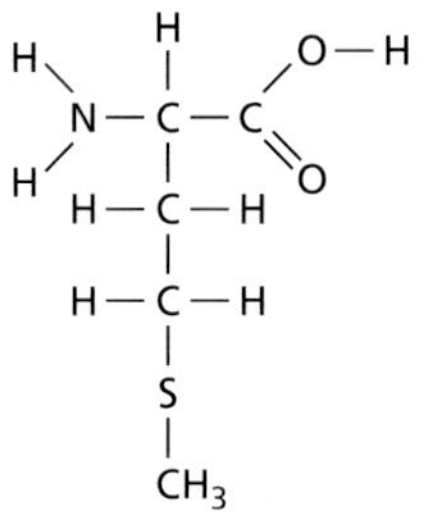

Figure 7.22 The structure of the amino acid methionine. Methionine has a hydrophobic side chain containing three carbons, a sulfide, and seven hydrogens.

Methionine is at the start of all translation

Methionine (**Figure 7.22**) has a large hydrophobic side chain and, as a result, it is frequently found buried in proteins. It contains a sulfur, like cysteine, but with a methyl group that may be activated. Methionine is abbreviated as Met or M and is encoded by a single codon, AUG.

The first amino acid of any new peptide chain generated by translation is a modified form of methionine, ***N*-formylmethionine** (**fMet**) (**Figure 7.23**). The start codon of a gene always codes for fMet in bacteria. AUG is the most common initiation codon and encodes methionine. However, the alternative initiation codons, GUG and UUG, still encode fMet at the start of translation and it is fMet that is incorporated at the beginning of the polypeptide chain. There are different tRNAs that start translation, and these are loaded with fMet. The genes of the *lac* operon, for example, use the alternative initiation codons GUG for *lacI* and UUG for *lacA*. Sometimes the *N*-formylmethionine is removed after translation and is not part of the final protein sequence.

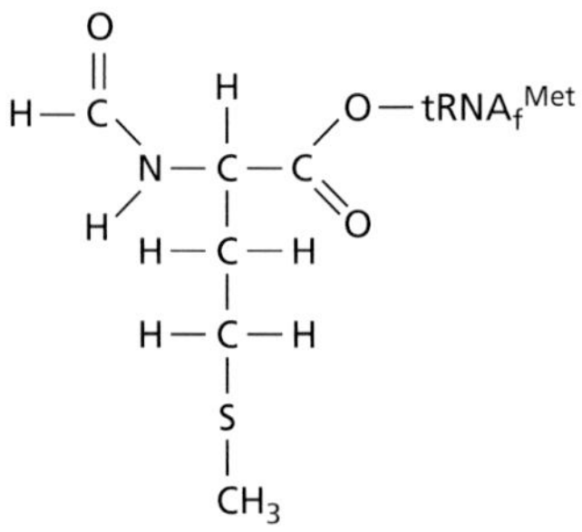

Figure 7.23 The structure of *N*-formylmethionine, a modified methionine. A formyl group (COH) is added to the amine group of the methionine, at the left of the structure. The carboxyl group is then bound to the $tRNA_f^{Met}$, at the right of the structure.

Phenylalanine has a rigid ring structure side chain

The side chain of phenylalanine is hydrophobic and includes a rigid aromatic group that forms a ring structure (**Figure 7.24**). These ring structures can stack, with the phenylalanine side chains forming a stable structure. Phenylalanine is one of the largest proteinogenic amino acids, along with tyrosine and tryptophan. Due to its hydrophobic nature, phenylalanine is almost always buried within the protein. Phenylalanine is abbreviated as Phe or F and is encoded by two amino acids: UUU and UUC.

The side chain for proline loops back to the amine group

L has a side chain that forms a ring structure; however, unlike other proteinogenic amino acids, the side chain ring for proline includes the amine group (**Figure 7.25**). Proline is the only cyclic amino acid. This cyclic form fixes the structure of the amino acid, which is unable to be flexible at the α carbon. The presence of proline causes a bend in the polypeptide chain; this can disrupt protein folding, making kinks in the structure. As a result of its structure, proline tends to be in the *cis* configuration when bound to other amino acids. Due to its structure, it is actually classified as an imino acid. In a peptide bond, the proline amino group has no H; therefore, it cannot donate H to stabilize protein structure. Proline is abbreviated as Pro or P and is encoded by four codons: CCU, CCC, CCA, and CCG.

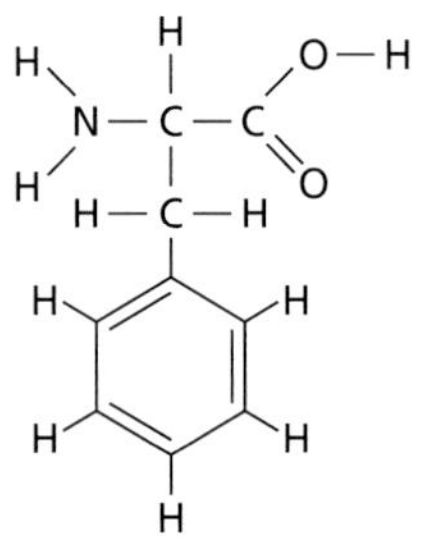

Figure 7.24 The structure of the amino acid phenylalanine. Phenylalanine has a hydrophobic side chain containing seven carbons, six of which are in a ring structure, and seven hydrogens.

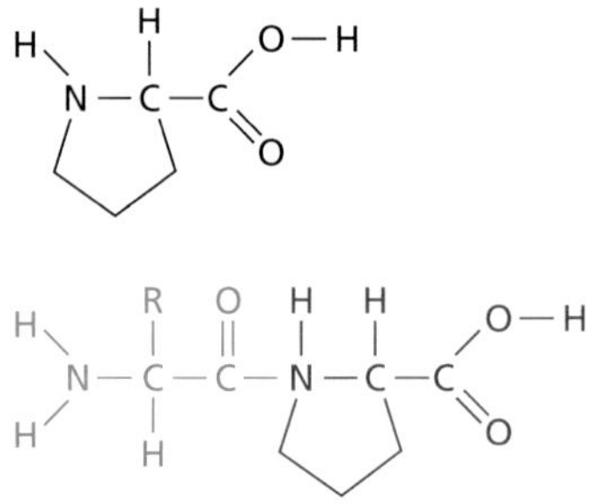

Figure 7.25 The structure of the amino acid proline. Proline has a side chain that loops back to the amine group (top). The proline amino acid (red) tends to be in a *cis* configuration with other amino acids (blue) due to its structure (bottom).

Serine has an uncharged side chain that readily donates hydrogen

The side chain of serine has a hydroxyl group (**Figure 7.26**) and the hydrogen from this side chain feature is readily donated. The side chain is uncharged and polar. The features of serine mean that it is often found on the outside of hydrophilic proteins. Serine is abbreviated as Ser or S and is encoded by six codons: UCU, UCC, UCA, UCG, AGU, and AGC.

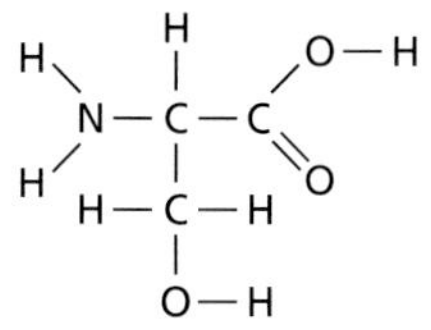

Figure 7.26 The structure of the amino acid serine. Serine has an uncharged side chain containing one carbon, one oxygen, and three hydrogens.

Threonine is similar to serine with an uncharged polar side chain

The side chain of threonine has two carbons, one oxygen, and four hydrogens (**Figure 7.27**). These form an uncharged side chain that is polar and hydrophilic. Side chain feature similarities mean that threonine is therefore similar to serine. Threonine is abbreviated as Thr or T and is encoded by four codons: ACU, ACC, ACA, and AGG.

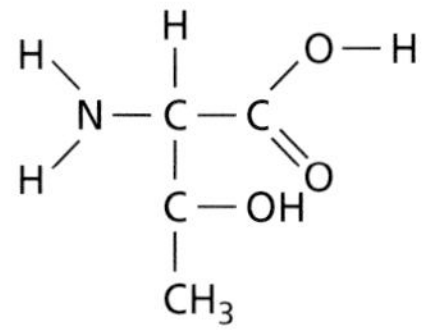

Figure 7.27 The structure of the amino acid threonine. Threonine has an uncharged side chain containing two carbons, one oxygen, and four hydrogens.

Tryptophan has a large side chain with a double ring structure

The hydrophobic side chain of tryptophan is large, incorporating a double ring structure (**Figure 7.28**). Tryptophan is the largest amino acid and is similar to phenylalanine and tyrosine. Tryptophan is abbreviated as Trp or W and is encoded by just one codon, UGG.

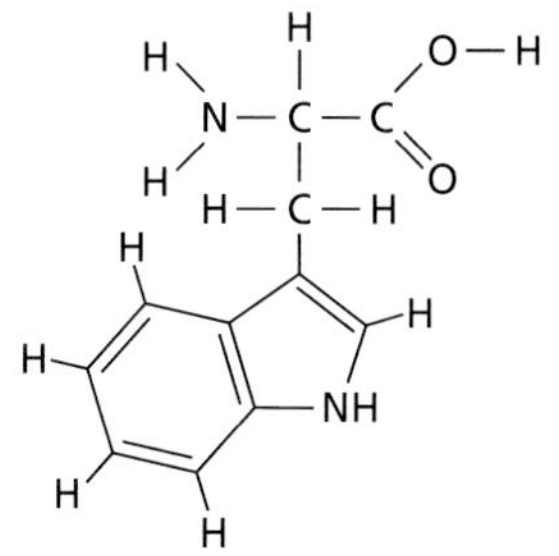

Figure 7.28 The structure of the amino acid tryptophan. Tryptophan has a hydrophobic side chain containing nine carbons, one nitrogen, and eight hydrogens forming a double ring structure.

Tyrosine has a hydrophobic ring structure side chain

The side chain of tyrosine is large and hydrophobic. It includes a ring structure (**Figure 7.29**), making it similar to phenylalanine and tryptophan. Tyrosine is abbreviated as Tyr or Y and is encoded by two codons: UAU and UAC.

Valine has a branched hydrophobic side chain, similar to isoleucine and leucine

Valine is one of the few proteinogenic amino acids with a branched side chain, the others being isoleucine and leucine. The side chain has three carbons and seven hydrogens (**Figure 7.30**). Due to its hydrophobic nature, valine is usually found inside proteins. Valine is abbreviated as Val or V and is encoded by four codons: GUU, GUC, GUA, and GUG.

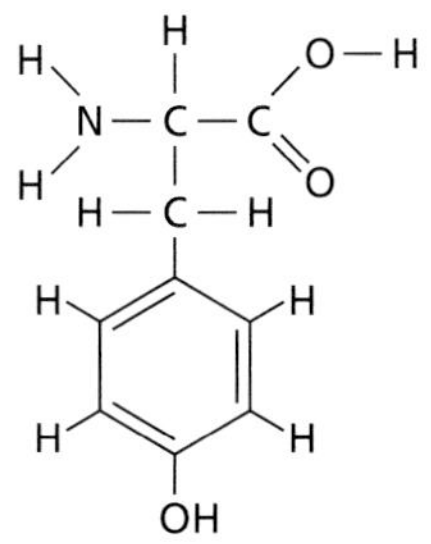

Figure 7.29 The structure of the amino acid tyrosine. Tyrosine has a hydrophobic side chain containing seven carbons, one oxygen, and seven hydrogens.

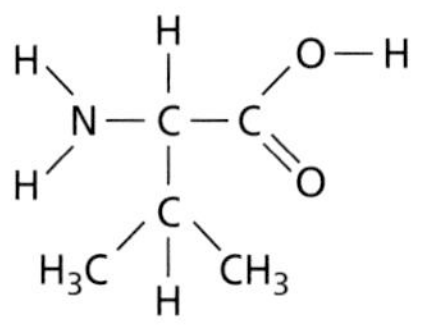

Figure 7.30 The structure of the amino acid valine. Valine has a hydrophobic side chain containing three carbons and seven hydrogens.

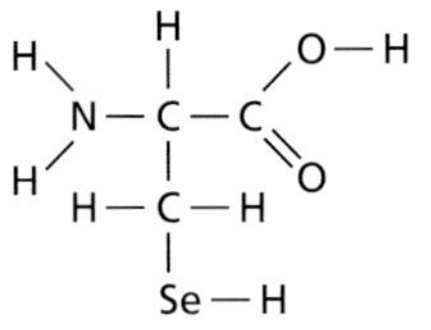

Figure 7.31 The structure of the amino acid selenocysteine. Selenocysteine has a side chain containing one carbon, one selenium, and three hydrogen.

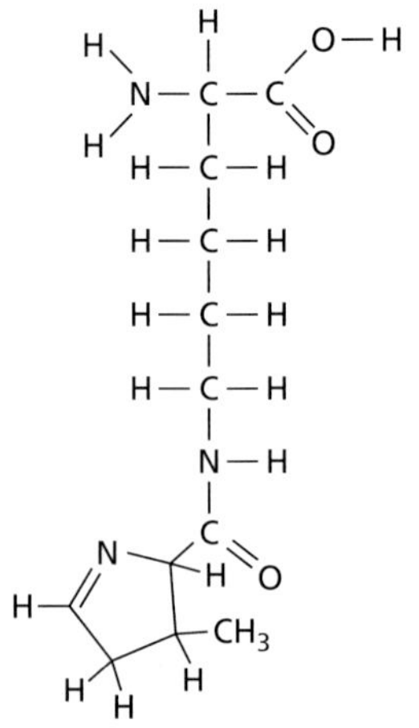

Figure 7.32 The structure of the amino acid pyrrolysine. Pyrrolysine has a side chain containing 10 carbons, two nitrogen, one oxygen, and 17 hydrogens.

Bacterial proteins can include other amino acids beyond the 20 with codons

In bacteria, in addition to these 20 amino acids, for which there are codons, there are other amino acids that can be incorporated into proteins during translation or added post-translationally. During translation, some bacteria can incorporate the amino acid **selenocysteine** and a few are able to incorporate **pyrrolysine**. These 21st and 22nd amino acids are both encoded by codons that are otherwise termination codons, signaling the end of translation. However, these termination codons become amino acid codons by nature of other features in the mRNA sequence that trigger the incorporation of selenocysteine and pyrrolysine, rather than inducing translational termination.

Selenocysteine is encoded by the UGA termination codon and the selenocysteine insertion sequence (SECIS) of 60 nucleotides. The presence of the SECIS element in the mRNA results in the UGA codon recruiting a selenocysteine tRNA rather than triggering translational termination. Just after the UGA on the mRNA, SECIS makes a stem-loop and this causes the UGA codon to be recognized as the selenocysteine codon, rather than the stop codon. Selenocysteine codons can be identified in the genome sequence data by the conserved bases and secondary structure of the SECIS. Selenocysteine is similar to cysteine, although selenium is in the place of sulfur (**Figure 7.31**), and is abbreviated as Sec or U.

Pyrrolysine is encoded by the termination codon UAG, associated with the PYLIS sequence on the mRNA. Pyrrolysine has a long side chain of 10 carbons (**Figure 7.32**) and is abbreviated as Pyl or O.

In addition, amino acids that do not occur naturally in bacterial proteins can be introduced artificially. **Alloproteins** are engineered proteins that do not occur naturally and that can include non-proteinogenic amino acids.

Key points

- Amino acids come together to make proteins.
- Amino acids have an amine group, a carboxyl group, and a side chain.
- The different side chains have different features that influence the structure and function of the amino acid within the protein.
- The peptide bond is double-bond like, which restricts the flexibility of the peptide chain at this bond.
- There are 20 amino acids encoded by the 61 standard three-base codons.
- A 21st and 22nd amino acid are encoded by what is otherwise a stop codon, with the added influence of another sequence within the mRNA.
- Most amino acids have more than one codon.
- The usage of particular codons is species specific.
- Charge, hydrophobicity, and polarity are features of amino acid side chains that influence the function and structure of the protein.

Key terms

Define the following terms introduced in this chapter. Check your answers using the definitions in the Glossary. These terms are also available as Flashcards online.

α-Amino acids
Amine group
Branched chain amino acids
Carboxyl group
Disulfide bond
fMet
N-formylmethionine
Post-translational modification
Primary amino acid structure
Proteome
Secondary amino acid structure
Selenocysteine
Stereoisomers
Steric hindrance
Tertiary amino acid structure
Zwitterion

Questions and discussion topics

Self-study questions

Answer each question using 50–100 words or a table or labeled diagram. Advice on where to find answers to these questions is available online.

1. What are the units that make up proteins? How many different types of these units are there and how does this compare to the units that make up DNA?
2. Draw the general structure common to all amino acids. Label the amine group and the carboxyl group.
3. Explain why the peptide bond that joins amino acids together is considered to be inflexible compared with the other covalent bonds within the amino acids.
4. Draw a *trans* dipeptide made up of one cysteine and one isoleucine.
5. Draw a *cis* dipeptide made up of one serine and one alanine.
6. What is the zwitterionic state of a peptide and why is this the usual state in physiological conditions?
7. If there are hundreds of different amino acids, why are only some referred to as proteinogenic amino acids?
8. Which amino acids form covalent disulfide bonds?
9. Which amino acid can be easily converted into another amino acid? Draw the conversion, including the intermediate stages in this three-step process.
10. What is the unusual feature of the structure of proline and how does this influence the structure of proteins containing proline?
11. What are the 21st and 22nd amino acids used by bacteria and how do the codons for these work?
12. How might amino acids with branched side chains and/or ring structures contribute to steric hindrance?

Discussion topics

These topics are presented for discussion in study groups, as part of class discussions, or on your own. These questions go beyond what is directly covered in this part of the book. Use the research literature and other reading to explore these topics in more depth. Tips to help prepare for topic discussions are available online.

1. This chapter has focused upon the amino acids that make up proteins, which are joined together through translation. Discuss the journey of an individual amino acid through the process of translation, starting from its association with tRNA, including the enzymes and protein complexes involved in joining this amino acid to the growing peptide chain, ending with its ultimate release from the ribosome.
2. In Chapter 3, mutations were discussed, including single-nucleotide polymorphisms where a single DNA base is changed to another base. Explore the ways in which the codons that encode the various amino acids could be changed to encode a different amino acid by changing just one base. How does this influence the characteristics of the amino acid that would be added into the protein by the mutated codon?
3. Make a table to classify the various amino acids by their common characteristics and explore how this might impact the role of these amino acids within proteins. If a mutation resulted in a change of one amino acid for another with the same characteristic, would the protein still function? Can you identify examples of this from research?

Online quiz questions

To further self-assess your understanding of the chapter material, please visit the following link, where you can participate in a range of interactive quiz questions:

www.routledge.com/cw/snyder

Further reading

Almeida P. *Proteins: Concepts in Biochemistry*. 1st ed. Garland Science, CRC Press, Taylor & Francis Group, 2016. ISBN 9780815345022.

Branden CI, Tooze J. *Introduction to Protein Structure*. 2nd ed. Garland Science, CRC Press, Taylor & Francis Group, 1999. ISBN 9780815323051.

Kessel A, Ben-Tal N. *Introduction to Proteins: Structure, Function, and Motion*. CRC Press, Taylor & Francis Group, 2018. ISBN 9781498747172.

Williamson M. *How Proteins Work*. 1st ed. Garland Science, CRC Press, Taylor & Francis Group, 2012. ISBN 9780815344469.

Chapter

Protein Folding and Structure 8

Proteins are assembled by the ribosomes as strings of amino acids, joined by peptide bonds. To become functional proteins, whether structural proteins or enzymes, the linear string of amino acids must fold into a three-dimensional protein structure. This structure may then undergo further modification, through the addition of non-amino acid components of the protein. For some proteins, conformational changes in the protein can be associated with the protein's function and location within the cell. In this chapter, the various characteristics of protein folding, structure, and modification are explored.

Primary amino acid structure is the linear sequence of amino acids joined by peptide bonds

As will be discussed in Chapter 9, the structure of a protein as it comes out of the ribosome during translation is the **primary amino acid structure**. It has the amine end of the first amino acid as the N-terminus. This $-NH_2$ is usually a $-NH_3^+$ due to the zwitterionic characteristic of peptide chains. At the other end of the peptide chain is the carboxyl end of the last amino acid at the C-terminus. This is a –COOH end that is generally $-COO^-$, as the zwitterionic form.

While the amino acid chain is how the protein is first formed, this is not how it remains within the cell. It is able to fold into a structure by virtue of the amino acids, their side chains, and features of the primary amino acid sequence. With structure then comes function.

From a genomics perspective it is relatively easy to predict the primary amino acid structure. This is based on the codons that are encoded in the DNA and transcribed into mRNA to be translated at the ribosome. However, what happens next to the primary amino acid structure is less easy to predict and relies upon an array of factors, not only in the amino acid chain itself, but also in the environment of the cell.

Secondary amino acid structure is a folding of the primary sequence of amino acids

An amino acid chain is able to fold into **secondary amino acid structures**. These are simple structures that are stabilized by the formation of hydrogen bonds between amino acids within the peptide chain. There are two basic secondary amino acid structures, the **α helix** and the **β sheet**.

An α helix structure forms from a linear peptide chain and is stabilized due to strong hydrogen bonds between the -CO of internal carboxyl groups and the -NH of internal amine groups along the chain (**Figure 8.1**). The hydrogen bond forms between an amino acid and the amino acid that is four along in the peptide chain. These hydrogen bonds stabilize the peptide chain into a helix structure. The side chains of the amino acids involved in the α helix structure are pointing out from the helix. The secondary structure of a protein contributes to its function and cellular location. For example, an α helix of 20 amino acids can span a bacterial membrane, making it an integral membrane protein.

DOI: 10.1201/9781003380436-11

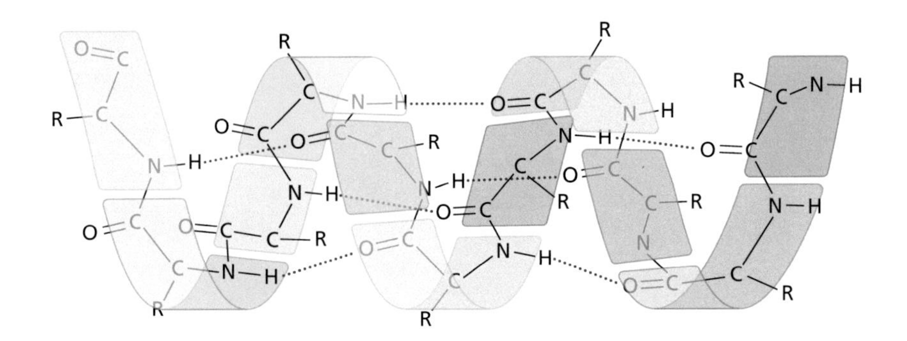

Figure 8.1 The α helix secondary amino acid structure is formed when a linear chain of amino acids forms into a spiral helix that is stabilized by hydrogen bonding. Amino acids that are internal to the peptide chain hydrogen bond to other amino acids that are four amino acids farther along in the chain. The –CO carboxyl group of one amino acid will hydrogen bond to the –NH amine group of another, giving stability to the helix structure.

A β sheet structure includes hydrogen bonds to stabilize the structure as well. In the case of β sheets, the hydrogen bonds are between neighboring parts of the peptide chain. The structure of β sheets comes in two forms. An antiparallel β sheet has neighboring hydrogen-bonded peptides in opposite directions (**Figure 8.2**). One strand is orientated with its carboxyl group to amine group and the other goes from its amine group to carboxyl group. In the antiparallel β sheet, the hydrogen bonding is symmetrical with one-to-one hydrogen bonding. A parallel β sheet has peptide chains that are orientated in the same direction, with the hydrogen bonding being non-symmetrical with one amino acid hydrogen bonding to two other amino acids (see Figure 8.2). As for an α helix, the β sheet secondary structure can also contribute to the function and location of the protein, wherein β sheets can also span the bacterial membrane.

Bulky side chains tend to interfere with protein secondary structures, so generally α helix and β sheet structures include glycines and alanines, due to

(a)

```
H R O     R      H R O     R      H R O     R
| | ||    |      | | ||    |      | | ||    |
N—C—C—N—C—C—N—C—C—N—C—C—N—C—C—N—C—C
      |    ||          |    ||          |    ||
      H    O           H    O           H    O

H R O     R      H R O     R      H R O     R
| | ||    |      | | ||    |      | | ||    |
N—C—C—N—C—C—N—C—C—N—C—C—N—C—C—N—C—C
      |    ||          |    ||          |    ||
      H    O           H    O           H    O
```

(b)

```
H R O     R      H R O     R      H R O     R
| | ||    |      | | ||    |      | | ||    |
N—C—C—N—C—C—N—C—C—N—C—C—N—C—C—N—C—C
      |    ||          |    ||          |    ||
      H    O           H    O           H    O

  R    O R H     R    O R H     R    O R H
  |    || | |    |    || | |    |    || | |
C—C—N—C—C—N—C—C—N—C—C—N—C—C—N—C—C—N
||   |       ||   |       ||   |
O    H       O    H       O    H
```

Figure 8.2 The β sheet secondary amino acid structure is formed when a linear chain of amino acids forms into an aligned sheet of amino acids that is stabilized by hydrogen bonding. Amino acids that are internal to the peptide chain hydrogen bond to other amino acids that are part of another part of the peptide chain. The –CO carboxyl group of one amino acid will hydrogen bond to the –NH amine group of another, giving stability to the helix structure. (a) In a parallel β sheet, both parts of the chain run in the same orientation. (b) The parts of the β sheet can be antiparallel, with the two parts of the peptide chain running from amine group to carboxyl group and the other from carboxyl group to amine group.

the simplicity of their side chain structures. Proline disrupts some secondary structures by virtue of its side chain structure that loops back upon itself. Therefore, proline is often at the end of a helix, bend, turn, or loop, rather than within these structures.

Tertiary amino acid structures form when secondary structures come together

The most complex structure that can form from a string of amino acids is a **tertiary amino acid structure**. Tertiary structures are formed when secondary structures come together with the assistance of several different forces. Hydrogen bonds contribute not only to secondary structures, but also to tertiary structures. For example, β sheets can stack with one another and these stacks of β sheets can form a **β barrel**. A β barrel protein structure forms a channel through the center of the β barrel that is frequently seen in membrane proteins, particularly outer membrane proteins in Gram-negative bacteria.

In addition to further interactions through hydrogen bonding, beyond that seen in secondary structures, the side chains of amino acids can interact and form tertiary structures. The forces involved inside chain interactions include ionic interactions and hydrophobic interactions that may result in hydrophobic amino acids being interior to the structure. Tertiary protein structures are also brought about through the disulfide bonds that can form between cysteines, creating bridges within tertiary amino acid structures (Figure 8.3). **Oxidative folding** is the process of formation of disulfide bonds between cysteines.

Quaternary amino acid structures form when tertiary structures come together

A single strand of amino acids can form tertiary structures. When two or more tertiary structures come together, they form a **quaternary amino acid structure**. Quaternary structures are influenced by hydrogen bonding, hydrophobic amino acids, ionic interactions, and disulfide bonds between two or more tertiary structures. Therefore, the same forces that contribute to tertiary structures are

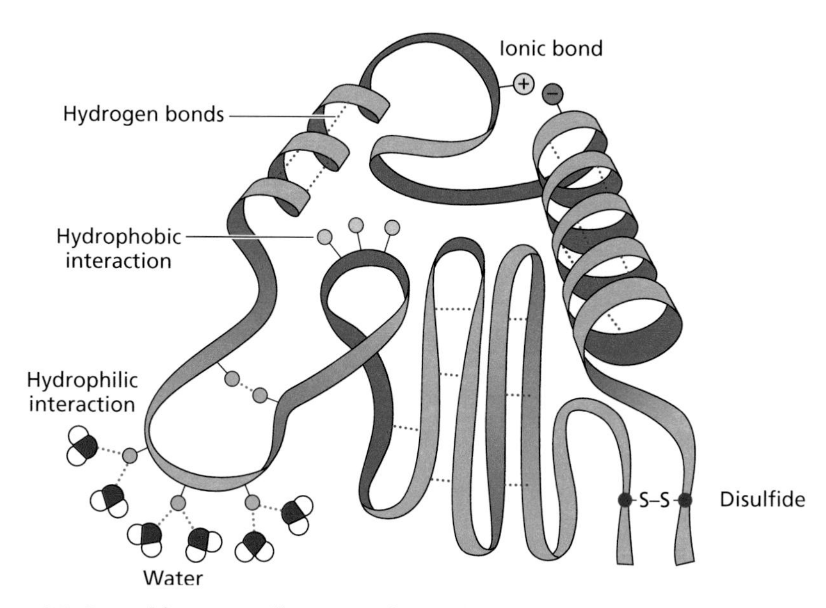

Figure 8.3 Several forces contribute to tertiary amino acid structures. Hydrogen bonding can bring a peptide chain together into secondary structures and can also contribute to tertiary structures. The side chains of the amino acids involved can cause ionic and hydrophobic interactions to occur, contributing to tertiary structure. The presence of cysteines in the protein means there is the potential for disulfide bonds to form within the structure.

involved in quaternary structures, the difference being that quaternary structures involve more than one protein. Peptides, as folded proteins, come together as dimers, trimers, or more with the resulting structures stabilizing each other further.

There are several examples of proteins that complex with other proteins to form quaternary structures. In each case it may be that these are two copies of the same protein or that different proteins form together into a final structure. For example, some regulatory proteins form dimers to bind DNA, made up of two identical polypeptide chains that themselves form tertiary structures, which come together in the dimeric quaternary structure. Some membrane proteins are multimers that together make a β barrel to form membrane pores.

If proteins come together in a quaternary structure, the individual proteins are referred to as subunits. For example, RNA polymerase core enzyme is made of five proteins, including the two α, subunits and the β, β′, and ω subunits, which are each translated as separate polypeptides. The polypeptides then come together in a quaternary structure and together with the sigma subunit will associate into the holoenzyme quaternary structure during the initiation of transcription, adding another subunit to RNA polymerase (**Figure 8.4**).

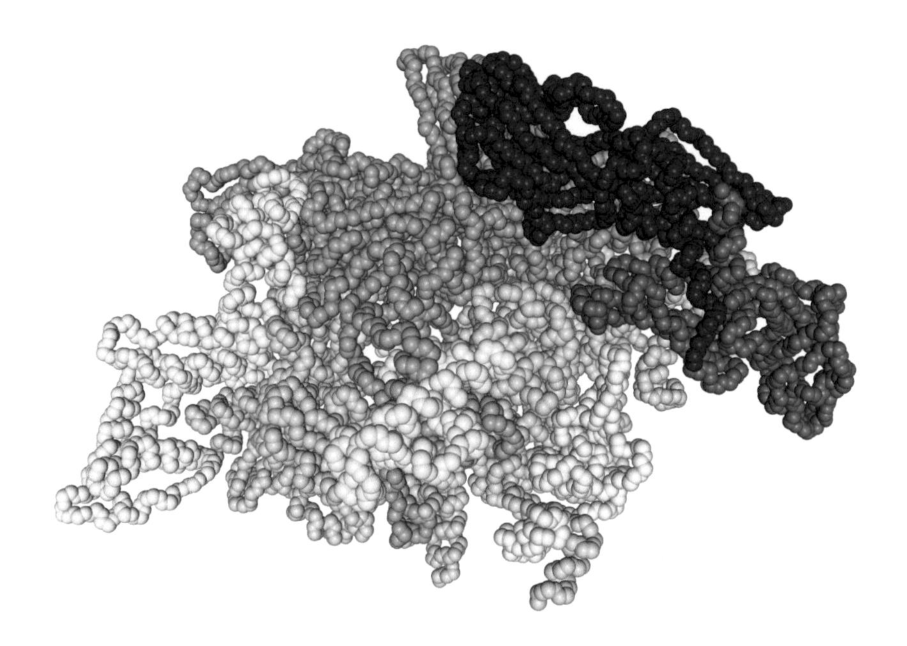

Figure 8.4 The quaternary amino acid structure of RNA polymerase. RNA polymerase is made up of six different proteins, shown here in six different shades of blue that come together in a quaternary structure. The core enzyme has five proteins, two α subunits, the β and β′ subunits, and the ω subunit. These are joined by the σ subunit. This structure is the RNA polymerase holoenzyme from *Thermus aquaticus* (DOI: 10.2210/pdb1L9U/pdb).

Proteins are assisted in folding by chaperones

Chaperones are proteins that assist in the folding of another protein. The complexity of protein folding means that the polypeptide chain may need some outside assistance in achieving the correct final structure. There are also forces that act within and upon the bacterial cell that may work against the formation of the correct protein structure. For this reason, many chaperone proteins are referred to as heat-shock proteins because increases in temperature and other cellular stresses result in protein misfolding. As a result, many chaperones are highly expressed in stress conditions.

Disulfide bonds between two cysteines form through oxidative folding; however, in some circumstances, the formation of the disulfide bond needs help. Some disulfide bonds form in the cytoplasm; however, for proteins that need to pass through the membrane to get to their final destinations, the disulfide bond may be formed later. For example, disulfide bonds in secreted and membrane proteins of Gram-negative bacteria will form in the periplasm. The process of forming disulfide bonds in the periplasm requires the proteins DsbA and DsbB. The disulfide bonds in these two proteins are able to pass this characteristic on

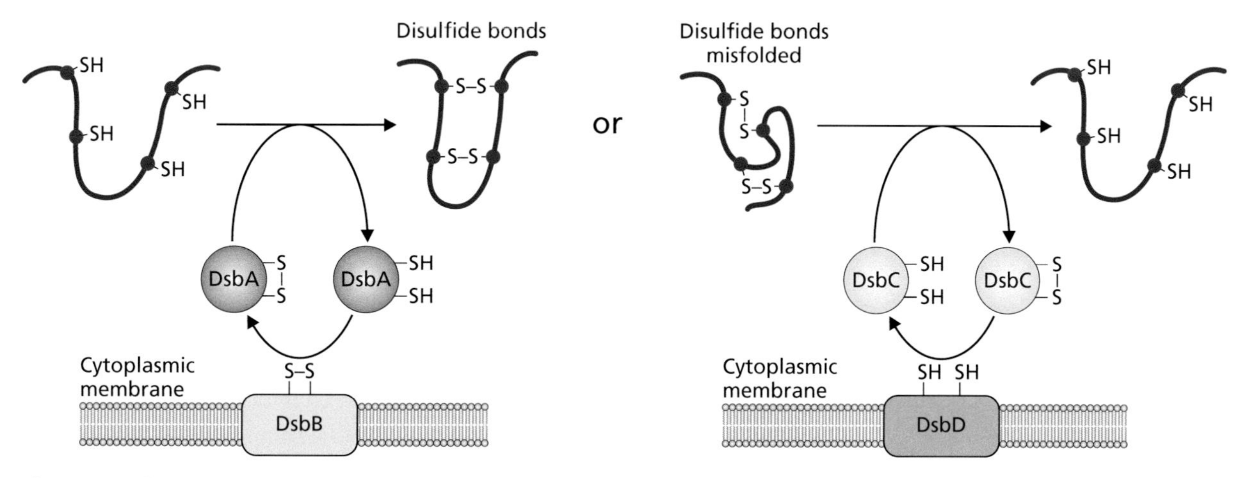

Figure 8.5 Chaperones assist in folding proteins. In the periplasm of Gram-negative bacteria, the chaperone proteins DsbA and DsbB aid in the formation of disulfide bonds. If incorrect disulfide bonds occur in the periplasm, then the chaperone proteins DsbC and DsbD will remove the erroneous bonds.

to the unfolded protein, forming disulfide bonds between the cysteines of the protein chaperoned by DsbA and DsbB. Incorrectly formed disulfide bonds in the periplasm are removed by the proteins DsbC and DsbD, which, similarly to DsbA and DsbB, are able to form disulfide bonds within their own structure and use these to break the disulfide bonds in the misfolded protein (**Figure 8.5**).

Some proteins include more than just amino acids

Not all of the proteins made by bacterial cells contain only amino acids; some undergo post-translational modifications to add other components to the protein or to modify the protein in other ways. Modifications can add on to the side chains of the amino acids or onto the N-terminal or C-terminal ends of the protein.

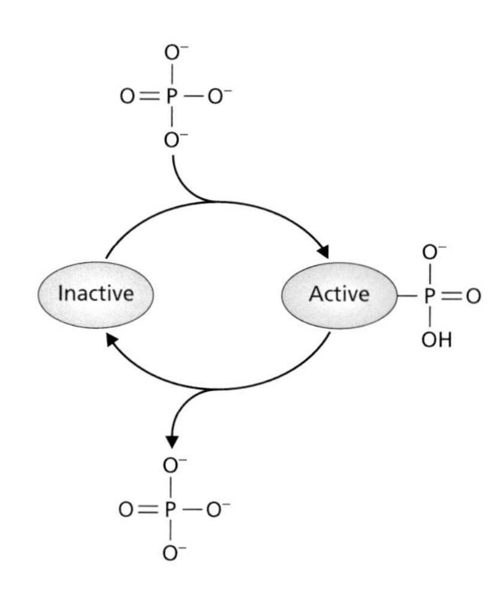

Figure 8.6 Phosphorylation can modulate the activity of proteins. Some regulatory proteins and enzymes are only active when they are phosphorylated. This reversible process provides a mechanism for switching between active and inactive proteins.

Phosphorylation adds a phosphate group to a protein, often activating the protein

Two-component regulatory systems and phosphorelays (see Chapter 5) rely on the **phosphorylation** of their components to function. Phosphorylation is when a phosphate group $(PO_3)^-$ is added to a protein. This addition of a phosphate can modulate the regulatory activity of regulatory proteins and the enzymatic activity of enzymes (**Figure 8.6**). Because proteins can be phosphorylated and dephosphorylated, the function and activity of proteins within the bacterial cell can be modulated.

Phosphorylation can modulate events in the cell through the reversible addition and removal of a phosphate group. This process is key to switching systems in the bacterial cell responsible for gene regulation, metabolism, and development. Phosphorylation involves kinase enzymes to add phosphates and phosphatase enzymes to remove them. These enzymes are able to modify tyrosine, arginine, and serine, and in some species histidine, lysine, aspartate, and cysteine (**Figure 8.7**).

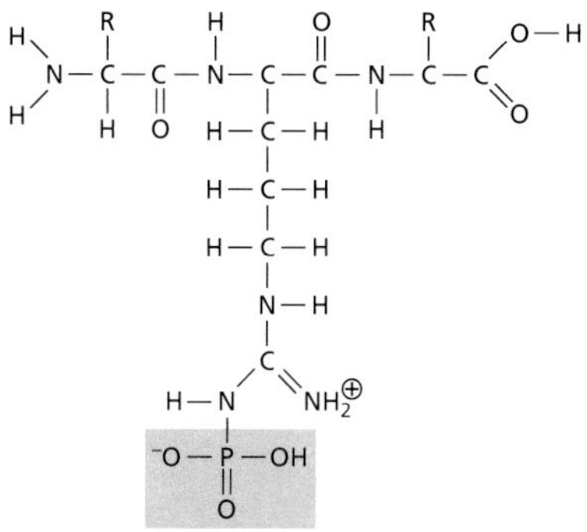

Figure 8.7 Phosphorylation of a protein adds a phosphate group. A phosphate, shown in purple, has been added onto the arginine in this peptide chain.

Lipids are added to proteins post-translationally, adding a hydrophobic region

Lipoproteins are proteins that include a lipid. The addition of the lipid is a post-translational modification, called **lipidation**. The lipid is a fatty acid derived from phospholipids, the components of the membrane. The addition of the lipid will make that portion of the protein hydrophobic. **Lipoproteins** have a lipid added

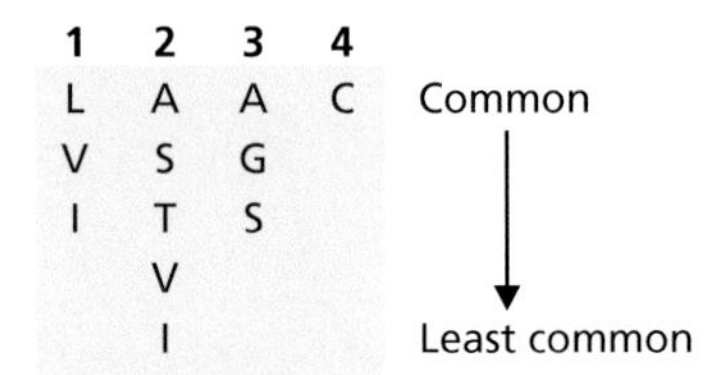

Figure 8.8 Sequence of the lipobox present in lipoproteins. Proteins that are to be lipidated contain a consensus sequence of four amino acids. The last of these is the conserved cysteine that is modified through the addition of the lipid. There are three amino acids that can be found in the first position of the lipobox, with leucine being the most common. In the second position, there are five possible amino acids, with arginine and serine being the most common. The third position, just before the conserved cysteine, is either arginine or glycine, and sometimes serine.

to them in the cytoplasmic membrane, which enables the protein to be anchored to the membrane via the hydrophobic nature of the lipid. The lipoprotein then remains associated with the membrane. The amino acid sequence of proteins that will be lipidated possess a lipobox, a sequence of amino acids at the end of the signal sequence including a cysteine that is recognized by the lipidation machinery (**Figure 8.8**).

The addition of a lipid to a protein increases its stability by providing it with an anchor in the membrane and enables it to interact with other membrane components. The addition of a lipid is a reversible process that can be a regulatory mechanism that alters the function and localization of a protein.

Glycoproteins have a sugar added to the protein

It is possible for a polypeptide chain to include a linked polysaccharide. The addition of the sugar will add a hydrophobic character to that portion of the protein, as a result of this post-translational modification. The process of covalently adding a polysaccharide is **glycosylation** and results in a **glycoprotein**. Glycoproteins are involved in adhesion and biofilm formation, including being involved in the proteins that make up pili and flagella.

Glycosylation can be the addition of an *O*-linked or an *N*-linked glycosyl group (**Figure 8.9**), that is, the addition of a polysaccharide, also referred to as a glycan. *N*-linked glycosylation occurs at the amine of asparagine; however, *O*-linked glycosylation adds a sugar to the hydroxyl groups of serine and threonine. While *O*-linked glycosylation in bacteria has been known for some time, *N*-linked glycosylation was once believed to be limited to eukaryotes. There are both *O*-linked and *N*-linked systems in bacteria, although *N*-linked systems are less common and found in only a few bacterial species. For example, *Campylobacter jejuni* has an *N*-glycosylation system.

Secreted and surface expressed glycoproteins are folded in the cytoplasm and glycosylation happens on the outer surface of the bacteria once the protein has passed through the membrane(s). Surface expressed glycoproteins include those that are included in the structure of the bacterial flagella and adhesins on the surface of the bacteria that interact with host cells and environmental surfaces.

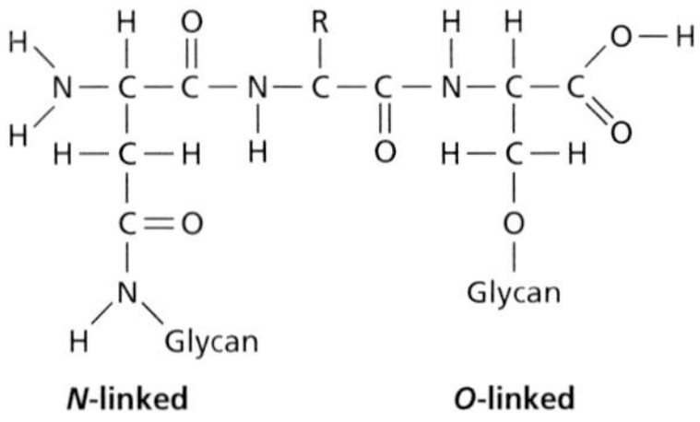

N-linked *O*-linked

Figure 8.9 Glycosylation adds a glycosyl group to a protein, making it a glycoprotein because of the presence of the glycan. The sugar can be added to asparagine for *N*-linked glycosylation or to serine or threonine for *O*-linked glycosylation.

Some proteins are modified through the addition of an oxygen

The addition of an oxygen as a hydroxyl group (-OH) to the side chain of proline or lysine results in **hydroxylation**, a post-translational modification of a protein. In particular, specific residues within the ribosome are hydroxylated by ribosomal oxygenases (**Figure 8.10**). The hydroxylation of elongation factors, termination factors, and ribosomal proteins contributes to the accuracy of translation by ensuring that the termination codon is properly recognized. For example, the 50S ribosomal protein Rpl16 participates in hydrogen bonding that stabilizes the structure of the aminoacyl-tRNA binding site within the ribosome. Altering the hydroxylation of Rpl16 therefore impacts translation and cell growth.

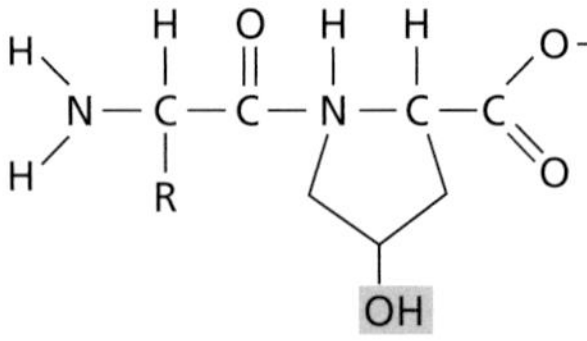

Figure 8.10 Hydroxylation of protein residues. Hydroxylation adds an oxygen group to the side chain of an amino acid. Hydroxylation of a proline adds an oxygen, shown in blue, to the ring structure of the side chain.

Acetylation adds an acetyl group to a peptide chain

Proteins can be post-translationally modified through the addition of an acetyl group (–CO–CH_3). **Acetylation** occurs when an acetyl group is added to the newly formed peptide chain. Frequently, this post-translational modification occurs at the N-terminus of the peptide; however, lysines within the amino acid sequence can also be acetylated.

N-terminal acetylation is an irreversible process, whereby an **N-terminal acetyltransferase** enzyme adds an acetyl group to the amine group at the N-terminus. The acetyl group comes from **acetyl coenzyme A (acetyl CoA)** (**Figure 8.11**). Acetyl CoA delivers acetyl groups to several processes within the bacterial cell, including the acetylation post-translational modification of

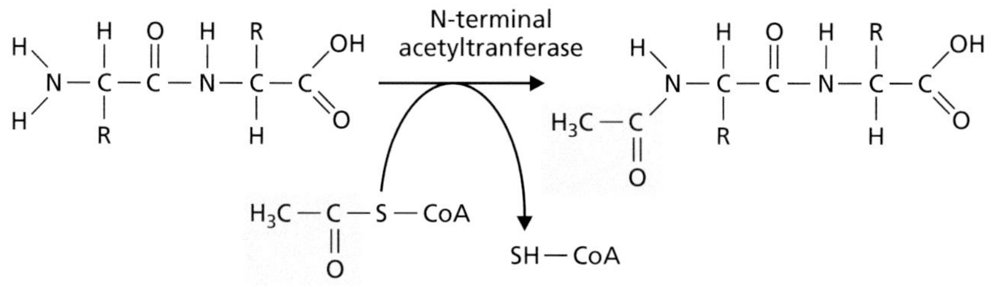

Figure 8.11 N-terminal protein acetylation. Acetylation of a protein transfers the acetyl group from acetyl CoA, shown in yellow, to the N-terminus of a protein.

proteins. The addition of the acetyl group neutralizes the zwitterionic positive charge of the amine end of the peptide chain. Because N-terminal acetylation is not reversible, it also blocks any further modification of the N-terminus. Some ribosomal subunits within *Escherichia coli* have been identified as having N-terminal acetylation. Proteomic technologies have discovered other proteins within bacterial cells that are acetylated on the N-terminus as well.

The acetylation of lysine is a reversible reaction; the acetyl group is added using **lysine acetyltransferase** and it can be removed by a **lysine deacetylase** enzyme (**Figure 8.12**). Due to its reversible nature, this form of acetylation can be involved in the modulation of protein function. Likewise, those proteins that are N-terminally acetylated can be present in the bacterial cell in both the acetylated and the non-acetylated form. The acetylation of a protein can influence the location of the protein within the bacterial cell. This post-translational modification can also change the DNA-binding activity, ATP-binding activity, protein-protein activity, and stability of the protein; therefore, the nature of the proteins can change due to the acetylation state. Acetylated proteins have also been demonstrated to be involved in bacterial chemotaxis and motility, in metabolism and cellular processes such as DNA replication, and in the virulence of pathogenic bacteria.

Succinylation and acetylation can happen at the same amino acid, but not both at the same time

Succinylation is a form of protein **acylation**, the addition of an acyl group, like acetylation. Both acyl groups share the characteristic oxygen double bonded to a carbon and an alkyl group; however, in succinylation a succinyl group ($-CO-CH_2-CH_2-CO_2H$) is added to the protein rather than an acetyl group ($-CO-CH_3$). Like acetylation, succinylation modifies a lysine. Succinylated proteins are involved in metabolism and cellular production of proteins. The same lysine amino acid in a protein can alternatively be subject to acetylation or succinylation, generating three different potential states for the same protein: acetylated, succinylated, or unmodified (**Figure 8.13**).

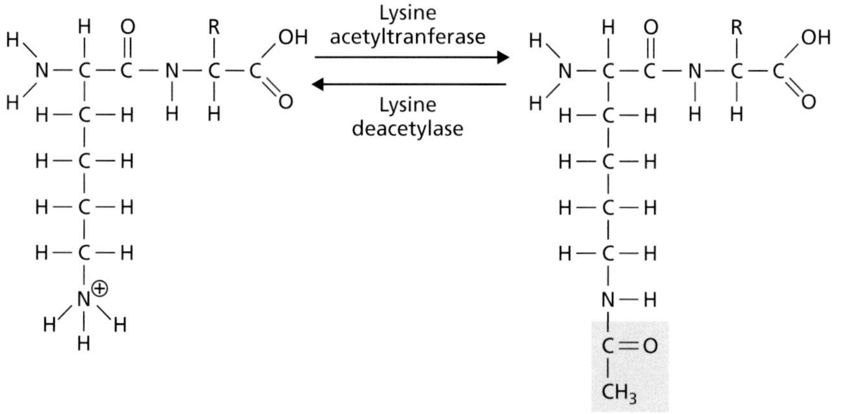

Figure 8.12 Reversible acetylation of lysine. Lysine acetylation, unlike N-terminal acetylation, is a reversible process. Lysine acetyltransferase adds an acetyl group (green) to the side chain of a lysine residue. The acetyl group can be removed by the enzyme lysine deacetylase.

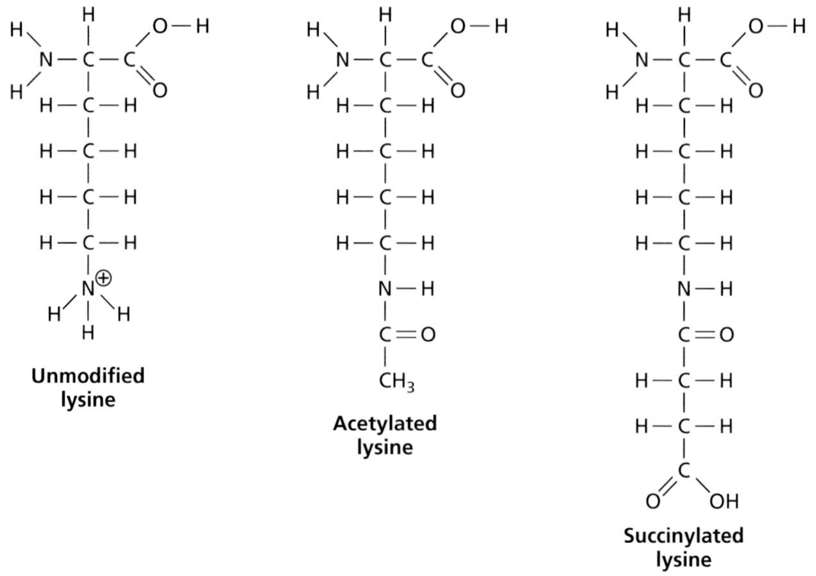

Figure 8.13 Acylation of lysine. The lysine residue in a protein can be acylated in one of two ways, either through the addition of an acetyl group, generating acetylated lysine, or through the addition of a succinyl group, generating succinylated lysine.

Unmodified lysine

Acetylated lysine

Succinylated lysine

CH_3

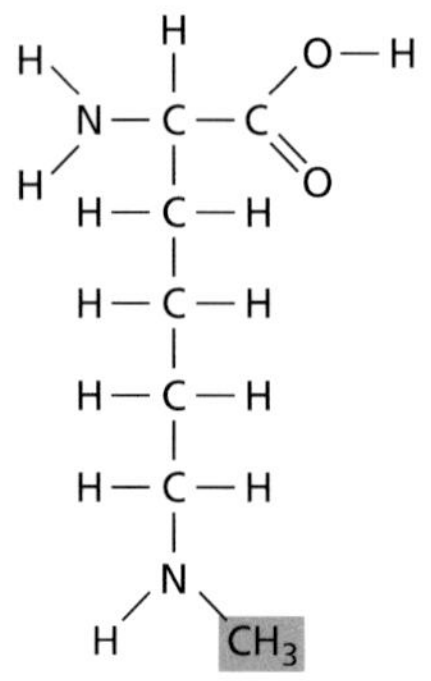

Figure 8.14 Lysines can also be methylated. The methylation of a lysine amino acid adds a methyl group (orange) to the end of the side chain.

Methylation post-translationally adds a methyl group to a protein

The addition of a methyl group ($-CH_3$) to a protein, **methylation**, occurs at a lysine amino acid (**Figure 8.14**). The covalent methylation of the side chain of lysine can alter the function and activity of the protein. For example, the first protein to be identified as being methylated was in the bacteria flagellar proteins of *Salmonella enterica* serovar *Typhimurium*. Changes in the methylation of pilin proteins can regulate cell mobility. Methylation of surface proteins is involved in adhesion. The methylation of lysines seems to play a role in virulence; for example, in some species, changes in methylation of surface proteins change pathogenicity.

Nitrosylation of bacterial proteins can modify regulatory networks

Cysteine residues on bacterial proteins can be ***S*-nitrosylated** through the addition of a thiol group (–SNO; **Figure 8.15**). For example, in *E. coli*, cells grown anaerobically in the presence of nitrate will *S*-nitrosylate some of its cellular proteins. As with other post-translational modifications, this addition can play a role in the regulation of protein activity and function. In *E. coli*, for example, the OxyR regulator becomes *S*-nitrosylated in anaerobic nitrate conditions, causing OxyR to operate on a different regulon than it does in aerobically grown bacteria. Because the cysteine is *S*-nitrosylated, it cannot participate in the formation of a disulfide bond. Therefore, this post-translational modification may have an impact upon the secondary structures that the protein can form.

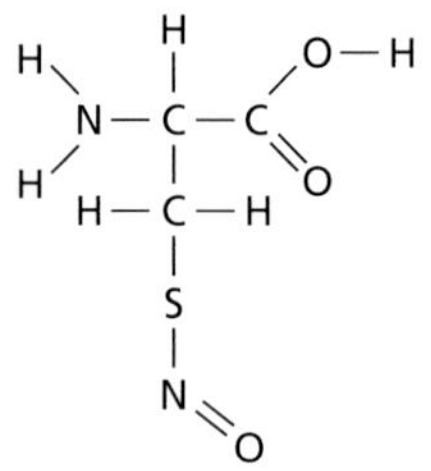

Figure 8.15 Cysteines can be *S*-nitrosylated. Cysteine residues in proteins can be post-translationally modified to be *S*-nitrosylated. A cysteine that is *S*-nitrosylated cannot form a disulfide bridge with another cysteine due to the modification at the sulfur. This inability to form a disulfide bond can cause a change in the structure of the protein.

Modification can remove the fMet at the start of the peptide chain

The first amino acid of any peptide chain is *N*-formylmethionine (fMet), which is a methionine amino acid with an added formyl group (–COH). It is removed for many proteins in the bacterial cell. First, the formyl group is removed by a **deformylase** enzyme (**Figure 8.16**) that is bound to the ribosome near where the peptide exits. For some proteins, only the formyl group is removed, with the starting methionine of the peptide chain being retained, now with the usual amine at the N-terminus.

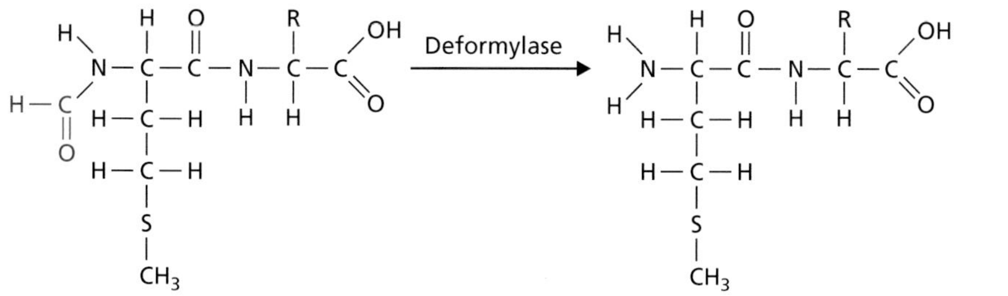

Deformylase

Figure 8.16 Removal of the formyl group from the start of a peptide chain. Many mature proteins in the bacterial cell do not retain the first amino acid as *N*-formylmethionine. A deformylase enzyme removes the formyl group, changing the first amino acid from *N*-formylmethionine to methionine.

Some mature proteins do not have a methionine at the N-terminus; therefore, after the removal of the formyl group from fMet these proteins are further modified. For these proteins, a **methionine aminopeptidase** removes the N-terminal methionine (**Figure 8.17**). This enzyme does not act to remove the first methionine on peptide chains where the second amino acid is an amino acid that is larger than a valine. It is therefore possible to predict those proteins that will have their N-terminal methionine removed.

Proteins are made in the bacterial cytoplasm, but may be transported elsewhere

Translation occurs in the bacterial cytoplasm, where the machinery of the ribosome, mRNA, and tRNAs work to build the peptide chain. Many proteins can be found in the cytoplasm, including those that make up the ribosome itself, the chaperone proteins that aid in protein folding, and enzymes responsible for the metabolism of the bacterial cell. If proteins do not localize appropriately, they are degraded by proteases.

Not all proteins stay within the cytoplasm. There are proteins that are embedded within the bacterial membrane, some of which are involved in transportation of molecules across the membrane. For example, the **Sec translocase**, also known as the **Sec pathway**, is a protein system within the cytoplasmic membrane that moves proteins across the membrane and inserts them into the cytoplasmic membrane. Unfolded proteins can be targeted for transport across the membrane by Sec translocase in two ways, both relying on a **signal sequence** of specific amino acids within the peptide chain. The growing peptide chain can be targeted

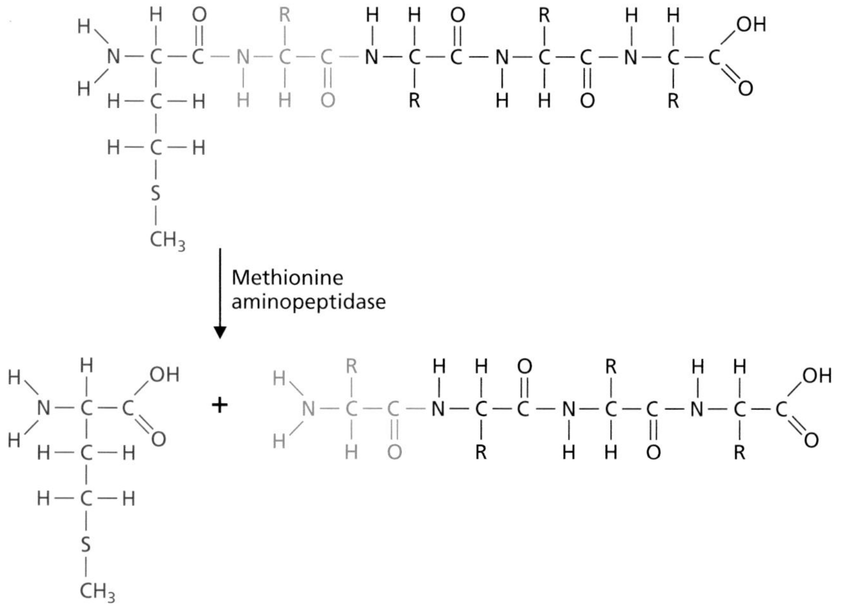

Methionine aminopeptidase

Figure 8.17 Removal of the first amino acid from a peptide chain. As it is made, the first amino acid in a peptide chain is *N*-formylmethionine. The formyl group can be removed from this by a deformylase enzyme (see Figure 8.16). However, some mature proteins have the resulting first methionine removed as well (red). This is accomplished through the action of a methionine aminopeptidase enzyme. The second amino acid that was incorporated into the peptide chain (blue) then becomes the N-terminal amino acid of the protein.

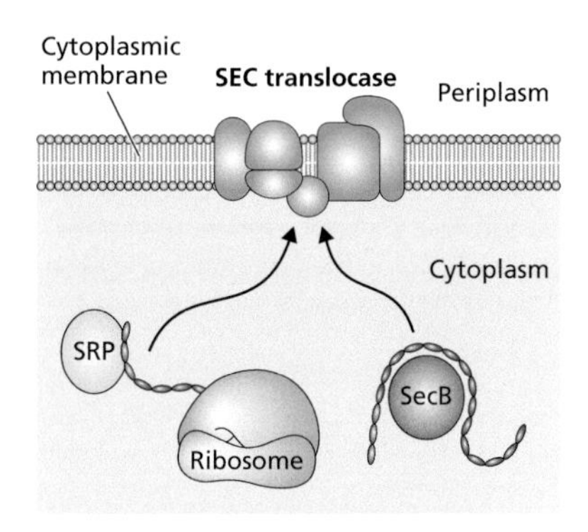

Figure 8.18 There are two mechanisms for recognition of the protein to be transported by the Sec translocase. This system transports unfolded proteins that carry a specific amino acid sequence (orange section of peptide) across the cytoplasmic membrane. The signal sequence can be recognized via two mechanisms. In the first, the peptide chain is recognized co-translationally, with the SRP binding to the signal sequence in the peptide chain. Alternatively, a complete yet unfolded peptide chain can be bound by SecB. Both mechanisms then target the peptide for transfer across the cytoplasmic membrane by the Sec translocase.

for transport co-translationally through recognition by the **signal recognition particle (SRP)** as it emerges from the ribosome. Alternatively, after the completion of translation, the signal peptide on the unfolded peptide chain can be recognized by SecB. The unfolded proteins are then transported by the Sec translocase across the membrane and then fold after translocation (**Figure 8.18**).

Proteins can also be translocated across the cytoplasmic membrane in their native folded state through the **Tat pathway** (**Figure 8.19**), also known as a **twin-Arg pathway**. Folded proteins destined to be outside the cell are translocated across the cytoplasmic membrane of Gram-positive bacteria by these pathways. Likewise, folded proteins destined for the periplasmic space in Gram-negative bacteria can be transported across the inner membrane by the Tat pathway.

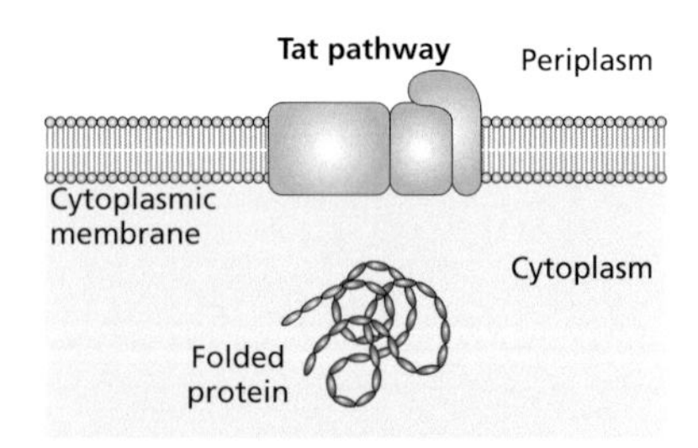

Figure 8.19 The Tat pathway transports folded proteins across the cytoplasmic membrane. Proteins that are folded in the cytoplasm can be transported across the cytoplasmic membrane by the Tat pathway, also called the twin-Arg pathway. A portion of the protein, the signal sequence (orange section), is recognized by the Tat pathway and targets the protein for transport.

Secreted proteins carry a signal to aid in their transport out of the cell

In bacteria such as *E. coli*, nearly half of the proteins are embedded in or cross the bacterial membrane(s). Whether they are transported via the Sec pathway, the Tat pathway, or another mechanism discussed in Chapter 9, most secreted proteins carry a specific amino acid sequence that is used to target the protein for transport. For example, there is a signal sequence of 18–30 amino acids present within the sequence of some proteins that are to be secreted out of the bacterial cell. Generally, secreted proteins have a positively charged N-terminus, usually from lysine or arginine amino acids. These secreted proteins also tend to have a nonpolar hydrophobic core and a polar cleavage region containing a cleavage site. The cleavage site, which breaks the amino acid chain following secretion, is at a cysteine (**Figure 8.20**). The sequence of the features of these proteins is not very well conserved, but the general structure is conserved, with some exceptions.

There are some secreted proteins with no charge or that are negative at the N-terminus and yet they are still translocated. There are also secreted proteins that are not cleaved, which results in them being anchored to the membrane by the transmembrane signal sequence. Therefore, while some secreted proteins can be predicted by sequence data and predicted amino acid features of the protein, others can only be determined experimentally.

Several types of secretion systems have been identified, which also contribute to the transport of proteins, and other components, across the bacterial membrane(s). Some work with the Sec and Tat pathways and some work on their own. These secretion systems are multi-protein complexes that work in a coordinated manner to secrete components out of the bacterial cell. Some activate secretion in response to contact with host cells and other stimuli. The various types of secretion systems will be explored in greater detail in Chapter 9.

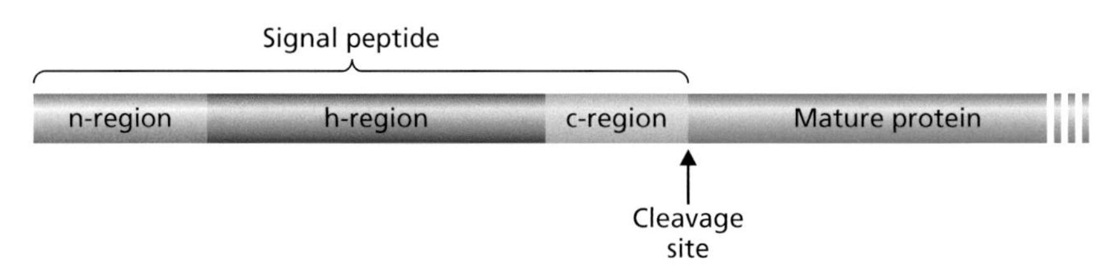

Figure 8.20 The features of the signal peptide region of a secreted protein. Although the amino acid sequence may vary, the general features of the N-terminus of proteins that are secreted are conserved. First there is an n-region that contains positively charged residues, followed by a hydrophobic h-region that can form an α helix, and lastly a c-region. The c-region is the location of the cleavage site, where the signal peptide is cleaved from the mature protein during transport across the membrane.

Key points

- The primary amino acid structure is the chain of amino acids as they come out of the ribosome during translation, but this is not their final form in the bacterial cell.
- Peptide chains can form α helices and β sheets as secondary amino acid structures, both of which rely on hydrogen bonding to form the structure.
- Three-dimensional tertiary structures of peptide chains vary, but can be the result of β sheets stacking into β barrels, or from amino acid side chains interacting with one another.
- Two or more tertiary protein structures can come together in a quaternary structure, such as seen in RNA polymerase with its subunits.
- Protein folding can be aided by chaperone proteins, which aid in the proper folding of proteins and the resolution of misfolding.
- Many post-translational modifications can add molecules to the amino acids that make up a bacterial protein.
- Modifications to proteins such as phosphorylation or lysine acetylation are a reversible means to activate and inactivate protein functions.
- The first amino acid in a peptide chain is *N*-formylmethionine; post-translationally the formyl group can be removed or the whole *N*-formylmethionine can be removed.
- Proteins can be transported across the cytoplasmic membrane unfolded by the Sec pathway or folded by the Tat pathway.

Key terms

Define the following terms introduced in this chapter. Check your answers using the definitions in the Glossary. These terms are also available as Flashcards online.

α Helix	Hydroxylation	Quaternary amino acid structure
β Barrel	Lipidation	*S*-nitrosylated
β Sheet	Lipoprotein	Sec pathway
Acetyl coenzyme A	Lysine acetyltransferase	Sec translocase
Acetylation	Lysine deacetylase	Secondary amino acid structure
Acylation	Methionine aminopeptidase	Signal recognition particle
Chaperones	Methylation	Signal sequence
Deformylase	N-terminal acetyltransferase	Succinylation
Glycoprotein	Phosphorylation	Tat pathway
Glycosylation	Primary amino acid structure	Tertiary amino acid structure

Questions and discussion topics

Self-study questions

Answer each question using 50–100 words or a table or labeled diagram. Advice on where to find answers to these questions is available online.

1 When an amino acid emerges from the ribosome during translation, which end is first, the amine group or the carboxyl group, and how does this reflect the terminus name of the peptide chain?

2 What are the two basic secondary structures that can be formed by amino acids and what stabilizes the formation of each of these structures?

3 Which tertiary amino acid structure forms from the stacking of secondary amino acid structures? How does the nature of this structure contribute to its function when part of a membrane protein?

4 What are the key differences between tertiary amino acid structures and quaternary amino acid structures?

5 Tertiary structures can include bridges within them formed by disulfide bonds. Which amino acids are involved in these bonds and what is the name of the process of formation of these bonds? Which proteins assist in the correct formation of disulfide bonds?

6 Which processes within a bacterial cell are influenced by phosphorylation of proteins? Which amino acids within proteins can be modified to add a phosphate?

7 What does the addition of a lipid to a protein do to the nature of the protein? How is it possible to recognize a protein that can be lipidated?

8 What addition do glycoproteins have added to the amino acid(s) and how does this influence the nature of the protein? What biological processes are glycoproteins involved in for bacteria?

9 Draw the structures involved when acetyl CoA acts on a peptide chain, resulting in N-terminal protein acetylation.

10 Draw a lysine residue and then the two potential acylations of lysine.

11 Lysine is also modified by methylation. Add to the drawing above a lysine that has been methylated. What cellular processes in bacteria include methylated proteins?

12 What modifications can occur to the first amino acid that is added to a peptide chain?

Discussion topics

These topics are presented for discussion in study groups, as part of class discussions, or on your own. These questions go beyond what is directly covered in this part of the book. Use the research literature and other reading to explore these topics in more depth. Tips to help prepare for topic discussions are available online.

1 In this chapter the structures of proteins have been explored, including internal bonds, modifications, and interactions between proteins and subunits. Using protein structural databases, investigate the structure of a protein of interest, identifying the presence of any α helices, β sheets, β barrels, hydrogen bonds, disulfide bonds, and post-translational modifications. Is there evidence in the research literature that chaperones are involved in the folding of this protein or other proteins involved in its modification?

2 Post-translational modifications, such as phosphorylation, can play a vital role in gene regulation, as seen with two-component regulatory systems, and influence such traits as virulence. Discuss the role of any post-translational modification of a specific protein or set of proteins that directly influences the virulence of a bacterial pathogen.

3 Protein transport and secretion is essential for the biological functions of the bacterial cell. Using a species of interest as an example, investigate those proteins that are transported via the Sec pathway and those that are transported via the Tat pathway. Compare and contrast the types of proteins transported by each pathway, the features of each protein, their final destinations, and the presence of any signal sequences.

Online quiz questions

To further self-assess your understanding of the chapter material, please visit the following link, where you can participate in a range of interactive quiz questions:

www.routledge.com/cw/snyder

Further reading

Proteins are assisted in folding by chaperones

Balchin D, Hayer-Hartl M, Hartl FU. Recent advances in understanding catalysis of protein folding by molecular chaperones. *FEBS Lett.* 2020; *594*(*17*): 2770–2781.

De Geyter J, Tsirigotaki A, Orfanoudaki G, Zorzini V, Economou A, Karamanou S. Protein folding in the cell envelope of *Escherichia coli*. *Nat Microbiol.* 2016; *1*(*8*): 16107.

Some proteins include more than just amino acids

Ambler RP, Rees MW. Epsilon-*N*-methyl-lysine in bacterial flagellar protein. *Nature.* 1959; *184*: 56–57.

Bastos PA, da Costa JP, Vitorino R. A glimpse into the modulation of post-translational modifications of human-colonizing bacteria. *J Proteomics.* 2017; *152*: 254–275.

Buddelmeijer N. The molecular mechanism of bacterial lipoprotein modification—How, when and why? *FEMS Microbiol Rev*. 2015; *39*(*2*): 246–261.

Eichler J, Koomey M. Sweet new roles for protein glycosylation in prokaryotes. *Trends Microbiol*. 2017; *25*(*8*): 662–672.

Lanouette S, Mongeon V, Figeys D, Couture J-F. The functional diversity of protein lysine methylation. *Mol Syst Biol*. 2014; *10*(*4*): 724.

Ouidir T, Kentache T, Hardouin J. Protein lysine acetylation in bacteria: Current state of the art. *Proteomics*. 2016; *16*(*2*): 301–309.

Ren J, Sang Y, Lu J, Yao YF. Protein acetylation and its role in bacterial virulence. *Trends Microbiol*. 2017; *25*(*9*): 768–779.

Chapter

9

Multiprotein Systems and Proteomes

The proteins produced by the bacterial cell can work independently; however, some work in coordinated multiprotein systems. These systems are responsible for secretion of proteins, synthesis of the bacterial peptidoglycan layer, cell division, uptake of nutrients, and efflux of hazards from the cell. Some multiprotein systems come together to form a functional part of the bacterial cell, while others come together to biosynthesize the non-protein components of the cell. Multiprotein systems will be explored in this chapter, along with the knowledge that has been gained through the study of bacterial proteomes. Like the genome or the **transcriptome**, the **proteome** is all of the proteins in the bacterial cell made and being expressed at a particular time and under the conditions in which the bacterial cell is grown. Although all proteins are encoded in the DNA and are translated from the mRNA, different levels of information can be obtained by exploring the proteins directly using **proteomics**.

Some cellular structural components are not directly encoded by genes

The genes of the genome encode the proteins found in a bacterial cell; however, there are components of bacterial cells that are not made of proteins. The bacterial membrane, for example, is made of **phospholipids**, not proteins. In order to grow, divide, and multiply in numbers, the bacterial cell must be able to make phospholipids, despite such non-protein components not being encoded in the DNA.

Non-protein components of bacterial cells, instead of being directly encoded in the genome, are made by proteins. Several protein enzymes come together in biosynthetic pathways to make the non-protein components. Although the components themselves are not encoded in the DNA, the proteins that manufacture the bacterial membrane phospholipids, the Gram-negative lipopolysaccharide, and the peptidoglycan cell wall are all produced by proteins. Therefore, the presence of particular genes for biosynthetic proteins in the genome can be indicative of the presence of various non-protein cellular components. However, for the genotype to be reflected in the phenotype, all of the proteins in a biosynthetic pathway need to be expressed.

Genetics of lipopolysaccharide production

The outer surface of the outer membrane of Gram-negative bacteria is dominated by **lipopolysaccharide** (**LPS**) or the shorter **lipooligosaccharide** (**LOS**) found in some Gram-negative species (**Figure 9.1**). As with the phospholipid membrane components, the LPS is not made of proteins. It is therefore not encoded in the genome; however, the proteins to make the various parts of this large molecule are not only present in the DNA but also well conserved between bacterial species.

DOI: 10.1201/9781003380436-12

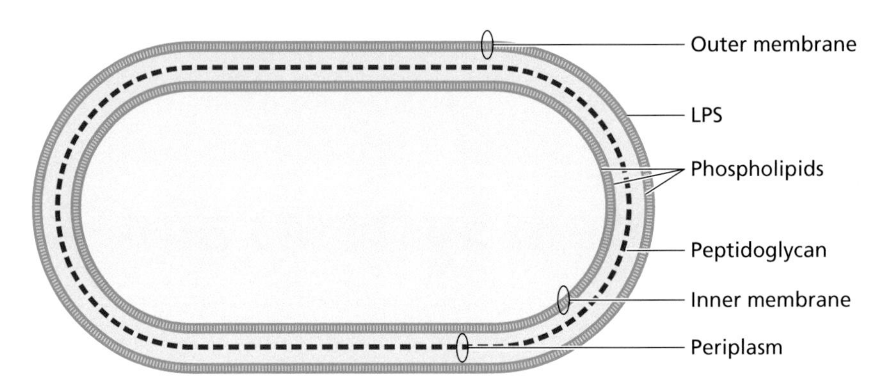

Figure 9.1 LPS makes up the outer leaflet of the outer membrane of Gram-negative bacteria. It is one of several non-protein components of the Gram-negative bacterial cell, including also the phospholipids of the inner membrane and inner leaflet of the outer membrane and the peptidoglycan of the cell wall, present in the periplasm.

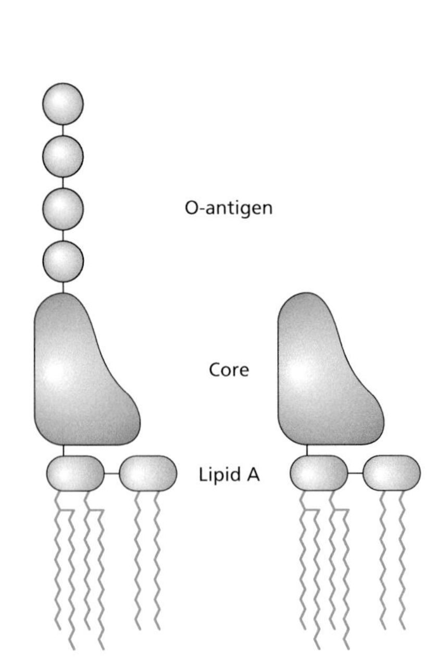

Figure 9.2 The three-part structure of LPS. LPS has three components: lipid A (beige), core (orange), and O-antigen (blue). The lipid A portion anchors the large LPS molecule in the outer surface of the Gram-negative outer membrane. The core and O-antigen portions of the molecule extend out from the bacterial cell into the surrounding environment. Most Gram-negative species have the longer LPS (left), while a few have shorter LOS on their surface (right), which does not have O-antigen.

LPS is a large molecule that has three distinct parts that are embedded in the outer membrane and extend into the extracellular environment: **lipid A**, **core**, and **O-antigen** (**Figure 9.2**). The lipid A part of LPS is made of two phosphorylated glucosamines and fatty acids. The hydrophobic nature of the fatty acids anchors the lipid A into the outer membrane. Therefore, the lipid A portion of the molecule is essential for LPS, as this is the part that keeps the whole of the molecule at the surface of the bacteria.

The **lipid A** portion of LPS is highly conserved between the various Gram-negative bacterial species. As such, the genes encoding the production of lipid A are likewise highly conserved. The genes involved in the synthesis of lipid A are *lpxA*, *lpxC*, *lpxD*, *lpxH*, *lpxB*, *lpxK*, *kdtA*, *lpxL*, and *lpxM* (**Figure 9.3**). Some of these genes are present together in clusters, which aids in the coordination of expression of the gene products and therefore the coordinated production of lipid A. The *lpxD-fabZ-lpxA-lpxB* cluster includes the *lpx* genes involved in lipid A biosynthesis and also the gene *fabZ*, which is involved in the production of phospholipids. The presence of genes for these two different biosynthetic processes together in one gene cluster may perhaps balance the production of these two essential non-protein membrane components.

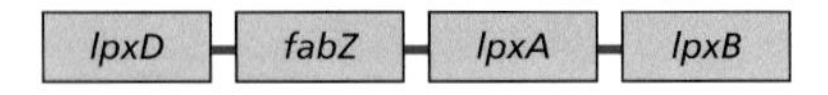

Figure 9.3 Lipid A genes. Some of the genes responsible for the production of the lipid A portion of LPS are present together in clusters, such as *lpxD*, *lpxA*, and *lpxB* shown here, facilitating the coordinated expression of the various enzymes needed for biosynthesis of lipid A.

The **LPS core** is an oligosaccharide made up of 10–15 sugars that is connected directly to the lipid A part of the LPS. The core itself is made up of the inner core, the outer core, and the kdo portion. These parts of the core are made using the enzymes encoded in three separate operons. The *waaD-waaF-waaC-waaL* gene cluster (**Figure 9.4**) is involved in the biosynthesis of the inner core oligosaccharide. The *waaQ* operon (see Figure 9.4) is adjacent to the other cluster and has 7 to 10 genes, the products of which are responsible for production of the outer core. Finally, the *kdtA* operon gene products add the kdo residue to the LPS core.

The portion of the LPS that is farthest from the bacterial cell is the **O-antigen**, which extends out into the environment. The O-antigen is made up of a repetitive glycan polymer of one to eight sugars that is attached to the LPS core. The structure, length, and characteristics of the O-antigen varies, dependent upon the bacterial species and strain. In *Escherichia coli*, for example, there are over 160 known different O-antigens in the various strains of this species. The phenotypic

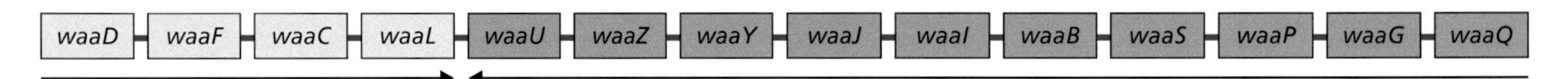

Figure 9.4 The operons for LPS core production. The LPS core *waaD* and *waaQ* operons are adjacent to one another yet transcribed in opposite orientations (arrows). The *waaD-waaF-waaC-waaL* genes of the *waaD* operon (light orange) encode the proteins that make the inner core. The convergently transcribed *waaQ* operon (dark orange) is responsible for the outer core and includes up to 10 genes. The kdo residues of the LPS core are added by the products of the *kdtA* operon (not shown).

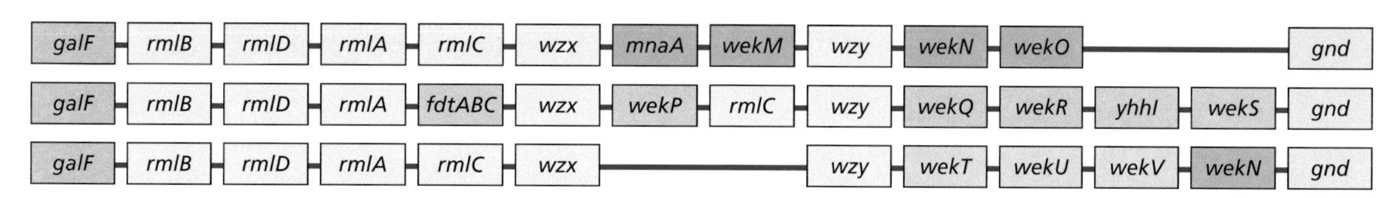

Figure 9.5 Genetics of the LPS O-antigen. The genes responsible for production of the O-antigen of LPS in *E. coli* are present between *galF* and *gnd*. The genes within this region vary between strains of *E. coli*, as shown in these three example regions. In the top region are *mnaA*, *wekM*, *wekN*, and *wekO*, which are not in the other two strains. In the middle, genes *fdtACB* replace *rmlC* between *rmlA* and *wzx* in the order of the region. The *rmlC* gene is present in this strain; however, it is at a later stage, after the unique gene *wekP*. In the bottom region there are no genes between *wzx* and *wzy*; however, like the other strains there are different genes between *wzy* and *gnd*.

characteristics that the LPS confers upon a bacterial colony growing on solid agar media has resulted in full-length LPS possessing strains being called "**rough**," while LPS with shorter O-antigen is referred to as "**smooth**." The length and features of the O-antigen of LPS can have an impact upon the bacteria's resistance to serum killing and antibiotics. Because it is a differentiating characteristic between strains, the O-antigen is the basis of serotyping in some species, enabling different strains with different LPS structures in the O-antigen to be identified, including strains of *E. coli*, *Yersinia*, and *Vibrio*. The variation in O-antigen means there is a great deal of variation in the genes that will be involved in its biosynthesis; however, most are part of the *rfb* gene cluster (**Figure 9.5**). Some species, such as *Neisseria* and *Haemophilus*, have no O-antigen in their LPS, where it is often referred to as LOS, reflecting its shorter structure with no O-antigen (see Figure 9.2).

The LPS has to be assembled and translocated to the outer surface of the cell

Components of the LPS are made within the bacterial cell, yet their final place is on the outer surface of the bacteria. LPS synthesis starts in the cytoplasm with lipid A, which must therefore be translocated across the membranes. Lipid A is hydrophobic; therefore, it begins embedded in the inner membrane. It is passed across to the periplasmic side of the inner membrane. It is here that the core and O-antigen are added. Subunits of O-antigen are also made in the cytoplasm where they are polymerized by the Wzy protein. The chain length of O-antigen repeats is determined by Wzz, an inner membrane protein. The O-antigen subunits are then translocated across the inner membrane by the Wzx membrane protein. Repeat chains of O-antigens are ligated to the LPS core and lipid A on the outer surface of the inner membrane by the protein WaaL, which is also present in the inner membrane. Then the complete LPS is translocated from the periplasmic aspect of the inner membrane to the outer leaflet of the outer membrane by the proteins LptA, LptB, LptC, LptD, LptE, LptF, and LptG. As with the biosynthesis of LPS, the genes responsible for translocation of LPS tend to be present on the genome in gene clusters (**Figure 9.6**).

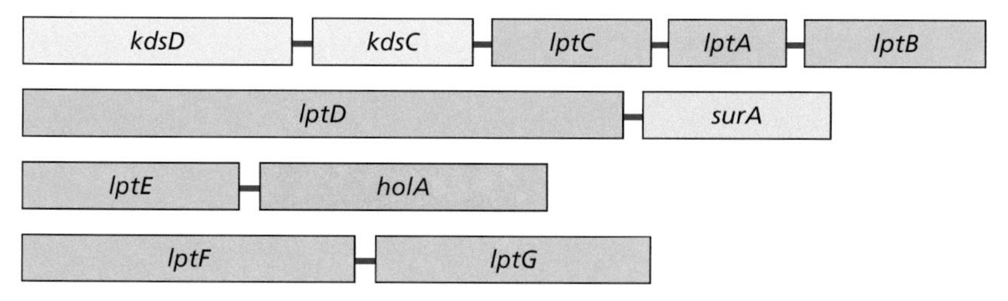

Figure 9.6 Genes that are responsible for translocation of LPS. Several of the genes encoding proteins involved in the translocation of LPS components from the cytoplasm to the periplasmic aspect of the inner membrane are present together in the chromosome. These include the *kdsD-kdsC-lptC-lptA-lptB* genes, the *lptD-surA* genes, the *lptE-holA* genes, and the *lptF-lptG* genes. The products of these genes come together to form the LptABCDEFG multi protein system responsible for the translocation of LPS from the inner membrane to the outer leaflet of the outer membrane.

Peptidoglycan is built by proteins encoded in a cluster of genes

Similarly to LPS, the components of the peptidoglycan cell wall are made in the cytoplasm. The products of the genes *murABCDEF*, *mraY*, and *murG* act, in this order, to generate peptidoglycan precursors in the cytoplasm (**Figure 9.7**). These precursors are then translocated to the periplasm where the repeating units of ***N*-acetyl glucosamine** (**GlcNAc** or **NAG**) and ***N*-acetyl muramic acid** (**MurNAc** or **NAM**) are assembled. The MurNAc components include peptide chains that add to the structural integrity of the peptidoglycan by cross-linking the layers of the cell wall.

Peptidoglycan is specific to bacteria. The component GlcNAc is not specific to bacteria, being found in the chitin of insects and fungi. However, the peptidoglycan component MurNAc is found in nature only in the cell walls of bacteria. The cell wall also contains both L-amino acids and D-amino acids. Most bacterial species share the same make-up of the peptidoglycan; however, in *Mycobacteria* the

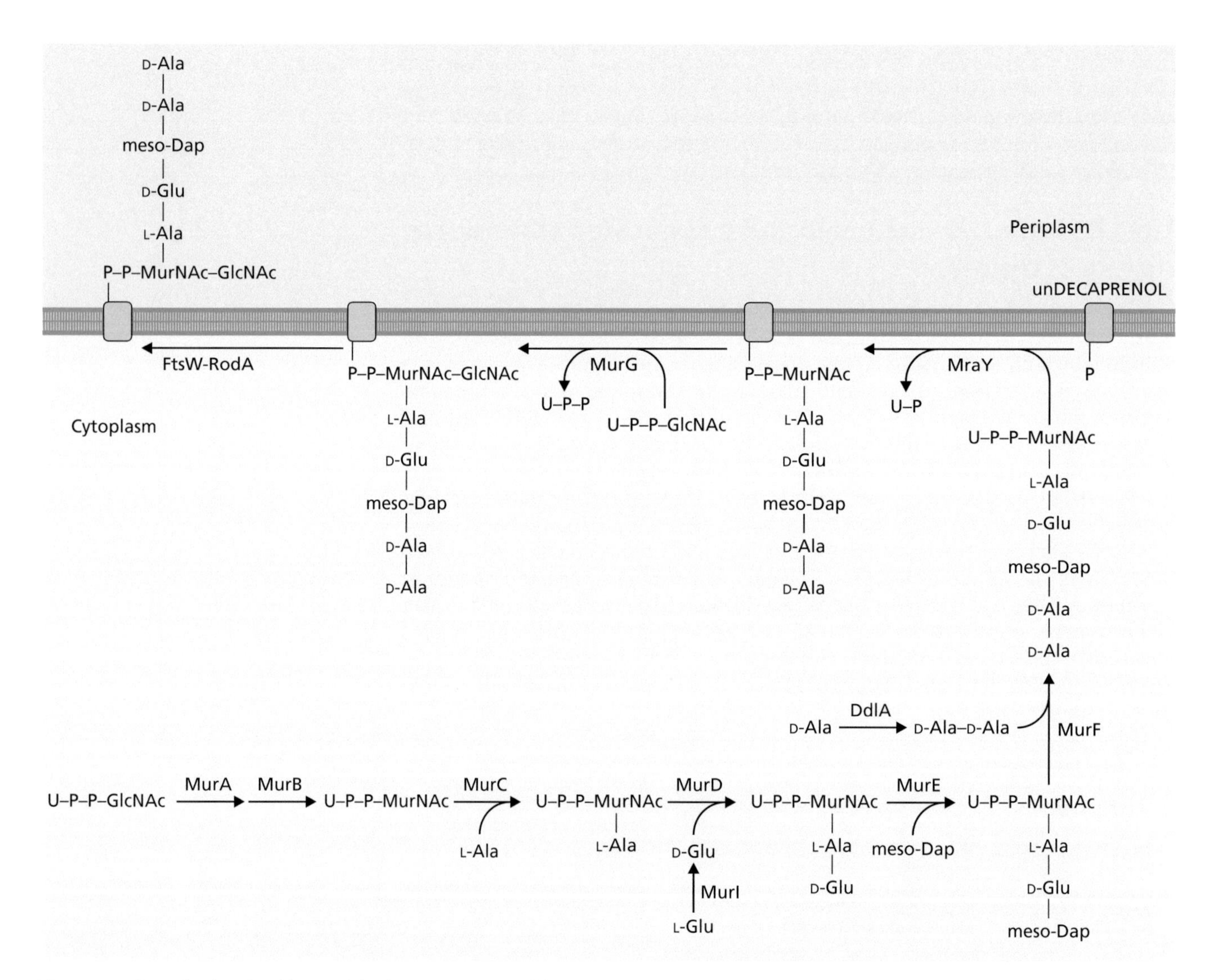

Figure 9.7 Proteins involved in peptidoglycan component biosynthesis. Many proteins are involved in the biosynthesis of the peptidoglycan components GlcNAc and MurNAc. Shown here are the actions of MurA, MurB, MurC, MurI, MurD, MurE, DdlA, MurF, MraY, MurG, and FtsW-RodA. The GlcNAc and MurNAc components are assembled in the cytoplasm and then flipped across the membrane into the periplasm. The process of biosynthesis starts with uridine diphosphate *N*-acetylglucosamine (U-P-P-GlcNAc). Amino acids and meso-Dap (diaminopimelic acid), a derivative of lysine, are added to MurNAc. At the membrane, LPS components are attached to the carrier undecaprenol.

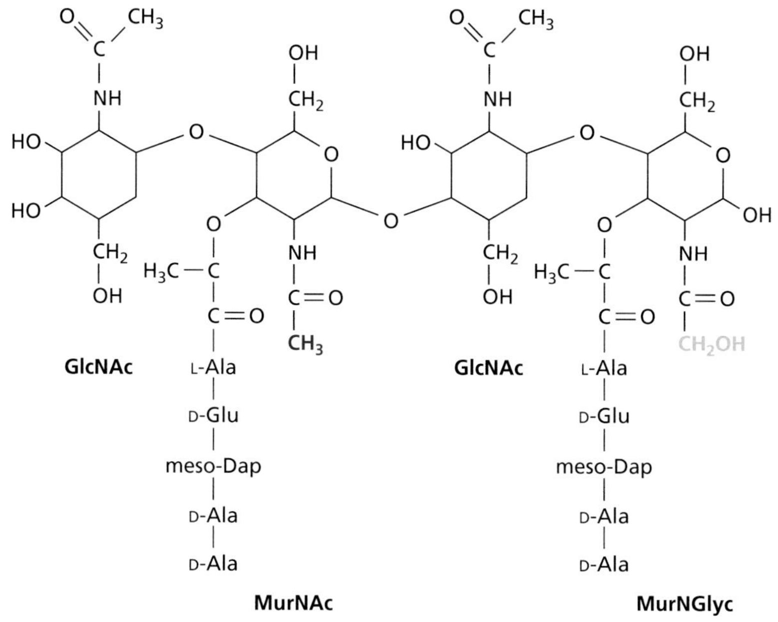

Figure 9.8 Structural components of *Mycobacteria* peptidoglycan. The cell wall of *Mycobacteria* can be different from the cell walls of other bacteria. While most species have cell walls with repeating GlcNAc–MurNAc structures, the *Mycobacteria* species can have GlcNAc–MurNAc or GlcNAc–MurNGlyc. Note that the key difference between these structures is the $-CH_3$ in MurNAc (purple) and the $-CH_2OH$ in MurNGlyc (green).

peptidoglycan is made of alternating GlcNAc and MurNAc (*N*-acetylated) as in other bacteria or GlcNAc and MurNGlyc (*N*-glycosylated) (**Figure 9.8**).

Many bacterial genes have been identified through the generation or discovery of knockout mutants, where a gene has ceased to function. The loss of phenotype then gives an indication as to the function of the gene. Peptidoglycan is essential to bacterial cell survival; therefore, deletion or mutation of peptidoglycan biosynthesis genes is lethal. Some of the genes responsible for the biosynthesis of peptidoglycan were therefore identified by creating **temperature-sensitive mutants**. These mutations allow the bacterial cell to have a normal phenotype and grow normally at one temperature, but to stop expressing the gene when the temperature changes, thereby causing a phenotypic change. In addition, **overexpression mutants**, where expression of gene products is increased, can reveal the function of essential genes. Since the gene is never lost, the mutation is not detrimental to the cell in the same way as a knockout mutation. Using these types of strategies, the roles of genes in the essential processes involved in peptidoglycan production have been determined. The shape of bacteria is determined by its peptidoglycan; therefore, mutations in some of these genes involved in its biosynthesis have an impact upon the cell morphology.

Many of the peptidoglycan biosynthesis genes are present within a cluster of genes alongside genes involved in cell division. Making peptidoglycan and dividing the cell into two cells are coordinated processes; therefore, the genes involved are found together in the genome. A large cluster of genes involved in division and cell wall synthesis, the *dcw* cluster, is highly conserved between bacterial species (**Figure 9.9**). As a bacterial cell grows, it must expand the peptidoglycan in a coordinated manner to maintain the differential in pressures inside and outside the cell.

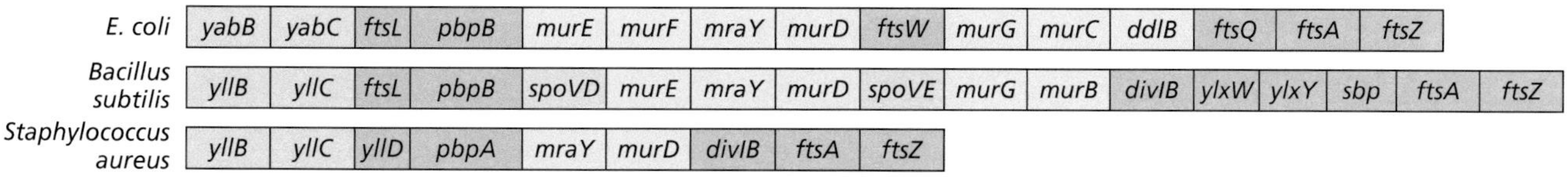

E. coli	*yabB*	*yabC*	*ftsL*	*pbpB*	*murE*	*murF*	*mraY*	*murD*	*ftsW*	*murG*	*murC*	*ddlB*	*ftsQ*	*ftsA*	*ftsZ*		
Bacillus subtilis	*yllB*	*yllC*	*ftsL*	*pbpB*	*spoVD*	*murE*	*mraY*	*murD*	*spoVE*	*murG*	*murB*	*divIB*	*ylxW*	*ylxY*	*sbp*	*ftsA*	*ftsZ*
Staphylococcus aureus	*yllB*	*yllC*	*yllD*	*pbpA*	*mraY*	*murD*	*divIB*	*ftsA*	*ftsZ*								

Figure 9.9 The *dcw* cluster. A large cluster of genes involved in division and cell wall synthesis in bacteria is highly conserved across several species. The processes of making the peptidoglycan (green genes) and of dividing the bacterial cell (purple genes) are both essential to bacterial life. In addition, these processes are linked, with more peptidoglycan being needed as the cell divides. The genes involved in these processes are found together in the *dcw* (division cell wall) cluster.

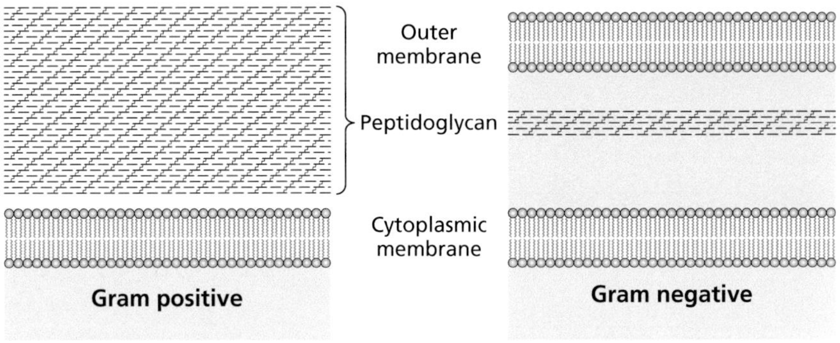

Outer membrane
Peptidoglycan
Cytoplasmic membrane
Gram positive
Gram negative

Figure 9.10 Different thicknesses of peptidoglycan in Gram-positive (left) and Gram-negative (right) bacteria. The Gram-positive peptidoglycan layer, which is on the outside of the cell, is very thick and can make up to 90% of the weight of the cell. There is a thin layer of peptidoglycan in Gram-negative bacteria, where the cell wall makes up about 20% of the weight of the cell.

Peptidoglycan is present in both Gram-negative bacteria and Gram-positive bacteria; however, the amount of cell wall varies between the two. In Gram-negative bacteria, the peptidoglycan layer makes up about 20% of the total weight of the cell, being a thin layer sandwiched between the inner and outer membranes. In Gram-positive bacteria, the exterior peptidoglycan layer is much thicker and can be up to 90% of the total weight of the cell (**Figure 9.10**).

Bacterial membrane phospholipids are made by proteins

Cellular membranes are made up, for the most part, of phospholipids, with their hydrophilic polar heads and hydrophobic non-polar tails (**Figure 9.11**). In order for the bacterial cell to grow and divide, more phospholipids must be generated. To adapt to changing environmental conditions, bacteria respond by changing their membrane phospholipids. As discussed in the previous chapter, such changes to the membrane can result in changes in exposure and activity of membrane proteins. Changes in the membrane can protect the bacterial cell from stress and may change during biofilm formation. Genomics has revealed that many of the genes involved in making non-protein cellular components of the bacterial cell, such as the biosynthesis of LPS and the generation of peptidoglycan synthesis, are conserved between species. For these, *E. coli* is seen as the representative species for such genetic investigations. Other cellular processes and the genes involved, such as genes responsible for the production and modification of phospholipids, differ between species. Regardless of whether the genes are conserved or unique to a specific bacterial species, the genes encode proteins, which are enzymes that in turn produce the phospholipids that make up the bacterial membrane. The genes themselves do not directly encode the phospholipids, yet without the genes there would be no phospholipids.

The lipid bilayer of a bacterial cellular membrane is 6–8 nm thick and is made of two layers of phospholipid molecules (**Figure 9.12**). The bacterial membrane in Gram-positive bacteria and the inner membrane in Gram-negative bacteria are made of mostly phospholipids. Within this membrane are α-helical proteins as well; however, greater than 95% of the membrane lipids are the phospholipids making up the lipid bilayer.

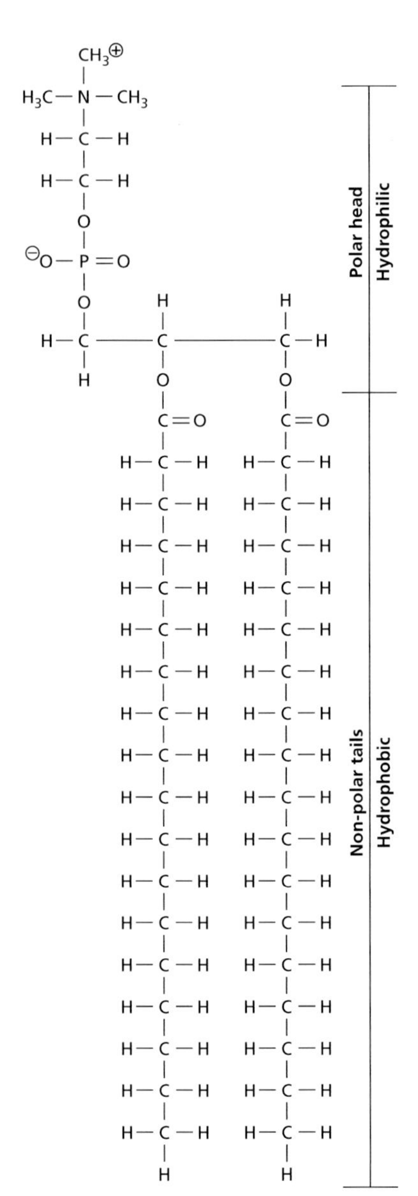

Figure 9.11 Structure of a phospholipid. The phospholipid has two parts: a hydrophilic polar head and a hydrophobic non-polar tail. The head includes a phosphate group. The bonds between the carbons in the tails can change from single bonds as shown here to double bonds (not shown), causing bends in the tail chains.

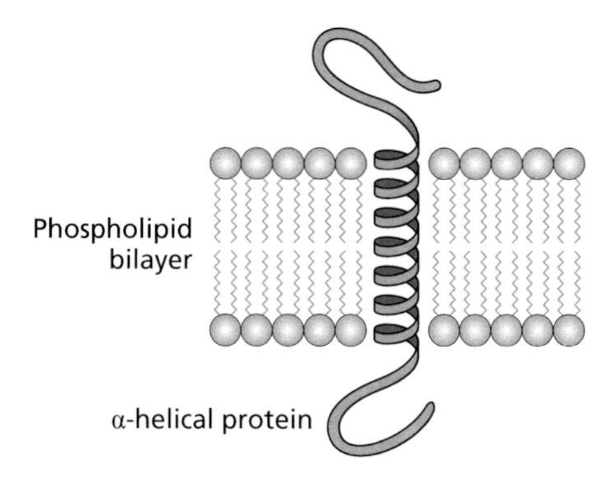

Figure 9.12 Structure of a phospholipid bilayer. The bacterial cytoplasmic membrane is made almost entirely of phospholipids. The hydrophilic heads form the surface of the membrane, while the hydrophobic tails form the center of the membrane. The membrane can include embedded α-helical proteins.

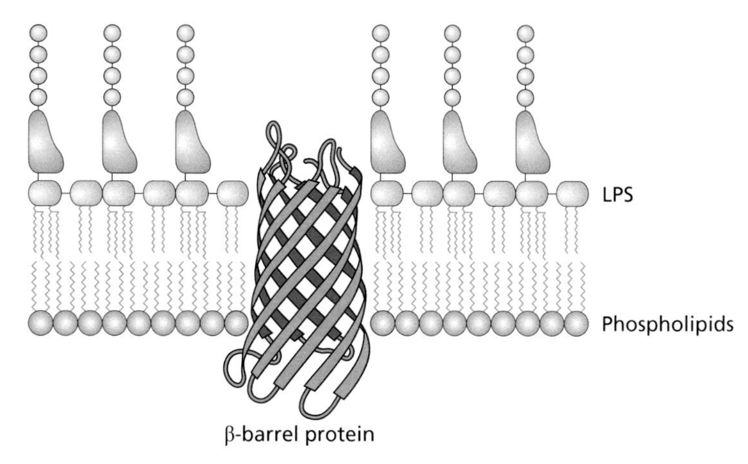

Figure 9.13 The outer membrane of the Gram-negative bacterial cell. The outer membrane of Gram-negative bacteria is made up of both phospholipids and LPS. Most of the outer leaflet of the outer membrane is LPS. Embedded proteins tend to be β-barrel proteins.

Gram-negative bacteria have a second cellular membrane, which also contains phospholipids. The outer membrane has both phospholipids and LPS making the membrane bilayer, rather than a bilayer of phospholipids as in the inner membrane (**Figure 9.13**). The proteins in the Gram-negative outer membrane tend to have β-barrel structures, unlike the α-helical proteins of the inner membrane. Lipoproteins are anchored by N-terminal acyl modifications and can be found in both membranes.

Extracellular polysaccharides make up the bacterial capsule

Some bacterial cells are covered in a **capsule**. The presence of a bacterial capsule is often important for virulence and survival within the host, enabling the bacteria to avoid being killed by the immune response. Capsules can also protect the bacterial cell from drying out and therefore are an important feature of species like *Staphylococcus epidermidis* that live on the surface of the skin. Previously, the experiments conducted that transferred virulence to non virulent bacteria were discussed, having been used to demonstrate that DNA is the genetic material. The feature that differentiated these virulent and non virulent bacteria was the capsule, present in the virulent bacteria.

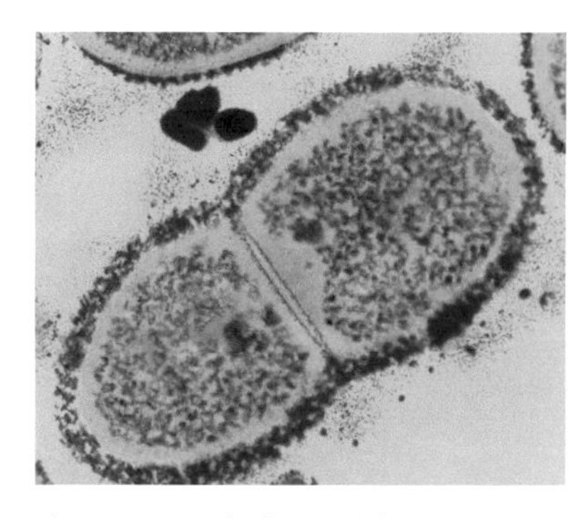

Figure 9.14 The bacterial capsule. Some bacterial cells are surrounded by a layer of extracellular polysaccharides that form the bacterial capsule. The capsule is visible here as the dark blue layer surrounding the bacterial cells. The capsule can help the bacteria avoid the immune system and resist drying. (Courtesy of Dr. Kari Lounatmaa/Science Photo Library.)

Bacterial capsules are made of extracellular polysaccharides that are anchored to the cell membrane through the lipid component of glycolipids. The capsule is present outside the outer surface of the Gram-negative outer membrane, outside the LPS, or beyond the thick peptidoglycan of the Gram-positive bacteria (**Figure 9.14**).

As with many of the other non-protein features of the bacterial cell, the genes responsible for the biosynthesis of the capsule tend to be present in clusters, close to each other on the genome. Changes in the repertoire of these genes, such that different genes are found in otherwise similar strains, influences the formation and structure of the capsule and therefore influences the strain-to-strain variation in the capsule. These differences in capsule types are often reflected in differences in serogroups, which are used to differentiate bacterial strains phenotypically. It has long been known that the DNA encoding capsule genes is able to transform bacteria of the same species that either do not have capsule genes, resulting in capsulate strains, or do have capsule genes to change the characteristics of the capsule produced. Horizontal gene transfer has been recognized with regard to bacterial capsule genes for quite some time, due to the easily identifiable change in phenotype.

Proteins make up bacterial cell structures

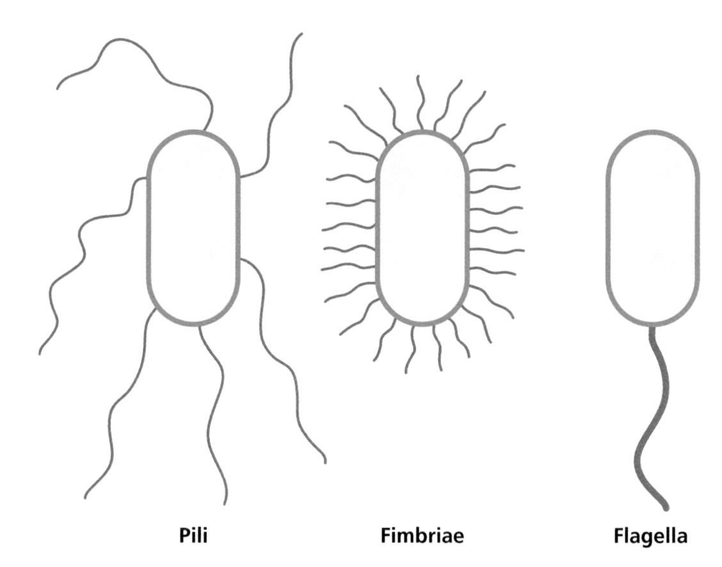

Figure 9.15 Bacterial cellular structures pili, fimbriae, and flagella. There are protein structures that protrude out of the cell for some bacterial species. They include the long thin pili (left) and the shorter, but more numerous fimbriae (middle), as well as the thicker flagella that confer motility to the bacterial cells (right).

Bacterial proteins are more than just the enzymes involved in the biosynthesis of non-protein cellular structures and cellular metabolism. Proteins also themselves form bacterial cellular structures, some of which can be visible microscopically, such as **pili**, **fimbriae**, and **flagella** (Figure 9.15). Pili are long thin structures that protrude from the surface of the bacterial cell, made up of protein subunits. These subunits are assembled together into the long pilus strand by pilus machinery, which is also made of protein. Pili can then be modified by enzymes, often changing the **antigenicity** of the pili, that is, its ability to be recognized by the immune system.

Fimbriae are similar to pili in that they are long thin protein structures extending from the bacterial cell. The difference between pili and fimbriae tends to be in the length and number that are on the cell; pili tend to be longer and there tend to be fewer of them, compared with fimbriae. Both structures enable the bacteria to attach to surfaces where they may allow invasion and penetration of host cells and where they may aid biofilm formation. Retraction of pili, through loss of protein subunits, causes **twitching motility**, enabling bacteria to move along surfaces. Some pili participate in bacterial conjugation, as discussed previously (Chapter 3).

Flagella also extend from the surface of bacteria and are also long structures made of proteins and assembled by protein machinery. Flagella are thicker than pili and fimbriae and tend to be fewer in number. The primary function of flagella is motility, achieved through rotation of the flagella. Each flagellum has three parts, a filament, motor complex, and hook, which give this cellular structure its characteristic look and function. The flagellar proteins and assembly machinery are all encoded by flagellar genes that are generally highly conserved in sequence and genetic organization.

Some bacterial proteins are enzymes that actively cause change

Many proteins within the bacterial cell have enzymatic activity. These proteins cause change within the cell, whether through action in metabolism, degradation of proteins and nucleic acids within the cell, phosphorylation and other modifications, or synthesis of non-protein components of the bacterial cell. When studying the proteins within a cell, the main roles for proteins are either an enzyme or a structural protein. Research involving knockout mutants and overexpression mutants has identified the functions of many of the enzymes within the bacterial cell, but not all. Metabolic maps and schematics of biosynthetic pathways can help reveal stages where enzymes are known to function, but for which the genes have not been identified. Homology searches, to identify genes similar to known enzymes in another species, may help fill in the gaps; however, some pathways and the enzymes involved are different between the species, limiting the identification of enzymes through homology alone.

Bacterial secretion systems move proteins across the bacterial membranes

Movement of proteins across the bacterial membranes and their secretion from the bacterial cell allows the bacteria to introduce proteins into their environment and surrounding cells. Secreted proteins can attach to eukaryotic cells, can scavenge resources from the environment, and can damage and disrupt surrounding cells with their toxic effects. Secreted proteins are made in the cytoplasm from the code in mRNA via the ribosome translational machinery. They are then secreted out of the cell via a **secretion system**.

There are different classes of secretion systems based on structural and functional differences; therefore, there are differences in genes encoding different secretion systems. Some secretion systems take secreted proteins across one membrane, two membranes, or even three membranes, when the proteins are secreted directly across a eukaryotic membrane and into the host cell. The Sec and Tat secretion systems were discussed in Chapter 8 as being involved in translocation of proteins across the bacterial cytoplasmic membrane. Most Sec and Tat transported proteins remain associated with the bacterial cell, either embedded in the membrane or present in the periplasm of Gram-negative species.

To secrete the protein out of the cell, another secretion system can either do all of the transport or take up where the Sec and Tat pathways have left off. Such secretion systems are prevalent in the Gram-negative bacteria, where secretion systems transport secreted proteins outside of the cell. There are currently nine recognized, distinct secretion systems that have been identified in Gram-negative bacteria.

The Type 1 Secretion System takes proteins across both membranes in one step

The **Type 1 Secretion System (T1SS)** of Gram-negative bacteria is made up of three components: **ABC transporter** in the inner membrane, a **membrane fusion protein** that crosses the inner membrane and bridges to the third component of the T1SS, and the outer membrane protein that makes a channel in the outer membrane through which the protein passes (**Figure 9.16**). The ABC transporter uses ATP as the energy source for transport of substrates across the membrane. The ABC transporter interacts with the membrane fusion protein component of the secretion system and recognizes the substrate for secretion. The membrane fusion protein aids the ABC transporter in recognition of the substrate to be secreted and bridges the periplasm to the outer membrane component of the secretion system. The outer membrane protein is a pore in the outer membrane, through which the secreted substrate passes upon leaving the bacterial cell. Often TolC is the outer membrane protein of a T1SS, being a protein that is used by several systems as a general outer membrane pore.

Vibrio cholerae uses a T1SS to secrete a toxin that contributes to the virulence of the species. Uropathogenic *E. coli* secrete hemolysin via a T1SS. The proteins that are transported by the T1SS pass across the membranes as an unfolded peptide. Folding then occurs following secretion.

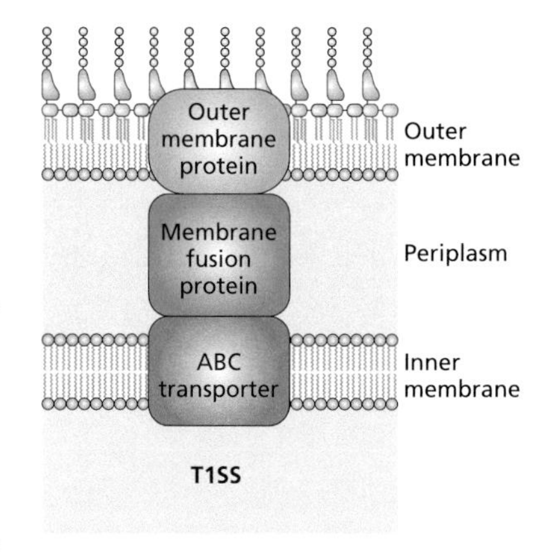

Figure 9.16 T1SS. A T1SS has three parts that cross the inner membrane, periplasmic space, and outer membrane. Embedded in the inner membrane is the ABC transporter (purple), which powers the secretion through ATP. The membrane fusion protein (blue) bridges the component in the inner membrane and outer membrane, spanning the periplasm. Finally, the outer membrane protein (light blue) forms a channel in the outer membrane through which the protein is secreted.

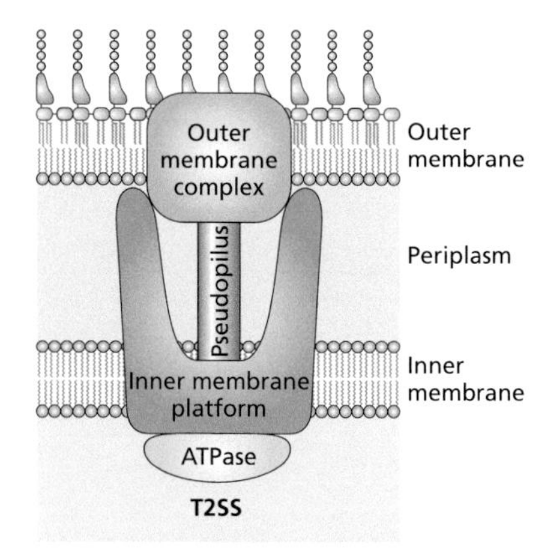

Figure 9.17 T2SS. A T2SS has four parts that cross the inner membrane, periplasmic space, and outer membrane. Embedded in the inner membrane is the inner membrane platform (purple) with the secretion ATPase (yellow), which powers the secretion through ATP. The outer membrane complex (green) forms a channel in the outer membrane through which the protein is secreted with the help of the pseudopilus (orange).

Type 2 Secretion Systems take a protein from the periplasm out of the cell

In Gram-negative bacteria, a protein can be translocated through the inner membrane by the Sec or Tat pathway. These proteins have an N-terminal signal sequence recognized by the Sec or Tat secretion systems. Following transport by Sec or Tat, the protein is then in the periplasm, where the **Type 2 Secretion System (T2SS)** can take it out of the cell. T2SSs act on folded proteins. Proteins transported by the Tat pathway are folded; therefore, they can be directly transported out of the cell by the T2SS. Unfolded proteins are transported by the Sec pathway; therefore, these must be folded in the periplasm before they are recognized by the T2SS for secretion.

The T2SS is made up of an **outer membrane complex**, an **inner membrane platform**, a secretion ATPase, and a **pseudopilus**. The outer membrane complex is a multimeric secretin that extends into the periplasm to contact the inner membrane platform. The inner membrane platform coordinates secretion through the T2SS following the arrival of a protein via the Sec or Tat pathways. The secretion ATPase is present in the cytoplasm of the bacterial cell and provides energy for transport from ATP within the cell. The T2SS pseudopilus is genetically related to pili, the long thin protein structures present in some species that extend from the surface of the bacterial cell. It is believed that the pseudopilus retracts and then pushes the folded protein through the T2SS outer membrane complex (**Figure 9.17**), in a manner similar to the retraction of pili in twitching motility.

T2SSs tend to secrete proteases, lipases, and phosphatases. *Vibrio cholerae* uses a T2SS to secrete cholera toxin, characteristic of this species. Another pathogen, *Pseudomonas aeruginosa*, secretes exotoxin A via a T2SS.

The Type 3 Secretion System can inject proteins like a syringe

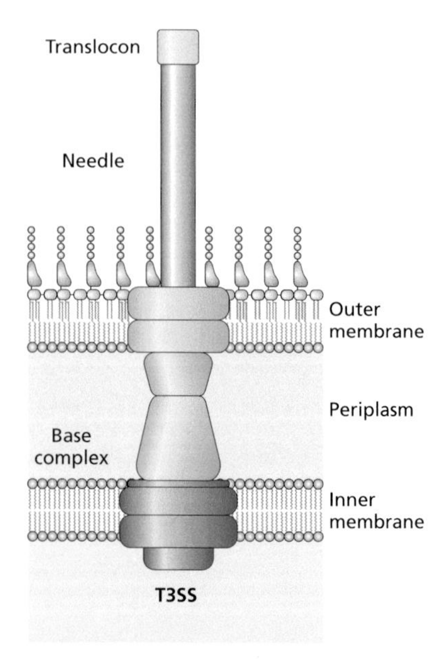

Figure 9.18 T3SS. A T3SS has three parts that cross the inner membrane, periplasmic space, outer membrane, and extend out of the cell. Embedded in the inner membrane, crossing the periplasm, and embedded in the outer membrane is the base complex, made up of several proteins that have homology to flagellar proteins. Extending out of the cell is the T3SS needle, which has at its end the translocon.

The **Type 3 Secretion System (T3SS)** is particularly prevalent in Gram-negative pathogens and symbionts. These bacteria have a needle and syringe structure that is similar to the bacterial flagella. The T3SS is able to secrete proteins across the inner membrane and the outer membrane. Some T3SSs transport proteins across both of the Gram-negative bacterial membranes and also across the eukaryotic membrane. In this way, T3SSs are able to inject proteins directly into host cells, delivering bacterial proteins directly to their targets. Proteins secreted by T3SSs are called **effectors** and have a secretion signal for type 3 secretion within their N-terminus.

The process of secretion via a T3SS is ATP dependent, as with the T1SSs and T2SSs. T3SSs transport unfolded proteins, which pass as peptide chains through the needle and syringe structure of the transport system. There is a core of nine proteins that are highly conserved in all T3SSs; eight of these are evolutionarily similar to proteins in flagella. There are an additional 10–12 other proteins that contribute to the T3SS.

Operons with T3SS genes are typically found in pathogenicity islands, that is, horizontally transferred DNA of the genome or plasmid. There are three parts to a T3SS: the **base complex**, the **needle**, and the **translocon** (**Figure 9.18**). The base complex, also called the **basal body**, has both cytoplasmic and outer membrane parts. The structure of the base complex includes several rings and a central rod, made up of about 15 proteins. The T3SS needle comes out of the basal body. The needle has a hollow core that is big enough for the passage of an unfolded protein through it. At the end of the needle is the translocon, which is able to sense contact with the host cell and regulates secretion of the effectors based on contact with the host cell. In species that transport effectors directly into host cells, the translocon assembles on contact with the host cell and forms a pore in the host cell membrane.

The Type 4 Secretion System includes conjugation systems and dna uptake systems

The **Type 4 Secretion System** (**T4SS**) includes the conjugation systems for transfer of DNA between bacterial cells; thus, there is homology between the T4SS and pili. While the conjugation system is involved in transfer of DNA involving cell-to-cell contact, the T4SSs can transport proteins, protein–protein complexes, and DNA–protein complexes either between cells or between the cell and the environment. T4SSs can secrete these substrates into other cells, be they bacterial or eukaryotic cells, and can secrete them out of the cell or be involved in uptake. Transport via T4SSs is through the inner and outer membrane and can also go through another membrane, like the T3SSs (**Figure 9.19**).

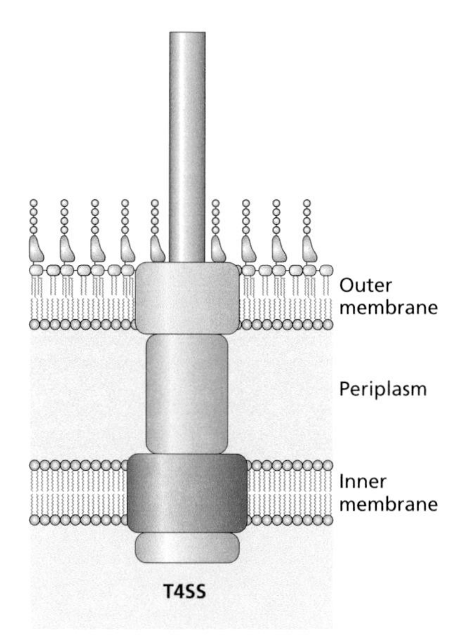

Figure 9.19 T4SS. A T4SS crosses the inner membrane, periplasmic space, and outer membrane. Several proteins make up the structure that spans and connects both membranes, providing a channel through which secretion substrates can pass. The T4SS proteins have homology to proteins of the pili and pilus assembly machinery, with those involved in conjugation making up the conjugation pilus.

Type 5 Secretion Systems are proteins that secrete themselves

In a **Type 5 Secretion System** (**T5SS**), there is no separate membrane channel or secretion apparatus. A T5SS protein secretes itself out of the cell. T5SS proteins have their own β-barrel domain that inserts into the outer membrane to form a pore through which the protein is secreted. A T5SS is only able to provide transport for a protein through the outer membrane. The T5SS proteins are brought into the periplasm as unfolded proteins secreted by the Sec pathway and, therefore, they have N-terminal Sec signal sequences.

There are three types of T5SS. The first involves proteins that are called **autotransporters** because they can transport themselves. The autotransporters are the most simplistic of the T5SSs. The transporter domain is in the C-terminal end of the protein and forms the outer membrane channel, made up of a 12-stranded β-barrel. It is possible that multiple copies of the autotransporter protein may come together to make the β-barrel pore. An example of a T5SS autotransporter is the immunoglobulin A (IgA) protease from *Neisseria gonorrhoeae*, which cleaves human IgA antibodies. The second type of T5SSs is the **two-partner secretion**, where a pair of proteins are involved. In two-partner secretion, one protein forms the β-barrel outer membrane channel and the other is the secreted protein. Lastly, **chaperone-usher secretion** involves three proteins. The usher component of chaperone-usher secretion forms the β-barrel in the outer membrane. The chaperone in the periplasm folds the Sec pathway transported secretion protein and delivers it to the usher. The third component of the chaperone-usher secretion T5SS is the secreted protein itself (**Figure 9.20**).

Type 6 Secretion Systems transport proteins into other cells, including other bacteria

A **Type 6 Secretion System** (**T6SS**) is also present exclusively in Gram-negative bacteria, although not in all species or all strains. Proteins can be transported from the bacterial cell into other cells, including other bacteria, fungi, and eukaryotic cells via the T6SS. The proteins that make up the T6SS are encoded within a gene cluster of up to 21 genes. These T6SS proteins are structurally homologous

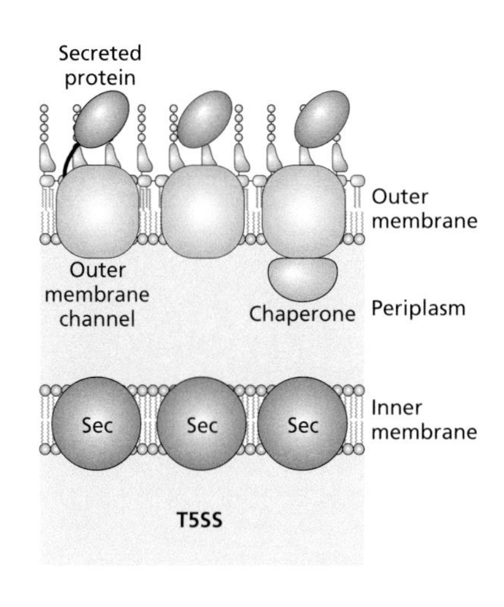

Figure 9.20 T5SS. A T5SS describes a protein that transports itself out of the periplasm. There are three types of T5SS, each relying on the Sec pathway (Sec) in the inner membrane to deliver the secreted protein to the periplasm. Autotransporters (left) form an outer membrane channel (green) in the outer membrane, through which the secreted portion (orange) of the autotransporter protein passes. In two-partner secretion (middle), two different proteins are the outer membrane channel and the secreted protein. Chaperone-ushered secretion (right) has a third component (green) that chaperones the folding of the secreted protein in the periplasm.

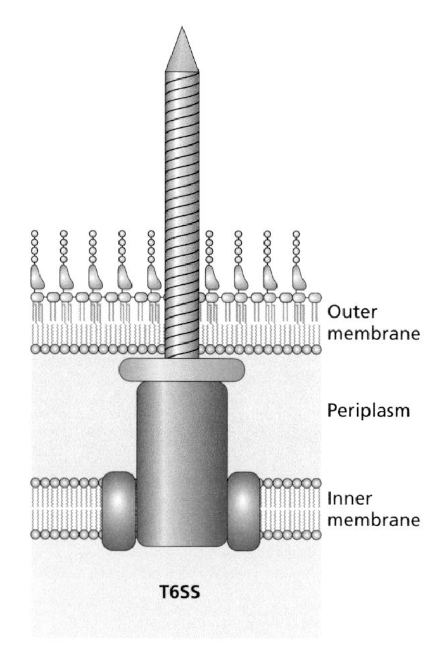

Figure 9.21 T6SS. A T6SS crosses the inner membrane, periplasmic space, and outer membrane. Several proteins that are homologous to bacteriophage make up the structure that spans and connects both membranes, providing a channel through which secretion substrates can pass.

to bacteriophage tails (**Figure 9.21**). The T6SSs contribute to interspecies competition, where one bacterium can inject another with proteins that will kill it.

The T6SS is a contractile nanomachine with a puncturing structure at the tip. The contractile mechanism of the T6SS is similar to the tail and spike of bacteriophage. The tip of the T6SS is made of the VgrG protein, which associates with various specialized effector proteins that are toxic to competitor cells found in the bacterial cell environment. In addition, there are cargo effector proteins that are not at the tip of the T6SS. Effectors include peptidoglycan hydrolases that break down bacterial cell walls, phospholipases that degrade phospholipids, DNases that break down DNA, and pore-forming toxins that compromise the integrity of competitor cell membranes. For each effector protein, the bacterial cell has an immunity protein to protect it from the antimicrobial effects of its own T6SS toxins. For example, the Ssp1 and Ssp2 antibacterial toxins are encoded in the T6SS gene cluster, as are their corresponding immunity proteins Rap1a and Rap2a.

Gram-positive Type 7 Secretion Systems can aid protein transport across the thick peptidoglycan layer

Gram-positive bacteria have the Sec and Tat pathway to translocate proteins across the membrane. To achieve more efficient protein secretion, many Gram-positives have two *secA* genes encoding the Sec pathway protein. The second of these genes encodes a protein that participates in the transport of only select proteins. Despite these systems, it is often not enough for a protein to escape the cell. In part this is due to the thick peptidoglycan layer, although proteins can eventually be secreted out by passive diffusion.

Type 7 Secretion Systems (**T7SS**) have been described in acid fast bacteria and Gram-positive bacteria. These systems are able to transport proteins through both the membrane and the peptidoglycan layer, reducing the reliance on passive diffusion to secrete the protein out of the cell (**Figure 9.22**). The T7SSs were first identified in *Mycobacteria tuberculosis* and *Staphylococcus aureus*, where the genes encoding the T7SS are present in clusters.

T7SSs are important for host–microbe and microbe–microbe interactions, such as those between *M. tuberculosis* and other Gram-positive bacteria. Although features and components are similar, the T7SSs form into two subtypes, T7SSa in acid fast bacteria and T7SSb in Gram-negative bacteria. T7SSa is involved in host–microbe interactions, iron acquisition, nutrient uptake, contact-dependent signalling, intracellular replication, immune suppression, and other processes.

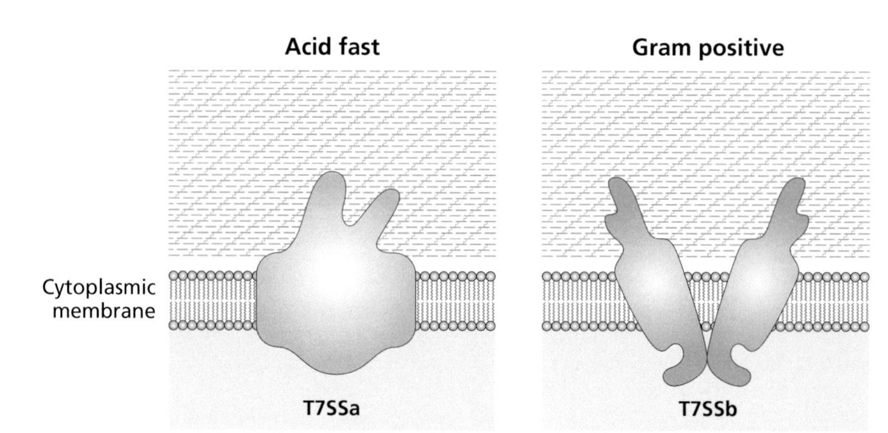

Figure 9.22 T7SS. A T7SS crosses the Gram-positive membrane and the thick peptidoglycan layer, helping to efficiently secrete proteins out of the cell. There are two types of T7SSs: T7SSa found in acid fast bacteria (left) and T7SSb found in Gram-negative species (right).

T7SSb targets both eukaryotic and prokaryotic cells and is involved in virulence and bacterial competition. There are theories that the T7SS may eventually be found in some Gram-negative bacteria as well as in *Mycobacteria* and Gram-positives; it is possible that the Gram-negative version of the T7SS has not yet been discovered.

Type 8 Secretion Systems are responsible for Curli fiber formation

The Gram-negative **Type 8 Secretion System** (**T8SS**) is responsible for **Curli fiber** biogenesis and bacterial cell aggregation, and also contributes to biofilm formation. The T8SS is also known as the nucleation secretion-assembly pathway and works together with the Sec general secretory pathway.

The extracellular Curli fibers are found in Proteobacteria and Bacteriodetes. These coiled fibers are 2 to 5 nm thick. The fibrous Curli surface structures were identified in *E. coli* in 1989. It is believed that Curli fibers are involved in pathogenesis.

Curli fiber biogenesis involves two operons, as found in *E. coli*. The *csgBAC* operon encodes the two structural Curli proteins, CsgA and CsgB, and a chapterone-like CsgC. The second operon, *csgDEFG*, is involved in secretion and assembly of the Curli fibers and the transcriptional regulation of these genes. To form the Curli surface structures, transport of the Curli components uses first the SecYEG system to pass through the Gram-negative inner membrane. Next, components cross to the cell surface via the CsgGEF multiprotein outer membrane transporter (**Figure 9.23**), forming the T8SS. CsgG forms the pore through the outer membrane. CsgE forms a cap on the periplasmic side of the CsgG pore and is responsible for the specificity of secretion of CsgA and CsgB through CsgG, crossing the outer membrane to reach the bacterial cell surface. The CsgF protein is also part of the T8SS multiprotein structure, but it is not required for secretion itself. Instead, CsgF is essential for formation of Curli fibers by the secreted CsgA and CsgB proteins.

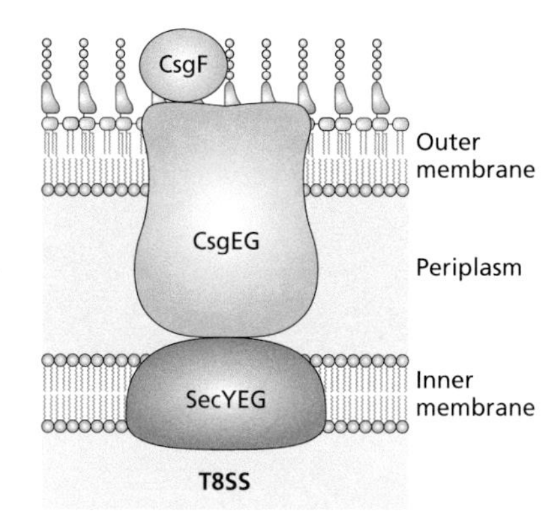

Figure 9.23 The T8SS transports proteins that make coiled surface structures called Curli. To cross the inner membrane, Curli components CsgA and CsgB are transported by the Sec general secretory pathway. This leaves the CsgA and CsgB proteins needing to cross the Gram-negative peptidoglycan layer and the outer membrane, which is achieved through CsgGEF.

Type 9 Secretion Systems have two different roles that depend on bacterial lifestyle

Type 9 Secretion Systems (**T9SSs**) have been identified in some Bacteroidetes species. They are involved in gliding mobility in environmental bacteria. Alternatively, in pathogens the T9SS is involved in pathogenesis.

In environmental bacteria *Flavobacterium johnsoniae* it was discovered that the core T9SS proteins are those responsible for gliding motility. In addition, the secretion of chitinase by *F. johnsoniae* uses the T9SS.

Porphyromonas gingivalis, the cause of periodontitis, uses the T9SS to move proteins across the Gram-negative outer membrane. *P. gingivalis* secretes virulence proteins gingipains, which combat the innate immune system and damage host tissues. However, there are no genes present in this species for T1SS, T2SS, T3SS, T4SS, T5SS, T6SS, or T8SS. Therefore, researchers investigated what is present and discovered the T9SS in *P. gingivalis*.

Proteins transported by T9SSs have a conserved domain at the C-terminus that is cleaved during secretion. T9SS effectors are either secreted into the environment or attached to the LPS.

The function of the T9SS relies on specific genes that encode the component proteins. Cytoplasmic and membrane proteins are encoded by *porLMXY*, those in the periplasm by *porDKNW*, and the outer membrane and surface by *porPQRTUVZ*. PorXY are a two-component regulatory system (**Figure 9.24**). PorLM together interact with PorXY and span the inner membrane, connecting cytoplasm and periplasm for the purposes of transferring the two-component signal to the rest of the T9SS complex. Interestingly, PorLMXY is not involved in the translocation of T9SS effector proteins; this is achieved by the Sec pathway (see Figure 9.24). PorDKNW are periplasmic proteins. The protein to be secreted

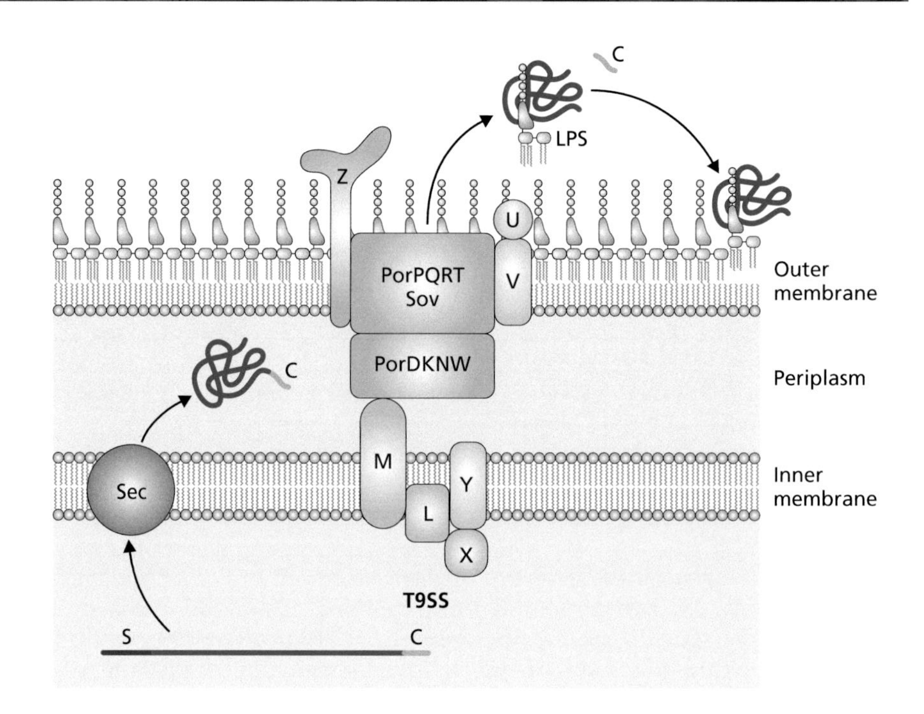

Figure 9.24 T9SS proteins. The PorXY two-component regulatory system signals to the rest of the T9SS through the membrane-spanning PorLM. The effector protein has two conserved sequences: a Sec signal peptide that is recognized by the Sec pathway and lost on passage through the inner membrane (S) and a characteristic C-terminal domain that is recognized by the T9SS and cleaved (C). The T9SS proteins PorKLMNQRTUVWXYZ are indicated via letter.

has within it a conserved sequence of amino acids that is recognized by the T9SS proteins. Upon transport, this C-terminal domain is cleaved. A multiprotein complex is believed to form from PorLM and PorKN. The remaining components are part of an outer membrane surface protein complex (PorPQRTUVZ and Sov; see Figure 9.24).

The Type 10 Secretion System is similar to bacteriophage lysis proteins

The secretion system designated as the **Type 10 Secretion System (T10SS)** has a member protein, holin, and a protein that edits the peptidoglycan layer, a peptidoglycan hydrolase, and is used for transport across the Gram-negative membranes and periplasmic space. Together with these two proteins are two other proteins that form a spanin, a protein system that spans from the inner membrane to the outer membrane.

Lambda bacteriophage lyze bacterial host cells using a multiprotein system. With these bacteria-specific viruses, one of the proteins involved in the process of bacterial cell lysis is a holin, responsible for movement across the inner membrane of another lysis-mediating protein, a peptidoglycan endopeptidase. The disruption of the peptidoglycan layer lyzes the bacterial cell, releasing the mature bacteriophage lambda particles. Also included in this process are two proteins that are spanins, which connect the inner and outer membrane of the bacteria.

The T10SS (**Figure 9.25**) was identified following analysis of chitinase secretion in *Serratia marcescens*. The *S. marcescens* genome sequences of the strains being studied did not contain T2SS genes, which are the recognized mechanism for secretion of chitinase in other bacteria. Instead, a cluster of previously unexplored genes was identified, *chiWXYZ*, which bears sequence similarity to bacteriophage lambda lysis genes.

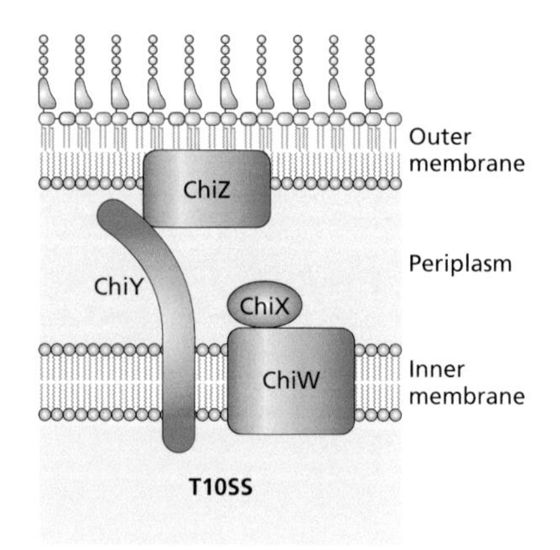

Figure 9.25 The T10SS. The T10SS is similar to protein complexes created by bacteriophage lambda for the purposes of lysis of the bacterial cell and release of the lambda bacteriophages that have replicated inside it. The T10SS, however, is involved in protein secretion from the bacterial cell. It involves four proteins, which work together in pairs. The first pair, ChiWX, are a holin and peptidoglycan cleavage enzyme. The second pair, ChiYZ, form a spanin that bridges the inner membrane and outer membrane of Gram-negative bacteria. Together the four proteins pass proteins for secretion across the inner membrane, through the peptidoglycan layer, and across the outer membrane to the exterior of the bacterial cell.

In *S. marcescens*, ChiW, the holin, is in the inner membrane where it generates a hole or passage for other proteins from the cytoplasm to the periplasm. ChiX is an L-Ala-L-Glu endopeptidase, an enzyme that cleaves the structure of the peptidoglycan layer. ChiY is also in the inner membrane and interacts with ChiZ, which has a signal peptide that is characteristic of spanin lipoproteins. Together, ChiYZ span both membranes as a spanin complex. The *chiWXYZ* genes are homologous to the genes involved in lambda bacteriophage-mediated bacterial cell lysis. Homologues of these have been identified in species other than *Serratia*. The specific activity of the peptidoglycan modifying enzyme appears to be flexible, yet the outcome is the same, resulting in cleavage of the peptidoglycan to permit protein translocation through the periplasm.

Efflux pump systems transport harmful substances out of the bacterial cell

Efflux pump systems are multiprotein systems that remove substances from the periplasm and/or cytoplasm that are damaging to the bacterial cell. Efflux pumps can play an important role in antibiotic resistance through actively removing antibiotics from the bacterial cell. The transport involved in efflux pump systems is powered by ATP or by the **proton motive force** (**PMF**). PMF results from cellular metabolism. Protons that are not used in the bacterial cell are exported to the surface and interact with either the LPS or the cell wall. These protons on the surface of the cell form an electrochemical gradient across the membranes that can be harnessed as an energy source for efflux pump transport.

There are five families of efflux pumps. The first are the **RND family** of efflux pumps. RND stands for resistance nodulation division, representing the cellular processes in which the RND proteins were believed to be involved. RND efflux pumps are only present in Gram-negative bacterial species. The classic RND efflux pumps in *E. coli* are AcrAB-TolC and AcrEF-TolC. RND family efflux pump systems are made up of three proteins: AcrB, which is in the inner membrane; AcrA, which is a membrane fusion protein that aids movement through AcrB; and TolC, which is the outer membrane channel that connects with AcrB (**Figure 9.26**). The protein TolC is a general outer membrane pore protein that is involved with other protein complexes, such as the T1SS, as discussed earlier.

The second family of efflux pump systems is the **MFS family**. MFS stands for major facilitator superfamily. This type of efflux pump is present in both Gram-negative and Gram-positive bacterial species. The efflux pump EmrD in *E. coli* is a member of the MFS family. It has 12 transmembrane helices in the inner membrane that form a channel for transport of hydrophobic substrates out of the bacterial cell using the proton motive force (**Figure 9.27**).

The third efflux pump family is the **ABC family**. In this case ABC stands for ATP-binding cassette. In this efflux pump system, there is a protein that binds to and uses ATP as the energy source for transport of substrates out of the cell (**Figure 9.28**). The ABC family efflux pumps are present in both Gram-negative and Gram-positive bacterial species.

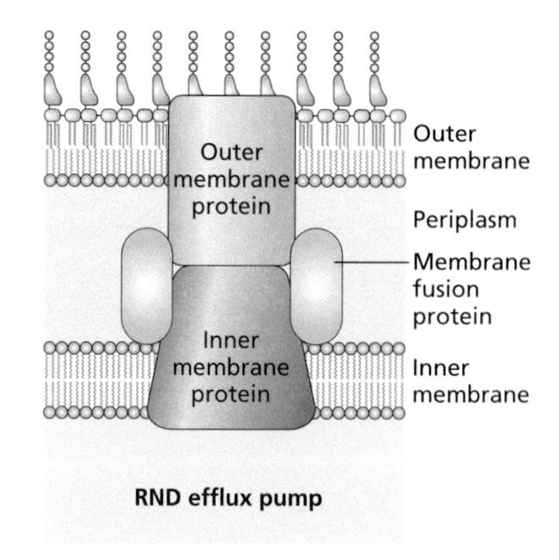

Figure 9.26 RND efflux pump. The RND efflux pump of Gram-negative bacteria crosses the inner membrane, periplasmic space, and outer membrane. It is made up of an inner membrane protein, a membrane fusion protein, and an outer membrane protein. These proteins work together using the PMF to remove substances from the cytoplasm and periplasm of the bacterial cell.

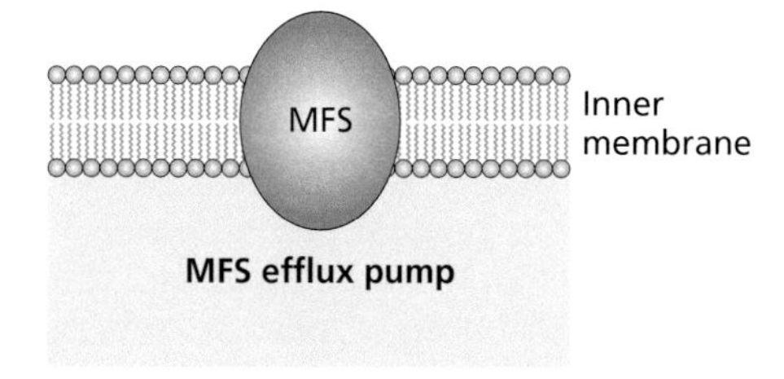

Figure 9.27 MFS efflux pump. The MFS efflux pump of Gram-negative bacteria crosses the inner membrane or the cellular membrane in Gram-positive bacteria.

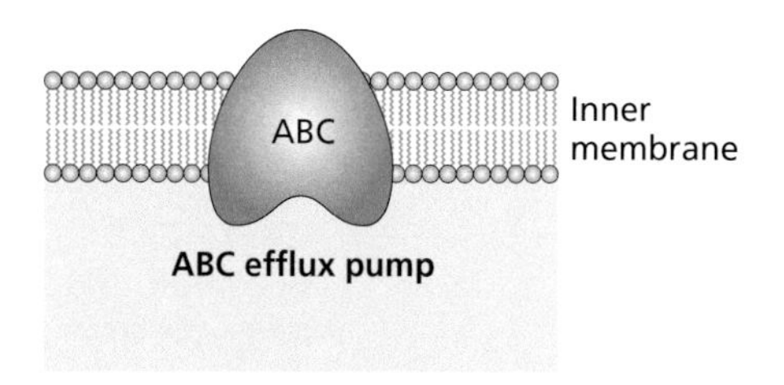

Figure 9.28 ABC efflux pump. The ABC efflux pump of Gram-negative bacteria crosses the inner membrane or the cellular membrane in Gram-positive bacteria.

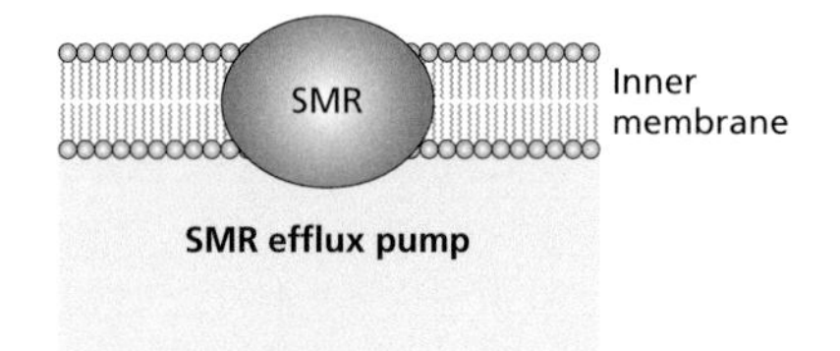

Figure 9.29 SMR efflux pump. The SMR efflux pump of Gram-negative bacteria crosses the inner membrane or the cellular membrane in Gram-positive bacteria.

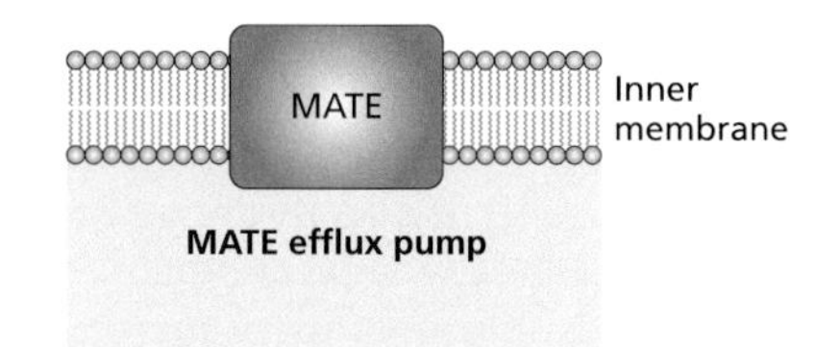

Figure 9.30 MATE efflux pump. The MATE efflux pump of Gram-negative bacteria crosses the inner membrane or the cellular membrane in Gram-positive bacteria.

The fourth family of efflux pump systems is the **SMR family**, which uses the PMF. These are classified as small multidrug resistance, SMR, and are present in both Gram-negative and Gram-positive bacterial species. The efflux pump system EmrE in *E. coli* is part of the SMR family. This efflux pump system includes eight EmrE proteins, each with four transmembrane helices (**Figure 9.29**).

The fifth and final efflux pump system family is the **MATE family**, where MATE stands for multidrug and toxic compound extrusion (**Figure 9.30**). MATE family efflux pumps are found in both Gram-negative and Gram-positive bacterial species and make use of sodium influx to drive efflux of harmful substances.

All of the expressed proteins are the proteomes

The bacterial genome is relatively stable. Every protein that can be made is there, encoded in the genes, but the genome does not represent which proteins are expressed and when. The bacterial transcriptome indicates the potential of protein production more than the genome. The mRNA within the bacterial cell gives an indication of protein expression. Yet, only the proteome shows the proteins that are currently present and presumably active in the bacterial cell at the time of assessment.

Proteomics analyzes all of the proteins. Like a transcriptome, the proteome changes over time and due to changes in cellular conditions. When conditions shift, the change in the proteome may be either more or less rapid than the changes in the transcriptome. This is due to the relative half-life and enzymatic degradation of the proteins and the mRNA molecules in the bacterial cell.

Through proteomics, it is possible to identify the protein products of genes annotated in the genomes as hypothetical. Comparative proteomics, where the proteins expressed in one condition are compared to the proteins expressed in another, can determine differences in the patterns of protein expression in various conditions.

Whereas sequence technologies made genomics and transcriptomics possible, **mass spectrometry** made proteomics possible. Previously, the study of proteins had been accomplished as standard using **2D gels**. The resolution of proteins in 2D gels works through the separation of proteins by the **isoelectric point** in the first dimension and then the protein **molecular weight** in the second dimension. Bacterial cellular proteins are resolved on 2D gels as spots that can be extracted from the gel and identified by mass spectrometry. Although the combination of 2D gels and mass spectrometry of spots is powerful and has revealed a great deal about the proteome, the resolution of 2D gels makes it difficult to identify low abundance proteins and those with high or low molecular weights. In addition, each individual spot of protein must be selected and extracted from the gel for mass spectrometry analysis, which is a labor-intensive process prone to introduction of contaminants. If there are few changes between two conditions and therefore relatively few spots to be extracted and prepared for mass spectrometry, this can be achieved. However, as the number of changes in protein expression increases, so does the difficulty in using this method for proteomic investigations.

Mass spectrometry technology enables the study of proteomes

Gel-free mass spectrometry technology, aka shotgun proteomics, has expanded the ability to study proteomes in much the same way as next-generation sequencing has expanded the scope of genomics and transcriptomics. By removing the reliance on 2D gels, the method was simplified and enabled the identification of proteins not previously resolved in the gels. Bacterial cells are complex mixtures of proteins; therefore, mass spectrometry technology needed to advance to be able to decipher this complexity.

The combination of mass spectrometry with **high-performance liquid chromatography** (**HPLC**) can identify and quantify entire proteomes. Using this method, bacteria are grown and lyzed to release the proteins. The extracted proteins are then digested to produce peptides. Fractionation of the peptides via HPLC will divide up the mixture of peptides to reduce the complexity of the peptide mixture. It is then possible to enrich the peptide pool for features such as post-translational modifications, if required, by capturing proteins carrying specific modifications. The peptides are then separated on HPLC and measured on high-resolution mass spectrometry. When the **mass spectra** are analyzed, the proteins will be identified (**Figure 9.31**).

The availability of genome sequences is particularly useful for mass spectrometry analysis because the peptides that are identified can be mapped against the genomic data to reveal the protein that generated the peptide fragment. Proteomics can identify proteins that can be important in understanding bacterial biological processes, which is particularly useful when new processes are explored, or new proteins are identified in genomics. The technology can be quantitative, enabling researchers to identify the abundance of proteins at different conditions and at different times, where transcriptomic quantitation is a less reliable measure of protein quantity. Despite advances in technology, only a subset of proteins is identified by proteomics. As with genomics and transcriptomics, it is the analysis and interpretation of the vast amount of data that is the challenge.

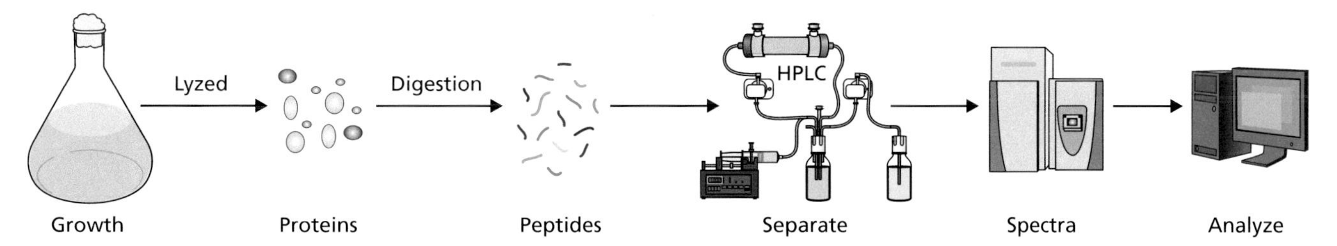

Figure 9.31 Proteomics workflow. A bacterial culture is lyzed, releasing the bacterial proteins. The proteins are digested, producing peptides, which are separated by HPLC and their mass spectra determined by mass spectrometry. The resulting data are then analyzed to identify proteins of interest.

Proteomics aids in identification of the core genome

The core genome is a set of genes that are in common between strains and species. The expression of the core genome can be seen in its encoded proteins. It is possible to use proteomics to identify a core set of bacterial genes that are essential for life for a diverse set of species. Using proteomics, this core set of bacterial genes identified in the genome is backed by evidence that those genes result in expressed proteins.

Liquid chromatography tandem mass spectrometry (**LC/MS/MS**) is a commonly used proteomics tool that separates peptides first using liquid chromatography and then identifies them using a tandem mass spectrometry approach. Despite its usefulness as an advance in proteomics, LC/MS/MS has limitations, especially for low expressed proteins. Therefore, it is likely that new methodologies will be developed.

Using LC/MS/MS it has been possible to determine if the predicted core genome is expressed and therefore likely to be a true representation of the core set of genes. Proteomics has found that 74% of the core genes were found to be expressed as proteins. Of these expressed core genes, it was found that 55% are involved in protein synthesis. This result mirrors transcriptome data where there is an overwhelming amount of rRNA, corresponding to the proteomic data being dominated by ribosomal proteins.

Just as genomics cannot determine if a protein is functional when expressed, proteomics cannot determine if all of the proteins that are detected are functional. Function of a protein depends on the protein conformation, location, and in some cases modification, presence of cofactors, substrates, and inhibitors.

Mass spectrometry is being used diagnostically to identify bacteria

MALDI-TOF-MS-based mass pattern and fingerprinting (**matrix-assisted laser desorption ionization-time of flight mass spectrometry**) is being used for the typing of bacteria, for the identification of bacterial toxins, and for the identification of antimicrobial resistance. This technology investigates the use of mass fingerprinting, creating a spectrum that is characteristic of particular bacterial species, as opposed to LC/MS/MS peptide sequencing, where the goal is to identify the peptides as corresponding to specific proteins.

The identification of bacteria using MALDI-TOF-MS is rapid, taking a fraction of the time required to identify bacteria using traditional laboratory techniques that take time and labor to complete. The speed of identification of bacteria can be important in an outbreak and hospital situations. To determine the identity of the bacteria, the bacterial culture is treated with a strong solvent and the extracted molecules are placed onto the MALDI-TOF-MS system. A computer takes the signal from the mass spectrometry and compares it against a database of patterns or fingerprints that have been produced previously, representing a range of bacterial species (**Figure 9.32**). This removes human error from the identification process. The process can be conducted on bacterial culture and can even be done directly upon blood culture. The results are reproducible and reliable. Use of MALDI-TOF-MS reduces human error that can be introduced into bacterial identification when making judgment calls about culture characteristics and results.

As with next-generation sequencing, there is a high initial cost for the MALDI-TOF-MS instrument. In addition, a robust database of MALDI-TOF-MS patterns is needed for accurate identification. Due to variations and similarities in patterns, it can often be difficult to identify close relatives; therefore, MALDI-TOF-MS remains a good tool for the identification of bacteria at the genus level, but not at the species or strain level.

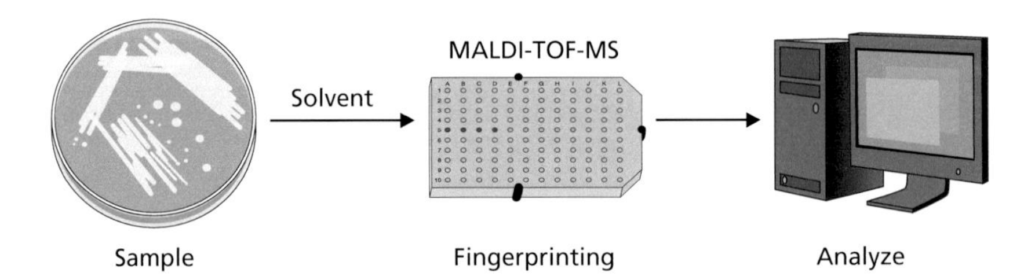

Figure 9.32 MALDI-TOF-MS identification of bacteria. A bacterial sample is subjected to a strong solvent. The material is then put onto the MALDI-TOF-MS system. This system will produce a mass pattern that is like a fingerprint, which can be analyzed to identify the bacteria.

Proteomics can be used to investigate antibiotic resistance

Using proteomics technologies, it is possible to investigate the differences between the proteomes of antibiotic-resistant and antibiotic-susceptible strains. First, the bacterial cultures are grown in the presence of sublethal antibiotics, where the antibiotic is present at a concentration just below what would be needed to kill the bacteria. Therefore, both the antibiotic-resistant and antibiotic-susceptible strains will grow.

Usually, the proteomics results are specific to antibiotic resistance, but also seen are proteins that are important for energy and metabolism, protein and nucleic acid synthesis, and stress responses. These proteins that are not directly involved in antibiotic resistance are presumed to be identified due to the presence of the antibiotic within the bacterial culture and the resulting stresses upon the cell.

The β-lactam antibiotics interfere with the synthesis and stability of peptidoglycan. The main mechanism of resistance to these antibiotics is through the expression of the β-lactamase enzyme, a protein that can be detected via proteomics. Other mechanisms of β-lactam resistance include the expression of efflux pump systems, the production of porins, and changes to the penicillin binding proteins involved in peptidoglycan biosynthesis that are the targets of penicillin. These protein components of resistance may be identified in proteomics data. Proteomics has identified the expression of porins involved in ampicillin resistance in *P. aeruginosa*. Such investigations have made it clear that in several species there are multiple overlapping resistance mechanisms that may all be at work within the same strain.

MALDI-TOF-MS is less expensive and more rapid than plate culture-based resistance determination, which takes several additional days and additional labor and human decision-making. MALDI-TOF-MS produces results based on expressed proteins unlike molecular methods relying on DNA or RNA, which can only identify the presence of the gene or transcript, not the protein that is ultimately the resistance determinant. The advantage in identifying antibiotic resistance using proteomics is that it is known that the protein is expressed. The bacterial cell both has the gene conferring resistance and is using it. Unfortunately, proteomics alone, like molecular tests for DNA or RNA, does not give minimum inhibitory concentration data. The minimum inhibitory concentration of the antibiotic to kill the particular bacterial strain contributes to epidemiology data about changes in resistance levels of bacterial species over time. In addition, these data are important for monitoring of resistance locally and globally and for making recommendations about prescriptions for effective management of disease with antibiotics.

Key points

- Some parts of a bacterial cell are not made of protein, such as phospholipids, LPS, peptidoglycan, and the bacterial capsule.
- The biosynthesis of non-protein components of the bacterial cell relies on proteins.
- Pili, fimbriae, and flagella are bacterial structures made of proteins that extend from the surface of the cell.
- Nine types of secretion systems have been identified in Gram-negative bacteria and one type has been identified in Gram-positive bacteria, which facilitate the passage of proteins out of the cell.
- Some secretion systems have evolved from other bacterial systems, such as pili, flagella, and bacteriophage.
- There are five families of efflux pump systems that actively remove substances from bacterial cells, including antibiotics.

- Proteomics is the study of the proteins within a bacterial cell, revealing all of the proteins encoded by the DNA that have been transcribed and translated into protein.
- Improvements in mass spectrometry have led to improvements in the results that can be generated in proteomic investigations.
- Proteomic technology can be used to identify bacteria.

Key terms

Define the following terms introduced in this chapter. Check your answers using the definitions in the Glossary. These terms are also available as Flashcards online.

2D gel	MALDI-TOF-MS	T5SS
Autotransporter	Mass spectrometry	T6SS
Capsule	Pili	T7SS
Curli fiber	Proteomics	T8SS
Efflux pump system	Proton motive force	T9SS
Flagella	T1SS	T10SS
HPLC	T2SS	Twitching motility
LC/MS/MS	T3SS	
Lipopolysaccharide	T4SS	

Questions and discussion topics

Self-study questions

Answer each question using 50–100 words or a table or labeled diagram. Advice on where to find answers to these questions is available online.

1. In what ways does the proteome relate to the genome and the transcriptome?
2. What are some of the non-protein components of bacterial cells? How are these biosynthesized within the bacterial cell?
3. What kind of bacterial cells have LPS and what are the three parts of the LPS molecule? Briefly describe the ways LPS can vary between strains of the same species.
4. What are the components that make up peptidoglycan for most bacterial species? Which of these is specific to bacteria, not being found elsewhere in nature?
5. In what circumstances might temperature-sensitive mutants or overexpression mutants need to be used to study the function of a gene, rather than a knockout mutant?
6. What are the features of phospholipids that allow them to form biological membranes? What joins phospholipids in making the outer membrane of Gram-negative bacteria?
7. Briefly, how do capsules help bacterial cells survive?
8. Using a drawing if you wish, differentiate between pili, fimbriae, and flagella.
9. What is the main difference between a secretion system and an efflux pump system, since both types of system are involved in transport out of the cell?
10. Of the 10 secretion systems, all except the T7SS involve both the outer and inner membranes. Why is the T7SS the exception and how else does it differ from the others?
11. How do efflux pump systems contribute to antibiotic resistance? Is this trait acquired or an innate feature of species that express efflux pumps?
12. Contrast the difference in resolution of proteins on a 1D gel that resolves only by molecular weight and a 2D gel that resolves by isoelectric point and then by molecular weight.

Discussion topics

These topics are presented for discussion in study groups, as part of class discussions, or on your own. These questions go beyond what is directly covered in this part of the book. Use the research literature and other reading to explore these topics in more depth. Tips to help prepare for topic discussions are available online.

1 Create a table listing in each row the secretion systems thus far identified in bacteria. Make columns across the components within the system, the location of the system, whether it is ATP-dependent or not, and what the secretion system transports. Fill in the table with information on example secretion systems that have been well researched.

2 Create a table listing the different types of efflux pump systems in different rows. In each column, specify the membranes that the components span, whether the efflux pump system is ATP-dependent, and the substrates that the efflux pump transports. Include for each examples of well researched efflux pump systems.

3 Investigate the ways in which MALDI-TOF-MS is being used diagnostically and in investigation of antibiotic resistance in bacterial populations. Discuss its use with regard to a particular bacterial species, by a certain hospital, or in addressing a specific health concern.

Online quiz questions

To further self-assess your understanding of the chapter material, please visit the following link, where you can participate in a range of interactive quiz questions:

www.routledge.com/cw/snyder

Further reading

Some cellular structural components are not directly encoded by genes

Rowlett VW, Mallampalli VKPS, Karlstaedt A, Dowhan W, Taegtmeyer H, Margolin W, Vitvac H. Impact of membrane phospholipid alteration in *Escherichia coli* on cell function and bacterial stress adaptation. *J Bacteriol*. 2017; *199*(*13*): e00849-e00915.

Wang X, Quinn PJ. Lipopolysaccharide biosynthetic pathway and structure modification. *Prog Lipid Res*. 2010; *49*: 97–207.

Bacterial secretion systems move proteins across the bacterial membranes

English G, Trunk K, Rao VA, Srikannathasan V, Hunter WN, Coulthurst SJ. New secreted toxins and immunity proteins encoded within the Type VI secretion system gene cluster of *Serratia marcescens*. *Mol Microbiol*. 2012; *86*(*4*): 921–936.

Green ER, Mecsas J. Bacterial secretion systems—an overview. *Microbiol Spectr*. 2016; *4*(*1*): 10.

Bhoite S, van Gerven N, Chapman MR, Remaut H. Curli biogenesis: bacterial amyloid assembly by the Type VIII Secretion Pathway. *Eco Sal Plus*. 2019; *8*(*2*): 10.1128.

Lasica AM, Ksiazek M, Madej M, Potempa J. The type IX secretion system (T9SS): highlights and recent insights into its structure and function. *Front Cell Infect Microbiol*. 2017; *7*: 215.

All of the expressed proteins are the proteome

Callister SJ, McCue LA, Turse JE, Monroe ME, Auberry KJ, Smith RD, Adkins JN, Lipton MS. Comparative bacterial proteomics: Analysis of the core genome concept. *PLOS ONE*. 2008; *3*(*2*): e1542.

Proteomics aids in identification of the core genome

Pérez-Llarena FJ, Bou G. Proteomics as a tool for studying bacterial virulence and antimicrobial resistance. *Front Microbiol*. 2016; *7*: 410.

Part IV

Genetics, Genomics, and Bioinformatics

Humans have been doing genetic experiments for thousands of years through selective breeding of plants and animals to select for traits we desire in crops, flowers, pets, animals, and even each other. Although humans did not know the molecular basis for inheritance, it was known that traits and features passed from generation to generation. In the modern era, bacteria have been at the heart of genetics and genomics research. Genomics brings together genetics, molecular biology, biochemistry, statistics, computer science, and disciplines such as engineering and biotechnology in developing technology to study genomes. In this part, the field of bacterial genetics will be further explored and expanded upon. Genetics and genomics have made important contributions to our understanding of the bacterial world, which will be explored in this part, as will the subject of **bioinformatics**, which uses computer-based tools to analyze genomic sequence data. Information in this section will build upon previous topics, adding depth of understanding of genetics and genomics and introducing bioinformatics.

DOI: 10.1201/9781003380436-13

Chapter

Genetics 10

This chapter will explain the field of bacterial genetics, starting from the terms and concepts particular to the discipline. Genetics is the foundation from which larger investigations, such as those in genomics, can be explored. Also presented are the origins of our understanding of genetics. Some of the key important insights gained by humans into genetics are highlighted here, demonstrating how we have arrived at our current state of understanding in the field. This chapter will begin to explore the genetic tools and techniques that have been used in the laboratory in the past and are still used today to unravel the information contained within DNA. Details on how these technologies work will be addressed in Chapter 18.

Terms and conventions in the field of bacterial genetics are straightforward

By now, terms like gene, genetics, genotype, phenotype, genome, and chromosome should be quite familiar. Unlike disciplines such as immunology, bacterial genetics and genomics do not have a lot of abbreviations to try to remember. DNA and RNA are relatively commonplace abbreviations, which often do not need definition. The abbreviation CDS has been used in previous chapters and refers specifically to the coding sequence from the start codon to the stop codon within an open reading frame, or ORF (see Chapter 2). A gene is a CDS, but not all CDSs are genes. This is because a sequence with a start codon and a stop codon that has been identified in sequence data is only speculated to be encoding a protein, based on the sequence evidence. To be a gene, that CDS has to have been shown to actually generate a protein via transcription and translation. CDS is therefore a designation assigned from computational analysis of genetic data, whereas genes are defined by laboratory experimentation. This nuance is often overlooked, with sequences being called genes without evidence to support the designation, and CDSs referred to as ORFs when an ORF is actually defined as a region between two stop codons (see Chapter 2).

In bacterial genetics, the name of the gene matches the name of the protein. The gene *dnaA*, for example, encodes the protein DnaA (**Figure 10.1**). For the most part, bacterial genes are named with a three-letter code that may correspond to a function or assumed function of the gene. This is followed by a capitalized letter. Different letters can help differentiate genes that are all involved in the same process. Gene names are always in italics and always start with a lower-case letter. Protein names, however, are not italicized and the first letter is capitalized. In this way, the reader can differentiate between the gene and the protein. Care should be taken when writing to ensure which form is used and whether the topic actually refers to the gene or its encoded protein.

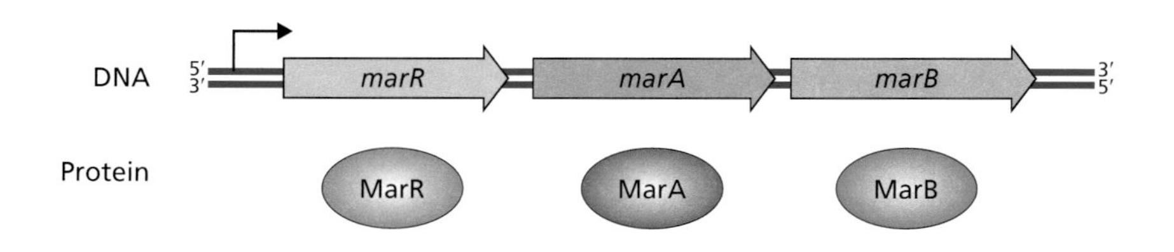

Figure 10.1 Nomenclature for genes and proteins. The DNA includes many genes, which are identified with names that reflect their function, such as these three genes involved in multiple antibiotic resistance: *marR*, *marA*, and *marB*. Note the differences in how the gene names are written (*marR*, *marA*, and *marB*) versus how the protein names are written (MarR, MarA, and MarB). In this way, it is possible to recognize whether a gene or protein is being referred to in writing. These names are based on the functions of the genes, so here *mar* comes from *m*ultiple *a*ntibiotic *r*esistance.

DOI: 10.1201/9781003380436-14

Humans understood about traits and inheritance long before the term genetics

Evidenced in the food in our markets and the pets with whom we share our homes, humans have had an understanding of the inheritance of traits for several thousand years. Modern dog breeds are very different from their ancestral wild relatives. Over time, through breeding practices in which favorable traits were selected, humans were able to alter the genetic make-up of these animals. The same is true of the crops that are grown for our food, with humans having selectively bred and planted those that were high producers, resistant to stresses, and tasted best. Although we did not call these selective breeding practices genetics, they were the earliest forms of genetically modified organisms, with humans directly influencing the genetic make-up of another species.

It was not until 1858 that Charles Darwin described the evolution of species and natural selection of traits. His concepts made possible a better understanding of inheritance of traits and could be directly applied to not only phenomena seen in the natural world, but also the changes that humans had been making for over 10,000 years to the species around them. Gregor Mendel's famous experiments that looked at the traits carried by pea plants, including careful records and detailed drawings of crosses between plants with different traits, was published in 1866. Mendel's observations led to the concept of there being units of heredity that were responsible for traits and that these units are unchanging particles. It was Hugo de Vries in 1889 who named the units of inheritance pangenes, which later became genes. It wasn't until 1905 that William Bateson coined the term genetics (**Figure 10.2**).

DNA was ignored and believed to be too simple to be the genetic material of inheritance

Friedrich Miescher published an account in 1869 of his discovery of a weak acid present in pus recovered from hospital bandages. This weak acidic substance was DNA, although at the time it was called nuclein because it was isolated from the nucleus. DNA is a very simple molecule and at the time was not believed to be the genetic material.

Although it may be hard for us to understand today, it was known that genes code for protein and that genes are on the chromosome, but it was not known that it was the DNA that made up the gene. Early in the 1900s, Mendel's work was rediscovered and applied to other organisms, including *Drosophila melanogaster* (fruit flies), which became a model organism for genetic research due to the ease of crossbreeding, their easy to identify phenotypes like eye color and wing structure, and short generation times. The concept of genes, Mendel's units of inheritance, became well established and genetic experiments were happening well before it was known that genes are made of DNA.

Indeed, in 1911, Alfred Sturtevant created the first physical map of a chromosome, which was expanded to a genetic map of the chromosome in 1913,

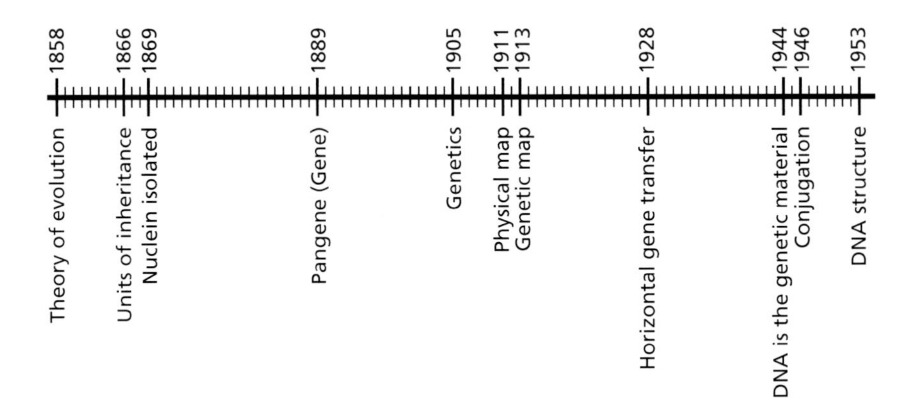

Figure 10.2 Timeline of genetics up to the structure of DNA. Key important dates over the century from Darwin's theory of evolution to Watson and Crick's structure of DNA.

highlighting where key genes could be found. Genetic transfer by the Griffith experiment, previously discussed (see Chapter 1), demonstrated that traits from one bacterial strain could be transferred to another strain. These results showed that chromosomes were the genetic material, supporting the knowledge and understanding of the time. These experiments also showed that bacteria participate in horizontal gene transfer. Some bacterial species, such as the streptococci used in the Griffith experiments, are able to take up DNA from other bacterial cells and incorporate it into their own genome, thereby expressing traits from the acquired DNA.

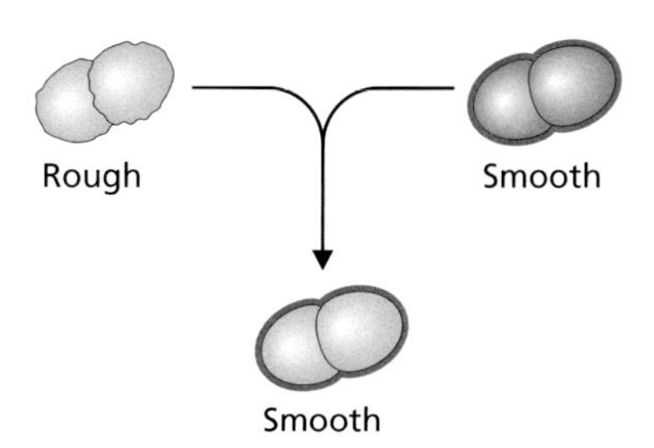

Figure 10.3 Horizontal gene transfer of the smooth trait in *S. pneumoniae*. Rough appearing bacteria (R strains) are not virulent (left, green cell). Smooth appearing bacteria (S strains) are virulent (right, pink cell), causing disease in humans. When rough and smooth strains are combined (arrows), the smooth trait can transfer to the rough strain (bottom, green cell), making them smooth. This demonstrates that the smooth trait can be transferred horizontally.

Bacterial genetics was the key to demonstrating the importance of DNA

As is highlighted throughout this book, several fundamental concepts in the fields of genetics and genomics have come about through research done in bacteria. One of the most important is the discovery made in the Avery, MacLeod, and McCarty experiment, which showed that DNA is the genetic material. These pivotal experiments demonstrated that the weak acidic substance first identified by Miescher from pus-filled hospital bandages was a critically important component of the cell and the key to genetics.

The rapid growth and short generation times of bacteria have been useful in studying their evolution and genetics, concepts that could be extrapolated to other organisms. The genetic passage of traits between bacteria, as observed by Griffith, Avery, MacLeod, and McCarty (see Chapter 1), has made important contributions to our understanding of genetics. In 1928, the Griffith experiment showed that transformation, the passage of genetic material from one bacterium to another, was possible. These experiments used *Streptococcus pneumoniae,* which caused important clinical infections at the time. Some *S. pneumoniae* strains had differences between strains that could be readily observed in the laboratory. Those agar cultures in which the bacterial colonies appeared rough were R strains, while S strains were smooth. The R strain *S. pneumoniae* did not cause disease and these non-virulent strains were able to be changed into virulent S strains in the course of the Griffith experiments (**Figure 10.3**).

There also seemed to be different virulent strains of *S. pneumoniae* that were causing bacterial pneumonia. Patients were able to make antibodies against one virulent S strain of *S. pneumoniae* but these antibodies were not necessarily able to recognize the antigens on other virulent S strains of *S. pneumoniae* in the laboratory (**Figure 10.4**). Oswald Avery studied the bacterial capsule that surrounded virulent S strains of *S. pneumoniae*. He established that the bacterial capsule is made of polysaccharide and discovered that different strains of *S. pneumoniae* had different and distinct polysaccharide capsules. Avery proposed that the capsules were antigenic and that this was the reason behind the differences in antibody responses seen in patients. The genes encoding the ability for *S. pneumoniae* to make distinct polysaccharide capsules are present in the DNA. Therefore, it is possible to not only transform a strain from having no capsule to expressing capsule, but also transform strains from one capsule to another.

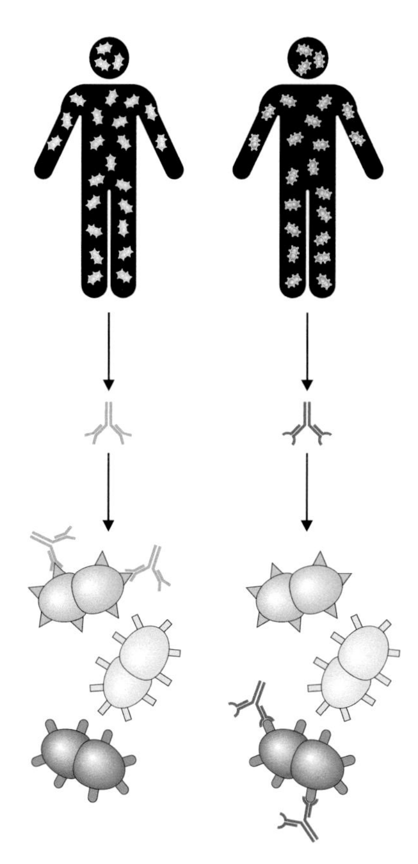

Figure 10.4 Antigenicity of S strains of *S. pneumoniae*. A patient (top left) infected with an S strain of *S. pneumoniae* (green bacteria) makes antibodies specific to the antigens on the surface of that S strain (middle left). These antibodies cannot recognize other S strains, only the one that infected the patient (bottom left). Different patients infected with different S strains, produce different antibodies, each only able to recognize the S strain that infected the patient (right panel).

Insights following the recognition of DNA as the genetic material led us to where we are today

The experiments of Griffith, Avery, MacLeod, and McCarty were early transformation experiments. Transformation is still used today to conduct genetic experiments on bacteria and for other research. There are, however, few bacteria that are naturally competent for transformation, meaning that there are not many species of bacteria that transfer traits in this way. Griffith was fortunate in his choice of bacterial species demonstrating transformation, a fortunate choice that Avery, MacLeod, and McCarty followed in their modifications of Griffith's experiments.

Later, other mechanisms for horizontal gene transfer were discovered, including the identification of conjugation in 1946. This involves the transfer of genetic material from one bacterial cell to another via a conjugation pilus and was described by Joshua Lederberg and Edward Tatum (see Chapter 3). The understanding of conjugation and the ability to replicate and manipulate this natural process in the laboratory was to be vital to understanding the structure of the bacterial genome, as discussed later in this chapter.

The study of genetics in bacteria led to two key discoveries in the 1960s. In 1961, a series of experiments began with contributions by several laboratories that deciphered the three-letter triplet code of the codons (see Chapter 2). The concepts of the control of gene expression also came in 1961 with the work of Jacob and Monod (see Chapter 3). Their work on the operon model in *Escherichia coli* has been expanded to other operon systems in many bacterial species. These fundamental experiments formed the basis of our understanding of DNA as the genetic material, of transformation, of horizontal gene transfer, of the genetic code, and of how that code is used by bacterial cells, which guide our investigations today.

Bacterial genetics is the cornerstone of all genetics

The key roles that bacterial experiments have played in the history of genetic research emphasize the contribution of bacterial genetics to our overall understanding of genetics today. The influence of bacterial genetic research on the field of genetics in general continues in the research that is being done currently. Today, bacteria are used not only to study the genetics and genomics of themselves and as models for other organisms, but also as genetically engineered bacteria. These bacteria, modified by humans, are able to express insulin, interferon, human growth hormone, clotting factors, and other human proteins that are missing or non-functional in some human diseases. Genetically modified bacteria also produce enzymes that are used in food production and reagents in diagnostic tests and laboratory research.

E. coli is the workhorse of **molecular biology**, having been adapted for use as a genetic-engineering tool. Molecular biology uses a variety of techniques to understand nucleic acids and proteins and the processes involving these biological molecules; thus, the application of molecular biology is important in the field of genetics. As an important tool in genetic research, *E. coli* is the most grown bacterial species in laboratories worldwide and is often grown in laboratories that are not interested in microbiology. Instead, the *E. coli* bacteria are used to carry introduced genetic material from other organisms and to express proteins from a range of other sources. From *E. coli* are also derived plasmids that are used for cloning of pieces of DNA and for the expression of proteins from these cloned DNA fragments. *E. coli* is also the source of enzymes such as polymerases and restriction enzymes that either are native to this species or have been introduced to the *E. coli* so that it can make the enzymes efficiently.

The identification and isolation of restriction enzymes is important for genetics research

Restriction enzymes are also known as **restriction endonucleases** because these proteins enzymatically cut nucleic acids in the middle, as opposed to exonucleases that digest at the ends. Restriction enzymes cleave both backbones of the DNA double helix and do so based on the enzyme's recognition of a specific sequence within the DNA. An important development in molecular biology, where we investigate biological systems at a molecular level, and our ability to understand and manipulate genetics in the laboratory was the discovery, isolation, and production of restriction enzymes as **recombinant proteins** from molecular biology generated **recombinant DNA**. Recombinant DNA is produced when DNA from two different sources are combined into one DNA molecule in the laboratory. In this way, recombinant proteins can be generated in *E. coli*

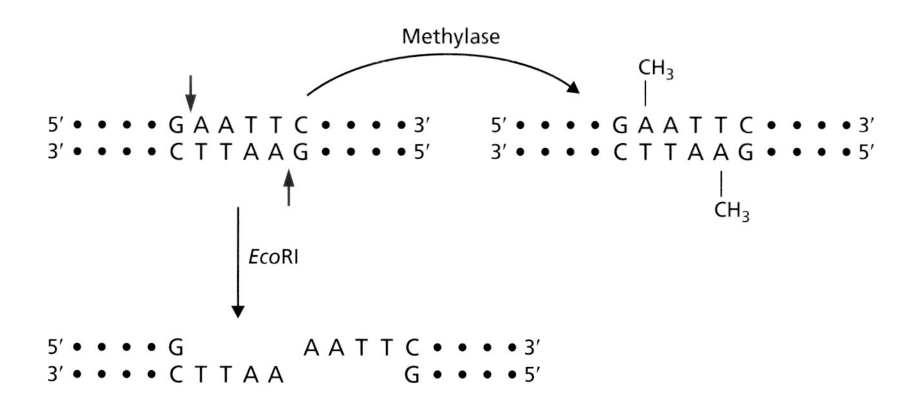

Figure 10.5 Restriction modification systems. Each restriction modification system includes a restriction enzyme, which cleaves DNA based on a specific DNA sequence, and a methylase, which methylates the DNA within the specific DNA sequence to protect it from cleavage by the restriction enzyme. Shown here, the recognition site for the *Eco*RI restriction enzyme is cleaved by the enzyme at the red arrows or is acted upon by the methylase to block the action of the *Eco*RI enzyme.

regardless of the bacterial species of origin. Restriction enzymes continue to play a vital role in genetics research, by virtue of the specificity of their enzymatic cleavage of DNA.

The discovery of restriction enzymes themselves supports the earlier observations of horizontal gene transfer. A restriction enzyme will digest any DNA that contains the specific recognition sequence for that enzyme. Bacterial cells therefore use restriction enzymes to digest and eliminate foreign DNA that enters the cell, thus restricting integration of DNA that may be harmful to the cell. Restriction enzymes have evolved to protect bacteria from the introduction of harmful DNA, such as bacteriophages. Bacteriophage DNA is introduced into the bacterial cell, where it subverts the bacterial cell for the manufacture of new bacteriophage particles, leading to the lysis of the bacteria. Therefore, digestion and elimination of this foreign DNA by restriction enzymes is beneficial. The bacteria's own DNA is protected from digestion, not by the absence of the enzyme's recognition sequence, but by modification of these sequences by a methylase that blocks recognition and digestion through methylation of the recognition sequence in the bacterial DNA (**Figure 10.5**). Together, the restriction enzyme and the methylase form a **restriction modification system**. Foreign bacterial DNA, introduced into the cell by horizontal gene transfer, may or may not be protected from digestion by the cell's restriction enzymes, depending on the methylation state of the DNA.

Unlike other enzymes that digest DNA, the specificity of the recognition site and therefore the specificity of the DNA ends makes restriction enzymes important in molecular biology and therefore in the study of genetics. The sequence specificity of the restriction enzymes has meant that DNA fragments could be generated by precision cutting of the DNA, either creating a straight cut across the DNA double helix, generating "**blunt ends**," or a staggered cut, generating "**sticky ends**" (**Figure 10.6**). The isolation of restriction enzymes from bacterial cells has meant that they could be used as molecular biology tools to specifically cut DNA, generating ends of DNA fragments with known sequences and enabling DNA fragments to be cloned into plasmids with complementary ends. Initially, restriction enzymes had to be extracted from bacterial cells and isolated for use in genetics experiments. Today, most restriction enzymes available commercially are recombinant proteins overexpressed in *E. coli*, which has simplified the process of isolation of the protein from bacterial cells. This reduced the cost per unit of restriction enzymes, making them more widely available molecular biology tools for investigations in the field of genetics.

Figure 10.6 Restriction enzymes generate either blunt or sticky ends to the DNA. A restriction enzyme cuts both of the DNA backbones of the double helix. Some cut the DNA straight across, resulting in no overhangs of nucleotides, which generates what is referred to as a blunt end (top). Some create a staggered cut in the DNA, making sticky ends that have nucleotide overhangs. These overhangs can be either at the 3′ end (middle) or at the 5′ end (bottom), depending on the action of the restriction enzyme.

There are four types of restriction enzymes, with type 2 being used most in laboratories

Restriction enzymes come in four different types, based on the make-up of the restriction modification system and distance between the cut site and the recognition site. Type 1 restriction enzymes cleave the DNA away from the recognition site. This means that, although the sequence of the recognition site is known and specific, the ends generated by the digestion are of an unknown sequence. In a type 1 restriction modification system, there is one protein that has both restriction and modification activity. This protein is made up of three subunits: HsdR does the restriction digestion, HsdM has the modification activity, and HsdS confers the sequence specificity for the recognition site.

Type 2 restriction enzymes cleave the DNA at the recognition site (see Figure 10.5). This makes them ideal for molecular biology and **molecular cloning**, the generation of recombinant DNA, because the sequence of the recognition site remains at the end of the DNA after it is cleaved. In the type 2 restriction modification systems, there are separate restriction and modification enzymes. The restriction enzyme forms a **homodimer**, where two copies of a protein come together into one quaternary structure that is then capable of specifically binding to the DNA. This homodimer structure influences the nature of the recognition site, which is a palindromic sequence of four to eight nucleotides. Cleavage of the DNA occurs at this site, which is the key feature making these enzymes the most commonly used and commercially available restriction enzymes.

The first type 2 restriction enzyme that was identified and isolated was the *Hind*II enzyme from *Haemophilus influenzae*. Restriction enzyme nomenclature comes from the species in which it was first identified, making up the first three letters of the enzyme name, which is written in italics. The remaining letters identify the strain of the bacteria and then Roman numerals based on the order in which restriction enzymes were identified in this strain. So, the type 2 restriction enzyme *Eco*RI comes from *E. coli* strain R and was the first enzyme identified in this strain. *Hind*I is a type 1 restriction enzyme, with *Hind*II, the first type 2, discovered later.

Type 3 restriction enzymes are a complex of restriction and modification enzymes that cleave DNA a short distance from the recognition site. The recognition sequence for a type 3 restriction enzyme is actually two different sequences that are present in inverted orientation relative to one another.

Unlike the other types of restriction enzymes, which are blocked by methylation generated by the associated methylase, the type 4 restriction enzymes specifically cleave methylated DNA. Type 4 restriction modification systems are otherwise similar to type 2 systems. Some classify the enzymatic activity within a CRISPR system as a fifth restriction enzyme type; this will be explored in more detail in Chapter 18.

The genetics of bacteria was unraveled using conjugation

In 1957, Jacob and Wollman interrupted conjugation in the laboratory to determine the location of genes on the *E. coli* chromosome. Bacteria with the F plasmid integrated into the chromosome as Hfr (see Chapter 3) were used to transfer the bacterial chromosome from one *E. coli* cell to another. As the chromosome crosses from one bacterial cell to another during conjugation, the gene locations on the chromosome were mapped based on the time the bacteria are attached by the conjugation machinery. During conjugation, the pilus retracts, pulling the donor and recipient cells closer together. This enables this Type 4 Secretion System (see Chapter 9) to transfer the DNA between the bacterial cells. Interruption of the conjugation at different times would allow a different amount of the chromosome to transfer over to the recipient. Conjugation experiments were interrupted quite simply - using a standard blender to separate the bacterial cells. Because the F plasmid can integrate into different locations in the chromosome in different Hfr strains, repeated experiments would produce results around the chromosome identifying the relative locations of genes.

The conjugation mapping experiments relied upon **auxotrophy**, the requirement for particular nutrients in the media for growth. **Auxotrophs** require specific nutrients in the media to grow; therefore, they can be easily identified through the use of media with different nutrients. Auxotrophies can differentiate bacteria that are otherwise identical, but which need different nutrients for growth.

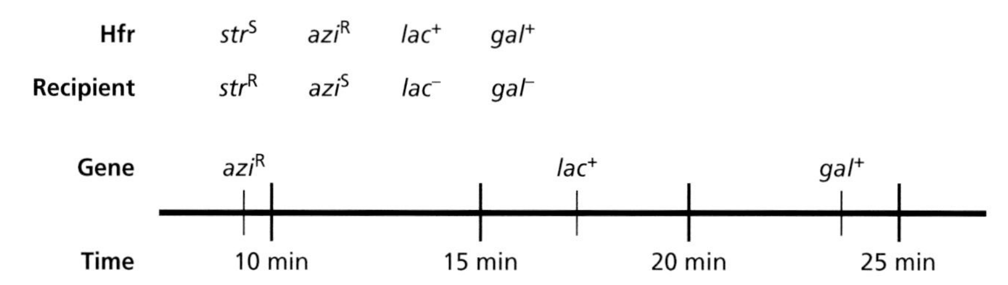

Figure 10.7 Conjugation mapping of *E. coli* using auxotrophies and antibiotic susceptibility. The Hfr strain is susceptible to streptomycin (*str*), resistant to azithromycin (*azi*), able to use lactose (*lac*), and able to use galactose (*gal*). The recipient strain is resistant to streptomycin, susceptible to azithromycin, and unable to use lactose or galactose. Conjugation is allowed to happen for a set time and then disrupted to stop conjugation. After conjugation, bacteria are cultured on media containing streptomycin, to select for the recipient cells. Only the recipient cells will grow, as the Hfr strain is susceptible to streptomycin. The time points at which these streptomycin-resistant recipient strains gain resistance to azithromycin and the ability to grow on lactose and on galactose produce a map of the genome with the relative distances of these genes indicated in the minutes of conjugation required to transfer the trait.

Genetic mapping experiments can track the locations of auxotrophy genes. The transfer of genes between cells in conjugation can reverse the auxotrophy, highlighting the genetic basis of the auxotrophic phenotype. In addition, genes conferring antibiotic resistance enable bacterial cells to grow on media containing the antibiotic. Timing this transfer of genes identifies the location of the gene on the chromosome. Through tracking of auxotrophy genes and antibiotic resistance genes, it was possible to generate a 100-minute map of *E. coli* that represented the complete conjugative transfer of the whole *E. coli* genome (**Figure 10.7**). These experiments were difficult; the connection between conjugating bacteria is very easy to break. Maintaining a connection for a few minutes is relatively easy, but maintaining connections for up to 100 minutes is technically challenging.

Through these timed conjugation experiments the relative positions of readily identifiable genes were determined. This produces a **genetic map** that gives the relative locations of genes and genetic features on the chromosome, such as auxotrophy genes, antibiotic resistance genes, and other genes of known function. In this way, the genetic features on the bacterial chromosome were first explored. While valuable, and certainly a cornerstone of understanding the *E. coli* genome for many years, conjugation does not happen in all bacterial species; therefore, generation of these genetic maps was not possible for every species of interest.

Physical maps can be made for any bacterial species using restriction enzymes

A **physical map** does not have gene location information like a genetic map. Instead, the base pair distances between restriction enzyme recognition sites around the chromosome are determined. The genetic features on physical maps are the type 2 restriction enzyme recognition sequences. Physical maps are made by digesting the chromosome with restriction enzymes and separating the **restriction fragments** by electrophoresis. The pattern of fragments on a gel is used to identify the lengths of the fragments. When several different restriction enzymes are used, singly and in **double digests** using two enzymes at the same time, the locations of these different recognition sites in relation to one another can be determined by determining the size of the restriction fragments and analyzing how they would go together. Because restriction enzymes are used to generate the map, these are also referred to as **restriction maps**. Restriction maps can be made for chromosomes and also for plasmids. Plasmid restriction maps are useful when planning to digest the plasmid with a restriction enzyme for the purposes of cloning (**Figure 10.8**).

In a physical map, the genes and other genetic features on the chromosome or plasmid are not identified. It is possible to add genetic map information onto

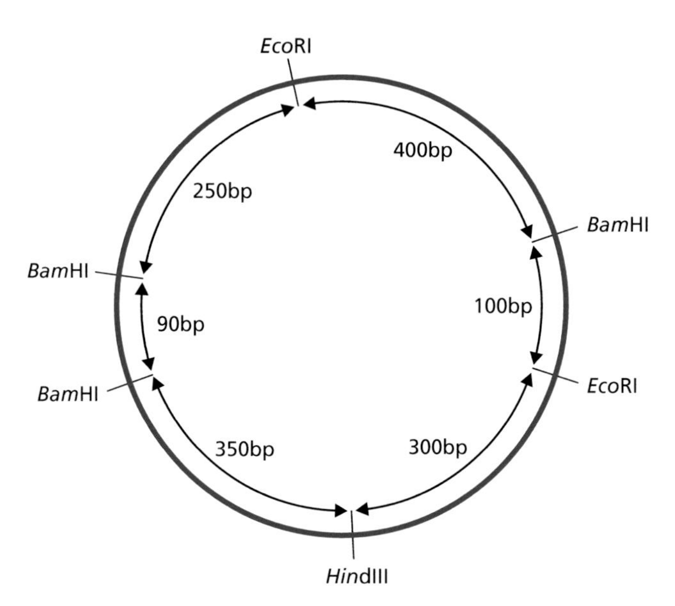

Figure 10.8 A physical map of a plasmid. Physical maps use restriction enzymes to generate DNA fragments, which are used to plot the distances between restriction enzyme recognition sites. Using several different enzymes can create a detailed map of the locations of the various enzyme recognition sites. Shown here are the locations of the restriction enzyme recognition sites for *Eco*RI, *Bam*HI, and *Hin*dIII.

a physical map by experimentally determining the genes within a restriction fragment. A physical map can be used, when overlaid with a genetic map, to identify restriction fragments that contain genes of interest and then use this information to isolate the fragment of DNA containing the gene. Because the restriction enzyme recognition sites are specific known sequences and the distances between them are determined in the physical map, this resource can be used to help in assembly of genome sequences and then to check that the assembly is accurate. The availability of a physical map was deemed to be important in the selection of strains for the first genome sequences. Although new physical maps using restriction enzymes are often not determined today, the methods behind their creation can still be used to verify genomic data. Many of the known functions of genes were determined through experiments that developed genetic and physical maps and experiments that used these data.

Experimentation reveals whether a CDS is a gene and what its function may be

While in the past mapping and molecular biology techniques were used to identify genes, today the abundance of sequencing data means that potential genes tend to be identified from DNA sequence, rather than experimental data. Although potential genes can be identified as CDSs through analysis of sequence data, only laboratory experiments can truly reveal the expression of the CDS as a gene and its function. Although there are many examples in a bacterial genome of regions where there is a start codon, a segment of uninterrupted DNA, and then a stop codon, identification from the sequence is based on sequence information, not on expression of that sequence as a protein. Even in CDSs with homology to known genes in other species, the function of the potential protein is only assumed to be the same as that of the homologous gene; it is not known to be the same.

Experiments in which a CDS is deleted or mutationally inactivated can demonstrate that this sequence includes a gene and can reveal its function. If, in the absence of the CDS sequence, a phenotypic difference is observed in the bacteria, the CDS can then be designated as a gene. If the phenotype gives a clear indication of the function of the gene, this is used to name the gene, using the conventions discussed earlier in this chapter. CDSs should not be given gene names when identified from sequence data alone. For this reason, CDSs are given other designations, often based on their relative locations around the chromosome, so that they can be identified and discussed in the literature prior to their status as a gene being known.

Library generation and library screening can identify genes and their functions

Transposon mutagenesis uses the insertion of transposons (see Chapter 3) into a bacterial chromosome to create a **library** of nearly identical bacterial **clones** where each has one transposon somewhere in the genome, disrupting the chromosomal sequence. Transposons used for the construction of libraries are specifically chosen to have antibiotic resistance markers that can be readily selected, to randomly insert in the genome, and to be small in size, sometimes being referred to as mini-transposons. The disruption of the genome is not targeted, but relies on the promiscuous integration of the transposon to semi-randomly generate mutations across the chromosome, causing **random mutagenesis**. The process is semi-random due to the nature of the target sites for the particular transposon used. The library is a collection of strains that can be used to look for changes in phenotype based on the different insertions of the transposon within the library.

It is possible to use **selection** to select library clones with a particular phenotype through the creation of a specific medium on which only those clones will survive. Auxotrophs and antibiotic resistance markers can easily be identified in this way. Library clones that grow can then be analyzed in more depth to see where the transposon has inserted in that particular bacteria. The transposon may have interrupted a gene directly involved in the selected phenotype. It may also be that the transposon has inserted into a promoter region that is important for expression of the phenotype or into a regulatory protein gene needed for expression of the phenotype.

Libraries can also be interrogated for specific phenotypes using **screening**. Unlike selection, all of the bacteria in the library are grown and then each individual clone is experimentally investigated for the presence of the phenotype, using a variety of molecular biology methods. This is a labor-intensive process because each library clone must be investigated individually for the phenotype of interest. The larger the library, and therefore the more extensive the transposon mutagenesis of the genome, the more experiments will be needed to identify the clone of interest. Like with selection, once the clone or clones with the desired phenotype are identified, the location of the transposon is then determined, using **polymerase chain reaction** (**PCR**) or sequencing, to identify genes involved.

Random mutagenesis identifies genes that have non-essential functions

A transposon library is a group of clones that are all identical except for the location of insertion of the transposon. Each clone has a transposon somewhere in the chromosome, interrupting the DNA sequence. If this mutation is detrimental to the bacterial cell, it will not grow and will not be part of the transposon library. This means that only non-essential genes will be disrupted by the transposon.

Locations in the chromosome that do not have transposon insertions in the transposon library will correspond to essential genes. When the locations of the various transposon insertions across the chromosome are assessed, the lack of any transposon insertions in a region suggests that there is an essential gene there, or another genetic feature that is essential, such as a promoter region driving essential genes or an ncRNA that the bacterial cell needs for growth.

If the growth conditions are changed for the culture used to grow the transposon library and some mutants are lost, the insertion locations of the transposons in the lost clones may be involved in growth in the new condition.

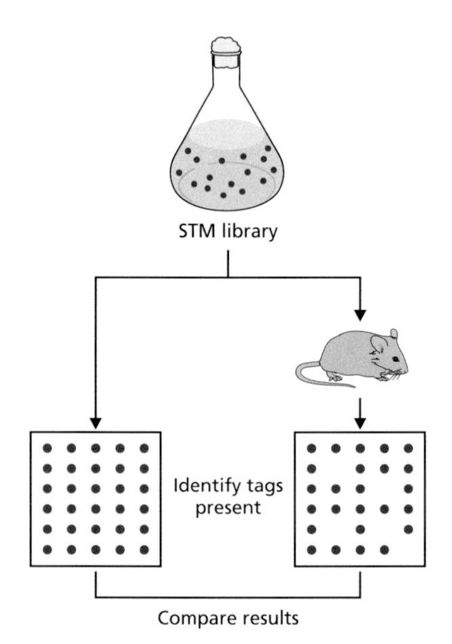

Figure 10.9 STM to identify genes needed *in vivo*. A signature-tagged library is grown in laboratory conditions. A portion of this culture is injected into mice (or another *in vivo* model system) to assess which library clones are lost *in vivo*. These clones will have disrupted genes that are essential for survival *in vivo*. By virtue of the tags, the clones from the laboratory culture and from the *in vivo* model can be assessed and those results compared to identify missing clones from the *in vivo* conditions.

By identifying the lost clones from the original culture conditions, it is possible to identify the insertion location and associate the phenotype with the genetic feature.

A specialized form of transposon library is called **signature-tagged mutagenesis** (**STM**). When the transposon library is made, specialized transposons are used that contain specific sequences, genetically engineered tags, within the transposon. Using molecular techniques such as PCR, the specific clones within the library can be identified. In experiments using STM, the transposon mutants are pooled and subjected to different growth conditions to see which can and which cannot survive in the different experimental growth conditions (**Figure 10.9**). Often STM has been applied to growth in laboratory ***in vitro*** conditions versus ***in vivo*** conditions in a host organism to identify genes required for growth *in vivo*. The signature tags enable the identification of clones more readily in the library. High throughput, inexpensive sequencing means that mutants identified from either standard transposon libraries or STM can be whole-genome sequenced to identify the location of transposon insertion.

The functions of genes can be determined using knockout technologies

To find the phenotype for a gene, a specific **knockout** of that gene may reveal the function. Disruption of the gene, the knockout, should result in a loss of function of that gene product. Libraries are particularly useful when the gene for a specific phenotype is sought, or when the functions of several genes need to be investigated. The identified loss of function should relate to the gene, although it should be noted that it is not always that simple.

In the same way that a transposon interrupts a gene sequence, a knockout mutation specifically gets rid of a gene. This can be achieved by specifically inserting a different sequence into the gene, disrupting the sequence of the desired gene, and rendering it unable to encode the protein. Unlike transposon mutagenesis, the gene is disrupted not by random process, but rather in a carefully designed experiment. Disruption of the gene sequence is often achieved with an antibiotic resistance cassette, which means that clones carrying the insertion in the gene can be selected for on media containing the antibiotic. Homologous recombination occurs between the chromosome and a fragment of engineered DNA containing the antibiotic resistance cassette flanked by sequences homologous to the chromosome. The flanking sequences are chosen so that recombination deletes the gene of interest. Bacterial cells that have not incorporated the antibiotic cassette into their chromosome will not be resistant to the antibiotic and will fail to grow on the plate, leaving only the mutants to grow. This knockout mutant can be used in experiments to try to determine the function of the disrupted gene.

Advances in molecular biology and molecular cloning mean that genes can be put into a plasmid and manipulated there. This **construct** enables the manipulation of the genetic sequence of interest. Once inserted into the plasmid, it is possible to insertionally inactivate the gene, delete all of the gene sequence, or delete part of the gene for the purposes of creating a knockout. A construct can also be generated using PCR techniques, which will be explored in detail in Chapter 18. Recombination between the genetically modified sequence and the bacterial chromosome will bring the changes to the gene into the bacterial

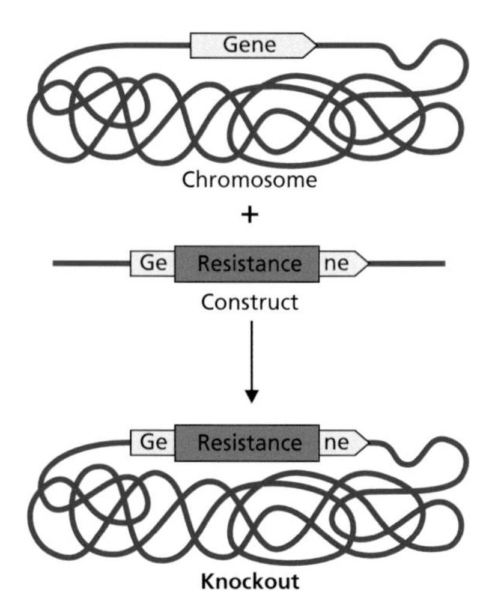

Figure 10.10 Knockout mutant generation. To knock out a gene of interest on the chromosome (top), a construct is generated, either on a plasmid or using PCR. A fragment of DNA containing the gene is created that has an antibiotic resistance cassette inserted into the gene, disrupting the sequence of the gene and thus preventing its function (middle). When this fragment recombines with the chromosome, the homology between the portion of the gene sequence in the construct and that in the chromosome means the construct ends up in the chromosome, generating the knockout of the gene.

genome, generating a knockout mutant (**Figure 10.10**). Molecular cloning techniques will be discussed in more detail in Chapter 18.

Knock out a gene and complement it back to check the phenotype is caused by the knocked out gene

Since the observed phenotype in a knockout mutant could be due to something else having changed in the genome, it is important to add back the gene and show a return of function. This process is called **complementation** and is often achieved by providing an intact copy of the gene on a plasmid (**Figure 10.11**), which is introduced into the bacterial cell. During the process of generating the knockout, a mutation could have appeared elsewhere in the genome, and this could cause the phenotype observed and presumed to be attributable to the knocked-out gene. Complementation provides additional supporting evidence that the phenotype associated with the knockout has actually been caused by the disrupted sequence of the knockout.

In the era of inexpensive genome sequencing, it is also possible to check for mutations and deletions of the genome that may have arisen during creation of the knockout mutant through whole-genome sequencing. Invariably there are changes seen in the whole-genome sequencing data. Even if two colonies from the same culture plate are sequenced, there can be differences due to errors introduced during replication and sequencing errors from the sequencing technology. Therefore, sequence data from knockout mutants have to be carefully assessed to see if there are sequencing errors or random mutations that do not impact the phenotype.

Through either targeted knockouts or random mutagenesis, like transposon libraries, many gene functions are identified. However, these techniques cannot be used on genes that are essential to bacteria. Deletion or disruption of a gene that is essential for bacterial growth and survival will result in dead bacteria that cannot be investigated in the laboratory. For essential genes, other strategies are needed.

Bacterial research has helped shape the field of genetics

In recent decades several key events involving bacteria have occurred that have contributed to the general field of genetics. In the 1950s, the DNA code was determined, and the principles of transcription, translation, replication, and operons were developed, as discussed briefly here and in earlier chapters. In the 1970s, key enzymes were identified, including restriction enzymes and reverse transcriptase, which are routinely used in molecular biology and have contributed to our understanding of genomes and transcriptomes. Recombinant DNA was first generated by Paul Berg in 1972, with 1973 seeing the creation of recombinant *E. coli* expressing non-*E. coli* genes, including a gene from *Xenopus laevis*, a toad. The first gene sequenced was from bacteriophage MS2, with both

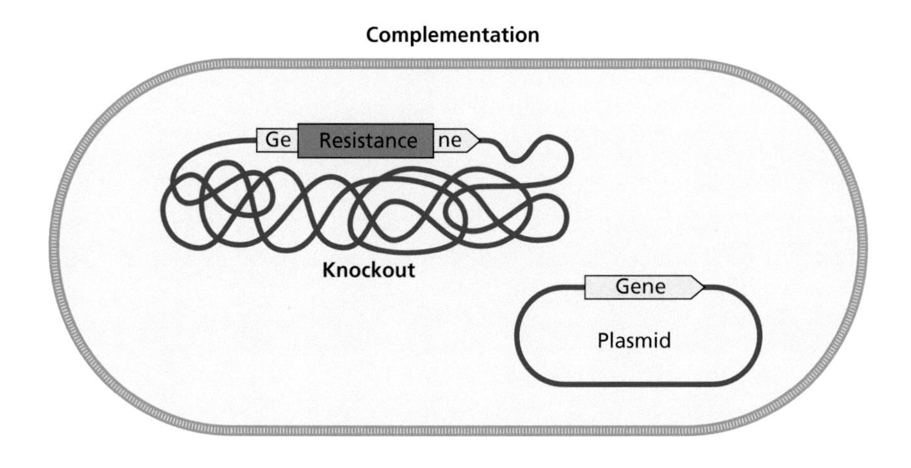

Figure 10.11 Complementation of a knockout mutant. To add to the evidence that a knockout mutant is responsible for an observed phenotype, the gene is added back into the bacterial cell, causing complementation. Often this is achieved by putting a functional copy of the gene on a plasmid and introducing it into the bacterial cell.

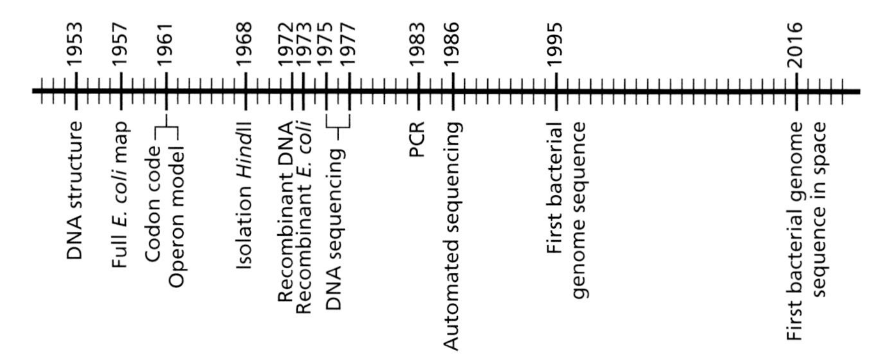

Figure 10.12 Timeline of events in the field of bacterial genetics from the structure of DNA to the present.

Sanger sequencing and Maxim and Gilbert sequencing being applied to bacterial sequences soon thereafter. Building upon earlier work and understanding of replication in the bacterial cell, Kary Mullis introduced the world in 1983 to the PCR, enabling researchers to copy specific sequences of DNA in the laboratory. PCR will be explored in depth in Chapter 18. The first automated sequencing machine was available in 1986, an enhancement of Sanger sequencing methodology that enabled the first bacterial genome sequence in 1995 of *H. influenzae*. Indeed, advances in technologies are often first applied to bacteria, due to the small size of their genomes and relative ease for genetic manipulation. As such, in 2016, *E. coli* was sequenced in space using a nanopore MinION system by astronaut Kathleen Rubins (**Figure 10.12**).

Key points

- Bacterial genetics is the study of how the nucleic acids within the bacterial cell contribute to its characteristics.
- Bacterial gene names are always in italics and the proteins they encode have the same name, but capitalized, making it easy to identify the gene and product in writing.
- Many of the advances in understanding the wider field of genetics have been made through experiments in bacteria, including identifying DNA as the genetic material.
- *E. coli* is often used in molecular biology for molecular cloning and expression of recombinant proteins, making it an important microorganism.
- The specificity of restriction enzymes in their cleavage of DNA has made possible the generation of physical maps of bacterial chromosomes and the creation of recombinant DNA, key advances in genetics.
- Timed conjugation experiments created genetic maps of bacterial genomes before it was possible to sequence the genome.
- From sequence data, we can predict where a gene might be and what it might do, but only through experimentation can we demonstrate which sequences are truly genes and what they do.
- Transposon mutagenesis, and variants of this method, randomly generates mutants that can contribute to understanding the functions of genes.
- Targeted gene knockouts and complementation can be used to investigate the function of a specific gene of interest.

Key terms

Define the following terms introduced in this chapter. Check your answers using the definitions in the Glossary. These terms are also available as Flashcards online.

Auxotroph	*In vitro*	Recombinant protein
Bioinformatics	*In vivo*	Restriction enzymes
Blunt ends	Knockout	Screening
Clone	Molecular cloning	Selection
Complementation	Polymerase chain reaction	Sticky ends
Construct	Recombinant DNA	Transposon mutagenesis

Questions and discussion topics

Self-study questions

Answer each question using 50–100 words or a table or labeled diagram. Advice on where to find answers to these questions is available online.

1. Describe the distinction between a gene, a CDS, and an ORF.
2. How is it possible to distinguish in writing between a gene name and a protein name? What is the general convention for naming bacterial genes and the proteins that they encode?
3. Early experiments in genetics used the fruit fly, *D. melanogaster*, to study phenotypes. Why was this model organism used and why does it continue to be used today?
4. What mechanism was involved in the horizontal gene transfer that occurred in the experiments by Griffith, Avery, MacLeod, and McCarty with *S. pneumoniae*? What phenotype was horizontally acquired?
5. What are bacterial capsules made from and are all capsules within a species the same? How do capsules impact the immune response to bacteria?
6. What does a restriction endonuclease do to a piece of DNA containing its recognition sequence? Draw the different outcomes for the DNA.
7. Compare the differences between a genetic map and a physical map.
8. What can be done experimentally to demonstrate the function of a CDS?
9. What is the difference between selection and screening? Which do you think is quicker or easier to do?
10. What sort of genes cannot be investigated using knockout mutants and transposon libraries? Why can they not be investigated in this way? Can essential genes be identified from library data?
11. What is the counterpart experiment to the generation of a knockout, which helps to verify the phenotype of the loss of function seen in the knockout? How would these experiments show the function of an unknown gene?
12. When was recombinant DNA first generated? Which species were involved in these experiments?

Discussion topics

These topics are presented for discussion in study groups, as part of class discussions, or on your own. These questions go beyond what is directly covered in this part of the book. Use the research literature and other reading to explore these topics in more depth. Tips to help prepare for topic discussions are available online.

1. Just as *D. melanogaster* was chosen as a model organism for the study of eukaryotic genetics, *E. coli* became the model organism for the study of bacterial genetics. Discuss the reasons why this was chosen as a readily available bacterial species for laboratories to investigate and to manipulate as molecular biology developed. How have discoveries made in *E. coli* been applicable to a broad range of other bacterial species, both Gram-negative and Gram-positive?
2. Identify a recent research article that has explored the function of a bacterial gene. Discuss how the researchers determined the function of the gene, including the genetic experiments conducted and the tests run to demonstrate the function. Was a knockout mutant made and loss of function shown? What were the controls in the experiment? What more needs to be done in future research?
3. Several important advances in the wider field of genetics and genomics have been based on work in bacteria. Within the last year, what vital research discoveries have been made due to investigations in bacterial species? Discuss the implications of these discoveries for the wider field of genetics and genomics and potential applications for the future.

Online quiz questions

To further self-assess your understanding of the chapter material, please visit the following link, where you can participate in a range of interactive quiz questions:

www.routledge.com/cw/snyder

Further reading

Bacterial genetics was the key to demonstrating the importance of DNA

Avery OT, Goebel WF. Chemoimmunological studies on the soluble specific substance of pneumococcus I: The isolation and properties of the acetyl polysaccharide of pneumococcus type I. *J Exp Med.* 1933; *58*: 731–755.

McCarty M. Discovering genes are made of DNA. *Nature.* 2003; *421*: 406.

Steinman RM, Moberg CL. A triple tribute to the experiment that transformed biology. *J Exp Med.* 1994; *179*: 379–384.

Insights following the recognition of DNA as the genetic material led us to where we are today

Crick FH, Barnett L, Brenner S, Watts-Tobin RJ. General nature of the genetic code for proteins. *Nature.* 1961; *192*: 1227–1232.

The identification and isolation of restriction enzymes is important for genetics research

Pingoud A, Wilson GG, Wende W. Type II restriction endonucleases – A historical perspective and more. *Nucleic Acids Res.* 2014; *42*(*12*): 7489–7527.

Ren J, Lee HM, Shen J, Na D. Advanced biotechnology using methyltransferase and its applications in bacteria: A mini review. *Biotechnol Lett.* 2022; *44*(*1*): 33–44.

van der Oost J, Patinios C. The genome editing revolution. *Trends Biotechnol.* 2023; *41*(*3*): 396–409.

Library generation and library screening can identify genes and their functions

Lin T, Troy EB, Hu LT-Gao L, Norris SJ. Transposon mutagenesis as an approach to improved understanding of Borrelia pathogenesis and biology. *Front Cell Infect Microbiol.* 2014; *4*: 63.

Random mutagenesis identifies genes that have non-essential functions

Cummins J, Gahan CG. Signature tagged mutagenesis in the functional genetic analysis of gastrointestinal pathogens. *Gut Microbes.* 2012; *3*(*2*): 93–103.

Bacterial research has helped shape the field of genetics

Castro-Wallace SL, Chiu CY, John KK, Stahl SE, Rubins KH, McIntyre ABR. Nanopore DNA sequencing and genome assembly on the international space station. *Sci Rep.* 2017; *7*(*1*): 18022.

Chapter

11

Genomics

Bacterial genomics, in which whole-genome sequence data are generated, interrogated, and explored, has only been possible since 1995. In that year, the first bacterial complete whole-genome sequences became available, and the era of genomics truly began. Previous genomic endeavors had to rely upon incomplete information, gained from techniques such as mapping and mutagenesis, as described in Chapter 10, and short targeted sequencing. Since the first bacterial genome sequences were completed, many thousands of bacterial samples have been sequenced, leading to a new understanding of the genes and sequence features within bacterial species and how they have diverged and evolved over time. This recent history of genomics will be explored in this chapter, highlighting key advances in technologies, discoveries that have been made, and how bacterial genomics has impacted our understanding of bacteria.

Automation of Sanger sequencing launched the era of bacterial genome sequencing

Sanger sequencing and Maxam and Gilbert sequencing, developed in the mid-1970s, allowed the specific base sequences of DNA to be determined using biological or chemical techniques, respectively. These methods were technically challenging and, therefore, the automation of Sanger sequencing made it possible to produce more sequence data more efficiently than ever before. This brief overview here will put into context the advances in Sanger sequencing that have led to genome sequencing.

Although Maxam and Gilbert sequencing was initially more popular, due to its chemical basis of sequence determination, ultimately Sanger sequencing became dominant, with its **sequencing by synthesis** approach. Sequencing by synthesis reproduces the process of replication, taking a single strand of DNA and using polymerase to add nucleotides. Through monitoring the addition of the four nucleotides, the sequence is able to be determined. Sanger sequencing initially used radioactively labeled nucleotides incorporated into the synthesized second strand of DNA to produce fragments and resolved these on large polyacrylamide slab gels. The radiolabeling of the four different nucleotides required four reactions to be set up and then resolved on four different lanes of the slab gel. After the gel was run, it was exposed to X-ray film and the radiation revealed the location of labeled nucleotides. To obtain the DNA sequence, the bands on the X-ray film across the four lanes were read by eye to reveal the order of the four nucleotides in the sequence (**Figure 11.1**).

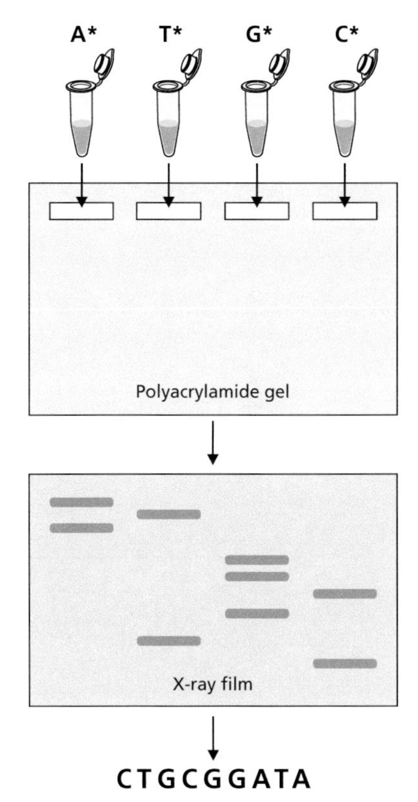

Figure 11.1 Sanger sequencing. The original Sanger sequencing method required radioactive nucleotides (indicated with a *), one for each base within DNA. These were used in four separate reactions to label DNA (top). Each reaction was run in a single lane of a polyacrylamide gel (top middle), which was then exposed to X-ray film to reveal the radioactive bands, each representing a base in the sequence (bottom middle). This X-ray film was read by eye, from bottom to top, to determine the sequence of the DNA (bottom).

To automate Sanger sequencing, the radiolabeling had to be replaced with something more easily detected by a computer-driven automated system, rather than relying on human interpretation from bands on a gel. **Automated Sanger sequencing** incorporated fluorescently labeled nucleotides. This change resulted in three main benefits. Using fluorescently labeled nucleotides, the bands on the polyacrylamide gel are detectable by a sensor and recorded automatically by a computer, rather than needing to be read and recorded by a human. Also, by using different fluorescences for each nucleotide, the four sequencing reactions needed in Sanger sequencing are able to be combined into one reaction in one tube, reducing set-up time for the sequencing reaction. Finally, since there is one sequencing reaction, containing information on all four of the nucleotides, the

DOI: 10.1201/9781003380436-15

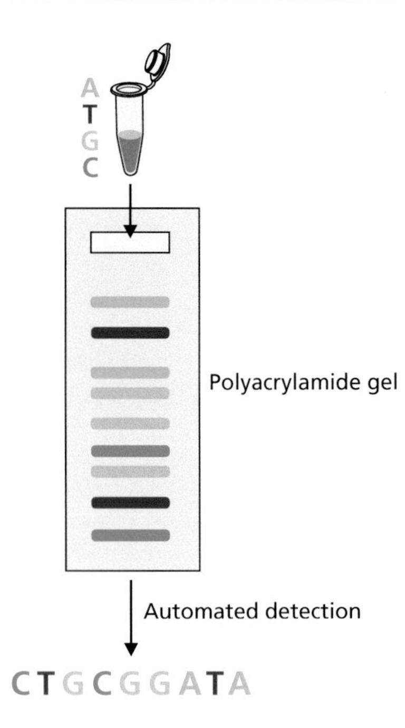

Figure 11.2 **Sanger sequencing with fluorescent nucleotides.** Rather than using four separate radioactive nucleotides in the sequencing reactions, use of nucleotides each labeled with a different type of fluorescence reduced both the number of reactions (top) and the number of lanes on the polyacrylamide gel (middle). Fluorescence also enabled automated detection (bottom), reducing human error in reading bands on an X-ray film as previously (see Figure 11.1).

number of lanes needed for each stretch of sequence on the polyacrylamide slab gel is reduced from four lanes to just one lane. More efficient use of the slab gel means that four times more sequence data can be gathered for each gel poured and run. This dramatically reduced preparation time for both the reaction and the gel, since fewer reactions and fewer gel lanes were needed to achieve the same amount of sequence data (**Figure 11.2**).

This stepwise advance in sequencing methodology still required researchers to pour large awkward polyacrylamide slab gels; therefore, these were replaced in the early 1990s with capillary gels. By containing each gel lane within a glass tube, the process was simplified in a way that could be completely automated. From this stage, other advances were based on increasing the capacity of the sequencer, such that more capillaries could be run at the same time and such that more sequence information could be gained from each capillary. These developments and the potential seen in early Sanger sequencing launched the sequencing of bacterial genomes. As groups embarked on large-scale sequencing efforts, new insights into the processes involved in sequencing led to these higher throughput methods, automating Sanger sequencing and transitioning to capillary gels.

The first genome sequence of a free-living organism was bacterial

The first genomes that were sequenced were the relatively small genomes of bacteriophages. Scaling up to a bacterial genome sequence was a major undertaking, going from a few hundred bases to several thousand. The first bacterial species for which genome sequencing was undertaken was the model microorganism *Escherichia coli*, which has 4.6 megabases in its genome. This project took an ordered and methodical approach to bacterial genome sequencing. The *E. coli* chromosome was physically and genetically mapped previously, providing a reference against which sequence data could be placed. The genome was then ordered into 40,000 base segments that overlapped one another. Each 40,000 base segment became the focus of sequencing efforts. This facilitated the assembly of the 200–800 bases of sequence data from each Sanger sequencing reaction into the correct order in each segment. The overlapping segments were then connected to each other to form the complete *E. coli* genome. The first 1.9 megabases of sequence data from this project was deposited into GenBank in 1992 and the project was finished in 1997.

E. coli was the first genome-sequencing project initiated, but it was not, however, the first to be finished. The first bacterial genome sequence, completed in 1995, was *Haemophilus influenzae*. This involved determining the sequence of the 1,800,000 bases of the *H. influenzae* genome using a new **shotgun sequencing** approach and took about a year to finish. In this method, the genome is randomly divided, cloned into a randomized library, and sequenced. Because the sequencing data are then produced in a random order, assembly of the genomic sequence relies heavily on computation and upon there being enough overlapping sequence fragments to determine where each 200–800 bases of sequence data join to one another (**Figure 11.3**). Although shotgun sequencing is more computationally intense, it is

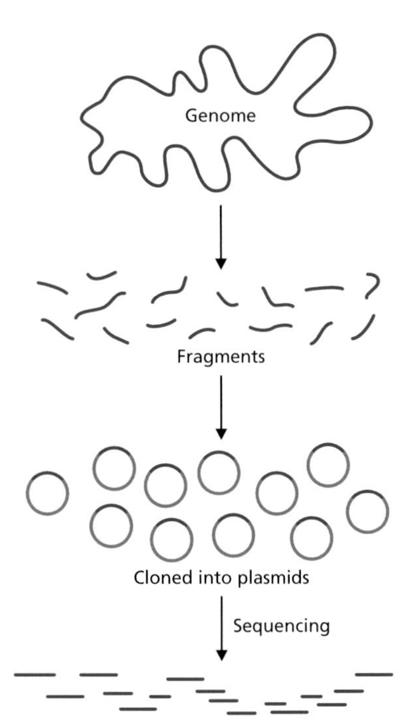

Figure 11.3 **Shotgun sequencing.** This methodology, applied to sequencing whole-bacterial genomes, involves first randomly dividing the genome into fragments that are cloned into a library. The order to the fragments as they were in the genome is lost. As the library is sequenced, it produces overlapping sequence data that is joined computationally to generate contiguous sequence data.

less laborious than making ordered libraries, reducing both cost and time. In the end, about 24,000 fragments of DNA sequence data became the *H. influenzae* genome sequence.

In the early years of bacterial genomics, most genome-sequencing projects were completed to one contiguous sequence, with all of the generated sequence fragments joined up as they would be within the bacterial cell. The goal in each case was to fully sequence all of the bases in the genome and fully annotate each coding sequence of the bacterial genome. There were, however, many limitations in the early days of bacterial genome sequencing. The processes of sequencing used for the first bacterial genomes were laborious and expensive, which in turn drove progress toward automating and increasing throughput of the Sanger sequencing methodology. Another limitation discovered during the course of the sequencing projects was that the libraries generated to facilitate all sequencing projects did not contain the entire genome. Regions of DNA that contained genes that were toxic to *E. coli* when these regions were cloned into plasmids and put into *E. coli* resulted in gaps in the sequence data. Because plasmids with these regions were toxic, no *E. coli* containing them were able to grow and therefore these regions could not be recovered in the resulting sequence data. Cloning into plasmids enables sequencing to use primers designed to anneal to the plasmid and sequence the inserted fragment. However, reliance on the propagation of the fragments within plasmids *in vivo* risks the loss of sequence data due to viability of clones. In order to fill in these gaps and complete the genome sequence, additional laboratory work and additional automated Sanger sequencing, targeting the missing regions, was required.

Early genome sequencing of bacteria provided opportunities for new insight and innovation

The availability of a bacterial genome sequence enabled researchers to explore the potential within the DNA, identifying features that might be genes and hypothesize as to their functions. As more whole-genome sequences were determined, new types of analyses could be done. **Comparative genomics**, in which two or more genome sequences are compared to one another, meant that it was possible to analyze similar sequences and identify **regions of difference**, where genomes contain different genetic features that are not in common between them (**Figure 11.4**).

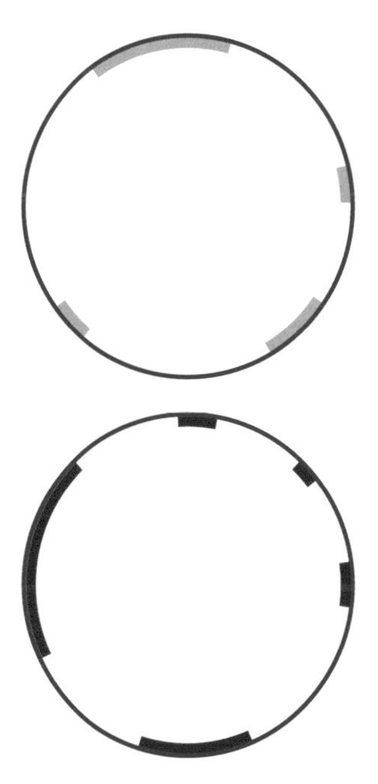

Figure 11.4 Regions of difference between bacterial genomes. Comparative genomics, evaluating differences and similarities between two genome sequences (top and bottom), shows that there are some regions in one genome that are not in the other (blue), and vice versa (red).

The first comparative genomics study was published in 1996, evaluating the differences and similarities between two genomes from different *Mycoplasma* species: *Mycoplasma genitalium* and *Mycoplasma pneumoniae*. Due to the small size of *Mycoplasma* spp. genomes, *M. genitalium*, with a 580 kb genome, was chosen as the second target of the new shotgun sequencing technique in 1995. In the following year, the 816 kb genome of *M. pneumoniae* was also sequenced. This provided a unique opportunity to conduct comparative genomics, for the first time comparing two complete genome sequences of species that were within the same genus and therefore closely related. All of the CDSs in *M. genitalium* were found to be within the *M. pneumoniae* genome sequence data, although the order of the CDSs was different, indicating chromosomal rearrangements between the two organisms. Additional CDSs in *M. pneumoniae* included new unique CDSs as well as some duplications of CDSs present as single copies in *M. genitalium*. This first comparative genomics study showed that the genome sequence data were able to reveal the depth of variation between species and enabled researchers to suggest how changes may have occurred during evolution.

Following on from this, a comparative genomics project comparing two strains of the same species provided even greater insight into bacterial genomic variation. *Helicobacter pylori* became the first comparative genomics study to be published within a single species. This analysis revealed that each of the two *H. pylori* strains contained strain-specific CDSs, many of which were present together in regions of difference between the strains. The value of these comparative genomics results in understanding strain diversity and bacterial genome evolution, coupled with growth of sequencing technology and reduction of sequencing costs, encouraged investigators to sequence multiple species of the same genus and multiple strains of the same species in the years since.

Bacterial genome sequencing required and fueled innovation

In 1997, the first genome sequences of the model bacterial organisms *E. coli* and *Bacillus subtilis* were completed, adding to the growing list of bacterial genome sequence data. The growth in size of sequence data from database entries of a few kilobases to a few megabases proved to be a challenge when it came to interrogating these data in available software and when it came to storage and retrieval of the data for analysis. The data therefore drove innovation into new ways in which the vast (in comparison to computer storage at the time) and growing amount of sequence data could be stored and the ways in which they could be analyzed. Without these innovations, investigations into the data would be very limited. These sequencing efforts into bacterial genomes were occurring at the time when the Human Genome Project was being pursued by a collection of sequencing centers around the world. Using the relatively smaller bacterial genomes, developers were able to test the ability of new software and data retrieval methods to enable them to ultimately analyze the larger human genome sequence.

By the year 2000, bacterial genome sequencing and comparative genomics brought about another innovation, **reverse vaccinology**. Using the genome sequence data, analysis was done to identify all of the potential targets for a new *Neisseria meningitidis* vaccine. It was known that the targets for a vaccine needed to be expressed on the surface of the bacteria and that they needed to be unique to bacteria. Theoretically, all of the proteins expressed on the surface of bacteria are encoded in the genome sequence, it was just a matter of doing the correct analyses to identify them. Following extensive sequence analysis and years of laboratory research, new vaccines were developed using this reverse vaccinology approach. Reverse vaccinology will be discussed in more depth in Chapter 20.

The emergence of next-generation sequencing technologies greatly increased sequence data

Early sequencing efforts using Sanger and automated Sanger methods demonstrated that bacterial genome sequencing could be achieved and would reveal important information about the bacterial species being studied. However, limitations to the application of these sequencing methods to large numbers of bacterial genomes remained, including the time, effort, and costs involved. High-throughput **next-generation sequencing** technologies dramatically reduced the time, effort, and cost per genome.

Several high-throughput technologies emerged in the mid- to late-2000s, most of which were based on similar principles. **Sequencing platforms**, the technologies and equipment for sequencing, by Roche (developed by 454), Illumina (developed by Solexa), and Life Technologies (developed by Ion Torrent) all used modifications of the sequencing-by-synthesis method pioneered by Sanger sequencing. Next-generation sequencing takes an approach that is massively parallel, with many individual sequencing reactions happening and being recorded within the sequencing equipment all at the same time. For each of these three next-generation sequencing technologies, the DNA is fragmented, similarly to shotgun sequencing. However, for next-generation sequencing these fragments are not cloned into plasmids. This addressed the previous methodology's issues with gaps in the final sequencing data due to non-viable clones. Instead, next-generation sequencing ligated specific sequencing adapters to the ends of each fragment. These sequencing platform-specific oligonucleotide adapters enabled the fragments to be captured, separated, amplified *in vitro*, and sequenced, taking the place of plasmids used in shotgun sequencing. The adapters attached to the fragments are captured onto the surfaces

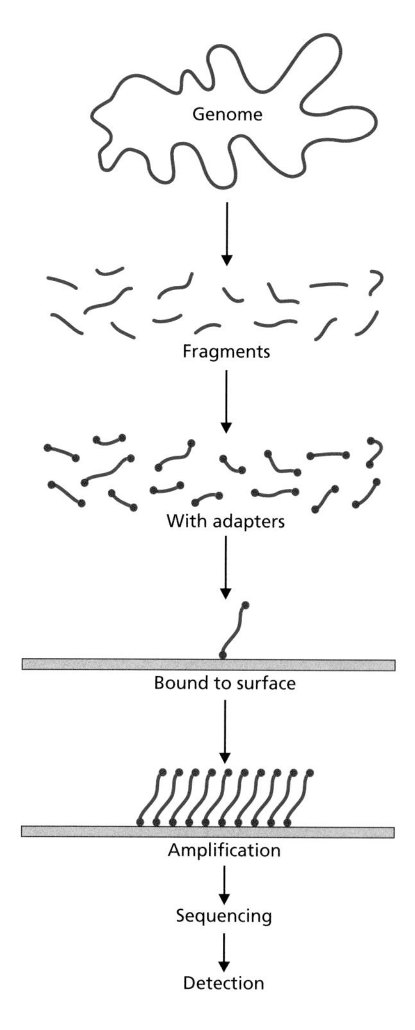

Figure 11.5 Next-generation sequencing principles. Most next-generation sequencing technologies use sequencing by synthesis, as established in the Sanger sequencing method, but with modifications. DNA is fragmented and sequencing technology-specific adapter sequences are ligated to the ends of the fragments. The adapters are used to attach the fragments to a solid surface (bead or flowcell), as primers for amplification of the fragments, and to initiate sequencing of the fragments. As nucleotide bases are incorporated into the DNA, the identity of each incorporated base is determined via fluorescence or changes in pH within the reaction vessel.

of microscopic beads (Roche and Life Technologies) or flowcells (Illumina) to keep each fragment discrete during the sequencing process. On these surfaces, the fragment is amplified *in vitro* to enhance the strength of the sequencing signal, be it fluorescence (Roche and Illumina) or pH-based (Life Technologies) (**Figure 11.5**). With the elimination of the cloning steps and a streamlining of the workflow, next-generation sequencing provided a means of obtaining bacterial genome sequence data that was quicker, easier to perform, and less expensive.

Although next-generation sequencing has been a major advance in the field of genomics, and is ideally suited to bacterial genome sequencing, there are some limitations of these technologies. Sanger sequencing has been able to reliably produce continuous stretches of sequence of 1,200 bases or more. Next-generation sequencing produces much shorter stretches of continuous sequence data. Although some technologies have increased the sequence information from a single sequencing-by-synthesis reaction (the **read length**), by 10 times their initial capacity, the longest reads from next-generation sequencing are still a third of what is possible with Sanger sequencing, in the range of 250–400 bases. As a trade-off, the amount of data produced from each genome is far greater in next-generation systems, with the **data depth** reflecting a significant increase in the number of times each base is sequenced, increasing the reliability of the data and also producing the needed increase in data to achieve enough overlaps of the shorter read lengths to assemble the sequences. Increased data depth also means that sequence variations within the sequenced sample can be detected. If a base has changed in just 10% of the bacterial population, this change may not be seen if that base has only been sequenced eight times. However, when it has been sequenced 100 times, the change in that fraction of the population is evident in the sequence data due to the depth of the data available on the identity of each base.

Next-generation sequencing has limitations

Despite the *in vitro* amplification steps eliminating the loss of clones and the resulting sequencing gaps, it is quite common for next-generation sequencing to be unable to resolve the data into whole-bacterial genomes. In part, this is due to the shorter read lengths of next-generation sequencing, which can only be partially overcome by the increased data depth. This is particularly true of those genome sequences that contain repetitive regions or that have regions of sequence similarity scattered in the genome. Common repeated features such as the copies of IS elements and prophages within the genome and the rRNA loci that are often present in more than one copy are also problematic regions for assembly. When sequenced as fragments, it is difficult to work out the location of the fragment on the chromosome, since the same DNA sequence is present in more than one location (**Figure 11.6**). This is akin to having a jigsaw puzzle where some of the pieces are identical in shape and color. Regions of the chromosome that have the same, or very similar, DNA sequences confound computational assembly systems that are looking for overlaps in sequence similarity in order to assemble the sequence data into longer and longer stretches of sequence. The sequences cannot be placed properly in the genome and are often seen as the same region rather than separate copies in different parts of the chromosome. When read lengths are shorter than repeated features in the genome, increasing the number of times that the genome is sequenced cannot overcome the inability to sequence fully across the repeated feature (see Figure 11.6).

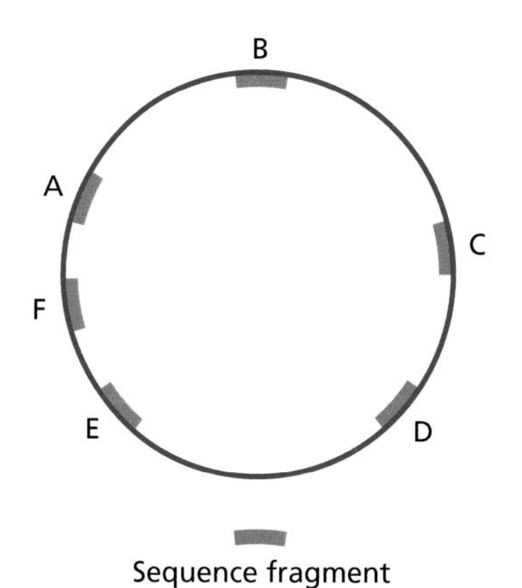

Figure 11.6 Repetitive regions complicate assembly. If a genome contains, for example, six repetitive regions that are identical to one another (A, B, C, D, E, and F), this can confound attempts to assemble sequencing fragments. The fragment shown could belong in any of the six regions of the genome.

Bacterial genome-sequencing projects shift focus due to next-generation sequencing limitations

Rather than one circular bacterial chromosome, next-generation sequencing data are often in several **contigs**: segments of contiguous sequence data that have been assembled together. For instance, if a genome sequence has been assembled into 50 contigs, it is not known in what order these 50 contigs belong in the chromosome because there is insufficient overlapping sequence data to complete the assembly. Next-generation sequencing genome projects therefore often leave the data in contigs, rather than pursuing the added laboratory work to manually sequence more information to complete the assembly. These partially assembled genome sequences make it difficult, if not impossible, to assess chromosomal inversions, gene orientation across the chromosome, and other differences in the chromosome that rely on data being assembled into one contig.

Likewise, duplications within the genome are a concern when attempting to assemble whole-genome data, even when looking at Sanger shotgun sequencing. Assembly relies on finding overlaps of identical stretches of sequence. If a region of the genome has been duplicated then there are two stretches of the genome that are identical to one another, or nearly so. When looking for similar reads, assembly systems will put these together, such that what is actually two copies in the genome looks as though it is only one copy. It should be that this one assembled region has twice as much sequence data as the rest of the genome, since it is in the DNA twice, so it is possible to spot duplications using data depth information, particularly if the region is large enough. Invariably, assemblies will therefore create one contig with the rRNA locus, although the sequence data has come from the two or more rRNA loci in the chromosome (**Figure 11.7**). To assemble across the rRNA loci, the duplication must be recognized in order to join the remaining other contigs to either side of the rRNA region.

Because next-generation sequencing is rapid, simple, and inexpensive, and because limitations in next-generation sequencing create only partially

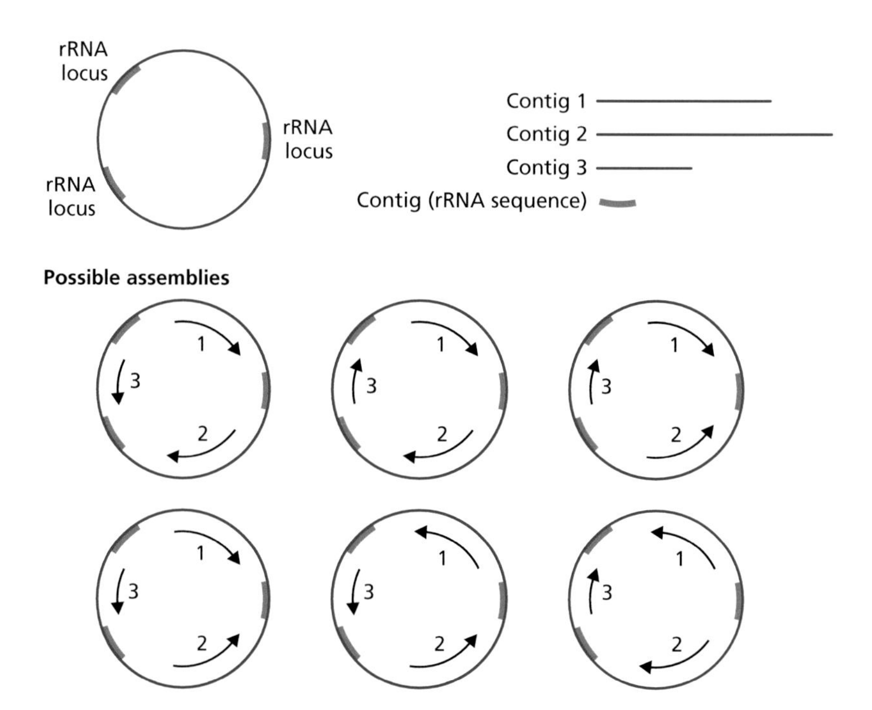

Figure 11.7 Incomplete assemblies and duplicated regions. A circular bacterial chromosome that has three identical rRNA loci will generate one contig representing all three regions, albeit with three times the coverage of the other contigs. The remainder of the genome can be assembled into three contigs, but no fewer, because the read length of the sequence data does not provide data across the whole of each rRNA locus. The genome could therefore be assembled in several different configurations, with the three non-rRNA contigs in different orientations and relative locations.

assembled bacterial genomes, there has been a shift in focus when it comes to analysis of sequence data. Rather than just sequencing one bacterial genome of interest, projects have undertaken to sequence hundreds of related genomes, with an aim to look at variations in the sequences of core genes and differences in the presence of accessory genes. Rather than sequencing one bacterial genome, as was typical in the mid-1990s and early-2000s, a decade later sequencing centers were pursuing projects with tens and sometimes hundreds of bacterial strains, all genome sequenced using the next-generation technologies. Next-generation genome sequences were often assembled to a point that reduced the number of contigs, but did not fully join together the bacterial genome. Nevertheless, valuable insight and information could be gained from these larger comparative genome projects. For newly identified species, draft genome sequences can be produced that may reveal some insight into the biology of the organism. Partially assembled next-generation sequencing data can be compared to an already assembled genome, making those bacterial genome sequences that are completely assembled a valuable resource.

Next-generation sequencing enables a massive expansion of comparative genomics

As mentioned earlier in the chapter, the first comparative genomics study investigated the differences and similarities between *M. genitalium* and *M. pneumoniae*. As comparative genomics progressed from two genome sequences to hundreds of genome sequences, it became clear that there is far more genetic diversity in the bacterial world than previously believed. Even strains of the same species that otherwise were typed as being identical were shown to contain variation at DNA level. While evolution was once believed to be a slow process relying on mutation of single bases, the bacterial genome sequences demonstrated that a great number of other factors than simple point mutations and SNPs are altering the genomes. These included the movement of mobile genetic elements like IS elements and transposons not only between different genomes, but also within genomes as well. Bacteriophage genomes integrated into bacterial chromosomes as prophages were excised, inserted in different locations, duplicated, and degraded through mutation to their associated CDSs. Large pathogenicity islands, horizontally transferred into some strains of a species, contributed to some of the major differences seen in genomic sequence data.

Next-generation sequencing and epidemiology

Bacteria of the same species, but which have different phenotypes and genotypes, can be differentiated from one another using various typing methods, which identify phenotypes such as the type of bacterial capsule. However, typing may not give a complete picture of the relationship between different isolates and therefore may give an inaccurate picture of the spread and epidemiology of the species. Methicillin-resistant *Staphylococcus aureus* (MRSA) isolates from around the world were genome sequenced, revealing relationships within and between lineages of the bacteria that were previously unknown. A global pattern of transmission of MRSA emerged, with isolates from different geographic locations having different genetic features and changes. This application of sequencing technology demonstrated the introduction of new isolates to new geographic areas, enabling targeted infection control measures and treatments to be put into place based on previous experience.

Bacterial genome-sequencing identification of the source of outbreaks

In 2010, a devastating earthquake hit Haiti and the international aid community responded by sending aid workers and supplies to the small island nation. In the aftermath of the earthquake, a cholera outbreak occurred. Haiti had not experienced cholera for nearly 100 years; therefore, the emergence of cases of cholera was a major concern. Bacterial genome sequencing provided an opportunity to identify the specific DNA features of the *Vibrio cholera* from Haiti and to potentially determine the origin of the outbreak. The sequence data, interestingly, did not show that the Haiti isolates were closely related to other Latin American cholera strains. Instead, there were several distinct sequence features in the genomic data that were most similar to Asian isolates. The location of the first outbreak cases, the location of UN aid workers, and the set-up of the sanitation in the area were also evaluated. This information, along with more targeted sequencing data, revealed that the *V. cholera* infecting earthquake survivors in Haiti had been brought from Nepal by UN aid workers from Nepal who had come in response to the environmental catastrophe.

In the summer of 2011, thousands of cases of bloody diarrhea, hundreds of cases of hemolytic-uremic syndrome, and dozens of deaths were reported in Germany. Cases of similar disease were then reported around Europe. The *E. coli* isolated from patients was indistinguishable by standard typing methods from other *E. coli* that did not cause this severity of disease. To address the question of why these bacteria were causing an outbreak with so many fatalities, next-generation sequencing of the genome was conducted. This rapidly produced genomic data that were released for the scientific community to interrogate and crowdsource their analysis. In less than a week, the isolate was identified as an enteroaggregative *E. coli* that had acquired genes for Shiga toxin and antibiotic resistance. Using the genomic data, a molecular test was developed to identify the outbreak strain and provide the recommended treatment.

In 2014, 3,000 genomes were sequenced from isolates of *Streptococcus pneumoniae* that were taken from a refugee camp in Thailand over a period of 4 years. This large study allowed unprecedented analysis of a large number of isolates coming from a common location, which showed how these pathogens had evolved over time. Changes to antigens present on the bacteria were seen over time, particularly several examples of switching events where recombination between bacteria switched the type or presence of capsule. Changes in antibiotic resistance in the isolates are also evident in the genomic data and were able to be mapped chronologically to the use of various antibiotics at the refugee camp.

Quick, easy sequencing means bacterial genomes can be given a second look

When bacterial genome sequencing began, it was the goal of sequencing projects to sequence new and unique samples, not to sequence the same samples again and again. Yet, as bacterial genome sequencing developed, it became evident that repeating some sequencing, **resequencing**, not only was practically possible, but also could uncover previously uninvestigated aspects of bacterial genomic biology.

The creation of transposon libraries, as discussed in Chapter 10, can help identify the function of CDSs. A transposon library can be grown in the laboratory and then genome sequenced to reveal the areas of the genome where no transposons have been inserted. This resequencing of the genome, where the genome is sequenced again, highlights any sequence changes, in this case the presence of the inserted transposons around the chromosome. When two conditions are compared, such as growth of the bacteria in the laboratory and then in an animal model, it is possible to identify the regions of the genome disrupted by transposons that are required for growth *in vivo*. This method has become known as **Tn-Seq, transposon sequencing**, and has been used to identify essential genes and those needed in particular environments (**Figure 11.8**).

Regions of the chromosome that interact with proteins can also be identified through resequencing in a process known as **ChIP-Seq**, where **chromatin immunoprecipitation** is done to reveal protein–DNA interaction and then the DNA regions are identified via sequencing. Proteins, such as regulatory proteins, interact with DNA and ChIP-Seq sequencing reveals where these interactions happen on the genome (**Figure 11.9**). The binding of DNA can be stabilized by

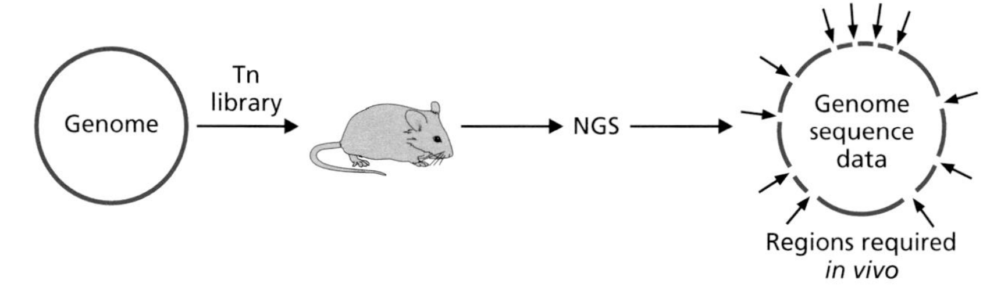

Figure 11.8 Tn-Seq can identify regions required in specific conditions, such as growth *in vivo*. A transposon library is made of the genome, which has previously been sequenced. The transposon library is subjected to the growth conditions, such as *in vivo* growth shown here. The bacteria that survive are genome sequenced by additional next-generation sequencing (NGS), resequencing the genome after the library has passed through *in vivo* conditions. Clones in the transposon library that have knocked out genes essential for growth *in vivo* will not be recovered and will not be present in the resequencing data (arrows).

cross-linking protein to DNA. Using an antibody that recognizes the protein, the fragment of DNA bound by the protein can be isolated and sequenced, identifying the protein binding site. Both Tn-Seq and ChIP-Seq use combined techniques to provide greater insight into bacterial genomics.

Sequencing genomes that had already been sequenced previously can resolve sequencing errors in the original data and could also reveal mutations that had accumulated within the genome since the first genome sequencing. A new investigative method has been established, referred to as **evolve and resequence**, in which bacteria are experimentally evolved *in vitro* or *in vivo* and the genome sequences before and after can then be compared. The famous experiments by Richard Lenski, in which *E. coli* has been continuously cultured in a laboratory for three decades, are now being genome sequenced at regular intervals to reveal how the *E. coli* genome has changed over time. Samples taken from patients with persistent bacterial infections are being sequenced to show how the bacteria mutate and evolve within the host. This has identified key factors involved in long-term survival *in vitro* and *in vivo*. These genomic studies looking at long-term growth of *E. coli* and persistent chronic bacterial infections will be discussed in Chapter 15.

Bacterial genome sequencing can uncover bacteria never before studied

Plague is caused by the bacteria *Yersinia pestis* and has caused infectious outbreaks in humans since the Bronze Age. Using DNA extracted from archeological samples collected from grave sites known to be used for plague victims, the genome sequences of *Y. pestis* from the past have been determined and analyzed. These samples included those from victims of the Black Death plague in Europe who died between 1348 and 1350. This revealed how plague

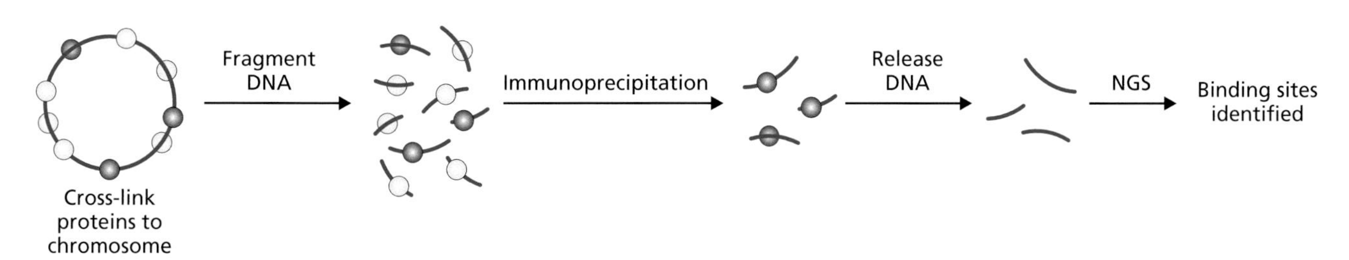

Figure 11.9 ChIP-Seq can identify protein binding sites in the genome. Proteins can be cross-linked to the DNA, preventing the protein from releasing the bound DNA. Fragmentation of the DNA and immunoprecipitation using an antibody against the protein separate out the bound DNA from the rest of the genome. Isolated fragments can be sequenced (next-generation sequencing, NGS) to reveal the protein binding sites.

bacteria were transmitted from Asia and across Europe. This historical perspective that can be achieved using ancient DNA from pandemics would otherwise be impossible to investigate. This has been useful in understanding how plague might be controlled today in areas where *Y. pestis* infections remain a public health concern. The sequencing of historic plagues and outbreaks can also enhance our understanding of how new infectious diseases emerge. Uncovering the genomics of plague will be discussed further in Chapter 15.

Metagenomic sequencing has uncovered a wider array of bacterial species than were previously appreciated. The vast majority of bacteria on the planet cannot be cultured in the laboratory because the needed culture conditions and nutrient requirements are not known. Therefore, the identification of life on this planet is far from complete, although metagenomic sequencing is revealing more through sequencing of DNA from an environment all together, without separating species or culturing samples. One of the first attempts to sequence a mixed pool of environmentally collected DNA was in 2004. Samples of water from the Sargasso Sea were collected, and sequencing of the DNA contained within them found over a million new CDS sequences. These presumably have originated from never before cultured and identified bacterial species.

Genome-sequencing data can be used on bacteria that we are unable to culture in the laboratory. *Tropheryma whipplei* could only be cultured in the laboratory in association with human cells in cell culture. No broth or agar media was known to be able to support growth of *T. whipplei*. When *T. whipplei* was genome sequenced, the data as to which metabolic pathways that the bacteria did and did not possess were used to engineer a medium that was shown to be capable of supporting growth, without the need for human cells. Culture-independent sequencing of DNA from samples has the potential to enable the identification of bacteria that cannot be grown in the lab, including identification of pathogens not previously known. Genome sequencing and analysis of metabolic pathways then make it possible to design new types of growth media, increasing the number of culturable bacteria.

Single-molecule sequencing is more sensitive and produces longer read lengths

Following on from experiments revealing the principles of genetics, the concept of genes and their place on DNA, and the structural determination of DNA, there came a desire to know the sequence of that DNA. This led to Sanger sequencing, which then led to the process being automated. When the desire to know the DNA sequence challenged the capacity of automated Sanger sequencing, high-throughput next-generation sequencing came about. The achievements you have read about in this chapter thus far have been made using these technologies. However, technology and the desire to generate more sequences did not stop there. The next challenges were to increase the sensitivity of the sequencing so that the need to amplify the DNA could be eliminated and to increase the read length. Both goals were achieved through **single-molecule sequencing**.

In all genome-sequencing technologies that came before, the genomic DNA was amplified to ensure there was enough DNA and enough signal to be detected. In shotgun sequencing, the fragmented genome pieces are cloned into plasmids, enabling their amplification within *E. coli*. In next-generation technologies, the fragments of the genome that are generated are amplified *in vitro* on the surface used by the sequencing technology. Single-molecule sequencing overcame the need to boost the sequencing signal with amplification by scaling down the process to focus on determining the base sequence of one single strand of DNA.

Two different technologies have emerged that are able to sequence a single molecule of DNA and as a result both produce much longer reads than previously achieved by Sanger sequencing or next-generation sequencing. The starting DNA material does not need to be heavily fragmented and the resulting data do not require as much assembly.

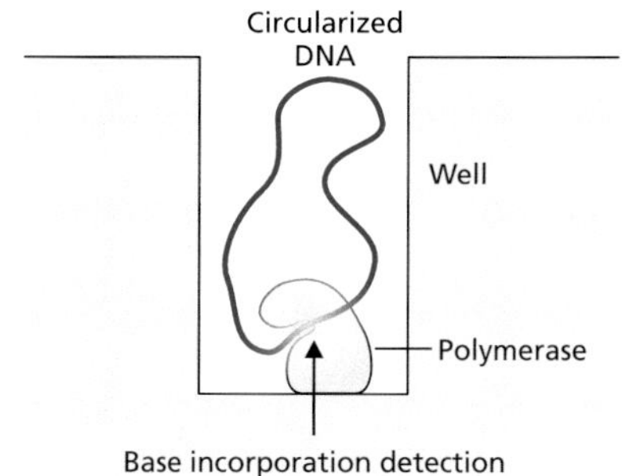

Figure 11.10 Pacific Biosciences single-molecule sequencing. To achieve sufficient fluorescence sequencing from a single molecule, this sequencing platform immobilizes the DNA polymerase used in the sequencing by synthesis reaction at the bottom of the sequencing well. Sensitive detection in this area of the well enables the circularized DNA to be sequenced. Circularization means that the same piece of DNA can be sequenced several times, thereby resolving sequencing errors by sequencing the same molecule several times.

Two single-molecule sequencing technologies have emerged, including physical reading of the DNA

The Pacific Biosciences single-molecule sequencing technology continues to use a sequencing-by-synthesis approach; however, unlike earlier technologies, this system immobilizes the DNA polymerase in a discrete location. The precise location where nucleotide incorporation and the resulting fluorescent signal is known, enabling the signal from one molecule of DNA to be detected (**Figure 11.10**). Single-stranded DNA (ssDNA) is threaded through the immobilized DNA polymerase, which is provided with labeled nucleotides for synthesis of the second strand. The area of anchored polymerase is monitored for incorporation of nucleotides and read as sequence (see Figure 11.10). The DNA fragment is circularized; therefore, the same fragment can be sequenced over and over again, to increase the accuracy of the sequencing data.

Nanopore single-molecule sequencing technology does not use sequencing by synthesis. Instead, the sequence of bases in the DNA is physically read as the single-stranded molecule of DNA passes through a **nanopore**. A nanopore has a very small opening that is created so that it only allows through ssDNA. As the ssDNA passes through the nanopore, the characteristics of the DNA are monitored and in this way the sequence of the DNA is determined. Oxford Nanopore technology uses nanopores embedded in a membrane, through which the ssDNA passes and individual bases are detected. To slow the passage of the ssDNA through the nanopore, a motor protein sits on top, making resolution and detection of data achievable (**Figure 11.11**).

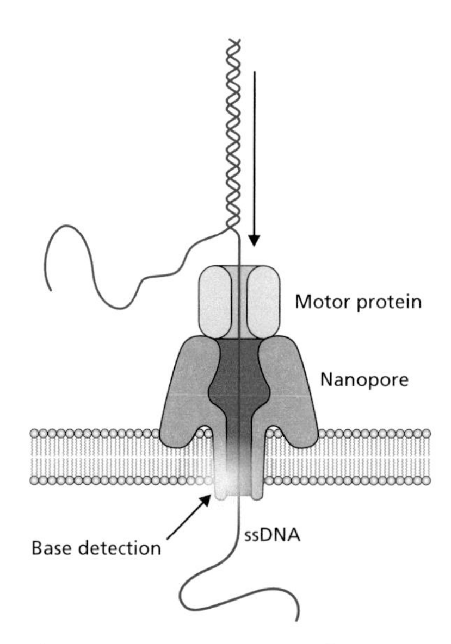

Figure 11.11 Nanopore single-molecule sequencing. This sequencing platform uses physical detection of the bases of the DNA as the ssDNA passes through the nanopore, slowed in its passage by a motor protein.

Both single-molecule sequencing technologies have lower accuracy than previous methods. This means that there is a higher likelihood that single-molecule technologies will incorrectly identify a base than if Sanger or next-generation sequencing is used. These inaccuracies can be overcome by resequencing the single molecule of DNA and getting a consensus identification of each base. However, this adds to the time involved. The trade-off for the loss of accuracy is the generation of much longer sequences that are 10–100 times longer than next-generation sequencing. Single-molecule sequencing technologies are therefore especially useful to use as a **scaffold** against which to assemble more accurate next-generation sequencing data. The scaffold is a rough sequence known to have errors, but against which more robust sequencing data can be assembled into the correct order and then the original genome can be configured.

Key points

- Bacterial genome sequencing became possible with advances in the Sanger sequencing-by-synthesis method that automated the process.
- Comparisons between sequence data identify differences at the DNA level that may influence the biology of the organism.
- Knowledge gained from the first bacterial genome sequences was motivational in the development of improved technologies and methods for rapid, inexpensive, accurate genome sequencing.
- Analysis of bacterial genome sequences can reveal information that can inform the development of vaccines and diagnostic tests and can be used in epidemiology.

- Genome sequence data are generated as fragments that must be pieced back together and assembled into the order of the original chromosome.
- Rapid, inexpensive genome sequencing has resulted in an abundance of sequencing data, including repeated sequencing of the same bacterial species.
- Resequencing of the same strain can reveal information about bacterial evolution and mutation and can be coupled with other techniques to reveal functional information.
- Metagenomic sequencing can be used to obtain information on bacterial presence in environmental and biological niches by sequencing all DNA from a sample area.
- Single-molecule sequencing is able to generate DNA sequence data without amplification.

Key terms

Define the following terms introduced in this chapter. Check your answers using the definitions in the Glossary. These terms are also available as Flashcards online.

Automated Sanger sequencing	Metagenomic sequencing	Reverse vaccinology
ChIP-Seq	Nanopore	Scaffold
Comparative genomics	Next-generation sequencing	Sequencing by synthesis
Contig	Read length	Sequencing platform
Data depth	Regions of difference	Shotgun sequencing
Evolve and resequence	Resequencing	Tn-Seq

Questions and discussion topics

Self-study questions

Answer each question using 50–100 words or a table or labeled diagram. Advice on where to find answers to these questions is available online.

1 What improvements have been made in Sanger sequencing since the technique was first established that have improved safety and ease of use for researchers?

2 What was the strategy that meant that although the *E. coli* genome-sequencing project was begun first, it was not the first bacterial genome sequence completed? What is the basic approach to this strategy?

3 What is comparative genomics and what can it reveal about the genome sequences that are being compared?

4 In next-generation sequencing, what is meant by the read length and how does this differ between sequencing platforms? Does data depth also differ?

5 How might read length and data depth impact upon the analysis of genome-sequencing data?

6 What kind of sequences can complicate assembly of genome sequence data and why?

7 How has the availability of next-generation sequencing caused a shift in the way genome sequence data are being used and the choice of bacteria to be sequenced?

8 Although genome sequencing is now inexpensive, it seems counterproductive to sequence the same bacterial strain again. What could justify resequencing a genome that has already been sequenced once?

9 What is the difference between genome sequencing and metagenomic sequencing? What might be learned from a metagenome?

10 Two different kinds of single-molecule sequencing have emerged. What is the main advantage of these methods over other sequencing technologies?

11 Of the sequencing methods described in this chapter, which do not use sequencing by synthesis? How do they work instead?

12 What is the difference between a contig and a scaffold? How are each formed?

Discussion topics

These topics are presented for discussion in study groups, as part of class discussions, or on your own. These questions go beyond what is directly covered in this part of the book. Use the research literature and other reading to explore these topics in more depth. Tips to help prepare for topic discussions are available online.

1. Describe how next-generation sequencing platforms such as 454, Illumina, and Ion Torrent built upon and developed what had previously been done by Sanger sequencing methods. Discuss common features between these next-generation sequencing technologies and also common features shared with Sanger sequencing. How do these relate to biological systems and processes?
2. Explore the ways in which next-generation sequencing has been used recently in bacterial epidemiology, either on a global scale or locally within a community or hospital. Assess the increased resolution possible using genomic data to demonstrate relationships between isolates and form timelines and theories of transmission events.
3. Explore the ways in which next-generation sequencing has been used recently in bacterial epidemiology, either on a global scale or locally within a community or hospital. Assess the increased resolution possible using genomic data to demonstrate relationships between isolates and form timelines and theories of transmission events.

Online quiz questions

To further self-assess your understanding of the chapter material, please visit the following link, where you can participate in a range of interactive quiz questions:

www.routledge.com/cw/snyder

Further reading

The first genome sequence of a free-living organism was bacterial

Fleischmann RD, Adams MD, White O, Clayton RA, Kirkness EF, Kerlavage AR. Whole-genome random sequencing and assembly of *Haemophilus influenzae* Rd. *Science*. 1995; *269(5223)*: 496–512.

Loman NJ, Pallen MJ. Twenty years of bacterial genome sequencing. *Nat Rev Microbiol*. 2015; *13*: 787–794.

Early genome sequencing of bacteria provided opportunities for new insight and innovation

Himmelreich R, Plagens H, Hilbert H, Reiner B, Hermann R. Comparative analysis of the genomes of the bacteria *Mycoplasma pneumoniae* and *Mycoplasma genitalium*. *Nucleic Acids Res*. 1997; *25*(*4*): 701–712.

Next-generation sequencing and epidemiology

Harris SR, Feil EJ, Holden MTG, Quail MA, Nickerson EK, Chantratita N. Evolution of MRSA during hospital transmission and intercontinental spread. *Science*. 2010; *327*(*5964*): 469–474.

Żukowska L, Zygała-Pytlos D, Struś K, Zabost A, Kozińska M, Augustynowicz-Kopeć E, Dziadek J, Minias A. An overview of tuberculosis outbreaks reported in the years 2011–2020. *BMC Infect Dis*. 2023; *23*(*1*): 253.

Quick, easy sequencing means bacterial genomes can be given a second look

Card KJ, Thomas MD, Graves JL Jr, Barrick JE, Lenski RE. Genomic evolution of antibiotic resistance is contingent on genetic background following a long- term experiment with *Escherichia coli*. *Proc Natl Acad Sci USA*. 2021; *118*(*5*): e2016886118.

Churchward CP, Calder A, Snyder LAS. Mutations in *Neisseria gonorrhoeae* grown in sub-lethal concentrations of monocaprin do not confer resistance. *PLoS ONE*. 2018; *13*(*4*): e0195453.

Good BH, McDonald MJ, Barrick JE, Lenski RE, Desai MM. The dynamics of molecular evolution over 60,000 generations. *Nature*. 2017; *551*(*7678*): 45–50.

Bacterial genome sequencing can uncover bacteria never before studied

Feres M, Retamal-Valdes B, Gonçalves C, Cristina Figueiredo L, Teles F. Did Omics change periodontal therapy? *Periodontol 2000*. 2021; *85*(*1*): 182–209.

Quince C, Walker AW, Simpson JT, Loman NJ, Segata N. Shotgun metagenomics, from sampling to analysis. *Nat Biotechnol*. 2017; *35*(*9*): 833–844.

Chapter

12

Bioinformatics

Bioinformatics is a complex, multidisciplinary field that focuses on the analysis and interpretation of biological data. In the context of bacterial genetics and genomics, bioinformatics refers to the investigation of nucleic acid sequence data and amino acid sequence data. Bioinformatics brings together computer science, mathematics, statistics, engineering, physics, chemistry, and biology. Primarily, bioinformatics is computer-based; however, the investigations pursued on the computer can then be brought into the laboratory for verification or further experimentation. In this chapter, the power of bioinformatics in interpreting bacterial data will be explored, including examples of how bioinformatics can be applied to various research questions.

A lot can be learned from looking at strings of A's, T's, G's, and C's

At its most basic, a bacterial genome sequence is a few thousand to a few million adenines (A's), thymines (T's), guanines (G's), and cytosines (C's). Taken as one continuous sequence, these four letters represent the chemical make-up and order of molecules present in the genomic DNA. Hidden in the nucleic acid order are all of the genes that encode the bacterial proteins, all of the sequences involved in regulation of expression of those proteins, and the sequences of all RNA molecules present within the bacterial cell. Using bioinformatics, the locations of the features within the genome can be revealed.

There are no markers that set out definitively where genes occur in the genome based on the DNA sequence of A's, T's, G's, and C's alone. To find the genes among the bases, bioinformatics makes a best guess at gene location. These guesses are based on laboratory evidence regarding features such as the nucleic acid sequence of the initiation codons (ATG, GTG, TTG) and the termination codons (TGA, TAG, TAA); however, these sequences can be present in the genome without being initiation or termination codons. In addition, laboratory experiments have revealed the consensus sequences of promoter regions, particularly those recognized by common sigma factors. This can help **bioinformaticians**, those who do bioinformatics, in predicting the locations of promoters and matching them up to predicted locations of genes. These predictions can be further strengthened by identifying the potential ribosome binding sites between the predicted promoter and the predicted CDS (**Figure 12.1**).

The DNA binding sites of some regulatory proteins have been determined in the laboratory, which can also be fed into bioinformatics analysis to predict possible binding site locations. This may reveal which CDSs are involved in the **regulon** controlled by the transcriptional regulatory protein. The various CDSs that are co-regulated by the same regulatory protein can give insight into the functions they conduct together or the conditions under which expression is beneficial. Research into RNAs within the bacterial cell has also enabled bioinformatics predictions of the locations of the ribosomal RNAs, the transfer RNAs, and the non-coding RNAs in the genomic sequence data.

DOI: 10.1201/9781003380436-16

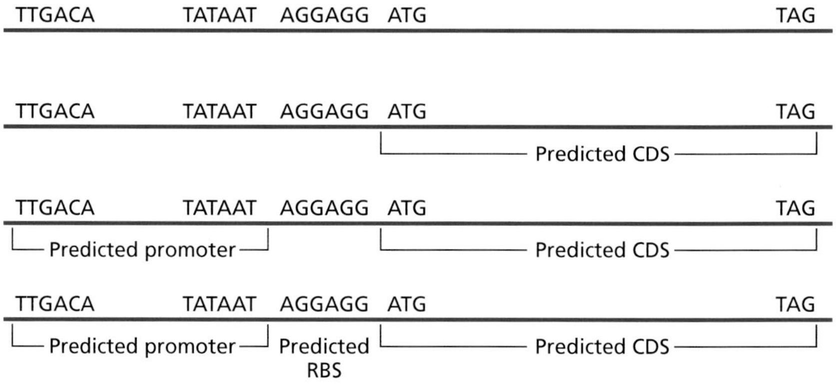

TTGACA TATAAT AGGAGG ATG TAG

TTGACA TATAAT AGGAGG ATG TAG
Predicted CDS

TTGACA TATAAT AGGAGG ATG TAG
Predicted promoter Predicted CDS

TTGACA TATAAT AGGAGG ATG TAG
Predicted promoter Predicted RBS Predicted CDS

Figure 12.1 Bioinformatics can predict potential genes and their associated sequences. Using information on initiation and termination codons, it is possible to predict the location of potential genes, as CDSs. Starting with just the DNA sequence data (top), potential initiation and termination codons can be identified. From this a predicted CDS is identified (second from top). Additional support for the predicted CDS comes from identification of a predicted promoter sequence (third from top) and from a predicted ribosome binding site (RBS; bottom).

Bioinformatics is essential for interpreting sequence data

Once sequence data are created, the goal is to interpret the data and make them available in public databases in a format that is accessible to other researchers so that the biological meaning of the sequence data can be understood in the context of the bacterial organism. To achieve this, bioinformatics is essential. The output of sequencing methods must first be processed before the search for genes and RNAs can occur. The **raw sequencing data** are the results of the experimental method for determining the order of nucleotides in a molecule of DNA. All sequencing methods are prone to errors and are less robust in their assessment of the sequence of the DNA as the read length (see Chapter 11) approaches the limits of the technology. There are also instances where a particular sequencing reaction will fail and not produce good quality data. Therefore, a first step for analysis of sequence data is to address the quality of the data and to trim the data down to sequence information that is believed to be correct. If bad data are left in the sequence, this may result in bad predictions later on down the line of investigation; therefore, it is important that the trimming and assessment of data quality is done well (**Figure 12.2**).

There are cases where the sequence of interest is shorter than the length limit of the technology and the sequence can be determined with one sequencing reaction. In these cases, generally an automated Sanger approach is taken, and, to increase the robustness of the sequence data achieved, two sequencing reactions will be used, one for each strand to sequence each strand of the DNA. This provides a check of each base determination using two different sequencing reactions that are sequencing the **reverse complement** sequence of one another

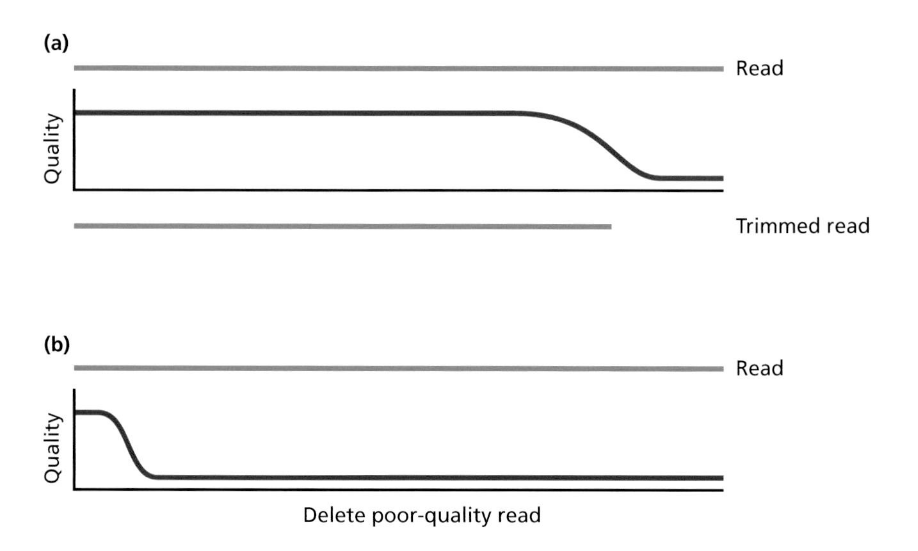

Figure 12.2 Raw sequencing data need to be trimmed. Raw sequencing data contain errors due to the limits of the sequencing technology and the chance that some sequencing reactions do not work well. The quality of the interpretations made later depends on the quality of the data; therefore, poor-quality sequences are removed before the sequencing data are analyzed. The top read (a) has a poor-quality sequence at the end; therefore, it is trimmed, removing the poor-quality sequence and producing a trimmed read that is included in later analyses. The bottom read (b) is poor quality throughout; therefore, the whole read is deleted and not included in further analyses.

Figure 12.3 Double-strand sequencing for small areas of interest increases accuracy. For a small region of interest that can be sequenced within the length limits of the sequencing methodology, it is possible to collect the needed sequence data with a single reaction. However, sequencing the DNA in both directions (sequencing reaction 1 and sequencing reaction 2) to sequence each of the strands of DNA will produce more robust sequence data (double-stranded data), having used two reactions that run in opposite directions to one another.

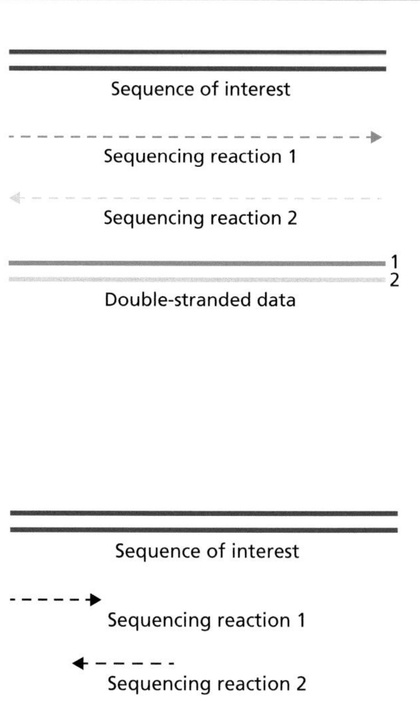

(**Figure 12.3**), being the sequence on the opposite strand where it is both in reverse and the complementary bases. In these cases, after trimming, a quick comparison of the two sequences and resolution of any differences between them is all that is needed.

For most sequencing efforts, more than one sequencing reaction, or double-stranded sequencing, is needed to achieve the identification of all of the bases of the relevant DNA. This may be because the region of interest for Sanger sequencing is longer than the read length that can be achieved. It may also be that the goal is the generation of a complete genome sequence. Either way, some **assembly** of the nucleic acid sequence data fragments into longer pieces of sequence is required. This will involve sequencing data that overlap with other sequencing data from the same sample, such that the overlaps can build upon each other and increase the length of known sequence information (**Figure 12.4**). The resulting assembled sequence information is a contig.

Sequence of interest

Sequencing reaction 1

Sequencing reaction 2

Sequencing reaction 3

Sequencing reaction 4

Sequencing reaction 5

Assembled contig

Figure 12.4 Assembly of sequence data relies on overlaps in the sequence information from each reaction. Each individual sequencing reaction generates a string of bases that are known to follow one after the other, regardless of whether they are on the forward or reverse strand. However, these may not cover the whole of the sequence desired. Therefore, several overlapping sequencing reactions are needed. These generate data where the overlaps allow assembly of the data into longer stretches of sequence. Once assembled, the contiguous sequences from different sequencing reactions are referred to as contigs.

Annotation predicts features in sequence data and notes their locations

An **annotation** is the labeling of features. Annotations are used in books, computer programs, and sequence data. In the context of DNA, the annotation describes the sequence, usually including the bacterial species and strain, as well as information as to the nature of the sequence, be it a short segment, plasmid, or a whole genome. An annotation will include the base numbers of particular features in the sequence. Some features, like CDSs, will be specific to one strand of the DNA, information that is also included in the annotation. The names of identified genes and RNAs and notations as to the attributes of the features in the sequence data will be annotated. These can include the amino acid sequence of the predicted protein product of a CDS, the predictions of subcellular locations of potential proteins, and other key pieces of information of relevance to the features within the sequence. Annotation is not just for genomes. Short segments of DNA and plasmids are also annotated to point out the features in the DNA or those believed to be in the DNA based on analysis (**Figure 12.5**).

A bacterial sequence annotation seeks to provide information on the CDSs and genetic features in the DNA. CDS annotations will include both genes with experimentally defined functions and sequences that look like they may be genes. Annotation of a feature such as a CDS can be as simple as noting its position within a stretch of DNA, but can also provide more information. The predicted protein sequence is often provided for CDSs and is based on the DNA sequence using a codon table to generate the amino acid sequence. If the sequence data suggest that there may be a signal peptide as part of a protein, then this can be noted in the annotation. The predicted amino acid sequence enables prediction of the molecular weight, isoelectric point, cellular location, and domains within the encoded protein. For CDSs that are known genes with experimentally derived functions, then the gene name can and should be included in the annotation. If the function or predicted function is enzymatic, an enzyme classification number can be included in the annotation. This is a numerical identifier for an enzyme based on its activity, identified as being similar in function regardless of feature names or sequence similarity. Annotations should include reference to experimental work, particularly where a feature is a gene with functions described in the literature. Some annotations also include GO (gene ontology) function, a set of terms for describing gene products, and COG (clusters of orthologous groups) designation,

Sequencing length 329

5–319 coding sequence

5–319 gene *expL* (an example gene)

5–319 amino acid sequence MRGLKNRG···

Figure 12.5 Gene annotation. This annotation is for an example DNA sequence of 329 bases, where the coding sequence starts at the fifth base and ends at base 319. This is a known gene, *expL*, with a predicted amino acid sequence (truncated here).

a database of similar protein sequences and their assigned functions. If the CDS described is a paralogue or orthologue of other annotated CDSs, this information may be valuable to include. It is possible that the sequence data may include **pseudogenes**, which are partial CDSs or those with internal stop codons. Like CDSs, these partial features can be included in the annotation and may be useful when comparing sequences.

Features in addition to CDSs can and should be included in an annotation. The identification of tRNAs and rRNAs is the most straightforward. Due to the conserved nature of these RNA sequences, the annotation of these features is quite robust. IS elements, transposons, and prophages are mobile genetic elements to include in annotations. Some of these are conserved between strains and species; therefore, the sequence similarities make them quite easy to identify and annotate. The sequences that include ncRNAs can be harder to annotate. Prediction programs for these may over-predict or under-predict the number and locations of ncRNAs in bacterial genome sequence data. The addition of experimental data on RNA expression, such as RNA-Seq, can enhance the annotation of ncRNAs and the source of such experimental data should be included in the annotation. Other sequence features within sequence data that can be included in annotations include repeats and the origin of replication.

The process of creating an annotation starts with the DNA sequence data

Annotation starts with the sequence information, generally as a **FASTA** format file. These data will be from the assembly of the sequence information into contigs. FASTA formatting of these data is common and is a generally accepted input for most sequence analysis tools. Information about the identity of the sequence is set out in the first line of the text in the FASTA file, set apart from the sequence information with a ">" symbol. This ">" is followed by some free text that should describe the data. This line ends with a hard return followed by the sequence data. The sequence data are generally just the A's, T's, G's, and C's; however, there can be ambiguous bases in the sequence data, such as Ns for any nucleotide, if there are issues in the data (**Figure 12.6**).

Sequence data can be written as plain text, and in many respects the FASTA file format is similar to plain text, but the first line contains important information about the sequence. This tends to be general information, such as the bacterial genus, species, and strain and perhaps some identifying features. In genome-sequencing projects, the FASTA files have information in the header that includes the number of the contig. Several contigs can be together in one file, which is sometimes referred to as a **multi-FASTA** format file. In these files, there is more than one ">" symbol, with each one differentiating the separate contigs containing sequence information (**Figure 12.7**).

```
>Example FASTA format sequence

ATTGGCCACGCTAATGCCCGTGTGGATTGCCATC
```

Figure 12.6 FASTA file format. In the first line of a FASTA file format, the ">" symbol is followed by a line of text that describes the sequence. In the second line, the sequencing data start and continue on until the end of the sequence.

```
>Example multi-FASTA format sequence 1

ATTGGCCACGCTAATGCCCGTGTGGATTGCCATC

>Example multi-FASTA format sequence 2

TCGGGAACTGTACG

>Example multi-FASTA format sequence 3

GCGGATGCGAAGGATGCCCACTGTTGACCGTTATTAAA
```

Figure 12.7 Multi-FASTA file format. Multi-FASTA is similar to the FASTA file format; however, there are multiple pieces of separate sequence information. Each sequence has a first line that starts with the ">" symbol and a description of the sequence, followed by the sequence in the next line. The end of one sequence is indicated by a hard return and a new line beginning with the ">" symbol. Multi-FASTA files can contain any number of separate sequences.

Nothing in the FASTA format points out the features within the DNA sequence. Although the header line may mention a key gene or genes within the sequence, it is not common for these to specify the base positions of genes, CDSs, or other genetic features. This is where annotation is needed and other formats of sequence data files are used.

Multiple lines of investigation into the sequence data features support the annotation

Adding either predicted or experimentally determined ribosome binding sites and transcriptional terminator locations can enhance CDS annotations, particularly for CDSs based solely on sequence-based predictions. It might be assumed when there are multiple potential initiation codons that the first should be selected, indicating the largest possible coding sequence, but this may not necessarily be the best choice. Identification of a ribosome binding site and transcriptional terminator around a potential coding sequence will better define the CDS location than just using the potential initiation and termination codons (**Figure 12.8**).

The inclusion of **evidence tags** in an annotation can provide an important trail back to the data used to formulate the final annotation. Evidence tags give researchers using the annotation the background that the annotators used to assign the particular locations, function, and other information in the annotation to a region of nucleic acid sequence. The basis of the decisions made when writing an annotation is not always clear, but can be included within the annotation itself. Experimental data can be cited. Tools that contributed to predictions can be cited. Empowered with this information, end users can use the evidence behind the annotation to decide for themselves if it is reliable (**Figure 12.9**).

Annotations tend to start with potential genes

The first step in the annotation of sequence data tends to be the identification of potential genes as CDSs. This makes use of the relatively easy identification of termination codons in each of the six potential reading frames, defining the ORFs, and then the identification of an initiation codon close to the start of each ORF, defining the CDS. Finding CDSs in this way can generate false positives, with sequences that look like CDSs but are not. Generally, these are short, being less than 100 bases. There could be a wrong start codon predicted, creating a false prediction of a CDS. In addition, evolution can create gene fragments, where what was a gene has been altered by frameshifts or in-frame termination codons to create a pseudogene. At its most basic, to submit a sequence to a public database where the data are accessible to other researchers, it must have the locations of CDSs annotated. Sequence data that support a journal article must be submitted to a public database; therefore, the sequence must have at least the most basic of annotations. Other features, including the predicted functions of the CDS, are not required, but are often added.

It is preferable for an annotation to indicate the function or potential function of the identified CDSs. To achieve this, the nucleic acid and predicted amino acid sequence of each CDS is compared to sequence databases to identify homology. If there is a hit, where the CDS sequence is similar to one or more sequences in

TGCCGTTACTTGATGGGTTTAAGGAGGTGAATG

Figure 12.8 Adding other features helps in determining CDS annotation. This sequence contains two potential initiation codons, underlined. Adding in the ribosome binding site, highlighted in yellow, it becomes clear that the second initiation codon is a better choice for the start of the coding sequence, since the ribosome binding site must come before the initiation codon.

TGCCGTTAGGAGGTTGATGGTGTTTAAGGTGCAT

Figure 12.9 An annotation with an evidence tag to justify a GTG start. In a sequence with two potential initiation codons, underlined, the inclusion of experimental evidence can clarify the choice. Here the region revealed through N-terminal protein sequencing, highlighted in yellow, supports annotation of the GTG annotation codon over the more standard ATG codon.

the databases, this homology is used in the annotation. Homology can reveal the identity of a known gene in the bacterial species, providing a wealth of annotation information. Homology searches may identify an **orthologue** in another bacterial species, a specific term for homologues of a common origin found in different species. The presence of orthologues suggests that the CDS is conserved between the two species and therefore may share a conserved function. **Paralogues** may also be identified, which are homologues contained within the same genome. These are likely the result of gene duplication events and each paralogous CDS may have the same function or may have diverged in its function. Likewise, there may be no homology with other sequences in the databases, or the homologues are of unknown function. This results in annotation of the CDS as a hypothetical gene encoding a hypothetical protein, which is either unique to the sequence being investigated or conserved among bacteria.

Homology and conserved protein domains can help identify the potential function of a CDS

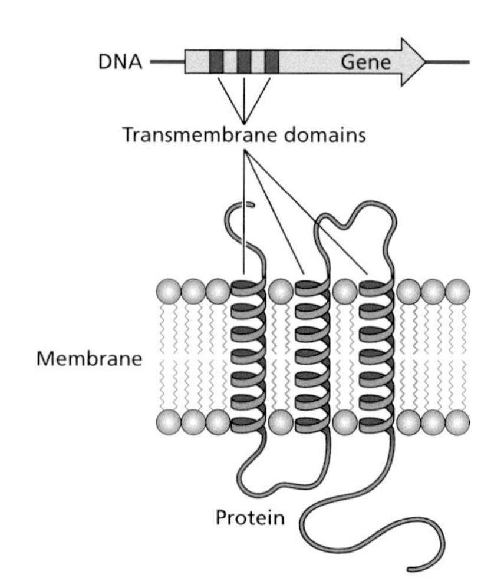

Figure 12.10 Protein domains. This integral membrane protein has three transmembrane α helices. Structurally, the α helices form in the protein and generate the transmembrane parts of the protein. These domains can be identified in the gene sequence and could help identify a CDS with an otherwise unknown function as a membrane protein.

DNA sequence homology between a CDS and a gene of known function is not as good as direct experimentation upon the CDS, but can give an indication of the potential function of the gene. This use of homology is relied upon extensively when annotating bacterial genome sequence data because demonstrating gene function in the laboratory for every potential gene is laborious and time consuming. Because CDS annotations rely upon a varying degree of sequence homology, it is possible that the annotated function is incorrect, but it is a starting point for further investigations. There are, however, a number of CDSs that do not correspond to gene sequences of known function. These either have no homology to other sequences or only have homology to other CDSs that have no known function. These are often annotated as encoding **hypothetical proteins**; the CDS sequence looks like it could be a gene, but there is no supporting experimental evidence. If hypothetical proteins have homologues in several different species, they can be referred to as **conserved hypothetical proteins**.

The functional units of proteins are **domains**, which have independently folding tertiary structures. Each domain has a distinct structure and function and most proteins have one or several domains (**Figure 12.10**). With the rise of DNA sequencing, protein domains are now often identified based on sequence data and not protein structure. Unlike structure-based identification of domains, which relies on the similarity of three-dimensional structures, sequence-based identification of domains can be automated due to the nature of the data. The identification of conserved domains within coding regions can help identify potential functions of a CDS, even when homology using the whole protein sequence cannot. The protein domain sequence is based on homology between protein fragments, rather than the whole protein. For example, a transmembrane domain suggests that a CDS is a membrane protein. The α helix or β barrel part of the protein identified in the sequence data suggests that the CDS has a transmembrane domain and is potentially encoding a membrane protein.

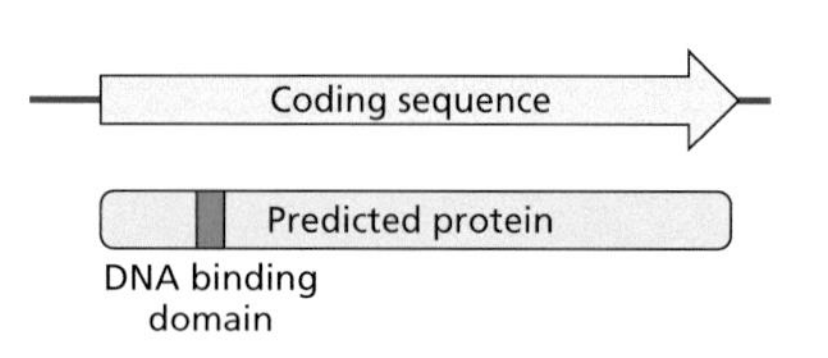

Figure 12.11 Domains within hypothetical proteins can improve annotations. A coding sequence translated into the predicted amino acid sequence and searched for protein domains reveals that there is similarity to DNA binding domains in a portion of the predicted protein, shaded blue. The coding sequence can be annotated as encoding a potential DNA binding protein.

When it comes to annotation, a predicted gene sequence might have an unknown function, but might carry a domain that is known to bind to DNA or be involved in transmembrane transport. The presence of a domain within a CDS can therefore aid in the annotation of the CDS, even when the whole of the CDS does not have an overall homology to other sequences. A CDS that might otherwise have been annotated as a hypothetical protein could be annotated as a DNA binding protein or transport protein, based on the domains identified (**Figure 12.11**).

About 20% of protein domains are identified as domains of unknown function. There are about 2,700 domains of unknown function in bacteria. These are identified as parts of predicted proteins that are common between predicted proteins of different species, but for which a function has not been identified. Some domains of unknown function are not only conserved, but also essential, being in CDSs identified as essential via techniques such as signature-tagged mutagenesis (see Chapter 10). When a protein structure for a hypothetical protein is determined, this may show that identified domains of unknown function are actually divergent versions of known domains. It is possible for proteins to be very different at the DNA and protein sequence levels but are very similar structurally.

Automated annotations rapidly produce an annotation that needs manual curation

Many annotations, particularly those associated with large-scale bacterial genome sequencing projects, are generated through automated processes with little or no manual curation of the resulting annotation. This is, in part, due to the requirement for a basic annotation to submit sequences to the databases, which is required for publication and dissemination of research findings. In addition, a preliminary annotation can help researchers on the large-scale project to achieve their analytical aims.

Any sequence annotation can have errors and all annotations should be looked at with a critical eye, whether generated through an automated system or months of meticulous manual annotation. Most automated annotation systems combine a variety of computational programs to find the different features that can be present in DNA. Some use more than one program to find the same features and then select the best, based on agreement between the predictions made. However, it is always best to manually curate an annotation to achieve the best results because there can be more than one ATG, GTG, and/or TTG toward the beginning of an ORF, because annotations in databases used can be incorrect, and because evidence trails supporting choices made during other annotations are not always recorded, which can all result in incorrect annotations.

The automated annotation of the tRNA and rRNA loci is often robust due to the conserved nature of these RNAs. The predictions for these features can be relied upon and form a solid set of data with regard to the number of copies of the rRNA loci and the types of amino acids and codons used by the bacterial strain being assessed.

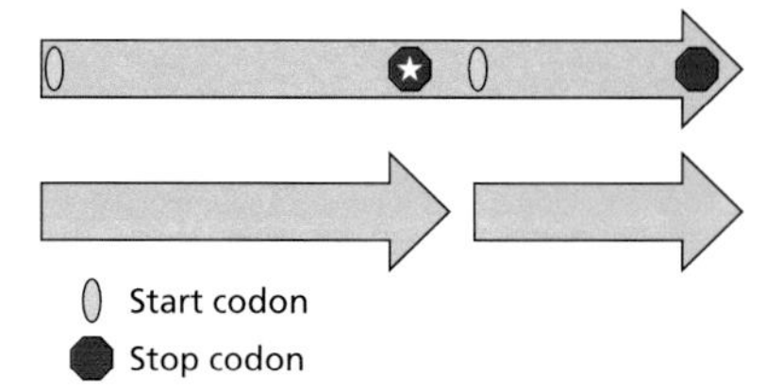

Figure 12.12 One gene becomes two. The introduction of a termination codon within a gene can sometimes result in the sequence forming two genes. One gene runs from the original initiation codon to the new termination codon (stop with *). The second gene uses an ATG, GTG, or TTG initiation codon that was otherwise within the original gene and ends with the original termination codon. The two genes generated will be in-frame with each other.

Some features in sequence data and annotation data can confuse new annotations

An annotation relies on the rules for a CDS being followed, but as biological organisms, bacteria do not always follow the rules. For instance, what was one gene can be split into two genes through mutation (**Figure 12.12**). This is, essentially, one gene with an internal termination codon that then looks like two genes from the sequencing data and, following the mutation, may actually be functional as two separate genes. Alternatively, the two fragments may now be non-functional.

Mutation can also fuse what was two genes into one. In this case, there are two genes that lose the termination codon between them and then become one gene, which again may or may not be functional (**Figure 12.13**). Some such genes encode two functional domains and bacteria are able to switch between them being either in one gene and therefore both domains in one protein or two genes and therefore one domain for each of the two proteins. Using only the sequencing data, it can be difficult to identify such switching because the sequencing process captures the DNA at one specific time point. Dynamic changes to the genome can be revealed through comparative sequence analysis and laboratory experimentation.

What is perceived as a mutation that has turned one gene into two or two genes into one could actually be an error in the sequencing data. However, even

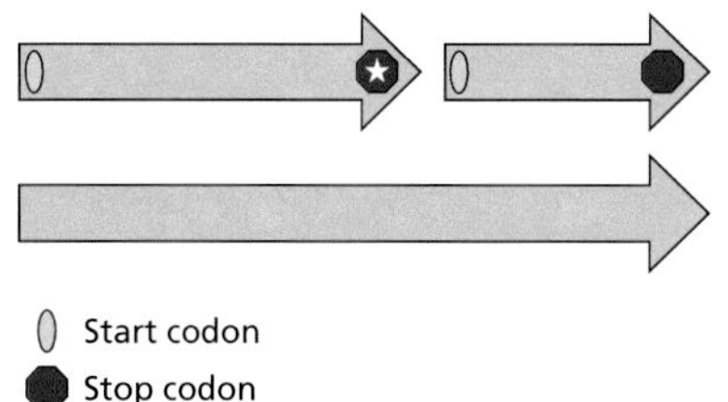

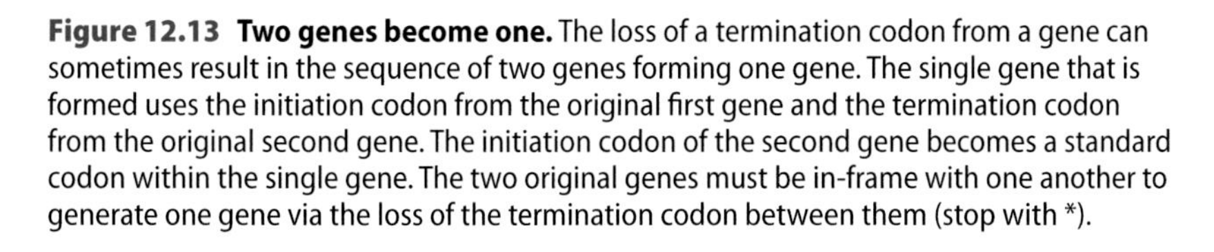

Figure 12.13 Two genes become one. The loss of a termination codon from a gene can sometimes result in the sequence of two genes forming one gene. The single gene that is formed uses the initiation codon from the original first gene and the termination codon from the original second gene. The initiation codon of the second gene becomes a standard codon within the single gene. The two original genes must be in-frame with one another to generate one gene via the loss of the termination codon between them (stop with *).

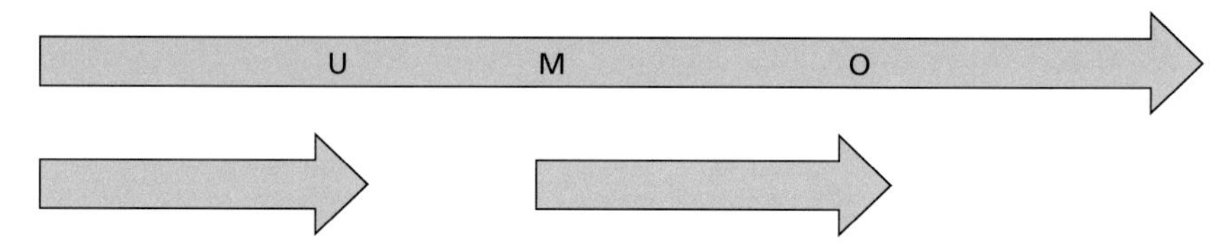

Figure 12.14 Selenocysteine and pyrrolysine codons and sequence annotation. A coding sequence (top line) containing a selenocysteine codon (U) and a pyrrolysine codon (O) could easily be annotated incorrectly. Identification of initiation codons and termination codons to define the coding sequences in the region incorrectly identifies two coding sequences, both much shorter than the original and neither containing the amino acids selenocysteine or pyrrolysine. The first uses the correct initiation codon, but takes the selenocysteine codon as a termination codon. The second uses an internal methionine codon (M) as an initiation codon and takes the pyrrolysine codon as a termination codon. Neither uses the true termination codon, which is after the pyrrolysine codon.

if potential sequencing errors that could generate such a situation are eliminated, the annotation in these situations can be tricky. Either of the annotations, as one gene or two, may be correct or both could be if the species switches between one gene and two genes. In such situations, a detailed annotation of the potentials is warranted.

As discussed in Chapter 7, there are genes that make use of the SECIS and PYLIS sequences to change a termination codon into a selenocysteine or pyrrolysine, respectively. Annotations often overlook these due to the nature of the progression of the annotation, which starts with ORF definition. Since the selenocysteine and pyrrolysine codons are stop codons, altered to encode these amino acids via the SECIS and PYLIS sequences, the ORF will be incorrectly defined at the start of the annotation process (**Figure 12.14**).

Naming genes is not straightforward, with some genes having more than one name

During the course of research into the functions of genes, it has been very easy for those that are orthologues to end up with different gene names in different species or even different gene names to be used by different laboratories to refer to genes in the same species. In some annotations, gene names are absent and only functional data for CDSs are included; however, others attempt to place gene names with sequences where possible. This can be problematic when there is more than one name for the gene. In a good annotation, where a gene is known in the literature by more than one name, the synonym will be listed in the annotation.

Complicating annotation of gene names are instances of the same name being used for different genes that have different functions. During the course of research, a research project will name a gene being investigated and use that name in reporting their findings in the literature. It may be that this name is already in use in this species or that the name is used in another species. When the whole of the genome is assessed for homologues and annotated, CDSs are assigned the same name using information from different sources, which can result in genomes having more than one gene with the same gene name.

Annotation errors, including spelling mistakes, can spread from one annotation to many others

Annotations rely on data from other annotations in the databases. Homology searches identify similar sequences in other annotated data and these other annotations contribute to new annotations. The result is that each annotation has the potential to be propagated across many other bacterial sequences. Errors in one annotation can expand to be errors in many, with the incorrect information spreading. This is most obvious in the propagation of spelling errors within annotations. While it may seem trivial to worry about spelling errors when some annotations are simply wrong in their identification of sequences and their function, spelling errors present a problem for analysis. For example, if the function of a CDS is not spelled correctly, this makes simple text searches for genes of that particular function impossible. It is possible for internet search engines to identify potential spelling errors and suggest to the user that they have misspelled something. Incorporation of such programming into the tools used in genetics and genomics would be useful for the identification of annotation spelling errors.

Gene locus identifiers are handy for labeling features in annotations, but reveal nothing about function

The original annotation of *Escherichia coli* strain K-12 assigned gene names to all of the annotated CDSs. Genes of known function with previously assigned names were annotated as such, but those that were not previously named were given names starting with *y*. This system was used in the early days of bacterial genomics to identify features and relative locations in the genome. These *y* gene names, however, tell us nothing about the functions of the CDSs.

CDS identifiers used in genome annotations are not strictly gene names. Gene names are based on the function of a gene and should not be applied to CDSs. Therefore, the *y* names should not be used in any annotation outside *E. coli* strain K-12. In such an instance where homology with one of these CDSs is noted, it is acceptable for the annotation to note the homology, but the *y* name should not be used as a gene name in other annotations. In *E. coli*, when a function is identified experimentally, the *y* gene designation should be replaced with a name related to the gene's function.

Later annotations of bacterial genomes used **locus identification numbers**. These are combinations of letters, to specify the bacterial genome, and numbers, to indicate the relative location on the chromosome. Locus identification numbers are simply markers for CDSs and unambiguously refer to a single feature in the genome (**Figure 12.15**). In most annotations, features will have both the locus IDs and the gene names, where applicable. As with the *E. coli y* names, the locus identification numbers should not be transferred across to another annotation's gene name. The identifiers from other genome sequence annotations can go into the notes in an annotation to indicate homology.

During the course of annotation, identified CDSs are numbered sequentially along with a specific tag to identify the genome sequence itself (see Figure 12.15). These locus identification numbers are not gene names, but can clearly identify a particular CDS from a particular genome sequence.

There are three major public databases for sequence data

Sequence data, either nucleic acid or amino acid, are available via public databases. Sequences that are published in the research literature are required to be deposited into one of these databases so that they are freely available to other researchers. There are three major public databases for sequence data, which together form the International Nucleotide Sequence Database Collaboration: the National Center for Biotechnology Information (NCBI) GenBank; the European Bioinformatics Institute European Nucleotide Archive (EBI ENA); and the DNA

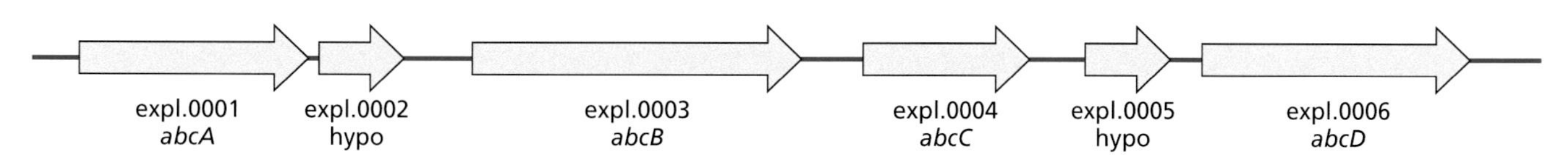

Figure 12.15 Locus identification numbers indicate relative locations of annotated CDSs. Annotation assigns a locus identification number to each identified coding sequence in numerical order along the sequence. To these locus identification numbers, additional annotation information can be added, such as gene names or identification as encoding a hypothetical protein (hypo). The locus identification numbers allow the sequence to be unambiguously referred to, whereas there may be multiple sequences annotated as *abcB* in the databases.

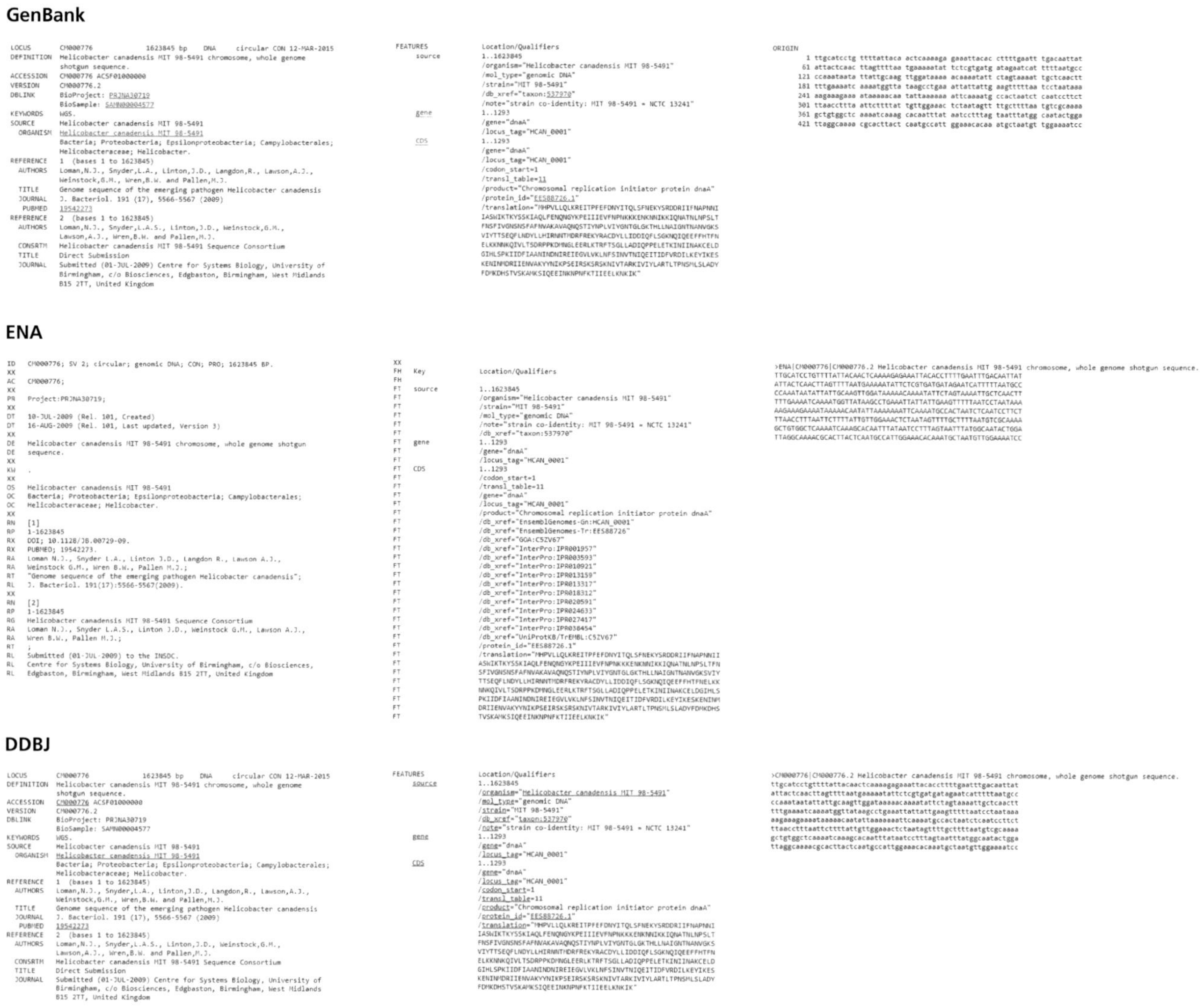

GenBank

```
LOCUS       CM000776               1623845 bp    DNA     circular CON 12-MAR-2015
DEFINITION  Helicobacter canadensis MIT 98-5491 chromosome, whole genome
            shotgun sequence.
ACCESSION   CM000776 ACSF01000000
VERSION     CM000776.2
DBLINK      BioProject: PRJNA30719
            BioSample: SAMN00004577
KEYWORDS    WGS.
SOURCE      Helicobacter canadensis MIT 98-5491
  ORGANISM  Helicobacter canadensis MIT 98-5491
            Bacteria; Proteobacteria; Epsilonproteobacteria; Campylobacterales;
            Helicobacteraceae; Helicobacter.
REFERENCE   1  (bases 1 to 1623845)
  AUTHORS   Loman,N.J., Snyder,L.A., Linton,J.D., Langdon,R., Lawson,A.J.,
            Weinstock,G.M., Wren,B.W. and Pallen,M.J.
  TITLE     Genome sequence of the emerging pathogen Helicobacter canadensis
  JOURNAL   J. Bacteriol. 191 (17), 5566-5567 (2009)
   PUBMED   19542273
REFERENCE   2  (bases 1 to 1623845)
  AUTHORS   Loman,N.J., Snyder,L.A.S., Linton,J.D., Weinstock,G.M.,
            Lawson,A.J., Wren,B.W. and Pallen,M.J.
  CONSRTM   Helicobacter canadensis MIT 98-5491 Sequence Consortium
  TITLE     Direct Submission
  JOURNAL   Submitted (01-JUL-2009) Centre for Systems Biology, University of
            Birmingham, c/o Biosciences, Edgbaston, Birmingham, West Midlands
            B15 2TT, United Kingdom
```

```
FEATURES             Location/Qualifiers
     source          1..1623845
                     /organism="Helicobacter canadensis MIT 98-5491"
                     /mol_type="genomic DNA"
                     /strain="MIT 98-5491"
                     /db_xref="taxon:537970"
                     /note="strain co-identity: MIT 98-5491 = NCTC 13241"
     gene            1..1293
                     /gene="dnaA"
                     /locus_tag="HCAN_0001"
     CDS             1..1293
                     /gene="dnaA"
                     /locus_tag="HCAN_0001"
                     /codon_start=1
                     /transl_table=11
                     /product="Chromosomal replication initiator protein dnaA"
                     /protein_id="EES88726.1"
                     /translation="MHPVLLQLKREITPFEFDNYITQLSFNEKYSRDDRIIFNAPNNI
                     IASWIKTKYSSKIAQLFENQNGYKPEIIIEVFNPNKKKENKNNIKKIQNATNLNPSLT
                     FNSFIVGNSNSFAFNVAKAVAQNQSTIYNPLVIYGNTGLGKTHLLNAIGNTNANVGKS
                     VIYTTSEQFLNDYLLHIRNNTMDRFREKYRACDYLLIDDIQFLSGKNQIQEEFFHTFN
                     ELKKNNKQIVLTSDRPPKDMNGLEERLKTRFTSGLLADIQPPELETKINIINAKCELD
                     GIHLSPKIIDFIAANINDNIREIEGVLVKLNFSINVTNIQEITIDFVRDILKEYIKES
                     KENINMDRIIENVAKYYNIKPSEIRSKSRSKNIVTARKIVIYLARTLTPNSMLSLADY
                     FDMKDHSTVSKAMKSIQEEINKNPNFKTIIEELKNKIK"
```

```
ORIGIN
        1 ttgcatcctg ttttattaca actcaaaaga gaaattacac cttttgaatt tgacaattat
       61 attactcaac ttagttttaa tgaaaaatat tctcgtgatg atagaatcat ttttaatgcc
      121 ccaaataata ttattgcaag ttggataaaa acaaaatatt ctagtaaaat tgctcaactt
      181 tttgaaaatc aaaatggtta taagcctgaa attattattg aagtttttaa tcctaataaa
      241 aagaaagaaa ataaaaacaa tattaaaaaa attcaaaatg ccactaatct caatccttct
      301 ttaaccttta attcttttat tgttggaaac tctaatagtt ttgcttttaa tgtcgcaaaa
      361 gctgtggctc aaaatcaaag cacaatttat aatcctttag taatttatgg caatactgga
      421 ttaggcaaaa cgcacttact caatgccatt ggaaacacaa atgctaatgt tggaaaatcc
```

ENA

```
ID   CM000776; SV 2; circular; genomic DNA; CON; PRO; 1623845 BP.
XX
AC   CM000776;
XX
PR   Project:PRJNA30719;
XX
DT   10-JUL-2009 (Rel. 101, Created)
DT   16-AUG-2009 (Rel. 101, Last updated, Version 3)
XX
DE   Helicobacter canadensis MIT 98-5491 chromosome, whole genome shotgun
DE   sequence.
XX
KW   .
XX
OS   Helicobacter canadensis MIT 98-5491
OC   Bacteria; Proteobacteria; Epsilonproteobacteria; Campylobacterales;
OC   Helicobacteraceae; Helicobacter.
XX
RN   [1]
RP   1-1623845
RX   DOI; 10.1128/JB.00729-09.
RX   PUBMED; 19542273.
RA   Loman N.J., Snyder L.A., Linton J.D., Langdon R., Lawson A.J.,
RA   Weinstock G.M., Wren B.W., Pallen M.J.;
RT   "Genome sequence of the emerging pathogen Helicobacter canadensis";
RL   J. Bacteriol. 191(17):5566-5567(2009).
XX
RN   [2]
RP   1-1623845
RG   Helicobacter canadensis MIT 98-5491 Sequence Consortium
RA   Loman N.J., Snyder L.A.S., Linton J.D., Weinstock G.M., Lawson A.J.,
RA   Wren B.W., Pallen M.J.;
RT   ;
RL   Submitted (01-JUL-2009) to the INSDC.
RL   Centre for Systems Biology, University of Birmingham, c/o Biosciences,
RL   Edgbaston, Birmingham, West Midlands B15 2TT, United Kingdom
```

```
XX
FH   Key             Location/Qualifiers
FH
FT   source          1..1623845
FT                   /organism="Helicobacter canadensis MIT 98-5491"
FT                   /strain="MIT 98-5491"
FT                   /mol_type="genomic DNA"
FT                   /note="strain co-identity: MIT 98-5491 = NCTC 13241"
FT                   /db_xref="taxon:537970"
FT   gene            1..1293
FT                   /gene="dnaA"
FT                   /locus_tag="HCAN_0001"
FT   CDS             1..1293
FT                   /codon_start=1
FT                   /transl_table=11
FT                   /gene="dnaA"
FT                   /locus_tag="HCAN_0001"
FT                   /product="Chromosomal replication initiator protein dnaA"
FT                   /db_xref="EnsemblGenomes-Gn:HCAN_0001"
FT                   /db_xref="EnsemblGenomes-Tr:EES88726"
FT                   /db_xref="GOA:C5ZV67"
FT                   /db_xref="InterPro:IPR001957"
FT                   /db_xref="InterPro:IPR003593"
FT                   /db_xref="InterPro:IPR010921"
FT                   /db_xref="InterPro:IPR013159"
FT                   /db_xref="InterPro:IPR013317"
FT                   /db_xref="InterPro:IPR018312"
FT                   /db_xref="InterPro:IPR020591"
FT                   /db_xref="InterPro:IPR024633"
FT                   /db_xref="InterPro:IPR027417"
FT                   /db_xref="InterPro:IPR038454"
FT                   /db_xref="UniProtKB/TrEMBL:C5ZV67"
FT                   /protein_id="EES88726.1"
FT                   /translation="MHPVLLQLKREITPFEFDNYITQLSFNEKYSRDDRIIFNAPNNII
FT                   ASWIKTKYSSKIAQLFENQNGYKPEIIIEVFNPNKKKENKNNIKKIQNATNLNPSLTFN
FT                   SFIVGNSNSFAFNVAKAVAQNQSTIYNPLVIYGNTGLGKTHLLNAIGNTNANVGKSVIY
FT                   TTSEQFLNDYLLHIRNNTMDRFREKYRACDYLLIDDIQFLSGKNQIQEEFFHTFNELKK
FT                   NNKQIVLTSDRPPKDMNGLEERLKTRFTSGLLADIQPPELETKINIINAKCELDGIHLS
FT                   PKIIDFIAANINDNIREIEGVLVKLNFSINVTNIQEITIDFVRDILKEYIKESKENINM
FT                   DRIIENVAKYYNIKPSEIRSKSRSKNIVTARKIVIYLARTLTPNSMLSLADYFDMKDHS
FT                   TVSKAMKSIQEEINKNPNFKTIIEELKNKIK"
```

```
>ENA|CM000776|CM000776.2 Helicobacter canadensis MIT 98-5491 chromosome, whole genome shotgun sequence.
TTGCATCCTGTTTTATTACAACTCAAAAGAGAAATTACACCTTTTGAATTTGACAATTAT
ATTACTCAACTTAGTTTTAATGAAAAATATTCTCGTGATGATAGAATCATTTTTAATGCC
CCAAATAATATTATTGCAAGTTGGATAAAAACAAAATATTCTAGTAAAATTGCTCAACTT
TTTGAAAATCAAAATGGTTATAAGCCTGAAATTATTATTGAAGTTTTTAATCCTAATAAA
AAGAAAGAAAATAAAAACAATATTAAAAAAATTCAAAATGCCACTAATCTCAATCCTTCT
TTAACCTTTAATTCTTTTATTGTTGGAAACTCTAATAGTTTTGCTTTTAATGTCGCAAAA
GCTGTGGCTCAAAATCAAAGCACAATTTATAATCCTTTAGTAATTTATGGCAATACTGGA
TTAGGCAAAACGCACTTACTCAATGCCATTGGAAACACAAATGCTAATGTTGGAAAATCC
```

DDBJ

```
LOCUS       CM000776               1623845 bp    DNA     circular CON 12-MAR-2015
DEFINITION  Helicobacter canadensis MIT 98-5491 chromosome, whole genome
            shotgun sequence.
ACCESSION   CM000776 ACSF01000000
VERSION     CM000776.2
DBLINK      BioProject: PRJNA30719
            BioSample: SAMN00004577
KEYWORDS    WGS.
SOURCE      Helicobacter canadensis MIT 98-5491
  ORGANISM  Helicobacter canadensis MIT 98-5491
            Bacteria; Proteobacteria; Epsilonproteobacteria; Campylobacterales;
            Helicobacteraceae; Helicobacter.
REFERENCE   1  (bases 1 to 1623845)
  AUTHORS   Loman,N.J., Snyder,L.A., Linton,J.D., Langdon,R., Lawson,A.J.,
            Weinstock,G.M., Wren,B.W. and Pallen,M.J.
  TITLE     Genome sequence of the emerging pathogen Helicobacter canadensis
  JOURNAL   J. Bacteriol. 191 (17), 5566-5567 (2009)
   PUBMED   19542273
REFERENCE   2  (bases 1 to 1623845)
  AUTHORS   Loman,N.J., Snyder,L.A.S., Linton,J.D., Weinstock,G.M.,
            Lawson,A.J., Wren,B.W. and Pallen,M.J.
  CONSRTM   Helicobacter canadensis MIT 98-5491 Sequence Consortium
  TITLE     Direct Submission
  JOURNAL   Submitted (01-JUL-2009) Centre for Systems Biology, University of
            Birmingham, c/o Biosciences, Edgbaston, Birmingham, West Midlands
            B15 2TT, United Kingdom
```

```
FEATURES             Location/Qualifiers
     source          1..1623845
                     /organism="Helicobacter canadensis MIT 98-5491"
                     /mol_type="genomic DNA"
                     /strain="MIT 98-5491"
                     /db_xref="taxon:537970"
                     /note="strain co-identity: MIT 98-5491 = NCTC 13241"
     gene            1..1293
                     /gene="dnaA"
                     /locus_tag="HCAN_0001"
     CDS             1..1293
                     /gene="dnaA"
                     /locus_tag="HCAN_0001"
                     /codon_start=1
                     /transl_table=11
                     /product="Chromosomal replication initiator protein dnaA"
                     /protein_id="EES88726.1"
                     /translation="MHPVLLQLKREITPFEFDNYITQLSFNEKYSRDDRIIFNAPNNI
                     IASWIKTKYSSKIAQLFENQNGYKPEIIIEVFNPNKKKENKNNIKKIQNATNLNPSLT
                     FNSFIVGNSNSFAFNVAKAVAQNQSTIYNPLVIYGNTGLGKTHLLNAIGNTNANVGKS
                     VIYTTSEQFLNDYLLHIRNNTMDRFREKYRACDYLLIDDIQFLSGKNQIQEEFFHTFN
                     ELKKNNKQIVLTSDRPPKDMNGLEERLKTRFTSGLLADIQPPELETKINIINAKCELD
                     GIHLSPKIIDFIAANINDNIREIEGVLVKLNFSINVTNIQEITIDFVRDILKEYIKES
                     KENINMDRIIENVAKYYNIKPSEIRSKSRSKNIVTARKIVIYLARTLTPNSMLSLADY
                     FDMKDHSTVSKAMKSIQEEINKNPNFKTIIEELKNKIK"
```

```
>CM000776|CM000776.2 Helicobacter canadensis MIT 98-5491 chromosome, whole genome shotgun sequence.
ttgcatcctgttttattacaactcaaaagagaaattacaccttttgaatttgacaattat
attactcaacttagttttaatgaaaaatattctcgtgatgatagaatcatttttaatgcc
ccaaataatattattgcaagttggataaaaacaaaatattctagtaaaattgctcaactt
tttgaaaatcaaaatggttataagcctgaaattattattgaagtttttaatcctaataaa
aagaaagaaaataaaaacaatattaaaaaaattcaaaatgccactaatctcaatccttct
ttaacctttaattcttttattgttggaaactctaatagttttgcttttaatgtcgcaaaa
gctgtggctcaaaatcaaagcacaatttataatcctttagtaatttatggcaatactgga
ttaggcaaaacgcacttactcaatgccattggaaacacaaatgctaatgttggaaaatcc
```

Figure 12.16 Differences in the file formats for sequence databases GenBank, ENA, and DDBJ. Although all three entries show excerpts for the same sequence data, the genome sequence data for *Helicobacter canadensis* strain MIT 98-5491, the presentation of those data is slightly different between the three public sequence databases. (Images are from National Library of Medicine [NLM], European Molecular Biology Laboratory-European Bioinformatics Institute [EMBL-EBI], and DDBJ web pages).

Data Bank of Japan (DDBJ). Each of these centers provides a repository for sequence data and searchable databases for DNA and RNA data, as well as protein sequence data and other information. The data contained at each of the three centers are synchronized daily to ensure that they carry the same information.

While the data in GenBank, ENA, and DDBJ are the same, each database presents the data in its own format. One of the features of the synchronization is to transfer the databank entry from one database's format to the format of the database taking in the data (**Figure 12.16**).

Comparative genomics finds that there are commonly shared genes and unique genes

In comparative genomics, the genomic features of different organisms are analyzed against each other. This can include the similarity between and the presence of the sequences of the CDSs, ncRNAs, and other features. Slight changes in regulatory sequences between two organisms can reveal differences in regulatory responses. Comparative genomics can also include analysis of the

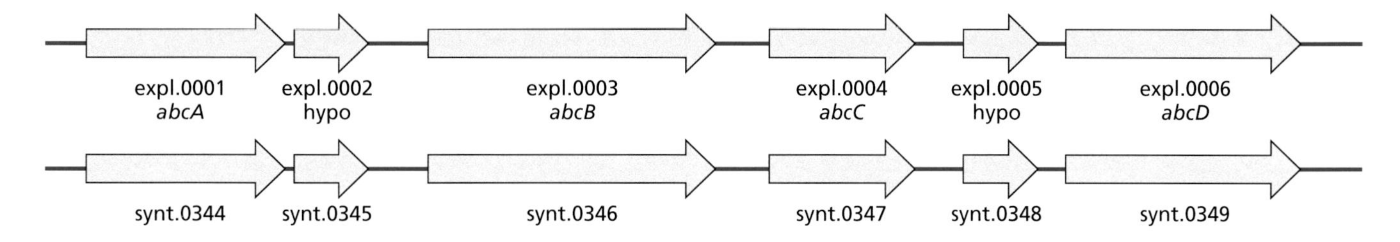

Figure 12.17 Synteny between genomes. In two syntenic regions from two different genome sequences, the first has locus identification numbers starting with expl, while the second has locus identification numbers starting with synt. Although they are from different genomes, the order of coding regions is the same between the two sequences, demonstrating synteny.

order of the CDSs and features in the genome, which is referred to as **synteny**. The positions of features such as genes, IS elements, transposons, and prophages may be quite different between two genomes that are otherwise **syntenic**, with the same order of features in the rest of the genome (**Figure 12.17**).

Similarities, differences, and evolutionary relationships between bacteria can be revealed via comparative genomics. Alignment of sequence data facilitates the identification of similarities between sequences and highlights regions of difference. From such analyses, the **core genome** can be identified. The core genome describes the genes, CDSs, and other features that are present in all strains within a taxonomic group. Core genomes can be described for a bacterial species or the term can refer to the genes in common within a genus or higher.

The opposite of the core genome is the **accessory genome**, referring to the genes that are present only in a single strain or only in some strains. Often it is the accessory genome that differentiates a virulent bacterial strain from an avirulent one. The acquisition of genes through horizontal transfer, such as antibiotic resistance genes, contributes to the accessory genome. Mobile elements like prophages and genetic islands are often part of the accessory genome, being present in some strains within a species, but not all. When all of the potential **accessory genes** and all of the **core genes** are considered together, this is the **pan-genome**, being all of the genes known within a taxonomic group (**Figure 12.18**). There are, for example, about 89,000 genes in the *E. coli* pan-genome.

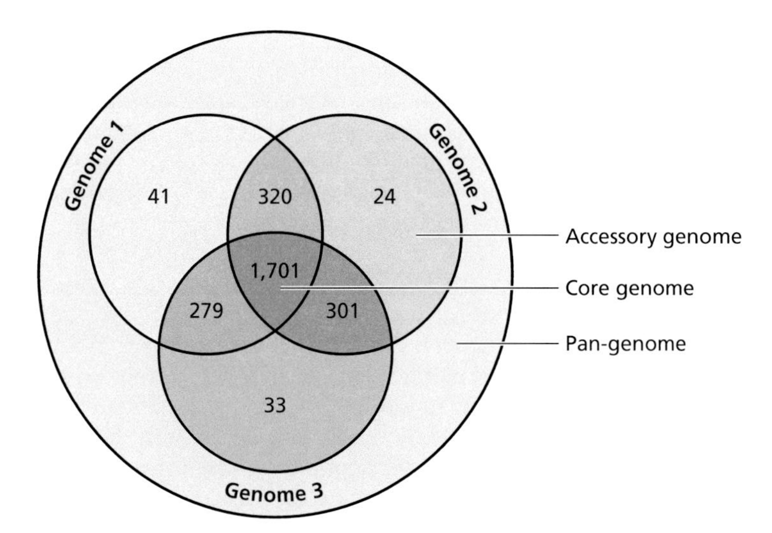

Figure 12.18 The core genome, accessory genome, and pan-genome. When comparing three genomes, shown here as different colored circles, there will be some genes that are shared, shown in the overlaps, and some that are unique to each genome. Here, there are 1,701 genes that are shared among all three bacterial genomes, making the core genome. In genome 2 (blue), there are 24 genes that are not in either of the other genomes, making the accessory genome. When all of the genes from all three genomes are taken together (gray-shaded circle), this is the pan-genome.

```
TTGACATTACCGTACTAAACCGGTATAATGCAGGAGGTTGA
         TTACCGTACTAAACCGGTATAAAG
   GACATTACCGTACTAAACCGGTATAAAGCA
                  AAACCGGTATAAAGCAGGAGGTTG
              GTACTAAACCGGTATAAAGCAGGAGG
```

Figure 12.19 Mapping of short unassembled sequence read data to a reference genome can reveal changes in the genetic sequence. The top line is the reference genome sequence. Below are aligned sequencing reads. Analysis reveals that one base has changed. The T highlighted in the reference genome is an A in the sequencing reads. This will change what might have been an ATG initiation codon into an AAG.

Comparative genomics can be done without assembly or annotation of sequencing data

With the advent of rapid and inexpensive genome sequencing, more sequence data are being produced than can be completely analyzed. It has therefore become essential to be able to extract from these data interesting bits of information about the bacteria. To this end, comparative analyses between genomes are done without assembly and annotation.

To start, the sequencing read data are collected from the sequencing platform, trimmed to remove poor-quality sequences, and quality checked. These data, without assembly, are then aligned against a fully annotated reference genome sequence; this is referred to as **mapping**. Similarities and differences between the complete reference genome and the unassembled sequences mapped against it are noted. This method can be used to readily identify variants like single base changes (SNPs), short indels, and short sequence changes, such as substitutions that add or delete repeats (**Figure 12.19**).

Mapping of the sequence reads against a reference genome can reveal small-scale genetics differences, but it is harder to identify larger regions of difference using this technique. Large differences will not align with the reference genome; therefore, to reveal these, it is necessary to look at the unaligned sequencing data to see genetic features that are not in the reference genome.

Horizontally transferred gene sequences tend to carry a signature that can identify them

Horizontal gene transfer is a key characteristic of bacteria, enabling the exchange of genetic material between strains and species. Horizontally acquired genes and DNA can also be called foreign genes and foreign DNA. DNA signatures of horizontally transferred sequences may be different from the rest of the genome in features like %G+C and codon usage (Chapter 3), as well as perhaps in amino acid usage, where using an amino acid in a protein sequence is rare for a particular species. Using bioinformatics, DNA signatures of the chromosome are analyzed to identify regions that are potentially horizontally transferred. A few bacterial species are naturally competent for transformation, like *Neisseria meningitidis* and *Neisseria gonorrhoeae*. For these species, there is extensive bioinformatics and experimental evidence of horizontal gene transfer in genome sequence data. Evolution in such species is heavily influenced by the ability to acquire DNA.

Recently acquired sequences are more likely to be different from the rest of the genome (**Figure 12.20**). Over time, sequences **ameliorate**, which means they change to match the genome with regard to %G+C and codon usage. It is therefore easier to identify recently horizontally acquired sequences than it is to identify those that have been within the bacterial genome for some time. There are some regions of genomes that may appear to be foreign when they are not. These regions are native to the genome, but due to the nature of the features in a region, it may look horizontally transferred based on its %G+C and codon usage.

Horizontally acquired regions can also be identified using phylogeny. Identification through phylogeny relies on orthologues being identified between

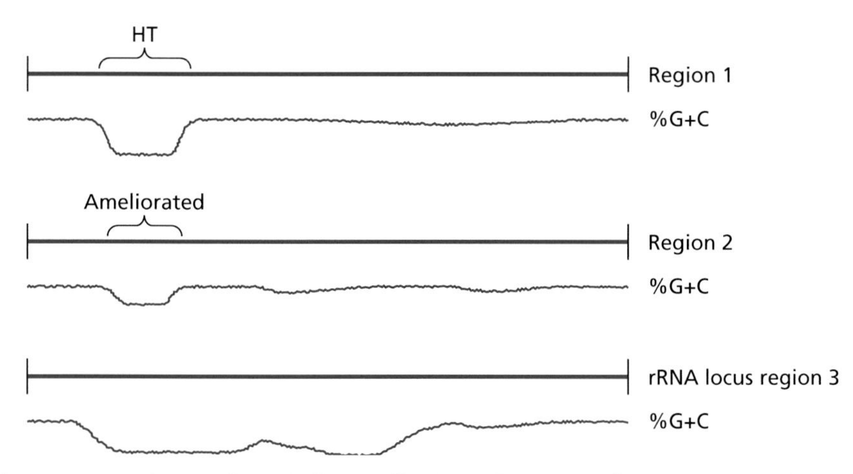

Figure 12.20 %G+C can be an indicator of horizontal gene transfer. Region 1 has a clear change in the %G+C for part of the sequence, which corresponds to some horizontally transferred DNA (HT). In region 2, the horizontally transferred DNA has been present in the genome for some time and has ameliorated to more closely match the %G+C of the rest of the region. Region 3 includes the rRNA locus, which has a %G+C signature that is different from the rest of the genome, by nature of the function of the sequence.

sequences. Those that have no orthologues could be foreign (**Figure 12.21**). Alternatively, the identified genes may be strain-specific because they were generated via duplication and change or deletion rather than horizontal gene acquisition.

With the focus on genomes and genomic sequence analysis, it could be easy to overlook the role that plasmids play as sources of horizontally transferred genes; plasmids can, in whole or in part, have originated from other bacteria. As extra-chromosomal DNA, plasmids are capable of self-replicating and of carrying with them the regulatory features needed to express the genes carried on the plasmid. Therefore, plasmids are efficient transmitters of horizontally transferred genes. Bacteriophages are also effective transmitters of horizontally transmitted genetic material (see Chapter 3). In addition to the prophage sequences themselves being horizontally acquired, bacteriophages may bring foreign genes with them into the genome, contributing to horizontal gene transfer between bacteria.

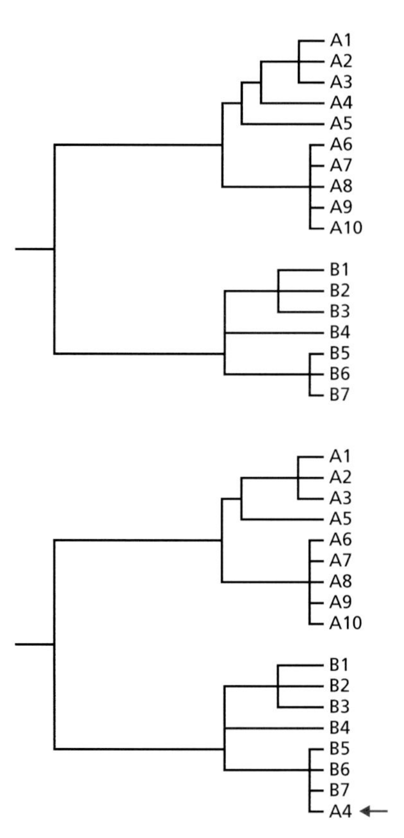

Figure 12.21 Phylogeny can reveal horizontally acquired sequences. This phylogenetic tree (top) shows two species, labeled A and B, with several strains representing each, all numbered. When a single gene is considered (bottom), it is revealed that the gene from A4 clusters with the genes from species B rather than with the other species A genes.

Genome sequence analysis can find unexpected features in the sequence data

Prior to whole-genome sequencing, many assumptions were made about the nature of genetic material within bacterial cells and our understanding of it. Long held beliefs such as bacteria having only circular chromosomes were overturned with the sequencing of linear chromosomes from *Borrelia burgdorferi* and other bacteria, confirming previous evidence. Flagellar gene systems were found within intracellular bacteria such as *Buchnera aphidicola* that do not express flagella yet use some of the flagellar proteins to form structures that transport proteins across the bacterial cell surface. Bacterial species were found to contain horizontally acquired eukaryotic DNA and vice versa. With important implications in our understanding of bacterial pathogens, virulence genes have been found in non-pathogens. These and other genomic analysis-based results have brought to light ecological and evolutionary contexts for gene presence in bacteria. Findings such as these have required a re-examination of the potential phenotypic influence that the presence of a gene or set of genes within a genome may have upon the characteristics of the microorganism.

Genome sequence data have enabled the identification of signatures for uptake signal sequences enhancing horizontal gene transfer that have not previously been identified. For example, in the *Apl* subclade of *Pasteurellaceae* including *Actinobacillus pleuropneumoniae*, *Mannheimia haemolytica*, and

Haemophilus ducreyi, horizontal gene transfer is enhanced between one another when the ACAAGCGGT sequence is present in the DNA. This sequence signature was identified through analysis of the sequence data for short sequences present in several copies around the chromosome at a frequency well above what might be expected by chance. It is different in both sequence and specificity from the AAGTGCGGT uptake signal sequence in the *Hin* subclade, which includes *Haemophilus influenzae*. Likewise, analysis of the genome of *Helicobacter pylori* revealed that there is no uptake signal sequence in this species, the assumption being that the acidic niche of the bacteria acts as a restriction to horizontal acquisition adequately.

As the understanding of bacterial genomes and genomic features has progressed, it has been possible to identify within the genome sequence data features that were never before recognized. Perhaps the most apparent of these is the identification of small non-coding RNAs in regions of the chromosome previously believed to be non-functional junk. In addition, genomic analysis has become better at finding repeat regions that might have regulatory functions, might be hotspots for recombination, or otherwise might be of interest, such as the CRISPR repeat regions that will be discussed in Chapter 13. With the inclusion of more sequence data, it has become easier to identify prophages and prophage fragments within bacterial genomes, particularly those that share homology with prophages in other genome sequence data, even from unrelated species. Other integrated features such as IS elements, transposons, and fragments of these features in the genome, as well as other regions that are potentially related to these elements, can now be readily identified.

Comparative genomics on closely related strains can reveal biologically important information

A key experiment in bacterial genetics is to knock out a gene and identify the change in phenotype. This requires laboratory investigation and it may be that the phenotype involved is not readily apparent. It is possible to compare the phenotypes of very closely related strains to achieve similar, although not perfect, results. Genome sequencing makes it possible to examine closely related strains that differ by only a few genes, but which have phenotypic differences such as the disease outcome for the host or the ability to tolerate antibiotics or environmental stress. Comparative analysis of the genome sequence data will identify the regions of difference, SNPs, indels, and other changes between the closely related strains. These are likely to be involved in the differences in phenotype and can inform directed laboratory experiments to investigate the functions and phenotypic implications of the genes.

Differences within strains can also be within systems that have evolved to differ. Some bacterial species use **phase variation** to switch ON and OFF certain genes within the genome. While the vast majority of genes are still regulated through standard regulatory systems, some are switched through **stochastic** phase variation mechanisms, a programmed randomness. The phase variable feature is associated with the phase variable gene; in this way it is not random because only those genes with the phase variable feature can randomly switch ON and OFF. Of the set of phase variable genes, their expression ON or OFF is a random process, which is why it is referred to as stochastic. Phase variation enables a portion of the bacterial population to express different combinations of phase variable ON and OFF genes, enabling the population as a whole to rapidly take advantage of the right combination of gene expression at any one time. The phase variable switch can be different between species, and species may employ more than one genetic feature as a phase variable switch. Generally, these include changes in the lengths of homopolymeric tracts or simple sequence repeats, inversion of short sequences, or changes in the methylation state of the DNA. All of these features can be readily identified from genome sequence data using bioinformatics, making it possible to identify never before known phase variable genes

(**Figure 12.22**). Phase variation can impact the transcription, translation, or protein sequence of a gene.

Bacteria also undergo **antigenic variation**, where the amino acid sequence or structure of a surface component of the bacteria is altered. Generally, this is to overcome the antigenicity of the surface feature and its recognition by the host immune system. By dynamically changing the antigens on the surface of the bacteria, the microorganisms are able to survive longer in the host through evading the immune system. Some key sequence features of antigenic variation have been identified and can be found through examination of genome sequence data. These include the presence of an expressed version of a gene along with shorter fragments of the gene sequence within the chromosome. Recombination with the silent cassettes generates mosaic genes that encode different antigens (**Figure 12.23a**). Recombination between paralogous genes within the genome that encode surface antigens can generate mosaic gene sequences, thus contributing to antigenic variation as well (**Figure 12.23b**). Other antigenic variation mechanisms use repeats within the gene, much as for phase variation, although rather than switching ON or OFF the expression, they change the protein lengths and structure, thus creating antigenic variation (Figure 12.22c).

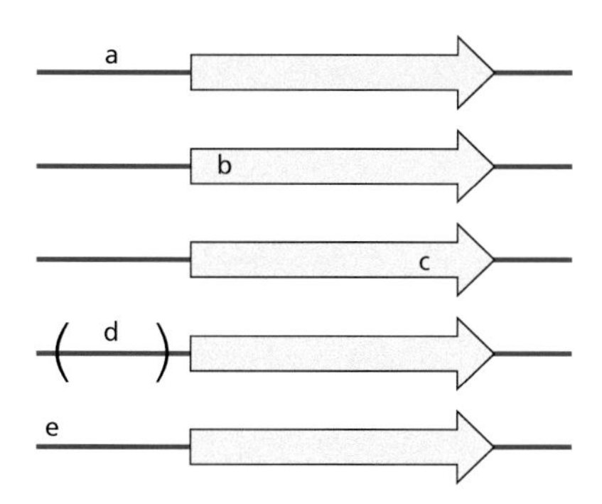

Figure 12.22 Phase variation mechanisms switch expression of associated genes. In transcriptional phase variation, changes in a repeat associated with the promoter region (a) alter the transcription of the gene. In translational phase variation, changes in a repeat within the first part of the coding region (b) alter translation of the gene due to the resulting frame-shift. In C-terminal phase variation, changes in a repeat toward the end of the coding region (c) alter the C-terminal sequence of the encoded protein. In inversion phase variation, a portion of the DNA, shown with parenthesis, inverts at its location (d) often resulting in a flip in the orientation of the promoter region or genes. In methylation phase variation, changes in the methylation state of the DNA can alter the binding of proteins, such as an activator or RNA polymerase (e).

Comparative genomics between non-related species gives insight into bacterial evolution

The availability of a vast quantity of genomic data enables the exploration of the diversity and similarity between non-related species, both within phyla and between phyla. Using bioinformatics to compare the sequence data, the genes that are conserved and present within all bacteria can be identified. These are common features that represent important key mechanisms for bacterial life such as DNA replication, transcription, and translation. Also conserved are genes involved in cell division, ATP-dependent transport, and stress responses. Through analysis of similarities in sequences, it is possible to identify and explore the implications of genes present in all bacteria and of genes present in all representatives of a particular branch of the phylogenetic tree. Given the number of bacterial species being revealed through non-culture-based metagenomic sequencing, it is interesting to consider the number of genes that are missing from our understanding of the non-culturable bacteria.

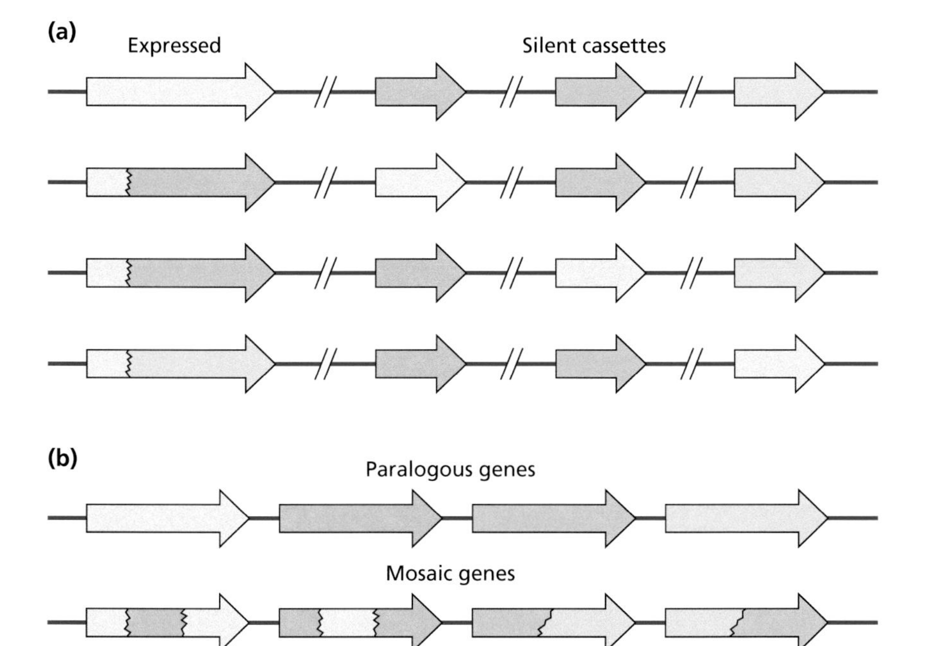

Figure 12.23 Antigenic variation through changes in sequences. (a) Antigenic variation can occur through recombination between the sequence of the expressed gene and the sequences of silent cassettes present elsewhere in the genome. (b) Antigenic variation can also be generated through recombination between paralogous genes, generating mosaic gene sequences that are different from the originals.

Strain identification using sequencing data is a powerful tool for tracking bacterial transmission

Next-generation sequencing and SNP analysis of the genomic data has enabled rapid identification of differences between sequence data from different bacterial samples. This allows the identification of bacteria that have come from a common source with resolution and certainty of association between isolates never achieved before with phenotypic or short sequence methods used for strain typing. As the bioinformatics involved in this type of analysis becomes more streamlined, with less user intervention in the generation of interpretable results, it becomes easier to use sequencing as a typing method because the output data are refined into a readily interpretable result. This enables fine detail identification of bacteria involved in outbreaks and transmission events, differentiating them from cases of infection occurring at the same time by random chance infection with unrelated strains. The origins of several outbreaks have been identified using next-generation sequencing methods analyzed by bioinformatics. The rapid, nearly real-time nature of the technology and analysis has meant that the origins of outbreaks can be identified and measures put in place to stop the current and future spread of infection.

It is important to successfully stop infections in a hospital setting or in an organism of agricultural importance. It is essential from a treatment and prevention standpoint to know if all or most of the infections come from the same strain (**Figure 12.24**). An example is the economic problem of bovine tuberculosis in the UK, where there has been a history of culls of European badgers (*Meles meles*), which are believed to harbor the *Mycobacterium bovis* bacteria and pass them on to cattle. Culling of badgers in outbreak areas has failed to reduce the cases of bovine tuberculosis. Indeed, culling of badgers seems to contribute to the spread of *M. bovis* to those badgers remaining in the area. In New Zealand, it is the brushtail possum (*Trichosurus vulpecula*) that is the wildlife reservoir blamed for spreading the disease to cattle. Genome sequencing of *M. bovis* from cattle and wildlife has been demonstrated to be far superior than traditional typing methods, providing greater resolution, sometimes revealing that the genetic signatures of cattle strains and local wildlife are quite different. Evidence is building, through use of genomics, which suggests that bovine tuberculosis is a complex situation in which wildlife play a role as reservoirs of the bacteria, but where spread between cattle is the major contributor to bovine disease.

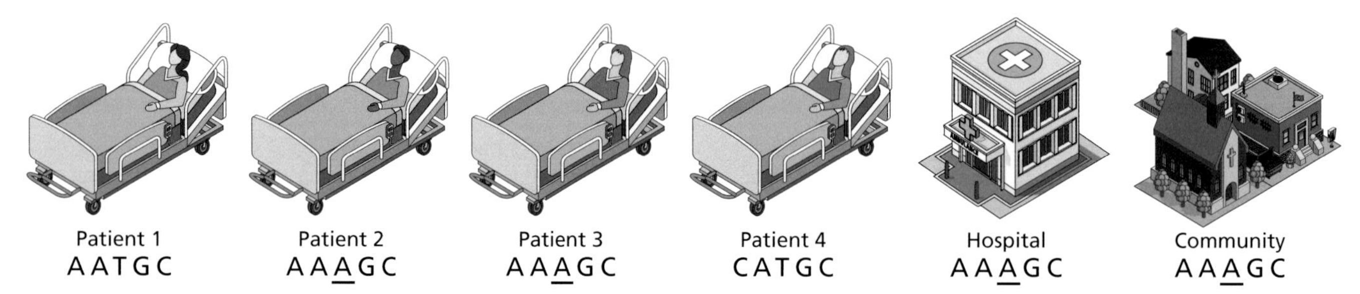

Figure 12.24 Sequencing and bioinformatics can identify bacterial transmissions. Sequence data show that patient 1 has an AATGC sequence, patient 4 has a CATGC sequence, and patients 2 and 3 have AAAGC sequences. Comparison of the sequences from patients 2 and 3 suggests that they acquired the infection from the same source or that one patient infected the other. Further sampling and sequencing finds that the same sequence signature, AAAGC, is present in an environmental sample from the hospital and also from a person in the community. This information may help contain the infection, which may have come from the community into the hospital and spread to the patients (same sequence as patient 2 and patient 3) or may have left the hospital for the community via a discharged patient or been carried by a visitor or hospital worker. Alternatively, the two patients may have been infected in the community before entering the hospital and then been the cause of the hospital environment sample. (Illustrations courtesy of Freepik.com.)

Bioinformatics analysis of sequence data from patients with chronic and persistent bacterial infections has enabled researchers to see the changes within the bacteria genome as they occur. In particular, ongoing infections in patients with cystic fibrosis, such as *Pseudomonas aeruginosa*, *Burkholderia dolosa*, and *Burkholderia multivorans*, have been genome sequenced, revealing changes in key genes contributing to within-host adaptation. **Convergent evolution** of genes in the genomes of these species between different patients suggests that similar evolutionary changes can occur in different bacterial lineages. Studies such as these investigations of genomic changes within patients have shown that SNPs can arise rapidly and therefore SNPs seen in outbreaks need to be considered in the context of the species and their rate of SNP accumulation. Outbreak sequence data that may look to be different enough to be from different originating bacteria may actually be from a similar source, if the species dynamically changes its sequences in a short period of time.

Function predictions can be made based on sequence similarity

Sequencing is able to resolve the A's and T's, and G's and C's, but it is bioinformatics that works out what it all means or what it might mean. Much of this is accomplished through the investigation into the similarity between one sequence and those other sequences in the public databases, such as GenBank. Algorithms such as **BLAST** (**basic local alignment search tool**) compare the query sequence to either the whole of the database or a portion of it, as selected by the user. Similar sequences are reported as **BLAST hits**, with each being given a score indicating the statistical significance of the similarity between the sequences, the **E value**. When assessing BLAST results it is important to look at both the percent identity between the query sequence and the subject sequence found in the database and also at the percent of the query that matches the subject (**Figure 12.25**). It is possible for a short sequence or a portion of a sequence to have 100% identity with another sequence, yet due to the short nature of the identity, this may reveal nothing about the query sequence. Indeed, it is possible to type in a random 15-base DNA sequence and have a BLAST hit of 100% to something in the database.

There are several different types of BLAST searches that have been developed over the years. There are BLAST searches that use a nucleotide query to investigate the nucleotide databases and those that search protein databases with a protein sequence. There are also BLAST searches that will translate the nucleotide sequence first, generating the predicted amino acid sequence that is then used in a search of the protein databases. Opposite to this is a BLAST that uses a protein search query to search the nucleotide database, by first generating a translation of all of the possible amino acid sequences from the nucleotide database. There are also a number of specialist BLAST searches available for assisting in specific tasks to try to identify similarity between nucleotide and amino acid sequences.

BLAST results can suggest the function of the sequence used as the query and its evolutionary similarity to other sequences in the database. Identification of similar sequences assists in the annotation of unknown sequences, such as bacterial genome sequence data. However, it can also lead to erroneous assumptions as to function, due to either errors in annotation of sequences in the databases or assignment of functions based on limited similarity. Even when two sequences are very similar to one another and the function of one has been experimentally shown, it is possible that differences in a single amino acid may be enough to give the proteins slightly different functions. BLAST and other bioinformatics tools will be explored in more detail in Chapters 16 and 17.

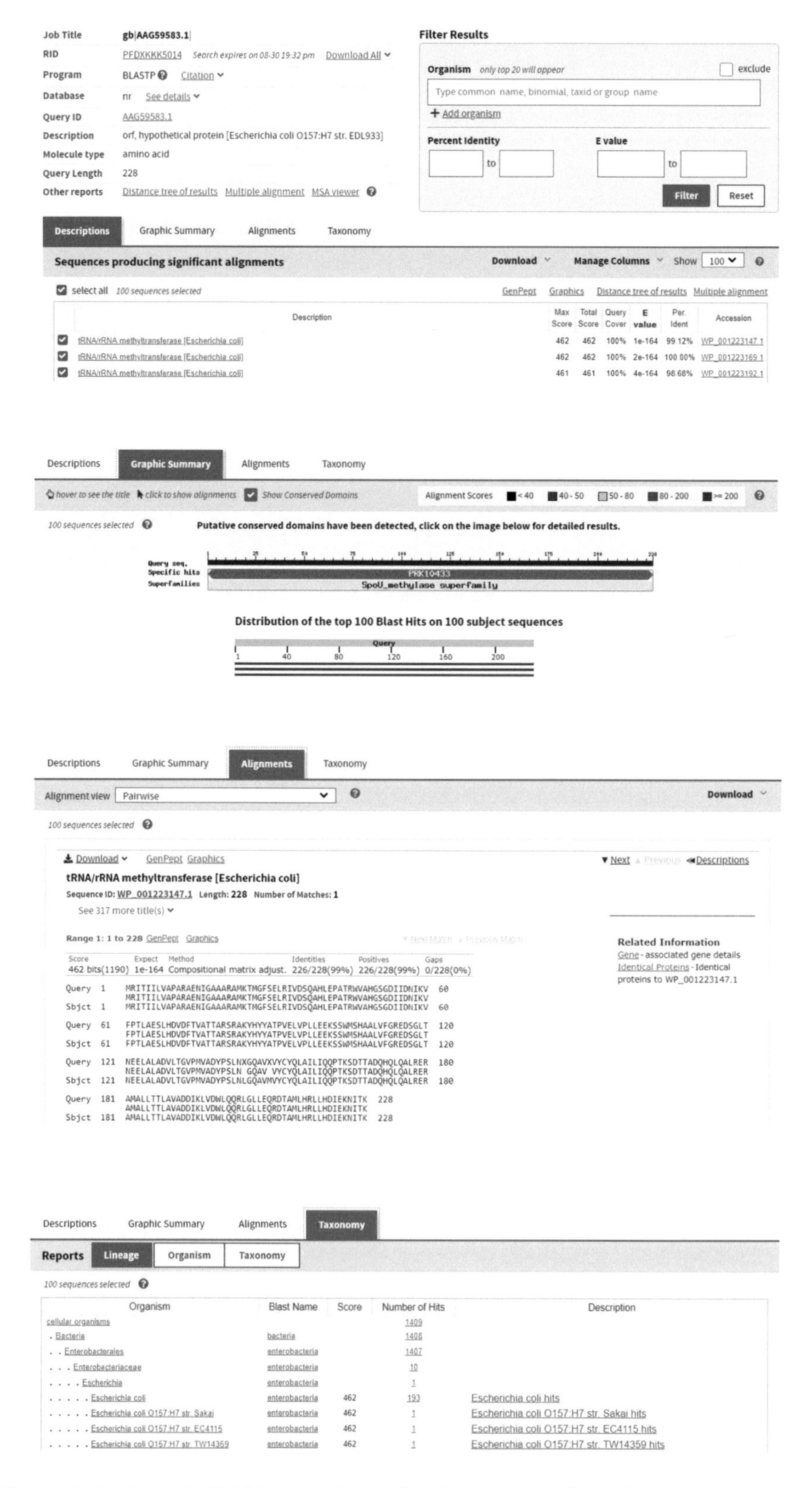

Figure 12.25 Example BLAST output, here using the sequence of LasT from *E. coli* 157:H7 strain EDL933. The query sequence is entered into the BLAST search, where it is used to interrogate all of the sequences in the database it has been assigned to search. The results are shown across several tabs. Under "Descriptions", the amount of similarity and the length of that similarity results in an E value. The "Graphic Summary" and "Alignments" show the similarity between the query sequence and sequences in the database, called the subject. "Taxonomy" shows the relatedness of the subjects to the query. (Images from the National Library of Medicine [NLM] web pages.)

Key points

- Bioinformaticians investigate the features within sequence data using computational programs.
- The quality of sequencing data is variable, and poor sequences need to be trimmed before they are investigated.
- Identified features in sequence data, such as potential genes, are noted in an annotation.
- An annotation becomes more robust as different pieces of evidence support the proposed function of the region of sequence.
- Most annotations focus on the identification of potential genes, called coding sequences, and the assignment of potential functions to these coding sequences.
- It is possible to generate an annotation automatically; however, some manual curation of the automated annotation is usually needed.
- Comparative genome sequencing analysis between two or more bacterial genome sequences identifies the shared features and different features.
- Bioinformatics analysis can provide useful information, based on computational analysis, which can be the basis of experimental laboratory investigations.

Key terms

Define the following terms introduced in this chapter. Check your answers using the definitions in the Glossary. These terms are also available as Flashcards online.

Accessory genome	Core genome	Pan-genome
Annotation	Domain	Paralogue
Antigenic variation	FASTA	Phase variation
Assembly	Hypothetical protein	Pseudogene
BLAST	Multi-FASTA	Regulon
Convergent evolution	Orthologue	Synteny

Questions and discussion topics

Self-study questions

Answer each question using 50–100 words or a table or labeled diagram. Advice on where to find answers to these questions is available online.

1. What features define a predicted CDS and what additional features can help support the evidence that the CDS may be a potential gene?
2. Should all sequencing data generated be included in analysis? Explain why or why not.
3. What is sequence annotation, what does it entail, and who is responsible for doing annotation?
4. What features other than CDSs might be included in a sequence annotation?
5. Write out an example of a sequence in FASTA format.
6. Expand the FASTA formatted sequence in question 5 to be a multi-FASTA.
7. What is the difference between an orthologue and a paralogue and what can each type of homologue tell us about the evolution of bacteria?
8. What are some of the features than can cause problems when attempting to annotate sequence data?
9. What is meant by the core genome, accessory genome, and pan-genome of a bacterial species and what type of genetic features can be found in each?

10 Briefly, what are the different mechanisms of phase variation and how can they switch gene expression in bacteria?

11 What is antigenic variation, how do bacteria use it to avoid the immune system, and what are some of the mechanisms bacteria use for antigenic variation?

12 How is BLAST used to identify sequences within a database, such as one of the International Nucleotide Sequence Database Collaboration databases that match a sequence of interest?

Discussion topics

These topics are presented for discussion in study groups, as part of class discussions, or on your own. These questions go beyond what is directly covered in this part of the book. Use the research literature and other reading to explore these topics in more depth. Tips to help prepare for topic discussions are available online.

1 Discuss what defines a CDS as being annotated as a hypothetical protein and at what point it becomes a conserved hypothetical protein. How might the identification of domains within annotated hypothetical proteins assist in identification of functions? Explore a research paper in which the function of a CDS was unknown at the start of the study and explore the steps involved in the study to determine a function.

2 Large sets of genome sequence data are often generated and comparatively analyzed by mapping the data against an annotated reference genome sequence. Select a recent research paper that has taken this approach and discuss the research question that the authors addressed with their investigation, how they generated the sequence data, the methods used for mapping and analysis, and the results obtained.

3 In this chapter, a few features are discussed that were revealed from the sequencing of bacterial genomes that were not expected in the species. Either explore these in more depth or investigate other research findings where novel discoveries have been made through analysis of genome sequence data. What is now known about the species from the sequence data that was not known previously from laboratory investigations alone and how has this led to new laboratory-based research?

Online quiz questions

To further self-assess your understanding of the chapter material, please visit the following link, where you can participate in a range of interactive quiz questions:

www.routledge.com/cw/snyder

Further reading

Automated annotations rapidly produce an annotation that needs manual curation

Richardson EJ, Watson M. The automatic annotation of bacterial genomes. *Brief Bioinform.* 2013; *14*(*1*): 1–12.

Comparative genomics finds that there are commonly shared genes and unique genes

Edwards DJ, Holt KE. Beginner's guide to comparative bacterial genome analysis using next-generation sequence data. *Microb Inform Exp.* 2013; *3*: 2.

Sun S, Xiao J, Zhang H, Zhang Z. Pangenome evidence for higher codon usage bias and stronger translational selection in core genes of *Escherichia coli. Front Microbiol.* 2016; *7*: 1180.

Genome sequence analysis can find unexpected features in the sequence data

Anderson MT, Steven Seifert H. Opportunity and means: Horizontal gene transfer from the human host to a bacterial pathogen. *mBio.* 2011; *2*(*1*): e00005-11.

Fraser CM, Casjens S, Huang WM, Sutton GG, Clayton R, Lethigra R. Genomic sequence of Lyme disease spirochaete, *Borrelia burgdorferi. Nature.* 1997; *390*(*6660*): 580–586.

Maezawa K, Shigenobu S, Taniguchi H, Kubo T, Aizawa S, Morioka M. Hundreds of flagellar basal bodies cover the cell surface of the endosymbiotic bacterium *Buchnera aphidicola* sp. Strain APS. *J Bacteriol.* 2006; *188*(*18*): 6539–6543.

Salzberg SL, White O, Peterson J, Eisen JA. Microbial genes in the human genome: Lateral gene transfer or gene loss? *Science.* 2001; *292*(*5523*): 1903–1906.

Sieber KB, Gajer P, Dunning Hotopp JC. Modeling the integration of bacterial rRNA fragments into the human cancer genome. *BMC Genomics*. 2016; *17*: 134.

Comparative genomics on closely related strains can reveal biologically important information

Snyder LAS, Loman NJ, Linton JD, Langdon RR, Weinstock GM, Wren BW, Pallen MJ. Simple sequence repeats in *Helicobacter canadensis* and their role in phase variable expression and C-terminal sequence switching. *BMC Genomics*. 2010; *11*: 67.

Zelewska MA, Pulijala M, Spencer-Smith R, Mahmood HA, Norman B, Churchward CP, Calder A, Snyder LAS. Phase variable DNA repeats in *Neisseria gonorrhoeae* influence transcription, translation, and protein sequence variation. *Microb Genomes*. 2016; *2*(*8*): e000078.

Strain identification using sequencing data is a powerful tool for tracking bacterial transmission

Robinson ER, Walker TM, Pallen MJ. Genomics and outbreak investigation: From sequence to consequence. *Genome Med*. 2013; *5*: 36.

Simar SR, Hanson BM, Arias CA. Techniques in bacterial strain typing: past, present, and future. *Curr Opin Infect Dis*. 2021; *34*(*4*): 339–345.

Tümmler B. Molecular epidemiology in current times. *Environ Microbiol*. 2020; *22*(*12*): 4909–4918.

Function predictions can be made based on sequence similarity

Altschul SF, Gish W, Miller W, Myers EW, Lipman DJ. Basic local alignment search tool. *J Mol Biol*. 1990; *215*: 403–410.

Part V

Bacterial Response, Adaptation, and Evolution

In order to survive in a changing world, all organisms must themselves be able to change, to respond to the ways in which their environment changes, to adapt to those changes, and to evolve to take best advantage of the resources available to them. Bacteria have been responding, adapting, and evolving for millions of years as our planet has changed and as new niches have developed. Over time, eukaryotic organisms developed, presenting bacteria with host organisms upon which they could be symbionts, pathogens, or **saprophytes**, that is, living on dead organic matter. In this section, the concepts of bacterial response, adaptation, and evolution will be explored in greater depth. These three concepts are very much overlapping, and some aspects will be covered in more than one chapter, such as quorum sensing, which is addressed in Chapters 13 and 14. This section will explore how bacteria respond to their environment, contributing to their phenotype and how this contributes to their abilities to adapt. The evolution of these responses and adaptations will then be discussed in the final chapter of the section.

DOI: 10.1201/9781003380436-17

Chapter

13

Bacterial Response

Although bacteria are single-celled organisms that are less complex in some ways than eukaryotic cells, they are not inert cells. Bacterial cells are dynamic, undergoing physical changes in their composition and make-up of cellular components and proteins in response to their surroundings. For some bacteria, this involves responding to being in the soil, ocean, or other environments. Other bacteria adapt to being within host organisms, including making use of mechanisms to modulate host immune responses used by either animals or plants to control the presence of bacteria. Bacterial responses to their environment enable the bacterial cells to enhance their survival, to persist in the environment, and to potentially transmit to a new environment (**Figure 13.1**).

This chapter will explore the mechanisms used by bacteria, both as individual cells and as populations, to respond to their environments. Signals received from the environment can include differences or fluctuations in temperature or pH, the presence of various nutrients or chemicals, and a whole host of other environmental cues. Based on these signals, the bacteria will respond, perhaps by forming a biofilm or expressing flagella and becoming motile or expressing genes to be able to take advantage of the presence of nutrients now available.

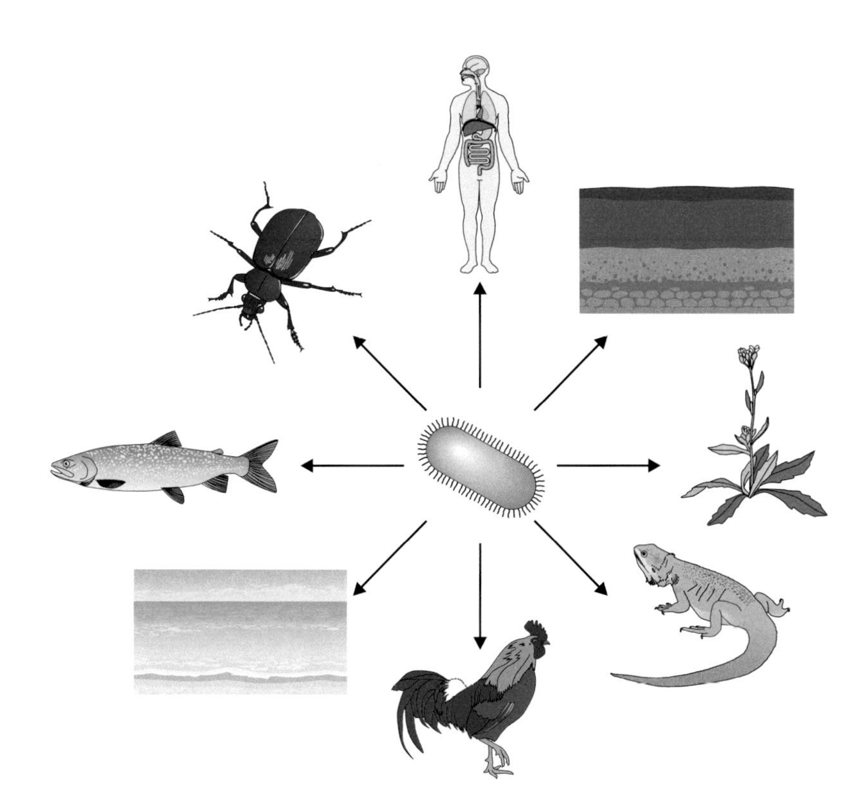

Figure 13.1 Bacteria respond to their environment. The same bacterial cell can find itself in several different environments. As shown here, the bacterial cell in the center could colonize a human, requiring a response to the temperature, nutrients, immune response, and environment of that niche. Likewise, these bacteria could end up (clockwise from the top right) in the soil, in a plant, in a lizard, in a chicken, in the ocean, in a fish, or in an insect. Each would be at a different temperature, would have different nutrients available to the bacteria, would contain different competitor organisms, and would provide other different stress challenges for the bacteria.

DOI: 10.1201/9781003380436-18

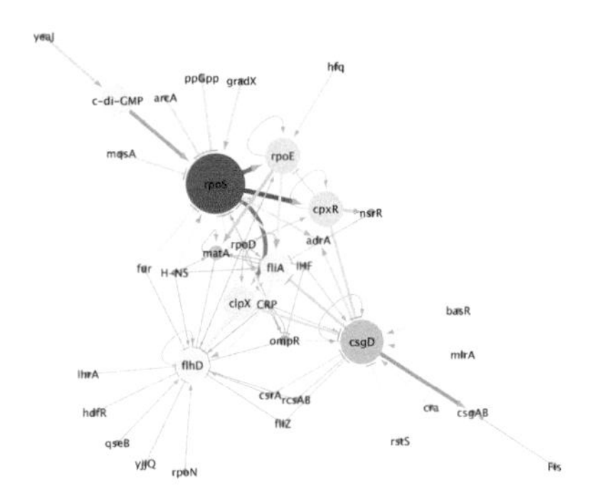

Figure 13.2 Regulatory network map of *Escherichia coli.* (Adapted from Amores GR *et al. Sci Rep*. 2017; 7(1): 16768. Published under CC BY 4.0.)

Studying responses often happens in pure bacterial cultures

Most of our experimental studies of bacterial responses to different conditions use pure laboratory cultures. Pure cultures enable the bacteria to be studied in a controlled manner, in which researchers can specifically design the experiments and examine the outcome of the results. This is not how the bacteria would exist in nature. Bacteria could grow alongside other bacterial species, mixed with various bacteriophages, eukaryotic microorganisms, biological surfaces, and an influx and efflux of various biological and inorganic material. However, through laboratory experimentation, it is possible to articulate the various regulatory networks that exist within bacteria that enable them to respond to changes in their local environment.

In a pure bacterial culture, it is possible to manipulate the bacterial environment and control the experiment so that the outcome can be carefully monitored and understood. The regulatory pathways in the bacterial response to the change in the environment can be elucidated and resolved. Using such methods, genetic pathways and regulatory networks have been mapped for many bacterial regulatory systems. What started as a simple understanding of the *lac* operon has developed into more complex maps with interconnecting repressors, activators, and regulatory RNAs that orchestrate the bacterial response to change (**Figure 13.2**).

Two-component regulatory systems enable bacteria to respond to their environment

As discussed in Chapter 5, two-component regulatory systems have a histidine kinase and a response regulator. The histidine kinase autophosphorylates then transfers the phosphate to the **receiver domain** of the response regulator, the protein domain that receives the phosphate. Phosphorylation of the response regulator activates its **output domain**, the protein domain that regulates gene expression when activated. Some histidine kinases both phosphorylate and dephosphorylate their response regulator, activating and inactivating its regulatory activity. Response regulators can also become activated via a phosphorelay system, described in Chapter 5. Response regulators generally have an N-terminal response regulator receiver domain and a C-terminal effector domain that binds DNA when active. The response regulator receiver domain undergoes a distinct conformational change when phosphorylated and tends to favor the formation of dimers. DNA-binding protein regulators tend to be dimers. In general, DNA-binding protein domain structures include helix–turn–helix structures where two α helices are joined by a short stretch of amino acids, with one sitting in the major groove and the other positioning the protein (**Figure 13.3**). Dimers of helix–turn–helix domain-containing proteins are therefore more robust in their association and specificity in binding to DNA, making them ideal regulatory proteins. There are many variations on the helix–turn–helix structure, which stabilize and enhance the DNA–protein interaction.

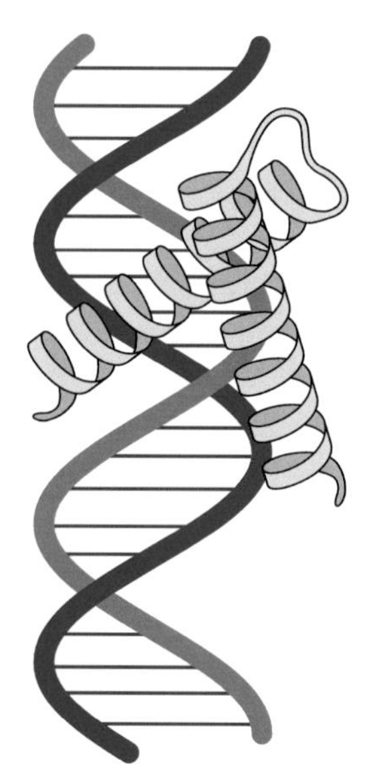

Figure 13.3 Helix–turn–helix DNA protein binding domain structure. The major protein structure in DNA-binding proteins is the helix–turn–helix domain, made up of an α helix, a short stretch of amino acids, and another α helix. One α helix is able to bind to the DNA within the major groove, while the other α helix positions the structure.

Receiver domains can also be present in proteins without DNA-binding activity in effector domains. The C-terminus can then have enzymatic activity that is activated and inactivated by the phosphorylation of the N-terminal receiver domain. The response regulator RNA-binding domain can prevent transcriptional termination at rho-dependent terminators. Response regulator enzymatic domains can include methylesterase, diguanylate cyclase, protein phosphatase, and phosphodiesterase. Some response regulators are regulated by anti-activator proteins, which alter the ability of response regulators to activate, regardless of phosphorylation. Anti-activators of response regulators stop the DNA-binding domain from binding to the promoter or stop it interacting with RNA polymerase. Due to the high degrees of homology between response regulators, they can be erroneously annotated as responding to environmental signals and controlling

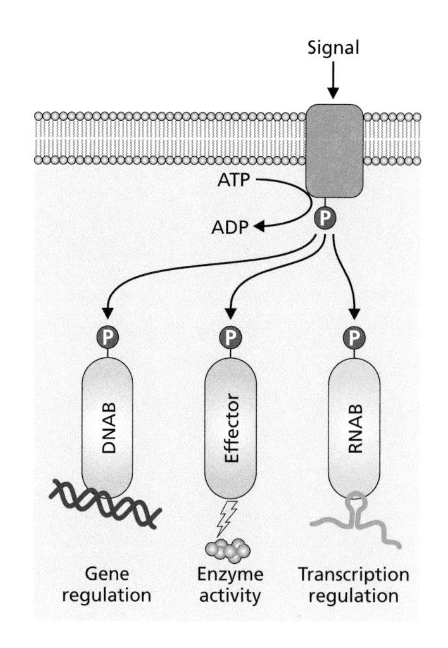

Figure 13.4 Two-component regulatory systems are involved in more than just gene regulation. As discussed in Chapter 5, response regulators of two-component regulators can have DNA-binding domains (DNAB) that are responsible for gene regulation (blue) in response to environmental signals. Additionally, response regulators can have other functions. Rather than DNA-binding domains, some of these proteins can have enzymatic domains, effectors (green), that are activated upon phosphorylation from the histidine kinase in the membrane (purple). Other response regulators have RNA-binding domains (RNAB) that are involved in transcriptional regulation (yellow) by modulating rho-dependent transcriptional termination.

genes due to functions of other proteins when their actual active function is different (**Figure 13.4**).

Mycobacterium tuberculosis is an intracellular pathogen that has 11 different two-component regulatory systems. The long-term survival of *M. tuberculosis* within the host requires the ability to respond to changing and adverse conditions. Some *M. tuberculosis* two-component systems regulate virulence, including regulation of production of cell wall lipases, phosphate uptake, aerobic respiration, and factors needed for intracellular infection and survival. They are able to scavenge nutrients within human host cells and avoid detection by the immune system. Fine tuning of the different regulatory networks within *M. tuberculosis* through the use of 11 different sensor kinases orchestrating their associated response regulators enables these bacteria to persist within the human body.

Bacteria decrease host glucose levels to impair the host immune response

As explored in earlier chapters, changes in the environment can be as simple as the differences in glucose levels, with the *lac* operon enabling the bacteria to make use of other available sugars when glucose levels are low (see Chapter 5). Bacteria can thrive in niches with high glucose levels. Indeed, diabetic patients with high glucose levels can have higher bacterial loads than those where their glucose levels are regulated. Bacteria can also respond to being within a host by lowering the hosts' glucose levels, thereby impairing the hosts' immune system.

When bacteria infect a host, there are consequences for the host. The host reacts to bacterial infection, in general, by increasing antimicrobial activities and decreasing the availability of essential nutrients to the bacteria, including the host itself decreasing the availability of glucose. In this way, the host works to try to prevent the proliferation of the bacteria. At the same time, the bacteria work to circumvent these processes, so that they can persist and survive, including using means to acquire nutrients and avoid antimicrobial actions. A reduction in host glucose levels, both by the host and by some bacterial species, causes an impaired host immune response, which exacerbates and prolongs the infection.

Bordetella pertussis infections, for example, are associated with low blood sugar. This is due to the bacteria targeting the beta cells in the host pancreas, inducing increased release of insulin relative to glucose. Controlled by a two-component regulatory system, *B. pertussis* secretes pertussis toxin, PTX, which is made of the A-protomer with the S1 subunit and the B-oligomer with the S2–S5 subunits. The B-oligomer binds to cell surfaces, causing various effects to host cells, including A-protomer-dependent binding to G-protein-coupled receptors (GPCRs) expressed on beta cells to stimulate insulin secretion. This increase in insulin reduces blood glucose levels, which is advantageous to the persistence of the bacteria (**Figure 13.5**). PTX is a complex toxin with various functions, causing a wide variety of biological activities in the human host.

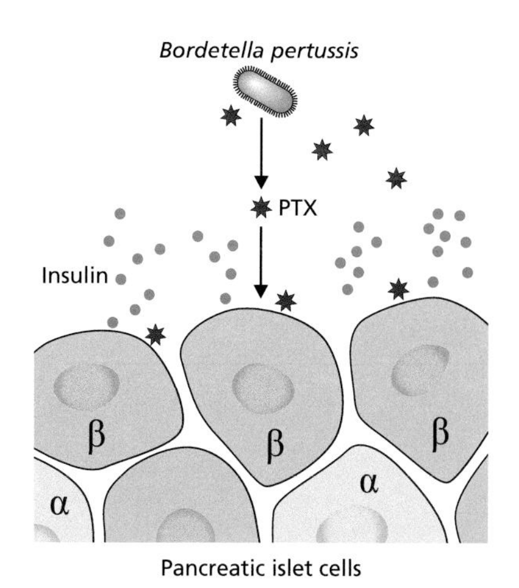

Figure 13.5 *Bordetella pertussis* pertussis toxin, PTX, induces the release of insulin. The PTX toxin has many effects within the host, one of which is its effect on the pancreas beta cells (β), where the PTX toxin stimulates the release of insulin. The increase in insulin reduces blood glucose levels, suppressing the immune response.

Bacteria modulate the immune response using the Type 3 Secretion System

Within the host, it is important for the bacteria to be able to evade the host immune system in order to survive. To do this, some bacteria use their Type 3 Secretion

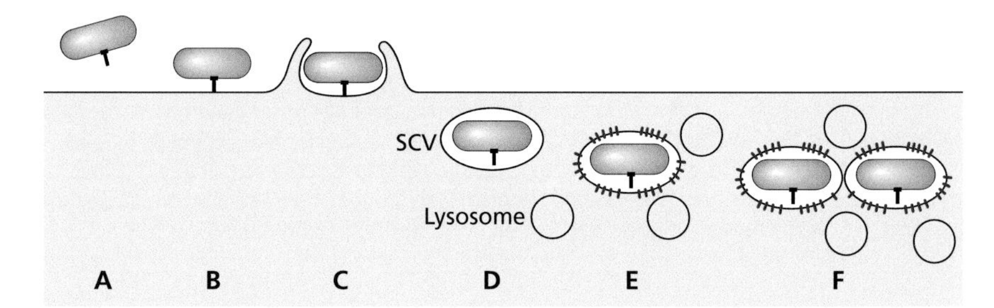

Figure 13.6 *Salmonella* use T3SS effectors to become engulfed by intestinal epithelial cells and resist lysosomal fusion. (A) The T3SS expressed on the surface of the *Salmonella* (B) inject effector proteins into the host intestinal epithelial cell, (C) causing the surface to surround the bacterial cell. (D) The bacterial cell is taken into the host cell within an SCV. (E) The T3SS releases other effectors within the SCV, altering its properties and preventing the lysosomes of the host cell from fusing with the SCV. (F) In some cases, vacuole division occurs at the same time as cell division of the *Salmonella*, increasing the number of SCVs within the host cell.

System (T3SS) to modulate the immune system (see Chapter 9). The T3SS at the surface of the bacteria is able to inject bacterial effector proteins into the host cells to change the way the immune system responds to the bacteria. This benefits the local bacterial population, including those that are not actually expressing the T3SS and its effector proteins.

Salmonella uses its T3SSs, as presented in Chapter 9, to inject effector proteins into host intestinal epithelial cells, those lining the intestine. These effector proteins cause the host cell membranes to extend outward and engulf the *Salmonella* cells. The bacterial cells are then taken within the intestinal epithelial cells, where the bacteria are within a vacuole of host cell membrane, called the ***Salmonella*-containing vacuole (SCV)**. The *Salmonella* T3SSs then inject other effector proteins into the vacuole, altering its structure and preventing the vacuole from fusing with the host cell lysosomes. The lysosomes contain lytic enzymes that are used to lyse bacteria; however, the *Salmonella* effector proteins block the fusion of the lysosomes with the vacuole, protecting the bacteria from lysis. The *Salmonella* can then safely undergo cell division within its vacuole in the host cell (**Figure 13.6**). Other studies have shown that it is possible for the vacuole itself to divide along with the *Salmonella* cell, increasing the number of vacuoles within the host cell and therefore increasing the number of lysosomes that would be needed to eradicate the engulfed bacteria, if they were able to fuse with the SCVs.

Some bacteria cheat and let others do all the work with their Type 3 Secretion Systems

The benefit of T3SS immune system modulation is enjoyed by the whole of the local bacterial population, not just those that are expressing the T3SS. The immune system in that area is suppressed, the bacteria in that area are safe from being killed by immune cells, and all of the bacteria enjoy the benefit. This means that there are bacteria in this local area that can "cheat." This situation has been exploited by what has been called "**cheater**" mutants that lack the T3SS: either they completely lack the genes for the system or they are not expressing the proteins for the system. Therefore, these cells do not have the metabolic expense of making the T3SS and the effector proteins that are being injected into the host cells. They are reaping the benefits of the T3SS-mediated immune modulation by other cells without having the energy cost (**Figure 13.7**). Cheaters arise in the population by chance mutation. However, they can thrive only when they are in the minority. They are reliant on the rest of the bacterial population to modulate the immune system, so they can survive.

It has been shown that *Pseudomonas aeruginosa* T3SS cheaters thrive in a mouse model of infection when co-infecting with the T3SS expressing *P. aeruginosa*. Non-virulent *Escherichia coli* can colonize the lungs of mice when these bacteria co-infect with T3SS *P. aeruginosa*, which suppress the mouse immune response and enable the non-virulent *E. coli* to colonize. Similarly, *Yersinia pestis* is able to cause infection only when it is expressing its plasmid-encoded T3SS, unless it is able to make use of the positive effects of another bacterial infection that is expressing T3SS-mediated immune suppression.

In contrast, *Salmonella enterica* typhimurium induces inflammation with its T3SS to eliminate competitor microbes. *Salmonella* typhimurium cheaters also exist and can establish infection in mice when they are co-infected into a mouse model with T3SS-positive strains. In this case, these T3SS cheaters gain by growing faster than the T3SS expressing wild-type and by not being targeted by antibodies against the T3SS proteins and its effectors. In this way, the *Salmonella* T3SS cheaters avoid T3SS activation of the innate immune response.

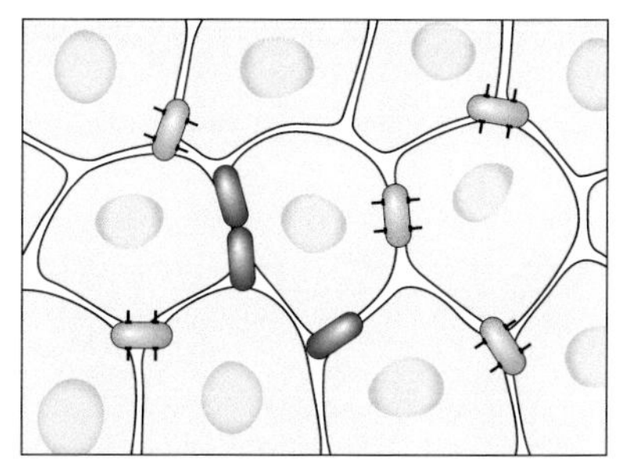

Figure 13.7 Cheaters take advantage of other bacteria expressing T3SS. In the area of the host shown here, the host immune system has been suppressed by the effectors being secreted through the T3SS on the orange bacterial cells, allowing the bacteria present to grow. The purple cells are also growing, although they are not expressing T3SS. They are enjoying the benefit of their neighbors having modulated the immune system, without having to expend any energy themselves.

A response might only be appropriate when the population is large

One bacterium expressing a toxin might not make a lot of difference. A large population of bacteria expressing a toxin is going to make a big impression. It is therefore important that there is a system in place for the bacteria to be able to tally the size of their population in a given location and tailor gene expression according to population size. This is referred to as quorum sensing, which was briefly discussed in Chapter 5. Quorum sensing phenotypes include biofilm maturation, competence for natural transformation, virulence, production of secondary metabolites such as antibiotics, motility, and sporulation, the process of forming bacterial spores resistant to environmental stress, and perhaps most spectacularly, bioluminescence (**Figure 13.8**).

Bioluminescence is most commonly seen in aquatic organisms and is a key case of why scientific experimentation and exploration into all fields is critical to gaining an understanding that can lead to insights into human health. It could not have been predicted at the time that research into why the bobtail squid glows

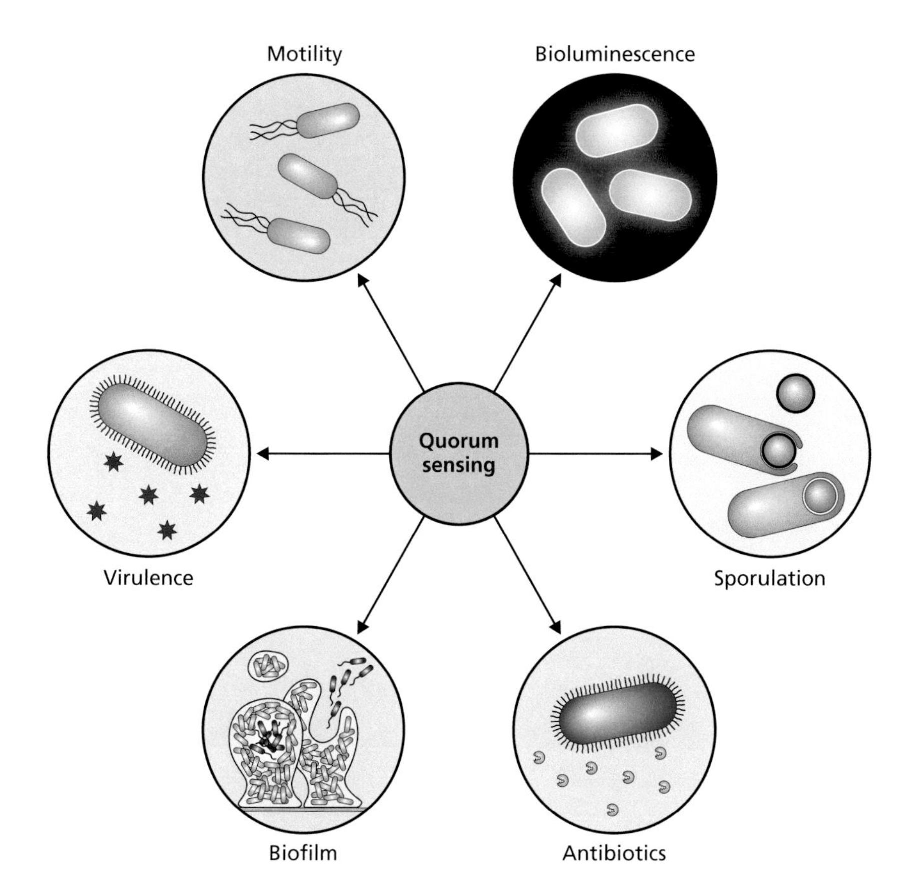

Figure 13.8 Quorum sensing. Quorum sensing is responsible for a variety of bacterial responses at the population level. Along with bioluminescence, there are an increasing number of traits that are attributed to quorum sensing. These phenotypes are advantageous when expressed as a large population, but can be energetically wasteful when expressed as lower numbers of cells.

at night would lead to insights into why catheters and other medical devices can become colonized with life-threatening infections. Both are cases where bacteria are responding and expressing certain genes only in cases where their populations are large. Both are very closely related quorum-sensing systems, yet the connection between them could not have been predicted when the research began.

Figure 13.9 Bioluminescence of the bobtail squid by *A. fischeri*. The squid has within its body a hospitable niche where the bacteria can grow and divide, reaching a large population size. When the bacterial quorum sensing system detects that there are enough bacteria present, the genes for bioluminescence are switched on, producing the characteristic glow of the bobtail squid. (From: wildestanimal/Shutterstock.com, with permission.)

Quorum sensing makes a beautiful bioluminescent glow in the ocean and in the lab

Some sea creatures, such as the bobtail squid, produce a beautiful bioluminescent glow. These squid have special organs called **photophores** or light organs in their bodies that become colonized by bioluminescent bacteria such as *Aliivibrio fischeri* (formerly *Vibrio fisheri*). The squid have evolved to have this organ as a hospitable environment for the bacteria so that the sea creature can then harness the light production of the bacteria, which prevents the squid from casting a shadow at night that might give away its presence to predators. These two organisms therefore live in **symbiosis**, a long-term interaction between two different species. The *A. fischeri* bacteria do not produce their bioluminescent glow all of the time. Making light is metabolically expensive and there is little to no benefit of doing so when there are only a few bacterial cells, since there will be little to no light produced. Indeed, during the day when the light is not needed, the squid purges its light organs of most of the bacteria, thereby turning off bioluminescence. When quorum is reached, when there is a large population of bacteria present in the photophore organ, then the *A. fischeri* bacteria glow (**Figure 13.9**).

Quorum sensing is a process to tally the bacterial cells

Quorum sensing allows bacteria to tally the number of bacterial cells in the vicinity and discover how many other bacterial cells of the same species there are, assessing the presence and concentration of bacterial cells. This process has, essentially, three main parts. In the first, the bacteria make and secrete an **autoinducer**, which is a molecule that the bacteria can recognize. The types of autoinducers used vary by species and some bacterial species produce more than one type of autoinducer. The second part is to detect the changes in the external concentration of autoinducer. As more bacteria are present, there should be more autoinducer detected in the environment. In some species, autoinducers from other species can be detected, even when they themselves do not make and secrete autoinducers. In the third and final part, the bacteria regulate their gene expression based on the detection of the autoinducer. The regulatory proteins and genes regulated in response to quorum sensing differ by species and the requirements for the expression of genes as the bacterial population grows. Once the amount of autoinducer reaches a certain level, this indicates that the bacterial population has reached the threshold for gene expression regulation (**Figure 13.10**).

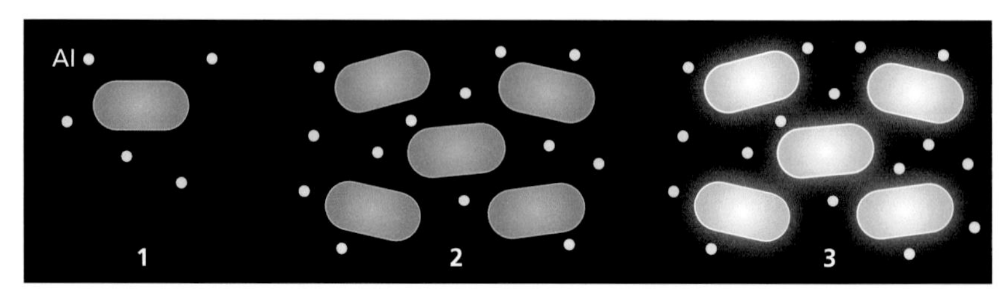

Figure 13.10 Quorum sensing has three parts. (1) The cells make autoinducer (AI). (2) The cells sense AI. (3) The cells respond to a threshold amount of AI that indicates there are enough other bacterial cells around to turn on advantageous genes, such as those for bioluminescence.

Autoinducers are abbreviated as AIs and most in Gram-negative bacteria tend to be molecules like acyl-homoserine lactones, synthesized by LuxI (**Figure 13.11**). These types of AI-1 molecules are often able to diffuse freely across the bacterial membrane, allowing them to leave the cell, accumulate in the environment, and diffuse into other bacterial cells. Some larger variants of AI are actively exported from cells by efflux pumps. There are specific receptors for Gram-negative AIs that are in either the inner membrane or cytoplasm, which recognize the AI structure. When the Gram-negative AIs bind to these receptors, they are able to trigger regulation of gene expression of the Gram-negative quorum sensing regulon. This regulon is the set of quorum sensing regulated genes that are orchestrated together as one system, with some genes being activated and some being repressed. In addition, the quorum sensing regulon often includes the genes encoding proteins for biosynthesis of the signaling molecule.

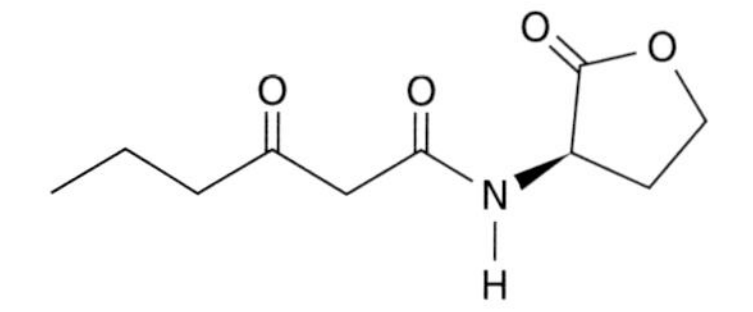

Figure 13.11 Acyl-homoserine lactone from *A. fischeri*. Gram-negative AI-1 autoinducer molecules can vary in their structure, particularly the length of the side chain and substitution on the acyl chain.

In Gram-positive species, the quorum sensing autoinducers are peptides of various lengths and modifications, autoinducing peptides abbreviated as AIPs (**Figure 13.12**). When there is high cell density of Gram-positive bacteria, the AIP binds to a two-component histidine kinase, which autophosphorylates and then phosphorylates and activates the response regulator. This response regulator activates transcription of the quorum sensing regulon. Some Gram-positive AIPs can enter back into cells as well as being present in the environment and in this way the AIP can directly interact with the regulation of gene expression, as well as through the two-component system.

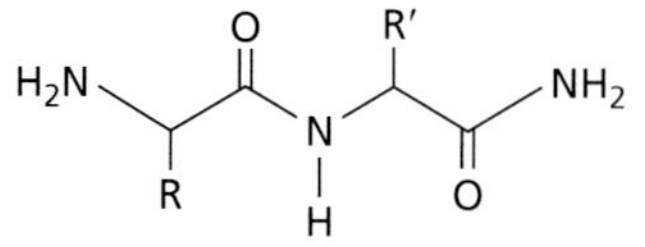

Figure 13.12 Gram-positive autoinducer molecules are usually short peptides. Unlike Gram-negative bacteria, which make organic molecules based on acyl-homoserine lactones, the AIP autoinducers used by Gram-positive bacteria are short peptide molecules. They can be of various sizes and amino acid sequences, which may be further modified, and are recognized by two-component histidine kinases on the surface of the bacteria.

Quorum sensing can involve not just monitoring the growth of the bacteria's own population, but also assessment of how many other bacteria there are around of other species. Interspecies quorum sensing involves a different autoinducer, AI-2, which is made by LuxS. Although there can be some chemical isomers and variations on the base structure, generally AI-2 is S-2-methyl-2,3,3,4-tetrahydroxy tetrahydrofuranborate (**Figure 13.13**). The cross-species communication possible through AI-2-mediated quorum sensing enables multi-species biofilms to develop and mature; however, other gene systems can be controlled by AI-2 signals as well. Indeed, AI-2 was first discovered through research into bioluminescence in *Vibrio harveyi*. It was observed that supernatants from cultures of other bacterial species were able to induce bioluminescence in *V. harveyi*.

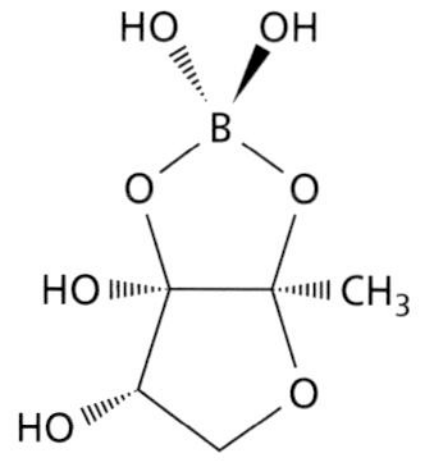

Biofilms mature due to quorum sensing signals

Biofilms are dense bacterial populations that form on surfaces in a matrix that the bacteria generate of extracellular polysaccharides, DNA, and proteins. Perhaps one of the most important realizations to come from the discovery of quorum sensing was that the development of biofilms is orchestrated by the signals sent between bacteria via autoinducers within the growing population. The maturation and development of the bacterial biofilm relies heavily on quorum sensing signals and its disruption can be influenced by disrupting quorum sensing. In several species, experimental studies have demonstrated that without functional quorum sensing systems, bacterial populations cannot form effective biofilms. Due to their placement within the biofilm, some cells will experience limited oxygen and nutrients; therefore, careful regulation of genes is necessary for the survival of the population.

Figure 13.13 AI-2, the molecules of communication between bacterial species. AI-2 is made by LuxS. Different bacterial species can receive AI-2 from one another, thereby communicating as one microbiome community.

One of the issues with biofilms is that they are notoriously difficult to eradicate, being **tolerant** to antibiotics. Bacteria have evolved biofilms to enable their survival, and this also happens to confer **tolerance** to the antibiotics that humans have started using against them. This means that although there is no genetic basis for resistance, there are physiological properties of the biofilm that are protective of the population as a whole that prevent killing of all of the bacteria by the antibiotic. Due to their properties, biofilms just happen to confer antibiotic tolerance. Regardless, it is believed that the density of the biofilm itself contributes to the observed tolerance of biofilms to antibiotics, as does perhaps the complexity of the extracellular matrix. However, antibiotics can be seen to penetrate through the biofilm, so this cannot be the whole of the reason that biofilms are tolerant to antibiotics. It is likely that the density of the biofilm and the extracellular matrix provide protection against the host immune response, rather than antibiotics. Additionally, the bacterial cells within the biofilm itself are in the stationary phase of growth or are very slow growing, which also makes them tolerant to antibiotic killing, referred to as a **persister** phenotype. This subset of cells within the biofilm is in a stationary phase-like state and is phenotypically distinct from the other bacterial cells in the biofilm population, resulting in antibiotic tolerance. Interestingly, bacterial cells that are released from biofilms tend to have higher tolerance than bacterial cells that have not been in the biofilm.

Persisters were first recognized in the 1940s, when it was observed that there was a small portion of some bacterial populations that could survive penicillin. These population variants were called persisters and were believed to be phenotypically different from the rest of the population, perhaps withstanding penicillin by the cell going dormant and not growing. Persisters are not mutants, but they are transiently and reversibly different phenotypically. They arise by stochastic events and enable the population of bacteria to adapt to fluctuating environments. In *E. coli*, the gene *hipA* encodes a toxin that inhibits translation. The HipA toxin therefore contributes to the formation of persisters by inhibiting cell growth, through inhibiting translation and putting the cells into a dormant state. About 0.1%–1% of cells in biofilms are persisters.

Quorum sensing has cheaters

As described for T3SS, there are also cheaters for quorum sensing systems. These are bacteria that do not participate fully in the quorum sensing system, yet which benefit from it. There are costs to individual cells cooperating in a population level activity, such as creating a biofilm or producing light through bioluminescence, but as a whole quorum sensing-induced gene expression benefits the bacterial population. The biofilm population is protected from the immune system and has a stable niche within the host. The bioluminescent population has a beneficial relationship with the host organism and a niche within the photophore where it is supplied with nutrients.

However, there have been mutants discovered within quorum sensing communities that benefit from these sorts of activities even though they do not participate. Such cheaters in *P. aeruginosa* lack quorum sensing and therefore do not receive signals and do not turn on the quorum sensing regulon genes, but they benefit from the production of beneficial factors by the rest of the bacterial community. It is believed that these *P. aeruginosa* cheaters would have a growth advantage in the host, particularly in the lungs of cystic fibrosis patients and on the mucosa, when colonizing alongside quorum sensing wild-type *P. aeruginosa*.

Quorum sensing plays an important role in the switch between life within the human host and within the aquatic environment for *Vibrio cholerae*. The genes regulated by this process enable the bacteria to cope with the stresses of changing from an infectious state to that within the environment, using quorum sensing to identify the number of other *V. cholerae* in the vicinity. Despite the importance of quorum sensing in the natural transitions of this species, there are cheaters in this system, which make use of the secreted extracellular proteases released by quorum-sensing bacteria.

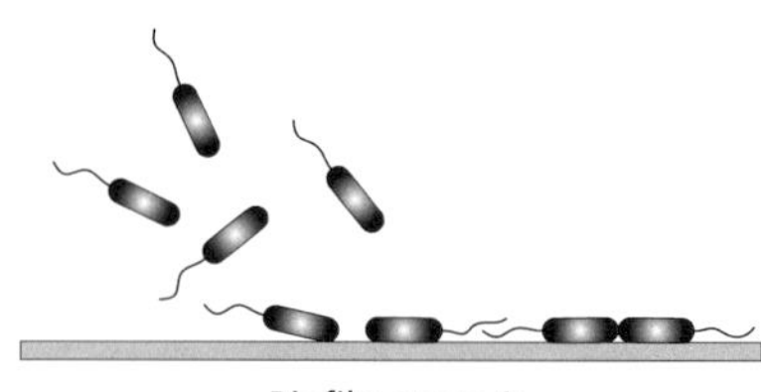

Figure 13.14 The first stage of biofilm formation: attachment to a surface. Free-living bacteria make contact with a surface and attach, using surface adhesins, as well as other surface structures such as pili or flagella to establish and maintain contact with the surface. Once in contact with the surface, cell division happens here, rather than as free-living organisms.

Going from free living to biofilm involves changes in gene expression

Bacterial biofilms develop in three stages, starting from free-living bacteria, which then attach to a surface. This attachment can make use of specific adhesins on the bacterial cell surface, as well as pili or flagella to adhere to the surface. Surfaces to which these bacteria attach can be biological or inorganic (**Figure 13.14**). Biofilms are a particular human health issue in medical devices, where they can form on catheters and indwelling artificial prosthetics. They can also form on living tissues, including in the intestines and on teeth as dental plaque, as well as on aquatic surfaces, such as rock, plants, and on the hulls of ships. Cell division then occurs here, on the surface.

In the second stage of biofilm formation, the cells grow into microcolonies that are attached to the surface. Exopolymeric substances hold the bacterial cells together, firmly attaching them to the surface and protecting the biofilm from host defenses and the environment (**Figure 13.15**). The extracellular matrix includes proteins, polysaccharides, and DNA. In addition to secretion of these, as the bacterial cells within the biofilm die and lyse, the cellular components such as DNA are released and contribute to the extracellular matrix. This matrix makes a slime on medical devices from *Staphylococcus epidermidis* and makes *P. aeruginosa* from the lungs of cystic fibrosis patients appear mucoid. Due to the three-dimensional structure of the biofilm, bacterial cells in different locations experience differences in gradients of nutrients, pH, oxygen, and quorum sensing signals. Some biofilms will form structures that extend out from the surface of the biofilm, away from the attachment surface, further varying the conditions to the different cells in the biofilm population.

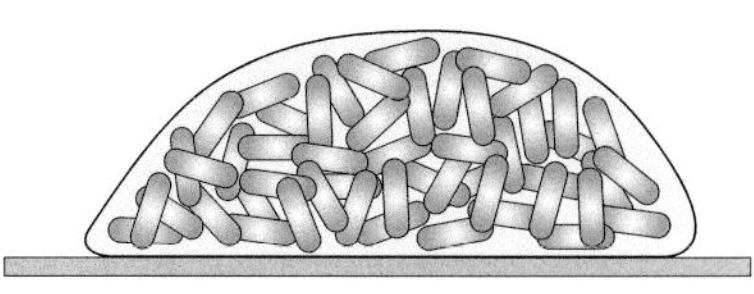

Figure 13.15 The second stage of biofilm formation: the development of the biofilm. The bacteria grow and divide, the colony spreading across the surface. The cells secrete proteins, polysaccharides, and DNA into the extracellular space, forming a matrix in and around the biofilm. The structure of the biofilm that develops has bacterial cells in very different locations within the biofilm that differ in their availability of nutrients and oxygen depending on whether they are close to the supporting surface or the outside of the biofilm.

In the third and final stage of the biofilm, some of the cells detach from the colony into the surrounding medium. Often this occurs due to limitations in the availability of nutrients for the biofilm, as well as restrictions on the growth of the biofilm (**Figure 13.16**). Detachment can happen in one of two ways. Some of the cells can return to a motile, free-living form and be released from the biofilm through the breakdown of the exopolymeric matrix. These motile cells are then capable of going on to form new biofilms, starting from the first stage with attachment to a new surface. Alternatively, sections of the biofilm, still surrounded by the matrix, can break away from the established biofilm and seed new areas of the surface, allowing the biofilm to spread.

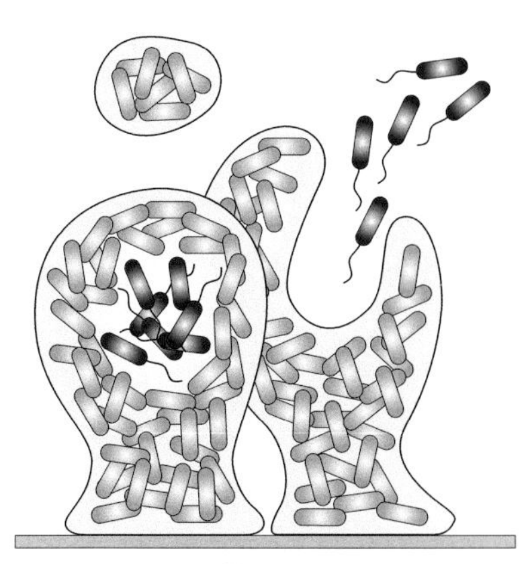

Figure 13.16 The third stage of biofilm formation: detachment of cells from the biofilm. As the biofilm grows and develops, the size of the biofilm population can become unsustainable based on the nutrients available; therefore, some of the cells switch back to being able to return to a free-living lifestyle. Parts of the extrapolymeric matrix break down, enabling these motile bacteria to be released, where they can go on to colonize new areas. Additionally, parts of the biofilm, surrounded by matrix, can break free and seed new biofilms elsewhere.

Biofilm dispersal is regulated by different elements between different bacterial species

In *Staphylococcus aureus*, the *agr* system is responsible for quorum sensing and biofilm dispersal. The genes of this system are present as one operon on the genome as *agrB*, *agrD*, *agrC*, and *agrA*, with a divergently transcribed RNAIII locus (**Figure 13.17**). The product of *agrD* is the precursor of AIP. This peptide is processed and secreted by the membrane protein AgrB. The autoinducer peptide is sensed by the two-component system AgrCA, where the sensor kinase is the membrane protein AgrC. The autophosphorylation of AgrC upon receipt of the AIP signal results in the transfer of the phosphate to the response regulator AgrA, which activates the transcription of the *agr* operon, increasing production of the AIP and the system to detect it, as well as activating virulence factors both directly and via RNAIII activation. RNAIII also represses the *S. aureus* regulatory protein Rot, which represses virulence factors, such as toxins. At low cell density, the *agr* system, working with RNAIII and Rot, permits biofilm formation, allowing biofilms to establish and reach high cell density before expressing virulence factors. As cell density increases, AIP levels increase, and biofilm dispersal is triggered in cells where virulence factor gene expression has been turned on.

In *Acinetobacter*, nutrient abundance triggers cell release from the biofilm. When there is low carbon, the biofilm is more compact. Cyclic-di-GMP (c-di-GMP) levels induce transition from the biofilm state to a planktonic lifestyle in response to nutrients, low levels of nitric oxide, or low levels of oxygen. Nitric oxide is an important biofilm dispersal signal in several bacterial pathogens, including *E. coli*, *S. aureus*, and *V. cholerae*. Some cells are released from biofilms due to the actions of other bacterial species. For example, *cis*-2-decenoic acid is a short-chain fatty acid signaling molecule made by *P. aeruginosa* that disperses biofilms made by *E. coli*, *S. aureus*, *Bacillus subtilis*, and *Klebsiella pneumoniae*, thereby being involved in interspecies competition for niches and resources.

The cells that are released from biofilms are distinct from free-living planktonic cells and from biofilm cells. For example, in *P. aeruginosa* the small regulatory RNAs RsmY and RsmZ are down regulated in dispersed cells and secretion genes are up regulated. RsmY and RsmZ bind to the regulatory protein

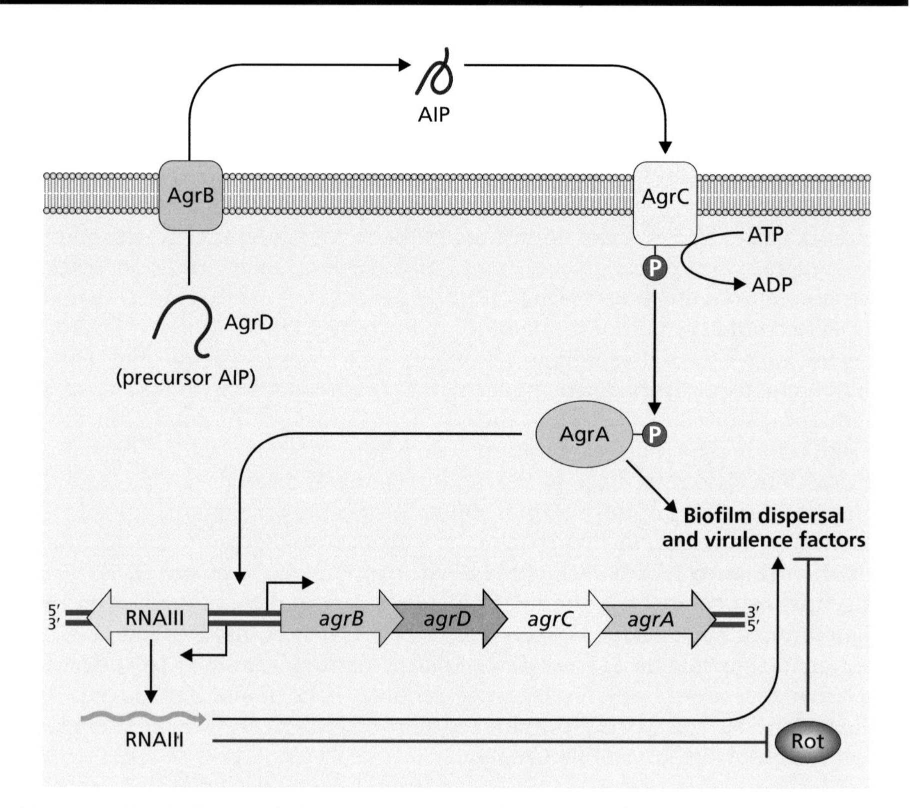

Figure 13.17 Biofilm regulating *agr* quorum sensing system of *S. aureus*. The AIP of *S. aureus* that controls biofilm dispersal and virulence factor expression is encoded by *agrD*. The precursor peptide is processed and secreted by AgrB and detected by the two-component system AgrCA, which activates expression of the *agr* operon and RNAIII. This activates biofilm dispersal and virulence factors and inhibits the repressor protein Rot.

RsmA, which results in expression of genes associated with *P. aeruginosa* chronic disease, whereas, when RsmY and RsmZ are not present, RsmA positively regulates genes for acute disease. The RsmF regulatory protein also interacts with RsmY and RsmZ and regulates genes that switch between acute and chronic disease. It is likely the differences in affinity of the proteins RsmA and RsmF for the RNAs RsmY and RsmZ fine tune which genes are expressed in each of the *P. aeruginosa* disease states. Therefore, reduced levels of RsmY and RsmZ in cells dispersed from the biofilm would result in RsmA and RsmF positively regulating acute disease gene expression. In addition to differences in ncRNA expression, the antibiotic susceptibility of cells released from biofilms is known to be different from their free-living counterparts and more akin to the cells seen within the biofilm itself, adding further credence to the belief that it is not the structure of the biofilm that is aiding the bacteria in survival.

c-di-GMP plays a key role in biofilm regulation in *P. aeruginosa*

P. aeruginosa has three key receptors that sense levels of bis-(3′–5′)-cyclic diguanosine monophosphate (c-di-GMP): WspA, YfiB, and RocS1. To make c-di-GMP, *P. aeruginosa* has **diguanylate cyclases**, which make the c-di-GMP from two GTPs, and *P. aeruginosa* has at least five well-characterized diguanylate cyclases: WspR, YfiN, SadC, RocA, and SiaD. In addition, c-di-GMP is broken down by **phosphodiesterases** into pGpG (5′-phosphoguanylyl-(3′-5′) guanosine); *P. aeruginosa* has at least five of these that have been well characterized: BifA, DipA, RocR, MucR, and NbdA. In addition to the role quorum sensing plays in maturation of a biofilm, other systems are involved in biofilm formation. For example, in *P. aeruginosa*, the protein WspA senses growth on surfaces, which then triggers autophosphorylation of WspE. WspE phosphorylates WspR to activate this diguanylate cyclase enzyme to make c-di-GMP. YfiB senses the cell membrane stress and activates YfiN to make c-di-GMP. The c-di-GMP levels regulate biofilm functions, such as synthesis of exopolysaccharides, adhesive pili and adhesins, secretion of extracellular DNA, and control of cell death and motility.

High levels of c-di-GMP induce the biosynthesis of adhesins and exopolysaccharides. *P. aeruginosa* makes three secreted polysaccharides, Pel, Psl, and alginate. Motility is inhibited and therefore the formation of a biofilm is encouraged. Low levels of c-di-GMP induce quorum sensing systems and other regulators within the cells, causing down regulation of adhesins and exopolysaccharides, therefore enhancing motility and contributing to biofilm dispersal. Other environmental cues can induce biofilm dispersion, including starvation and low oxygen. Biofilms can also break up or shed cells that can colonize new areas.

Working in concert with the c-di-GMP controlled systems to control biofilms and virulence in *P. aeruginosa* is regulation by the quorum sensing systems. An estimated 10%–12% of the *P. aeruginosa* genome is controlled by quorum sensing, using several different quorum sensing AI-1 molecules. LasI makes *N*-(3-oxo-dodecanoyl)-L-homoserine lactone ($3OC_{12}$-AHL; OdDHL), which is detected by LasR. Once activated, LasR then activates elastase and endotoxin A. The enzyme RhlI makes the autoinducer *N*-butyryl-L-homoserine lactone (C_4-AHL; BHL), which is detected by the sensor RhlR. RhlR in turn activates the production of lectins (LecA and LecB), rhamnolipid, elastase, pyocyanin, chitinase, and pyoverdine. The proteins PqsA, PqsB, PqsC, PqsD, and PqsH make the autoinducer 2-heptyl-3-hydroxy-4-quinolone (PQS), which is detected by PqsR, in addition to directly interacting with other proteins in the bacterial cell. PQS is also incorporated into the cell membrane and there induces the production of **outer membrane vesicles** (**OMVs**) that contain active PQS capable of signaling cells. These OMVs are blebs of membrane that contain outer membrane protein and may also contain some cellular material, such as DNA fragments, in small parts of the membrane that have separated away from the bacterial cell.

Bacteria have their own immune system to protect them from bacteriophages

There has been a great deal in the popular media about the CRISPR systems and their potential capacity to edit the human genome, but for the purposes of bacterial genetics and genomics, we are fundamentally interested first in their origin, as a bacterial immune system. The **clustered regularly interspersed short palindromic repeats** (**CRISPRs**) and the **CRISPR-associated genes** (**Cas**) together help protect the bacterial cell from bacteriophages through the processing of RNAs and interaction of these RNAs and proteins. Bacteria have evolved this CRISPR–Cas system, which uses sequence memory, to protect the bacteria against bacteriophage infection and also against plasmids and conjugative elements.

Within the bacterial genome there are CRISPR regions, with short direct repeats and spacer sequences between them. Preceding these elements is an AT-rich region. Associated with these repeats are the CRISPR-associated genes (Cas). To generate immunity to bacteriophages for the bacteria, the Cas protein complex incorporates bacteriophage sequences into the CRISPR locus. The bacteriophage sequences are put between the repeats of the CRISPR system in the genome as the spacers. Each spacer is added sequentially, thus the spacers are essentially a history of the bacteriophages encountered by the bacteria. When the Cas complex is activated, it destroys the recognized bacteriophage DNA that enters the cell (**Figure 13.18**).

The CRISPR array (crRNA array) of repeats is transcribed from the genome and processed into individual CRISPR RNAs (crRNAs) (**Figure 13.19**). These crRNAs with spacer sequences and partial repeats are available to hybridize to complementary nucleic acids coming into the bacterial cell. If there is a complementary sequence in an invading bacteriophage or other invading DNA, then the spacer hybridizes to the DNA and this binding triggers degradation of

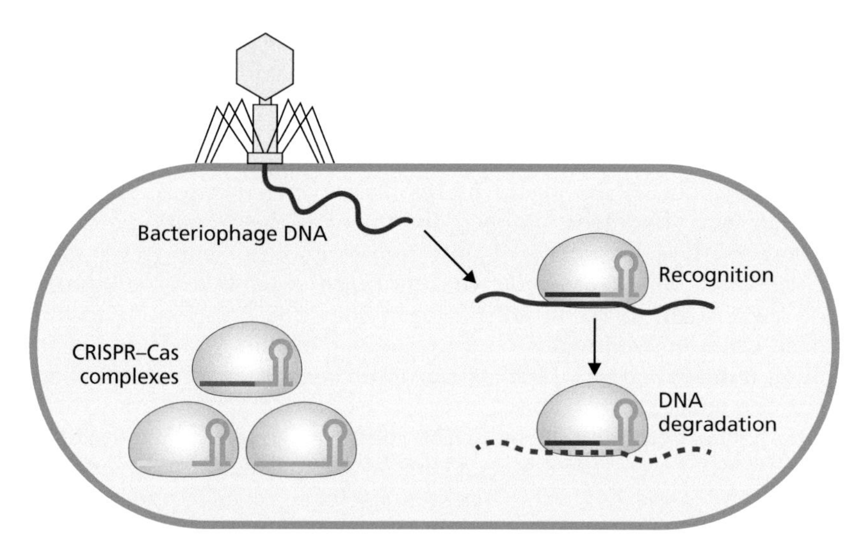

Figure 13.18 The CRISPR–Cas system protects the bacteria from invading bacteriophages through recognition of the bacteriophage sequence. The segments of the CRISPR sequences and Cas proteins form CRISPR–Cas complexes within the bacterial cell. When a recognized bacteriophage invades the bacterial cell, the complementary sequence is hybridized by the CRISPR sequence within the CRISPR–Cas complex and targets the bacteriophage DNA for degradation. This protects the bacterial cell from the bacteriophage attack.

the bacteriophage DNA bound by the Cas protein complex. The Cas1 dsDNA endonuclease and Cas2 dsDNA/ssRNA endonuclease are able to integrate new spacer sequences into the crRNA array, enabling the bacteria to recognize new bacteriophage sequences.

Some of the Cas proteins appear to have secondary or alternative functions, as well as their roles in immunity against bacteriophages. In *Legionella pneumophila*, CRISPR–Cas gene *cas2* is involved in the acquisition of new spacers from bacteriophages, yet *cas2* is also required for survival in amebae hosts. In *Francisella novicida*, CRISPR–Cas gene *cas9* is a dsDNA nuclease that is required to repress production of bacterial lipoprotein BLP that induces an innate immune inflammatory response; therefore, *cas9* is involved in immune evasion. Deletion

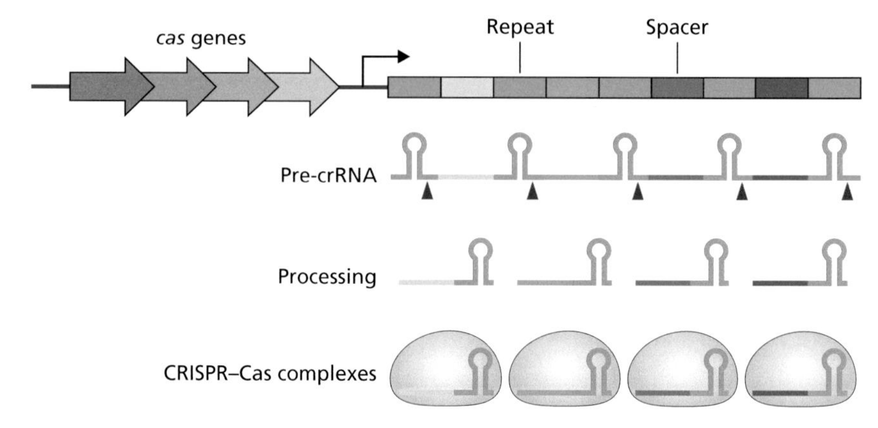

Figure 13.19 The CRISPR array is processed into crRNAs that form CRISPR–Cas complexes. The CRISPR array of repeats in the bacterial genome is transcribed into the pre-crRNA, a strand of RNA that has repeats and spacer sequences. This is cut at the repeats, such that each resulting piece has one spacer containing a sequence unique to a bacteriophage that previously invaded the bacterial cell. These crRNAs complex together with a Cas protein, ready to defend the bacterial cell from future bacteriophage invasion.

of *cas*9 in *Neisseria meningitidis* affects adhesion to and invasion of host cells. Deletion of *cas*9 in *Campylobacter jejuni* reduces virulence. CRISPR–Cas systems have been demonstrated to inhibit biofilm formation and swarming motility in *P. aeruginosa*.

Key points

- Bacterial cells are dynamic, changing in response to changes around them.
- Regulatory networks enable bacteria to respond by switching gene expression on and off in complex and overlapping ways.
- Two-component regulatory systems are the major means for bacteria to sense their environment and enact gene expression changes, including sensing the number of other bacteria in the vicinity as well as changes in nutrients and stresses.
- Two-component systems can also trigger activation of enzymes and transcriptional termination regulation, as well as gene regulation, through the use of response regulators with different effector domains.
- Bacteria can use T3SS to modulate the host immune system, allowing them to colonize and survive within the host; other bacteria can take advantage of this and cheat without having to make a T3SS themselves.
- Some bacterial behaviors are only advantageous when the bacterial population in the area is large; therefore, a quorum sensing system is used to tally the bacteria present.
- Quorum sensing is responsible for the spectacular displays in nature where sea creatures are able to glow with bioluminescence, making use of light production by bacteria that they host within special photophore organs within their bodies.
- Gram-negative bacteria commonly use acyl-homoserine lactone-based autoinducers, AI-1, for quorum sensing, while Gram-positive bacteria use short peptides, AIP, and interspecies quorum sensing uses AI-2.
- Quorum sensing regulates biofilm gene expression, switching on genes for the maturation of the biofilm community when there are sufficient bacteria present and having a role in dispersal.
- CRISPR–Cas systems have evolved as a sequence-based system to protect bacteria from invasion by bacteriophages; CRISPR–Cas complexes of crRNA segments and Cas proteins within the bacterial cell hybridize with bacteriophage DNA and target it for degradation to protect the bacterial cell.

Key terms

Define the following terms introduced in this chapter. Check your answers using the definitions in the Glossary. These terms are also available as Flashcards online.

Autoinducer	Microbiome	Receiver domain
Cas	Outer membrane vesicles	Saprophyte
Cheater	Output domain	Symbiosis
CRISPR	Persister	Tolerance

Questions and discussion topics

Self-study questions

Answer each question using 50–100 words or a table or labeled diagram. Advice on where to find answers to these questions is available online.

1. Using one example bacterial species, describe the different environmental changes encountered by this species to which it must respond.
2. Briefly, how do the pure bacterial cultures used in laboratory experiments differ from how the bacteria live in nature?
3. Building on knowledge of two-component regulatory systems from Chapter 5, draw an example system, including the different domains of the response regulator, how they function, and the structures involved.
4. Describe or illustrate how *Salmonella* can use T3SS effector proteins to evade the host immune system.
5. Briefly describe how "cheaters" are able to survive within the host without possessing functional genes needed for survival like those for the T3SS.
6. Describe, perhaps with illustrations, how bioluminescence occurs within aquatic organisms and contrast that with how it can be experimentally induced in a laboratory.
7. Give an example of a gene or set of genes that it is important for a bacterial species to express only when a bacterial population has reached a critical threshold size. Explain why this gene or genes should not be expressed in a smaller population.
8. How many different types of autoinducers, in general, have been described? Which types of bacteria, for the most part, make each of these autoinducers and sense them?
9. What bacterial system is important for the development and maturation of biofilms? How could targeted inhibition of this system be beneficial?
10. What is the difference between antibiotic resistance and the antibiotic tolerance exhibited by persister cells in biofilms?
11. When a bacterial cell is infected by a bacteriophage, what system protects the cell and how does it do this?
12. What can we learn about the history of the bacterial cell by the spacer sequences between the repeats in the CRISPR array?

Discussion topics

These topics are presented for discussion in study groups, as part of class discussions, or on your own. These questions go beyond what is directly covered in this part of the book. Use the research literature and other reading to explore these topics in more depth. Tips to help prepare for topic discussions are available online.

1. Two-component regulatory systems are often thought of in terms of regulation of gene expression via a DNA-binding protein that is phosphorylated and then binds to promoter regions in DNA. However, response regulators can have other functions as well. Explore and discuss the role of two-component regulatory systems in one bacterial species, where the response regulators are involved in cellular functions other than transcriptional regulation.
2. List some bacterial responses that are regulated by quorum sensing. Associate each with a bacterial species that uses quorum sensing to regulate this response, the quorum sensing genes involved, and the genes regulated. Indicate why this is beneficial to the bacterial population for that species.
3. Regulation of biofilm dispersal is important for bacterial cells, but also is the subject of research into the control of infectious diseases. Using examples of current research as a guide, explore and discuss why it is believed that understanding and perhaps modulating biofilm dispersal signals may be important in controlling biofilm infections.

Online quiz questions

To further self-assess your understanding of the chapter material, please visit the following link, where you can participate in a range of interactive quiz questions:

www.routledge.com/cw/snyder

Further reading

Two-component regulatory systems enable bacteria to respond to their environment

Francis VI, Stevenson EC, Porter SL. Two-component systems required for virulence in *Pseudomonas aeruginosa. FEMS Microbiol Lett.* 2017 *364*(*11*). doi: 10.1093/femsle/fnx104.

Bacteria decrease host glucose levels to impair the host immune response

Freyberg Z, Harvill ET. Pathogen manipulation of host metabolism: A common strategy for immune evasion. *PLoS Pathog.* 2017 *13*(*12*): e1006669. doi: 10.1371/journal.ppat.1006669.

Some bacteria cheat and let others do all the work with their T3SSs

Czechowska K, McKeithen-Mead S, Al Moussawi K, Kazmierczak BI. Cheating by type 3 secretion system-negative *Pseudomonas aeruginosa* during pulmonary infection. *Proc Natl Acad Sci USA*. 2014; *111*(*21*): 7801–7806.

Price PA, Jin J, Goldman WE. Pulmonary infection by *Yersinia pestis* rapidly establishes a permissive environment of microbial proliferation. *Proc Natl Acad Sci USA*. 2012; *109*(*8*): 3083–3088.

A response might only be appropriate when the population is large

Rutherford ST, Bassler Bonnie L. Bacterial quorum sensing: Its role in virulence and possibilities for its control. *Cold Spring Harb Perspect Med*. 2012; *2*(*11*): a012427.

Striednig B, Hilbi H. Bacterial quorum sensing and phenotypic heterogeneity: How the collective shapes the individual. *Trends Microbiol*. 2022; *30*(*4*): 379–389.

Quorum sensing makes a beautiful bioluminescent glow in the ocean and in the lab

Verma SC, Miyashiro T. Quorum sensing in the squid-vibrio symbiosis. *Int J Mol Sci*. 2013; *14*(*8*): 16386–16401.

Quorum sensing is a process to tally the bacterial cells

Thoendel M, Kavanaugh JS, Flack CE, Horswill AR. Peptide signaling in the *Staphylococci. Chem Rev*. 2011; *111*(*1*): 117–151.

Warrier A, Satyamoorthy K, Murali TS. Quorum-sensing regulation of virulence factors in bacterial biofilm. *Future Microbiol*. 2021; *16*: 1003–1021.

Biofilm dispersal is regulated by different elements between different bacterial species

Davies DG, Marques CN. A fatty acid messenger is responsible for inducing dispersion in microbial biofilms. *J Bacteriol*. 2009; *191*: 1393–1403.

Fedtke I, Gotz F, Peschel A. Bacterial evasion of innate host defences – the *Staphylococcus aureus* lesson. *Int J Med Microbiol*. 2004; *294*: 189–194.

c-di-GMP plays a key role in biofilm regulation in *P. aeruginosa*

Chua SL, Liu Y, Li Y, Ting HJ, Kohli GS, Cai Z. Reduced intracellular c-di-GMP content increases expression of quorum sensing-regulated genes in *Pseudomonas aeruginosa. Front Cell Infect Microbiol*. 2017; *7*: 451.

Bacteria have their own immune system to protect them from bacteriophages

Egido JE, Costa AR, Aparicio-Maldonado C, Haas PJ, Brouns SJJ. Mechanisms and clinical importance of bacteriophage resistance. *FEMS Microbiol Rev*. 2022; *46*(*1*): fuab048.

Makarova KS, Haft DH, Barrangou R, Brouns SJ, Carpentier E, Horvath P. Evolution and classification of the CRISPR-Cas systems. *Nat Rev Microbiol*. 2011; *9*(*6*): 467–477.

Watson BNJ, Steens JA, Staals RHJ, Westra ER, van Houte S. Coevolution between bacterial CRISPR-Cas systems and their bacteriophages. *Cell Host Microbe*. 2021; *29*(*5*): 715–725.

Chapter

14

Bacterial Adaptation

The world around us is constantly changing and, in order to survive, living organisms have to be able to change as well. The availability of nutrients fluctuates and, as explored in Chapter 5, bacteria can adapt by altering the regulation of genes such as the *lac* operon in response to the presence of nutrients like glucose and lactose. There are many other nutrients needed by bacterial cells and they must adapt to the presence or absence of these nutrients, as well as the presence of competitors in the environment that need the same nutrients.

Bacteria must also adapt to the complexity of their environmental niche, which may change in pH, temperature, or other physiological conditions. The adaptation to change is more than just response to the environment, it is also the ability to respond to specific and particular environments that are part of the organism's lifestyle. These are the niches in which the species will most frequently find itself and within which it has evolved to thrive. The transitions between niches necessitate responses; therefore, adaptation and response go hand-in-hand, and adaptation to various niches is a part of evolution. Therefore, the whole of Part V is very much interrelated. In this chapter, some of what was explored in responses to Chapter 13 will be revisited in the context of adaptation and some of what will be explored in the context of evolution in Chapter 15 will be foreshadowed. Niche adaptation, cell cycle regulation, phase variation, antigenic variation, and other mechanisms that enable bacterial cells to adapt to and survive their surroundings will be explored.

Within a niche, bacteria have to adapt to their peers and other bacteria

Bacteria live in complex communities beside other microorganisms, sharing niches and competing for nutrients in dynamically changing environments. As a result, bacteria have to respond and adapt to the presence of these other bacteria, bacteriophages, fungi, and parasites. Depending on the niche, there may also be host cells present, including immune response cells in some cases, and potentially beneficial or harmful biomolecules in the environment. Some of the other organisms in the niche are their peers, daughter cells of the same bacterial population. Many are competitors for the same nutrients, which may ebb and flow from the environment due to physical conditions, as well as use by biological organisms. All of this variation in the microenvironment of the bacterial niche requires adaptation for survival, through either regulated responses or variations within the bacterial population such that some cells are able to thrive even when others of the same population are lost.

Bacteria that have similar nutrient requirements will be in competition with one another when those nutrients are in limited supply. As the populations of the bacteria grow, the nutrients become depleted and competition increases. For example, within a host niche, one of the limiting nutrients that is essential for bacterial growth is iron; however, free iron in this niche is rare. To acquire iron in the host, many bacteria use iron-scavenging **siderophore** molecules, which are iron-binding molecules made by the bacteria and secreted into the environment. Some bacteria can make use of siderophores that are made by other bacteria, reducing the energy cost of the siderophore biosynthesis and robbing their

DOI: 10.1201/9781003380436-19

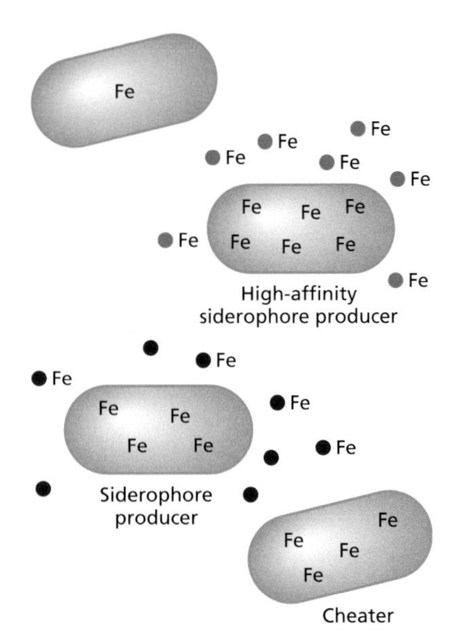

Figure 14.1 Iron acquisition within a host niche. Free iron (Fe) in the niche is rare and it can be difficult for bacterial cells to acquire enough to survive. Therefore, some bacteria make siderophores, molecules that are able to bind iron and scavenge it from the host environment (red and orange spots). Different siderophores may have higher affinities for iron, giving some bacteria an advantage over others (orange spots). Some bacteria, "cheaters," may make use of the siderophores made by other bacterial cells, saving themselves the energy of making their own siderophores.

competitors of iron. Additionally, different siderophores have different affinities for iron that may make some species, or even some strains within species, more effective competitors (**Figure 14.1**).

Bacteria can also adapt to the presence of other bacteria in their niche by trying to kill them. This reduces competition for nutrients and releases the nutrients within the dead bacterial cells. Bacteria, including lactic acid bacteria such as *Lactobacillus*, produce **bacteriocins**. These are antimicrobial peptides of 30–60 amino acids in length and are encoded by the bacteriocin operon. Immunity to the bacteriocin is conferred by an adjacent gene on the operon, protecting the bacteria that make the bacteriocin from their own antimicrobial peptides. Bacteriocins act on the cytoplasmic membrane, generally of Gram-negative bacteria, disrupting the proton motive force essential for bacterial cellular processes, including transport of nutrients across the membrane. It is possible for bacteria to evolve resistance to bacteriocins, generally through efflux or degradation of the antimicrobial peptide; however, bacteriocins are a highly effective means for bacterial cells to reduce or eliminate competitor bacterial cells.

GlcNAc has a role as a signaling molecule as well as being part of the bacterial cell wall

The bacterial cell wall, peptidoglycan, is made up of the sugars *N*-acetylglucosamine (GlcNAc) and *N*-acetylmuramic acid (MurNAc). During bacterial cell growth and division, GlcNAc is released from the cell due to the necessary remodeling of peptidoglycan that has to happen in order for the cell to expand and for the process of division to occur. The release of GlcNAc into the environment is therefore a signal that bacterial cells are dividing and can trigger neighboring bacterial cells to adapt to the presence of growing and dividing cells in their proximity. Quorum sensing was previously discussed in Chapters 5 and 13, whereby a signaling molecule is made, released, and sensed as an indicator of population size. In the case of GlcNAc, this is a molecule that is released as part of the process of bacterial cell growth and division. GlcNAc is used as a signaling molecule for bacteria to adapt to signals of growth from each other and other organisms.

For example, the presence of GlcNAc stimulates antibiotic production in soil bacteria. Although we use antibiotics today to combat infectious diseases, these antimicrobial agents originally evolved as mechanisms to enable bacterial cells to kill neighboring bacterial cells that were their competitors. The signal that there are other bacteria around, the presence of released GlcNAc, therefore turns on the production of antibiotics. Likewise, the GlcNAc released from growing Gram-positive bacteria induces the production of toxins and virulence factors by *Pseudomonas aeruginosa*. Some bacteria take the GlcNAc signal as a sign to leave the niche and find a new place to colonize; *Escherichia coli* decreases the production of its type 1 fimbrial adhesins thereby increasing transmission to new hosts in response to the presence of GlcNAc. The presence of GlcNAc also reduces the production of Curli fibers and fimbrial adhesins, important for adhesion and biofilm formation, enhancing dissemination of the bacteria away from the competitors.

Competitor bacteria can be killed with specialized Type 6 Secretion Systems

As seen in Chapter 9, some bacteria use Type 6 Secretion Systems to target competitor bacterial cells; killing them reduces competition for nutrients and releases nutrients from the dead bacterial cells. A Type 6 Secretion System is a multiprotein surface structure made by some Gram-negative bacterial species that is able to directly inject toxins into host cells and bacterial competitors. In some ways this is similar to Type 3 Secretion Systems that inject effector proteins into host cells; however, the Type 6 Secretion System can also be used against other bacterial cells and competitor fungal cells, as well as against host cells.

Type 6 Secretion Systems inject toxins into competitor bacteria that degrade bacterial cell walls, disrupt bacterial cell membranes phospholipids, form pores in bacterial membranes, and degrade nucleic acids. As well as the effector proteins that are secreted by the system, bacterial cells with Type 6 Secretion Systems produce an immunity protein that protect the bacteria from its own toxins. Most Type 6 Secretion System effectors target Gram-negative competitor bacteria, which can include related bacterial species.

For example, many of the bacterial species in the Order *Bacteroidales* colonize the human gut at the same time and compete for space and resources. By diversifying their metabolisms they can make use of different nutrients within the niche. However, *Bacteroidales* can also use Type 6 Secretion Systems and secrete toxins to eliminate the competition.

Caulobacter differentiate between motile and sessile cells

Caulobacter are a genus of bacterial cells that have a distinct cell cycle, which generates both motile **swarmer cells** that have flagella and stalked **sessile cells** that adhere to surfaces and are non-motile. Cells differentiate from being motile cells to sessile cells, with each cell type providing an adaptive advantage to the bacterial population in the aquatic environment in which *Caulobacter* are generally found. When a sessile cell divides, a sessile daughter cell remains attached to the surface and a motile daughter cell is released to the environment (**Figure 14.2**). The stalked sessile cells are optimized for adhesion to surfaces, nutrient uptake, DNA replication, and cell division, while the motile swarmer cells are optimized for motility via flagella and are able to disperse the bacteria, enabling them to colonize new niches. The swarmer cell can either grow the population in the local area when nutrients are abundant or go to new areas when nutrients become depleted. The swarmer cells also have adhesive pili associated with the flagella; as the swarmer cell transitions to becoming a sessile cell in its site of colonization, the flagella is lost and these pili attach to a surface and retract, bringing the cell close to the surface so that the stalk can form (see Figure 14.2).

The cell cycle of *Caulobacter* is controlled by coordinated expression of the regulatory proteins CtrA, DnaA, GcrA, and CcrM. These are master regulator proteins that control regulons involved in the differentiation of the *Caulobacter* cells. The overlapping of the expression of the different master regulator proteins

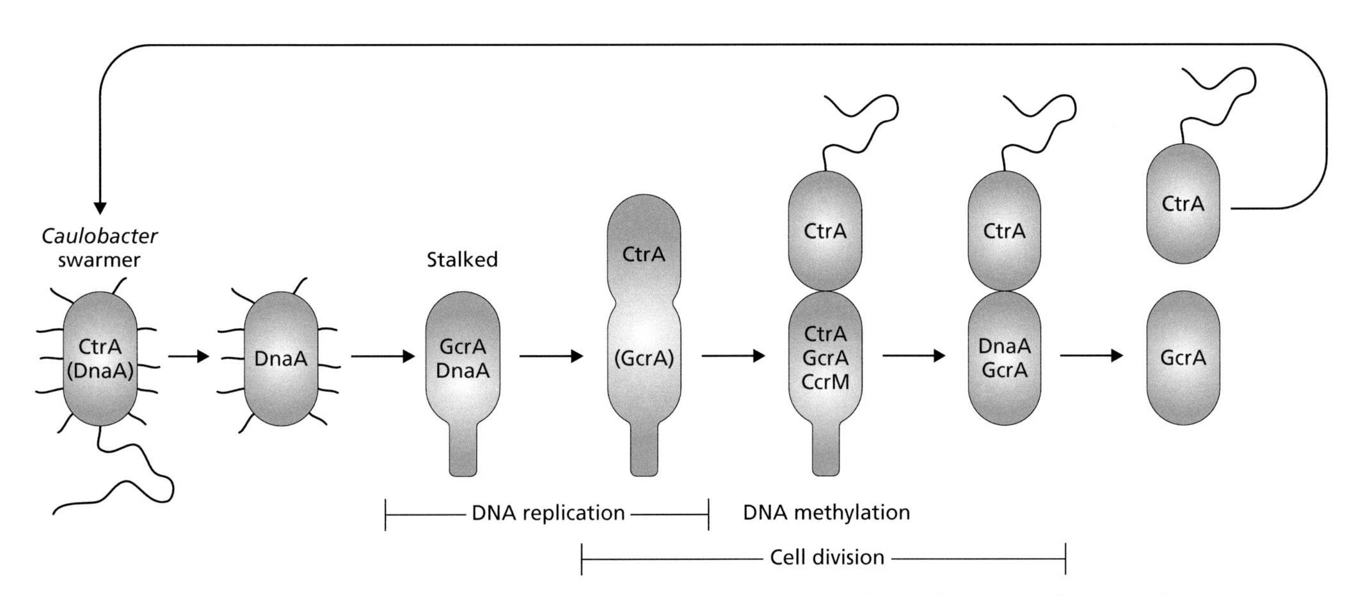

Figure 14.2 The cell cycle of *Caulobacter* is controlled by master regulatory proteins. The motile swarmer cell expresses the master regulator CtrA and starts to express DnaA. The cell loses its flagella, as shown in the second cell in the sequence. As DnaA is joined by GcrA, the motile cell becomes a stalked sessile cell and DNA replication begins. Levels of GcrA diminish as CtrA returns and cell division begins, with a flagellated daughter cell forming. The stalked cell remains, able to divide again to produce more flagellated daughter cells. The motile daughter cell is able to move to new areas before becoming sessile and starting a new cycle.

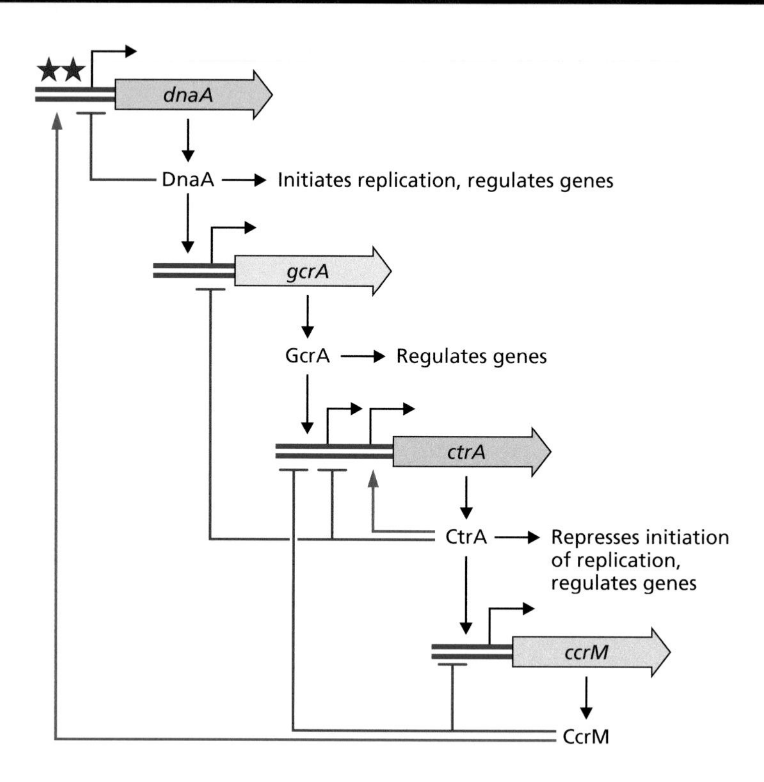

Figure 14.3 *Caulobacter* cell cycle master regulator gene expression. The master regulatory protein genes of *Caulobacter* activate and inhibit the expression of each other, providing feedback loops that orchestrate their expression and influence the timing of their presence within the swarmer and stalked cells of *Caulobacter*. The ☆☆ indicates the methylation site targeted by CcrM in the *dnaA* promoter. The arrows indicate interactions where the protein activates expression. The red blunt-ended lines indicate interactions where the protein inhibits expression.

determines the expression of flagella, the adhesion of the sessile stalked cell to the surface, and the process of DNA replication and cell division (see Figure 14.2). Division generates one sessile stalked daughter cell and one swarmer cell. The stalked daughter cell remains sessile and it is here that CcrM, a DNA methyltransferase, acts. Swarmer cells are motile initially, but then become sessile cells as part of the cell cycle, due to the influence of the master regulator proteins (see Figure 14.2).

The control of the genes that regulate the expression and degradation of DnaA, GcrA, CtrA, and CcrM is regulated in an interconnected manner (**Figure 14.3**). In general, proteolysis ensures that the activity of proteins will stop at the right time, through degradation of proteins and therefore their inactivation. Proteins are synthesized by the cell and are cleared away by proteolysis, orchestrating the expression of gene products. In the *Caulobacter* cell cycle, DNA replication only occurs after CtrA is cleared from the cell and DnaA accumulates. In the gene expression cascade, the *dnaA* gene is transcribed and translated into DnaA, which initiates replication and regulates genes, including activating expression of *gcrA*. Accumulation of DnaA will also inhibit *dnaA* expression. As GcrA is expressed it regulates genes, including the expression of *ctrA*. CtrA represses the initiation of replication and regulates genes, including *ccrM* and *ctrA* itself. CcrM inhibits the expression both of itself and of *ctrA* and also methylates the promoter of *dnaA*. In this way, there are feedback loops within the expression of these key proteins involved in the *Caulobacter* cell cycle that orchestrate the development of the swarmer and sessile cell phenotypes and ensure efficient and tight regulatory control of the cell cycle stages.

Staphylococcus aureus secretes several proteins to inhibit host defenses as part of adapting its niche to its needs

Sometimes niche adaptation involves bacteria adapting a niche to their own needs. In the case of *Staphylococcus aureus*, the bacteria secrete a number of different proteins that change the host, making it more hospitable for the bacterial cells. *S. aureus*-secreted virulence factors inhibit **opsonization**, a host cell process of coating of the bacteria with antibodies or complement proteins to target the bacteria for **phagocytosis**. Immune cells, such as macrophages, are able to recognize opsonized bacterial cells, engulf them, and neutralize the threat to the host from the pathogens. Specifically, *S. aureus* protein A, which is present on the surface of these bacterial cells, binds to the Fc region of host antibodies and therefore inhibits appropriate antibody opsonization and phagocytosis of the bacteria (**Figure 14.4**).

Complement activation is also inhibited by secreted *S. aureus* virulence proteins. The *S. aureus* collagen adhesion protein, Cna, inhibits complement component C1q, preventing its interaction with complement component C1r. In addition, staphylococcal clumping factor A, ClfA, targets complement component C3b for degradation by host complement control protein factor I. Taken together, *S. aureus* has evolved a set of proteins that are able to inhibit the host innate immune response and reduce complement-mediated killing and bacterial cell opsonization by complement. This enables the bacterial cells to control their host niche and make it more hospitable.

S. aureus-secreted proteins also inhibit the immune response through lysis of neutrophils, thereby directly attacking the cells of the immune system. Specifically, staphopain protease ScpA inhibits the activation of neutrophils, while the SspB staphopain protease appears to target neutrophils for phagocytosis. In addition, the *S. aureus* aureolysin Aur is able to neutralize the host antimicrobial peptide LL-37 and complement component C3, providing further protection for *S. aureus* from host defenses. This wide range of factors listed, and others, all contribute to making the niche more hospitable for *S. aureus*, in a sense adapting the host niche to the bacteria, rather than the bacteria adapting to fit the niche.

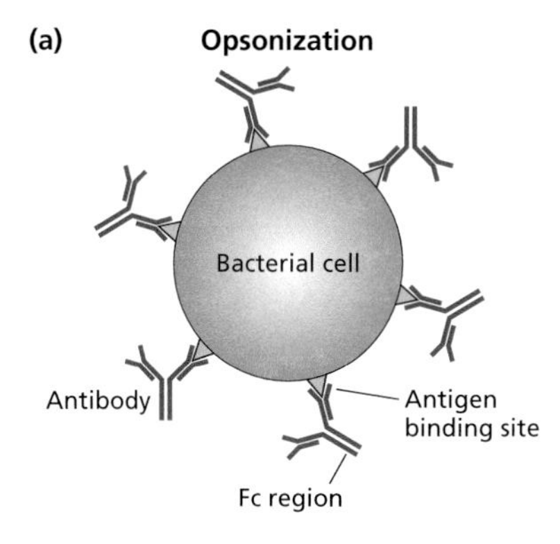

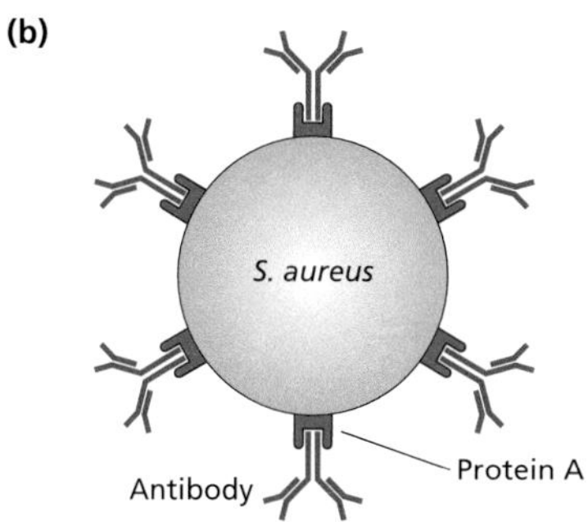

Figure 14.4 *S. aureus* opsonization and protein A. (a) The host immune system makes antibodies against bacterial surface antigens. These antibodies have an antigen binding site at one end and an Fc region at the other end. Opsonization coats the bacteria with antibodies so the Fc region can facilitate phagocytosis and elimination of the bacteria. (b) *S. aureus* protein A on the surface of the bacteria binds the Fc region of host antibodies, preventing the antigen binding site of the antibody from properly binding to the bacteria and opsonizing the bacterial cells.

Intracellular bacteria adapt to life inside the cells of the host

Many bacteria live within eukaryotic host organisms. Often these bacteria are living within the organism, but not within the organism's individual cells. However, there are some bacteria that actually enter into the eukaryotic cells themselves, as **intracellular** bacteria. As mentioned in Chapter 3, these often have lower %G+C than extracellular bacteria. Some bacteria are **facultative intracellular** bacteria, which means they can live outside host cells the way the vast majority of bacteria live or they can live within host cells. Switching between intracellular and extracellular states requires adaptation, including switching between different mechanisms of nutrient acquisition, survival against host defenses, and stress responses to different niches. Some very specialized bacteria are **obligate intracellular** bacteria, which can survive only within host cells. Obligate intracellular bacteria often lack some metabolic and biosynthetic pathways and instead rely on the host cell for nutrients and ATP. These bacteria are so reliant upon other cells for their survival that they have to be grown in host cells to culture them in the laboratory, making experiments involving obligate intracellular bacteria such as *Mycobacterium leprae, Coxiella burnetii,* and *Rickettsia* spp. a bit more complicated. These cells still do adapt; however, the stimuli are now changes within the intracellular niche.

Being intracellular means that these bacteria cannot be detected by host antibodies. **Antibodies** are proteins that are secreted by host immune cells, specifically targeting foreign **antigens**, such as the surface proteins and surface structures of bacteria. Antibodies circulate in the host and will bind to bacterial antigens they recognize, targeting the bacterial cells for elimination. Antibodies cannot enter host cells; therefore, they cannot find bacteria within host cells to target them for clearance by the immune system. Bacteria can hide from the antibodies by going inside the host cells, which is a distinct advantage for intracellular bacteria.

Host cells are not without means of protecting themselves, however. The intracellular bacteria must protect themselves from host cell lysosomal enzymes and they must be able to scavenge nutrients away from the host cell in order to survive being inside the cell while hiding from the antibodies and the immune response. *Legionella pneumophila*, as an example, prefer to live within

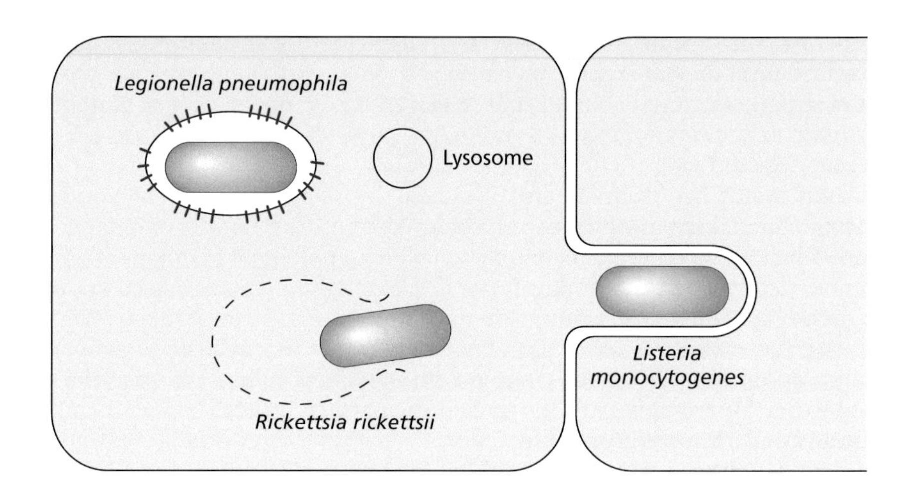

Figure 14.5 Survival strategies of intracellular bacteria. Intracellular bacteria use various strategies to avoid being killed within the host cell. *Legionella pneumophila* blocks fusion of the lysosome with the phagosome it inhabits. *Rickettsia rickettsii* and *Listeria monocytogenes* escape the phagosome into the cytoplasm before the lysosome fuses with the phagosome. *L. monocytogenes* is then able to spread to adjacent host cells.

macrophages. Macrophage cells are professional immune cells that engulf and destroy bacteria, yet *L. pneumophila* induce their own uptake into macrophages and, once inside, the bacteria block the fusion of the **lysosome** (containing the lysosomal enzymes), with the **phagosome** (containing the *L. pneumophila* to escape destruction). Other species avert degradation by lysosomal enzymes by other means. Rather than blocking the fusion of the lysosome and phagosome, *Rickettsia rickettsii* destroys the phagosomal membrane and escapes into the cytoplasm, thereby avoiding fusion with the lysosome and digestion by the host cell. *Listeria monocytogenes* also escapes the phagosome into the cytoplasm before fusion of the phagosome with the lysosome and from here is able to spread to nearby host cells (**Figure 14.5**). Each of these processes is regulated by an adaptive response, which senses the uptake of the bacterial cell by the host cell and the presence of the bacterial cell in the phagosome, triggering the bacterial intracellular survival strategies.

Mycobacterium tuberculosis adapts both itself and its host

The only known niche for *Mycobacterium tuberculosis* is the human host. The life cycle of these bacteria is a delicate balance because they must evade the host immune system to survive. Part of their time is spent in a non-replicative state where they are resistant to the immune system and antibiotics that act on actively dividing bacterial cells. This stage of the life cycle can last for decades. To transmit to a new host, the *M. tuberculosis* must replicate, with the bacterial cells being transferred via aerosol. Once in the lung alveoli, the bacteria are phagocytosed by alveolar macrophages and dendritic cells. Within the phagocytic cells, the *M. tuberculosis* replicate, migrate to the host lymph nodes, and spread to infect additional macrophages. When the host adaptive immune response begins, the *M. tuberculosis* cells enter the non-replicative stage of their life cycle, establishing the characteristic chronic tuberculosis infection.

M. tuberculosis have been shown to reside mostly within macrophage phagosomes, although they can be found in other cells and in the cytoplasm. There is also some heterogeneity in the nature of the phagosomes, suggesting that the bacteria are able to regulate their environment and adapt it to their needs, enabling the extreme longevity of this chronic infection. It is clear that *M. tuberculosis* can interfere with lysosome fusion and with phagosome acidification, as well as being able to modulate the phagosome environment according to the specifics of the host cell, length of infection, and location within the host. The *M. tuberculosis* gene *icl*, encoding enzyme isocitrate lyase, is strongly induced *in vivo* to enable the bacteria to utilize fatty acids as an energy source within the macrophages. The infection also induces elevated lipid metabolism in the human host. Therefore, the *M. tuberculosis* switches to a lipid metabolism as its preferred energy source and then induces the host to make more.

In some susceptible individuals, where the immune system is weakened or becomes weakened, the bacteria are able to replicate, sometimes only slowly. The macrophages containing these *M. tuberculosis* make granulomas that degrade into caseous centers with a characteristic cheesy appearance of tuberculosis disease. This caseating granuloma lung tissue progresses in the disease, breaking down and liquefying. The *M. tuberculosis* are able to rapidly replicate in the liquid environment and are transmitted to a new human host when the patient coughs, aerosolizing the liquefied disease matter.

Legionella adapt by knowing when not to grow

L. pneumophila has quite varied environments in which it can live, being found as an intracellular bacteria and as an extracellular bacteria. In freshwater ecosystems, *L. pneumophila* lives as biofilm or as free-living planktonic forms of the bacterial cells. However, these bacteria appear to replicate only within eukaryotic cells, such as *Acanthamoeba castellanii*, an ameba found in the environment. In the environment, the bacteria prefer to establish a parasitic relationship with protozoa, such as these ameba, because the *L. pneumophila* are then provided with a safe place to replicate, a source for nutrients, and protection from environmental stresses. *L. pneumophila* can also enter humans, inhaled in aerosols, where they infect and replicate in alveolar macrophages and cause Legionnaires' disease. Infection of humans is, in many ways, accidental. Human macrophages are evolutionarily similar to the protozoan phagocytes, and *L. pneumophila* that have been growing in protozoa are readily able to infect human macrophages, whereas those grown on agar plates are not. Human-to-human transmission is rare, with cases of Legionnaires' disease arising due to human-created environmental sources where protozoa infected with *L. pneumophila* can thrive. Intracellular *L. pneumophila* are within a vacuole and there use serine as a carbon source to fuel replication and metabolism. When this amino acid is depleted, the **stringent response** turns on expression of virulence traits and protects the bacteria so that it can survive long-term stress and starvation in the extracellular environment. As discussed in Part II, bacterial cells respond to stress through coordinated regulation of sets of genes to combat the encounters with stress. Alternative carbon sources, including glucose, are used and the bacteria are able to escape into the environment ready to find a new host cell.

The transition of the bacteria from its extracellular to its intracellular niches causes phenotypic changes in the bacterial cell. The intracellular *L. pneumophila* in the phagocytic protozoa and macrophages are replicative and metabolically active; however, the extracellular bacteria in the environment are motile, resistant to stress, and virulent. There are also some intermediate forms that have been reported, which presumably facilitate transition from one niche to another. Regulation of gene expression adapts the bacterial cell to the various hosts and niches.

These bacteria can be found in extreme environments, such as water sources where the temperatures are at 0 °C and those where they are over 60 °C, as well as acidic conditions. They are readily able to adapt to these extreme changes in their environment and to detect whether the situation is hospitable for replication or if the situation calls for the expression of genes for motility, stress resistance, and virulence.

Group A streptococci within the host experience adaptation, mutation, and death

Group A streptococci (*Streptococcus pyogenes*) have been reported to switch from a virulent phenotype, $SpeB^{hi}/SpeA^{-}/Sda1^{low}$, to a hypervirulent phenotype, $SpeB^{-}/SpeA^{+}/Sda1^{high}$, after about 3 days of growth *in vivo*. This switch is not reversible, and the original phenotype is gone from the population by day 7 of the infection. This change in phenotype is due to pressures from the immune system and selection for bacteria adapting to the host. Specifically, mutations in the CovRS two-component regulatory system are selected for, generating the hypervirulent phenotype. SpeB goes from being highly expressed in the virulent cells, $SpeB^{hi}$, to being absent in the hypervirulent, $SpeB^{-}$. The lack of SpeB in the hypervirulent group A streptococci preserves the activity of other virulence factors made by the bacteria. Sda1 goes from low-expression levels ($Sda1^{low}$) to high-expression levels ($Sda1^{high}$) in the hypervirulent bacteria. Sda1 is a potent DNase that protects the group A streptococci from neutrophil killing. SpeA is a toxin, which is not made in the virulent group A streptococci ($SpeA^{-}$), but is produced in the hypervirulent bacteria ($SpeA^{+}$). Therefore, the changes in expression of these three key virulence factors change the bacterial phenotype from virulent ($SpeB^{hi}/SpeA^{-}/Sda1^{low}$) to hypervirulent ($SpeB^{-}/SpeA^{+}/Sda1^{high}$).

This is an interesting example because transcription studies have shown that the virulent bacteria attempt to adapt and survive in the host environment, but ultimately fail. Changes are seen in metabolism and stress response gene expression, but in the end, these bacterial cells become extinct in the bacterial population within the host. Only those with the *covS* sensor kinase mutations that generate the hypervirulent phenotype survive to the 7th day of *in vivo* growth conditions in the host niche.

Adaptation of the host to enhance the spread of infection

The insect **endosymbiont** *Wolbachia* infects reproductive tissues and spreads from mother to offspring through the cytoplasm of the insect eggs. The bacteria are able to influence the behavior of the host insect, causing adaptation of their host environment for the advantage of the bacteria. Because the *Wolbachia* are endosymbionts, they rely on the insect hosts and it is essential that they spread to new insect hosts to continue to grow as a population; therefore, mechanisms that increase their spread are key. The infection causes a reproductive issue for insects referred to as cytoplasmic incompatibility, where embryos from infected males and uninfected females die. Insects that are infected with *Wolbachia* also exhibit characteristic changes in reproductive behaviors, including male feminization, parthenogenesis, or male killing, that can negatively impact the reproduction of the uninfected population of insects. Infected female insects are likely to be more reproductively successful, being able to mate with infected and uninfected males successfully, increasing the spread of the *Wolbachia*.

Listeria monocytogenes can adapt to an intracellular or soil niche

L. monocytogenes lives in the soil as a **saprophyte**. This niche gives it ready availability to the nutrients of the dead and decaying plants and enables it to live its saprophytic lifestyle. From this niche, *L. monocytogenes* can find itself ingested by humans or animals and it is at this point that it takes on a pathogenic lifestyle and becomes a facultative intracellular bacteria. Inside host cells there is an upregulation of factors that allow cell-to-cell spread of the bacteria and allow bacterial replication within the host cell (see Figure 14.5). The cell-to-cell spread of *L. monocytogenes* protects the bacteria from the host immune system because the bacterial cells never emerge from the host cells. They spread from being within one cell to being within the next cell and are never exposed to the host antibodies.

Within animals and humans, *L. monocytogenes* expresses proteins InlA and InlB that are **internalins** that help bacteria invade host cells. InlA binds E-cadherin on host cells. InlB binds the hepatocyte growth factor receptor on host cells. These two very specific host cell targets mean that the bacteria are able to bind their proteins to the surface of the host cell and then make use of the host cell's own mechanisms to internalize particles. *L. monocytogenes* highjacks mechanisms already in place and uses them to bring itself into the host cell. In this case, the *L. monocytogenes* ends up inside the eukaryotic cell inside a **vacuole**, a small membrane-bound space. Once inside the host cell, *L. monocytogenes* escapes from the phagosome using LLO (listeriolysin O), which forms pores in the vacuole membrane, and phospholipases.

The process of *L. monocytogenes* adopting an intracellular lifestyle is regulated by the protein PrfA and is based on the available carbon sources in the different niches as well as differences in the temperature. Almost all of the intracellular processes of *L. monocytogenes* are regulated by PrfA and its expression is transcriptionally regulated by an RNA thermosensor that detects the host body temperature. This provides a clear means of detection of the animal and human niche versus the soil niche, based on temperature. There are 10 genes directly regulated by PrfA, with another 145 shown to be within the regulon by microarray and proteomic assays. Without PrfA, *L. monocytogenes* is 100,000-fold less virulent in mouse infection. Evidence has shown that PrfA is also involved in extracellular biofilm formation, suggesting that it has roles in intracellular and extracellular niche adaptation.

Environmental bacteria like *Lactobacillus plantarum* can live in a wide variety of niches

Niches within the environment are many and varied, and bacteria such as *Lactobacillus plantarum* can adapt to live within a wide variety of plants and in the gastrointestinal tracts of animals and humans, and are associated with fermentation processes including the production of sourdough and sauerkraut. Of particular interest is *L. plantarum* growth in pineapple juice, which is normally inhospitable for bacterial growth because it is acidic and has a high concentration of carbohydrates, fibers, and phenols. The ability of this species to adapt to this and other diverse and inhospitable environments suggests that it has a high capacity for adaptation.

Transcriptomic and metabolomic investigations showed that *L. plantarum* is able to sense the plant host environment and adjust its carbohydrate metabolism accordingly. This allows the bacteria to not only adapt to the niche, but also metabolically outcompete other bacteria, enabling it to thrive.

Analysis of the genomic content of a collection of *L. plantarum* strains isolated from diverse environments showed no connection between genes and niches of isolation. This suggests that these bacteria have not restricted themselves to specific lineages adapted to certain niches, but rather that the bacterial species as a whole retain the ability to exploit a wide array of the available environment. This fits with the genus being the largest and most diverse of the lactic acid bacteria, a group of Gram-positive, acid-tolerant bacteria.

Pseudomonas aeruginosa adapts to live in a wide variety of environments

The human pathogen *P. aeruginosa* can adapt to live in a wide range of environments. In addition to humans, this species is able to infect nematode worms, insects such as fruit flies and moths, aquatic organisms including zebrafish, plants, and protozoa such as amebas, as well as residing in plant and environmental niches. Human infections tend to involve patients where the immune defenses are compromised in some way, such as the infection of burns and wounds where the protection of the epithelial layer is compromised; in contact lens wearers when the surface of the eye is scratched by the lens; and in cystic fibrosis patients, where the lungs accumulate mucus secretions.

The chromosome of *P. aeruginosa* is millions of bases larger, at about 6.3 Mb, than other bacterial species such as *E. coli* and *Bacillus subtilis* at 4.6 and 4.2 Mb, respectively. Additional genetic material in *P. aeruginosa* includes additional biosynthetic enzymes, regulatory systems, transcriptional regulators, transport systems, and metabolic pathways. Therefore, the larger genome increases the diversity of niches and environments that can be survived and explored by *P. aeruginosa*. Plasticity in the genome, through deletion, loss, and diversification of genetic material in the genome, generates population diversity across the species.

Mastitis-causing bacteria *Streptococcus uberis* can adapt to different niches within cows

Streptococcus uberis causes bovine mastitis in dairy herds and is able to colonize the gut, genital tract, and mammary glands of cows. It is an opportunistic pathogen; therefore, *S. uberis* is capable of living a non-pathogenic lifestyle and taking advantage of a situation for pathogenicity when the opportunity arises. Analysis of its genome suggests that it has genes that would allow it to adapt to different environmental niches with different metabolic options available.

Analysis of the genome shows that the metabolism of *S. uberis* is well adapted to make use of the products arising from rumination of plant materials in the bovine gut and it also well adapted to make use of nutrients available in bovine mammary glands during lactation. These genomic features make the species distinctly different from the other members of the *Streptococcus* genus and uniquely adapted to the niche in which *S. uberis* is found.

Several distinct types of *S. uberis* are often isolated from the same herd of dairy cows; therefore, cases of mastitis arising are not due to single outbreak strains. *S. uberis* can be found not only in body sites of the cattle, but also in feces, bedding, and samples from the pastures; however, it appears that the bacteria are not able to persist for very long in the environment. The disease therefore seems to arise as an opportunistic infection, with the bacteria remaining for a short time in the environment, until they are able to colonize a host and establish themselves in a permissive niche within the cow. If that niche is the mammary glands of a lactating cow, then mastitis can develop, but disease need not be the outcome for *S. uberis*, with disease depending upon the niche within the bovine that is colonized.

Bacteria adapt to avoid recognition by the host immune system through antigenic variation

Host immune systems learn how to recognize everything that is part of itself, therefore enabling them to identify invaders such as bacteria by their surface structures. Using antigenic variation, bacteria change their surface proteins and other structures to attempt to avoid recognition and evade the host immune system (see Chapter 12). Antigenic variation allows bacteria to stay within a host for an extended period and to re-infect a host that it has previously infected. The proteins and structures on the bacterial surface that can be recognized by the immune system are antigens and these are targeted by host antibodies that are specifically made to bind to the antigen. Once the host has made an antibody against an antigen, it will be able to quickly mount an immune response targeting that antigen again in the future. Therefore, it is essential for the adaptation of the bacteria that they are able to alter their antigens.

Antigenic variation is about adaptation of the population to avoid the immune system. The changes to the surface antigens enable the bacterial population to survive but not the individual bacterial cells, some of which will still be recognized by the immune system.

Through antigenic variation, the bacterial population continually alters its surface, making it difficult for the immune system to eliminate it. There are several different mechanisms by which bacteria can change their antigens, including **gene conversion**, where recombination within the chromosome alters the sequence of an expressed gene.

Several different species use gene conversion as a mechanism of antigenic variation

Anaplasma marginale, a cattle pathogen, generates thousands of outer membrane protein variants in its bacterial population. The major antigens on the surface of *A. marginale* are Msp2 and Msp3, which each have hypervariable regions that make them quite diverse antigenically. Elsewhere in the genome are loci that encode multiple alternative hypervariable regions that are not expressed; these are **silent cassettes** (see also Figure 12.23). Recombination of these sequences into the Msp2 or Msp3 expression sites generates antigenic variation of these surface structures (**Figure 14.6**).

Borrelia burgdorferi expresses different forms of the VlsE surface lipoprotein and avoidance of host immune system recognition involves gene conversion of *vlsE*. The *vlsE* gene is located close to the right telomere end of the linear plasmid lp28-1 of *B. burgdorferi* preceded by a 51 base pair (bp) inverted repeat and 15 silent *vls* cassettes. The host immune response is able to recognize the expressed VlsE and therefore kills off most of the bacterial population, with the exception of those bacteria within the population that can escape immune recognition by expressing a variant lipoprotein (**Figure 14.7**).

Neisseria gonorrhoeae pilin antigenic variation is the result of recombination between silent copies of the pilin sequence in the genome, *pilS*, and the expression locus, *pilE*. This system has been extensively studied in this species and in the related species *Neisseria meningitidis*. The changes to the *pilE* locus have been shown to be as small as one nucleotide or as large as the entire replacement of the *pilE* sequence with that of a *pilS* locus (**Figure 14.8**).

These alternative sequences, the silent cassettes present in these unrelated species that share a common mechanism of antigenic variation, do not simply swap with the expression locus. The recombination events between the expression locus and the silent cassettes occur progressively over many bacterial generations, which generates mosaic sequences, combining portions of sequences from different sources.

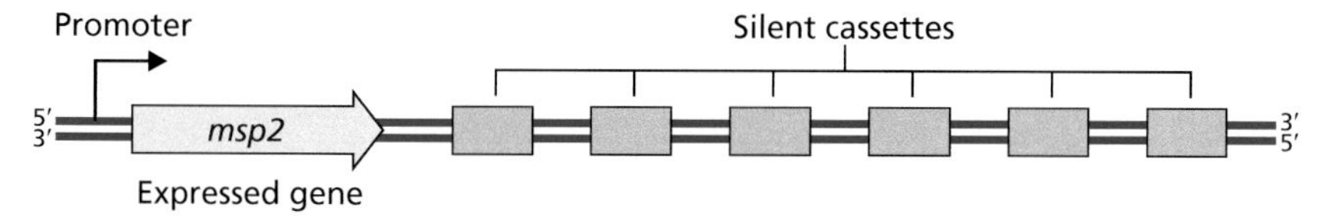

Figure 14.6 ***Anaplasma marginale msp2* expression locus and alternative hypervariable regions.** Antigenic variation of Msp2 occurs through recombination between the alternative hypervariable regions (silent cassettes) and the expressed *msp2* locus (expressed gene).

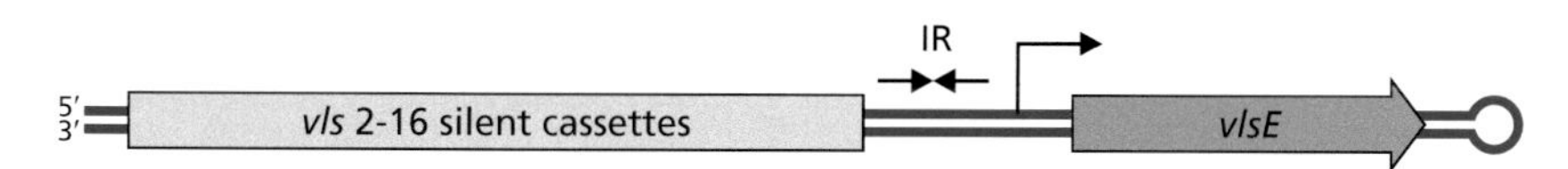

Figure 14.7 ***Borrelia burgdorferi vlsE* gene and *vls* cassettes.** At the right telomere end (loop) of the linear plasmid lp28-1 of *B. burgdorferi* is the *vlsE* gene encoding the VlsE lipoprotein. Antigenic variation is achieved through gene conversion with sequences from one of the 15 silent *vls* cassettes, separated from *vlsE* by a 51 bp inverted repeated (IR).

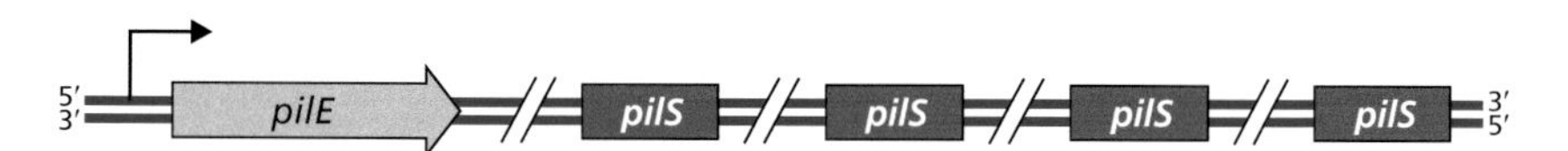

Figure 14.8 ***Neisseria gonorrhoeae pilE/pilS* antigenic variation system.** The PilE pilin protein is encoded by the *pilE* gene. Antigenic variation occurs through recombination with the *pilS* silent cassette sequences present elsewhere in the chromosome (indicated by //).

Phase variation is an important means of adaptation, but is not a means of response

The expression of some genes is switched on and off via the mechanisms of phase variation (see Chapter 12). Phase variable expression is not in response to a change in the environment or a stimulus and cannot be controlled by the cell. Phase variation is a programed random event, referred to as stochastic, which means that within the whole population of bacterial cells some of the phase variable genes will be switched on and some will be switched off and it is the fitness of the cells themselves that will determine whether they survive. Only a portion of a bacterial population needs to survive. If a particular phase variation combination provides those cells with an adaptive advantage, then they will be selected for, while others in the population are selected against. Some antigens can be switched on and off in their expression by phase variation, but critically, unlike mechanisms of antigenic variation, phase variation is always reversible (see Figure 12.22). Not all phase variation changes the expression of antigens; some phase variation is also involved in other aspects of adaptation, including stochastic switches in the bacterial population that confer selective advantages. For example, the *Helicobacter pylori* phase varies restriction-modification systems. This places sets of genes under the regulatory control of a single phase variable gene, the modification component of the restriction-modification system, forming a **phasevarion**. Just as a regulon is controlled by a regulatory protein, a phasevarion is controlled by the switching of a phase variable gene. The *H. pylori* phasevarions enable the bacterial population to take advantage of the microenvironments encountered within the host.

There are various genetic mechanisms of phase variation that switch gene expression transcriptionally or translationally via sequence inversions or simple sequence repeats (see Chapter 12). Phase variation can also occur due to epigenetic mechanisms, such as methylation of gene expression regulatory regions. Most phase variation does not involve transposition or recombination because these processes are not precise and therefore not reversible back to exactly the original sequence. However, inversion of segments of DNA and the insertion and excision of a few specific IS elements are examples of reversible and therefore phase variable transposition events.

In *Moraxella bovis* and *Moraxella lacunata*, DNA inversion switches expression of type IV pili. This is a phase variable event that also achieves antigenic variation by switching between expression of the TfpQ and TfpI pili. In *Bacteroides fragilis*, there are eight loci that encode eight different capsular polysaccharides that are each controlled by phase variation by an invertible switch. Mycoplasmas generate surface diversity through phase variable expression of the *vmcABCDEF* genes. This creates a dynamic surface with adaptive differences in expressed proteins. The population is therefore stochastically generated to be diverse with cells within the population variously expressing combinations of Vmc lipoproteins on their surfaces.

Small noncoding RNAs also have a role in enabling bacteria to adapt

Small noncoding RNAs (ncRNAs) regulate gene expression, as explored in Chapter 4. Their production requires less energy than the biosynthesis of proteins, and regulatory RNAs are able to act faster than protein regulators in altering expression. Small ncRNAs have an integral role in bacterial adaptation, as a result. As discussed in Chapter 4, small regulatory RNAs impact bacterial gene expression via several mechanisms.

ncRNAs have a key role in adaptation to the host immune response, to transition to an intracellular lifestyle, and to stress responses. For example, in the *Salmonella*-containing vacuole mentioned in Chapter 13, where the *Salmonella* are held within the host cell in a bacteria-modified vacuole, there are two ncRNAs within the bacteria called RfrA and RfrB that control expression of genes important for intracellular survival. Homologues of these ncRNAs are seen in *Brucella*, where they are called AbcR1 and AbcR2, which together contribute to growth in macrophages and chronic infection.

The small ncRNAs are also likely to be involved in adaptation to hostile environments introduced by humans, such as the use of antimicrobial agents against infectious diseases. Studies in *S. aureus* show a correlation between ncRNAs and their impact on gene regulation with the exposure of the bacteria to the antibiotics vancomycin, linezolid, ceftobiprole, and tigecycline. This suggests that the ncRNAs are able to influence the expression of genes within the bacteria that may be able to increase the resistance of the bacteria to these antimicrobial agents. The NrrF ncRNA in *N. gonorrhoeae* also influences antimicrobial resistance. NrrF has been shown to reduce the levels of *mtrF* mRNA in *N. gonorrhoeae*, which reduces the ability of these bacteria to achieve high-level antimicrobial resistance using the MtrCDE efflux pump, particularly in low iron conditions. In *E. coli*, the DsrA ncRNA triggers expression of the MdtEF multidrug efflux pump through positive control of the alternative sigma factor σ^S. The MdtEF efflux pump decreases the susceptibility of *E. coli* to a range of antibiotics and, in addition, the DsrA ncRNA regulates capsule biosynthesis and Type 3 Secretion System genes in the locus of enterocyte effacement (LEE).

Key points

- In a niche, bacteria not only have to adapt to the environment, but also to the other microorganisms that are present, plus any immune cells or antimicrobials that may be in the niche.
- Many Gram-negative bacteria have Type 6 Secretion Systems and use them to get rid of competitor bacteria by injecting them with effector proteins and killing them.
- Different bacterial species have different strategies for avoiding the host immune system, including antigenic variation, capture of host antibodies to prevent them properly binding, and hiding within host cells, among others.
- When bacteria adapt to a new niche, only a portion of the population needs to survive and thrive.
- Analysis of the genes within the bacterial genome can indicate whether the bacteria have the capacity to adapt to a wide variety of environments or if they have a restricted range.
- To avoid recognition by the immune system, some bacteria have antigenic variation mechanisms that have one expressed gene and several silent cassettes that can provide alternative genetic sequences.
- Phase variation is an important mechanism of adaptation used by some bacterial species to randomly switch on and off expression of a specific set of genes.
- Small noncoding RNAs (ncRNAs) have key roles in adaptation and are able to act more quickly than protein regulators in enacting expression change in the bacterial cell.

Key terms

Define the following terms introduced in this chapter. Check your answers using the definitions in the Glossary. These terms are also available as Flashcards online.

Antibody	Internalins	Phagocytosis
Antigen	Intracellular	Phagosome
Bacteriocin	Lysosome	Siderophore
Gene conversion	Opsonization	Vacuole

Questions and discussion topics

Self-study questions

Answer each question using 50–100 words or a table or labeled diagram. Advice on where to find answers to these questions is available online.

1. What kind of challenges might bacterial cells encounter to which they must adapt?
2. Why are siderophores needed by bacteria? What is the advantage of expressing high-affinity siderophores? What do siderophore "cheaters" do?
3. What two systems discussed in this chapter do bacterial cells possess to eliminate competitor bacteria? Briefly, how do they work?
4. Draw the *Caulobacter* cell cycle including cell morphologies, master regulatory protein control, and major cellular events.
5. Which are more difficult to culture in a laboratory, obligate intracellular bacteria or facultative intracellular bacteria? Why?
6. The terms antibody and antigen are frequently confused. Identify three ways that you can differentiate and distinguish these terms and the importance of the interactions between antibodies and antigens.
7. Where are *Legionella pneumophila* found? Where do they replicate? How are intracellular and extracellular *L. pneumophila* different?
8. What are the different ways various bacterial species can avoid digestion by host cells following phagocytosis?
9. *Listeria monocytogenes* is able to adapt to living in a human or animal host or living in soil. How does it sense which niche it is in and which regulatory mechanisms are involved in determining this location?
10. What evidence is there that *Lactobacillus plantarum* is able to grow in a variety of environments?
11. How does the genome size of *Pseudomonas aeruginosa* compare to other bacterial species? What accounts for this genomic difference and how does this benefit *P. aeruginosa* as a species?
12. Draw an example of a locus that generates antigenic variation through gene conversion. Show how the expression locus will change the antigenicity of the expressed protein following recombination.

Discussion topics

These topics are presented for discussion in study groups, as part of class discussions, or on your own. These questions go beyond what is directly covered in this part of the book. Use the research literature and other reading to explore these topics in more depth. Tips to help prepare for topic discussions are available online.

1. Explore in more depth the ways in which *Staphylococcus aureus* subvert the immune system and therefore make their niche more hospitable. Discuss the genes and expressed proteins involved and the regulation of expression as adaptation to the host and its immune response.
2. In this chapter, two different strategies were presented for avoiding recognition by the immune system, antigenic variation and survival intracellularly. Using examples of species that use these strategies, discuss the benefits of possessing both antigenic variation and the ability to hide within host cells.
3. Several examples of bacterial adaptation have been presented in this chapter. Select a bacterial species of interest and investigate how it adapts to changes in its surroundings, whether it is found in multiple environments that present the bacterial cells with very different physiological challenges or it adapts to changes that occur within its niche.

Online quiz questions

To further self-assess your understanding of the chapter material, please visit the following link, where you can participate in a range of interactive quiz questions:

www.routledge.com/cw/snyder

Further reading

Competitor bacteria can be killed with specialized Type 6 Secretion Systems

Mariano G, Trunk K, Williams DJ, Monlezun L, Stahl H, Pitt S J, Coulthurst S. A family of Type VI secretion system effector proteins that form ion-selective pores. *Nat Commun*. 2019; *10*: 5484.

Staphylococcus aureus secretes several proteins to inhibit host defenses as part of adapting its niche to its needs

Guerra FE, Borgogna TR, Patel DM, Sward EW, Voyich JM. Epic immune battles of history: Neutrophils vs. *Staphylococcus aureus*. *Front Cell Infect Microbiol*. 2017; *7*: 286.

Intracellular bacteria adapt to life inside the cells of the host

Oliva G, Sahr T, Buchrieser C. The life cycle of *L. pneumophila*: Cellular differentiation is linked to virulence and metabolism. *Front Cell Infect Microbiol*. 2018; *8*: 3.

Group A streptococci within the host experience adaptation, mutation, and death

Aziz RK, Kansal R, Aronow BJ, Taylor WL, Rowe SL, Kubal M, Chhatwal GS, Walker MJ, Kotb M. Microevolution of Group A Streptococci *in vivo*: Capturing regulatory networks engaged in sociomicrobiology, niche adaptation, and hypervirulence. *PLoS One*. 2010; *5*(*4*): e9798.

Bacteria adapt to avoid recognition by the host immune system through antigenic variation

Palmer GH, Bankhead T, Seifert HS. Antigenic variation in bacterial pathogens. *Microbiol Spectr*. 2016; *4*(*1*): 10.1128.

Phase variation is an important means of adaptation, but is not a means of response

Gor V, Ohniwa RL, Morikawa K. No change, no life? What we know about phase variation in *Staphylococcus aureus*. *Microorganisms*. 2021; *9*(*2*): 244.

Phillips ZN, Tram G, Seib KL, Atack JM. Phase-variable bacterial loci: how bacteria gamble to maximise fitness in changing environments. *Biochem Soc Trans*. 2019; *47*(*4*): 1131–1141.

Chapter

15

Bacterial Evolution

Evolution of bacteria encompasses the ability to respond and adapt. Evolution is the survival of the population by using what has already been obtained in the genetic material. Bacterial genomes can evolve slowly. They have the potential to be very stable, to be very slow-changing, and to have only a few base pair changes appear in their genome over thousands of generations. DNA replication can have very high fidelity, enabling the genome to be copied accurately from generation to generation. In contrast, bacterial genomes can also evolve very quickly in a very short time span, with bacterial genomes having the capacity for mutations to occur and to be tolerated, so long as they provide a selective advantage, or at least do not provide a disadvantage to the bacterial cell or the bacterial population. Changes in the bacterial genome can arise during evolution via mutation and horizontal gene transfer and recombination.

As genomes evolve and bacterial populations are acted upon by natural selection, it is possible to see how evolution has influenced the diversity of the bacterial phylogenetic tree. Due to their short generation times, it has also been possible to study the evolution of bacteria as it happens.

Evolution can be studied within bacterial cultures

Both the stability of the bacterial genome and its capacity to change have been the focus of genomic research for quite some time, particularly in the model organism *Escherichia coli*. Of note are the long-term evolution experiments conducted in Richard Lenski's laboratories at Michigan State University, where *E. coli* cultures have been maintained and studied for several years, including phenotypic and genomic studies of mutants that have spontaneously arisen over time (**Figure 15.1**). This research group grew the same 12 cultures of *E. coli* continuously from February 1988 to May 2022, equating to 75,000 bacterial generations, and this experiment is still ongoing. In June 2022, the LTEE (*E. coli* Long-Term Evolution Experiment) was passed on to Jeffrey Barrick's laboratory at the University of Texas at Austin. The LTEE started with two *E. coli*, strains REL606 and REL607, which are **isogenic** apart from REL607 having a spontaneous mutation enabling it to use the sugar L-arabinose. Isogenic strains have otherwise identical backgrounds, making them good candidates for comparative studies. In the Lenski LTEE, 12 cultures, six of each strain, are transferred into fresh media daily. After 75 days, cultures are preserved in freezer stocks and checked for contamination by plating (see Figure 15.1). Assessments of changes to the genome and phenotype can be done at any time during the experiment.

The relatively short generation times of bacteria enable researchers to study genomic changes over many generations of the organisms, which cannot be achieved in the same time frames in eukaryotes. What the Lenski experiment has achieved in 30 years would take 1,518,000–2,277,000 years to study in humans (generation time 22–33 years), 13,270 years to study in mice (generation time 10 weeks), and 1,324 years to study in *Drosophila* (generation time 7 days). Evolution experiments in bacteria have thus provided tremendous insight into evolution in general, not just for bacteriology. In addition to *E. coli*, evolution experiments have been conducted on several other bacterial species, including *Pseudomonas* spp., which have large complex genomes with several overlapping regulatory systems, which are important for adaptation to various niches, as discussed in Chapter 14.

DOI: 10.1201/9781003380436-20

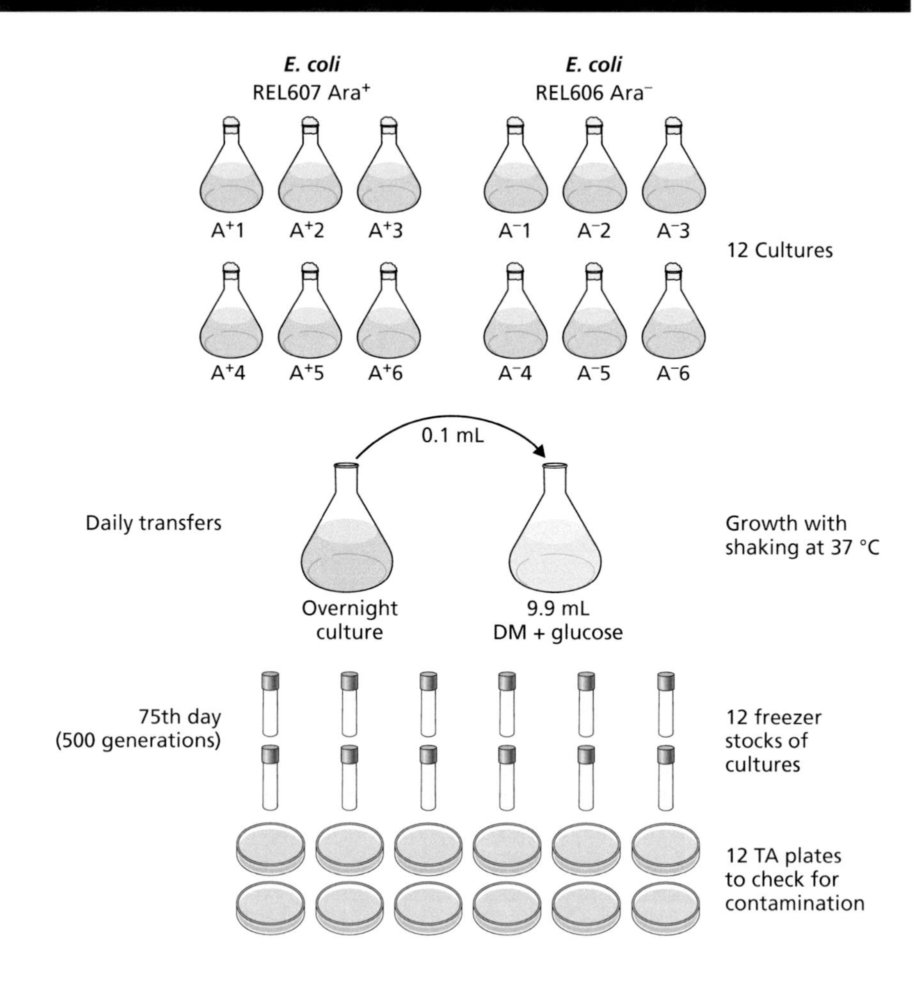

Figure 15.1 Experimental evolution experiments. Richard Lenski's research group has been growing *E. coli* cultures continuously since February 1988. These *E. coli* started from a single culture and have been through over 50,000 generations. The experimental design involves the growth of 12 cultures that are continuously maintained. There are six cultures of *E. coli* REL607, which is Ara^+, and six cultures of the isogenic strain *E. coli* REL606, which is Ara^-. Each day 0.1 mL of media containing growth from the previous day's culture is transferred into 9.9 mL of DM media with glucose and grown at 37 °C with shaking. Each 75th day, freezer stocks are made of each culture and the cultures are checked for contamination by inoculating some of the culture onto TA plates and incubating these at 37 °C to confirm that only *E. coli* are growing.

Much of what we know about the rate at which evolution occurs comes from laboratory evolution experiments and from observations of evolution happening in strains isolated from their natural environments. There is no one universal rate at which mutations will occur within a bacterial genome and therefore predictions of bacterial evolution need to be made on a species by species basis and in some cases a strain by strain basis. Variations in mutation rates within a species and the ability to control mutation rates can influence the evolution of different bacterial strains within a species.

Bacteria can evolve within the host and we can see this happen with sequencing technologies

Advances in genome sequencing have revealed two important and previously often overlooked aspects of microbiology. The first is that bacteria can evolve within the host. Several studies have been done that have shown the within-host changes to various bacterial genome sequences over time, revealing the mutations that occur and the phenotypic implications (**Figure 15.2**). In one study, the genome sequences of *Burkholderia pseudomallei* chronic infections from seven cystic fibrosis patients were analyzed over a period of time, with samples collected at about 4 months and 55 months apart. This revealed changes to the genomes including mutations resulting in antibiotic resistance, adaptations to survive in the host, and reduction in virulence leading to persistence of the infection.

The second important aspect of microbial infections often previously overlooked is that a host can be infected with two (or more) different isolates of the same species. Often in the diagnostic laboratory a single organism is identified as infecting the host, yet it is known that patients can be infected with more than one type of bacteria of the same species at the same time. This has become more

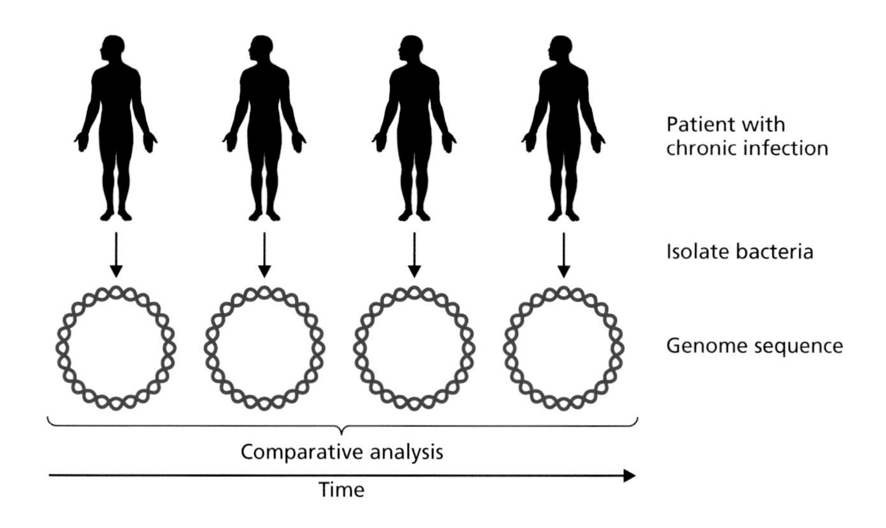

Figure 15.2 Within-host evolution of bacteria. Genome sequencing has enabled researchers to monitor the changes to the bacterial genome that occur within the host during the course of the infection. In patients with a chronic infection (or other long-term infection), the bacteria are isolated at various times during the course of the infection in a single patient and genome sequenced. Whether as complete circular genomes (shown here) or unassembled sequence reads, the data is comparatively analyzed to investigate the within-host evolution of the bacteria.

apparent with whole-genome sequencing (**Figure 15.3**). When we look at an infection in detail, often with genome sequencing, we can see that although the patient has the symptoms of one infection, it may actually be caused by more than one strain of the same species.

Together, these two important aspects of bacterial infection set up a tremendous opportunity for within-host evolution. The basic mutation rate of the bacterial species will provide a base rate at which the bacterial species can change within the host. The presence of another closely related bacterium of the same species presents the opportunity for horizontal gene transfer, particularly as they are occupying the same or closely associated niches. If these are bacterial species that are naturally competent for transformation, the opportunity for evolutionary change is even greater. Of course, bacteria also tend to be part of complex microbial communities from which there is the potential to acquire new genetic material. By capturing the genome sequence data at the start of an infection and then later on in a persisting infection, it is possible to track the evolutionary changes to the genome and to identify how the infecting bacteria have diversified within the host, perhaps forming into subpopulations that have and have not mutated certain genes or that have and have not acquired certain genes.

Antibiotic resistance is an easily observable evolutionary event

Most evolutionary events are slight changes for the bacterial cell that give it a slight advantage over other cells, thus selective pressures favor the new phenotype.

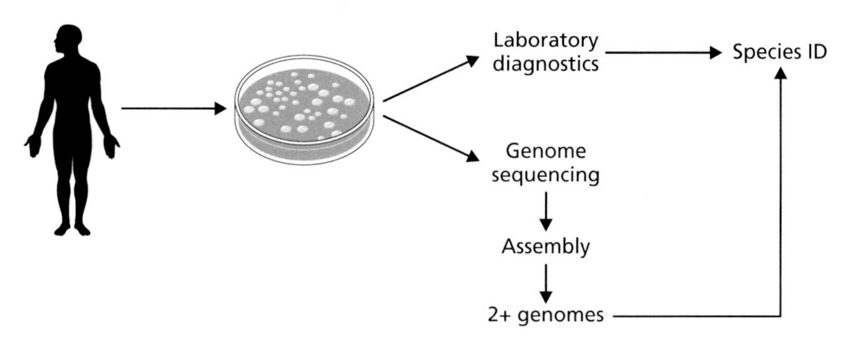

Figure 15.3 Infections can be caused by more than one strain of the same species. When the infection-causing bacteria are isolated from a patient, laboratory diagnostics can identify the species of the bacteria, however this will not identify the presence of more than one strain of the same species in the patient. Genome sequencing of the isolated bacteria can result in assembly of the data into two or more distinct genomes, making it clear that there are multiple infecting strains of the bacterial species.

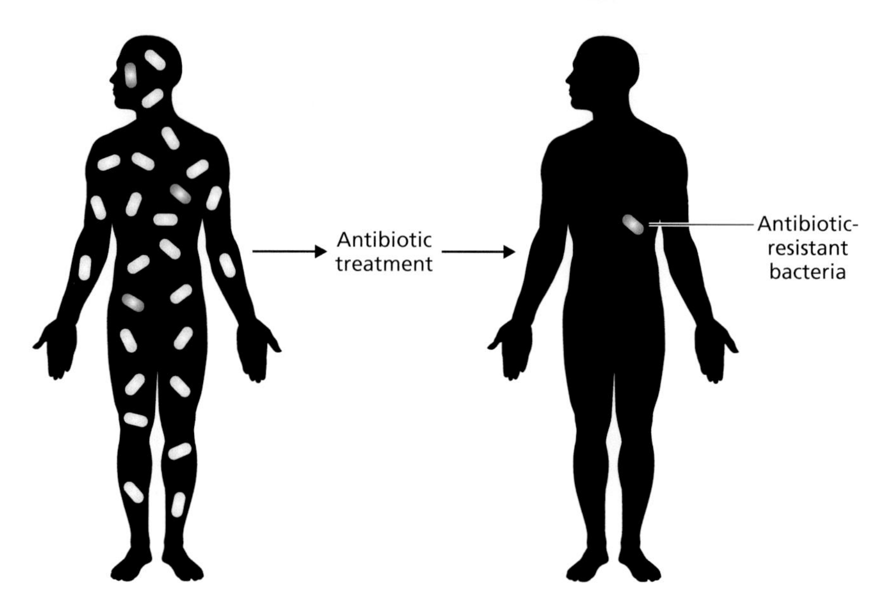

Figure 15.4 Antibiotic resistance evolution in microbiome bacteria. When antibiotics are used against an infection (yellow bacteria), there is a chance that resistance can arise in the bacteria that are part of the microbiome in the body (blue bacteria). Thus, the evolution of resistance can be observed.

Sometimes, evolution makes some striking changes, but most commonly, the changes are very small. Antibiotic resistance is a difference with a big impact that can evolve quite suddenly in a bacterial **lineage**, due to the strong selective pressures of the antibiotics that kill the bacteria without the resistance trait.

Unlike horizontal transfer of an already evolved resistance gene, evolution of antibiotic resistance generally occurs in the opportunistic pathogen and in the non-pathogen part of the microbiome. These are bacterial species that are inhabiting the host organism and otherwise not causing any harm through their colonization at the time of antibiotics exposure. The introduction of an antibiotic to a host organism exposes all of the bacteria within the body to selective pressures, regardless of the intended target of the antibiotic.

Exposure of the microbiome to antibiotics sets up a situation where resistance can evolve within any of the bacterial species within the host organism. Some microbiome species will be naturally resistant to the antibiotic. This could be because they do not have the target of the antibiotic or because the antibiotic is unable to reach the target within the bacterial cell. If the microbiome species is otherwise susceptible, some cells within the population may survive killing by the antibiotic due to selection of pre-existing mutations within the population. Antibiotic resistance can then evolve in these microbiome bacteria (**Figure 15.4**). Although they were not the intended targets to be killed by the antibiotic, because they were in the body treated by the antibiotic, these bacteria end up evolving resistance.

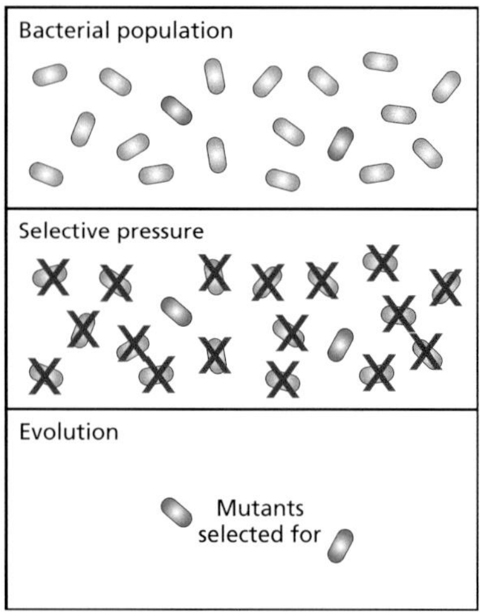

Figure 15.5 Evolution selects for mutations already present in the population; it does not cause mutations. Evolution does not cause mutations to occur in response to selective pressures. Rather, selection acts on the bacterial population and those cells with mutations that give them an advantage (blue) will outcompete the other cells in the population.

Mutations can be introduced into bacterial DNA by a variety of factors

The key factor to remember about evolution is that the situation does not cause mutations to happen. Mutations are there and are selected for by the situation. The populations of bacteria are exposed to a situation where some of the bacterial cells have a fitness advantage due to a mutation that is already present in those cells (**Figure 15.5**). The situation does not induce the mutation. Mutations are already present in the population. Perfect replication the way it was illustrated earlier in this book (see Chapter 1) almost never happens. Strand breaks occur during replication and have to be repaired; this is in addition to polymerase errors in replication. Any time a repair has to happen, there is a chance of mutation, despite the efficiency of repair mechanisms.

However, there are a variety of factors that can introduce mutations and some of these factors include situations where it can be beneficial for the bacteria to acquire a mutation in order to gain a selective advantage evolutionarily. For example, DNA can be damaged by UV light and by chemicals. It is essential that bacteria can repair such DNA damage, so that detrimental mutations do not accumulate. In some cases, these mutations can remain uncorrected in the genome and may provide a selective advantage in protecting the bacteria from UV or the chemicals to which the bacteria were exposed. Stress turns on stress responses including molecular mechanisms that cause mutations, which enables **evolvability**, the capacity to evolve adaptations. The general stress response involving RpoS (σ^S) SOS DNA-damage response, membrane stress response, and the stringent response can all result in mutations. These stress responses regulate mutagenesis and thus regulate the ability of the bacteria to evolve. It must be remembered, though, that the exposure did not create the mutation in specific locations to deal with the problem; evolution is a random process of mutation and selection. The mutations happen randomly in the DNA; it is selection that determines if those mutations have a benefit.

Sublethal exposure to some antimicrobials can damage DNA and cause genomic changes. For example, the antibiotic bleomycin binds to DNA and induces double-strand breaks, which may result in mutations. The quinolone class of antibiotics inhibits the DNA replication enzyme topoisomerase ligase domain without affecting the nuclease domain. It therefore cuts DNA without re-ligation, potentially introducing mutations. Production of reactive oxygen species, ROS, in response to antibiotics like beta-lactams, quinolones, and aminoglycosides can also damage DNA. ROS can cause DNA strand breaks and incorporation of oxidized guanine into the genome. Antimicrobials can also trigger the movement of mobile genetic elements present in the genome, which can cause mutations through their introduction into new regions of the chromosome. For example, in *Staphylococcus aureus* subinhibitory exposure to quinolones upregulates the SOS response, which increases IS256 transposition and movement of integrating conjugative elements.

Yersinia pestis, causing plague, has evolved from *Yersinia pseudotuberculosis*

Historically, there is great interest in the major infectious diseases that have had a significant impact upon the human population, including the Black Death of the fourteenth century. The bacterial species responsible is *Yersinia pestis*, which has actually caused three major pandemics in recorded history: the Justinian plague (541–542 AD); the Black Death (1346–1353 AD); and modern plague beginning in 1894. *Y. pestis* has been investigated at the genomic level to understand the loss and acquisition of genomic content from this species. These data suggest that the acquisition of a number of genomic islands and plasmids led to the evolution of *Y. pestis*, and therefore the emergence of plague, from another species, *Yersinia pseudotuberculosis*.

Since its emergence as a species, *Y. pestis* has lost DNA regions as well as acquired them. These changes have generated different lineages of *Y. pestis*, sometimes referred to as biovars and genomovars. Genomic analysis has determined that modern plague originated in the Yunnan Province of China as *Y. pestis* biovar orientalis. This modern lineage evolved from the *Y. pestis* lineage that caused the Black Death, as determined through genome sequence data obtained from DNA fragments extracted from victims buried in plague pits. Genome sequencing has also confirmed that *Y. pestis* was responsible for the Justinian plague, but that neither the Black Death plague nor modern plague had evolved from this earlier *Y. pestis* lineage.

To obtain the ancient DNA needed in order to investigate the genomics of ancient diseases, a source of DNA that is likely to have resisted degradation is

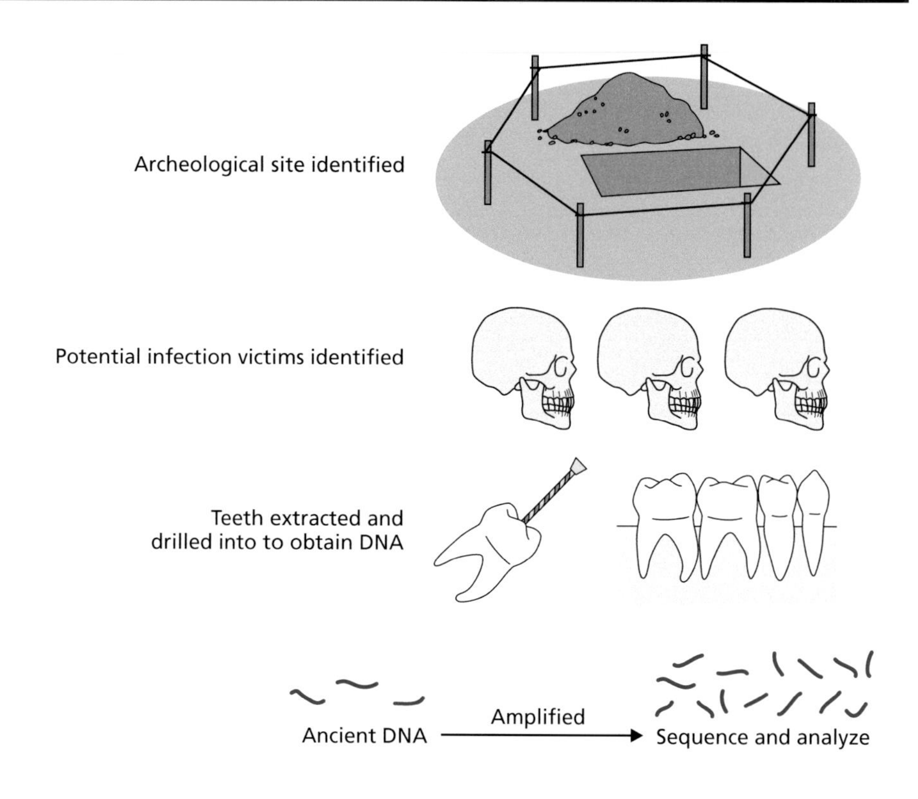

Figure 15.6 Genomic data can be obtained from *Yersinia pestis* historical plague cases. An archeological site with buried plague victims is identified and the skeletons exhumed. DNA is extracted by drilling into the teeth. The extracted fragments of DNA are amplified and then genome sequenced and analyzed.

needed. Researchers in this field attempt to obtain samples by drilling into the teeth of skeletons to extract DNA, in the hope that enough intact DNA fragments will remain that can be interpreted once they are sequenced. Since there is very little DNA recovered, and since the majority of what is recovered will generally be of human origin, the nucleic acids extracted are amplified before sequencing (**Figure 15.6**). This may introduce some errors and bias in the process, but increases the amount of starting material enough to obtain viable sequencing data. In this way, important genetic information has been obtained about historical infectious diseases, such as plague, which provides insight into the evolution of bacterial genomes over hundreds of years.

Neisseria meningitidis, causing meningococcal meningitis and septicemia, acquired its capsule fairly recently

Some pathogens, such as *Y. pestis* that causes plague, have been causing disease for hundreds of years. Other bacteria have evolved to be pathogens more recently. These are emerging infectious diseases, when they are only recently evolved or only recently recognized to be the **etiological agent**, the cause of a disease. Commensal bacteria have the capacity to acquire traits that mean they become pathogens, which may be what has happened with *Neisseria meningitidis*, the cause of meningococcal meningitis and septicemia.

The first historical accounts of meningitis diseases come from North America around 1805, while the closely related pathogen *Neisseria gonorrhoeae* causing gonorrhea is believed to be referred to in the Bible and other ancient texts. This fairly recent emergence of *N. meningitidis* was believed to be due to the horizontal acquisition of the capsule genes from *Pasteurellaceae*. The meningococcal capsule is a key virulence trait characteristic of this species, however genomic research has revealed that non-pathogenic *Neisseria* also possess capsule genes that may have been the source of the *N. meningitidis* capsule. The first pathogenic *N. meningitidis* are likely to have come from a lineage that is a common ancestor of *N. gonorrhoeae* and the non-pathogen *Neisseria lactamica*. *N. meningitidis* therefore horizontally acquired the 24 kb capsule locus as a whole relatively recently and,

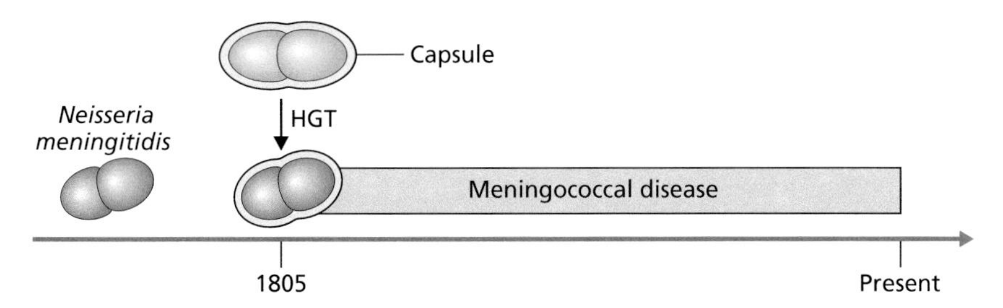

Figure 15.7 ***Neisseria meningitidis*** **evolved to have a capsule only about 200 years ago.** *N. meningitidis* causes potentially fatal diseases, yet genomic investigations show that the bacterial capsule, a vital virulence trait that contributes to its pathogenesis, has been recently acquired via horizontal gene transfer (HGT) around 1805. The bacterial cells, Gram-negative diplococci, were then able to express a capsule and cause meningococcal disease.

this combined with the other virulence determinants, resulted in these bacteria becoming the etiological agent of meningococcal disease (**Figure 15.7**).

The number of pseudogenes in a species can reveal how recently it has adapted to a new niche

As bacterial species evolve to a new niche, they can develop characteristic pseudogenes, followed by complete loss of genomic sequence regions no longer needed by the bacteria in their new environment. Since some parts of the genome are not needed, they can be eliminated from the genomic content, thus saving the bacterial species the energy of making the gene product and replicating the genetic information. When genes are not frequently transcribed and translated, mutations that randomly accumulate in these regions are less likely to be fixed by error correction mechanisms. Genome sequences from intracellular bacteria such as *Rickettsia prowazekii*, which causes typhus, and *Mycobacterium leprae*, which causes leprosy, are clear examples of **reductive genome evolution**. This is where isolated lineages become adapted to a discrete and restricted niche and their genomes reduce to only what is needed to survive in that niche (**Figure 15.8**).

Clavibacter michiganensis subspecies *sepedonicus* causes ring rot in potatoes, which has devastating agricultural consequences. Genome sequencing has revealed extensive chromosomal rearrangements by the IS elements present in over 100 copies in the genome. These rearrangements have generated over 100 pseudogenes. There has also been gene acquisition by horizontal gene transfer and gene loss by deletion of genetic material. Analysis of these genomic data suggests that the restriction to living exclusively within the plant niche was a recent adaptation. Because competition within the potato niche is low and the nutrients available are less diverse than other niches in the soil, selective pressures on some parts of the genome have been relaxed, thus the functions of these genes have been lost.

Recent endosymbiont of the tsetse fly *Sodalis glossinidius* has a large number of pseudogenes in its genome. Potentially as many as 40% of the *S. glossinidius* CDSs are pseudogenes. Many of these pseudogenes are still transcribed and some are translated. There may still be some residual effects for some of these truncated proteins where the functions are perhaps changing and evolving. What is being observed in the *S. glossinidius* genome is a snapshot in evolutionary time.

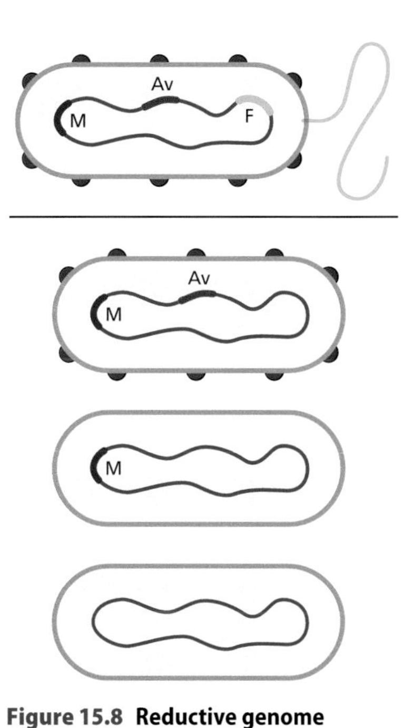

Figure 15.8 Reductive genome evolution eliminates genetic material that is no longer needed in a niche. Evolution eliminates the unneeded genetic material. Shown here is a bacterial cell with flagella genes (F, green), antigenically variable surface proteins (AV, red), and an assortment of metabolic pathways to take advantage of a range of environments (M, purple). When this cell enters a new environment where these traits offer no advantage, they can be lost over time and eliminated from the genome.

Evolution of the bacterial surface to cope with the immune system and vaccines

The surfaces of bacterial species that are exposed to a host immune system have evolved strategies to avoid being killed. Often the antigenic surface components of bacterial cells are regulated either by environmental signals or by phase variable switching. The bacterial surface presented to the host immune system is therefore changeable, which will help it avoid recognition.

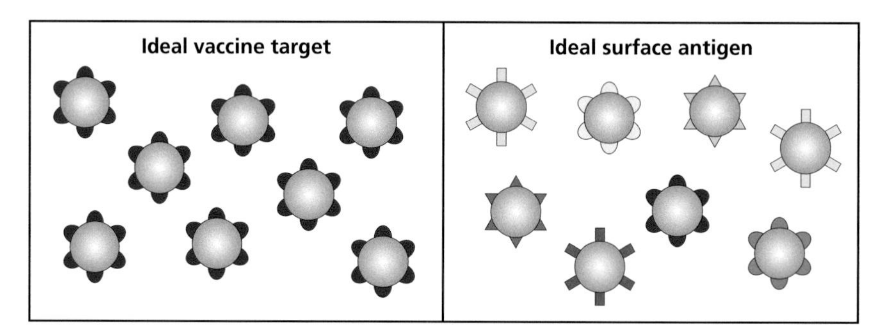

Figure 15.9 Ideal vaccine targets (for humans) versus ideal surface antigens (for bacteria). When humans design vaccines, an ideal vaccine target is an antigen present on all bacteria and that is identical on all bacteria (left). The ideal surface antigen for bacteria is changeable, to avoid recognition by the immune system (right). The opposite nature of these ideals complicates efforts at vaccine design.

Ideally, vaccine targets should be surface antigens. A vaccine candidate needs to be present on the surface of the bacterial cell at all times, needs to be present on all bacterial cells of a bacterial species, and needs to be similar enough in its protein sequence for the vaccine to offer protection against all strains of the species (**Figure 15.9**). Unfortunately, when it comes to vaccine design, because surface antigens are readily available for recognition by the immune system, they have evolved to be antigenically variable, regulated in expression via environmental signals, phase variable, and combinations of these. Such variable surface antigens do not make ideal classical vaccine targets. They are not always expressed on the surface of the bacterial cell. Their sequences are highly variable within the species.

However, there are some cases where antigenically variable surface components can be viable vaccine components. If the lack of expression of the antigen is otherwise detrimental to the bacterial cell, leaving it open to the host immune defenses, then the vaccine may be effective. When the antigen is being expressed by the bacteria, then the vaccine-primed immune system can recognize the bacteria and eliminate them. Without the antigen, the bacteria are not protected, and the immune system can eliminate them (**Figure 15.10**). As an example, *N. meningitidis* express a protein on their surfaces that can bind to the human complement protein Factor H and helps the bacteria avoid being killed by human complement. This Factor H binding protein is a part of one of the vaccines against *N. meningitidis* and is believed to be effective because cells that switch off expression of Factor H binding protein will have switched off their protection from complement killing.

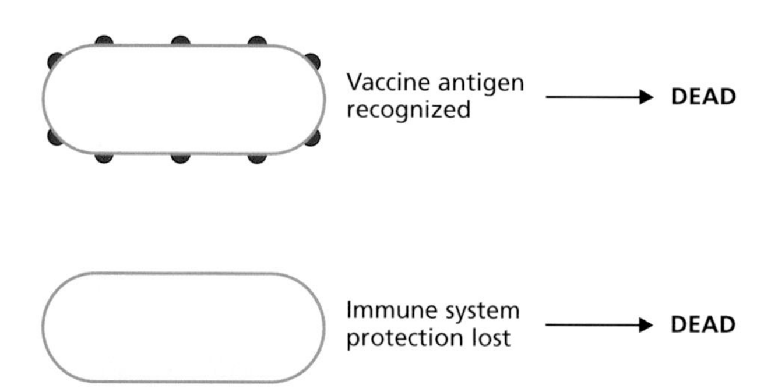

Figure 15.10 Vaccination against bacterial antigens is essential for survival. An alternative strategy for vaccine design is to find a vaccine target that the bacteria needs to express on its surface, where the vaccine-primed immune system can see it and target the cell for killing (top cell), or doesn't express it on its surface and the immune system can then target the cell for killing (bottom cell). In either state, the immune system eliminates the bacterial cell. Without vaccination, it is essential that the bacterial cell expresses this antigen to avoid killing by the immune system, therefore vaccination against this antigen means the bacterial cell is killed whether it expresses the antigen (by the vaccine) or not (by the immune system).

Horizontal gene transfer can bring new genes into a species, contributing to its evolution

Horizontal gene transfer is a mechanism whereby genetic material from one bacterial cell is transferred to another bacterial cell. This process can contribute to evolution through movement of bacterial genes or whole systems into a species that previously did not possess these traits. Striking examples of horizontal gene transfer can be observed in the spread of antibiotic resistance gene cassettes from their evolution in one bacterial species across to other species. Often these resistance genes are carried on plasmids or transposons, which facilitate their incorporation into the new genetic background.

Enterotoxigenic *E. coli*, or ETEC, is the main cause of travelers' diarrhea and diarrhea in young children. The bacterial cells cause disease because they bind to the epithelial cells of the small intestine using class 5 fimbriae and there secrete enterotoxins. *Burkholderia cepacia* can colonize the lungs of cystic fibrosis patients and attack respiratory epithelium. There is evolutionary evidence that the *E. coli* class 5 fimbriae have been horizontally transferred to *B. cepacia*, where the structure is referred to as the cable pili. In *B. cepacia*, the fimbriae/pili also mediate attachment to epithelial cells, although in this context they allow the bacteria to adhere to lung epithelial cells (**Figure 15.11**).

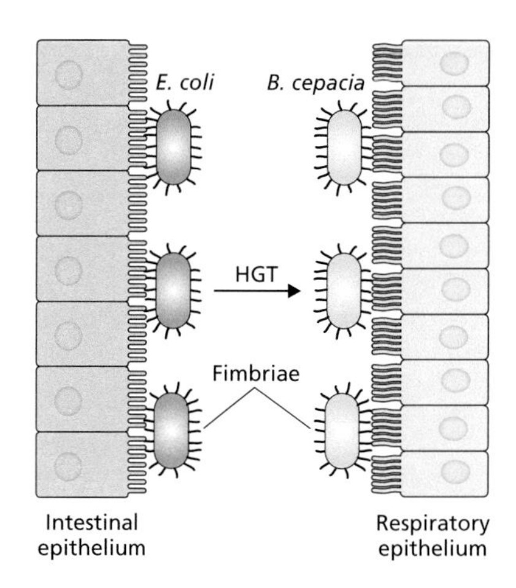

Figure 15.11 Class 5 fimbriae from *E. coli* aid *B. cepacia* in causing disease. The class 5 fimbriae of enterotoxigenic *E. coli* allow these bacteria to attach to the epithelial cells of the small intestine (left). Horizontal gene transfer of the class 5 fimbriae system from *E. coli* to *B. cepacia* enables these bacteria to bind to the epithelial cells of the lungs of cystic fibrosis patients (right).

Horizontal acquisition of toxins has contributed to the evolution of virulence in several bacterial species. For example, some strains of *S. aureus* have acquired through horizontal gene transfer the superantigen toxic shock syndrome toxin-1 (TSST-1). This toxin induces an inflammatory cascade causing staphylococcal toxic shock syndrome. TSST-1 is encoded by *tst* on staphylococcal pathogenicity island 1. This whole genetic island of 15 kb was horizontally acquired and can be horizontally transmitted. Also, aquaculture to provide food for the growing human population can be hampered by bacterial infections, particularly at the early stages of life. *Vibrio parahaemolyticus* causes acute hepatopancreatic necrosis disease in shrimp with up to 100% mortality. These particular strains of *V. parahaemolyticus* have acquired a 69 kb plasmid that encodes two toxin genes, evolving to be this devastating pathogen.

The particular nature of an environmental niche can create opportunities for evolution

Many traits of bacteria that we view as being virulence factors should be considered in an evolutionary context as characteristics that provided a specific advantage in a particular niche, which may have originally had no bearing on virulence whatsoever. These virulence factors evolved outside of the realms of bacterial-host interactions or in niches other than those ultimately inhabited by the pathogenic bacteria. In this context, it is therefore less surprising when "virulence genes" are found in non-pathogens because in these organisms the genes have evolved in a different ecological context or outside of the scope of causing disease. Virulence genes in one organism are simply genes for enhanced survival in another organism.

Bacteria have been evolving on the planet for far longer than we have been potential hosts for them. The Earth is dominated by cold environments where **psychrophilic** cold adapted microbial communities that have evolved to thrive in the cold survive. Antarctic bacteria *Pseudoalteromonas haloplanktis* grows at –2.5 to 25 °C. Proteomics suggests cold adaptation to growth at 4 °C involves protein *S*-thiolation and changes in nutrient uptake and metabolism. Cold environments also tend to be low in nutrients, high or low in pH, high in osmotic pressure, and may have excessive UV and/or radiation.

Some evolution in bacterial species has occurred in response to human intervention. Antibiotics are an obvious example, but vaccine use has also exerted selective pressures that have driven evolution. In *Bordetella pertussis*, there are few single-nucleotide polymorphisms (SNPs) but many chromosomal rearrangements are observed in comparisons between the genome sequence data of isolated strains. It is notable that there has been a resurgence in *B. pertussis* disease since the 1990s despite vaccination. This is possibly due to a change from use of whole-cell vaccines to acellular vaccines at about this time. It may also be, or perhaps in addition to the switch in vaccines, that there has been an adaptation and evolution in circulating *B. pertussis* strains, resulting in strains with increased toxin production and the observed escape from vaccine protection.

It is possible for completely new genes to evolve

While we tend to explore evolution to investigate the accumulation of mutations or the horizontal acquisition of genes, it is possible for entirely new genes to evolve, albeit relatively rarely. The origin of new genes is the result of one of three different processes: divergence of a gene sequence following duplication of the gene; *de novo* generation of a gene from noncoding DNA; and gene fusion and fission events from existing gene sequences (**Figure 15.12**).

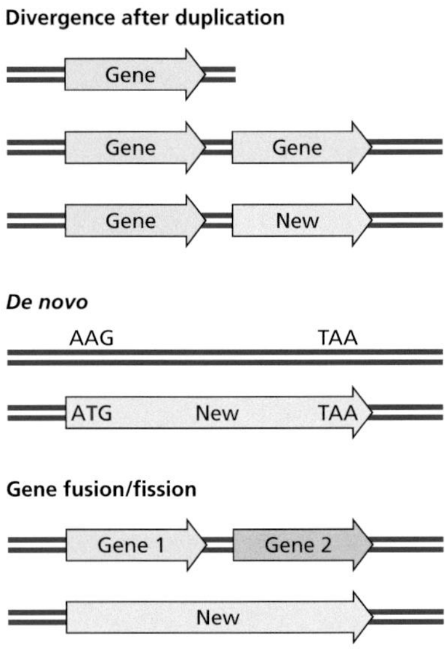

These processes all rely on recycling existing sequences. They make use of DNA sequences that already exist. In the first or the third example, this is a gene sequence that exists and codes for something else and is changed following its duplication or is changed in its function through fusion or truncation. In the second example, this is a noncoding sequence that acquires an initiation codon and thus in a rare event becomes a coding region.

In the case of gene fusions, it is possible to create new genes with starting material from sequences from across the microbial world. Horizontal gene transfer can bring in DNA from a variety of sources. Most commonly this is DNA from other bacteria, with that from closely related species being most compatible in codon usage and regulatory features. Natural competence for transformation may nonselectively take up environmental DNA from any source, including unrelated bacteria and other sources of DNA. Transduction brings in DNA from bacteriophages. Therefore, viruses, phages, fungi, bacteria, eukaryotes, and archaea are all potential sources of DNA for new genes. There is evidence of DNA from these sources in bacterial genomic data.

Divergence after duplication
Gene
Gene Gene
Gene New
De novo
AAG TAA
ATG New TAA
Gene fusion/fission
Gene 1 Gene 2
New

Figure 15.12 Evolution of new genes. New genes can arise by three different mechanisms. After duplication of a gene, the sequence can diverge into a new gene (top). A new gene can arise *de novo*, as new, from a noncoding region when a mutation generates an initiation codon (middle). Fusion of two gene sequences creates a new gene (bottom) and likewise fission dividing a gene into two can generate two new genes (bottom in reverse).

Key points

- Long-term evolution experiments make use of the short generation times of bacteria to explore evolution.
- Bacteria are able to evolve within a host organism, as selective pressures act upon the bacterial population.
- More than one strain of the same species can infect the same host or the same niche, setting up the potential for horizontal gene transfer and evolutionary changes.
- Antibiotic resistance can be acquired through horizontal gene transfer of a resistance trait or through selection of cells in the bacterial population that have mutations conferring resistance.
- Evolution selects for mutations that are already present in the bacterial population.
- The evolution of infectious diseases such as plague can be studied through extraction of ancient DNA from archeological sites.
- Some pathogenic bacteria have evolved fairly recently, through the acquisition of virulence factors by non-pathogenic bacteria.
- The presence of numerous pseudogenes in a genome can be an indicator that the bacterial species has recently adapted to living in a new niche and is in the process of evolving.
- Completely new genes can evolve via a number of mechanisms, arising from noncoding DNA, from the duplication and divergence of an existing gene, and from the fusion or fission of an existing gene or genes.

Key terms

Define the following terms introduced in this chapter. Check your answers using the definitions in the Glossary. These terms are also available as Flashcards online.

Etiological agent	Isogenic	Psychrophilic
Evolvability	Lineage	Reductive genome evolution

Questions and discussion topics

Self-study questions

Answer each question using 50–100 words or a table or labeled diagram. Advice on where to find answers to these questions is available online.

1 Why is it important in experimental design to compare isogenic strains of bacteria? Why not just compare any two bacterial strains?

2 How do *E. coli* generation times compare to humans, mice, and *Drosophila* (fruit flies)? What about other experimental models like rats and zebrafish?

3 Which two important aspects of bacterial infection provide for the opportunity of within-host evolution?

4 Using antibiotic resistance as an example, briefly describe how selection acts on a population of bacteria. When does the mutation occur, before or after selection?

5 Name some factors that introduce mutations into bacterial DNA.

6 Briefly, what has genome sequencing revealed about the evolution of the cause of the Black Death?

7 What is a pseudogene? What does the presence of pseudogenes reveal about the process of evolution of the bacteria?

8 In what type of bacteria do we tend to see reductive genome evolution? Why does this occur?

9 What features of a surface antigen make it an ideal vaccine candidate? How are these features not ideal for bacteria?

10 *Pseudoalteromonas haloplanktis* has evolved to grow in Antarctic conditions. What sort of environmental conditions does this species encounter?

11 What are the three ways new genes arise in bacteria?

12 Several examples are presented in this chapter of evolution due to selective pressures. List these and other examples of selection that can drive evolution.

Discussion topics

These topics are presented for discussion in study groups, as part of class discussions, or on your own. These questions go beyond what is directly covered in this part of the book. Use the research literature and other reading to explore these topics in more depth. Tips to help prepare for topic discussions are available online.

1 Experimental evolution experiments, such as what has been accomplished by the Lenski laboratory over 30 years, have made significant contributions to our understanding of evolution. Discuss the insights gained from experimental evolution research in another bacterial species. Does the experimental design differ from Lenski? How long was it conducted for? What has been learned?

2 Several studies have used comparative genomics to investigate the evolution of pathogen genomes within patients with chronic bacterial infections. Discuss one of these studies. Over what time period were the bacteria isolated from each patient? What changes were found in the genome? What does this reveal about the evolution of the bacteria within the human body?

3 Horizontal gene transfer (HGT) brings new genes into bacterial genomes and has made significant contributions to the evolution of many bacterial species. Explore the role of HGT in one bacterial species of interest and how the acquisition of genes has led to an evolutionary change for that species or a lineage of the species.

Online quiz questions

To further self-assess your understanding of the chapter material, please visit the following link, where you can participate in a range of interactive quiz questions:

www.routledge.com/cw/snyder

Further reading

Evolution can be studied within bacterial cultures

Callaway E. Legendary bacterial evolution experiment enters a new era (News Q&A). *Nature*. 2022; *606*: 634–635.

Lenski RE. The *E. coli* long-term experimental evolution project site, 2022. https://lenski.mmg.msu.edu/ecoli/index.html

Bacteria can evolve within the host and we can see this happen with sequencing technologies

Viberg LT, Sarovich DS, Kidd TJ, Geake JB, Bell SC, Currie BJ, Price EP. Within-host evolution of *Burkholderia pseudomallei* during chronic infection of seven Australasian cystic fibrosis patients. *mBio*. 2017; *8*(*2*): e00356–e00417.

Yersinia pestis, causing plague, has evolved from *Yersinia pseudotuberculosis*

Feldman M, Harbeck M, Keller M, Spyrou MA, Rott A, Trautmann B. A high-coverage *Yersinia pestis* genome from a sixth-century Justinianic Plague victim. *Mol Biol Evol*. 2016; *33*(*11*): 2911–2923.

Neisseria meningitidis, causing meningococcal meningitis and septicemia, acquired its capsule fairly recently

Bartley SN, Mowlaboccus S, Mullally CA, Stubbs KA, Vrielink A, Maiden MC, Harrison OB, Perkins TT, Kahler CM. Acquisition of the capsule locus by horizontal gene transfer in *Neisseria meningitidis* is often accompanied by the loss of UDP-GalNAc synthesis. *Sci Rep*. 2017; *7*: 44442.

Horizontal gene transfer can bring new genes into a species, contributing to its evolution

Arnold BJ, Huang IT, Hanage WP. Horizontal gene transfer and adaptive evolution in bacteria. *Nat Rev Microbiol*. 2022; *20*(*4*): 206–218.

Wiedenbeck J, Cohan FM. Origins of bacterial diversity through horizontal genetic transfer and adaptation to new ecological niches. *FEMS Microbiol Rev*. 2011; *35*(*5*): 957–976.

The particular nature of an environmental niche can create opportunities for evolution

Bart MJ, Harris SR, Advani A, Arakawa Y, Bottero D, Bouchez V. Global population structure and evolution of *Bordetella pertussis* and their relationship with vaccination. *mBio*. 2014; *5*(*2*): e01074–e01114.

Part VI

Gene Analysis, Genome Analysis, and Laboratory Techniques

The next three chapters look at the tools and techniques involved in doing the research that has been discussed in the previous five parts of this book. Here, the discussions will move away from theory. These chapters will delve into the application of skills to the various forms of analysis and experiments that are conducted on DNA, RNA, and proteins. Whether these are the sequence data being investigated on a computer or the physical molecules being investigated at the laboratory bench, the methods involved are included here. First gene analysis techniques will be explored and then the specialized analysis tools used when scaling up to investigate whole bacterial genomes. Links to web resources referred to in these chapters are available on the Instructor and Student Resources Companion Website. Finally, the last chapter in this section is dedicated to laboratory techniques that are key to investigating bacterial genetics and genomics.

DOI: 10.1201/9781003380436-21

Chapter

Gene Analysis Techniques

16

Back in Chapter 2, the way in which a coding sequence could be identified was presented and a gene was defined. Throughout this book there have been several discussions of CDSs and genes. In this chapter, the techniques for analyzing genes will be explored in depth, imparting the knowledge needed to be able to conduct such analyses.

Sequence searches are done to find out what else is similar to this gene

Having generated a gene sequence, it is often wondered whether this looks like any other gene sequences that are known to exist. This can be determined by doing a search of the public sequence databases for similar sequences. One of the most commonly used programs in sequence analysis is the Basic Local Alignment Search Tool, BLAST, mentioned in Chapter 12. This is a computer algorithm for comparing nucleotide sequences of DNA or RNA, or for comparing amino acid sequences of proteins to databases of nucleic acid or amino acid sequences. It is so widely used because it does for genetics and genomics research what a Google search does for helping us find stuff on the internet. BLAST takes the gene or protein you are interested in and finds other sequences similar to it, then displays them for you so you can see how close a match they are to your original.

The BLAST search program means it is possible to search a database for any sequences that match or are similar to the sequence you are interested in, your **query sequence** (**Figure 16.1**). Often this is the gene sequence of interest, but it could be any sequence data; it could possibly be a fragment of a gene, a promoter

```
Query  1        ATGAAACAATCCGCCCGAATAAAAAATATGGATCAGACATTAAAAAATACATTGGGCATT  60
                ||||||||||||||||||||||||||||||||||||||||||||||||||||||||||||
Sbjct  1299898  ATGAAACAATCCGCCCGAATAAAAAATATGGATCAGACATTAAAAAATACATTGGGCATT  1299839

Query  61       TGCGCGCTTTTAGCCTTTTGTTTTGGCGCGGCCATCGCATCAGGTTATCACTTGGAATAT  120
                ||||||||||||||||||||||||||||||||||||||||||||||||||||||||||||
Sbjct  1299838  TGCGCGCTTTTAGCCTTTTGTTTTGGCGCGGCCATCGCATCAGGTTATCACTTGGAATAT  1299779

Query  121      GAATACGGCTACCGTTATTCTGCCGTGGGCGCTTTGGCTTCGGTTGTATTTTTATTATTA  180
                ||||||||||||||||||||||||||||||||||||||||||||||||||||||||||||
Sbjct  1299778  GAATACGGCTACCGTTATTCTGCCGTGGGCGCTTTGGCTTCGGTTGTATTTTTATTATTA  1299719

Query  181      TTGGCACGCGGCTTCCCGCGCGTTTCTTCAGTTGTTTTACTGATTTACGTCGGCACAACC  240
                ||||||||||||||||||||||||||||||||||||||||||||||||||||||||||||
Sbjct  1299718  TTGGCACGCGGCTTCCCGCGCGTTTCTTCAGTTGTTTTACTGATTTACGTCGGCACAACC  1299659

Query  241      GCCCTATATTTGCCGGTCGGCTGGCTGTATGGTGCGCCTTCTTATCAGATAGTCGGTTCG  300
                ||||||||||||||||||||||||||||||||||||||||||||||||||||||||||||
Sbjct  1299658  GCCCTATATTTGCCGGTCGGCTGGCTGTATGGTGCGCCTTCTTATCAGATAGTCGGTTCG  1299599
```

Figure 16.1 BLAST search result from a 300-base query. The top line of this alignment was used to search a database for sequences that are similar. The sequence searched for is the query sequence and is 300 bases in length (Query). It has retrieved the sequence aligned below it, the target sequence, which was the target of the search (Sbjct). The numbers (1,299,898–1,299,599) are the position within the target sequence where the query sequence matches; in this case the numbers are descending because the match is to the reverse complement of the database sequence entry.

DOI: 10.1201/9781003380436-22

region, an operon, or an ncRNA, for example. The sequences found are each the **target sequence** of each BLAST search hit (see Figure 16.1). BLAST searches can be used to identify species, find protein domains, build phylogenetic trees, propose functions, and map and annotate previously unknown sequences.

The first BLAST program was published in 1990, by a team from the US National Institutes of Health. BLAST is what is called a **heuristic** technique, which means that the approach to sequence searching is practical and sufficient for getting the job done, but it might not be the most logical, or the most optimal, or the most rational way to do it. A logical, optimal, rational approach would take excessive time, and the end result is that BLAST does good enough alignments to find similar sequences, which is really the goal. There are better programs for finding alignments, discussed later, but BLAST's alignments are good enough for doing its searches. Another example of heuristics is when you make an educated guess. This uses past knowledge to make an informed, quicker decision than would otherwise be made by a more in-depth analysis.

Before there was BLAST, there was FASTA

The FASTA file format for sequence data was mentioned in Chapter 12, where the first line of the sequence data starts with a ">" symbol followed by some descriptive text. The sequence data then come in the second line. This file format comes from the program FASTA, a search tool that was a predecessor to BLAST.

FASTA was first designed for protein sequence similarity searches and released in 1985. In 1987, DNA and translated DNA searches were added, however FASTA was replaced by BLAST in the 1990s, the latter being superior in speed and scalability with the ever-increasing size of the databases. Being able to keep up with the needs of researchers as databases grew in size, through its scalability, has meant that BLAST is still used today.

The FASTA input format, however, remains in use today, leaving a legacy of the name FASTA, pronounced Fast A with a long A sound at the end. The FASTA file format is described and shown in Chapter 12, earlier in this book. Because the methods used are different, it may be interesting to see if your search results differ with a FASTA search versus a BLAST search.

BLAST quickly finds the most similar sequences

The breakthrough that came with BLAST was speed. Even with the vast increase in public databases as more and more sequence data are added each day, BLAST is able to keep up and produce results relatively quickly. In the original BLAST, there could be gaps in the alignment generated, where similarity slipped, but newer versions such as BLAST2 produce fewer gaps. BLAST can be used via a web browser interface, making it readily accessible to many researchers, or it can be downloaded for local use in the command-line of Unix. The program is open source, which means that anyone can have access to the code and can change it. As such, there are many versions of BLAST that have been developed since the original in 1990.

One way that speed is achieved is through not including any areas of poor similarity in the BLAST data returned from the search. These data do not, therefore, return to the user whole gene sequences, unless the whole gene has similarity to the query. To speed the BLAST search process and eliminate search hits to sequences that may not be robust to the query, all **low complexity regions** are excluded from the search (**Figure 16.2**). Sequences like repetitive stretches of DNA fall into the category of low complexity regions. Since completely unrelated sequences can share the same repeat and therefore confuse the BLAST search, these are excluded. The filter for low complexity regions can be manually turned off by the user, but this is likely to generate erroneous results, which must be taken into account when interpreting the data. In the next step of the BLAST search process, the search is divided into "words" of a set length. These are referred to as k-letter words, with protein searches generally being three-letter words and DNA

```
Query  1       ctcttctcttctcttctcttctcttcCGCAGCGCAGGCGGCAAGTGAAGACAATGGCCGC  60
               ||||||||||||||||||||||||||||||||||||||||||||||||||||||||||||
Sbjct  733538  CTCTTCTCTTCTCTTCTCTTCTCTTCCGCAGCGCAGGCGGCAAGTGAAGACAATGGCCGC  733479

Query  61      GGCCCCTATGTGCAGGCGGATTTGAACTACGCCTACGAACACATTACCCACGATTATCCC  120
               ||||||||||||||||||||||||||||||||||||||||||||||||||||||||||||
Sbjct  733478  GGCCCCTATGTGCAGGCGGATTTGAACTACGCCTACGAACACATTACCCACGATTATCCC  733419

Query  121     GCCGACAACGCCAAAGTCTTCGACGACTACCGCGACATCAAAACCCGCTCCACACACCCA  180
               ||||||||||||||||||||||||||||||||||||||||||||||||||||| ||||| 
Sbjct  733418  GCCGACAACGCCAAAGTCTTCGACGACTACCGCGACATCAAAACCCGCTCCACCCACCCC  733359

Query  181     CGCCTTTCCGTCGGCTACGATTTCGGCAACTGGCGCATCGCCCTCGATTACGCCCGCTAT  240
               ||||||||||||||||||||||||||||||||||||||||||||||||||||||||||||
Sbjct  733358  CGCCTTTCCGTCGGCTACGATTTCGGCAACTGGCGCATCGCCCTCGATTACGCCCGCTAT  733299

Query  241     AATAAGTGGAAACACGGCAAGCATATTCGCACAGAACAAAACAAATCAGTTCAAAACGGC  300
               |||||||||||||||||||||||||||| | ||||| | |  |||  |||  |||| |||
Sbjct  733298  AATAAGTGGAAACACGGCAAGCATATTCACGCAGAAAATACTAAAAAAGTCAAAAATGGC  733239
```

Figure 16.2 BLAST excludes low complexity regions from sequence searches. At the start of this 300-base search there is a region of repetitive DNA in the query sequence. BLAST ignores the CTCTT repeats when searching the database and indicates this in the output by displaying the letters in lowercase. Although these bases align between the query and the target, they were not used in the search for the target.

being 11-letter words. The query sequence then becomes a list of words. This list is used to search the database for the highest scoring matching words, including looking for data on matches to adjacent words to increase the score. The process continues until an alignment is returned to the user.

In addition to the alignment with similar sequences, BLAST output provides statistics about the matches from the database. One of these is the **E value** (Expect value), which represents the number of alignments to the query sequence that would be expected by chance (**Figure 16.3**). The more similar the query to the target sequence, the closer the E value is to 0. Identical sequences have an E value of 0; those that are nearly identical will also be 0 or approaching 0, based on how similar the query is to the target. These also have an **Ident value**, for identity, of 100% (see Figure 16.3). Note that these scores are only for the BLAST aligned segment of the sequence, not the whole gene or genetic region used in the search. Other values listed in the BLAST output are **Max score** (the maximum score), **Tot score** (the total score), and **Query Cover** (the query coverage; see Figure 16.3). The Max score is the highest score for this BLAST hit, taking into account all aligned and matched nucleotides (or amino acids) and taking into account all penalties for mismatches and gaps. Tot score sums the alignment scores for segments from the same subject, which means that a target sequence that is similar in more than one place to the query will have those separate places taken into account. Query cover addresses how much of the query has been found in the database; it might be that not all of the sequence has been found. There may be 100% identity to the query sequence being searched, but this only hits the first 150 bp of the query, which is actually 3,500 bp long, and the rest of the sequence remains unknown.

Description	Max score	Tot score	Query cover	E value	Ident value
Neisseria lactamica partial *opa* gene for Opa A protein, strain NL_4	1258	1258	100%	0.0	100%
Neisseria lactamica strain Y92-1009, complete genome	1076	2010	100%	0.0	94%

Figure 16.3 BLAST output information for Figure 16.2 showing Max Score, Total score, Query cover, E value, and Ident value. This information relates to the BLAST hit from Figure 16.2, where the first 300 bases are shown. In total, the query is 697 bases. The first row is the query sequence finding itself in the database, therefore all of the values are at their maximum for this sequence because the query and the target are identical. The target sequence in Figure 16.2 is shown in the second line. Although there is alignment over the whole of the 697 bases (Query cover 100%) there is only 94% identity (Ident value), influencing the Max score and Tot score. The similarity of the sequences is high, therefore the E value is 0.0.

There are five basic versions of BLAST, addressing different search tasks

BLASTN is a specialized form of BLAST search that takes a DNA query, a nucleotide sequence, and will find the most similar DNA sequences that match it from within a DNA database (**Figure 16.4**). This is a straightforward and simple way to show that a gene sequence is present as a gene sequence in another bacterial genome. This can be a genome from the same species or another species, and the similarity match may help identify the function of the gene. If the homologous sequence has an annotated function, then it is possible that the query sequence might share this same function.

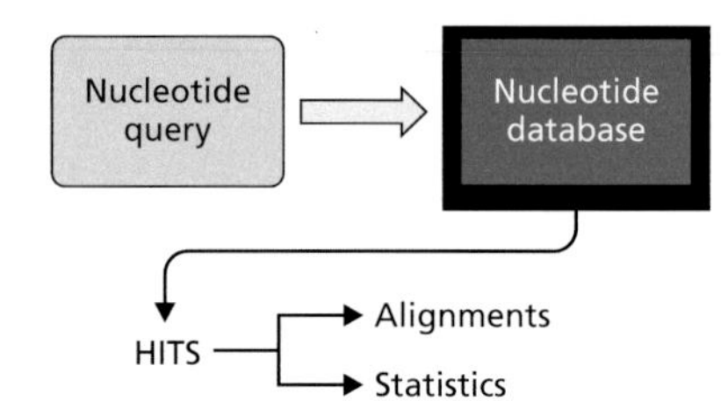

Figure 16.4 BLASTN starts with a nucleotide query and searches a nucleotide database. The results of the search, the hits, are presented to the user as alignments between the query sequence and the sequences that were found by the search, the target sequences. The statistics comparing the query and target sequences are also presented to the user.

BLASTP is similar to BLASTN, however it starts with a protein query, an amino acid sequence. Using this query, BLASTP finds the most similar protein sequence from a protein database (**Figure 16.5**). In deciding whether to use BLASTN or BLASTP, keep in mind that BLASTN will return more closely related sequences, since the DNA sequences will be more similar to one another than the protein sequences returned by BLASTP will have been at the genetic level. This is because of the variation in codons that can all encode the same amino acid; recall from Chapter 7 that most amino acids have two codons and some have as many as six, therefore the DNA sequence for two identical amino acid sequences can be quite different.

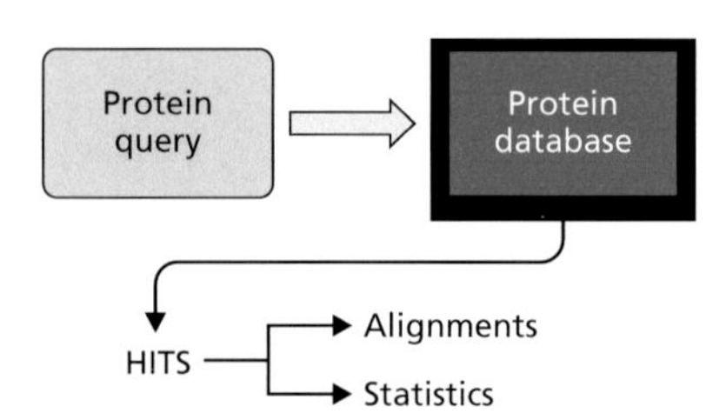

Figure 16.5 BLASTP starts with a protein query and searches a protein sequence database. The results of the search, the hits, are presented to the user as alignments between the query sequence and the sequences that were found by the search, the target sequences. The statistics comparing the query and target sequences are also presented to the user.

It is possible that you may have a DNA sequence and you want to see if there are any similarities between the proteins it could potentially encode and the proteins in a protein database. **BLASTX** helps with this. Rather than having to first translate the DNA sequence yourself, BLASTX does this part for you, in all six potential reading frames: three forward and three reverse. These six amino acid sequences are then used to interrogate a protein database as would be done in BLASTP (**Figure 16.6**).

Likewise, it is possible that you have a protein sequence and would like to see if it potentially matches the DNA database. First, BLAST needs to know what the DNA database potentially encodes as amino acids. Therefore, it does a complete six-frame translation of the nucleotide database and then uses these data to check for similarities to the query. This is the **TBLASTN** search (**Figure 16.7**).

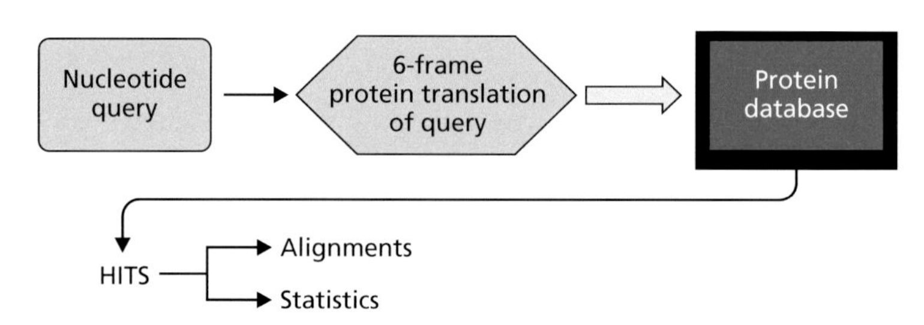

Figure 16.6 BLASTX starts with a nucleotide query and first conducts a six-frame protein translation of the query before searching a protein sequence database. The results of the search, the hits, are presented to the user as alignments between the query sequence and the sequences that were found by the search, the target sequences. The statistics comparing the query and target sequences are also presented to the user.

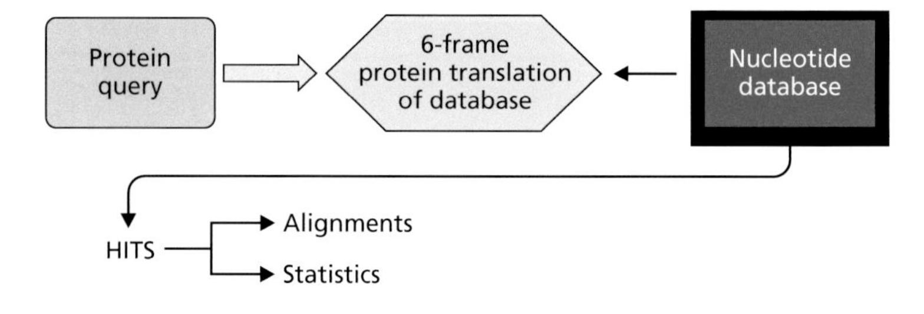

Figure 16.7 TBLASTN starts with a protein query, but before it is used to search a database, first a nucleotide database is six-frame protein translated and then this translated database is used for the search. The results of the search, the hits, are presented to the user as alignments between the query sequence and the sequences that were found by the search, the target sequences. The statistics comparing the query and target sequences are also presented to the user.

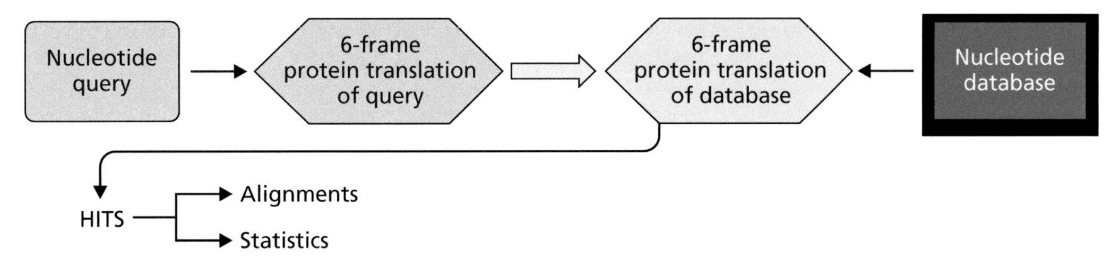

Figure 16.8 TBLASTX starts with a nucleotide query and a nucleotide database. Both the query and the database undergo six-frame protein translations before the translated query is used to search the translated database. The results of the search, the hits, are presented to the user as alignments between the query sequence and the sequences that were found by the search, the target sequences. The statistics comparing the query and target sequences are also presented to the user.

TBLASTX is the slowest of all BLAST searches because it does a six-frame translation of both the query (as in BLASTX) and the database (as in TBLASTN). A TBLASTX search is done when investigating a sequence to find distant relationships between nucleotide sequences (**Figure 16.8**). Sequences will come up on TBLASTX searches that are not revealed by BLASTN.

There are other versions of BLAST that do specialist searches

The five basic versions of BLAST are all about straight searches of nucleotide or protein databases with nucleotide or protein sequences. However, there are other versions of BLAST that have other functions. For example, PSI-BLAST is able to find more distant relatives of sequences, specifically proteins. It makes a list of all closely related proteins and forms a **profile sequence**. Rather than a single query sequence, this set of sequences combined in the profile sequence highlights significant sequence features of the group. The search finds a set of similar sequences to the profile sequence. This larger group forms a new profile sequence (**Figure 16.9**). Iterative searches, which repeat the search processes, are done by PSI-BLAST to identify more distant evolutionary relationships that might otherwise be missed by a simple BLASTP.

There is a lot of data in the sequence databases, and it grows every day. In an effort to generate BLAST results that may produce more basic information and therefore be more clearly informative for some researchers, **smartBLAST** was created. From a protein query sequence, the five best protein matches in a reference database, called the **landmark database**, are displayed (**Figure 16.10**). If there are hits from different organisms, these will be reported by preference. If five matches in the landmark database cannot be found, smartBLAST will then use the non-redundant protein database, the same database BLASTP would use. The output of smartBLAST combines a graphical view and a phylogenetic tree (see Figure 16.10). There are 27 genomes in the landmark database. Of these, 11 are bacterial species.

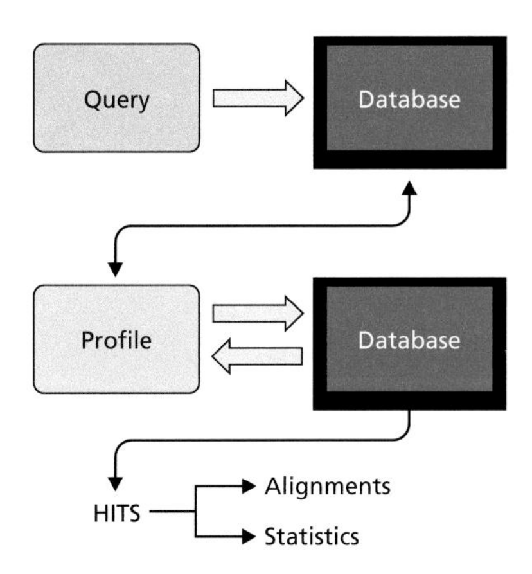

Figure 16.9 PSI-BLAST starts with a query and searches a database to find some similar sequences. With these, a profile is built. This profile of several sequences is used to search the database again. This can happen several times, represented by the reverse arrow. The results of the iterative search, the hits, are presented to the user as alignments between the query sequence and the sequences that were found by the search, the target sequences. The statistics comparing the query and target sequences are also presented to the user.

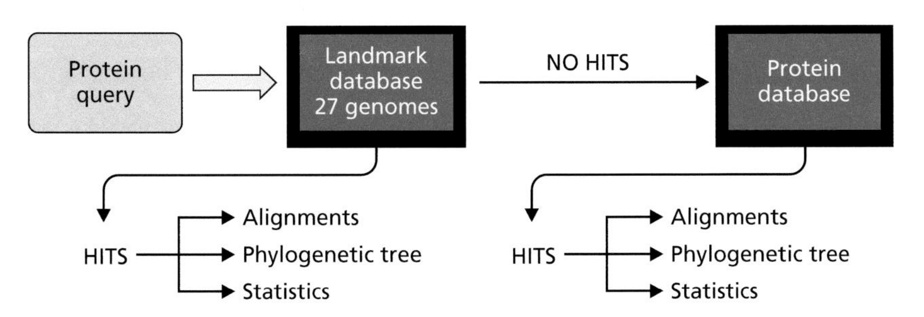

Figure 16.10 smartBLAST starts with a protein query and searches the landmark database of 27 specific genome sequences. If there are no results from this search, the query will be used to search a protein database instead. The results of the search, the hits, whether from the landmark database or protein database search, are presented to the user as alignments between the query sequence and the sequences that were found by the search, the target sequences. The statistics comparing the query and target sequences are also presented to the user.

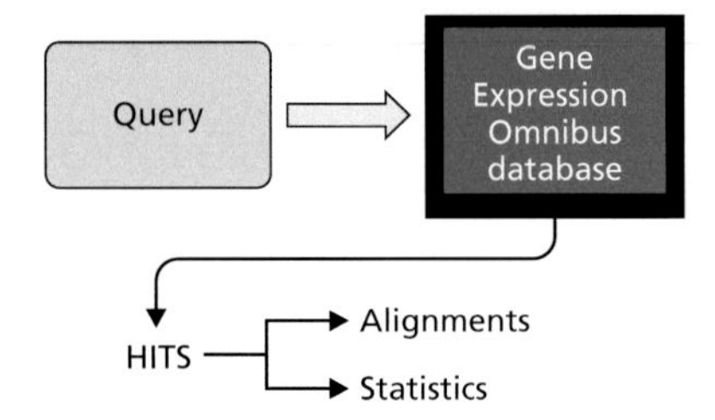

Figure 16.11 GEO-BLAST starts with a query and searches the Gene Expression Omnibus database. The results of the search, the hits, are presented to the user as alignments between the query sequence and the sequences that were found by the search, the target sequences. The statistics comparing the query and target sequences are also presented to the user.

Although many researchers access BLAST programs via a web browser, which makes it very accessible, there are also several command-line versions of BLAST. For instance, MEGABLAST is a command-line version of BLAST and is used when investigators have a lot of sequences to be compared. It is able to batch-load many searches via the command-line and generate the output data for later analysis. BLAST+ versions are command-line versions of BLAST. Users download BLAST+ source codes and use them locally from the command-line or in custom installations.

In addition to doing a search against the nucleotide or protein database, BLAST can be used to search against GEO, the Gene Expression Omnibus database. This is a repository for functional genomics data including microarray and RNA-seq expression gene information. Since this is gained from experimental evidence, the GEO-BLAST can identify if the query is a sequence that has been expressed in other bacteria and under what conditions (**Figure 16.11**).

Searches can look for more than just similarities

Embedded within BLASTP is a second search that looks for **conserved domains** within the amino acid sequences (**Figure 16.12**). Protein domains (Chapter 12) are amino acid sequences that have been identified as encoding functional features of proteins, and those that are conserved are found across proteins of different species. A conserved domain search can also be run as a stand-alone analysis of protein or translated DNA sequences.

Related to the conserved domain search is a tool that enables researchers to find proteins that share the same conserved domain architecture. This is CDART, Conserved Domain Architecture Retrieval Tool. This identifies protein sequences that have the same sequential order of conserved domains as the query sequence. This can identify proteins that are evolutionarily distant and may not have sequence similarity, but which may have related functions (**Figure 16.13**).

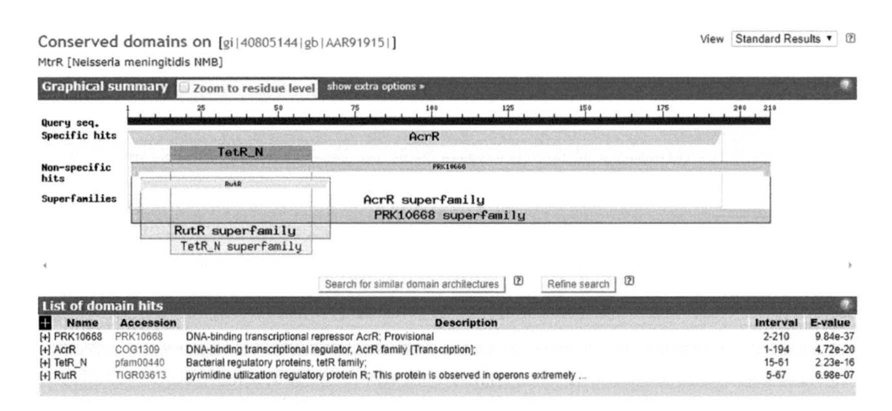

Figure 16.12 Conserved domain search. The results of a conserved domain search where the query is the MtrR protein sequence from *Neisseria meningitidis* strain NMB. Highlighted in the graphic are regions of the 210 amino acid protein sequence that share similarity to protein domains in the conserved domain database. Below the graphic is a list of the domain hits. This information can be expanded at the [+] to reveal more functional information on each domain, perhaps providing additional information about the function of the query protein.

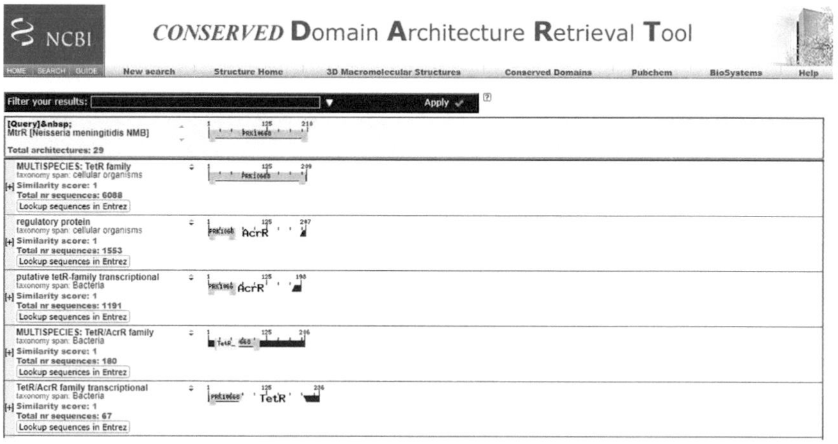

Figure 16.13 Conserved Domain Architecture Retrieval Tool. The results of a Conserved Domain Architecture Retrieval Tool (CDART) search where the query is the MtrR protein sequence from *Neisseria meningitidis* strain NMB. The graphical outputs show the similar architecture of conserved domains in other proteins between this 210 amino acid protein sequence and sequences in the conserved domain database. For each result, additional information is available by expanding the information at the [+] and following the available links.

Alignments of similar sequences are useful for further analysis

As can be seen in the preceding figures, the BLAST output produces an alignment between the query sequence and the target sequence, showing in the output the similarity between the two sequences. Visualization of the similarity between sequences is useful for further analysis, where it may be possible to see if sequences share common encoded domains, for example. The alignment generated by BLAST may not be ideal, however. The key to BLAST is that it is fast and that it is doing a search, not that it is carefully doing an accurate alignment of the two sequences. For an accurate alignment, other programs are used.

Alignments can be either local - as they are in the partial alignments used in BLAST searches - or global. A **local alignment** will only align those portions of the sequence with similarity. If there is no similarity, no alignment will be made by a local alignment program. In contrast, a **global alignment** is one that aligns the whole of the two sequences. In a global alignment, the computer will align the sequences whether or not there is similarity. Since a global alignment program will always produce a result no matter how poor or non-existent the similarity, it is really only useful once it is already known that there is some similarity, generally from a BLAST result.

Local alignments are used to compare the portions of the sequence that are similar

The Smith-Waterman algorithm, developed and released in 1981, produced a local alignment of two sequences in a **pairwise alignment**. All pairwise alignments involve two sequences. Comparisons are made between segments of the two sequences, such that all possible lengths of the sequences are compared to optimize the similarity in the final alignment (**Figure 16.14**). The Smith-Waterman local alignment is a variation of the Needleman-Wunsch global alignment algorithm. Unfortunately, Smith-Waterman requires more time to generate its alignments than most other alignment programs.

Available online is the EMBOSS suite of bioinformatics tools, the European Molecular Biology Open Software Suite. Among these is the EMBOSS Water alignment program, which uses the Smith-Waterman algorithm for the alignment

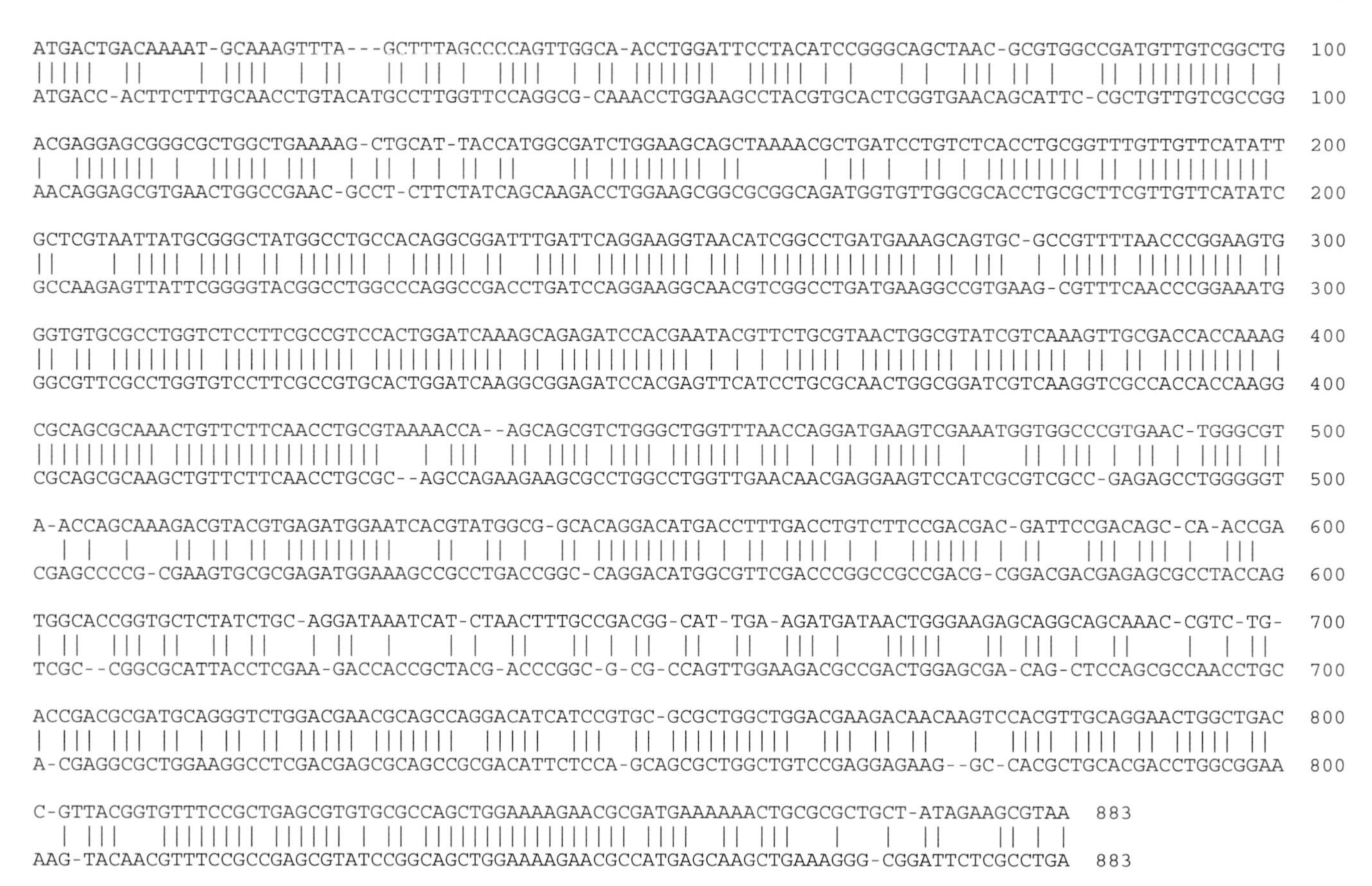

Figure 16.14 Smith–Waterman alignment. This Smith–Waterman alignment of the *rpoH* nucleotide sequences from *Escherichia coli* (top line) and *Pseudomonas aeruginosa* (bottom line) shows where the two sequences are similar along their lengths by aligning the bases and indicating identical bases with a vertical line. Where the insertion of spaces in one sequence would improve the matches between the pair of sequences, these are shown by a dash. Note that the lengths for each line overall include these introduced spaces; *rpoH* is 855 bases in both species, including the termination codon.

```
E_coli_RpoH      1 MTDKMQSL-ALAPVGNLDSYIRAANAWPMLSADEERALAEKLHYHGDLEA     49
                   ||..:|.: ||.|..||::|:.:.|:.|:||.::||.|||:|.|..||||
P_aeru_RpoH      1 MTTSLQPVHALVPGANLEAYVHSVNSIPLLSPEQERELAERLFYQQDLEA     50

E_coli_RpoH     50 AKTLILSHLRFVVHIARNYAGYGLPQADLIQEGNIGLMKAVRRFNPEVGV     99
                   |:.::|:|||||||||::|:||||.|||||||||:||||||:|||||:||
P_aeru_RpoH     51 ARQMVLAHLRFVVHIAKSYSGYGLAQADLIQEGNVGLMKAVKRFNPEMGV    100

E_coli_RpoH    100 RLVSFAVHWIKAEIHEYVLRNWRIVKVATTKAQRKLFFNLRKTKQRLGWF    149
                   ||||||||||||||||::|||||||||||||||||||||||..|:||.|.
P_aeru_RpoH    101 RLVSFAVHWIKAEIHEFILRNWRIVKVATTKAQRKLFFNLRSQKKRLAWL    150

E_coli_RpoH    150 NQDEVEMVARELGVTSKDVREMESRMAAQDMTFDLSSDDDSDSQPMAPVL    199
                   |.:||..||..|||..::|||||||:..|||.||.::|.|.:|...:|..
P_aeru_RpoH    151 NNEEVHRVAESLGVEPREVREMESRLTGQDMAFDPAADADDESAYQSPAH    200

E_coli_RpoH    200 YLQDKSSNFADGIEDDNWEEQAANRLTDAMQGLDERSQDIIRARWLDEDN    249
                   ||:|...:.|..:||.:|.:.:...|.:|::||||||:||::.|||.|:
P_aeru_RpoH    201 YLEDHRYDPARQLEDADWSDSSSANLHEALEGLDERSRDILQQRWLSEE-    249

E_coli_RpoH    250 KSTLQELADRYGVSAERVRQLEKNAMKKLRAAIEA    284
                   |:||.:||::|.|||||:||||||||.||:..|.|
P aeru RpoH    250 KATLHDLAEKYNVSAERIRQLEKNAMSKLKGRILA    284
```

Figure 16.15 EMBOSS Water. This EMBOSS Water alignment of the RpoH amino acid sequences from *E. coli* (top line) and *P. aeruginosa* (bottom line) shows where the two sequences are similar along their lengths by aligning the amino acids and indicating identical amino acids with a vertical line. Similar amino acids are shown with dots. Where the insertion of spaces in one sequence would improve the matches between the pairs, these are shown by a horizontal dash.

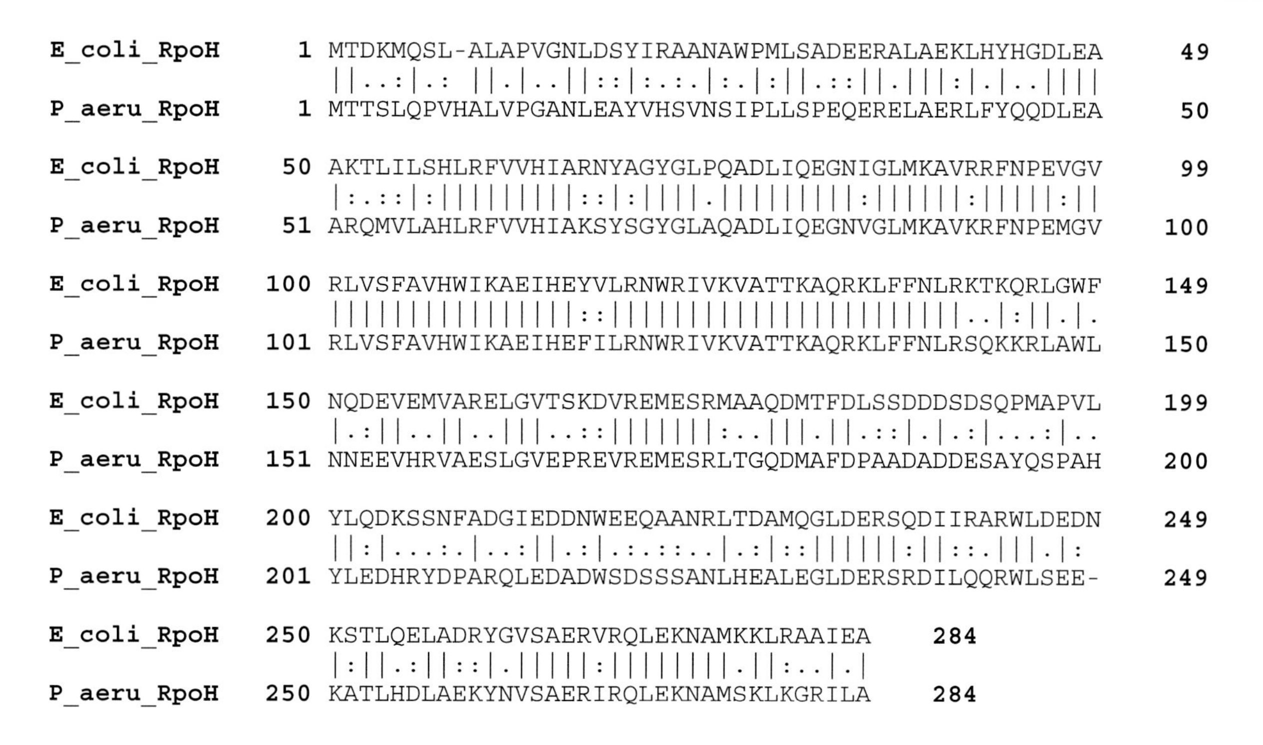

Figure 16.16 EMBOSS Matcher. This EMBOSS Matcher alignment of the RpoH amino acid sequences from *E. coli* (top line) and *P. aeruginosa* (bottom line) shows where the two sequences are similar along their lengths by aligning the amino acids and indicating identical amino acids with a vertical line. Similar amino acids are shown with dots. Where the insertion of spaces in one sequence would improve the matches between the pairs, these are shown by a dash.

of two sequences, either nucleotide or protein (**Figure 16.15**). These pairwise local alignments match up the similar regions in two sequences and may indicate where they share similar functions or evolutionary relationships.

Another EMBOSS program, EMBOSS Matcher, uses an alignment algorithm based on LALIGN to do a local alignment between two sequences. This is quite rigorous. Unlike many other alignment programs, LALIGN is able to find internal duplications within a sequence by doing local alignments with itself for protein or nucleic acid sequences (**Figure 16.16**).

Global alignments will align any sequences, similar or not

The Needleman-Wunsch global alignment algorithm, developed and released in 1970, was the first alignment program. Originally, it was used to find similarities in amino acid sequences between proteins (**Figure 16.17**). The Needleman-Wunsch

```
MTDKMQSL-ALAPVGNLDSYIRAANAWPMLSADEERALAEKLHYHGDLEAAKTLILSHLR  59
MT  +Q + AL P  NL++Y+ + N+ P+LS ++ER LAE+L Y  DLEAA+ ++L+HLR
MTTSLQPVHALVPGANLEAYVHSVNSIPLLSPEQERELAERLFYQQDLEAARQMVLAHLR  60

FVVHIARNYAGYGLPQADLIQEGNIGLMKAVRRFNPEVGVRLVSFAVHWIKAEIHEYVLR  119
FVVHIA++Y+GYGL QADLIQEGN+GLMKAV+RFNPE+GVRLVSFAVHWIKAEIHE++LR
FVVHIAKSYSGYGLAQADLIQEGNVGLMKAVKRFNPEMGVRLVSFAVHWIKAEIHEFILR  120

NWRIVKVATTKAQRKLFFNLRKTKQRLGWFNQDEVEMVARELGVTSKDVREMESRMAAQD  179
NWRIVKVATTKAQRKLFFNLR  K+RL W N +EV  VA  LGV  ++VREMESR+  QD
NWRIVKVATTKAQRKLFFNLRSQKKRLAWLNNEEVHRVAESLGVEPREVREMESRLTGQD  180

MTFDLSSDDDSDSQPMAPVLYLQDKSSNFADGIEDDNWEEQAANRLTDAMQGLDERSQDI  239
M FD ++D D +S   +P  YL+D    + A  +ED +W + ++  L +A++GLDERS+DI
MAFDPAADADDESAYQSPAHYLEDHRYDPARQLEDADWSDSSSANLHEALEGLDERSRDI  240

IRARWLDEDNKSTLQELADRYGVSAERVRQLEKNAMKKLRAAIEA  284
++ RWL E+ K+TL +LA++Y VSAER+RQLEKNAM KL+  I A
LQQRWLSEE-KATLHDLAEKYNVSAERIRQLEKNAMSKLKGRILA  284
```

Figure 16.17 Needleman–Wunsch alignment. This Needleman–Wunsch alignment of the RpoH amino acid sequences from *E. coli* (top line) and *P. aeruginosa* (bottom line) shows where the two sequences are similar along their lengths by aligning the amino acids and indicating identical amino acids in the middle line. Similar amino acids are shown (+). Where the insertion of spaces in one sequence would improve the matches between the pairs, these are shown by a dash.

algorithm is still used to generate high-quality, optimal global alignments of two sequences, however it is computationally one of the slower alignment programs, especially for long sequences. As mentioned, a global alignment will always produce an alignment, even when two sequences are not similar to each other, so it should not be used unless it is already known that sequences are similar. However, if sequences have been identified as similar from a BLAST search, for example, a Needleman–Wunsch alignment will generate a superior alignment compared to the alignment that is made during the BLAST search process. It is therefore worth doing this sort of global alignment when a high-quality alignment is desired.

Among the EMBOSS suite of programs is EMBOSS Needle, a global alignment tool that uses the Needleman–Wunsch algorithm to align two protein or nucleotide sequences. There is also a modification of Needle, called Stretcher, that can accommodate larger sequences (**Figure 16.18**).

More complex comparisons need multiple sequence alignment algorithms

For alignments of more than two sequences, different algorithms are needed. Clustal was one of the earliest programs used for **multiple sequence alignments** of more than two sequences. It was developed by Des Higgins in 1988 and could be used for amino acid or nucleic acid sequences. There have been many improvements and variations since that time. The current standard version is ClustalΩ. ClustalΩ progressively develops a multiple alignment from a series of pairwise alignments, using the pairwise alignments between each of the input sequences to be aligned to create the best final alignment of all of the sequences together (**Figure 16.19**). ClustalΩ is one of the fastest online multiple sequence alignment tools available and produces accurate and reliable results. The scalability of the ClustalΩ program allows it to process hundreds of thousands of sequences into a multiple alignment in a few hours. This tool enables researchers to identify regions of similarity that are in common across multiple protein sequences, often indicating key amino acids that may be conserved functional domains of the protein.

T-Coffee is another multiple alignment program, which is the Tree-based Consistency Objective Function for Alignment Evaluation. It generates pairwise alignments to guide the final multiple alignment. This alignment can use previously generated data to guide the alignment, being able to combine previously obtained multiple alignments, use structural information, and identify motifs.

Protein localization can be predicted from the amino acids

PSORT, Protein Subcellular Localization Prediction Tool, was the first bioinformatics program that was used on a wide scale for predicting the location of proteins in Gram-negative bacteria. Computational predictions of subcellular locations are based on known protein-sorting signals and how they direct proteins to these spaces in bacterial cells. PSORTb predicts cellular locations for proteins in Gram-negative and Gram-positive bacteria, as well as archaea. As discussed in Chapter 8, proteins are localized in the bacterial cell by specific mechanisms such as the Sec and Tat pathways. PSORT looks for the signals in the amino acid sequences recognized by these pathways and predicts the subcellular destinations of the protein based on signal sequence presence or absence (**Figure 16.20**).

An alternative program to predict the cellular location of a protein is SignalP, which also looks for telltale signal peptides within the amino acid sequence. SignalP software predicts whether there is a signal peptide cleavage site in an amino acid sequence. It is also able to differentiate signal sequences and N-terminal transmembrane regions. This program can help identify proteins that will be exported out of the bacterial cell or will end up within the membrane and

```
E_coli_RNaseE     1 MKRMLINATQQEELRVALVDGQRLYDLDIESPGHEQKKANIYKGKITRIE    50
                    ||||||||||.|||||||||||||:||||||...||||||||||:|||:|
P_aeru_RNaseE     1 MKRMLINATQPEELRVALVDGQRLFDLDIESGAREQKKANIYKGRITRVE    50

E_coli_RNaseE    51 PSLEAAFVDYGAERHGFLPLKEIAREYFPANYSAHGRPNIKDVLREGQEV   100
                    |||||||||:|||||||||||||:||||..  |..||.|||:||.|||||
P_aeru_RNaseE    51 PSLEAAFVDFGAERHGFLPLKEISREYFKK--SPEGRINIKEVLSEGQEV    98

E_coli_RNaseE   101 IVQIDKEERGNKGAALTTFISLAGSYLVLMPNNPRAGGISRRIEGDDRTE   150
                    |||::|||||||||||||||||||.||||||||||||||||||||::|.|
P_aeru_RNaseE    99 IVQVEKEERGNKGAALTTFISLAGRYLVLMPNNPRAGGISRRIEGEERNE   148

E_coli_RNaseE   151 LKEALASLELPEGMGLIVRTAGVGKSAEALQWDLSFRLKHWEAIKKAAES   200
                    |:|||..|..|..|||||||||:|:|.|.|||||.:.|:.|.|||:|:..
P_aeru_RNaseE   149 LREALNGLNAPADMGLIVRTAGLGRSTEELQWDLDYLLQLWSAIKEASGE   198

E_coli_RNaseE   201 RPAPFLIHQESNVIVRAFRDYLRQDIGEILIDNPKVLELARQHIAALGRP   250
                    |.|||||:||||||:||.||||||||||:|||:....|.|...|..: .|
P_aeru_RNaseE   199 RGAPFLIYQESNVIIRAIRDYLRQDIGEVLIDSIDAQEEALNFIRQV-MP   247

E_coli_RNaseE   251 DFSSKIKLYTGEIPLFSHYQIESQIESAFQREVRLPSGGSIVIDSTEALT   300
                    .::||:|||...:|||:.:|||||||:||||||:||||||||||.||||.
P_aeru_RNaseE   248 QYASKVKLYQDSVPLFNRFQIESQIETAFQREVKLPSGGSIVIDPTEALV   297

E_coli_RNaseE   301 AIDINSARATRGGDIEETAFNTNLEAADEIARQLRLRDLGGLIVIDFIDM   350
                    :|||||||||:||||||||..||||||:||||||||||:|||||||||||
P_aeru_RNaseE   298 SIDINSARATKGGDIEETALQTNLEAAEEIARQLRLRDIGGLIVIDFIDM   347

E_coli_RNaseE   351 TPVRHQRAVENRLREAVRQDRARIQISHISRFGLLEMSRQRLSPSLGESS   400
                    ||.::|||||.|:|||:..||||:|:..||||||||||||||.|||||:|
P_aeru_RNaseE   348 TPAKNQRAVEERVREALEADRARVQVGRISRFGLLEMSRQRLRPSLGETS   397

E_coli_RNaseE   401 HHVCPRCSGTGTVRDNESLSLSILRLIEEEALKENTQEVHAIVPVPIASY   450
                    ..|||||:|.|.:||.|||||:|||||||||||:.|.||.|.||..:|::
P_aeru_RNaseE   398 GIVCPRCNGQGIIRDVESLSLAILRLIEEEALKDRTAEVRARVPFQVAAF   447

E_coli_RNaseE   451 LLNEKRSAVNAIETRQDGVRCVIVPNDQMETPHYHVLRVRKGEETPTLSY   500
                    ||||||:|:..||.|... |..|:|:|.:||||:.|.|:|  :::|
P_aeru_RNaseE   448 LLNEKRNAITKIELRTRA-RIFILPDDHLETPHFEVQRLR--DDSP----   490

E_coli_RNaseE   501 MLPKLHEEAMALPSEEEFAERKRPE-QPALATFAMPDVPPAPTPAEPAAP   549
                           |.:|..:..|.|..:..| ||..:|..:.....|.....|..|
P_aeru_RNaseE   491 -------ELVAGQTSYEMATVEHEEAQPVSSTRTLVRQEAAVKTVAPQQP   533

E_coli_RNaseE   550 VVAPAPKAAPATPATPA-QPGLLSRFFGALKALFSGGEE--TKPTEQPAP   596
                      ||....||..||.|. :|.|......:|.:||:|.::  .||.|...|
P_aeru_RNaseE   534 --APQHTEAPVEPAKPMPEPSLFQGLVKSLVSLFAGKDQPAAKPAETSKP   581

E_coli_RNaseE   597 KAEAKPERQQDRRKPRQNNRRDRNERRDTRSERTEGSDNREENRRNRRQA   646
                    .|| :..||.:||..||.|||......:.|.|..:..:.|.|.:....:|
P_aeru_RNaseE   582 AAE-RQTRQDERRNGRQQNRRRDGRDGNRRDEERKPREERAERQPREERA   630

E_coli_RNaseE   647 QQQTAETREGRQQAEVTEKARTADEQQAPRRERSRRRNDDKRQAQQEAKA   696
                    ::...|.|..|::.|..|  |.|.|::.||..|..|.....|:.:|..:.
P_aeru_RNaseE   631 ERPNREERSERRREERAE--RPAREERQPREGREERAERTPREERQPREG   678

E_coli_RNaseE   697 LNVEEQSVQETEQEERVRPVQPRRKQRQLNQKVRYEQSVAEETVVAPVAE   746
                    ....|:..:...:|...||.:..|:.|: .::.|.|:...||.   ...|
P_aeru_RNaseE   679 REGREERSERRREERAERPAREERQPRE-GREERAERPAREER---QPRE   724

E_coli_RNaseE   747 ETVAAEPIVQEAPAPRTELVKVPLPVVAQTAPEQQEENNADN--RDNGGM   794
                    :..|.:....||.|         ||.......::|::.:.:.  |.:.|.
P_aeru_RNaseE   725 DRQARDAAALEAEA---------LPNDESLEQDEQDDTDGERPRRRSRGQ   765

E_coli_RNaseE   795 PRRSRRSPRHLRVSGQRRRRYRDERYPTQSPMPLTVACASPELASGKVWI   844
                    .|||.|..|...|||:.......:.  ..:|:....|.|:..:|.....:
P_aeru_RNaseE   766 RRRSNRRERQREVSGELEGSEATDN--AAAPLNTVAAAAAAGIAVASEAV   813

E_coli_RNaseE   845 RYPIVRPQDVQVEEQRE-----------QEEVQVQPMVTEVPVA--AAVE   881
                    ...:.:......|...|           .|.|:.|...:|....  |.:|
P_aeru_RNaseE   814 EANVEQAPATTSEAASETTASDETDASTSEAVETQGADSEANAGETADIE   863

E_coli_RNaseE   882 PVVSAPVVEEMAEVVEAPVPVA-EPQPEVVETTHP-EVIAAAVTEQPQVI   929
                    ..|:..||.:.|:.....|..| |..|...|:... |...:||....:..
P_aeru_RNaseE   864 APVTVSVVRDEADQSTLLVAQATEEAPFASESVESREDAESAVQPATEAA   913

E_coli_RNaseE   930 TESDVAVAQEVAEHAEPVVEPQEETADIEEVAETAEVVVAEPEV---VAQ   976
                    .|....|..|.|..:||....:...|.....|......:.:|..   :.:
P_aeru_RNaseE   914 EEVGAPVPVEAAAPSEPATTEEPTPAIAAVPANATGRALNDPREKRRLQR   963

E_coli_RNaseE   977 PAAPVVAEVAAEVETVTAVKPEI----TVEHNHATAPMT-RAPAPEYVPE  1021
                    .|..:..|.||..|......|.:    .|....|:|... .||..|.:.:
P_aeru_RNaseE   964 EAERLAREAAAAAEAAAQAAPAVEEVPAVASEEASAQEEPAAPQAEEIAQ  1013

E_coli_RNaseE  1022 A--PRHSDWQRPTFAFEGKGAAGGHTATHHASAAPARP--QPVE  1061
                    |  |..:|..:.....|.:.:....|.|.||.......  :|..
P_aeru_RNaseE  1014 ADVPSQADEAQEAVQAEPEASGEDATDTEHAKKTEESETSRPHA  1057
```

Figure 16.18 EMBOSS Stretcher alignment. This EMBOSS Stretcher alignment of the RNase E enzyme amino acid sequences from *E. coli* (top line) and *P. aeruginosa* (bottom line) shows where the two sequences are similar along their lengths by aligning the amino acids and indicating identical amino acids with a vertical line. Similar amino acids are shown with dots. Where the insertion of spaces in one sequence would improve the matches between the pairs, these are shown by a dash.

```
N_men_RpoH      MPQMNNAFALPAIQSGNGSLEQYIHTVNSIPMLSQEEETRLAERR-IKGDLNAAKQLILS   59
E_coli_RpoH     ---MTDK-MQSLALAPVGNLDSYIRAANAWPMLSADEERALAEKLHYHGDLEAAKTLILS   56
P_aeru_RpoH     ---MTTSLQPVHALVPGANLEAYVHSVNSIPLLSPEQERELAERLFYQQDLEAARQMVLA   57
                   *.            ..*: *:::.*: *:** ::*  ***:   : **:**: ::*:

N_men_RpoH      HLRVVVSIARGYDGYGLNQADLIQEGNIGLMKAVKRYEPGRGARLFSFAVHWIKAEIHEF   119
E_coli_RpoH     HLRFVVHIARNYAGYGLPQADLIQEGNIGLMKAVRRFNPEVGVRLVSFAVHWIKAEIHEY   116
P_aeru_RpoH     HLRFVVHIAKSYSGYGLAQADLIQEGNVGLMKAVKRFNPEMGVRLVSFAVHWIKAEIHEF   117
                ***.** **:.* **** *********:******:*::*  *.**.*************:

N_men_RpoH      ILRNWRLVRVATTKPQRKLFFNLRSMRKNLNALSPKEAQDIADDLGVKLSEVLEMEQRMT   179
E_coli_RpoH     VLRNWRIVKVATTKAQRKLFFNLRKTKQRLGWFNQDEVEMVARELGVTSKDVREMESRMA   176
P_aeru_RpoH     ILRNWRIVKVATTKAQRKLFFNLRSQKKRLAWLNNEEVHRVAESLGVEPREVREMESRLT   177
                :*****:*:***** *********. ::.*  :. .*.. :* .***   :* ***.*::

N_men_RpoH      GHDIAIMAD--NSDDEDSFAPIDWLADHDSEPSRQLSKQAHYALQTEGLQNALAQLDDRS   237
E_coli_RpoH     AQDMTFDLSSDDDSDSQPMAPVLYLQDKSSNFADGIEDDNWEEQAANRLTDAMQGLDERS   236
P_aeru_RpoH     GQDMAFDPAADADDESAYQSPAHYLEDHRYDPARQLEDADWSDSSSANLHEALEGLDERS   237
                .:*:::      ..:.   :*  :* *:  : :  :..       :  * :*:  **:**

N_men_RpoH      RRIVESRWLQDDGGLTLHQLAAEYGVSAERIRQIEAKAMQKLRGFLTEEAEAV   290
E_coli_RpoH     QDIIRARWLDEDNKSTLQELADRYGVSAERVRQLEKNAMKKLRAAIEA-----   284
P_aeru_RpoH     RDILQQRWLSEE-KATLHDLAEKYNVSAERIRQLEKNAMSKLKGRILA-----   284
                : *:. ***.::   **::** .*.*****:**:* :**.**:. :
```

Figure 16.19 ClustalΩ alignment. This ClustalΩ alignment of the RpoH amino acid sequences from *N. meningitidis* (top line), *E. coli* (middle line), and *P. aeruginosa* (bottom line) shows where the three sequences are similar along their lengths by aligning the amino acids and indicating identical amino acids (*) and similar amino acids (: and .) below the alignment. Where the insertion of spaces in one sequence would improve the matches between the alignments, these are shown by a dash.

PSORTb Results

```
SeqID: N_gonorrhoeae_MtrE
 Analysis Report:
  CMSVM-       Unknown       [No details]
  CytoSVM-     Unknown       [No details]
  ECSVM-       Unknown       [No details]
  ModHMM-      Unknown       [No internal helices found]
  Motif-       Unknown       [No motifs found]
  OMPMotif-    OuterMembrane [matched 3 rules (Rule64, Rule116, Rule137)]
  OMSVM-       OuterMembrane [No details]
  PPSVM-       Unknown       [No details]
  Profile-     Unknown       [No matches to profiles found]
  SCL-BLAST-   OuterMembrane [matched 11353796: Outer membrane integral membrane protein]
  SCL-BLASTe-  Unknown       [No matches against database]
  Signal-      Non-Cytoplasmic [Signal peptide detected]
 Localization Scores:
  Cytoplasmic            0.00
  CytoplasmicMembrane    0.00
  Periplasmic            0.00
  OuterMembrane          10.00
  Extracellular          0.00
 Final Prediction:
  OuterMembrane          10.00
--------------------------------------------------------------------------------
```

Figure 16.20 PSORTb results. This is the output from PSORTb for *Neisseria gonorrhoeae* MtrE, the outer membrane protein of the MtrCDE efflux pump system of this Gram-negative species. PSORTb has identified the outer membrane protein motifs in the amino acid sequence of MtrE (OMPMotif), matched an outer membrane integral membrane protein (SCL-BLAST), and detected the signal peptide present in the sequence (Signal). This results in a correct Final Prediction as an outer membrane protein (10 out of 10).

those where the N-terminal amino acids of the encoded protein are cleaved in the mature protein (**Figure 16.21**).

To gain additional information about a predicted protein, another tool that can be employed is the EMBOSS tool pepstats, which gives the statistics for peptide and protein sequences. Although it does not directly identify protein

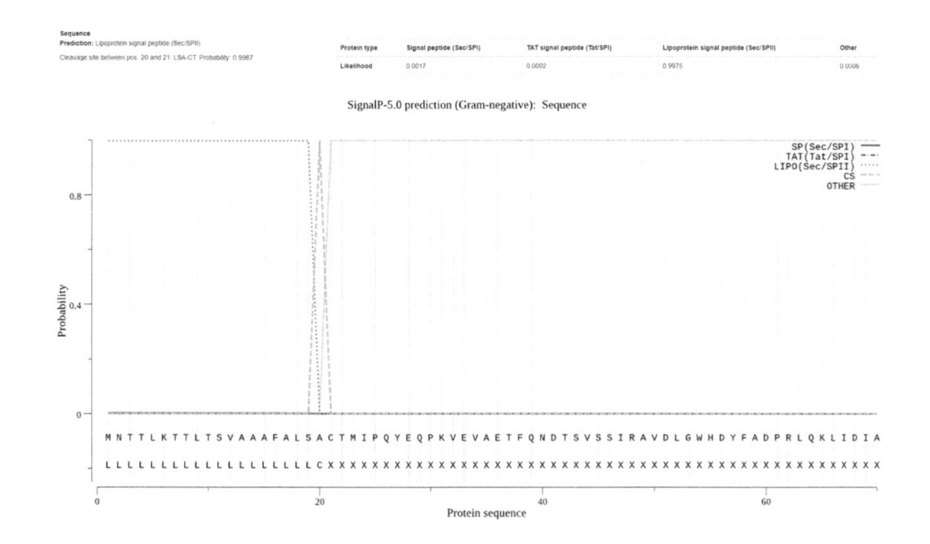

Figure 16.21 SignalP results. This is the output from SignalP for *N. gonorrhoeae* MtrE, the outer membrane protein of the MtrCDE efflux pump system of this Gram-negative species. SignalP assesses the probability of the presence of different signal peptides in the sequence, presenting these data in graphical and table format. Here it has predicted a lipoprotein signal peptide present in the sequence and indicates that the amino acid sequence will be cleaved between the 20th and 21st amino acids.

localization, the protein statistics information can contribute to an overall picture of the protein. The input to the program is an amino acid sequence. This generates various statistics about the protein including its molecular weight, the number of amino acids in the protein, the average weight of those amino acids, the charge of the protein, and its isoelectric point. This information on the physical qualities of the protein can be useful when planning laboratory experiments, since these predictions can help give the parameters of the experimental design. The information can also be added to other accumulated information, building a body of evidence supporting a location and/or function of a predicted protein.

DNA sequence to gene to amino acid sequence to 3D protein structure, ideally

Starting from the DNA sequence, it is possible to predict the CDS and to search with this sequence to identify any similar sequences, perhaps finding a proposed function for the potential gene. Certainly, translation of the DNA sequence into the predicted amino acid sequence is straightforward and generates the primary structure of the protein (Chapter 7). Ideally, to be able to predict the three-dimensional structure of a protein (Chapter 7) starting from the DNA sequence would be a real achievement for genetics research. It would be possible to then investigate how the protein structure might interact with other protein structures and other molecules within the bacterial cell, the environment, and the host.

From the features and order of the amino acids in the protein sequences, it is possible to predict a potential protein structure. Several different approaches are used to try to predict how the protein might fold within a biological system under certain conditions. Interestingly, there is an ongoing research community-wide experiment to test the effectiveness of protein structure prediction software. This massive project is called CASP, Critical Assessment of protein Structure Prediction. Since 1994, protein structure prediction methods have been tested every 2 years and the results made available.

Some online servers will integrate several structural prediction tools into one web-based interface, making it worth the effort to look for those that might provide more options, rather than just providing one tool. For example, PredictProtein integrates 16 different prediction tools and four different databases, all through one web interface. ExPASy, the Bioinformatics Resource Portal, also has a wealth of tools available for protein and structural predictions.

De novo protein structure predictions base structures just on the amino acids

It has long been a goal to be able to give a primary amino acid sequence to a computer and have a three-dimensional protein structure prediction determined as the output. However, even with only 1,300 protein folds occurring in nature, this problem is complex, especially when the nature of such folds can be influenced by the environment in which the protein folds in the cell and the presence of chaperones or co-factors.

Predictions for small proteins and those containing a single functional domain tend to be more accurate when compared to experimentally determined structures. As complexity increases in the protein, the accuracy of the structural prediction is less and the demands on computational resources are more.

Transmembrane helix prediction can find membrane proteins

Due to the hydrophobic nature of the interior of the phospholipid bilayer, any protein that spans the membrane will have regions of hydrophobic amino acids. It is possible to identify these and therefore identify transmembrane helices in proteins. In addition, although proline is hydrophobic, it is not internally part of alpha-helices due both to its bulky side chain and inability to contribute to hydrogen bonding (see Chapter 7), so transmembrane prediction software takes side-chain features such as this into account as well when searching for transmembrane alpha-helices. These predictions can reveal not only structural information about helices within proteins, but also localization information, being embedded in the membrane.

Homology modeling of protein structures based on known structures

Homology modeling methods create a three-dimensional structure of a protein based on an experimentally determined three-dimensional structure of a related homologous protein and the amino acid sequence of interest. This process is also called **comparative modeling**. Two homologous protein sequences are aligned and the amino acids of the protein of interest are mapped onto the experimentally determined structure to produce a predicted structure.

Protein threading can suggest a protein structure based on protein fold similarity

Protein threading is useful when there are no homologous proteins with an experimentally determined structure. In this case, there may be proteins that, although not homologous, share overall the same type of **protein fold**. This means that some of the internal structures of the protein are similar to other known protein structures in the way that they fold. This method is also called **fold recognition**. There are only about 1,300 unique folds that proteins are able to make in nature, therefore it is computationally possible to determine potential folds based on amino acid properties and known potential fold patterns. Most new protein structures determined experimentally are based on known folds.

Some gene tools are used to help design laboratory experiments

Primer3 is a program that is used to design PCR primers, sequencing primers, and hybridization probes. The parameters of design can be altered by the user to tailor the resulting primers, or probe, to the particular needs of the laboratory application (**Figure 16.22**). Poorly designed primers are often at the heart of failed PCR experiments, therefore the use of Primer3, with carefully selected design parameters, is an important first step. This program aids researchers in the design of primers for PCR by suggesting sequences for primers that are matched in their annealing temperatures (see Chapter 18) and will generate a PCR product of the desired size for the researcher. Further enhancements to this program have helped researchers to avoid some common primer-derived pitfalls in PCR that

OLIGO	start	len	tm	gc%	any_th	3'_th	hairpin	seq
LEFT PRIMER	35	20	59.11	55.00	25.81	0.00	0.00	cagttggcaacctggattcc
RIGHT PRIMER	510	20	58.94	55.00	19.94	4.76	0.00	ctcacgtacgtctttgctgg

SEQUENCE SIZE: 855
INCLUDED REGION SIZE: 855

PRODUCT SIZE: 476, PAIR ANY_TH COMPL: 0.00, PAIR 3'_TH COMPL: 0.00

```
    1 atgactgacaaaatgcaaagtttagctttagccccagttggcaacctggattcctacatc
                                        >>>>>>>>>>>>>>>>>>>>

   61 cgggcagctaacgcgtggccgatgttgtcggctgacgaggagcgggcgctggctgaaaag

  121 ctgcattaccatggcgatctggaagcagctaaaacgctgatcctgtctcacctgcggttt

  181 gttgttcatattgctcgtaattatgcgggctatggcctgccacaggcggatttgattcag

  241 gaaggtaacatcggcctgatgaaagcagtgcgccgttttaacccggaagtgggtgtgcgc

  301 ctggtctccttcgccgtccactggatcaaagcagagatccacgaatacgttctgcgtaac

  361 tggcgtatcgtcaaagttgcgaccaccaaagcgcagcgcaaactgttcttcaacctgcgt

  421 aaaaccaagcagcgtctgggctggtttaaccaggatgaagtcgaaatggtggcccgtgaa

  481 ctgggcgtaaccagcaaagacgtacgtgagatggaatcacgtatggcggcacaggacatg
                <<<<<<<<<<<<<<<<<<<<

  541 acctttgacctgtcttccgacgacgattccgacagccaaccgatggcaccggtgctctat

  601 ctgcaggataaatcatctaactttgccgacggcattgaagatgataactgggaagagcag

  661 gcagcaaaccgtctgaccgacgcgatgcagggtctggacgaacgcagccaggacatcatc

  721 cgtgcgcgctggctggacgaagacaacaagtccacgttgcaggaactggctgaccgttac

  781 ggtgtttccgctgagcgtgtgcgccagctggaaaagaacgcgatgaaaaaactgcgcgct

  841 gctatagaagcgtaa

KEYS (in order of precedence):
>>>>>> left primer
<<<<<< right primer
```

Figure 16.22 Primer3 results. This Primer3 output shows the selection of a pair of PCR primers for the *E. coli rpoH* sequence. Default primer Tm parameters of a minimum 57 °C and maximum 62 °C were used, producing here primers with Tms of approximately 59 °C. The size of the product was selected as 400–500 bases, producing here a product with this primer pair of 476 bases. Both the left primer (>>>>>) and right primer (<<<<<) are 20 bases in length and their positions in the sequence of *rpoH* are indicated.

will be discussed in Chapter 18, such as primer dimers and hairpins that can result in there being no PCR products.

Primer-BLAST is a combination of Primer3 and BLAST to find primers for PCR and check that they are specific through BLAST. The BLAST part of the program uses the Primer3 output as the query sequence and finds any other hits in the targeted genome that mean the primers would generate a PCR product

Primer pair 1	Sequence (5'->3')	Template strand	Length	Start	Stop	Tm	GC%	Self complementarity	Self 3' complementarity
Forward primer	TTAACCCGGAAGTGGGTGTG	Plus	20	278	297	59.89	55.00	4.00	0.00
Reverse primer	TCCTGCAACGTGGACTTGTT	Minus	20	764	745	60.11	50.00	4.00	2.00
Product length 487									

Figure 16.23 Primer-BLAST results. This Primer-BLAST output shows the selection of a pair of PCR primers for the *E. coli rpoH* sequence. Default primer Tm parameters of a minimum 57 °C and maximum 62 °C were used, producing here primers with Tms of approximately 60 °C. The size of the product was selected as 400–500 bases, producing here a product with this primer pair of 487 bases. Both the forward primer and reverse primer are 20 bases in length. This pair of primers was BLAST searched against the non-redundant nucleotide database entries for *E. coli* to confirm that the primers will specifically anneal to only these locations in the genome; primers that anneal elsewhere are not displayed. The output of this search is displayed to the user, who is able to select sequences that are the intended target for PCR primer annealing.

other than the one desired (**Figure 16.23**). Such non-specific PCR products can be a huge issue in PCR and waste time in the laboratory, therefore good primer design at the start can save aggravation later.

There are several programs that can help identify restriction enzyme digest sites within DNA from sequence data, which can aid in the design of experiments in the laboratory, such as the generation of recombinant DNA (see Chapter 18). Often the companies that supply restriction endonucleases have these tools on their web sites. For example, New England Biolabs has its NEBcutter to identify restriction sites in sequence data. These programs can often also suggest enzymes that have compatible sticky ends, facilitating the joining together of fragments of DNA (**Figure 16.24**).

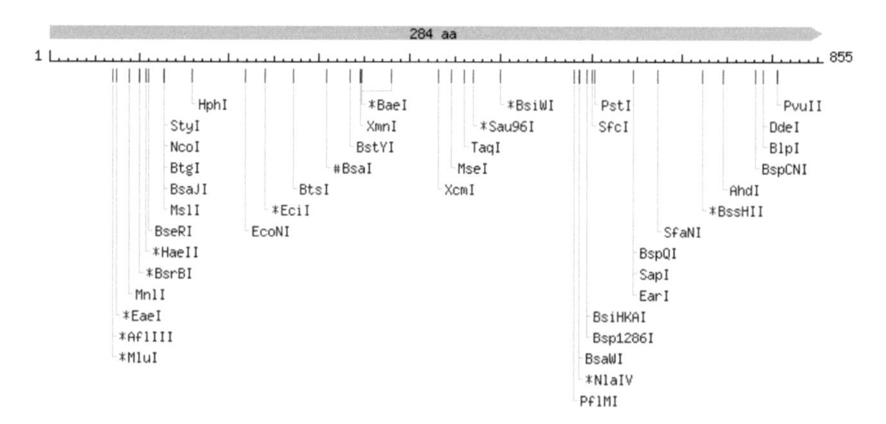

Figure 16.24 NEBcutter results. Using NEBcutter, the restriction enzyme recognition sequence sites within the *E. coli rpoH* gene sequence can be readily identified and displayed either graphically, as here, or in a list ordered by their position within the sequence.

Key points

- BLAST searches can identify similar sequences in the databases to the sequence being used for the search.
- Specialized BLAST searches like PSI-BLAST, GEO-BLAST, and Primer-BLAST use the BLAST search algorithm to find out more information about a sequence.
- Searches for conserved protein domains can identify functional regions within protein sequences.

- Alignments of sequences, either pairwise between two sequences or multiple alignments of several sequences, enable researchers to compare sequence data and features within DNA, RNA, and amino acid sequences.
- The signal sequence that is part of a secreted protein can be readily identified by programs like PSORT and SignalP and used to predict the cellular location of the protein from the amino acid sequence.
- There are several methods to attempt to predict protein structure from DNA sequence, based on homology to proteins with known structures or to protein folds, or based on the 1,300 potential ways amino acids are known to fold.
- Gene analysis programs can help design experiments in the laboratory, increasing the chances of success and efficiency.

Key terms

Define the following terms introduced in this chapter. Check your answers using the definitions in the Glossary. These terms are also available as Flashcards online.

BLASTN	Global alignment	Pairwise alignment
BLASTP	Heuristic	Protein threading
BLASTX	Homology modeling	Query sequence
Comparative modeling	Local alignment	TBLASTN
Conserved domain	Low complexity regions	TBLASTX
Fold recognition	Multiple sequence alignment	Target sequence

Questions and discussion topics

Self-study questions

Answer each question using 50–100 words or a table or labeled diagram. Advice on where to find answers to these questions is available online.

1. Create a table that compares and summarizes the different versions of BLAST, including their inputs and outputs (nucleotide or amino acid), the databases searched, and other information that you can use for reference.
2. Download the sequence for any bacterial gene of interest; note the bacterial species. Run the sequence through a BLASTN search and identify the top hit that has come from a different bacterial species. Record the alignment between the query and the target and the data for this BLASTN result.
3. Download the amino acid sequence for the bacterial gene used in question 2; this should be from the same bacterial species. Run the sequence through a BLASTP search and identify the top hit that has come from a different bacterial species. Record the alignment between the query and the target and the data for this BLASTP result.
4. During the BLASTP search for question 3, if the conserved domain search returned a result, record the information gained from that search, including the expanded and detailed information about the domains identified. If no domains were found, try a different protein sequence until a conserved domain is identified.
5. Use the DNA sequence that was used in the BLASTN search in question 2 in a BLASTX search. Does this identify the same protein hits as the BLASTP results from question 3? What differences are there and why?
6. Use the DNA sequence that was used in the BLASTN search in question 2 in a TBLASTX search. Are these results similar to any of the previous BLAST results? What differences are there and why?
7. Using any available online tool, do a pairwise local alignment of two of the homologous DNA sequences identified from your BLAST searches. These could be your query and one target sequence, or they could be two target sequences. Record your results, noting similarity scores and any gaps.

8 Using any available online tool, do a pairwise global alignment of two of the homologous DNA sequences identified from your BLAST searches. These should be the same two sequences used in question 7. Record your results, noting similarity scores and any gaps. Is there a difference between the local alignment from question 7 and this global alignment?

9 Using any available online tool, do a multiple alignment of five of the homologous amino acid sequences identified from your BLAST searches. Record the results. If there are conserved domains identified in question 4, highlight these in the alignment. Are there notable regions of conservation between the protein sequences?

10 Use the amino acid sequence that was used in question 3 in a PSORTb search. What does the output tell you about the potential cellular localization? Try some other protein sequences that might give different results.

11 Using the same amino acid sequence that went into PSORTb, run it through a SignalP prediction. Is a signal peptide cleavage site predicted? How about the other sequences you tried in question 10?

12 Use the gene sequence from question 2 to design PCR primers using Primer-BLAST. Adjust the parameters to make PCR products of three different lengths (at least 50 bp different) with Tms of 55 °C.

Discussion topics

These topics are presented for discussion in study groups, as part of class discussions, or on your own. These questions go beyond what is directly covered in this part of the book. Use the research literature and other reading to explore these topics in more depth. Tips to help prepare for topic discussions are available online.

1 PSI-BLAST can find potential distant evolutionary relationships between sequences. Starting with the sequence of a hypothetical protein of unknown function, run PSI-BLAST through several iterations until proteins of annotated functions are returned. Evaluate the credibility of these results and whether these hits might be reliable evidence to support experiments to investigate if the function of the hypothetical protein matches the annotated functions of these PSI-BLAST hits.

2 Discuss the ways in which the potential structure of the protein investigated in question 3 could be determined using online tools. Are there any experimentally identified structures for these proteins or for related proteins identified in your searches? Is it possible to predict a structure for one of these proteins using any of the techniques discussed here or others you have found?

3 Often in molecular biology a gene of interest is subject to restriction digestion. Using an online restriction enzyme recognition site identification program, such as NEBcutter, find the restriction digest sites in the gene used in question 2. Evaluate the restriction enzymes that recognize these sites. Identify those that cut the gene only once. Find a pair of restriction enzymes that work at the same temperature and in the same buffer that will cut the gene at the beginning and at the end, producing sticky ends.

Online quiz questions

To further self-assess your understanding of the chapter material, please visit the following link, where you can participate in a range of interactive quiz questions:

www.routledge.com/cw/snyder

Further reading

BLAST quickly finds the most similar sequences

Altschul S, Gish W, Miller W, Myers E, Lipman D. Basic local alignment search tool. *J Mol Biol*. 1990; *215*(*3*): 403–410.

Maryam L, Usmani SS, Raghava GPS. Computational resources in the management of antibiotic resistance: Speeding up drug discovery. *Drug Discov Today*. 2021; *26*(*9*): 2138–2151.

Protein localization can be predicted from the amino acids

Petersen TN, Brunak S, von Heijne G, Nielsen H. SignalP 4.0: Discriminating signal peptides from transmembrane regions. *Nat Methods*. 2011; *8*: 785–786.

Yu NY, Wagner JR, Laird MR, Melli G, Rey S, Lo R. PSORTb 3.0: Improved protein subcellular localization prediction with refined localization subcategories and predictive capabilities for all prokaryotes. *Bioinformatics*. 2010; *26*(*13*): 1608–1615.

DNA sequence to gene to amino acid sequence to 3D protein structure, ideally

Yachdav G, Kloppmann E, Kajan L, Hecht M, Goldberg T, Hamp T. PredictProtein - An open resource for online prediction of protein structural and functional features. *Nucleic Acids Res.* 2014; *42*(*W1*): W337–W343.

Homology modeling of protein structures based on known structures

Waterhouse A, Bertoni M, Bienert S, Studer G, Tauriello G, Gumienny R. SWISS-MODEL: Homology modelling of protein structures and complexes. *Nucleic Acids Res.* 2018; *46*(*W1*): W296–W303.

Some gene tools are used to help design laboratory experiments

Koressaar T, Remm M. Enhancements and modifications of primer design program Primer3. *Bioinformatics.* 2007; *23*(*10*): 1289–1291.

Ye J, Coulouris G, Zaretskaya I, Cutcutache I, Rozen S, Madden T. Primer-BLAST: A tool to design target-specific primers for polymerase chain reaction. *BMC Bioinf.* 2012; *13*: 134.

Chapter 17

Genome Analysis Techniques

Genome sequences are large, involving thousands to millions of individual A, T, G, and C bases. Therefore, analysis of the data is a significant task, requiring dedicated computer programs that are specifically designed to accommodate long strings of sequence data. In addition, it has to be remembered that the sequence is double stranded, which means that there can be information in either direction, so both must be analyzed for a thorough investigation. With the vast number of genomes available to researchers today, comparative analysis can reveal the regions of similarity and differences between strains and species, can aid in the annotation of genes and other genetic features, and can show phylogenetic relationships between bacteria. Presented in this chapter are computational tools and programs that are freely available. Other commercial tools are available, which are not discussed here.

A few things happen to the genome sequencing data before the search for genes

When the genome sequencing machine stops, it is time to get to work with the analysis, but before the search for interesting genes and genetic features in the bacterial genome can happen, the data need to be processed a bit first. No matter the type of genome sequencing platform used, the technology is in some way interpreting a signal as a nucleic acid, whether that is a signal from light being emitted or a pH change or a charge differential across a nanopore. Interpretation of these signals into the A's, T's, G's, and C's is **base calling** (**Figure 17.1**), and this step is important, is specific for each sequencing technology, and is constantly being improved. Data that were processed through previous versions of base calling software can often be re-analyzed with the software containing the improvements. This re-analysis can be worth doing, particularly for older data, but relies on keeping the original data from the sequencer needed for base calling.

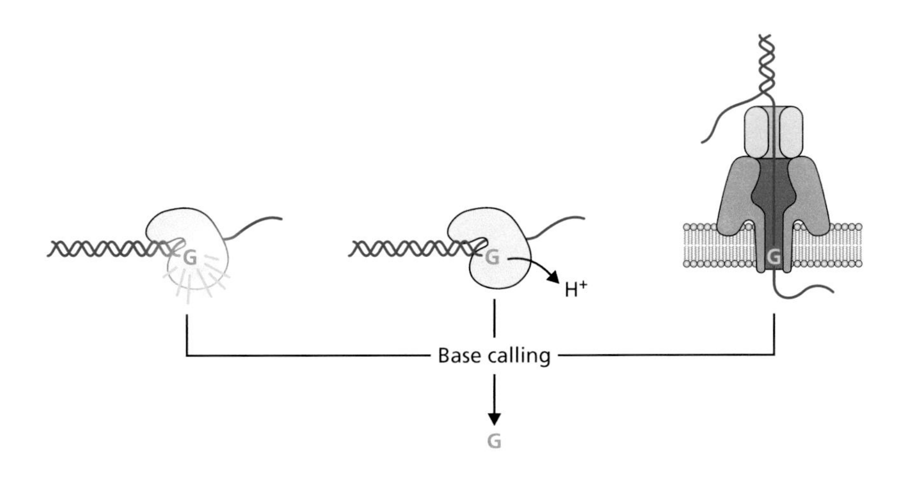

Figure 17.1 Base calling from DNA sequencing where bases are detected by light, pH, or nanopore signals and the interpretation of these signals into bases. In sequencing by synthesis-based technologies, the incorporation of nucleotides by polymerase is detected by the emission of light (left) or the release of hydrogen and the resulting change in pH (middle). Although hydrogen is always released with the incorporation of a nucleotide, it is only detected by some technologies, therefore no hydrogen is shown on the left. Nanopores detect the passage of nucleotides through the pore (right). In each instance, the signal from the technology is interpreted to generate the sequence data in a process called base calling.

DOI: 10.1201/9781003380436-23

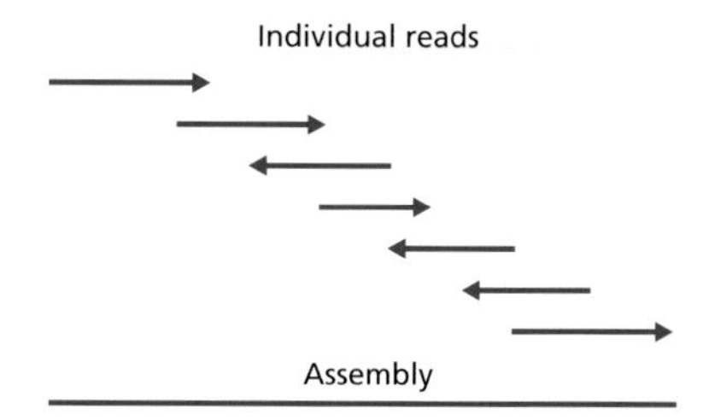

Figure 17.2 Assembly of sequencing reads. As discussed in Chapter 12, the individual sequencing reads generated by sequencing technologies need to be joined together. This produces an assembly, which is then ready for analysis.

The next stage is assembly, which was discussed in Chapter 12. At this stage, the separate sequences from the genome sequencing machine are pieced together in an attempt to reconstruct the chromosome as it is in the bacterial cell (**Figure 17.2**). There are various assembly programs and, like for base calling, these are often associated with the particular sequencing technology being used. If there is no specific assembly software that is packaged with the genome sequencing platform, there are likely to be particular assembly tools that are favored for the type of data produced by that technology. In addition, different genomes may work better with different assembly tools, by nature of features such as their %G+C. It may be worth trying different assembly programs to see what works for your data. Remember that for most bacterial genomes the GC skew around the circular chromosome can give an indication of whether the assembly is correct, although recent rearrangements, as well as incorrect assemblies, can alter GC skew (see Chapter 3).

Finally, before analysis of the genetic features, the post-assembly data should undergo **polishing**. The process of polishing improves the accuracy of the data by going back over it to compare the final data to the original base calling data. This can improve error rates in the sequence and can give greater confidence in changes identified later in the analysis as being true mutations and not sequencing errors.

Identification of features in genomic data is a key aspect of analysis

When a genome has been sequenced, the first questions asked are usually which genes are present and what features make it different from other bacteria that have already been sequenced before? Therefore, the identification of genetic features within the genome sequence data is a key aspect of analysis. The most numerous of these are the coding sequences, the best guess at the genes that are encoded by the bacterial DNA. Also, within the genomic data are all of the RNAs, including the rRNAs, tRNAs, and ncRNAs. The genome also contains, as have been described in the previous chapters, other features such as transposons, IS elements, repeats, prophages, and regulatory elements. Therefore, genome annotation (Chapter 12), finding these features and noting their locations within the genomic data, is an essential step in bacterial genome sequence analysis.

Because of all of those different features that are present in the DNA sequence, the process of bacterial genome data annotation needs to use different tools for identification of the various genetic features. It is possible to use tools individually to annotate the different features present in bacterial genomes, however most genomics researchers will start their annotations by using a pipeline that applies several tools at once, or in sequence, to the genome data, giving a single automated annotation output to the user.

A pipeline that combines several different tools into one resource is a great starting point because these annotation pipelines combine many of the individual feature identification tools to the whole of the genomic data at once, assign feature numbers consistently across the genome, and format the data into a GenBank, ENA, or DDBJ file format (Chapter 12), as needed (**Figure 17.3**).

Figure 17.3 An example pipeline for annotation of assembled sequence data. Once the sequencing data are assembled from the individual reads, the assembly can be analyzed for features. Using bioinformatics tools, simultaneously or sequentially, identify the CDSs, locate the rRNAs, find the tRNAs, and predict the ncRNAs, then annotate each of these. Further, it is possible to identify transposons (Tn's) and IS elements (IS's), locate prophages, find repeats, and predict regulatory elements to enhance the annotation. When complete, the annotation is uploaded to GenBank, ENA, or DDBJ in the format of the database.

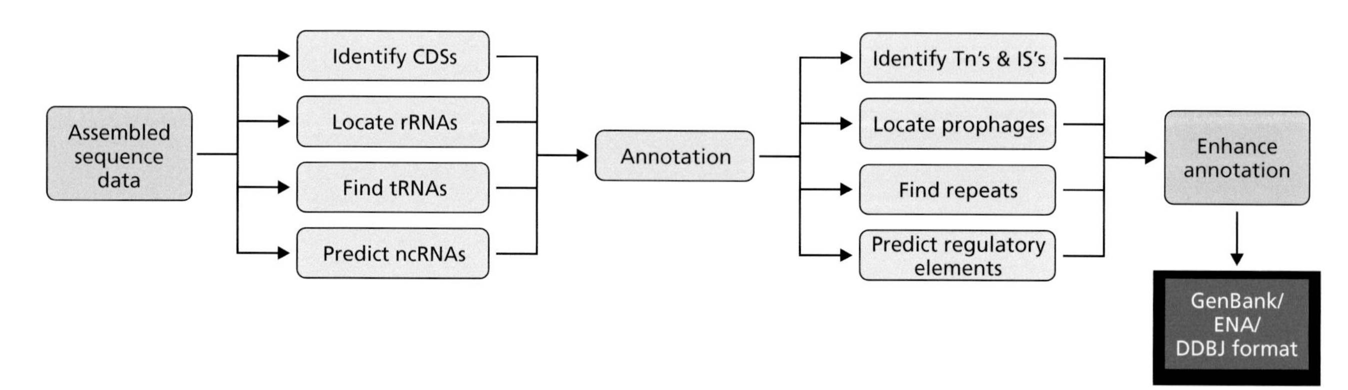

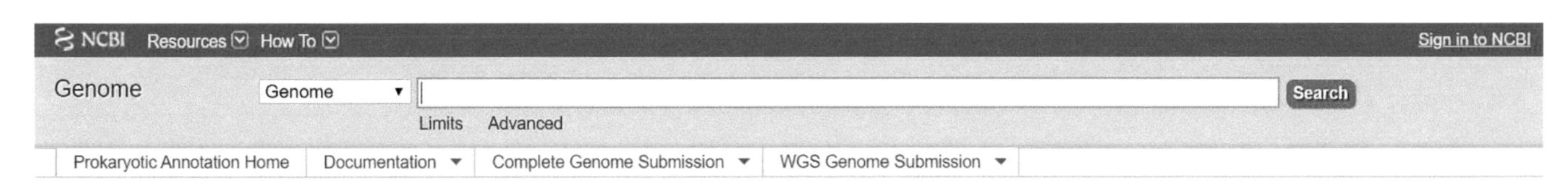

Figure 17.4 Screenshot of NCBI Prokaryotic Genome Annotation Pipeline (PGAP). This valuable tool is available from the National Center for Biotechnology Information (NCBI) and facilitates the depositing of prokaryotic genomes into GenBank. Genome sequence assembly data are uploaded to the web site, which generates an automated annotation in GenBank format. (From ncbi.nlm.nih.gov/genome/annotation_prok/.)

Automated annotation pipelines usefully combine feature identification tools

Three of the most commonly used genome annotation pipelines are the NCBI Prokaryotic Genomes Annotation Pipeline (PGAP) from the National Center for Biotechnology Information (the GenBank people), RAST (Rapid Annotation using Subsystem Technology), and Prokka. PGAP annotates bacterial and archaeal genomes, including chromosomes and plasmids (**Figure 17.4**). It predicts and annotates the locations and functions of protein-coding sequences, tRNAs, small RNAs, pseudogenes, direct and inverted repeats, and mobile elements, including transposons and IS elements. The functions of CDSs are based on homology and protein domains, making use of both the publicly available NCBI databases and several internal databases not available for outside interrogation.

RAST is another fully automated bacterial genome annotation service available for bacterial and archaeal genomes (**Figure 17.5**). It "calls" genes, meaning it identifies potential CDSs and makes functional predictions based on homology to sequences in the public databases. This service is free to use, and annotations are available usually within 24 hours of submission of a complete or draft genome sequence. RAST identifies CDSs and RNAs and annotates their functions. Recent improvements to the pipeline have made it more customizable

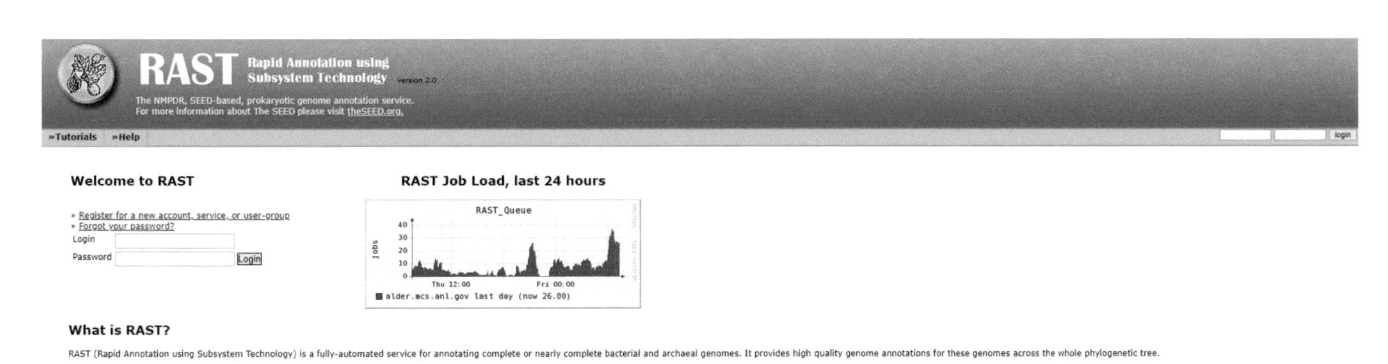

Figure 17.5 Screenshot of RAST (Rapid Annotation using Subsystem Technology). This valuable automated annotation tool provides a rapid annotation of bacterial genome assembly data, linked to functional data that can be interrogated and explored. (From rast.theseed.org/FIG/rast.cgi.)

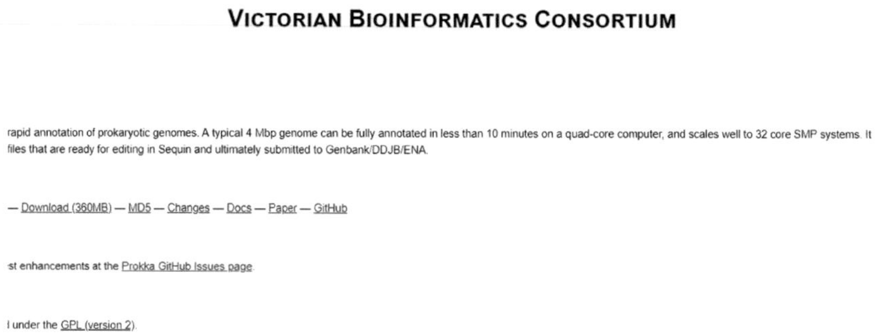

VICTORIAN BIOINFORMATICS CONSORTIUM

rapid annotation of prokaryotic genomes. A typical 4 Mbp genome can be fully annotated in less than 10 minutes on a quad-core computer, and scales well to 32 core SMP systems. It
files that are ready for editing in Sequin and ultimately submitted to Genbank/DDJB/ENA.

— Download (360MB) — MD5 — Changes — Docs — Paper — GitHub

st enhancements at the Prokka GitHub Issues page.

l under the GPL (version 2).

Figure 17.6 Screenshot of the original source of Prokka. Prokka (currently available at https://github.com/tseemann/prokka) is able to produce an automated annotation of a bacterial sequence assembly within a few minutes, locally on the user's own computer servers, rather than having to upload the data remotely. There are a number of local program requirements in addition to downloading Prokka (BioPerl, BLAST+, HMMER, Aragorn, Prodigal, tbl2asn, GNU Parallel, and Infernal), as well as recommended (Barrnap and MINCED) and optional programs (RNAmmer, HMMmmer, and SignalP). (From vicbioinformatics.com/software.prokka.shtml.)

as the RAST toolkit, RASTtk. The RASTtk version enables researchers to choose their tools used in the feature identification and annotation pipeline. In addition to providing an annotation file for the genomic data, RAST provides users with a dashboard to explore the details of the annotated features in the genome, including which functional categories they fall into, such as biosynthetic genes, structural proteins, or tRNAs. It also analyzes the various potential metabolic pathways possible for the organism, based on the CDSs found in the annotation pipeline process.

Prokka is another option for bacterial genome sequence annotation (**Figure 17.6**). It is a software tool freely available that rapidly annotates bacterial genomes, producing files suitable for submission to GenBank, ENA, or DDBJ. Unlike NCBI's pipeline and RAST, the Prokka pipeline does not require users to upload their data to a remote server for processing. With Prokka, users download the software to use it; this requires BioPerl and other local installations. Prokka uses the command-line to fully annotate a bacterial genome in about 10 minutes on a typical desktop computer. One of the local program installations required is Prodigal, the Prokaryotic Dynamic Programming Genefinding Algorithm, which finds CDSs in bacterial and archaeal sequences. Prodigal is rapid and accurate, and comparable to or better than other gene-finding programs such as Glimmer. The Prokka annotation pipeline option is particularly useful to researchers who need an annotation fast and already have the command-line and supporting software installed. In this situation it takes just minutes to get a high-quality annotation of a genome. Coupled with modern next-generation sequencing, a bacterial isolate can have its DNA extracted, sequenced, assembled, and annotated in a day.

The functional information is considered to be more robust from RAST than from Prokka; strengths and limitations of the three automated annotations are summarized in **Figure 17.7**. Keep in mind, as discussed in Chapter 12, that all annotations should be manually curated to ensure they are robust.

Pipeline	Strength	Limitation
PGAP	Embedded into NCBI	Less straightforward for manual assessment of annotations
RAST	On-line tools to interrogate annotation	Genome uploaded and queued onto a remote server
Prokka	Local installation is rapid	Functional information less robust

Figure 17.7 Summary of PGAP, RAST, and Prokka. Strengths and limitations of the PGAP, RAST, and Prokka automated annotation pipelines.

Visualization of an automatically generated annotation can aid manual curation

Once an annotation has been created, it should be checked to ensure that the automated choices for CDS locations are the best ones available and that the functional assignments fit with what is known in the literature for the bacterial species. It may be that there are errors in other annotations in the public databases and these can be replicated into new annotations through automated systems, however researchers with a critical eye and a knowledge of the bacterial species and the published literature can improve upon automated annotations through manual curation. This process is facilitated with a visualization tool, such as the Artemis Genome Browser. This program for visualizing genomic data was specifically designed for bacterial genomes and therefore takes the particular features of bacteria into account.

The Artemis Genome Browser displays the sequence, annotation, and six-frame translation in a graphical user interface (**Figure 17.8**). The user is able to use Artemis to then edit the annotation, which may have been generated by PGAP, RAST, or Prokka. Users can also add entirely new annotation features that were not part of the original automated annotation or delete features. Artemis

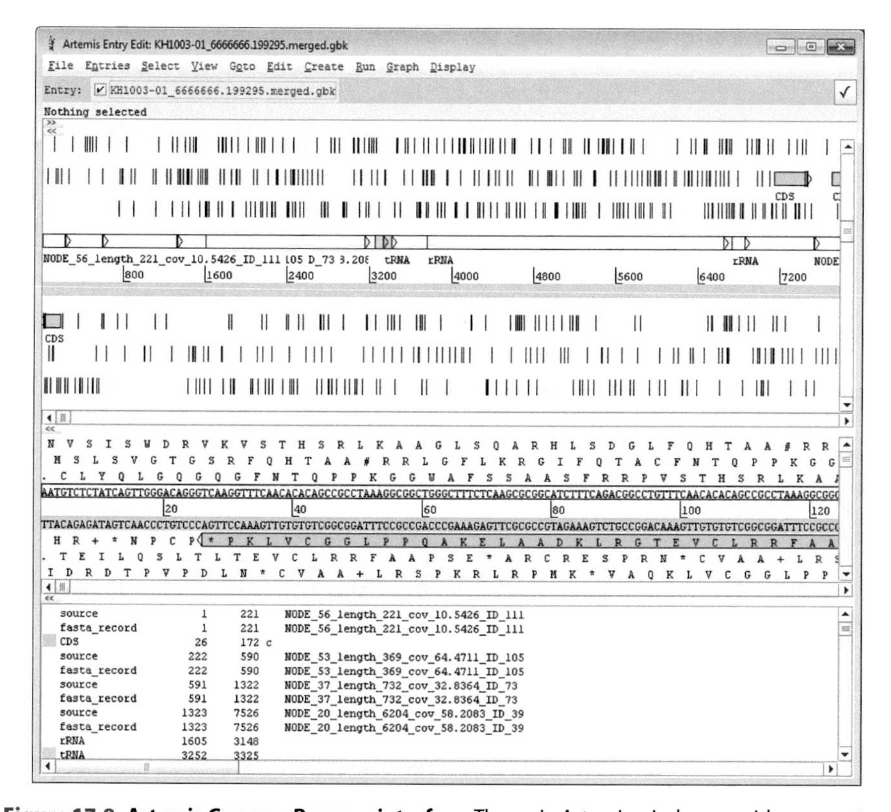

Figure 17.8 Artemis Genome Browser interface. The main Artemis window provides access to tools that aid with editing annotation along the top and three windows with different features. The top window is an overview of a large region, here several assembled contigs merged together (Nodes 56, 53, 37, and 20). The vertical lines in the top window represent termination codons in each of the six reading frames, three on the positive strand (above the features line) and three on the negative strand (below the features line). The middle window zooms in on a portion of the sequence, displaying individual bases and translated amino acids for all six reading frames, with the positive strand on top and the negative below the DNA sequence. Termination codons are represented by *, #, and + depending on their sequence. The bottom window shows the annotations for the region. This will indicate the contig that features are on for partial assemblies. Clicking on a coding sequence, such as the blue box above, will highlight the feature in the zoomed view and the annotation information. An added annotation will appear in the bottom window, as will any additions made to existing annotations. (From Carver T *et al. Bioinformatics*. 2012; *28(4)*: 464–469.)

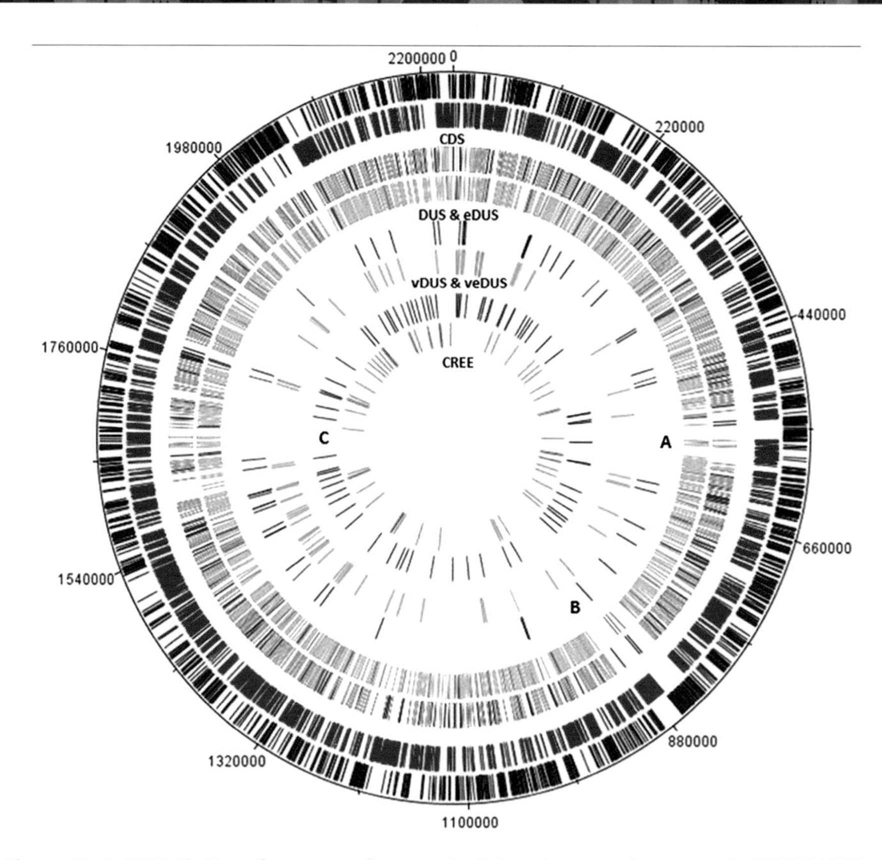

Figure 17.9 DNA Plotter of sequence features in *Neisseria gonorrhoeae* strain NCCP11945. DNA Plotter enables the visualization of sequence features around the circular chromosome of bacterial genomes. Here the CDSs on the positive and negative strands are shown in reds. The DNA uptake sequences (DUS) and a longer extended DNA uptake sequence (eDUS) are shown in blues, highlighting not only how frequent these features are, but also that there are some gaps. A few of the variant DNA uptake sequences (vDUS) and longer extended variant DNA uptake sequences (veDUS) from the non-pathogenic *Neisseria* spp. are present, shown in green. The Correia Repeat Enclosed Elements (CREE), unique to the *Neisseria* spp., are also mapped to the chromosome and shown in purple. This shows regions where these repeats are less frequent (A, B, and C), perhaps indicating regions of horizontal transfer or other unusual sequence features in the genome. (From Spencer-Smith *et al. Microb Genomics* 2016; *2*: e000069. With permission from the Microbiology Society.)

also displays sequence features that are not annotated, but which can be useful in sequence analysis and in assessing annotations, such as %G+C and GC skew, as well as showing the individual sequence reads that contributed to the assembled sequence data, if needed.

Artemis was improved upon in 2007, allowing multiple users to work on and change an annotation file simultaneously, making large genomic projects more efficient at the stage of manual curation. Built in to Artemis is DNA Plotter, which generates circular chromosome figures, popular in publications about genomes and genomic features (**Figure 17.9**). This gives a circular representation of the whole of the chromosome and allows the user to select features to display as concentric rings on the plot.

Another option for annotation visualization is GenDB, which is both a genome annotation system and a genome annotation visualization system for bacterial genomes, allowing automated and manual annotation. GenDB is modular and can be easily linked to other systems and tools. The GenDB system identifies and annotates CDSs using an array of tools as a first step, which can then be followed by the easy-to-use interface for manual curation and annotation.

Annotation with GenDB can be done by a team, even if the team members are present in different laboratories at different sites or in different countries. Each member of the team is able to sign into the data through a graphical web interface

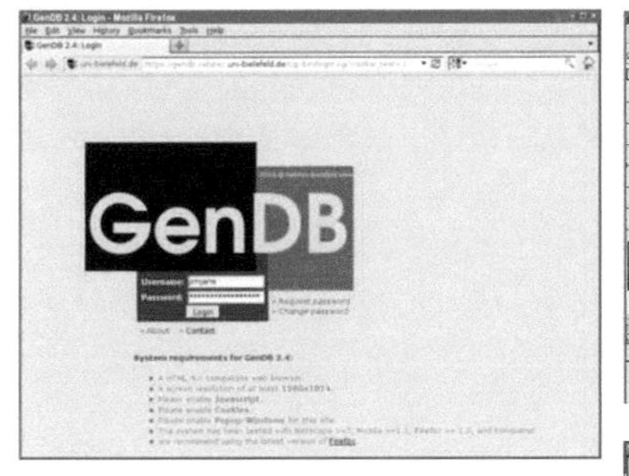

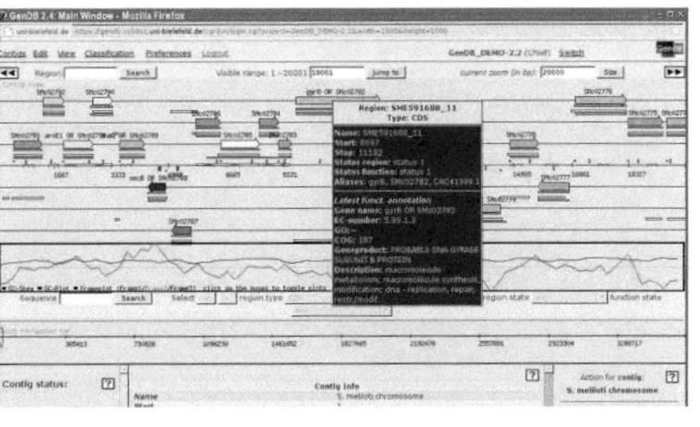

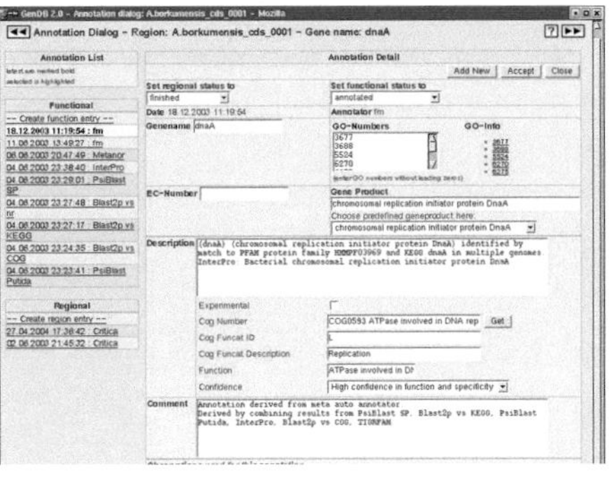

Figure 17.10 GenDB bacterial genome annotation via web browser interface. GenDB requires a local installation on a server, but then the genome data can be accessed for annotation using a web browser through a log-in page (left). The automated annotation generated using a pipeline of annotation tools can be accessed via the graphical interface (top right), which shows the evidence behind the annotations (bottom right). When the evidence has been assessed and the user is satisfied with the annotation, it can be exported in GenBank format. (Courtesy of GenDB.)

to access the data that are stored on a remote server (**Figure 17.10**). GenDB requires a Unix installation and Perl/C++ on a server, which is then accessed by users through the web interface, therefore GenDB needs some local support for use and implementation. However, once running, the web interface is intuitive for end users and links through to the sources of functional assignments from the automated annotations, facilitating manual curation within the graphical interface. In partially assembled genomes, each contig is presented separately.

Other visualization tools that allow users to see the annotated features in the context of the sequence data are discussed later in the chapter. These additional tools are integrated into programs that also enable comparisons of genomic data or that provide other additional functionalities, as will be discussed.

Comparisons show orthologues and paralogues, revealing evolutionary relationships between genes

Once annotations are completed, meaningful comparisons can be made between genome sequences. Even without an annotation, sequence data can be compared against fully annotated genomes and use the annotation data from the reference genome as a guide to the unannotated features in the new data. Comparisons between data can reveal homologous sequences that are orthologues, as discussed in Chapter 12. These orthologous sequences can give an indication of the relatedness between bacteria.

Tools such as the OrthoMCL gene clustering program use **reciprocal best hits** to identify orthologues. Reciprocal best hits are the top hit list in similarity searches for sequence identity both ways, between the genome and the database and between the database and the genome. This particular program forms groups of orthologues and paralogues, therefore it will also identify gene duplication events within the genomic data. Unfortunately, OrthoMCL doesn't scale up to hundreds of genomes, so it is not useful for very large comparative genomics projects.

Regardless of the tools used to investigate orthologues and paralogues, these investigations can uncover the evolutionary relationships between genes. The degree of similarity between orthologues may be able to suggest the divergence of the lineage of speciation of the genomes under investigation. Likewise, the similarity between paralogues can suggest how recently duplication events have created new genes.

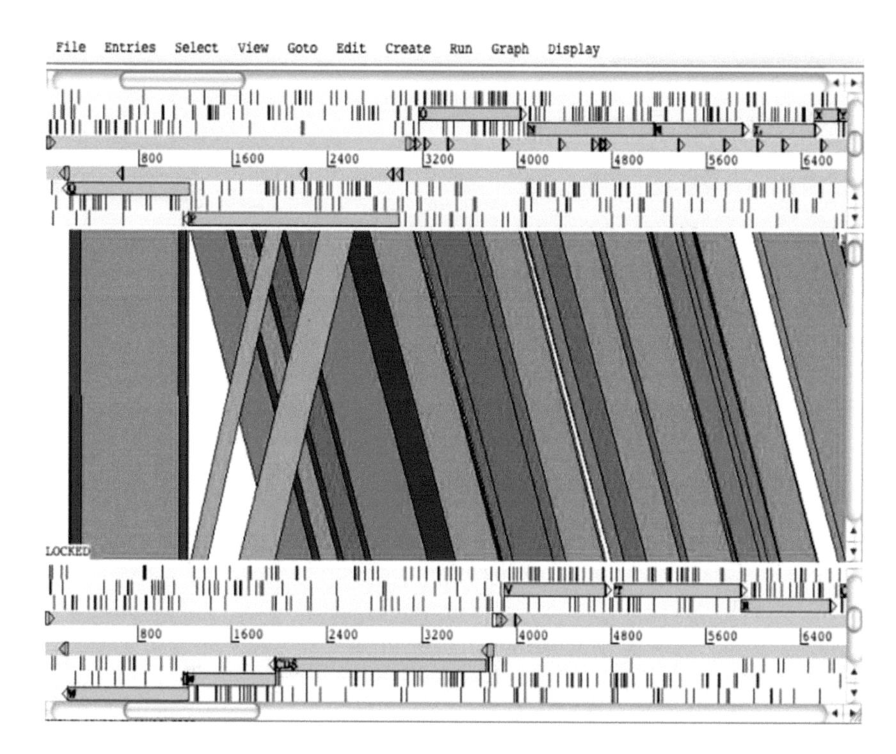

Figure 17.11 Screenshot of ACT showing red similarity regions between Artemis-visualized sequences. Red between each pair of sequences shows a region of similarity that is orientated in the same direction. Where the red lines have a slant, the region has some rearrangement interrupting the synteny. Although not shown here, blue between a pair shows a region that has inverted, but is similar. (From Carver TJ *et al. Bioinformatics* 2005; *21*: 3422–3423.)

Genomes can be aligned, just like genes can be aligned

Advances in sequencing technology have meant that it is necessary not only to be able to perform gene alignments, but also to be able to do whole bacterial genome alignments. Researchers want to be able to investigate at a whole genome level characteristics like synteny and to be able to find rearrangements and regions of difference. To achieve this, several genome alignment programs have been developed.

The Artemis Comparison Tool, ACT, is a Java-based application that displays pairwise alignments of two or more bacterial genomes. The two sequences are first compared by BLAST and then visualized in ACT. Regions of similarity between the genomic sequences are shown in red when in the same orientation or in blue when in the reverse orientation (**Figure 17.11**). Within the ACT user interface, it is possible to zoom in to the nucleotide level to be able to see the exact sequence of the DNA or to zoom out to see the whole of the genome on the computer screen at once. Using ACT, researchers can readily see regions of difference and connections between the DNA sequences that are being compared. The input files can be annotated GenBank/EMBL files or unannotated FASTA formatted files.

Features of Artemis are integrated into ACT, and analyses conducted in ACT can be used to inform annotations using these integrated Artemis annotation tools. Researchers can therefore dynamically add to or edit the annotations of the genomes that are being compared in ACT similarly to a genome being visualized in Artemis. The number of comparisons possible in ACT is limited by the practicality of visualization; recommendations are for a few genomes, 3–4, possibly up to about 7. WebACT used a web browser interface for ACT to compare researchers' own sequences or to look at pre-computed comparisons of genomes from the databases, however it went offline in December 2017.

Of course, most bacterial genomes are circular, not linear. ACT and other genome alignment programs are aligning a linear file of the bacterial genome,

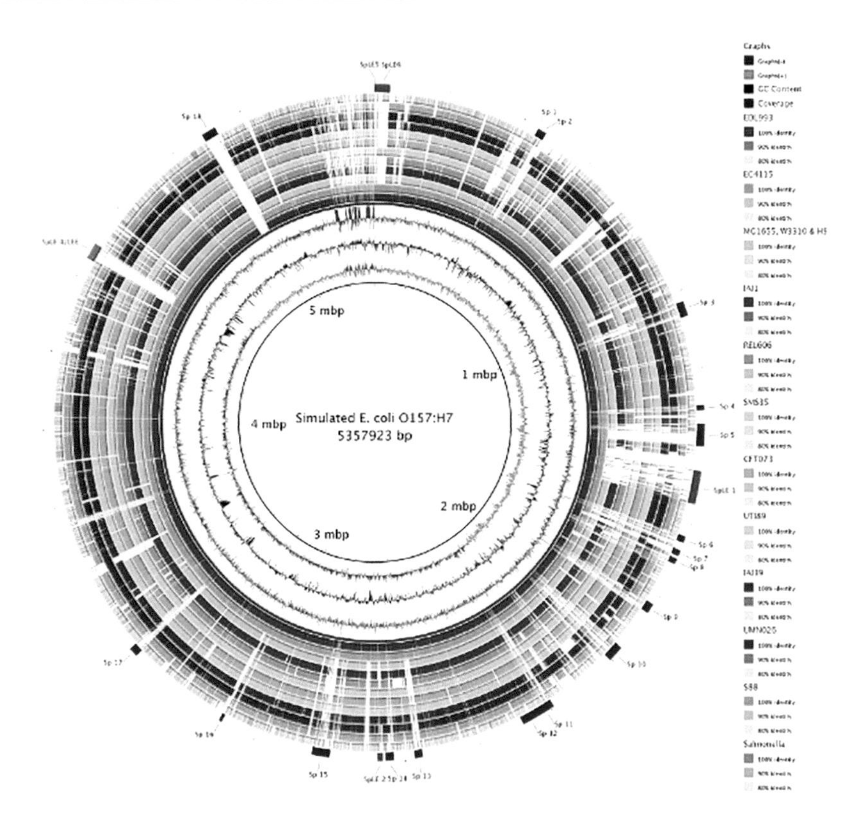

Figure 17.12 BRIG comparison of *E. coli* and *Salmonella* genome sequences BLAST searched against *E. coli* O157:H7. The key to the right shows the color coding used for the different genome sequences that have been compared to *E. coli* O157:H7. This visualization tool enables regions of difference between genomes to be readily identified as white areas of the circles. (From Alikhan NF *et al. BMC Genomics* 2011; *12*: 402.)

which is not how it actually occurs in nature. BRIG, the BLAST Ring Image Generator, addresses this problem. BRIG makes circular figures of BLAST comparisons between complete genomic data in a Java-based application. In this way, it enables visualization around the circular chromosome of comparisons of the reference sequence to a set of query sequences from the BLAST results, plotted as colored rings (**Figure 17.12**). BRIG can compare many sequences in this way, however each has to be entered into the system by hand and there is a visualization limit to the rings, after which it becomes difficult to interpret.

The use of BRIG can help to identify common elements across several genomes or regions missing from some. However, careful selection of the starting reference genome is important. It is against the reference that all of the query sequences are compared. If there are regions that are in the query sequences, but which are not in the reference, these regions will not show on the BRIG displayed result. Unfortunately, BRIG is no longer actively maintained, but it is still useful and still makes attractive and informative figures.

Mauve genome alignments make stunning figures, as well as being a useful research tool

Mauve is a freely available multiple genome alignment program that can be downloaded for use. It is similar to sequence alignment for genes, but scaled up, allowing researchers to look at regions of similarity between genomes. The Mauve genome alignment tool can generate alignments even in the presence of large-scale rearrangements, inversions, and other evolutionary changes. The ability to easily align multiple genomes makes it possible to do comparisons between the content of the genomes, looking at similarities and differences. Although doing these alignments requires computational resources, the creators of Mauve have

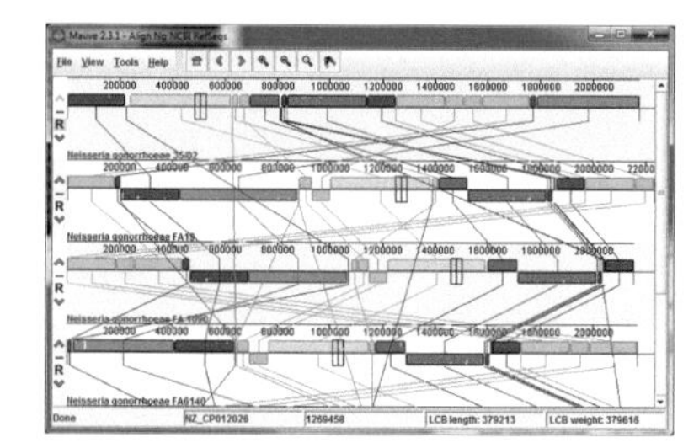

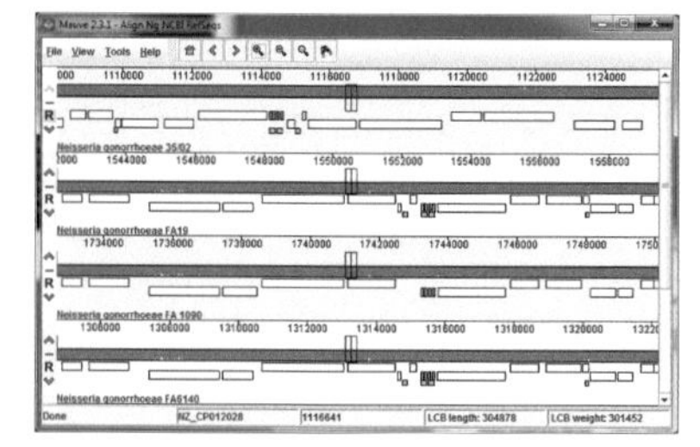

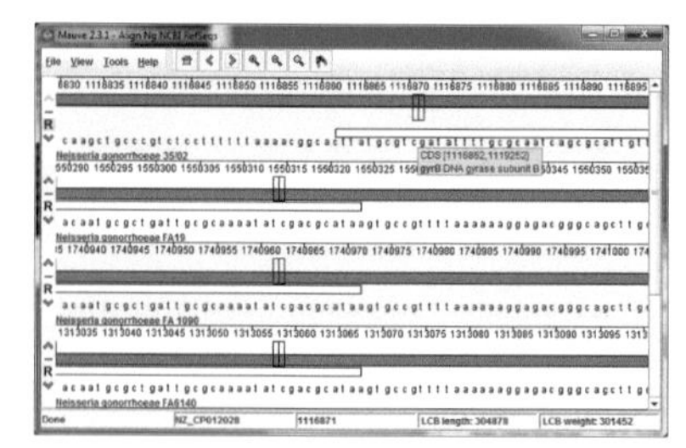

Figure 17.13 Mauve genome alignments show similar regions across the whole genome and can zoom in to the sequence level. A progressiveMauve alignment of four *Neisseria gonorrhoeae* strains shows LCBs (locally collinear blocks), regions of similarity, as colored blocks connected by lines (top panel). The same region across the four genomes is shown in the same color; regions inverted relative to the genome in the top line have an LCB below the line. Zooming in on a similar region in all four genomes reveals annotation information, with CDSs as white boxes, tRNAs as green boxes, and other features in pink (middle panel). Note this region is inverted in the other three genomes relative to the reference. It is possible to zoom in to the DNA sequence level (bottom panel), here showing the TAA termination codon of the annotated *gyrB* CDS, which is revealed when the mouse hovers over the white annotation box.

designed their program so that it scales well and will work on any length bacterial genomes of any reasonable number on nearly any laptop or desktop computer, given enough time. Their web site claims that two *Yersinia pestis* genomes can be aligned in under 2 minutes, while nine divergent Enterobacterial genomes can be aligned in a few hours. It is not feasible, however, to align more than about 50 genomes, both in terms of computing time and in terms of viewing and reasonably comparing the data.

The algorithms used today are in progressiveMauve, an improvement upon the original in terms of accuracy of the alignments and of being applicable to more genomes, including those that are less than 50% identical. Mauve's main strength is its ability to identify and align **orthologous regions** of the genomes. These are areas where two or more genomes have homology. These regions may be reordered or inverted relative to one another in their placement in the genome, and investigating these rearrangements of orthologous regions can provide insight into evolutionary events. Mauve identifies conserved segments of DNA during alignment. These are referred to within the software as locally collinear blocks (LCBs), and each is represented with a uniquely colored block that matches the corresponding orthologous region in the other genomes (**Figure 17.13**). These LCBs make it readily apparent to the researcher where there are genomic regions of similarity, areas that have been transposed to another place in the genome, and regions that have undergone inversion.

Regions of difference (see Chapter 3) can also be readily identified within Mauve. These display as gaps in the LCBs in the Mauve graphical output (**Figure 17.14**). Zooming in on the sequence data in the graphical output, it is possible to see the annotation of genomes being compared, perhaps revealing

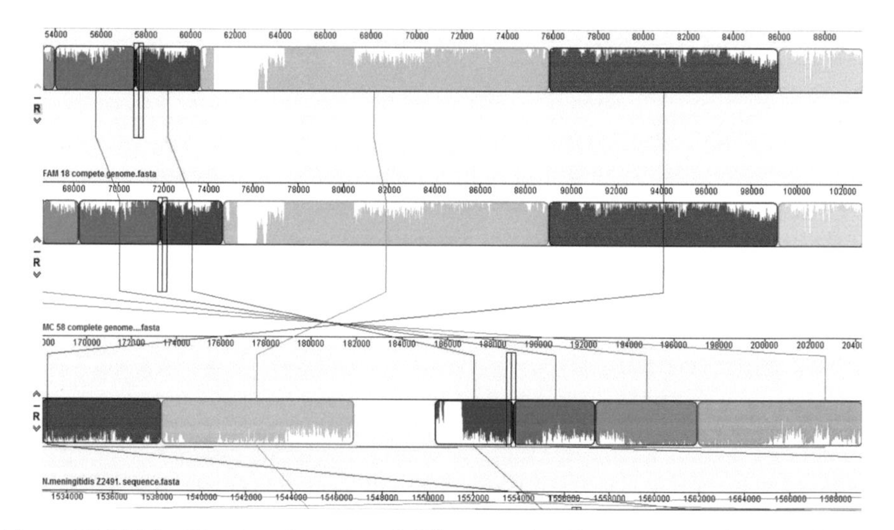

Figure 17.14 Identification of a region of difference using Mauve. LCBs can be set to show degrees of similarity within them, indicated here by the white within the colored blocks. Compared here are genome sequences of *Neisseria meningitidis* serogroup C strain FAM18 (top), *N. meningitidis* serogroup B strain MC58 (middle), and *N. meningitidis* serogroup A strain Z2491 (bottom), showing the capsule locus differences. The genes in this locus are responsible for the different serogroups of these strains.

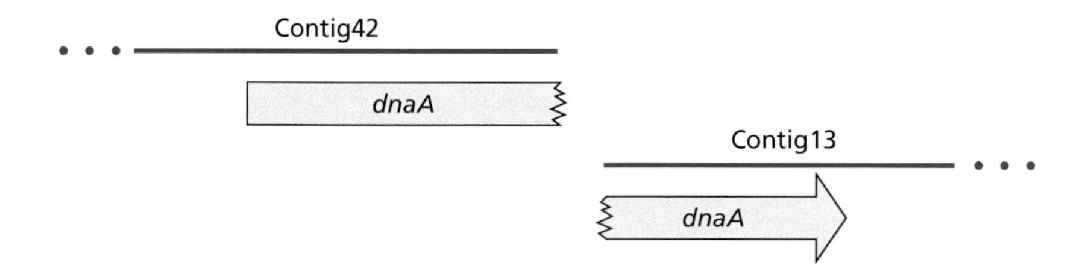

Figure 17.15 CDSs can span the ends of contigs. Close analysis of the features of the sequences at the ends of contigs can aid in joining together sequences that do not assemble, perhaps because there is not enough sequence overlap or due to some sequence error at the ends generating mismatches. Here, the beginning of the gene *dnaA* is identified at the end of Contig42 and the end of *dnaA* is identified at the end of Contig13. Although these two contigs have not assembled, the CDS sequence information suggests they should be joined. This discovery makes it possible to PCR across this portion of *dnaA* to fill in and join Contig42 to Contig13 with some automated Sanger sequence data.

that a bacteriophage sequence is present in one bacterial genome, but is not in the other, and therefore explaining the origin of the region of difference between the genomes.

There are other tools within Mauve that can facilitate the analysis of genomic data as well. For example, it is possible to search the genomic data and the annotation data in Mauve to further aid comparative genomic research. To determine if a gene or gene cluster is present in all of the aligned genomes in the comparison, the annotation could be searched for the gene name. Even if the gene is not annotated, or not correctly annotated, in all of the genomes, the alignment will reveal if there is homology at the sequence level between these sequences at the location of the CDSs. It is also possible to order and orient contigs, which can help with manual assembly of the genome once assembly software has done all that it can to assemble the data. It may be that in analyzing data it is recognized that the start of *dnaA* is present in Contig42 and the end of *dnaA* is present in Contig13 (**Figure 17.15**). Although these sequences were probably not annotated, being gene fragments, their alignment against other genomes may have revealed the fragments of *dnaA* and indicated the order and orientation for Contig42 and Contig13.

There is value in typing data, even in the genomics age

Prior to rapid and inexpensive bacterial whole-genome sequencing, several techniques were used to attempt to identify the relatedness of different bacterial isolates to one another with the goal of identifying common features, lineages, and outbreaks. One such technique is **multi-locus sequence typing** (**MLST**), which sequences discrete selected loci and uses these as a means to type and compare different bacterial isolates. Although in many ways MLST has been superseded by whole-genome sequencing, it remains a useful tool for large-scale population comparisons. Due to the size and historical value of data in MLST databases, which contain sequence data from carefully selected loci across bacterial chromosomes, it can be beneficial to extract the MLST information from genomic datasets and compare them to the MLST populations. This enables researchers to place these sequenced strains into the phylogenetic context of other established strains within known MLST **clonal complexes**, groups of genetically similar bacteria (**Figure 17.16**). An MLST-based phylogenetic tree of *Helicobacter pylori*, for example, has demonstrated that this bacterial species has genetic markers that are geographically distinct, aligning with the human population and its migration across the globe. The mother to child transmission of the *H. pylori* species has meant that its genome has evolved in lineages that trace the migration of our ancestors, grouping into the major regions of the world (see Figure 17.16).

Galaxy provides a full analysis suite for biological data

The tools described in this chapter thus far have come from a variety of sources. Although all freely available for use, they are not centralized in one location. Seeing a need for a one-stop resource that could analyze biological data and track the analyses that were done, Galaxy was developed and released in 2005. Galaxy is a widely used and continuously updated web-based platform for investigations of sequence data, although it should be noted that this is not specifically for bacterial data. Galaxy provides a graphical user interface for genome analysis programs that are otherwise working on the command-line. There is a menu of

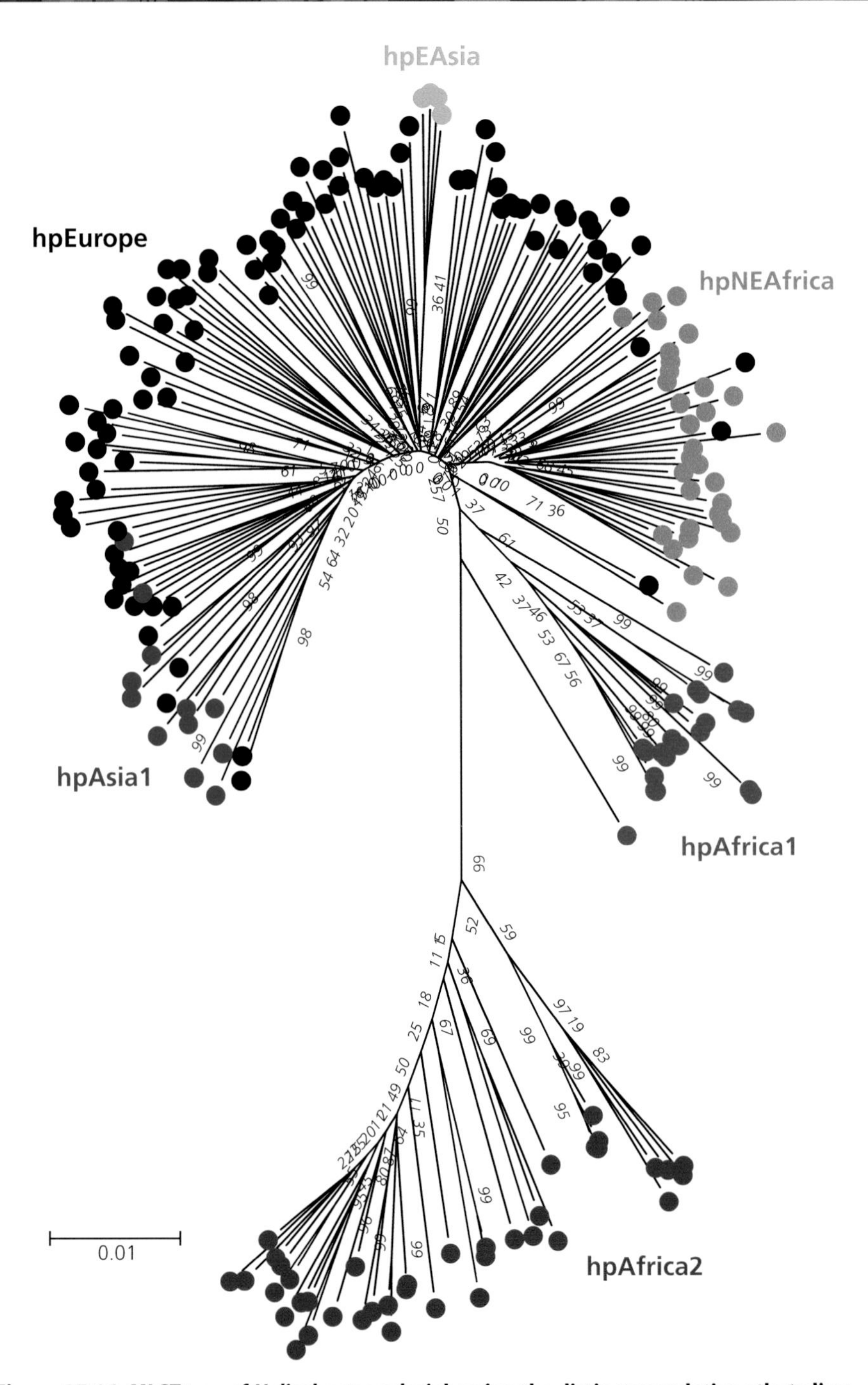

Figure 17.16 MLST tree of *Helicobacter pylori* showing the distinct populations that align with the migration of humans across the globe. Using specific conserved gene sequences, a phylogenetic tree is constructed that identifies the bacterial strains that are genetically similar. Because *H. pylori* tends to be transferred from mother to child, the strains have mostly remained with human lineages. This neighbor-joining tree was constructed by MEGA using aligned MLST sequences, identifying *H. pylori* groups based on geography.

various programs to select from with details of what each does, parameters set to default or that can be changed by the user, links to more information about the program, and the citations that should be used in publications when referencing (**Figure 17.17**). Users can upload their own data or access data from public databases. Originally designed for genomics, Galaxy is now also used, through the integration of additional bioinformatics tools, for transcriptomics and gene expression analysis, proteomics, and other applications involving biological data, such as drug design.

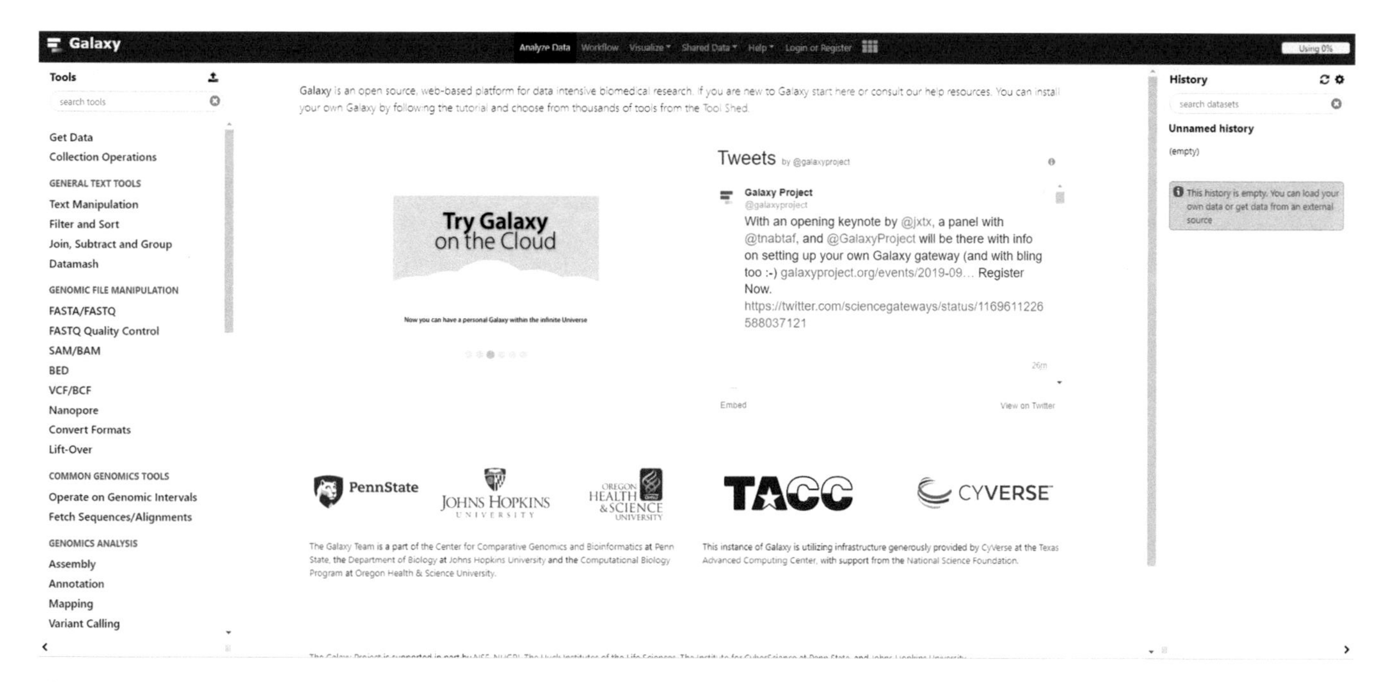

Figure 17.17 Screenshot of Galaxy. Galaxy is a web-based suite of tools for analysis of genomic sequence data and other data related to biomedical research. (From usegalaxy.org.)

Another strength of Galaxy is that it is available to anyone, anywhere, through the web interface, and that it tracks your workflow as you do it. Each process or analysis applied to your data is recorded. If you take your genome sequence data from your genome sequencing machine and trim the data to remove poor-quality data at the ends of reads and poor-quality reads, this will be clearly recorded in your workflow and the output of the trimming given a new identity, distinct from the original data. Therefore, when it comes time to publish, there is a clear record of all changes to the data, the software that was used to analyze the data, and the settings and parameters for the software (**Figure 17.18**).

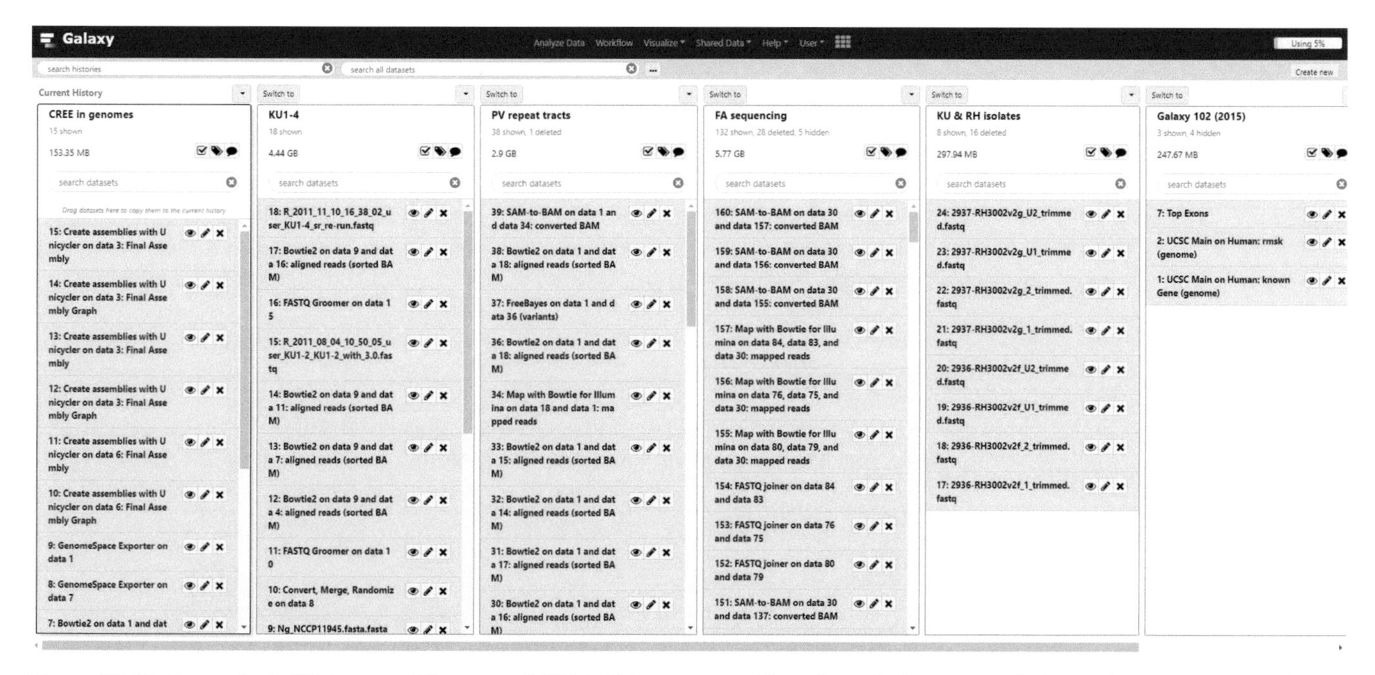

Figure 17.18 Screenshot of Galaxy workflow record. Within Galaxy, your workflow, the tools that were applied to analyze your sequence data, is recorded and kept track of for you. This is a valuable resource for accurate reporting and reproducibility.

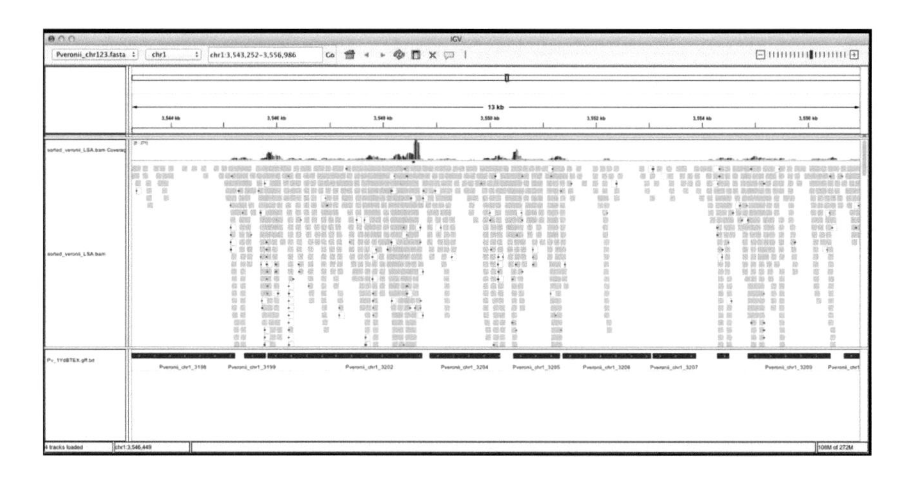

Figure 17.19 Screenshot of IGV showing the data and the various tracks. Here it is possible to visualize the assembled sequence data and also the individual sequencing reads that contributed to this assembled consensus. Using this information, individual SNPs can be investigated to determine the read coverage for them. For example, this might reveal that an SNP is present in 80% of the sequencing reads, with 20% of the bacterial population retaining the original sequence. (From https://wp.unil.ch/sequenceagenome/2015/05/06/bioinformatics-treatment-of-bacterial-rnaseq-data/.)

There is very little computer programing knowledge needed with Galaxy, making genomic bioinformatics more accessible to many researchers. Because it is web based, it does not rely upon someone setting up Unix and Perl systems to use the tools; these are run on the Galaxy host servers on behalf of the end user. Integral to Galaxy is its training session, Galaxy 101, which takes new users through an introduction to Galaxy tools and the workflow by performing example step-by-step tasks.

There are graphical interfaces available for some of the data outputs of Galaxy, although some are very biased toward eukaryotic projects and require a bit of extra time and setting up to coordinate with a chosen bacterial genome as the reference, rather than the human or mouse genome. The Integrative Genomics Viewer (IGV), for example, enables researchers to export datasets from Galaxy and explore them in a graphical interface (**Figure 17.19**). With IGV it is possible to conduct further manual analyses including zooming in on SNPs to look at them in genomic context and assessing read coverage of mapped reads against a reference genome.

Potential for use of artificial intelligence to analyze bacterial genome sequence data

With the advent and availability of web interfaces to interact with artificial intelligence (AI), some have speculated about how it can be used. AI can be particularly useful in the analysis of large datasets, such as might be desired when exploring the bacterial genome sequence data. The key for using AI in this manner is to have a good level of understanding about the types of tools that can be used to investigate genomic data, a clear research question to be investigated, and an understanding of the best ways to ask an AI for assistance in achieving the desired research outcome. For those who are less savvy in the use of bioinformatics programs or in developing new ways to interrogate the genome sequences, the use of AI may be a valuable tool.

Key points

- Base calling, assembly, and polishing of the genomic data from the sequencing machine all precede identification of the genetic features in the DNA.
- Genome sequence annotation to find all of the CDSs, RNAs, and other genetic features often starts with an automated annotation pipeline.
- The NCBI Prokaryotic Genomes Automatic Annotation Pipeline, RAST, and Prokka are three popular automated bacterial genome annotation systems.
- All automated annotations should be manually curated, which can be facilitated through use of a visualization program, such as Artemis, GenDB, or Mauve.

- Comparisons between genome sequences can reveal important similarities and differences, which is why programs such as ACT and Mauve that can align multiple whole bacterial genomes are beneficial research tools.
- Galaxy provides a web-based interface for bioinformatics tools that are otherwise only available via the command-line and tracks the workflow applied to data, making reporting of analysis steps in publications easier.

Key terms

Define the following terms introduced in this chapter. Check your answers using the definitions in the Glossary. These terms are also available as Flashcards online.

Base calling	Multi-locus sequence typing	Polishing
Clonal complexes	Orthologous regions	Reciprocal best hits

Questions and discussion topics

Self-study questions

Answer each question using 50–100 words or a table or labeled diagram. Advice on where to find answers to these questions is available online.

1 Outline the stages involved in assembly, which prepares genome sequence data for analysis.

2 What kinds of features can be identified in genome sequence data and annotated for inclusion with GenBank/ENA/DDBJ entries?

3 Download the complete genome sequence of any bacterial strain sequenced in the 1990s or early 2000s. Using the FASTA format of these data, generate a new automated annotation using PGAP, RAST, or Prokka. Note that this exercise will take some time to generate the annotations and if you choose to try Prokka, this requires local installations. Save this annotation.

4 Use the Artemis Genome Browser to view the annotation generated in question 3. Look through some of the genomic data and note at least three regions: one that looks like it could be an operon; one that includes rRNA or tRNA loci; and one where CDSs switch from one strand of the DNA to the other.

5 Use Artemis to generate a circular chromosome figure of the genome downloaded in question 3, displaying features from the annotation generated in question 4. Try to display the rRNA loci, %G+C, and GC skew. How does this compare to the original publication about this genome?

6 Use ACT to compare the genome sequence from question 3 to one or two other genome sequences from the same species. Are there any regions of rearrangement and/or inversion? Are there regions of difference, which are present in one genome sequence but not in the other?

7 Download and use Mauve to compare the original annotation of the bacterial genome sequence from question 3 to the automated annotation generated. Use progressiveMauve for the alignment. Scan through the annotations and note up to five differences between the annotations. Why do you think there are these differences?

8 Use progressiveMauve to compare the same genome sequences that were compared in question 6 with ACT. How are the differences between the genome sequences displayed in Mauve, including rearrangements, inversions, and regions of differences? Is anything notable in Mauve that was not in ACT?

9 Adjust the zoom and the settings in the Mauve alignment from question 8 to display a region of difference of your choice. Decide if you prefer the LCBs to be a solid color or to display the degrees of similarity by including white within the colored blocks. Export the image and save it.

10 Use the search function within Mauve to find *dnaA* in the genomes aligned in question 8. Zoom in to the DNA level and assess the sequence across all of the genome sequences to see if there are any differences in any of the aligned genome sequences.

11 Access Galaxy through the online web interface, register, and work through the Galaxy 101 tutorial. Note how the stages of analysis are recorded and saved, so they can be referred to later.

12 Using what you have learned from Galaxy 101, apply this knowledge to another sequence. This might be the sequence downloaded for use in question 3 or recently generated genome sequence data.

Discussion topics

These topics are presented for discussion in study groups, as part of class discussions, or on your own. These questions go beyond what is directly covered in this part of the book. Use the research literature and other reading to explore these topics in more depth. Tips to help prepare for topic discussions are available online.

1 Compare an early bacterial genome sequence publication, such as the first *Escherichia coli*, *Bacillus subtilis*, or *Haemophilus influenzae* genome sequence, to a recent genome sequencing publication. What are the striking differences about what is reported in these papers? What are the differences in the methods used in sequencing, assembly, annotation, and analysis?

2 Many of the bacterial genome sequence data that are generated are not assembled to a closed circular genome (for those species that have circular genomes). Discuss the reasons why this process is difficult and time consuming. What analyses can be done with an assembly that is still in several contigs (or scaffolds)? Which tools described in this chapter can be used on this kind of genomic data?

3 Discuss how MLST has contributed to understanding a specific bacterial population, paying particular attention to the methods used in the sequencing, sequence data assembly, annotation, and analysis. What insight was gained into the specific bacterial species, its geographic and/or chronological distribution, its antimicrobial resistance profile, and the ways it is/was transmitted?

Online quiz questions

To further self-assess your understanding of the chapter material, please visit the following link, where you can participate in a range of interactive quiz questions:

www.routledge.com/cw/snyder

Further reading

A few things happen to the genome sequencing data before the search for genes

Huang F, Xiao L, Gao M, Vallely EJ, Dybvig K, Atkinson TP, Waites KB, Chong Z. B-assembler: A circular bacterial genome assembler. *BMC Genomics.* 2022; *23*(*Suppl 4*): 361.

Land M, Hauser L, Jun S-R, Nookaew I, Leuze MR, Ahn T-H. Insights from 20 years of bacterial genome sequencing. *Funct Integr Genomics.* 2015; *15*(*2*): 141–161.

Automated annotation pipelines usefully combine feature identification tools

Brettin T, Davis JJ, Disz T, Edwards RA, Gerdes S, Olsen GJ. RASTtk: A modular and extensible implementation of the RAST algorithm for building custom annotation pipelines and annotating batches of genomes. *Sci Rep.* 2015; *5*: 8365.

Delcher A, Bratke K, Powers E, Salzberg S. Identifying bacterial genes and endosymbiont DNA with Glimmer. *Bioinformatics* 2007; *23*(*6*): 673–679.

Seemann T. Prokka: Rapid prokaryotic genome annotation. *Bioinformatics* 2014; *30*(*14*): 2068–2069.

Tatusova T, DiCuccio M, Badretdin A, Chetvernin V, Newrocki EP, Zaslavsky L, Lomsadze A, Pruitt KD, Borodovsky M, Ostell J. NCBI prokaryotic genome annotation pipeline. *Nucleic Acids Res.* 2016; *44*(*14*): 6614–6624.

Visualization of an automatically generated annotation can aid manual curation

Carver T, Harris SR, Berriman M, Parkhill J, McQuillan JA. Artemis: An integrated platform for visualization and analysis of high-throughput sequence-based experimental data. *Bioinformatics.* 2012; *28*(*4*): 464–469.

Meyer F, Goesmann A, McHardy AC, Bartels D, Bekel T, Clausen J. GenDB – An open source genome annotation system for prokaryotic genomes. *Nucleic Acids Res.* 2003; *31*(*8*): 2187–2195.

Spencer-Smith R, Roberts S, Gurung N, Snyder LAS. DNA uptake sequences in *Neisseria gonorrhoeae* as intrinsic transcriptional terminators and markers of horizontal gene transfer. *Microb Genomics.* 2016; *2*(*8*): e000069.

Genomes can be aligned, just like genes can be aligned

Alikhan NF, Petty NK, Ben Zakour NL, Beatson SA. BLAST Ring Image Generator (BRIG): Simple prokaryote genome comparisons. *BMC Genomics.* 2011; *12*: 402.

Carver TJ, Rutherford KM, Berriman M, Rajandream MA, Barrell BG, Parkhill J. ACT: The Artemis comparison tool. *Bioinformatics.* 2005; *21*(*16*): 3422–3423.

Mauve genome alignments make stunning figures, as well as being a useful research tool

Charles C, Conde C, Vorimore F, Cochard T, Michelet L, Boschiroli ML, Biet F. Features of *Mycobacterium bovis* complete genomes belonging to 5 different lineages. *Microorganisms*. 2023; *11*(*1*): 177.

Cho H, Song ES, Heu S, Baek J, Lee YK, Lee S, Lee SW, Park DS, Lee TH, Kim JG, Hwang I. Prediction of host-specific genes by pan-genome analyses of the Korean *Ralstonia solanacearum* species complex. *Front Microbiol*. 2019; *10*: 506.

Darling AE, Mau B, Perna NT. progressiveMauve: Multiple genome alignment with gene gain, loss and rearrangement. *PLoS One*. 2010; *5*(*6*): e11147.

Lee JK, Seong MW, Shin D, Kim JI, Han MS, Yeon Y, Cho SI, Park SS, Choi EH. Comparative genomics of *Mycoplasma pneumoniae* isolated from children with pneumonia: South Korea, 2010–2016. *BMC Genomics*. 2019; *20*(*1*): 910.

There is value in typing data, even in the genomics age

Falush D, Wirth T, Linz B, Pritchard JK, Stephens M, Kidd M. Traces of human migrations in *Helicobacter pylori* populations. *Science*. 2003; *299*(*5612*): 1582–1585.

Matsunari O, Shiota S, Suzuki R, Watada M, Kinjo N, Murakami K, Fujioka T, Kinjo F, Yamaoka Y. Association between *Helicobacter pylori* virulence factors and gastroduodenal diseases in Okinawa, Japan. *J Clin Microbiol*. 2012; *50*(*3*): 876–883.

Galaxy provides a full analysis suite for biological data

Afgan E, Baker D, Batut B, van den Beek M, Bouvier D, Čech M. The Galaxy platform for accessible, reproducible and collaborative biomedical analyses: 2018 update. *Nucleic Acids Res*. 2018; *46*(*W1*): W537–W544.

Chapter

18

Laboratory Techniques

In previous chapters, the theory of laboratory techniques has been described. In this chapter, the application of those theories will be explored in detail, with discussion of the specifics of the techniques including tips and recommendations for protocols and experimental design, as well as troubleshooting and alternative strategies.[1,2]

The study of bacterial genetics and genomics fundamentally focuses on DNA, therefore starting with lysis of bacterial cells for DNA extraction

To extract DNA, several laboratory methods have been developed and improved upon. For most extraction methods, bacterial cells are suspended in lysis buffer. This is usually made up of TE, which is 10 mM Tris-HCl (tris hydrochloride) and 1 mM EDTA (ethylenediaminetetraacetic acid) to which is added a detergent, such as SDS (sodium dodecyl sulfate) at a concentration of 0.5%. Alternatively, 1% Triton X-100 can be used as the detergent. Detergent agents not only aid in disruption of the cell membranes, but they also remove membrane proteins and lipids from the DNA extraction.

For Gram-negative bacterial cells, add 200 μg/mL proteinase K to the lysis buffer and incubate at 56 °C for 30 minutes to 1 hour. For Gram-positive bacterial cells, lysis is more difficult, largely due to the thickness of the peptidoglycan layer. The lysis of acid-fast bacteria can also be tricky. Addition of lysozyme can be beneficial in these cases, however lysozyme is less effective against staphylococci and many *Bacillus* strains. Lysozyme is especially effective when combined with EDTA, which is present already in TE (Tris-HCl and EDTA) and is effective at disrupting bacterial cell walls, provided the enzyme has access to the peptidoglycan layer, which it does not in Gram-negatives unless the outer

---

[1] This chapter is not a laboratory manual; however, several laboratory techniques are described in sufficient detail that they could be translated into laboratory use with minimal additional research. Such additional research must include completion of all locally required safety documentation, full awareness of the hazards involved in all experiments, and use of the appropriate personal protective equipment by all persons affected. Experiments should only be performed in an appropriate laboratory environment. The exception to this location restriction is the at-home activity described at the end of the chapter for extraction of DNA in the kitchen using common household items.

[2] The next-generation sequencing protocols are specific to their sequencing machine platforms and are being continuously improved upon; they are not discussed in this chapter. Rather, there are overviews of these in Part IV of this book. Additionally, many manufacturers have specific kits for some of the techniques described here in this chapter that may have enhanced features or success rates. Familiarity with the techniques here will provide a basis for understanding the scientific basis of such kits and therefore aid in optimization or troubleshooting of such kits. Whether using a long-standing laboratory protocol or one from a purchased kit, always read all the way through a protocol before starting an experiment. Be sure you understand the experiment completely before beginning the experimental work and if not, that you research the steps that you do not understand. Knowing the purpose of all of the steps of a protocol will help with troubleshooting should you obtain results that are unexpected. Being familiar with a protocol before starting it will enable you to prepare all of the solutions, reagents, equipment, and materials that will be needed for the experiment, including setting temperatures for incubators or water baths or having bacterial cultures ready for the next experiment.

DOI: 10.1201/9781003380436-24

membrane is otherwise compromised. Lysostaphin is effective at aiding in the lysis of staphylococci, but is only effective against staphylococci.

Often physical methods of breaking open Gram-positive cells work best, such as beating the cells in a tube with small sterile beads or **sonication**. Sonication applies sound to agitate the sample, thereby breaking open the bacterial cells and releasing the DNA. To be precise, this technique uses ultrasonic waves delivered via an ultrasonicating bath or ultrasonic probe. Spores can also be difficult to break open to be able to extract the DNA. Like Gram-positive bacteria, physical methods are generally used to get the spores open.

DNA extraction using phenol produces very pure, large quantities of DNA

When a large quantity of very pure DNA is required, an old and established method of extraction using phenol is employed, although this carries with it hazards. Phenol is acutely toxic and causes chemical burns, while at the same time numbing the skin so that exposure is not immediately detected. It is imperative that material safety data sheets are read for this reagent before use, the risks are assessed, and gloves are changed frequently; phenol goes through nitrile gloves rapidly, therefore these should be avoided for phenol work.

There are phenol DNA extraction methods, and RNA extraction methods discussed later in this chapter, that have eliminated the lysis step by shaking the cells in phenol either on its own or with glass beads. Either way, the cells need to be broken open for the DNA to be released from the bacterial cells. The extraction method starts with the DNA in an aqueous solution.

Phenol/chloroform/isoamyl alcohol (25:24:1) pH 7.8–8.2 is used for DNA extraction. The pH is important. If acidic phenol is used, the DNA will not be recovered; the DNA will be in the phenolic layer and the RNA alone will be in the aqueous layer. Acidic phenol is used for RNA extraction, which will be covered in a later section of this chapter. Check if the supplied bottle of phenol/chloroform/ isoamyl alcohol has a layer of water or buffer on the top. If so, do not remove this; pipette from below the water level to use the phenol/chloroform/isoamyl alcohol. Add an equal volume of phenol/chloroform/isoamyl alcohol to bacterial cells suspended in broth or PBS. Mix until a suspension is formed. Some protocols suggest vortexing this mixture for 1 minute, while others recommend gentle agitation for 10 minutes. The latter results in more intact DNA, whereas the former will lyse more cells. The goal of this step is to create an emulsion.

Phase separation and DNA precipitation in a phenol DNA extraction result in isolated DNA

Centrifuge the emulsion at the maximum speed allowable for the centrifuge and tubes for at least 1 minute to separate the aqueous and phenolic phases (**Figure 18.1**). This may require up to 10 minutes of centrifugation. Generally, the upper layer will be the aqueous phase and thus contain the DNA. However, if the sample is high in salt or otherwise alters the properties of the emulsion, the aqueous phase can be at the bottom. Using a colored phenol available from many suppliers can aid in the identification of the phenolic and aqueous phases. At this stage, the aqueous phase needs to be transferred to a new tube, however it is important to be careful not to disturb the interphase, the layer between the aqueous and phenolic phases (see Figure 18.1). It is also important not to disturb the phenol layer. The interphase contains proteins that will contaminate the DNA extraction, including potentially DNases that can degrade the DNA and other proteins that may inhibit downstream experiments. Phenol contamination of the DNA from the phenol phase at the bottom of the tube may also inhibit downstream experiments as well as being a hazard.

Transfer the aqueous phase to a new tube by pipetting with a 100–1,000 μL volume micropipette. Safely dispose of the phenolic phase, tubes, and tips in chemical waste according to hazardous waste protocols.

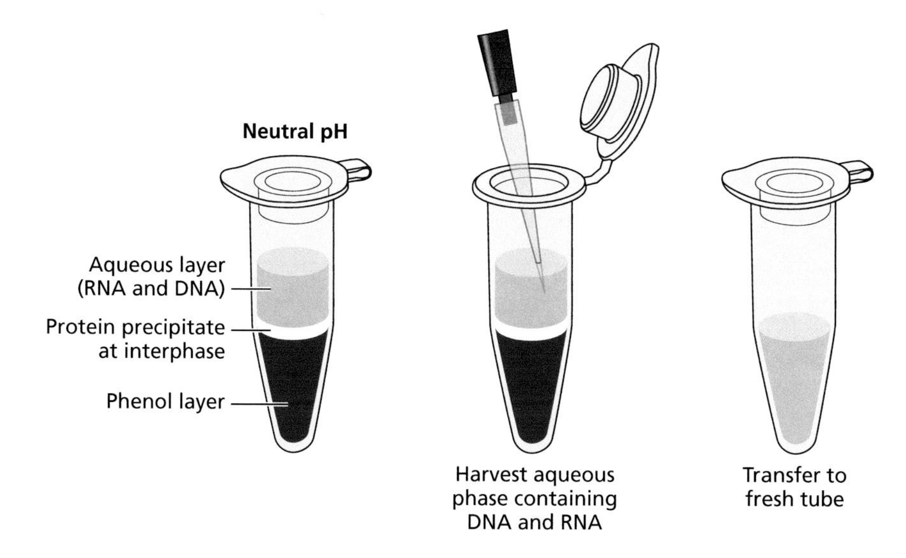

Figure 18.1 Phenol extraction phases. Following centrifugation, the phases separate into a top aqueous phase containing DNA and RNA, a bottom phase that is the phenol, and a middle interphase layer of protein. The top layer is transferred via pipette to a fresh tube, taking care not to disturb the interphase or phenol layers.

To precipitate the DNA out of solution, add 3 M NH_4OAc at one-tenth of the volume of recovered aqueous phase, pipetted under the liquid level of the aqueous phase in its new tube, and mix (**Figure 18.2**). Next, add 2.5 volumes of 100% EtOH to the aqueous phase and mix (**Text Box 18.1**). At this stage, it should be possible to start to see the DNA floating in the mixture, appearing as a wisp that may have bubbles attached to it (see Figure 18.2). Extraction can continue from here or the sample can be chilled, either at −80 °C for 1 hour or −20 °C overnight, which is believed to increase yield. To pellet the DNA to the bottom of the tube, centrifuge at a safe maximum speed for 2–5 minutes, ideally at 4 °C (see Figure 18.2). Carefully remove the ethanol supernatant from the DNA pellet. Wash the pellet with 70% EtOH by adding 200 µL to the pellet, repeating the centrifugation, and again carefully removing the ethanol supernatant from the DNA pellet. For increased purity, wash again with 70% EtOH.

Text Box 18.1

If you have 100 µL of aqueous phase, first add 10 µL of 3 M NH_4OAc and then 250 µL 100% EtOH.

The DNA pellet must be air dried, allowing the ethanol to evaporate. Leave the tube open on the bench in a clean location until all droplets are gone. To speed up this process, be sure all of the 70% EtOH liquid is removed carefully with a small pipette tip. DNA that is pure will be hard to see as a pellet. DNA that is contaminated with protein will appear as a white pellet, whereas pure DNA will be more transparent and can, at times, have a yellow tinge. Once the pellet is completely dry, suspend the DNA pellet in 10 mM Tris-EDTA pH 8.5 or ultra-pure water. Store the DNA at 4 °C short term or at −20 °C long term in aliquots to avoid

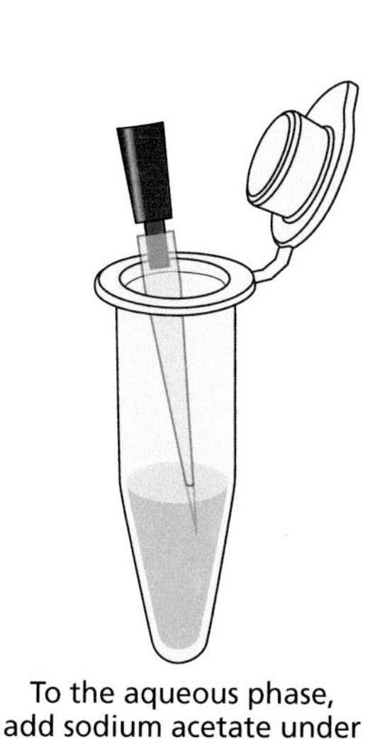

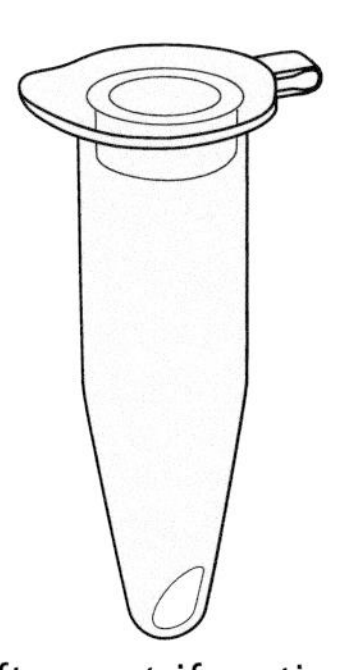

Figure 18.2 DNA precipitation. To the aqueous phase, add sodium acetate under the surface of the liquid. Then, add ethanol to precipitate the DNA. After centrifugation, a pellet will be evident at the bottom of the tube.

repeated freeze-thaw. Freezing and thawing nucleic acids breaks the backbone, particularly of genomic DNA, resulting in fragmented DNA, therefore repeated freeze-thaws should be avoided.

Additional considerations for phenol DNA extraction can improve the outcome

Some protocols recommend repeating the phenol extraction two to three times. This involves taking the aqueous phase off to a new tube and then adding an equal volume of fresh phenol/chloroform/isoamyl alcohol to it. This is again mixed, centrifuged, and the aqueous phase removed. The purpose of repeating these steps is to reduce protein contamination from the interphase layer, which should visibly reduce each time. If there are issues with protein contamination, this is recommended, but generally the additional steps are not needed, particularly if care and attention are taken when removing the aqueous phase.

Some protocols also recommend extracting the aqueous phase one to two times with equal volumes of chloroform/isoamyl alcohol. This reduces phenol contamination in the final DNA preparation, which can interfere with downstream experiments. If phenol contamination is a problem, this is recommended, but if the aqueous phase is carefully removed, phenol contamination can usually be avoided. At the point of taking the aqueous phase into a new tube, do not try to take too much of it. If as much as possible of the aqueous phase is taken in an effort to increase DNA yield, it is more likely that there will be both phenol and protein contamination of the DNA. In an attempt to get a better result, a bigger yield of DNA, the extraction will actually be compromised.

Ultra-pure water (18 MΩ) is best for all applications, but it is not buffering, so DNA doesn't necessarily last as long in ultra-pure water as it might in TE (Tris-EDTA). The inhibitory effects of EDTA on some downstream experiments can be avoided by taking a middle ground and using 10 mM Tris pH 8.5 as a DNA storage buffer, with no EDTA. DNA is more stable at the slightly basic pH of 8.5, thus TE and 10 mM Tris are made to pH 8.5. Pellets will suspend more readily in Tris solutions than in water. Prior to suspension of the pellet in water, TE, or 10 mM Tris, the pellet should be completely dry. Ethanol contamination of DNA cannot be detected by a spectrophotometer, however the presence of ethanol in the DNA extraction can impact downstream experiments. For example, the sample may float up from wells of an agarose gel, even with sufficient loading dye, or the sample may not freeze at –20 °C.

Text Box 18.2

If 70 mL is measured into a graduated cylinder and then the graduated cylinder is filled to the 100 mL mark with distilled water, this will not make 70% ethanol; the resulting solution is less than 70% and closer to 67%–68% ethanol. Instead, measure 70 mL of ethanol, then separately measure 30 mL of distilled water. When the 70 mL of ethanol and 30 mL of water are combined they will make 70% ethanol.

It should be noted that when mixing 70% ethanol it is not appropriate to measure both the ethanol and the water volumes in the same graduated cylinder at the same time (**Text Box 18.2**). If the ethanol being used is also old and may have absorbed water from the atmosphere, the starting 70 mL will not have been 100% ethanol so the resulting solution will be even farther off 70% than expected. The correct procedure is to use the graduated cylinder to measure 70 mL of ethanol, pour this into a bottle, then use the graduated cylinder to measure 30 mL of distilled water and pour this into a bottle (see Text Box 18.2). Having done this, now measure the total volume of the two. It will be less than 100 mL, but the solution made will accurately be 70% ethanol. This is because ethanol and water dissolve in each other, so the volumes cannot be accurately measured together in the way either two aqueous solutions or two alcohol solutions could be. If in doubt about the age of the ethanol and if it has taken up atmospheric water and is possibly not near 100% (or 99.8% or 99.5% as it was when purchased), make 75% or 80% ethanol to use in the later stage of the DNA extraction; this will wash off residual salts from DNA too and not risk losing the DNA in the aqueous portion.

Most DNA extractions use columns

Most DNA extractions make use of column matrix technology, whether they are for chromosomal DNA or plasmid DNA. There are also kits for RNA extraction, covered later in the chapter, and those for cleaning up DNA-based reactions.

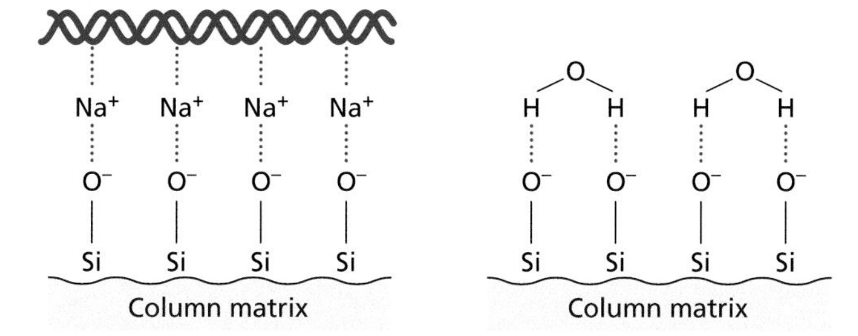

Figure 18.3 Extraction columns bind DNA using a silica resin matrix. Within spin columns used for DNA extraction, there is a filter containing a matrix of silica resin (Si), which binds to nucleic acids (left) at the optimal pH and salt conditions. The DNA is released from the column by elution with an aqueous-based solution (right), which can be water or the solution provided with the kit.

Spin columns or vacuum columns are available in commercial kits and most contain a silica resin that binds nucleic acids (**Figure 18.3**). Depending on the concentrations of salts and other components in the buffers that are included in the extraction kits, the extraction is specific for genomic DNA, plasmid DNA, or RNA.

In extraction column protocols the manufacturer's instructions need to be followed, but in general the principles are similar. Bacterial cells are mixed with kit lysis buffers, which generally contain salts such as guanidine thiocyanate or urea. These types of salts disrupt proteins that might be bound to the DNA and disrupt the interaction of nucleic acids with water, encouraging the binding to silica. Lysis buffers in the kits also contain detergents, such as sodium dodecyl sulfate (SDS), to solubilize proteins and lyse the bacterial cells, and EDTA to inhibit DNases present in the bacterial cell, similarly to the phenol extraction. Also, as with the phenol DNA extraction, addition of enzymes such as proteinase K (Gram-negative) or lysozyme (Gram-positive) may help. Once lysis is achieved, a binding buffer is added to adjust the buffer conditions from lysis to conditions that will permit the DNA to bind to the silica within the column.

The mixture is added to the extraction column and centrifuged. This forces the liquid through the column and provides the opportunity for the DNA to be bound by the silica in the column (**Figure 18.4**). The flow-through liquid mixture of lysis buffer and binding buffer is disposed of and the DNA on the column is then washed with a kit-provided wash solution (**Text Box 18.3**). Generally, the column is washed twice to remove residual salts from the first stage lysis and to clean the DNA that is bound to the silica column of protein. As for the DNA binding, the wash is applied to the top of the spin column, which is centrifuged to force the wash through the column, and the flow-through of the wash solution

Text Box 18.3

The wash solution is generally supplied without the needed essential ingredient, ethanol, which is added by the user. The wash solution is not 100% alcohol because this would result in an overabundance of small fragments and degraded DNA.

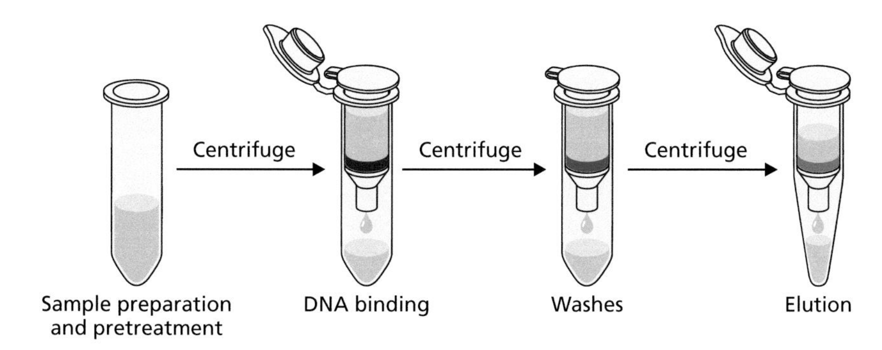

Figure 18.4 Steps of using an extraction column. To extract DNA with an extraction column, the sample is prepared for going onto the column using the supplied buffers. This mixture is put onto the column and centrifuged, so that it flows through the extraction column filter and the DNA binds to the column. Alternatively, a vacuum pump is used to draw the liquid through the column. The DNA on the column is washed with a supplied wash solution, before the DNA is eluted off the column, ready to use.

is disposed of before proceeding (see Figure 18.4). Following one to two washes, the spin column is centrifuged again to remove any residual wash solution. This ensures a cleaner elution of the DNA off the column, as residual ethanol on the column interferes with elution. Finally, 10 mM Tris pH 8.5, TE, a supplied elution buffer, or pure water is added to the column, causing the DNA to dissociate from the silica and enter the aqueous solution. Upon centrifugation, the flow-through now contains the DNA and should be retained (see Figure 18.4).

Troubleshooting DNA extractions can increase yield and quality of the DNA

If the yield of DNA from either a phenol or column extraction is low, this could be due to incomplete lysis of cells. It is important to research the options for breaking open the bacterial cells that are being used in your experiment, remembering the thick peptidoglycan of Gram-positive species and the resilient structure of spores can be problematic.

Use fresh high-quality EtOH when making up the wash solution in column kits and for washes in phenol/chloroform/isoamyl alcohol extractions. Poor-quality EtOH and old bottles that have been open may take up water from the atmosphere, reducing them from nearly 100%. An almost empty bottle is likely to be quite a bit less than 100%; it might be best to make 75% or 80% EtOH rather than 70% in this context or to purchase a fresh bottle for DNA extraction.

Protein contamination can have a negative impact upon the results of downstream experiments. Following the DNA extraction, the A_{280} reading on a spectrophotometer should be assessed and compared to the A_{260} reading. Absorbance at 280 nm on a spectrophotometer corresponds to proteins in aqueous solution, while DNA and RNA absorb at 260 nm. If the A_{280} reading is high compared to the A_{260}, such that the $A_{260/280}$ ratio is less than 1.7, this indicates that the DNA extraction is contaminated with protein. Protein contamination of a DNA extraction can be due to an excess of starting material, where too large a mass of bacterial cells are used at the start. Increasing yield is not simply a matter of starting with more sample, just as it is not possible to roast a chicken in a quarter of the time at four times the temperature. Each protocol has a sample limit. Exceed this limit and protein contamination occurs.

Proteins are not the only source of contamination. A poor $A_{260/230}$ ratio on the spectrophotometer is indicative of phenol contamination or salt contamination. Take care in removing the aqueous phase in phenol extractions and use adequate washes with both protocols.

Lastly, a quantity of DNA may be obtained, but it could be very fragmented and therefore unsuitable for the desired experiment. Degradation of DNA is generally due to the treatment of the sample at the lysis stage. Gentle agitation with rocking or rotary mixing is recommended rather than vortexing if samples are degraded. This can be important if the DNA is being extracted for long-range PCR, covered later in this chapter, or long-range sequencing.

A quick (and dirty) DNA extraction can be achieved by boiling

A simplified method of DNA extraction involves boiling the bacterial sample and can be effective for a quick, albeit generally highly protein-contaminated, extraction for both Gram-negative and Gram-positive bacteria.

Grow a culture of bacteria to 10^6 cfu/mL and collect a pellet of the bacterial cells by centrifugation, which are then resuspended in 200 μL TE. The TE bacterial suspension is boiled for 15 minutes, with the cap of the microfuge tube being secured with a locking clip or as a screw cap, so it does not pop open during the boiling process. This could be a safety hazard and should not be attempted with bacterial pathogens that may be an aerosolization risk; risk assessments for all experiments are important. Following the 15 minutes of boiling, the tube is centrifuged at maximum speed for 5 minutes and the DNA-containing supernatant is removed to a new tube. This method is akin to the method used for colony PCR (see page 325) and will not produce DNA that is very clean or pure, but it will rapidly release DNA from the bacterial cells and separate them from the cellular material by centrifugation, and the resulting DNA may be viable for some applications.

The first recombinant DNA experiments in the 1970s were made possible because of restriction enzymes, which are still used today

Once DNA is extracted from the bacterial cell, it can be used for a variety of molecular biology applications, including generation of recombinant DNA. As discussed in Chapter 10, recombinant DNA is DNA that originated from two different biological sources, such as putting a gene from one bacterial species into a plasmid from another species. This is most commonly done for the purposes of genetic manipulation of the gene sequence. The first recombinant DNA to be put into a biological organism was in 1973 by Herbert Boyer and Stanley Cohen. This work was preceded by Paul Berg and Janet Mertz's *in vitro* joining of DNA in 1971; they did not transform this into living cells because at the time there were concerns about whether this was safe. Similar methods to these are still used today to cut pieces of DNA with restriction enzymes so they have compatible ends and then join them with ligase to make DNA from two different original sources.

Restriction digestion uses restriction endonucleases (see Chapter 10), enzymes that are commercially provided in a glycerol solution. Due to the presence of the glycerol, the tube with the enzyme will not freeze, therefore keep the restriction enzyme in the freezer until you need to add it to the reaction and then put it back in the freezer right away. Strict discipline with enzyme storage and use increases the life of the enzyme, sometimes years after the manufacturer's stated expiration date.

The buffer used for the restriction digestion reaction is frozen and it is important that this is defrosted completely and mixed before use. If not, the ratio of components in the buffer that goes into your reaction will be wrong and the ratio of the components in the tube of buffer stock gets worse as more of the tube is used. Over time, with repeated improper defrosting of the restriction enzyme buffer, it is possible for the laboratory to end up with buffer that is the wrong composition, which means the restriction enzyme does not work and the digestions for the entire research group fail. This is more often due to poor care with the reaction buffer than expiration of the activity of the restriction enzyme itself.

Set up a restriction digestion with the optimal reaction conditions

Most restriction digestions are straightforward, but there are a few things to keep in mind when designing an experiment. Some variants, if forgotten, can result in a failed digest. For example, check if additives like bovine serum albumin (BSA) are needed for the enzyme or enzymes being used. This information will be available from the supplier of the restriction enzymes. Also, check the digest temperature. Most commercially available restriction enzymes work at 37 °C, so this tends to be the assumption for a restriction digestion temperature. However, some enzymes, such as *Apa*I and *Sma*I, work best at 25 °C, *Bsm*I at 65 °C, and *TspM*I at 75 °C. The temperature of the restriction enzyme reaction is dependent on the temperature that is optimal for the bacterial species from which the enzyme was isolated. It is therefore important that the assumption is not made that this is 37 °C.

If there is no digestion of the DNA, the first thing to do is to check that the DNA sequence has the expected restriction enzyme digest site using one of several online tools. Next, check the laboratory techniques. Check that the correct buffer was used for the restriction enzyme. If more than one enzyme is being used at the

Volume	Reagent
6.9 μL	dH_2O
1.0 μL	DNA (1 μg/μL)
1.0 μL	10x restriction enzyme buffer
1.0 μL	BSA
0.1 μL	Restriction enzyme (10 U/μL)
Total 10.0 μL reaction	

Figure 18.5 Restriction digestion reaction set-up. To set up a typical restriction digestion reaction, three things are needed: a restriction enzyme; the buffer for the restriction enzyme; and the DNA for the restriction enzyme to cut. Some enzymes require bovine serum albumin (BSA). The reaction needs to be in the proper final volume, made up with distilled water (dH_2O). Start with the largest volume, pipette smaller volumes into it, and add the enzyme last, as shown here in this table.

same time, check that the best buffer has been used for the double digest. That buffer might not be the best buffer for either restriction enzyme being used on its own, but will be one that allows both to function. Methylation-sensitive enzymes will either only cut or not cut depending on the methylation state of the DNA, therefore the lack of digestion could be due to the methylation state of your DNA. Check the specifics of the enzyme that is being used to see if methylation of the DNA, or lack of methylation, may alter the activity of the restriction enzyme.

To digest the DNA, combine DNA, restriction enzyme, buffer, BSA (if needed), and dH_2O (distilled water) to the desired volume. For molecular cloning, you will generally need 1 μg of DNA to have enough to work with for downstream manipulation. The reaction volume can be as small as 10 μL; this can be a cost saving on the units of restriction enzyme used, provided enough is added to fully digest the amount of DNA in the sample in the time allowed for the reaction. These parameters are available from the manufacturer of the product (**Text Box 18.4**; **Figure 18.5**).

Text Box 18.4

Generally, 1 unit of restriction endonuclease cuts 1 μg DNA in a 50 μL reaction in 1 hour. Therefore, set up a reaction with 1 μg DNA, 0.1–0.2 μL restriction enzyme (10 U/μL), 1 μL 10× buffer, 1 μL 10× BSA (where needed), and up to 10 μL volume with dH_2O.

There are a few additional considerations to remember when doing restriction digestions

Add the restriction enzyme last. In fact, add all enzymes last to reactions. These are the active components, often stored at –20 °C. Remove them just before use and put them right back at –20 °C. There is no need to defrost enzymes. Enzymes are supplied in solutions containing glycerol, so they do not freeze. This also means that they are denser than water, therefore it is important to mix enzymes such as restriction endonucleases and polymerases added to PCR (later in this chapter) into reactions. Do so gently, stirring with the pipette tip, so as not to shear the DNA and damage it.

Incubate restriction digestions at the correct temperature, usually 37 °C, for the correct time. The time for a digestion was usually 1 hour, however improvements in restriction digestion technology in 2006 reduced the time of the same reaction to 5 minutes for some restriction enzymes. Purchasing fast enzymes can be a time saving on protocols, enabling more research to be done in a day. Note that if the restriction enzyme being used has a warning about **star activity**, do not leave the reaction longer than needed to achieve digestion and do not use more enzyme than is needed to achieve digestion; leaving the reaction longer or using too much enzyme will make non-specific cuts.

Some enzymes can be **heat inactivated**, which stops the enzyme cutting DNA. This can be useful if, for example, the DNA needs to be digested with two different enzymes but there is no compatible buffer, so the two digestions must happen sequentially rather than together. Whether a specific enzyme can be heat inactivated and at what temperature is indicated by the supplier.

Restriction digestions are used to change DNA sequences and join sequences together

When joining together DNA sequences, such as when generating recombinant DNA, look for enzymes with **compatible cohesive ends**, if there isn't an enzyme that cuts both of the bits of DNA to be combined. For example, when wanting to join two pieces of DNA, it would be simple if both of them could be cut by the *Eco*RI restriction enzyme, since the sticky ends of the restriction fragments would hybridize together, facilitating ligation of the fragments. However, if there is no

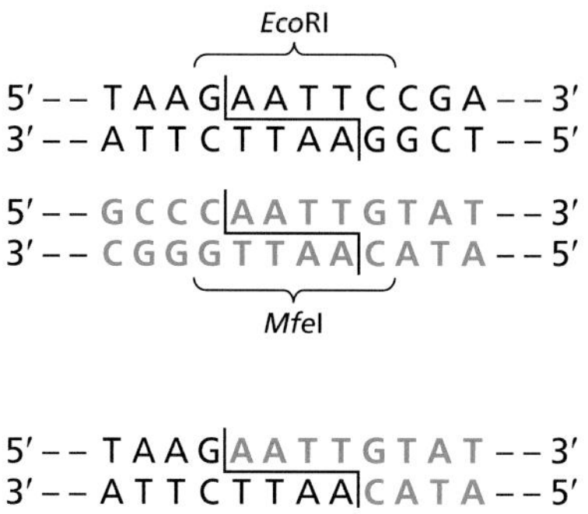

Figure 18.6 Compatible cohesive ends. The sequence shown on the top is cut by the restriction enzyme *Eco*RI, with the recognition sequence GAATTC. This cuts between the G and the A (staggered line), leaving a TTAA 5′ overhang, shown on the bottom strand. The sequence shown in the middle (in blue) is cut by the restriction enzyme *Mfe*I, with the recognition sequence CAATTG, which is slightly different. This cuts between the C and the A (staggered line), also leaving a TTAA 5′ overhang, shown on the bottom strand. Because both enzymes leave TTAA 5′ overhangs, they have compatible cohesive ends, which means that the fragments generated can be joined to one another, as shown in the bottom sequence. Here, the left part comes from the top sequence and the right (in blue) from the middle.

*Eco*RI

5′-- T A A G|A A T T C C G A --3′
3′-- A T T C T T A A|G G C T --5′

5′-- G C C C|A A T T G T A T --3′
3′-- C G G G T T A A|C A T A --5′

*Mfe*I

5′-- T A A G|A A T T G T A T --3′
3′-- A T T C T T A A|C A T A --5′

one enzyme that cuts both pieces of DNA, it may be possible to find two different enzymes that generate the same sticky ends. For example, *Eco*RI and *Mfe*I both generate TTAA 5′ overhangs (**Figure 18.6**). Once cut, the overhanging ends of the compatible cohesive ends will match (see Figure 18.6).

To prevent a plasmid from sticking back together spontaneously via hydrogen bonding between the bases after being cut with a single enzyme or enzymes with compatible ends, treat the plasmid with phosphatase. The phosphatase enzyme removes the terminal phosphate, preventing rejoining of the DNA by ligase enzymes (**Figure 18.7**). Alkaline phosphatase removes the 5′ phosphate groups from the plasmid vector so it will not ligate back together. In order for ligation to happen, the ligase enzyme needs to use the insert, the piece of DNA to be cloned into the plasmid, which has the 5′ phosphates. Alkaline phosphatase can be added to the restriction enzyme reaction when the plasmid vector is digested (see Figure 18.5); the water is reduced to allow for the added volume of the alkaline phosphatase enzyme. Before proceeding to the next stage in cloning, the DNA fragments need to be purified away from the alkaline phosphatase, restriction enzyme(s), and buffers. This can be achieved using a variety of commercially available reaction clean-up kits.

If the cloning strategy means that there is a need to make blunt ends, but the available restriction digest recognition sites are sticky ends, it is possible to fill in the overhangs with the Klenow enzyme. Both 3′ overhang removal and fill in of 5′ overhangs are accomplished using the DNA polymerase I (Klenow) fragment (**Text Box 18.5**).

Text Box 18.5

To remove 3′ overhangs and fill in 5′ overhangs, add 1 unit Klenow per µg DNA with restriction enzyme buffer or ligation buffer and 33 mM of each **deoxyribonucleoside triphosphate** (dNTP) for 15 minutes at 25 °C. To stop the reaction, add EDTA to a final 10 mM concentration and heat to 75 °C for 20 minutes.

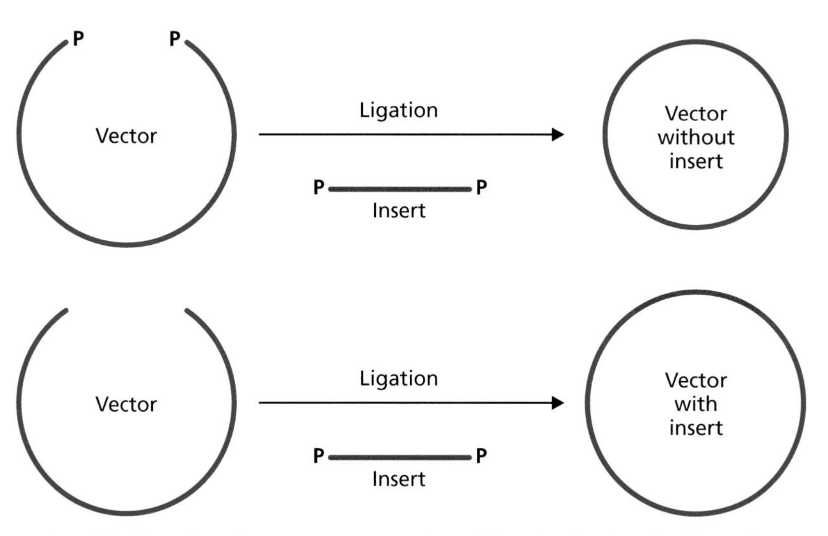

Figure 18.7 Alkaline phosphatase to prevent plasmid ligating back on itself. In a ligation of a vector and insert, the plasmid vector is most likely to ligate back together without the insert, if the ends of the vector are compatible (top panel). Alkaline phosphatase removes the 5′ phosphate from DNA (P). This phosphate is required for ligation. Using alkaline phosphatase on the plasmid vector prevents the plasmid from ligating back together. The plasmid will only circularize with the insert (red), which still has phosphate ends (bottom panel).

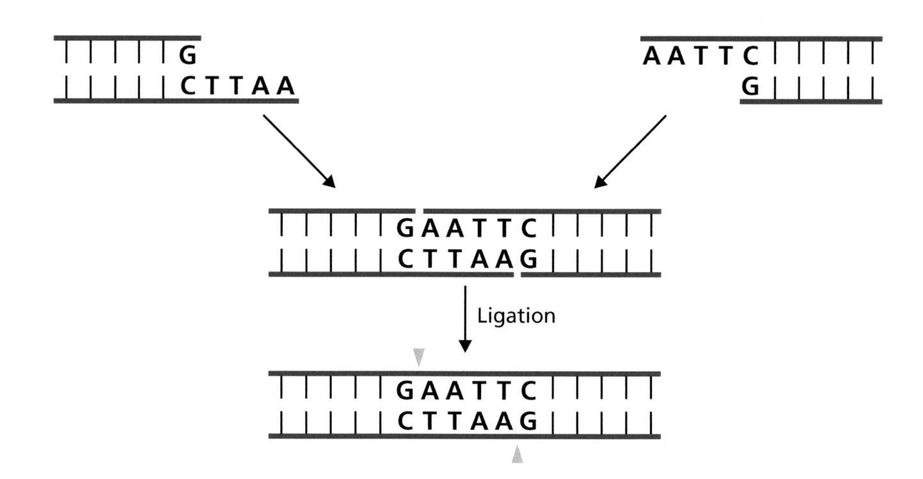

Figure 18.8 Ligation closes the backbone of DNA. Breaks in the backbone of DNA are joined by the ligase enzyme. For example, when two pieces of DNA cut by restriction enzymes come together with sticky ends, hydrogen bonding between the bases aligns the strands and keeps them together, but the backbone of the DNA is discontinuous. Ligase is able to join the 5′-phosphate and the 3′-hydroxyl groups of adjacent nucleotides, completing the helix.

Cut ends of DNA need to be ligated together to complete cloning

To generate recombinant plasmids commonly used when cloning genes or sequences that will be manipulated or transformed into bacteria, it is necessary to ligate the insert into the plasmid. Ligation covalently connects the DNA backbone of the restriction enzyme digested DNA fragments. T4 DNA ligase joins the 5′-phosphate and the 3′-hydroxyl groups of adjacent nucleotides (**Figure 18.8**).

The ratio of insert to vector in a ligation reaction is important and an online calculator that takes account of the length of the fragments to be ligated is recommended. Otherwise, if the insert is smaller than the vector, a molar ratio of 3 insert to 1 vector is usually adequate. In total, approximately 100 ng of DNA should be in the ligation reaction (**Figure 18.9**). The ligation reaction should be set up on ice, adding the ligase enzyme last (**Text Box 18.6**). Chill the tube on ice and transform 1–5 μL of the mixture directly into 50 μL of competent cells.

The optimal temperature for T4 DNA ligase enzyme activity is 25 °C, however this is not necessarily the best temperature for annealing the ends of the DNA fragments. Sticky end ligations have overhanging bases that can hydrogen bond and therefore efficiently ligate at room temperature, at about 22 °C in 3 hours or 4–8 °C overnight. Blunt end ligations are less efficient and generally are done at 15–20 °C for 4–18 hours. If you are unsure how long to leave a ligation or at what temperature, it is possible to transform the ligation into competent bacterial cells after a few hours with 1 μL of the ligation reaction and then continue with the rest of the reaction overnight, if desired. There are suggestions that ligations done on melting ice can be very successful. On ice, the cold temperature permits the hybridization of the DNA fragments together. As the ice melts, the temperature rises to what is more optimal for the ligase, yet the temperature rises slowly so the hybridization is not lost.

Text Box 18.6

Generally, a 10 μL ligation reaction is set up with 1 μL of 10× ligase buffer, 25 ng vector DNA, 75 ng insert DNA, enough dH_2O to take the final volume to 10 μL, and finally 1 μL DNA ligase (see Figure 18.9). Gently mix by stirring the reaction with the pipette tip. To ligate cohesive ends, incubate the ligation reaction overnight at 16 °C or at room temperature for 10 minutes. To ligate blunt ends, or single-base overhangs, incubate at 16 °C overnight or at room temperature for 2 hours. Alternatively, a thermocycler can be set to alternate between 30 seconds at 10 °C and 30 seconds at 30 °C. The ligase enzyme can be heat inactivated at 65 °C for 10 minutes.

Volume	Reagent
6 μL	dH_2O
1 μL	10x ligase buffer
1 μL	Vector DNA (25 ng/1 μL)
1 μL	Insert DNA (75 ng/1 μL)
1 μL	DNA ligase
Total 10.0 μL reaction	

Figure 18.9 Ligation reaction set-up. To set up a typical ligation reaction, four things are needed: ligase; the buffer for the ligase; and the two pieces of DNA for the ligase to stick together. The reaction needs to be in the proper final volume, made up with distilled water (dH_2O). Start with the largest volume, pipette smaller volumes into it, and add the enzyme last, as shown here in this table.

Important considerations when performing ligations

Ligase buffer contains ATP and is stored frozen at –20 °C. As with restriction endonuclease buffer, the ligase buffer must be thawed completely to preserve the correct ratios of the components of the buffer. It must not be used when partially thawed. The ATP in this buffer is an additional complication because it degrades with each freeze–thaw cycle. It is therefore recommended that the ligase buffer is completely defrosted when it arrives from the supplier and carefully aliquoted into smaller volumes of 5–10 μL. Since buffers are usually provided from suppliers in excess, a new aliquot tube can generally be used each time, preserving the activity of the ATP. The creation and use of aliquots also minimize the degradation of DTT (dithiothreitol), which is present in the ligation buffer. DTT can precipitate upon freezing. Vortex the thawed ligation buffer for 1–2 minutes to resuspend the DTT prior to use.

It is very easy to neglect doing controls for things like ligations, but when the ligation does not work, a control can help you to work out why. In this case, it is important to include alongside the ligation reaction a second tube that has only a vector, with no insert DNA. This will both confirm that the vector has completely digested, and that the phosphatase treatment has worked (when needed). If this control generates ligated circular DNA and any colonies on transformation, it means the vector was either not fully cut or the phosphatase did not fully work.

PEG, polyethylene glycol, is sometimes used in ligation reactions as an agent that promotes "molecular crowding," which means it brings the fragments together and so is particularly good for blunt end ligations. It is best, if this is done, to use PEG that is pure, and to adjust the molar ratio and check the resulting ligations because **concatemerizations** can occur. This is when two or more inserts ligate together, generating an unexpected construct. Some protocols that recommend PEG will use PEG 8000 at 10% final reaction concentration in a ligation of 3–6 hours. Other protocols recommend using final concentrations of 5% PEG 4000 in blunt reactions. There are commercially available products that provide more rapid ligation than standard ligase and buffer. These tend to make use of optimized ligation buffers that include PEG. One manufacturer's quick ligation reaction buffer differs from its standard buffer through the addition of a final concentration of 6% PEG 6000, for example.

Cloning of sequences is often important in bacterial genetics and genomics research

To understand the functions of bacterial sequences, often these are cloned into plasmids or other vectors for manipulation. Plasmids engineered for molecular cloning have a **multiple cloning site** (**MCS**) with many restriction enzyme recognition sites. This is sometimes also called the **polylinker**. To make use of the plasmid for cloning, it is cut at the MCS using a restriction enzyme that either also cuts either side of the sequence to be inserted or that has compatible cohesive ends with the enzyme that will be used to cut the insert. As mentioned earlier, the addition of alkaline phosphatase to the restriction digestion of the plasmid will prevent the ligation reaction from just joining the plasmid back together. Instead, ligation to produce a closed circular piece of DNA will need to incorporate the desired insert. The sequence to be inserted into the plasmid is cut with either the same restriction enzyme as used to cut the plasmid or an enzyme with compatible cohesive ends (**Figure 18.10**).

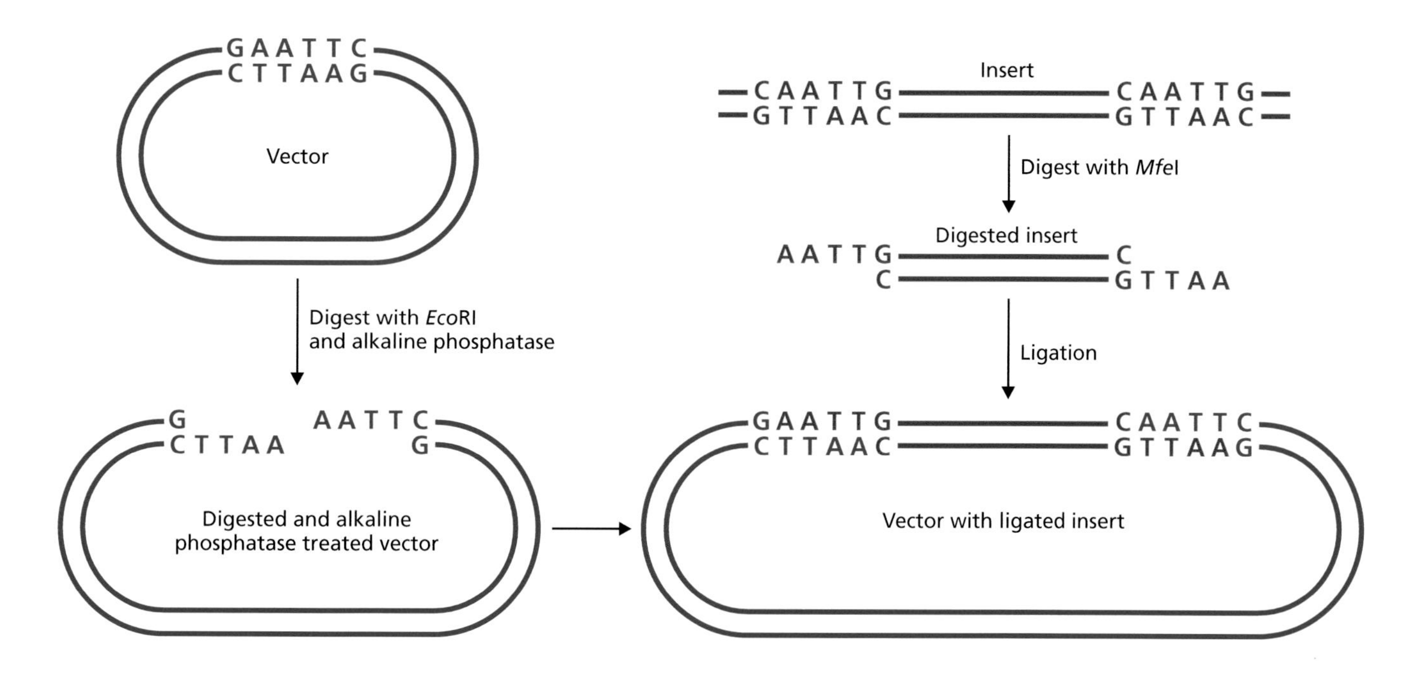

Figure 18.10 Cloning an insert into a vector using restriction enzymes. The plasmid vector is cut with a restriction enzyme, here *Eco*RI. Alkaline phosphatase is added to this reaction to prevent the plasmid ligating back upon itself. The insert is cut with *Mfe*I, which has compatible cohesive ends. The sticky ends of the insert and the vector join in the ligation reaction and the DNA backbone of the vector and insert are joined by ligase.

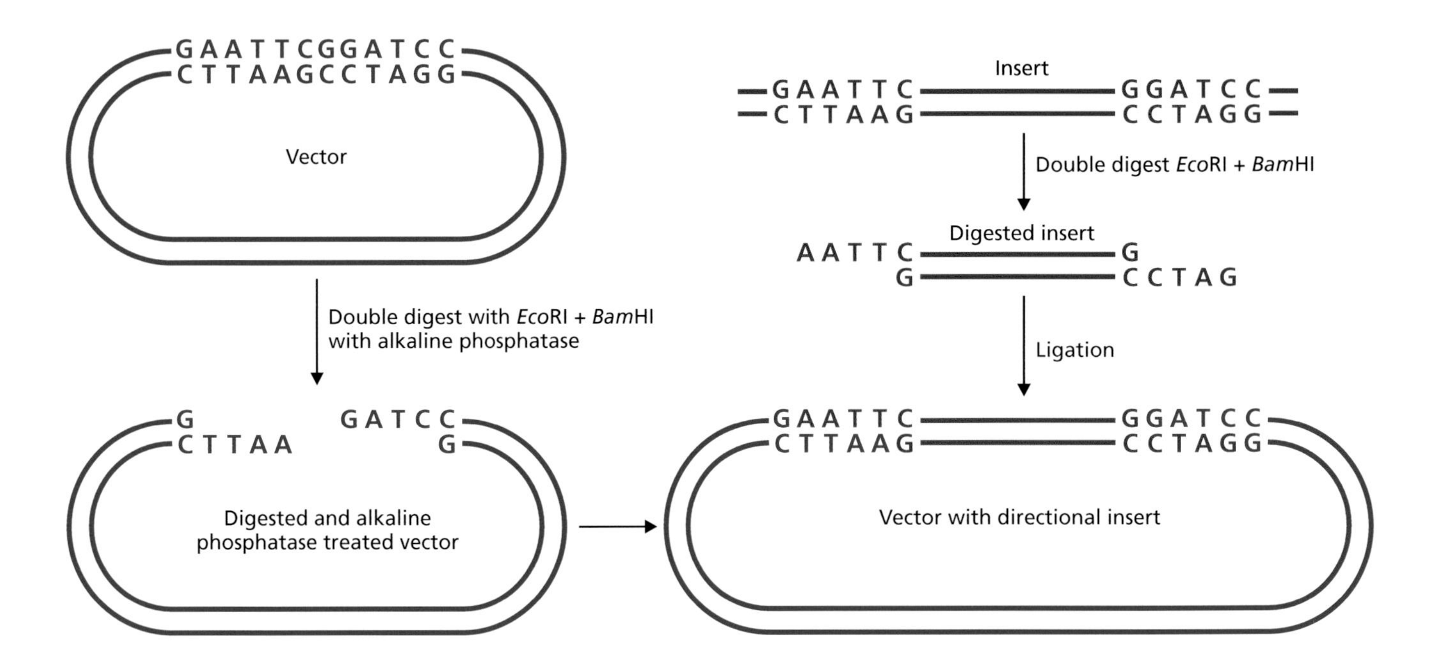

Figure 18.11 Directional cloning of an insert into a vector using restriction enzymes. The plasmid vector is cut with two restriction enzymes, here *Eco*RI and *Bam*HI. Alkaline phosphatase is added to this reaction to prevent the plasmid ligating back upon itself. The insert is also cut with *Eco*RI and *Bam*HI, which means that the sticky ends of the insert and the vector join in only one orientation. During the ligation reaction, the DNA backbone of the vector and insert are joined by ligase.

With the preceding strategy, the insert can be cloned into the plasmid vector in either orientation, which may not matter depending on the experiment. However, sometimes **directional cloning** is desirable, where the insert needs to be in the plasmid in a specific orientation. In this case, the cloning is designed so that the plasmid is cut at the MCS with two different restriction enzymes that have incompatible cohesive ends (**Figure 18.11**). This means that the insert will only go into the plasmid in one orientation, and it prevents the plasmid from being able to be ligated back together because the two ends of the plasmid will not be compatible with one another. It is, however, still a good idea to include alkaline phosphatase here, because any incomplete digestions that only cut once will be prevented from upsetting the cloning by re-ligating. The tiny piece of DNA that is removed by the cutting with the two enzymes also needs to be purified away so it does not interfere in the ligation process. This can be done using a clean-up column or by resolving the restriction digestion band on an agarose gel and isolating the DNA band from the gel. Either would generally be done with a column-based kit. The sequence to be inserted is cut either with the same two restriction enzymes or with enzymes that have compatible cohesive ends to those used for the plasmid.

Selection of the restriction enzymes to be used for the cloning can be aided by online tools. It may be necessary to use blunt cutting enzymes, but these are less efficient in cloning because the ends of the DNA will not make use of the hydrogen bonding between nucleotide bases to aid the ligation process. It is possible to engineer restriction digest sites onto the ends of the sequence to be inserted, if necessary. This can be helpful if none exists, to place restriction recognition sites where they are needed, or to make use of restriction endonucleases that are the best choice for the cloning. As in the PCR description that follows, primers can be designed with additional sequence information at the 5′ end, which then becomes part of the amplified DNA.

TA cloning exploits a feature of PCR to rapidly clone sequences

One of the most popular methods of cloning, TA cloning, uses a PCR-generated product and inserts it into a plasmid vector by virtue of T and A overhangs, thus its name. TA cloning is simpler and quicker than traditional cloning with restriction enzymes because it is not reliant on there being convenient restriction enzyme recognition sites in the sequence or engineering them by PCR. However, TA cloning is not directional, as can be achieved with careful design of a traditional cloning experiment.

Taq and other PCR enzymes lack 5′ to 3′ proofreading activity and as such add an additional adenosine triphosphate to the 3′ end of the product (**Figure 18.12**). Because of the 3′ A overhangs on PCR products, these pieces of DNA can be cloned into plasmids with complementary 3′ T overhangs (see Figure 18.12). The process can be optimized by designing the PCR product so that primers have G's at their 5′ ends; this increases the probability of addition of the A. Choice of polymerase for the PCR is also important; not all polymerases will add the A's, including most with proofreading activity, which tend to be preferred for making accurate PCR products with no introduced mistakes in the sequence. To ensure accuracy, it is possible to generate the amplicon with a proofreading polymerase and then to add some *Taq* polymerase to the end of a proofreading PCR just to have it add on the A's or to purchase a special proofreader that adds on the final A to the product.

The vector can be purchased in a TA cloning kit or created in the laboratory. To generate a TA cloning vector, the plasmid is linearized with a blunt cutting restriction enzyme and then 3′ Ts are added by terminal transferase from ddTTP (dideoxythymidine triphosphate). This is the same kind of T used in Sanger sequencing, which when incorporated will not allow another base to be added to it. In this way, only one T is added to each 3′ end of the cut plasmid (see Figure 18.12).

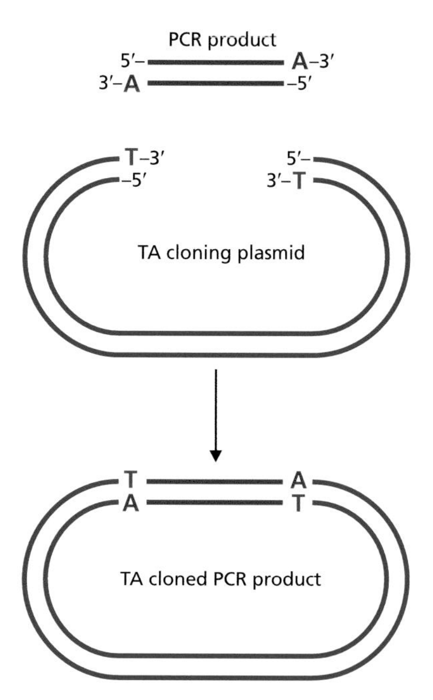

Figure 18.12 TA cloning. PCR products have 3′ A overhangs (top), therefore it is possible to clone them into plasmids that have 3′ T overhangs (middle). The A ends on the PCR product hydrogen bond to the T ends on the plasmid, circularizing the plasmid with the PCR product as an insert (bottom), resulting in a TA cloned PCR product in the plasmid vector.

Some commercially available kits augment ligation and cloning with accessory proteins and exploitation of other systems

Fragments can be cloned into vectors without DNA ligase using TOPO TA cloning, where the process of TA cloning is enhanced by using a TA cloning vector with DNA topoisomerase I attached (**Figure 18.13**). Topoisomerase acts in replication to cleave and rejoin supercoiled DNA. TOPO vectors have topoisomerase covalently bound to the cut ends of the plasmid. During cloning, the PCR product end associates with the vector end and topoisomerase covalently links their backbones together. Some TOPO vectors also clone in PCR products made with proofreading enzymes, which lack the additional A required for TA cloning. These blunt TOPO cloning reactions are made more efficient by the topoisomerase bound to the vector than would be a traditional blunt ligation.

There are other cloning methods that rely on DNA pairing and homologous recombination, needing compatible regions between vectors and DNA inserts. For example, ligation independent cloning (LIC) uses long overlaps between insert and vector to stabilize the hybridization of base pairs compared to traditional restriction enzyme cloning where this would be generally no more than 4 bp. Overhangs of 10–12 bp are made by T4 DNA polymerase and specifically designed primer sequences. The insert and vectors annealed together contain single-stranded nicks that are repaired once the plasmids are transformed into the *Escherichia coli* where they are replicated.

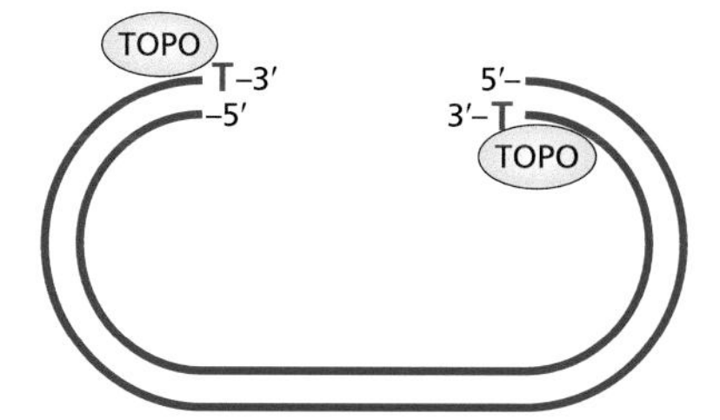

Figure 18.13 TOPO TA cloning. TA cloning (see Figure 18.12) through the use of a TOPO TA cloning vector, a plasmid that has 3′ T overhangs like a TA cloning vector and also has DNA topoisomerase I attached to these ends. When a PCR product with 3′ A ends hybridizes with the plasmid, the topoisomerase joins the DNA backbones, making TOPO TA cloning more efficient than TA cloning.

Gateway cloning is a proprietary system of recombination-based cloning owned by Invitrogen and designed to enable researchers to easily move a cloned piece of DNA between different plasmids. Each move retains the reading frame of the insert sequence by use of a set of recombination sequences, the Gateway *att* sites, and system-specific cloning enzymes. To use the system, the insert first goes into the Gateway entry clone by adding Gateway *attB1* 5′ and Gateway *attB2* 3′ via engineered PCR primers. The amplicon with *attB1* and *attB2* ends is recombined into the Gateway donor vector by a proprietary enzyme mix, creating the Gateway entry clone. The cloned insert can then go into Gateway destination vectors for a variety of experiments. Gateway is based on site-specific recombination of λ bacteriophage, which enables the prophage genome to integrate into the *E. coli* chromosome. The λ bacteriophage genome has a recombination site *attP* and *E. coli* has a corresponding recombination site *attB*. Two enzymes are involved in the prophage integration, integrase from λ bacteriophage and IHF from *E. coli*. Integration creates *attL* and *attR* sites flanking the prophage that enable reversible

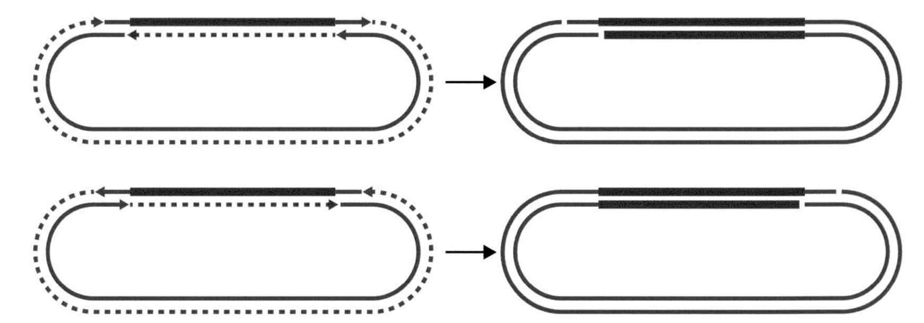

Figure 18.14 Circular polymerase extension cloning, CPEC. In this cloning process, polymerase is used to complete the synthesis of the second strand of DNA around a single strand of plasmid and a single strand of insert, thus joining them together. A denatured single-stranded insert with overlapping similarity to a denatured single-stranded plasmid vector hybridizes (left). The overlapping regions are used by polymerase to prime extension and generate the complementary strand (right).

excision of the prophage from the *E. coli* chromosome by the Xis phage protein. Gateway and other similar recombination-based cloning techniques are based on bacteriophage integration and excision mechanisms, which are covered in more detail in Chapter 21.

CPEC, circular polymerase extension cloning, uses polymerase to extend overlapping insert and vector sequences to make complete circular plasmids. The overlaps hybridize and act as primers for polymerase extension around the circular plasmid (**Figure 18.14**). As with LIC, the plasmids are transformed into the bacterial cells where the nicks are resolved during replication. It is possible with this process to include more than one overlapping insert in the design for CPEC in a design process similar to overlapping PCR, discussed later in this chapter.

Antibiotic resistance markers on plasmids help us find the transformed bacterial colonies

Antibiotic selection (see Chapter 10) is made possible due to the addition of an antibiotic resistance cassette, or two, to plasmid vectors used for molecular cloning. In this way, bacteria with the plasmid can be selected for by using the antibiotic in the media. To use antibiotic selection, use a plasmid vector encoding an antibiotic resistance cassette and ensure the bacterial strain that the vector is being transformed into is susceptible to killing by the antibiotic. For example, if using cloning vector pUC18 (or pUC19, which is identical apart from having an MCS in the reverse orientation), selection is with ampicillin. The plasmid encodes a β-lactamase that breaks down ampicillin. For the popular pCR TOPO TA cloning vector, selection is also ampicillin resistance, but in addition a resistance cassette for kanamycin provides flexibility for selection choice.

Test that the *E. coli* does not grow on LB agar plates with a final concentration of 100 μg/mL ampicillin (**Text Box 18.7**). Use these LB-amp plates to test the untransformed *E. coli*. So long as there is no growth, this *E. coli* and plasmid vector combination can be used for cloning. Following transformation, only those *E. coli* cells that have taken up the plasmid will be able to express the ampicillin resistance phenotype. A similar strategy can be used for other plasmid cloning vectors with other antibiotic resistance markers.

Text Box 18.7

Make LB-amp plates. Make 100 mL LB agar. Make a 100 mg/mL ampicillin stock solution by dissolving 1 g sodium ampicillin in 10 mL dH_2O. Filter sterilize the solution using a syringe-top 0.22 μm filter into a sterile 15 mL tube or other suitable vessel. This ampicillin antibiotic stock can be stored at 4 °C for up to 3 weeks or at −20 °C for up to 6 months; check the storage times and temperatures for other antibiotic stocks. Antibiotics should not be autoclaved, therefore they are added into the media (agar or broth) after autoclaving. To the 100 mL autoclaved LB agar, cooled to less than 65 °C, add 100 μL of the 100 mg/mL ampicillin stock solution. Pour the agar plates, which will have a final concentration of 100 μg/mL ampicillin.

Blue-white screening helps us find the colonies transformed with plasmids with the insert

Blue-white screening is used to identify the bacterial colonies that have been transformed with plasmids that have inserts (recombinants) versus those that do not. Selection only reveals whether the plasmid has entered the cells, not if the plasmid vector has the insert ligated into it. For this, screening is one solution.

Blue-white screening is a common strategy using specifically engineered vectors where the MCS is within the gene fragment *lacZ*α. The full *lacZ* gene encodes β-galactosidase, which can convert X-gal (5-bromo-4-chloro-3-indolyl-β-D-galactopyranoside) into 5-bromo-4-chloro-3-hydroxyindole that dimerizes and oxidizes to the blue pigment 5,5′-dibromo-4,4′-dichloro-indigo. When the engineered vector with the *lacZ*α, encoding just the N-terminal part of *lacZ*, is used with an *E. coli lacZ* mutant with a deletion in the *lacZ* N-terminus, together they can make a functional LacZ able to convert X-gal and make colonies appear blue, in a process called **α-complementation**. Without the vector, colonies appear white. By combining antibiotic selection, so the *E. coli* without the vector will die, with blue-white screening it is possible to see which cells have plasmids with and without inserts (**Figure 18.15**).

LB agar plates are made with both 100 μg/mL ampicillin and 20 μg/mL X-gal, plus 0.1 mM IPTG (isopropyl β-D-1-thiogalactopyranoside) to induce expression of the *lac* operon (**Text Box 18.8**). All colonies growing on these plates will have taken up the plasmids and be expressing the antibiotic resistance. Some colonies will be blue; these will be using the intact *lacZ*α from the plasmid vector to rescue the *E. coli* mutant by α-complementation, creating a functional β-galactosidase to break down the X-gal. The blue color means there is no insert in the vector. Insertion into the MCS disrupts the *lacZ*α and prevents the α-complementation resulting in white colonies, which is why it is important to have the antibiotic selection as well. Pick the white colonies from the LB-ampicillin-X-gal-IPTG agar plate into LB-ampicillin broth and grow overnight for plasmid extraction, generally using a column-based extraction kit, followed by restriction digestion

Text Box 18.8

X-gal is made in a stock at 20 mg/mL by dissolving 0.2 g X-gal in dimethylformamide (DMF) or dimethylsulfoxide (DMSO) and stored at −20 °C in the dark (generally with the tube wrapped in foil). IPTG is made in a stock of 0.1 mM by dissolving 0.238 g IPTG in 10 mL dH_2O and filter sterilizing into a sterile tube. Aliquots of IPTG stock should be stored at −20 °C for up to a year. LB-amp-X-gal-IPTG agar plates are made, for example, by adding to 100 mL of autoclaved LB agar cooled to less than 65 °C 100 μL of ampicillin stock (100 mg/mL), 200 μL of X-gal stock (20 mg/mL), and 100 μL IPTG stock (0.1 mM).

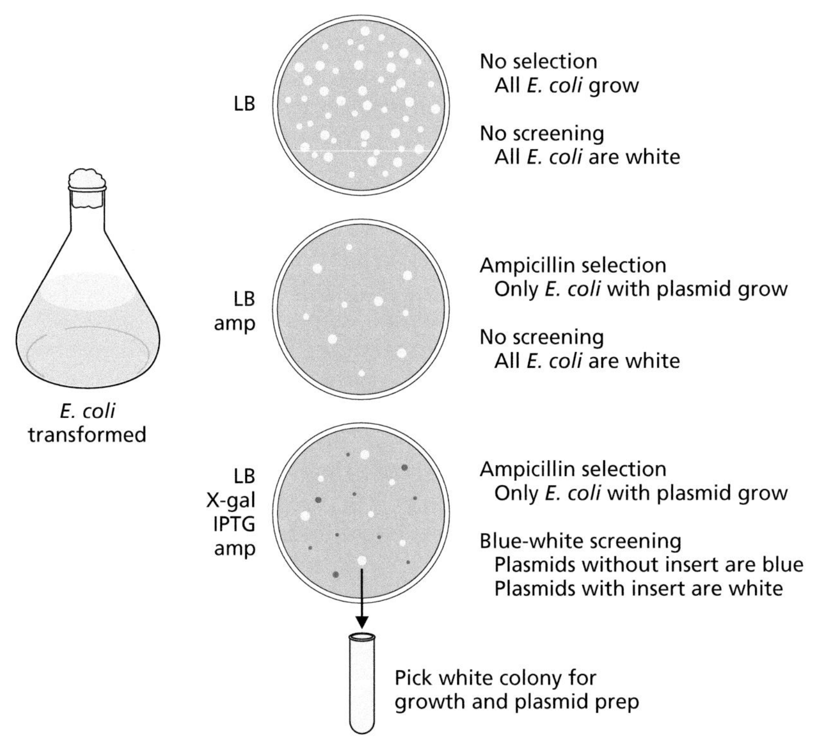

Figure 18.15 Selection and screening for plasmid uptake. After transformation of *E. coli*, the cells can be plated onto different media to reveal the results. Plating onto LB agar has no selection and no screening, so all of the bacterial cells will grow, those that have taken up the plasmid and those that have not. Adding ampicillin to the media, where the plasmid carries an ampicillin resistance marker, will select for those *E. coli* cells that have been transformed by the plasmid. However, this does not reveal which of those plasmids contain a cloned insert. Screening, through addition of X-gal and IPTG, shows through blue-white coloring of the ampicillin-resistant colonies that grow which have the insert (white) and which do not (blue). White colonies on LB-X-gal-IPTG-amp plates are picked for growth and plasmid preparation, having been selected as bacteria containing transformed plasmid with the cloned insert.

Text Box 18.9

To add X-gal and IPTG on top of already poured LB-amp plates, spread 40 μL IPTG stock and 40 μL X-gal stock on top of the agar with a sterile spreader. Let the liquids dry about 30 minutes and then use the plate for the transformation.

and resolution of the fragments on an agarose gel (or E-gel or agarose capillary chip) to confirm insert presence and size.

If LB-amp plates are already poured, it is possible to put the X-gal and IPTG on top of the agar (**Text Box 18.9**). The resolution of the blue-white coloring may not be as uniform as adding the IPTG and X-gal directly to the media, but this method will suffice if it is suddenly discovered that there are no plates, and the transformation needs to happen.

In some rare cases, very small inserts of <100–200 bp seem to be tolerated enough to generate partial α-complementation activity, therefore resulting in pale blue colonies. If the insert is small and no white colonies are apparent, consider whether the insert desired might reside in a pale blue colony. Colony PCR screening can also be used, although this is much more time-consuming than blue-white screening. See the section on PCR later in this chapter for more details.

Laboratory techniques of molecular biology are able to copy segments of DNA in processes similar to replication

A bacterial species that tolerates high temperatures, *Thermus aquaticus*, was isolated and its *Taq* polymerase enzyme was later used in PCR, the polymerase chain reaction, which is an *in vitro* process similar to replication, but for a small section of DNA rather than a whole chromosome. Because the native temperature for *T. aquaticus* is high, the DNA polymerase enzyme from within this bacterial species is active at higher temperatures, making it possible to develop PCR amplification.

PCR (see Chapter 10) has become a standard for amplification of DNA fragments, detection of DNA sequences, modifications of genes, for molecular cloning, and as a staple of molecular biology. The protocols for standard end-point PCR are generally similar, with some temperature variations for primer annealing and polymerase activity. The timing and number of PCR cycles are also variable, generally depending on constraints of time and equipment. *Taq* polymerase extends at 72 °C while proofreading enzymes may extend at 68 °C. Primers are generally designed to anneal between 50 °C and 65 °C. Denaturation of DNA happens at high temperature, so PCR needs a high temperature stage, although this range may be lower in some protocols to reduce damage to the polymerase. Therefore, cycles of temperature from 92 °C to 98 °C to denature the DNA double helix, then 50–65 °C to anneal the primers, then 68–72 °C for polymerase to extend from the primers for 20–35 cycles are typical (**Figure 18.16**). The speed of the reaction *in vivo* is far faster than the cycling of the thermocycler equipment can change temperatures within the tube *in vitro*, therefore although pauses at each temperature of 30 seconds to a minute are usual, 10 seconds is probably sufficient, certainly for denaturation and annealing. Extension may require longer times, depending on the length of the PCR product and the polymerase used. Indeed, some researchers set up a two-temperature cycle, rather than three, with the temperature alternating between 92 °C–98 °C and a lower temperature that permits annealing of primers and then extension. Polymerase can begin to act at temperatures lower than its optimum and will continue to extend as the PCR thermocycler ramps up to the denaturation temperature. In a two-step PCR with a fast thermocycler, it may be necessary to slow down the ramping speed, the speed at which the temperature changes, to allow the polymerase time to extend and generate the product. For a three- or two-step PCR, a final extension in the last cycle is recommended to finish incomplete extensions.

Temperature	Activity
92–98 °C	Denature DNA double helix
50–65 °C	Anneal PCR primers to DNA
68–72 °C	Extend primers by polymerase

Figure 18.16 Typical PCR cycling parameters. There are typically three PCR cycling temperatures, which each have a range of temperatures. The denaturation and extension temperatures are dependent upon the polymerase being used for PCR and both its tolerance for high temperature and its optimal extension temperature. The oligonucleotide primers being used inform the annealing temperature.

PCR primers must be designed well, avoiding potential for annealing in the wrong place, as discussed in Chapter 16. In the laboratory, failures in obtaining PCR products or where multiple products are amplified are generally resolved in one of two ways: adjusting annealing temperature or adjusting $MgCl_2$ concentration. Although primer annealing temperature calculations are fairly accurate, adjustments of even 0.5 °C can improve results. Specificity in PCR can be improved by adjusting the $MgCl_2$ concentration of the reaction buffer.

Depending on the properties of the water in your local area, even when using ultra-purified water of 18 MΩ, it may be necessary to include additional $MgCl_2$ to PCR to generate amplicons as standard or under certain conditions. If primers are designed well but amplification is problematic, setting up a matrix of reactions across different annealing temperatures and $MgCl_2$ concentrations may reveal an optimal condition (**Figure 18.17**). Often the reaction buffer included with the PCR polymerase from the manufacturer contains $MgCl_2$, making it possible only to increase concentration, however some polymerases are supplied with buffers without $MgCl_2$, making a wider range of adjustments possible.

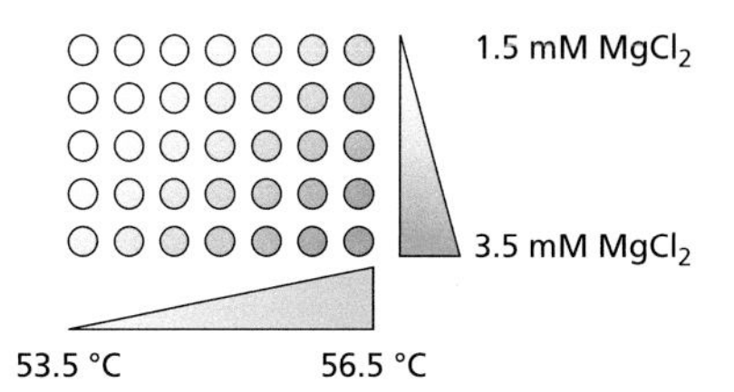

Figure 18.17 Gradient PCR. To address issues with obtaining optimal PCR products, one troubleshooting technique is to do a gradient PCR, where a gradient of temperatures, gradient of $MgCl_2$ concentrations, or both are conducted to find the ideal conditions. Shown here, across a PCR machine heat block, a temperature range of 53.5 °C to 56.5 °C is programmed for the annealing temperature. In addition, reactions will be set up where the reaction buffer has a range of $MgCl_2$ from 1.5 to 3.5 mM. Therefore, 35 PCR tubes will be run in total ranging from 53.5 °C in 1.5 mM $MgCl_2$ to 56.5 °C in 3.5 mM $MgCl_2$ and the combinations between.

PCR can be altered slightly to address experimental needs

Standard end-point PCR uses extracted DNA (see Figure 18.16). It is also possible to set up a rough PCR where the DNA is not extracted from the cells; instead, whole bacterial cells are used in the reaction. This can be done either by boiling a single bacterial colony in 10–50 μL TE or dH_2O (with a lid lock for safety) and using 1–2 μL in PCR (**Figure 18.18**) or by putting a colony directly into the PCR tube and setting the initial denaturation step for 5–10 minutes (see Figure 18.18). The burst bacterial cells release their DNA to be the template for the PCR. **Colony PCR** is useful for screening bacterial colonies following transformation to determine if the screened colonies have the desired genetic changes.

There are times when long PCR products may be needed, including amplifying across repetitive elements to determine the copy number of repeats and amplifying across regions to resolve genome assemblies. In general, standard PCR polymerases will amplify fragments of only a few kilobases. For longer products, there are special enzymes with better **processivity**, meaning that the polymerase stays on the DNA for longer, extending the DNA longer than *Taq* or other standard PCR polymerases. With long-range PCR there is a need to increase the extension time to at least 1 minute per kilobase. This technique is good for amplicons of 5 kb and generally works for those of 15 kb. In some cases, companies claim their long PCR enzymes can get products of up to 30 kb.

PCR can be used to engineer sequences through design of the sequences of the primers. Most engineering with PCR is to simply add sequences to the end of the PCR amplicons by adding them to the 5′ ends of the primers, such as the addition of restriction enzyme recognition sequences to facilitate cloning (**Figure 18.19**). Changes to the 5′ ends of primers do not interfere with the annealing of the primers, therefore it is straightforward to make these sorts of changes.

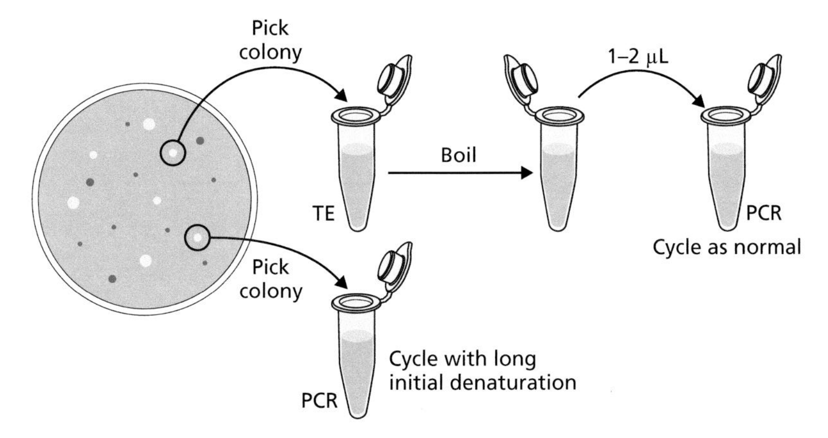

Figure 18.18 Colony PCR. To obtain a PCR product directly from a bacterial colony, two approaches can be taken. A single colony can be picked and suspended in TE, which is boiled, and then 1–2 μL of the crude lysate is used in a normal PCR as the template DNA. Alternatively, a single colony can be picked into a PCR to serve as the template DNA. Here the initial denaturation step should be set for 5–10 minutes to break open the bacterial cells, releasing the DNA to make it available for the PCR.

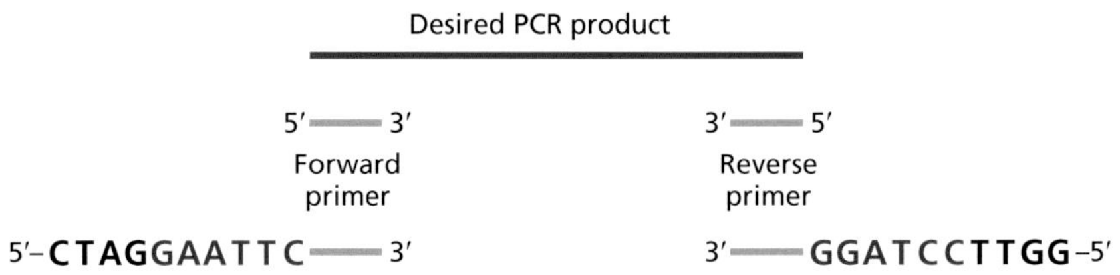

Desired PCR product

5′—3′ Forward primer

3′—5′ Reverse primer

5′-CTAGGAATTC—3′

3′—GGATCCTTGG-5′

Figure 18.19 Adding restriction enzyme digest sites to the ends of PCR products. Conveniently placed restriction digest recognition sites are not always easy to find; sometimes it is easier to engineer one. Once primers are designed to a desired PCR product, the recognition sequences for restriction enzymes can be added to the 5′ ends without interfering with the function of the PCR primers. Here the *Eco*RI site has been added to the forward primer (left, red) and *Bam*HI to the reverse primer (right, red) to facilitate directional cloning of the PCR product (see Figure 18.11). Often some additional bases 5′ of the restriction digest site aid in efficiency of cutting fragments such as PCR products (black); consult the manufacturer's recommendations.

Site-directed mutagenesis systems help researchers make specific changes to DNA

In investigating bacterial genetics and genomics, it is often desirable to mutate sequences so that their functions can be assessed. Frequently this is accomplished when the sequence is cloned, and a PCR primer is designed with the change in the middle. The primer is used to amplify around the plasmid, creating a new sequence with the change (**Figure 18.20**). A single base in the plasmid can be changed by designing a primer that is identical to the sequence of the plasmid aside from the desired change, which is made in the middle of the primer. The primer will otherwise anneal well to the template and there will be amplicons produced that will be synthesized to contain the change in the primer. Likewise, primers can be designed that include either small insertions, where the primer has a few additional bases relative to the original sequence, or small deletions, where the primer has fewer bases in the middle than the original sequence. Amplification results in site-directed mutagenesis, driven by the engineered primer (see Figure 18.20).

Using this purely PCR-based strategy, base changes are made in cloned sequences. The bases within the primer become an integral part of the plasmid following amplification and sequential changes can be made, if necessary. However, for larger insertions or deletions, modifications via restriction enzyme removal of sequence may be a better strategy. This type of mutagenesis may make use of existing restriction enzyme digest sites within the sequence, perhaps identified using one of the many online tools to locate recognition sequences. However, when none is present, they are not conveniently located, or do not work compatibly with other enzymes in the mutagenesis plan, it may be that PCR is needed to aid in the addition of restriction digest sites again. It may be possible to

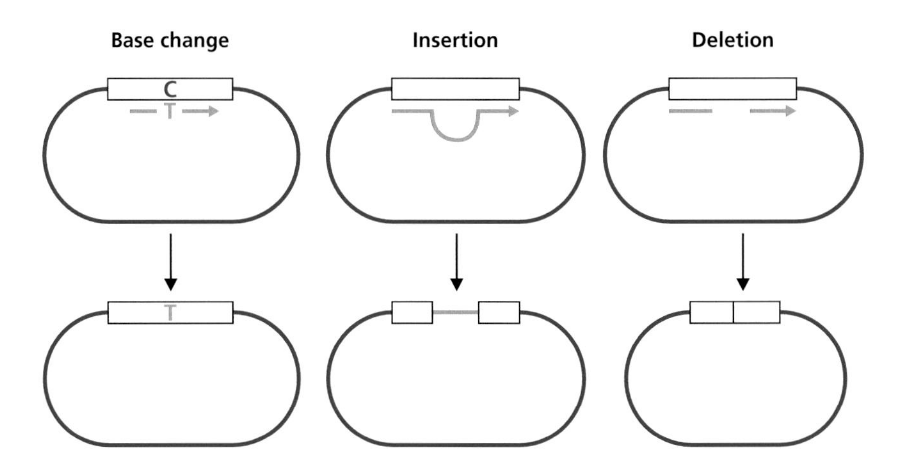

Figure 18.20 Site-directed mutagenesis. It is possible to make small changes to a cloned insert in a plasmid vector. To accomplish this, a primer is designed that is homologous to the insert apart from the small change to be made, a base change (left), small insertion (middle), or small deletion (right). This primer is used to amplify around the plasmid, therefore introducing the new sequence into the plasmid and insert (bottom row).

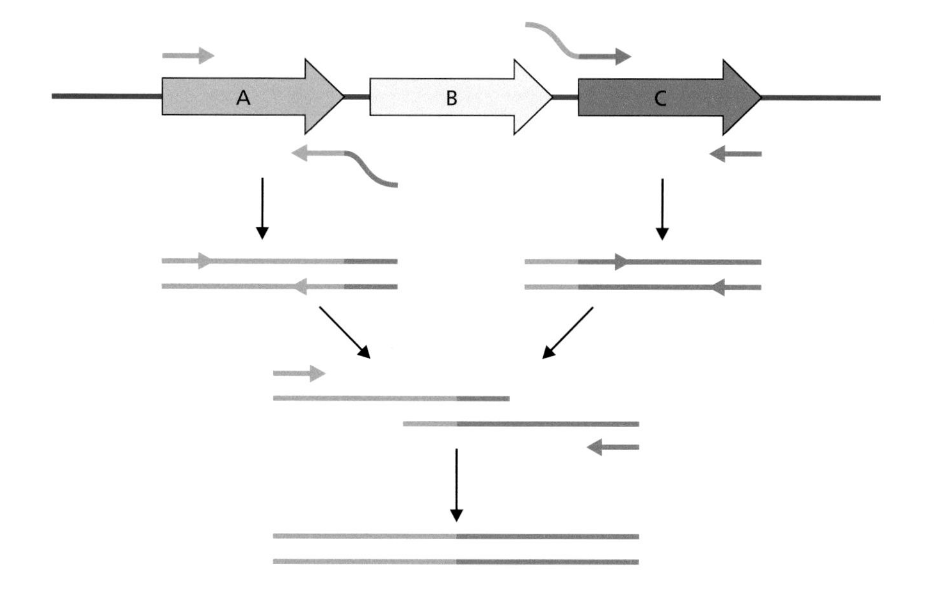

Figure 18.21 Overlapping PCR. To use overlapping PCR to delete the B gene from this operon, primers are designed to the A gene so that the reverse primer has some of the beginning of the C gene (orange) at its 5′ end and primers are designed to the C gene so that the forward primer has some of the end of the A gene (green) at its 5′ end (top). PCR with these primers generates two products: amplification of the A gene with a bit of the beginning of the C gene sequence joined to it and amplification of the C gene with a bit of the end of the A gene sequence joined to it. These products, plus the A gene forward primer and C gene reverse primer, are used in a final round of PCR to make one product that has a full A gene joined to a full C gene. The B gene sequence has been deleted out of the sequence by overlapping PCR.

generate an ideally placed *Mfe*I digest site through change of a single base within the insert, for example. Site-directed mutagenesis can therefore contribute to larger insertions or deletions.

Alternatively, **overlapping PCR** can be used to generate a modified sequence for cloning. This technique makes it possible to change sequences within a PCR product as it is being generated by designing overlapping PCR primers, so that two (or more) reactions are run and then assembled into one product with the change (**Figure 18.21**). Overlapping PCR connects sequences together; therefore, it can be used to generate deletions where distant sequences are brought together and the sequences between them are removed, insertions where new sequences are placed within existing sequences, or to bring together novel sequences. Two primers are designed as normal; these are the outer two primers and are used twice in the process. Two primers are designed with the overlapping sequences; these are only used in the first round of PCR. The 3′ end of the overlapping primers anneals to the sequence to be amplified and works with the normal primer that pairs with it in the first round of amplification (see Figure 18.21). The 5′ end of the overlapping primers does not anneal; however, these sequences match the sequence that will be joined later in the process. Because this 5′ end is part of the primer, it too is amplified in the first round of PCR and becomes part of the PCR product and thus the template for the subsequent cycles of PCR. The amplicons generated in this first round of overlapping PCR therefore have the original sequences fused to a bit of the sequence they will join to at the end of the process (see Figure 18.21). It is possible to design primers so that there are several overlapping PCR products, rather than just two. Following the generation of these products, a second round of PCR uses the products as templates with the outer primers only, to generate the final product.

Loop-mediated isothermal amplification (LAMP) quickly amplifies DNA at a single temperature

Since the advent of PCR there have been other methods of DNA amplification developed, some of which do not require complex thermocycling, making them good options for fieldwork, for example. **Loop-mediated isothermal amplification** (**LAMP**), unlike PCR, does not require expensive specialist equipment; it can be performed at one temperature in a simple water bath, heat block, or incubator. Amplification by LAMP is more rapid than PCR, often being complete in less than 15 minutes. LAMP is also more tolerant to inhibitors than PCR, which can be affected by proteins and impurities in the sample. LAMP primers are designed differently, but as for PCR there are online design programs to aid in optimal design.

The target sequence has six unique sequences targeted by the LAMP primers, three forward (F1, F2, and F3) and three backward (B1, B2, and B3), plus their complementary sequences on the opposite strand (F1c, F2c, F3c, B1c, B2c, and B3c; **Figure 18.22a**). There are four primers designed to these sequences: the forward outer primer containing sequence F3 (FOP); the forward inner primer containing sequence F2, but with F1c at its 5′ end (FIP); the backward outer primer containing sequence B3 (BOP); and the backward inner primer containing sequence B2, with B1c at its 5′ end (BIP; **Figure 18.22b**). At the start of LAMP, the FIP invades the strands of the DNA, displacing one strand to hybridize its F2 sequence to F2c (**Figure 18.22c**). Extension of this primer generates a strand that has the sequence F1c–F2–F1 at the 5′ end (**Figure 18.22d**). This strand is displaced by the FOP with primer sequence F3 (**Figure 18.22e**). While the FOP extends, the displaced strand forms a self-hybridizing loop with the complementary F1c and F1 sequences at its 5′ end (**Figure 18.22f**). This strand is targeted by the BIP, with its B2 sequence annealing to the B2c sequence toward the 3′ end of the strand (**Figure 18.22g**). Extension from the BIP results in a strand that has F1–F2–F1c at the 3′ end and B1–B2–B1c at the 5′ end (**Figure 18.22h**). Displacement of this strand by the BOP (**Figure 18.22i**) results in a strand that has self-hybridizing loops at both ends, where the F1 and F1c hybridize and the B1 and B1c hybridize (**Figure 18.22j**). LAMP rapidly generates such structures, which are able to self-prime further amplification from their 3′ end or can be further amplified by FIP or BIP primers, depending on their F2/F2c and B2/B2c sequences. Due to internal self-priming, LAMP generates complex concatemers, where the original dumbbell-like structure (see Figure 18.22j) becomes joined to others.

Some LAMP kits are designed with a visual detection system that produces a distinct color change or fluorescence in the reaction tube upon production of LAMP amplification products, making detection very rapid. LAMP can therefore eliminate the need for a thermocycler, electrophoresis equipment, and agarose gel visualization system, which would be needed for standard PCR detection. LAMP is primarily useful for diagnostics and detection. Due to the way in which primers have to be designed, with internal complementation, and the generation of concatemerized sequences, LAMP is not suitable for generation of products for cloning and genetic manipulations.

Following *in vitro* manipulation of DNA, it has to be transformed into a bacterial cell

Following ligation of a plasmid, the DNA is transformed into a bacterial cell, which involves getting DNA into somewhere it does not normally go. The workhorse of molecular biology is *E. coli*, but this bacterial species is not naturally competent for transformation. *E. coli* cells have to be manipulated in the laboratory to be transformed. Before transformation, it is important to remember that ligase itself inhibits transformation. Ligase can be heat inactivated at 65 °C for at least 10 minutes. Alternatively, clean up the ligated DNA on a column; this can be especially beneficial for electroporation as it will also reduce salt in the sample, which can inhibit electroporation. It has been noted that heat inactivation of PEG can inhibit transformation, therefore if PEG has been added to increase ligation efficiency or if a fast ligation kit with PEG was used, clean-up is recommended, rather than heat inactivation.

The first method published for *E. coli* to take up foreign DNA uses calcium chloride. It is simple and easy to use. *E. coli* are grown until they are at the early to intermediate log phase of growth as determined by the growth curve (**Figure 18.23**). Generally, bacteria are grown in liquid media over a period of time, with the increase in cells monitored by optical density (OD) using a spectrophotometer blanked on sterile media that has not been inoculated with bacteria. Graphing the OD readings every hour or half hour should reveal when the bacterial cells have left the lag phase of growth and entered the log phase of exponential growth, ideal

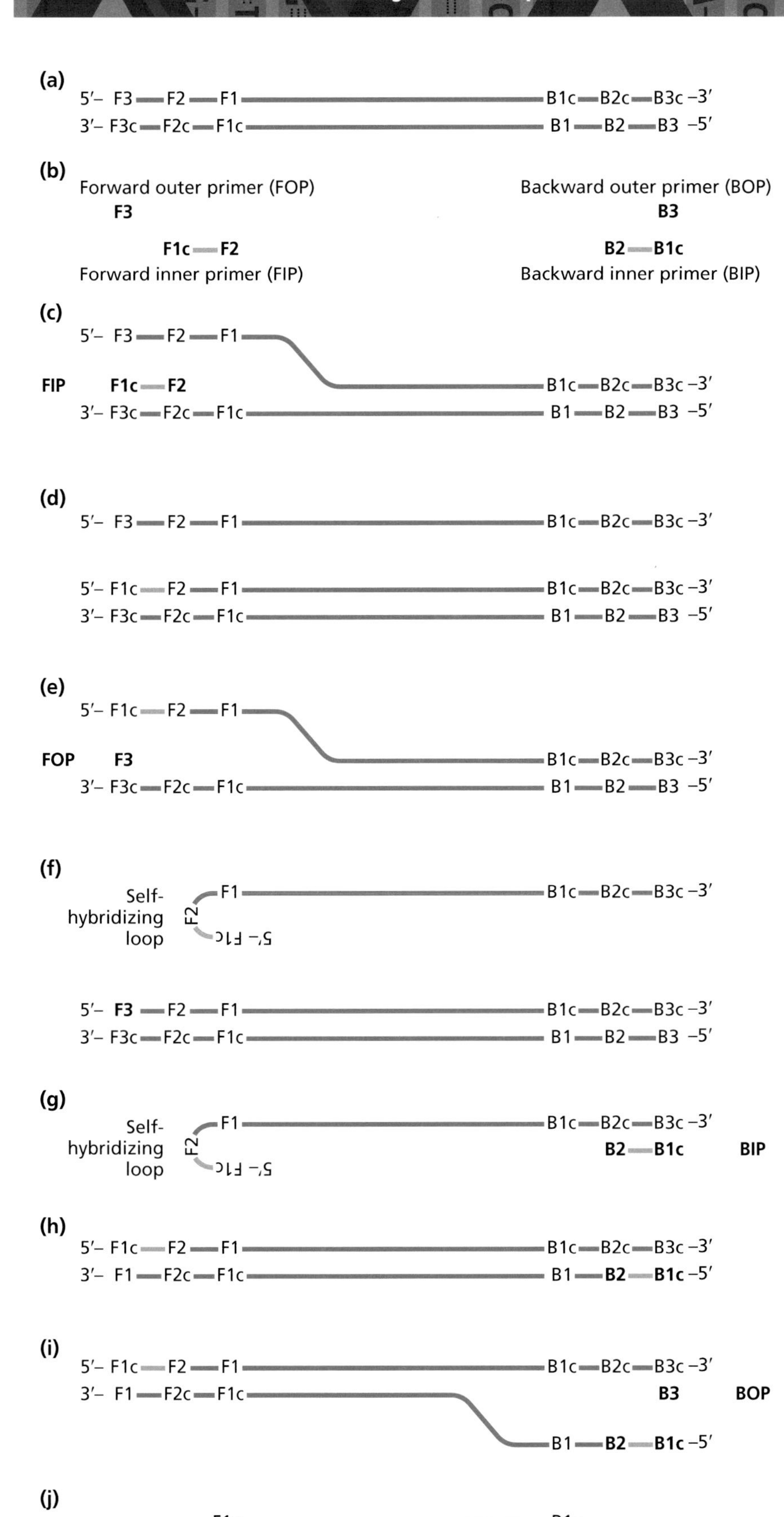

Figure 18.22 LAMP, loop-mediated isothermal amplification. Six unique sequences and the reverse complementary sequences are identified in the target sequence (a). The forward outer primer (FOP), forward inner primer (FIP), backward outer primer (BOP), and backward inner primer (BIP) are designed against these six sequences (b). The FIP anneals, displacing one strand (c), and extends, making a strand with the F1c–F2–F1 sequence at one end (d). The FOP anneals, displacing this strand (e). As the FOP extends, the displaced strand generates a self-hybridizing loop, characteristic of LAMP (f). BIP anneals to the other end of the strand with the self-hybridizing loop (g) and extends, generating a strand with F1–F2–F1c at one end and B1–B2–B1c at the other end (h). The BOP displaces this strand (i) and it generates self-hybridizing loops at both ends (j). Such structures can prime their own amplification at their 3′ end and can be amplified by annealing with FIP or BIP.

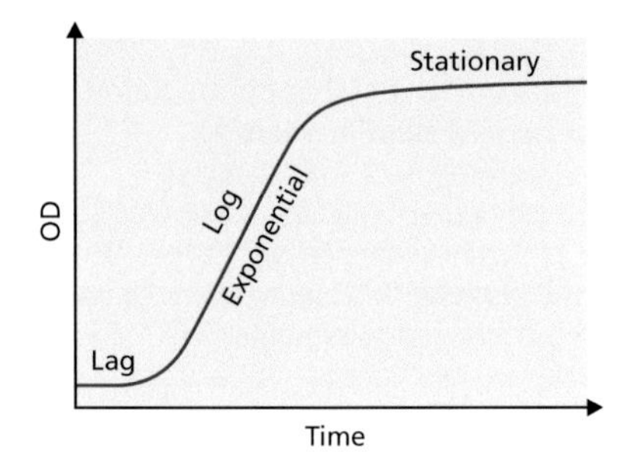

Figure 18.23 ***E. coli* growth curve.** The optical density (OD) of the bacterial cells in liquid culture is monitored over time to determine the number of bacteria in the culture. The OD will increase as the bacteria grow and divide. At the start of the culture, this process can be slow; this is the lag phase. As the population adapts to the culture and builds, its multiplication becomes more apparent in the log phase; this is exponential growth and is ideal for making competent cells for transformations. As the nutrients and resources within the culture become scarce, growth slows, and the OD stabilizes in the stationary phase.

for generation of cells for transformation. As a guideline, an OD_{600} of 0.6–0.8 is typical of exponential phase *E. coli*. The *E. coli* are then put on ice in a calcium chloride buffer. Details of this protocol are discussed in the next section. Cells from the stationary phase should not be used. Although these cells are not dead, nor are they "stationary" or otherwise not growing or dividing; many of the cells in this population are different phenotypically than those in the log phase. They will be experiencing stress in the culture deprived of nutrients and resources, which can adversely influence their ability to take up DNA in transformation.

It is important with chemically competent *E. coli* cells that the bacteria are kept on ice or chilled at all times. Refrigerated centrifuges should be used, set to 4 °C, and cold buffers should be used, either kept in the refrigerator or on ice. Even plastics such as tips and microfuge tubes are put in refrigerators or on ice to chill them for the protocol. Any cells not used immediately can be stored at −70 °C to −80 °C for 6 months to 1 year in 15%–50% glycerol. It is also important to include controls in the transformation. One of these should be to include an "uncut vector DNA control" to get the transformation efficiency of the cells per µg of DNA. Another control is a "no DNA control" to show no growth without transformation and no contamination of the cells. If there is growth on the selective media, then the cells are either contaminated or have the plasmid or a resistance to the selection antibiotic already. Each new batch of competent cells should also be transformed with a positive control plasmid to check the efficiency of transformation.

It is best to give cells some time to express antibiotic resistance genes on the plasmid before plating the bacteria on media containing the antibiotic. However, when using ampicillin, it is possible to plate immediately if time is short because it takes time for the ampicillin concentration to accumulate in the cells.

Text Box 18.10

Grow *E. coli* in LB broth to OD_{600} of 0.4 and put on ice for 20 minutes. Chill the needed tubes and reagents on ice as well, including 0.1 M $CaCl_2$ (11.1 g anhydrous calcium chloride in 100 mL dH_2O, filter sterilized by 0.22 µm filter for 1 M $CaCl_2$) and 0.1 M $CaCl_2$ + 15% sterile glycerol. Centrifuge the culture at 2,500 g for 10 minutes at 4 °C. Remove the supernatant and suspend the pellet in 20 mL of ice-cold 0.1 M $CaCl_2$. Leave the suspension of cells in the 0.1 M $CaCl_2$ on ice for 30 minutes. Centrifuge the cells at 2,500 g for 10 minutes at 4 °C. Suspend the pellet in 5 mL of 0.1 M $CaCl_2$ + 15% sterile glycerol. Use immediately or store at −70 °C/−80 °C.

Calcium chloride provides a quick method to obtain competent cells for immediate use

For the calcium chloride method, the bacteria are kept in a calcium chloride solution on ice to make the cells permissive to transformation (**Figure 18.24**; **Text Box 18.10**). Calcium chloride competent cells are fairly quick and inexpensive to make. They can be made and then used immediately for transformation. These cells can also be stored at −70 °C/−80 °C, however, they are often best used the day they are made.

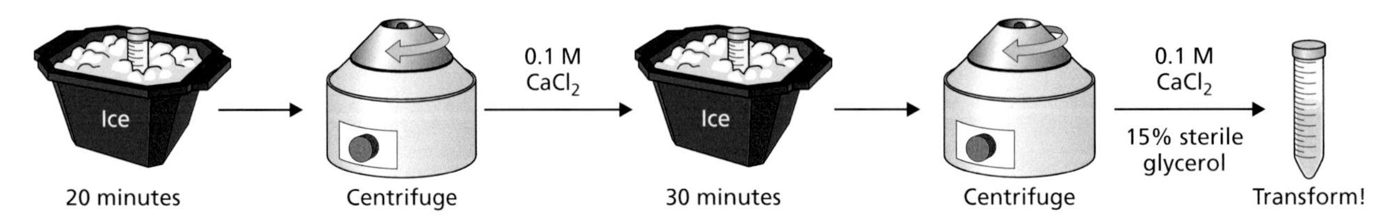

Figure 18.24 Calcium chloride generation of competent cells. A quick way to make competent cells for immediate use is to grow *E. coli* to mid-log phase and put the culture on ice for 20 minutes. Centrifuge to collect the cells, suspend the pellet in ice-cold 0.1 M $CaCl_2$, and leave on ice for 30 minutes. Centrifuge the cells again and suspend in 0.1 M $CaCl_2$ and 15% sterile glycerol. Use immediately for transformations.

Chemically competent cells with Inoue buffer have the best reputation for good rates of transformation and reliability

The Inoue method is a reliably reproducible method of laboratory production of chemically competent *E. coli* cells, which starts by growing the bacteria at 18 °C and then washing the cells in reducing volumes of Inoue buffer. Because the bacteria will grow more slowly, the cultures can be set up to grow overnight and the competent cells made early in the morning. To start, make the Inoue buffer (**Text Box 18.11**). In addition, make SOC (**Text Box 18.12**).

For the cells, grow *E. coli* in 100 mL LB broth overnight at 18 °C to OD_{600} of 0.55 (**Figure 18.25**). Alternatively, SOB media can be used; this is similar to SOC, but lacks the addition of glucose. Centrifuge the culture, split into two sterile 50 mL tubes at 2,500 g for 10 minutes at 4 °C, and remove the supernatant completely, first by pouring off the liquid and then removing the residual with a small pipette tip. Suspend cells in 32 mL ice-cold Inoue buffer gently using a swirling motion or stirring with a pipette tip. Centrifuge the cells at 2,500 g for 10 minutes at 4 °C. Remove the supernatant completely and suspend in 8 mL Inoue buffer and add 600 μL cold DMSO. Mix well, yet gently. Keep the cells on ice for 10 minutes, then aliquot as 50 μL into pre-chilled 1.5 mL tubes. Snap freezing is recommended in liquid nitrogen or a dry ice and ethanol bath prior to long-term storage at −70 °C/−80 °C.

Text Box 18.11

Inoue buffer: 55 mM $MnCl_2{\cdot}4H_2O$; 15 mM $CaCl_2{\cdot}2H_2O$; 250 mM KCl; 10 mM PIPES (0.5 M, pH 6.7). Filter sterilize with a 0.45 μm syringe-top filter, aliquot, and store at −20 °C.

Text Box 18.12

SOC: add 20 g tryptone, 5 g yeast extract, and 0.5 g NaCl to 950 mL dH_2O and shake until dissolved. Add 10 mL of 250 mM KCl and adjust the pH to 7.0, then make the final volume 975 mL. Sterilize by autoclaving, then cool to about 60 °C, which should be warm to the touch but not too hot to handle. Add 20 mL of filter sterilized 1 M glucose and 5 mL of autoclave sterilized 2 M $MgCl_2$ to complete the SOC.

Chemically competent cells can be made with TSS buffer

A similar technique to the Inoue method is one that uses a different buffer, TSS, transportation and storage solution. Some researchers prefer this method, while others favor Inoue. TSS does not require growth of the bacteria in different culture conditions, like Inoue, and there are fewer steps in the TSS method, however the efficiency of the cells obtained may be better with one method or the other. Therefore, it is worth trying a method and, if the cells are not as competent as you would like, perhaps trying another method of preparation of competent cells. To make TSS chemically competent cells, TSS buffer is needed (**Text Box 18.13**).

Grow the *E. coli* in 100 mL LB broth to OD_{600} of 0.4. Chill on ice the TSS and 50 1.5 mL microcentrifuge tubes. When the *E. coli* reaches the needed OD_{600}, put the culture into two 50 mL sterile tubes and put these tubes on ice (**Figure 18.26**). This will enhance the chilling of the culture and enable balanced centrifugation of the culture. Centrifuge the *E. coli* culture tubes at 2,500 g for 10 minutes at 4 °C. Remove the LB broth supernatant from the cell pellet by first pouring off the liquid and then carefully pipetting off all of the remaining broth. Suspend the cell pellet in one of the two tubes in 5 mL of the chilled TSS. Once suspended, transfer the liquid to the second tube and suspend the second pellet in the same TSS, combining the two pellets. Be careful with the cells when suspending them. Gentle vortexing can be used, but best is gentle agitation or mixing with a large

Text Box 18.13

TSS buffer: 5 g PEG; 1.5 mL 1 M $MgCl_2$; 2.5 mL DMSO in LB broth, adjusted to pH 6.5 prior to taking the final volume to 50 mL. Some syringe-top filters cannot tolerate DMSO; it is fine to add this after filtering.

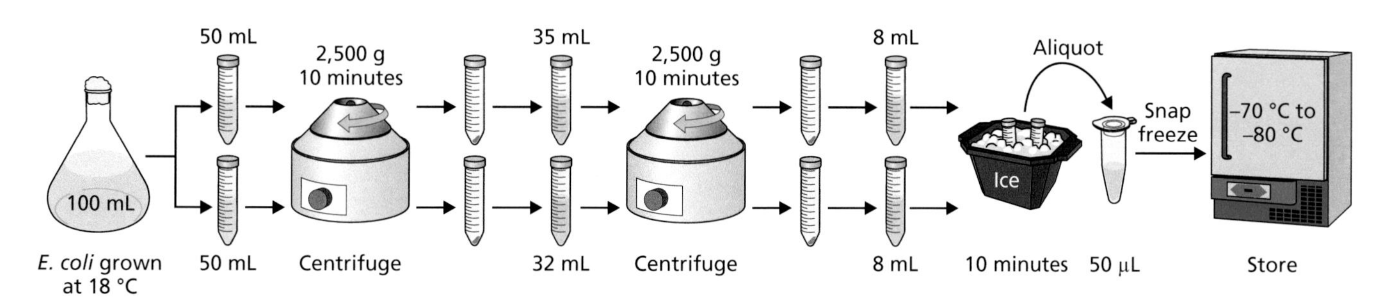

Figure 18.25 Inoue method of generation of competent cells. This method starts with a culture of *E. coli* that have been grown overnight at 18 °C. A 100 mL culture is split into two centrifuge tubes of 50 mL and centrifuged at 2,500 g for 10 minutes at 4 °C. The resulting pellet is resuspended in 32 mL of ice-cold Inoue buffer and then centrifuged again at 2,500 g for 10 minutes at 4 °C. The pellets are this time resuspended in 8 mL of ice-cold Inoue buffer and put on ice for 10 minutes before aliquoting in 50 μL portions, snap freezing, and storing at −70 °C to −80 °C.

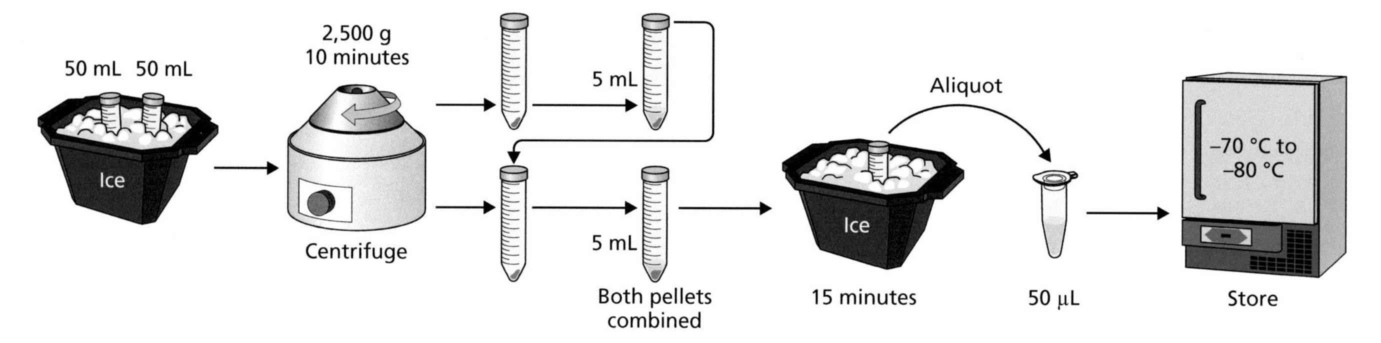

Figure 18.26 TSS method of generation of competent cells. The *E. coli* culture is split into two centrifuge tubes containing 50 mL and chilled on ice prior to centrifuging at 2,500 g for 10 minutes at 4 °C. The bacterial cell pellet of one of these is resuspended in 5 mL of ice-cold TSS buffer. This 5 mL with suspended bacterial cells is used to resuspend the second pellet as well, combining the two pellets into one 5 mL volume, which is chilled on ice for 15 minutes. This is aliquoted into 50 μL portions and stored at −70 °C to −80 °C.

pipette tip. Do not pipette up and down forcefully or with a small tip or the cells could lyse. Chill the suspended tube of cells on ice for 15 minutes.

Aliquot the TSS chemically competent cells at 50 μL volumes into the chilled 1.5 mL tubes and freeze at −70 °C/−80 °C or keep the cells on ice and use them immediately. Some protocols suggest snap freezing in a dry ice ethanol bath or liquid nitrogen before freezing at −70 °C/−80 °C. Aliquots of competent cells, TSS or Inoue, are measured for single use, so there is no need to freeze–thaw, however if you find you are running low, they can be split and 25 μL could be used per ligation/control.

Transformations using chemically competent cells use similar methods, regardless of how the cells were made

Chemically competent cells are transformed by putting the DNA and the competent cells together on ice, giving the cells a brief heat shock, and then letting the cells recover. For all chemically competent cells, start by thawing cells that have been stored at −70 °C/−80 °C. Some recommend briefly warming the tube in the palm of your hand, then placing it on ice, while others recommend just thawing the cells on ice. Add 1–2.5 μL of the DNA to be transformed into the chilled tube containing 25–50 μL of chemically competent cells and leave these cells and DNA on ice for 30 minutes. For just 90 seconds, put the tube at 42 °C for the heat shock (**Figure 18.27**). Generally, this is done in a water bath for good thermal transfer. This 42 °C heat shock is a critical step. Some protocols conduct the heat shock for a shorter period of 30 or 45 seconds. At the end of the heat shock, immediately transfer the tube to ice to chill for 2 minutes. Add 1 mL of LB or SOC

Figure 18.27 Transformation using heat shock. Competent cells are thawed on ice, then 1–2.5 μL of the DNA to be transformed into the bacterial cells is added. The DNA and cells are left on ice for 30 minutes. The tube is briefly heatshocked in a 42 °C water bath for 90 seconds and then immediately put on ice for 2 minutes. To prepare the cells for incubation, 1 mL of LB or SOC is added, and the tube is incubated at 37 °C for 45 minutes before plating 200 μL onto selective media (such as LB-amp-X-gal-IPTG) and incubating overnight at 37 °C. The remainder of the bacterial cells (800 μL) can be stored for plating later, if there are no colonies on the plate.

to the tube and incubate at 37 °C with shaking for 45 minutes. This incubation gives time for the resistance marker to be expressed, as mentioned earlier. SOC is sometimes preferred as a richer medium for the bacteria. Some protocols also suggest a 1 hour incubation, however this can run into a second round of bacterial cell division and artificially generate duplicate clones. At the end of 45 minutes, plate 200 μL of the cells onto LB agar containing the appropriate selection and screening supplements for the plasmid used in the cloning (i.e., ampicillin, X-gal, and IPTG). Incubate the LB plate overnight at 37 °C. The remaining 800 μL of the transformation should be stored on the bench or at 4 °C overnight. Should the LB selection plate have no colonies after overnight incubation, or no white colonies for blue-white screening, centrifuge this SOC tube and plate the remainder of the transformation.

Electroporation provides an alternative to chemically competent cells

Electroporation requires an electroporation cuvette and electroporator (**Figure 18.28**), as well as electrocompetent cells. To make electrocompetent cells, grow 100 mL of *E. coli* to OD_{600} of 0.6 and then put the culture on ice in two 50 mL sterile tubes. Centrifuge the culture at 1,000 g for 15 minutes at 4 °C, then remove the supernatant LB broth from the pellet completely. Suspend the cell pellet in 50 mL of ice-cold sterile dH_2O and centrifuge at 1,000 g for 15 minutes at 4 °C. Remove the supernatant, suspend the pellet in 25 mL of ice-cold sterile dH_2O, and centrifuge again at 1,000 g for 15 minutes at 4 °C. Remove the supernatant and suspend the pellet in 1 mL of ice-cold 10% sterile glycerol. Aliquot the cells as 50 μL portions into 20 pre-chilled 1.5 mL tubes, in which they can be stored for 6 months at −70 °C/−80 °C. If these cells repeatedly produce arcing during electroporation, as described in the next section, add an additional cold water wash step to this electrocompetent cell preparation process.

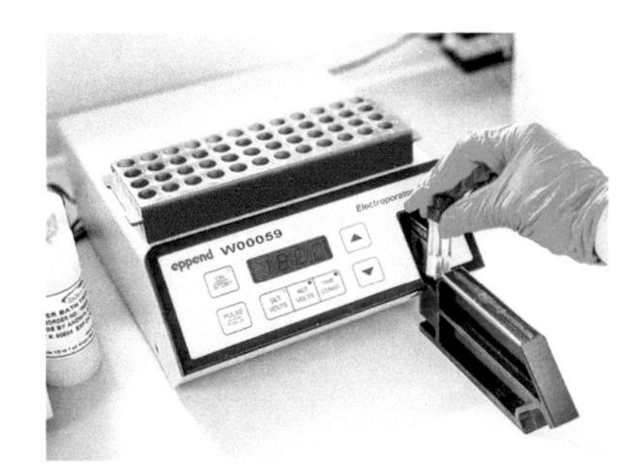

Figure 18.28 Transformation using electroporation. Electroporation requires electroporation equipment, including an electroporator and electroporation cuvettes. The electroporator delivers the electric shock through the electroporation cuvette, which holds the bacterial cells and the DNA. This shock moves the DNA into the bacterial cells.

The process of electroporation is sensitive to salts, but quick to perform

To electroporate, chill the electroporation cuvettes and DNA on ice. Warm LB plates to 37 °C. Thaw electrocompetent cells from −70 °C/−80 °C on ice; alternatively, freshly made electrocompetent cells kept on ice can be used immediately. For 1 mm cuvettes, set the electroporator to 1.25–1.5 V or the setting recommended for the cuvette size and electroporator combination by the manufacturer. Add 2 μL of purified DNA to 50 μL of electrocompetent cells, on ice, mixing gently with the pipette tip. Transfer the cells and DNA to the chilled electroporation cuvette, being sure there are no air bubbles. Wipe the moisture off the outside of the cuvette without warming it and place it in the electroporator. Pulse the cells with an electric shock. If there is a loud pop, repeat the process with a new cuvette, tube of cells, and less DNA. The pop noise is arcing of the electricity caused by too much salt in the DNA. If less DNA does not work, then the DNA needs to be cleaned up further with a spin column. When the electroporation is successful (there is no pop), add 1 mL of SOC immediately to the cuvette. SOC is a rich bacterial medium that helps the bacterial cells recover from the process. It is a good idea to include an electroporation control with no DNA to show that the zap of electricity does not kill the cells; this should be plated onto a plain LB agar plate without selection.

Put the SOC and cells in a microcentrifuge tube on ice for 2 minutes to recover, then at 37 °C with shaking for 1 hour. After an hour, plate 200 μL of the SOC and cells onto LB agar containing the appropriate selection and screening supplements (i.e., ampicillin, X-gal, and IPTG). Incubate the plate overnight at 37 °C. Leave the microcentrifuge tube with the SOC and cells on the bench overnight or at 4 °C. If there are no transformed colonies on the plate the next day or if none of the colonies has the desired insert, centrifuge the remaining 800 μL and plate the rest of the electroporated cells.

Expression studies rely on extraction of high-quality RNA, which means controlling RNases

This chapter has mostly been concerned with techniques related to DNA because these relate directly to bacterial genetics and genomics. However, to investigate gene or ncRNA function, it may be beneficial to explore RNA-based experiments. In such cases, starting with good-quality RNA is important. As with DNA, RNA can be extracted by one of two general methods, using either a phenol method, which produces high-quality RNA but uses hazardous chemicals, or a spin column method, which is quicker and safer but can be contaminated with proteins and DNA.

For many applications, a purchased spin column kit will work well, provided precautions are taken when handling RNA. Recall from Chapter 4 that RNases are everywhere and that RNA is structurally very unstable, being prone to pulling itself apart even when it is not degraded by the ever-present RNases. Use products that remove RNases from surfaces in the laboratory environment, such as *RNaseZap*, and wear gloves that are changed frequently and wiped with RNase surface remover frequently. Use a solution to preserve the RNA in the bacterial cells prior to the extraction of the RNA, such as RNA*Later*, which will stop production of new mRNA and prevent RNases within the bacterial cells from degrading the transcripts.

In addition, all plastics should be specifically purchased for RNA work as RNase-free, including use of pipette tips that include within-tip filters. These filters stop any contamination from within the body of the pipette entering your experiment and vice versa. RNase-free plastics come certified from the manufacturer as not having RNases at the time of arrival at your laboratory, but once the packaging is opened, there is a risk of contamination of tips and tubes. Take care in handling these to preserve their RNase-free quality. Despite the popular belief, autoclaving does not eliminate RNases, even when done twice.

RNA extraction columns work similarly to DNA extraction columns, with some slight variations

RNA extraction columns work in a similar manner to DNA extraction columns, relying on the features of nucleic acids and silica to bind to the column, and wash with an ethanol-based wash solution. The most striking variation when comparing most DNA and RNA commercial kits is that the RNA extraction kits have RNase-free components. Generally, the columns will come individually packaged to ensure they remain RNase-free during the life of the kit. Some kits may require the user to purchase their own RNase-free tubes and RNase-free water for the elution step, while other manufacturers will provide these.

Aside from needing to take care due to RNases, the other common issue with RNA extraction using the spin column-based methods is that they frequently result in RNA that is contaminated with DNA. For downstream experimental analysis, residual DNA can give erroneous results, therefore it is important to completely remove any DNA from the extracted RNA. Most kits will have an optional step to DNase treat the RNA while it is still bound to the column. This should be indicated as an essential step.

At the end of the protocol, elute the RNA with RNase-free water into an RNase-free tube and then aliquot the RNA into RNase-free tubes in single-use volumes to prevent freeze–thawing. Store the RNA at –70 °C/–80 °C and use it within a year.

Acidic phenol extraction of RNA makes high-quality, pure RNA

So long as your technique for removing the aqueous layer from atop the phenol layer is good, the acidic phenol extraction method for RNA is one of the best. The principles are similar to phenol-based DNA extraction, but the change in pH of the phenol to acidic means that the DNA from within the bacterial cells ends up

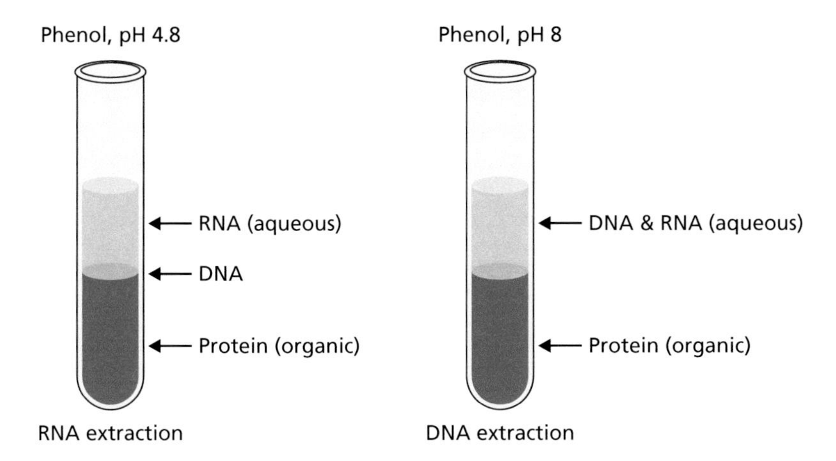

Figure 18.29 Phenol phases for RNA extraction compared to DNA. For RNA extraction, acid phenol is used at pH 4.8. This has the distinct advantage that only RNA will be in the aqueous phase, with the DNA being in the interphase or the organic (phenol) phase.

in the phenolic phase and only the RNA is in the aqueous phase following the first round of centrifugation (**Figure 18.29**). As with the DNA extraction using phenol/chloroform/isoamyl alcohol described earlier in this chapter, the bacterial cells are lyzed, and the aqueous mixture is combined with an equal volume of phenol/chloroform at pH 4.8. Acidic phenol is used because following emulsion and then centrifugation, the DNA and proteins are in the organic phase and interphase, while the RNA alone is in the aqueous phase. In addition, acidic phenol minimizes RNase activity.

Mix the equal volumes of bacterial cells in aqueous solution and acidic phenol/chloroform by inverting or gently agitating for 10 minutes. Centrifuge the mixture at the maximum speed allowable for the centrifuge and tubes for 10 minutes to separate the aqueous and phenolic phases. As with the DNA extraction, the top layer will generally be the aqueous phase and contain the RNA, however high salt content can invert the layers. Again, using colored phenol can help identify the phases and is a good safety precaution when using a highly toxic chemical like phenol. Likewise, it is essential to take only the aqueous phase off to a new tube to avoid protein contamination, phenol contamination, and in this case also DNA contamination from the interphase or organic phase (see Figure 18.29). RNA can also become contaminated with the guanidinium thiocyanate that is present in commercial acidic phenol preparations.

The steps in RNA precipitation are the same as for DNA, in that 1/10th volume of 3 M NH_4OAc is added directly to the liquid of the aqueous phase in its new tube and mixed before addition of 2.5 volumes of 100% EtOH and mixing (**Figure 18.30**). As previously, extraction can continue from here or the sample can be chilled, for 1 hour at –70 °C/–80 °C or overnight at –20 °C to attempt to increase yield. Centrifuge the tube again at maximum for 2–5 minutes to pellet the RNA, ideally at 4 °C (see Figure 18.30). Carefully remove the ethanol supernatant

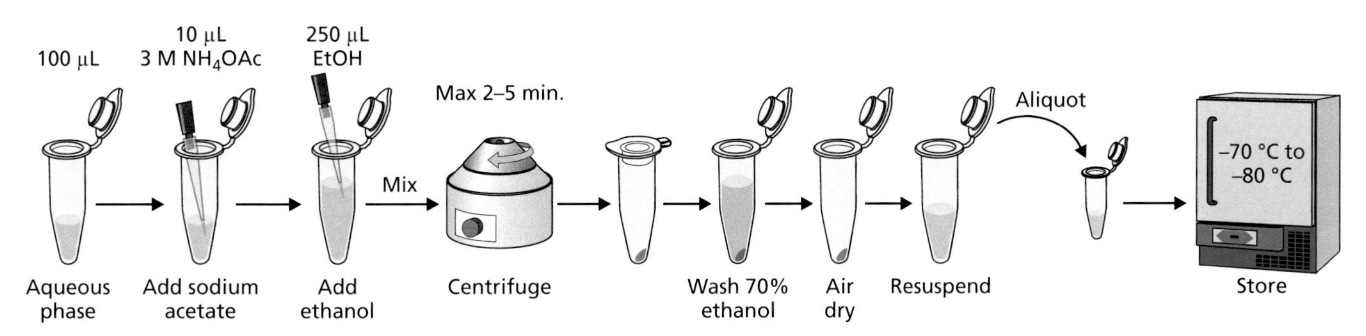

Figure 18.30 Precipitation of RNA. To precipitate RNA from the aqueous phase, first add 1/10th volume of 3 M NH_4OAc, here 10–100 µL of aqueous phase. Mix, then add 2.5 volumes ethanol, here 250 µL, and mix. Centrifuge to pellet the RNA at maximum for 2–5 minutes. Remove the supernatant from the pellet and wash with 70% ethanol. Remove the 70% ethanol and air dry the pellet. Once dry, resuspend the pellet, aliquot, and freeze to store at –70 °C to –80 °C.

from the RNA pellet, which is then washed with 70% EtOH and air dried before suspension in RNase-free water. RNA should always be stored in small aliquots at −70 °C/−80 °C. To further protect the RNA during storage, an add-in RNase inhibitor can be added to the RNA at the point of aliquoting, prior to freezing.

Key points

- DNA can be extracted using phenol, which is hazardous but gives high quality and high yield, or using spin columns, which is quick and easy.
- DNA contamination with phenol, ethanol, or proteins can compromise the results of downstream experiments, therefore care needs to be taken in extraction.
- Restriction endonuclease digestion of DNA makes it possible to join together pieces of DNA from different sources, creating recombinant DNA.
- Digestion of an insert and a plasmid vector with the same restriction enzyme or with restriction enzymes with compatible cohesive ends facilitates cloning and ligation.
- PCR products can be cloned directly into TA cloning vectors by virtue of the A overhangs left by *Taq* polymerase during PCR.
- Antibiotic selection and blue-white screening aid in the identification of bacterial cells containing plasmid vectors containing the desired insert sequence.
- PCR can be used to specifically modify cloned DNA and change its sequence.
- There are several methods to generate competent cells for transformation of DNA, either using chemicals or electroporation.
- Extraction of RNA is similar to extraction of DNA, however great care must be taken due to the ubiquitous presence of RNases.

Key terms

Define the following terms introduced in this chapter. Check your answers using the definitions in the Glossary. These terms are also available as Flashcards online.

α-Complementation	Heat inactivated	Polylinker
Colony PCR	LAMP	Processivity
Compatible cohesive ends	MCS	Sonication
Directional cloning	Overlapping PCR	Star activity

Questions and discussion topics

Self-study questions

Answer each question using 50–100 words or a table or labeled diagram. Advice on where to find answers to these questions is available online.

1. What needs to happen to a bacterial cell to extract the DNA or RNA? Name methods by which this can be achieved. Why do the methods sometimes need to differ for Gram-negative and Gram-positive bacteria?
2. Make a table listing at least three methods for DNA extraction from bacterial cells in the rows, with "pros" and "cons" in the columns. Note down the advantages ("pros") and disadvantages ("cons") of each extraction method.
3. What are some of the key things to remember when using a restriction endonuclease in the lab?

4 How would the reaction set-up differ if the restriction digestion from Figure 18.5 was to be made without BSA and as a double digest with two restriction enzymes?

5 How might the reaction you made in question 4 need to be set up if the two enzymes worked best at two different temperatures?

6 Describe two methods that could be used to clone a PCR product into a plasmid. What considerations need to be taken into account with regard to the generation of the PCR product and any processing of the product after PCR for each method?

7 Briefly describe how blue-white screening works and how it helps researchers identify which colonies to pick for plasmid extraction.

8 What are the advantages of LAMP compared to PCR? What are the disadvantages?

9 Compare and contrast the different methods for making *E. coli* competent cells. This could be briefly in text, visually in a diagram, or represented in a table format.

10 Electroporation is a rapid means of getting DNA inside of a bacterial cell, both in terms of preparing electrocompetent cells and in terms of doing the electroporation. What steps are involved in making the cells electrocompetent and what key factors lead to success in electroporation?

11 What special precautions need to be taken when working with RNA that are not as necessary when using DNA?

12 It is possible to extract DNA out of a bacterial cell with phenol or to selectively extract RNA. What change(s) in the protocol make it possible to obtain only RNA from a phenol-based extraction?

Discussion topics

These topics are presented for discussion in study groups, as part of class discussions, or on your own. These questions go beyond what is directly covered in this part of the book. Use the research literature and other reading to explore these topics in more depth. Tips to help prepare for topic discussions are available online.

1 Using an online tool and the sequence for a gene of interest, explore the restriction enzymes that will cut the gene sequence. Discuss suitable enzymes for directional cloning into the MCS (multiple cloning site) of the plasmid pUC18. Once cloned, how would you use restriction digestion to delete a portion of the gene sequence, generating a knock-out mutant of the gene? In your choice of enzymes, take care in selection, as some may cut more than once, may cut the plasmid as well as the gene, or may cut the new sequence generated by the join between plasmid and insert. For each restriction digestion reaction, include in your discussion whether two different restriction enzymes are needed and, if so, whether they digest at the same temperature and in the same buffer. Such information will be available online from the manufacturer.

2 PCR is a powerful tool not just for copying fragments of DNA, but also as seen here for manipulating DNA sequences and modifying them. Investigate research studies that have used one or more of the techniques described here to alter the sequence of bacterial DNA using PCR. Discuss how the technology was applied, including the materials and methods described in the research publication and the results outcome of this experimental work. Extrapolate from this study what further investigations could be done in this strain using this technique to learn more about the biology of the organism.

3 Some bacterial species are naturally competent for transformation, as discussed in Chapter 3. *E. coli* can be made chemically competent or electrocompetent, as discussed here. Other species can be made competent as well. Investigate bacterial species that are either made into chemically competent cells or made into electrocompetent cells. Discuss the similarities and differences between the methods for that species and for *E. coli*.

Online quiz questions

To further self-assess your understanding of the chapter material, please visit the following link, where you can participate in a range of interactive quiz questions:

www.routledge.com/cw/snyder

Activity: Extract DNA in your own kitchen

The techniques described in this chapter are for use in the laboratory, but this activity can be done in your kitchen at home using materials you may have around the house or can get in the store. This activity demonstrates the basics of DNA extraction: cell lysis; detergent separation of proteins and membranes; and alcohol precipitation of DNA. There are several videos available online demonstrating methods similar to this that are fun to watch or possibly to create a video of your own to share.

To start, squash up some fruit in a resealable sandwich bag. A banana, a few strawberries, or other soft fruit works well. Make some soapy water in a small glass by adding a spoonful of dish soap and a dash of salt to a quarter cup of warm water. Stir gently to mix.

Pour this soapy water into the bag of squashed fruit. Seal the bag. Mix thoroughly. Pour the liquid from the bag through a strainer/sieve/coffee filter into a clear glass, so that the liquid is retained in the glass. Dispose of the bag and chunks of fruit that have been strained off.

To the strained liquid in the clear glass, add ice-cold alcohol. This can be rubbing alcohol, vodka, or any clear alcohol. Try to pour the alcohol along the side of the glass, slightly tipped, so that it layers on top of the strained liquid. Set the glass on a table or counter and watch as the DNA strands separate out. The DNA will appear as wisps of milky white material, covered in bubbles that may look to have a mucus-like consistency. The DNA can be lifted out of the mixture using a glass or wooden stirring stick.

Further reading

Loop-mediated isothermal amplification (LAMP) quickly amplifies DNA at a single temperature

Notomi T, Okayama H, Masubuchi H, Yonekawa T, Watanabe K, Amino N, Hase T. Loop-mediated isothermal amplification of DNA. *Nucleic Acids Res.* 2000; *28*(*12*): e63.

Following *in vitro* manipulation of DNA, it has to be transformed into a bacterial cell

Chung CT, Miller RH. Preparation and storage of competent *Escherichia coli* cells. *Methods Enzymol.* 1993; *218*: 621–627.

Huang M, Liu M, Huang L, Wang M, Jia R, Zhu D, Chen S, Zhao X, Zhang S, Gao Q, Zhang L, Cheng A. The activation and limitation of the bacterial natural transformation system: The function in genome evolution and stability. *Microbiol Res.* 2021; *252*: 126856.

Chemically competent cells with Inoue buffer have the best reputation for good rates of transformation and reliability

Inoue H, Nojima H, Okayama H. High efficiency transformation of *Escherichia coli* with plasmids. *Gene.* 1990; *96*(*1*): 23–28.

Chemically competent cells can be made with TSS buffer

Chung CT, Niemela SL, Miller RH. One-step preparation of competent *Escherichia coli*: Transformation and storage of bacterial cells in the same solution. *Proc Natl Acad Sci USA.* 1989; *86*(*7*): 2172–2175.

Part VII

Applications of Bacterial Genetics and Genomics

The last part of the book brings together three chapters on very different topics, but which all relate to applications of bacterial genetics and genomics and associated fields. In the first, the field of biotechnology is explored, focusing on the aspects of biotechnology that make use of bacterial cells and systems. Infectious diseases are touched on in the following chapter. There are entire volumes dedicated to this topic; the material presented here is an overview of the topic to complement the rest of the chapters of this book and put them into the context of human bacterial disease. In Chapter 21, the bacteriophages are presented in more depth. Although bacteriophages are presented in Chapter 1 and mentioned throughout the book, this chapter focuses on them specifically as their own entity, rather than just in the context of the bacterial genome. These chapters together provide additional depth and enhanced understanding and appreciation of how bacterial genetics and genomics can be applied in the fields of biotechnology and infectious disease control, as well as in understanding the ubiquitous nature of bacteriophages.

DOI: 10.1201/9781003380436-25

Chapter

19

Biotechnology

The use of biological processes or systems, such as the use of biological organisms, for the benefit of improving human life is **biotechnology**. This benefit can be in the broadest of senses. Biotechnology can clearly be the use of organisms to create or produce medicines, but it can also be to clean up our environment so that we have a healthier planet on which to live. Biotechnology can also be applied to the creation of novel cosmetic products that enrich our lives and make us happier, thereby bringing us joy, or which improve efficiency in industry that enables processes to be safer for human workers or for workers to be more productive, perhaps improving work–life balance. The use of microorganisms, especially bacteria, is a major part of the field of biotechnology. While many biological systems are used in biotechnology, this chapter, in keeping with the theme of this book as a whole, will focus only on those that use bacteria in some way to achieve their end result.

Biotechnology is far older than genetic engineering

When we think of biotechnology, we tend to think of modern scientific technologies. However, the use of biologicals for human benefit is ancient, if we consider processes such as beer brewing, mead and wine making, and cheese making. While the fermentation processes require yeasts, which are microbes outside the scope of this book, cheese making uses enzymes that originally came from sources such as calves' stomachs. These enzymes often now come from bioengineered bacteria, being recombinantly generated. Although cheese making itself predates recorded history, today we are able to make it using modern scientific techniques, including recombinant enzymes expressed in bacteria.

Rennet used in making cheese is extracted from the fourth stomach of calves that have been slaughtered for veal. After processing, rennet contains chymosin, a protease, as well as pepsin and lipase. The role of the chymosin enzyme is to coagulate milk, which is beneficial to calf digestion and enables humans to manufacture cheese. In older cows, the level of chymosin drops, making the extraction from young calves' stomachs important for cheese making. Due to the expense and limited availability of rennet, cloning of the chymosin gene into bacteria and expression of the protein was an attractive solution. The fermentation-produced chymosin (FPC) from the bacteria is the same as the chymosin produced in the calf stomach. Today, most FPC is produced in either the fungus *Aspergillus niger* or the yeast *Kluyveromyces lactis*, but some is also made in *Escherichia coli*.

With the rise in awareness of lactose intolerance in humans, particularly adult humans, there has been an increasing market demand for lactose-free products. In infants, the lactase enzyme is made in the small intestine to digest human breast milk from the mother, but with age the expression of this enzyme can decrease or stop altogether. Consumption of lactose-containing foods by lactose-intolerant individuals results in abdominal cramps, gas, diarrhea, and discomfort. In addition to non-dairy alternatives to milk and cheese products, biotechnology has come up with some solutions. The lactase enzyme has been produced using recombinant technologies in a similar way to chymosin and added in the production of milk and cheese products to reduce lactose, often to levels where companies are labeling products as lactose-free.

DOI: 10.1201/9781003380436-26

Biotechnology impacts many aspects of our lives and of research

Biotechnology is a broad discipline that touches many aspects of science, technology, economics, and society. In medicine, biotechnology is used to develop new pharmaceuticals such as monoclonal antibody-based therapies and recombinant protein-based drugs. In agriculture and environmental sciences, biotechnology is involved in genetic modification of plants, often involving bacteria as intermediates within which to manipulate the genetic material, as well as aiding in the development of efficient biofuels and use of bacterial species for bioremediation. In industrial biotechnology, bacteria aid in production of enzymes that enhance products as diverse as food and laundry soap.

Biotechnology permeates our lives, being present in our homes and market shelves. Biological laundry soaps contain digestive enzymes, generally lipases, proteases, or both. Some products, including dishwasher detergents, also contain amylases. Lipase is able to digest fats and oils while protease is able to digest proteins and starch, thereby removing the two most common sources of stains on clothing. Cloning and expression of lipase, protease, and amylase enzymes in bacteria using biotechnology mean that large quantities of the enzymes can be made inexpensively. Indeed, around 25%–30% of enzymes are used by the detergent industry.

There is a fascinating ecosystem at the floor of the ocean that occurs when a whale dies and sinks to the deep ocean, below 1,000 m, called whale fall (**Figure 19.1**). This environment has created an ideal opportunity to explore microorganisms that are adapted for growth in cold water, less than 20 °C, and that possess cold water adapted enzymes that function at these temperatures. For manufacturers of laundry soap, this is an ideal environment, especially with the move to set washing machines to cold water washes to save on energy and preserve clothing. From studies of whale fall, some of the lipase fat-dissolving enzymes in laundry soap now come from organisms that have evolved to live in cold water, making them ideal for cold water wash conditions (see Figure 19.1).

Large quantities of bacteria are grown in bioreactors to yield large quantities of recombinant proteins

To harvest large quantities of recombinant proteins for biotechnology applications, **bioreactor** systems are frequently used. Bioreactors are culture vessels intended for industrial use or for highly controlled long-term laboratory experiments. These closed bacterial growth systems enable large quantities of bacteria to be grown, enabling recombinant protein harvesting on a much larger scale than would be achieved in a typical laboratory bacterial culture (**Figure 19.2**).

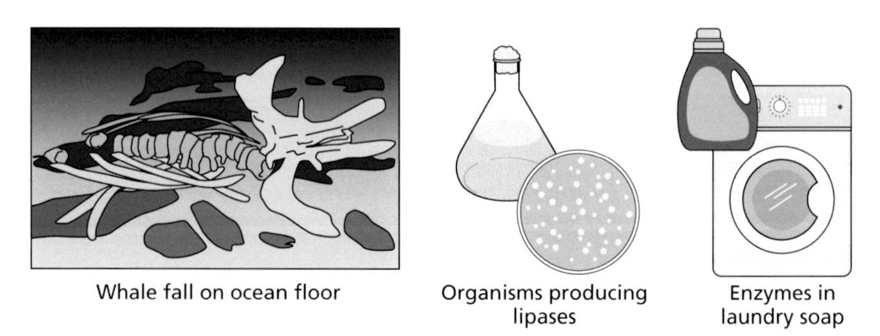

Figure 19.1 Whale fall ecosystems are sources of cold adapted enzymes. When a whale dies and its carcass falls to the deep ocean floor, it creates an ecosystem. Other organisms thrive on breaking down the nutrients of the whale body. These include organisms producing lipases adapted to function well at cold water temperatures. These have been collected, grown in laboratories, and used as enzymes in laundry soaps to break down fats in cold water laundry washes.

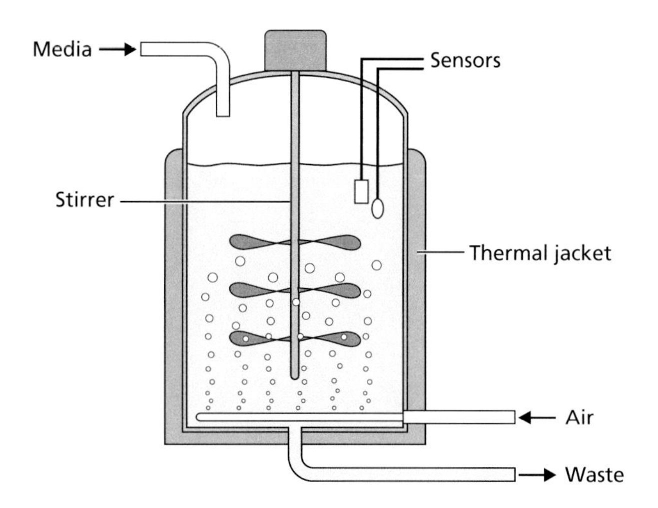

Figure 19.2 An example of a bioreactor. Bioreactors are contained systems for growth of bacteria under highly controlled conditions that can be carefully monitored. The culture is often stirred and aerated, as needed for the species, and temperature controlled, with the aid of the thermal jacket on the bioreactor vessel. As nutrients are depleted, more media can be added and waste removed.

At its most basic, a bioreactor is any engineered system that is able to support growth of microbes or biochemical reactions. Bioreactors can be vessels of less than a liter or of 50,000 liters, potentially more. Growth conditions within the bioreactor are carefully controlled and maintained, with new media added, temperature and pH controlled, oxygen levels adjustable, and the system, including stirring rate of the culture, fully monitored (see Figure 19.2). When growing large volumes of bacteria, such as *E. coli* and *Bacillus* spp. expressing recombinant proteins, the bioreactor maintains the optimal growth conditions to produce the best yield of protein. The bioreactor simulates the ideal conditions for the bacteria to grow and express the recombinant protein.

A typical liquid culture in a microbiology laboratory uses a broth medium in a shaking flask at a set temperature in an incubator. Under these conditions the temperature within the culture is not uniform, with the temperature within the flask differing from that toward the surface of the flask. The culture is shaken in an attempt to aerate the broth; however, this also is not uniform throughout. A bioreactor attempts to address this and other variations in culture conditions to produce more reproducible cultures each time, as well as optimal conditions for each bacterial cell within the culture. However, setting up bioreactor cultures is more complex than a simple liquid culture, which is why bioreactors are not used for routine bacterial growth.

Human insulin expressed in *Escherichia coli* is a classic example of biotechnology

Insulin is needed for patients with diabetes. Insulin was once isolated from the pancreas of pigs, cows, and sometimes salmon. This was a slow and expensive process that was unable to keep up with the demands of clinical need in a growing population. In addition, some patients were allergic to insulin harvested from animals and others were not comfortable with the use of medicines derived from animals. Overcoming these issues, the gene for human insulin was cloned into bacteria and expressed as a recombinant protein.

Humulin was licensed for human use in 1982. Two DNA sequences were introduced into *E. coli*: one for the insulin A chain of 21 amino acids and one for the B chain of 30 amino acids. In humans, insulin is made as one polypeptide chain, proinsulin, which has the A chain and B chain separated by the 33 amino acid C peptide (**Figure 19.3**). Proteolytic cleavage removes the C peptide. The A chain and B chain are then connected by two disulfide bonds; a third disulfide bond occurs within the A chain. Production of the A and B chain peptides in *E. coli* results in recombinant human insulin with the same structure (see Figure 19.3).

Today, millions of diabetics use synthetic insulin that is made mostly by bacteria or yeast in some cases. Recombinant insulin can be a perfect match for human insulin, or it can be changed slightly using genetic engineering to produce

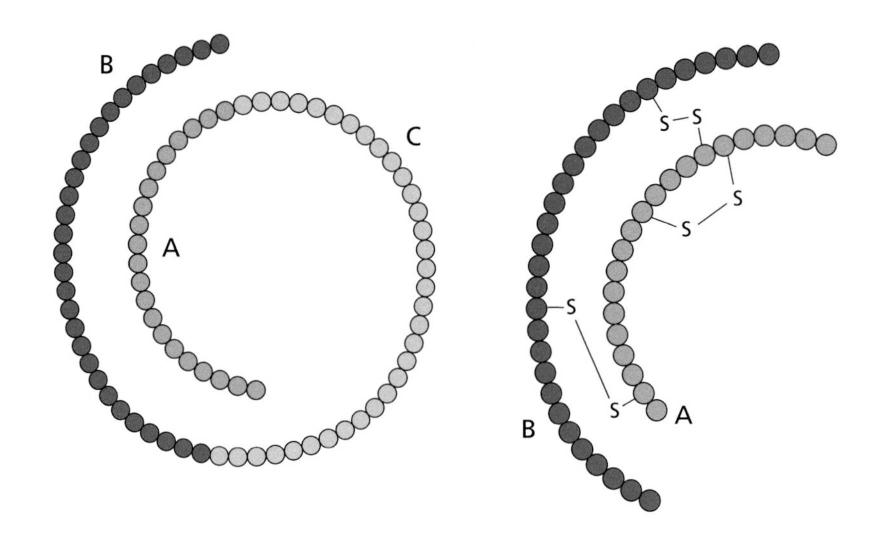

Figure 19.3 Human insulin. Insulin is made in humans as one continuous chain of amino acids, initially proinsulin, which is 84 amino acids (left). To generate insulin, the 33 amino acid C peptide is removed from proinsulin, releasing the 21 amino acid A chain and the 30 amino acid B chain. The A chain and B chain come together, connected by two disulfide bonds, with a third disulfide bond within the A chain (right).

insulin analogs, including insulin lispro, known as Humalog. Humalog was the first insulin analog, entering the market in 1996, and is a fast-acting insulin. The amino acid sequence of the B chain of Humalog differs from normal human insulin at the C-terminus, where the lysine and proline residues are reversed, which blocks the formation of insulin dimers and hexamers. Because these multimeric forms of insulin cannot form, there is more insulin available for fast absorption by the patient. This minor change gave patients more flexibility in managing their lives because they did not have to wait as long following an insulin injection to have a meal. There are several different analogs of insulin in production today by different pharmaceutical companies.

Many recombinant drugs have been made since insulin

In the 1980s, *E. coli* was used to express several recombinant drugs, including Humatrope and Protropin for hGH deficiency, Roferon A for hairy cell leukemia, and Intron A for cancer, genital warts, and hepatitis. The availability of these human proteins was an important contribution by biotechnology made possible by bacteriology. Human growth hormone (hGH) deficiency is the result of the body being unable to make enough growth hormone. This can manifest at birth or later in life, resulting in short stature, poor bone density, and decreased muscle mass. Therapeutic intervention involves replacement of the hormone. Until recombinant hGH was available, it was extracted from the pituitary glands of cadavers.

The cytokine-based drugs used in cancer chemotherapy Aldesleukin, Interleukin-2, Filgrastim, and Pegfilgrastim are produced in *E. coli.* A range of other anticancer biopharmaceuticals are also made in *E. coli*, including Interferon alpha-2a, Interferon alpha-2b, Interferon gamma-1a, and Endostatin.

Denileukin diftitox is a fusion protein made in *E. coli* as a diphtheria toxin fused to a human cytokine and is used to treat T-cell lymphoma. Two diphtheria toxin fragments, toxin A and toxin B, are fused to human interleukin-2 (IL-2; **Figure 19.4**). This biotechnology-generated drug uses two bacterial species in generation of the final drug product Denileukin diftitox. *E. coli* is used for expression and the diphtheria toxin comes from *Corynebacterium diphtheriae.*

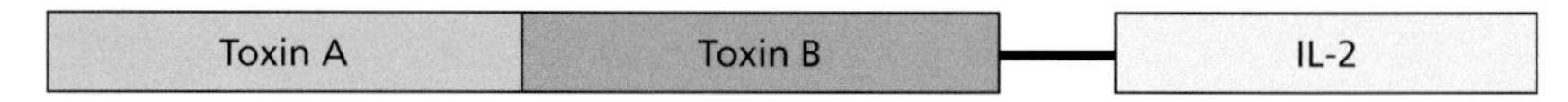

Figure 19.4 Denileukin diftitox fusion protein. Biotechnology has generated a drug that fuses two diphtheria toxin fragments from *C. diphtheriae* (left end) with human IL-2 (right end). The fusion protein is expressed in *E. coli*, producing a drug that specifically targets IL-2 expressing cancer cells for destruction by the toxin.

Diphtheria toxin is a potent exotoxin, which is selectively delivered to cancer cells expressing the IL-2 receptor by virtue of the fusion protein drug. Clinical use of this drug was discontinued in 2014, due to production and purification issues with the *E. coli* expression system, however yeast expression systems are being explored as replacements.

Recombinant production of influenza virus vaccines

The surface of the influenza virus is coated in the glycoprotein hemagglutinin (HA; **Figure 19.5**). The HA antigen can be recognized by the body's immune system and when bound by antibodies that recognize HA, the influenza virus becomes neutralized. This is the basis of the seasonal flu vaccine, the vast majority of which are produced in egg-based manufacturing systems (see Figure 19.5) that are limited in their flexibility and in capacity to scale up in times of epidemic and pandemic outbreaks of flu. In addition, some patients experience complications due to the egg-based nature of the propagation of the vaccine.

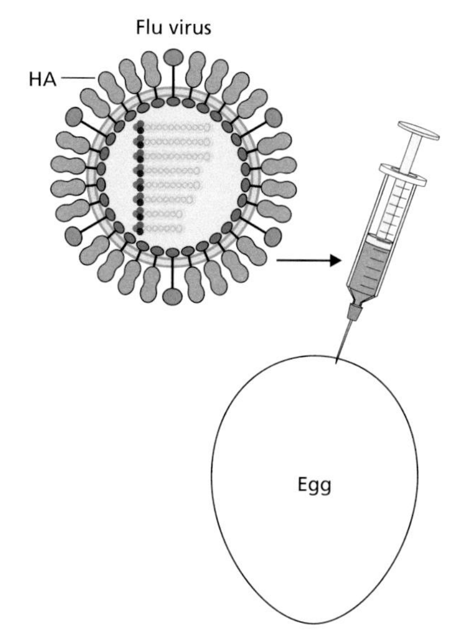

Figure 19.5 Flu virus vaccine made within eggs. One of the surface antigens of influenza virus is hemagglutinin (HA), which can be recognized by the immune system. Once bound by antibodies the flu virus is neutralized. It is therefore useful to be able to replicate flu viruses within eggs to be able to produce flu vaccines. Eggs are injected with the virus, which multiplies. The eggs are later harvested for the viral particles and the batch is tested before being approved for use.

Recombinant production of HA expressed in bacterial systems using biotechnology techniques is an attractive alternative to egg-based production. However, several attempts have resulted in HA that is less antigenic and that had lower yield than the established production methods. To address this issue, researchers at Cytos Biotechnology took an approach that combined viral and bacteriophage sequences. Rather than expressing the whole of the HA protein in *E. coli*, fragments of the protein were investigated and domains that were shown to be readily expressed and that folded into structures that were the same as the native flu virus protein structure were identified. This provided a portion of the HA that could be highly expressed in *E. coli*. This HA fragment was fused to the bacteriophage Qβ virus-like particle (VLP), a carrier that has been used in development of other potential vaccines to enhance antigenicity. Within the Qβ particles is RNA that is a ligand for TLR7/8; the fusion therefore enhances antigenicity of the HA or other antigens that are combined with it.

The HA Qβ VLP proteins expressed in *E. coli* are not glycosylated as they would be in humans. This may be a limitation to this technology, however the investigators discovered that the antigens expressed in their bacterial cultures were able to offer better cross-protection to other flu viruses in mouse model experiments than might be expected. This suggests that recombinantly generated flu vaccine may have better protective abilities as well.

Live recombinant vaccines use live bacteria to deliver antigens

While several recombinant vaccines have been developed where bacteria, yeast, or other cells express proteins, there have also been investigations into live recombinant vaccines, where live organisms are introduced into the human body to express antigens and trigger the immune response. The rationale is that the whole organism will generate a broader immune response than a single recombinant protein and that a live vaccine will be able to replicate within the body, unlike an inactivated vaccine.

Advances in bacterial genetics and genomics have enabled specific changes to be made in bacterial strains. Virulence factors can be completely removed from the genome, yet antigenic proteins can be retained. Additional antigens can be added and the strain that is to be introduced to the patient can be fully susceptible to antibiotics, so that it can be eliminated with antimicrobial chemotherapy if required.

One bacterial strain that has been extensively investigated as a live recombinant vaccine vector is *Mycobacterium bovis* BCG. Due to its previous use as a vaccine, it is known to be safe, having been administered to over 3 billion people. Recombinant BCG (rBCG) has been explored for several antigens, expressed by *M. bovis* BCG, which acts as an adjuvant for the immune system recognition of the

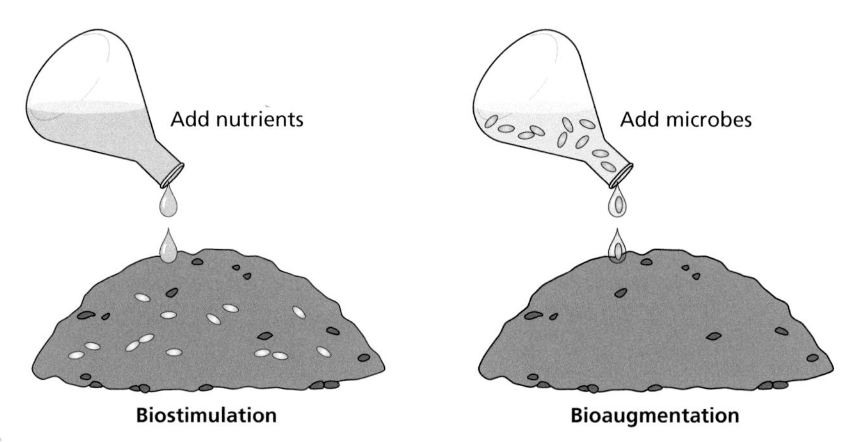

Figure 19.6 Biostimulation and bioaugmentation to tackle pollution. Bioremediation uses microbes to degrade the pollution. Two approaches can be taken to achieve this. Biostimulation uses microbes already present in the environment, introducing nutrients or other changes to give those microbes a boost (left). Bioaugmentation instead puts new microbes into the environment, which are specifically chosen for their ability to degrade the pollutants (right).

antigens. These include work on rBCG-HIV that produced a cellular and humoral immune response against HIV and rBCG-S1PT that produced a cellular immune response against *Bordetella pertussis* toxin.

Bioremediation uses the microbial world to correct the pollutants we have introduced into the natural world

Bioremediation seeks to restore contaminated sites using microorganisms to degrade the contaminants. Treatments can happen in place, *in situ*, or the contaminated material can be removed and treated elsewhere, *ex situ*. There are several approaches to bioremediation, including naturally occurring microbial communities, stimulated to break down contaminants. In **biostimulation**, changes in pH, moisture, nutrients, and other conditions are made to enhance the growth and bioremediation properties of the bacteria that are already present (**Figure 19.6**). Alternatively, **bioaugmentation** can be used, where specific microbes are chosen for their degradation capabilities (see Figure 19.6). It is possible to combine biostimulation and bioaugmentation. Generally, bioaugmentation is short-lived, since the introduced bacteria die quickly in their new environment. Where present, the organisms already adapted to the environment and able to help it recover are better able to achieve bioremediation.

Bioremediation is concerned, for the most part, with cleaning up historical sites or accidental sites of contamination. **Biotreatment** of wastewater and byproducts of industry does not strictly fall within the realms of bioremediation, as the goal of biotreatment is to treat these before they enter the environment.

Bioremediation using bacteria present in the environment can help us reclaim sites

In Medellin, Colombia, there is a landfill called "El Morro," which was in use from 1972 to 1984. Over 25 years later, the Colombian government decided to turn the area into a park. Something similar had been done at Mount Trashmore Park in Virginia Beach, Virginia, USA in 1973 (**Figure 19.7**), as well as several other landfill sites elsewhere. To investigate the feasibility of conversion of the Medellin site into a park, samples of the landfill and associated bacterial species were taken. This revealed that bioremediation was naturally occurring. Plans were therefore put in place to attempt to aid the bacteria in their work through the provision of additional nutrients. What was once a deprived area is now an urban park.

Figure 19.7 Mount Trashmore, Virginia, USA. Once a landfill in use until 1971, in 1973 it became the world's first park built on a waste landfill. This has become not only a popular park, but also a demonstration of landfill reuse and bioremediation that has inspired other sites worldwide. (With permission from Sherry V Smith/Shutterstock.com.)

Several useful bacterial species have been identified in various environmental contexts. However, *Dechloromonas aromatica* is an especially useful bioremediation species. It can oxidize benzoate, chlorobenzoate, and toluene causing reduction of oxygen, chlorate, and nitrate. It is uniquely able to anaerobically oxidize benzene, a contaminant of ground and surface water.

As an example of biostimulation, oil naturally exists in the Gulf of Mexico and therefore seeps onto the ocean floor as a natural part of that ecosystem. When there was the massive Deepwater Horizon leak, spilling 4.9 million barrels of oil into the Gulf in 2010, there were already communities of microbes existing in this environment capable of bioremediation of oil. Through the addition of nutrients to the water near an oil spill, the bacterial population can be assisted in its bioremediation. In addition, to aid the bacteria being able to access the oil, chemical dispersants were added to the water. The dispersants broke up the oil spill, creating more oil and water interface opportunities for the bacteria to gain access to the oil.

Genetic modification for bioremediation can provide organisms with new features

Using genetic engineering, bacteria for bioremediation can be enhanced. Genes can be added that degrade pollutants and/or genes can be added that can monitor pollution levels. There are, however, concerns about releasing genetically modified organisms into the environment and the potential for their genes to be horizontally transferred. Some of these concerns can be overcome by creating strains that will die once their bioremediation job is done and that can be detected easily, for example through expression of bioluminescence.

With bioengineering, new bioremediators have been created, such as *E. coli* that are able to remove mercury, a function not available in nature for any organism. These genetically modified bacteria would be contained within filters that are used at mercury-contaminated sites to sequester away the mercury, which is otherwise highly toxic in the environment as mercury or when transformed to methylmercury. There is some hope that bacteria will not only be able to remove heavy and rare metals from the oceans, where these metals are sources of contamination for wildlife and our food chain, but also that once these are sequestered by the bacteria, they will be recoverable. If we are able to remove the contaminating metals from the oceans and reuse them in industrial applications, rather than further damage the environment by trying to retrieve more from the Earth, this would be beneficial overall.

Production of materials by bacteria

Acetobacter naturally produce nanocellulose, as do other Gram-negative and some Gram-positive cells, but the *Acetobacter* is prolific in its generation. Biotechnology-based investigations have applied this bacterial biosynthesis to generate sustainable, growable items. For example, a mold can be made of a pair of shoes. The bacteria are then grown in the mold and in the process of them naturally producing nanocellulose, they make shoes. This method can even generate different colors. Once growth of the shoe is complete, the structure is sterilized to remove the bacteria, leaving behind only the cellulose. The resulting shoes are therefore biodegradable.

Bacterial nanocellulose has also found applications in medicine. These applications include topical use for wound dressings and implants for repair or replacement of structural defects in the body.

Bacteria can be a renewable source of bioenergy

Given the growing global demand for energy and issues with traditional sources of energy and their impact on the environment, there is interest in **bioenergy** and the role microbes can play. Bioenergy is energy from a biological source, in this case bacteriological. Microbial metabolism can create biohydrogen, biomethane, bioethanol, and bioelectricity, as well as addressing the problems of plastics from petroleum products through generation of microbial biopolymers.

One goal of bioenergy research is to use microbes to convert the biomass of the waste of our consumption into energy in a clean and renewable way. Most bacteria can convert sugars to ethanol, and some can metabolize plant matter to do so. *E. coli* have been genetically modified to generate biodiesel, through breaking down sugar and secreting high-grade biodiesel. Due to the properties of the fuel and its export from the bacterial cell, it floats to the top of the culture vessel, making harvesting easy. In addition, the *E. coli* are modified to express and secrete cellulase, able to break down plant matter that is indigestible by humans. The cellulase genes come from *Clostridium stercorarium* and *Bacteroides ovatus*, bacterial species that live in soil and ruminating animals. This addition was made to minimize the impact of production of the biodiesel on human food production. The secreted cellulase breaks down the inedible plant material to sugars that the *E. coli* metabolize into biodiesel (**Figure 19.8**).

The bacterial species *Geobacter sulfurreducens* and *Shewanella oneidensis* are able to transfer electrons to bioelectrochemical devices such as microbial fuel cells and microbial electrolysis cells (**Figure 19.9**), making them useful for generating biohydrogen and bioelectricity. This technology is not yet on a scale to create energy of commercial use, however there is potential for future development.

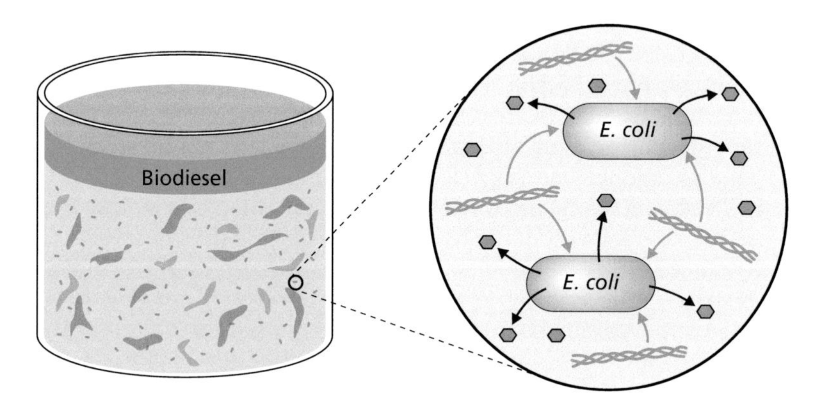

Figure 19.8 ***E. coli* breakdown of plant matter to make biodiesel.** Genetically modified *E. coli* can make high-grade biodiesel, which floats to the top of the culture vessel. Addition of genes for secreted cellulase into the *E. coli* enables them to break down plant matter into sugars (insert), which are then converted into biodiesel.

Figure 19.9 Bioelectrochemical devices. Microbial fuel cells are not yet at a scale where they can produce commercial levels of electricity, but there is potential for future development of microbes as power sources. (Courtesy of Volker Steger/Science Photo Library.)

Key points

- Biotechnology uses biological processes for the benefit of humans.
- Bacteria make several important contributions to biotechnology, both directly and indirectly.
- Biotechnology, putting biological systems to use for human benefit, is an ancient discipline, when techniques such as the production of wine, beer, bread, and cheese are considered.
- Modern biotechnology includes the expression of recombinant proteins by bacterial cells.
- Large volumes of bacteria are grown for industrial purposes in bioreactors, enabling large-scale production of recombinant proteins.
- Biotechnology has made a positive contribution to human health, including the expression of human insulin in *E. coli* as early as 1982, followed by several other recombinant drugs.
- Some vaccines are based on recombinant antigenic proteins and live recombinant vaccines are in development.
- Use of bacteria in bioremediation can help recover the natural environment from human-generated pollutants.
- Bacteria can be the source of renewable energy in bioenergy developments including generation of biodiesel and as bioelectricity generators.

Key terms

Define the following terms introduced in this chapter. Check your answers using the definitions in the Glossary. These terms are also available as Flashcards online.

Bioaugmentation	Bioremediation	Biotreatment
Bioenergy	Biostimulation	
Bioreactor	Biotechnology	

Questions and discussion topics

Self-study questions

Answer each question using 50–100 words or a table or labeled diagram. Advice on where to find answers to these questions is available online.

1. Using your own words, define biotechnology. Describe three things that are a part of your life that are a result of modern biotechnology.
2. Describe two ways in which biotechnology is being used in the dairy industry to improve production, reduce reliance on animal-derived enzymes, and market dietary-specific products.
3. What conditions can be altered in a bioreactor? What advantages does this have over growing a bacterial culture in a flask with liquid growth media in a standard shaking incubator?
4. In what year was human insulin cloned into *E. coli* licensed for use? How does this work? What modifications have been made since?
5. Many recombinant drugs are expressed in *E. coli.* What are the advantages of using *E. coli*? What are the disadvantages?
6. What are the options for production of vaccines against influenza? What are the advantages of each of the options?
7. Some vaccines are called "live recombinant vaccines." Why are they live? What is recombinant in them? What is the advantage of this type of vaccine?
8. Bioremediation can be achieved in a few ways. What are these and how do they differ?
9. What is the difference between bioremediation and biotreatment? Why is bioremediation necessary if biotreatment happens?
10. What do genetically modified bioremediators do? What additional functions would you design into a bioremediator?
11. What is bioenergy and why is it attractive as a form of energy?
12. What are microbial fuel cells and what are some potential advantages in their use?

Discussion topics

These topics are presented for discussion in study groups, as part of class discussions, or on your own. These questions go beyond what is directly covered in this part of the book. Use the research literature and other reading to explore these topics in more depth. Tips to help prepare for topic discussions are available online.

1. The cloning of human insulin and licensing of its use was a major breakthrough in biotechnology. Since then, many other drugs have been designed and used, including some that are discussed here. Explore and discuss one of these in more depth, or another of interest. How does the therapeutic aid in the disease it is fighting? Is it using *E. coli* or another expression system? Are there any issues?
2. Investigate an example of bioremediation in your local area. What form of bioremediation was used? Was just one approach attempted, or were different techniques combined, as needed? At what stage is the bioremediation project and how successful has it been thus far?
3. Generation of biodiesel may provide a solution for fuel, so long as the source does not threaten food security. What is being done by researchers to try to protect food crops for human consumption, yet utilize biomass for the production of biodiesel as well? Expand upon the example in this chapter, explore other examples, and discuss the advances in this area.

Online quiz questions

To further self-assess your understanding of the chapter material, please visit the following link, where you can participate in a range of interactive quiz questions:

www.routledge.com/cw/snyder

Further reading

Large quantities of bacteria are grown in bioreactors to yield large quantities of recombinant proteins

Habe H, Sato Y, Aoyagi T, Inaba T, Hori T, Hamai T, Hayashi K, Kobayashi M, Sakata T, Sato N. Design, application, and microbiome of sulfate-reducing bioreactors for treatment of mining-influenced water. *Appl Microbiol Biotechnol.* 2020; *104*(*16*): 6893–6903.

Obom KM, Magno A, Cummings PJ. Operation of a benchtop bioreactor. *J Vis Exp.* 2013; (*79*): 50582.

Priyadarshini BM, Dikshit V, Zhang Y. 3D-printed bioreactors for in vitro modeling and analysis. *Int J Bioprint.* 2020; *6*(*4*): 267.

Recombinant production of influenza virus vaccines

Jegerlehner A, Zabel F, Langer A, Dietmeier K, Jennings GT, Saudan P. Bacterially produced recombinant influenza vaccines based on virus-like particles. *PLoS One.* 2013; *8*(*11*): e78947.

Bioremediation uses the microbial world to correct the pollutants we have introduced into the natural world

Pushkar B, Sevak P, Parab S, Nilkanth N. Chromium pollution and its bioremediation mechanisms in bacteria: A review. *J Environ Manage.* 2021; *287*: 112279.

You W, Peng W, Tian Z, Zheng M. Uranium bioremediation with U(VI)-reducing bacteria. *Sci Total Environ.* 2021; *798*: 149107.

Bacteria can be a renewable source of bioenergy

Bokinsky G, Peralta-Yahya PP, George A, Holmes BM, Steen EJ, Dietrich J. Synthesis of three advanced biofuels from ionic liquid-pretreated switchgrass using engineered *Escherichia coli*. *Proc Natl Acad Sci USA.* 2011; *108*(*50*): 19949–19954.

Gomaa OM, Costa NL, Paquete CM. Electron transfer in Gram-positive bacteria: Enhancement strategies for bioelectrochemical applications. *World J Microbiol Biotechnol.* 2022; *38*(*5*): 83.

Chapter

20

Infectious Diseases

There are many other texts that have been written exclusively on infectious diseases and which focus on these exclusively throughout. This chapter provides an overview of the contributions that bacterial genetics and genomics have made to the control of infectious diseases including expanding our understanding of pathogenesis and evolution of bacteria. The examples presented here are not exhaustive and there are other stories embedded within the book in other chapters as well. Within this chapter the innovations that can be brought to infection control through exploration of bacterial genome sequences will be introduced.

It is useful to remember when reading this chapter that there is a Glossary of Bacterial Species at the end of this book. This provides additional information on all bacterial species mentioned in this book, which may be useful for putting species into context.

The study of bacterial pathogen genes has led to new drugs to control infectious diseases

One-half of the antibiotics still in use were discovered in the 1950s and 1960s (**Figure 20.1**). Today, these drugs are no longer protected by their original patents. The antibiotics can therefore be made by any company, which means they are inexpensive and readily available. However, overuse, misuse, and bacterial evolution have led to resistance to many of the established antibiotics by important human pathogens. For example, antibiotic resistance in pneumonia is a major problem for treatment. The Gram-negative bacteria *Pseudomonas aeruginosa*, *Klebsiella pneumoniae*, *Acinetobacter baumannii*, and Gram-positive methicillin-resistant *Staphylococcus aureus* are major causes of hospital-acquired pneumonia.

Efforts have been put into discovering novel antibiotics to overcome resistance. Through these efforts, key points of antimicrobial drug design were realized. First, due to genetic variation of bacteria, potent antimicrobials have been identified that kill just one strain, species, or genus, but do not have broader activity, being **narrow-spectrum antibiotics**. In a clinical setting, it is beneficial to have antibiotics that work against a range of species, being **broad-spectrum antibiotics**, because the **etiological agent**, the bacteria causing the infection, is not known. In many cases this has meant that antimicrobials that are deemed too

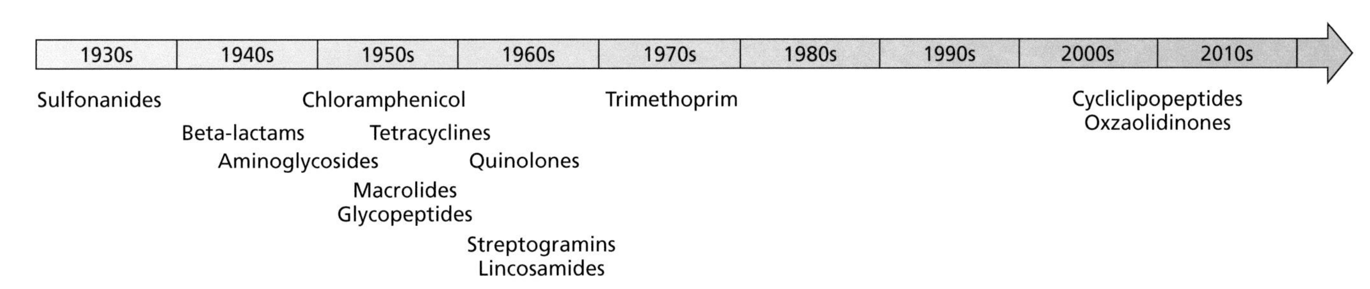

Figure 20.1 Antibiotics discovery timeline. Most antibiotics were discovered in the 1950s and 1960s. There had been no new classes of antibiotics since trimethoprim in the 1970s until cycliclipopeptides and oxazolidinones in the 2000s.

DOI: 10.1201/9781003380436-27

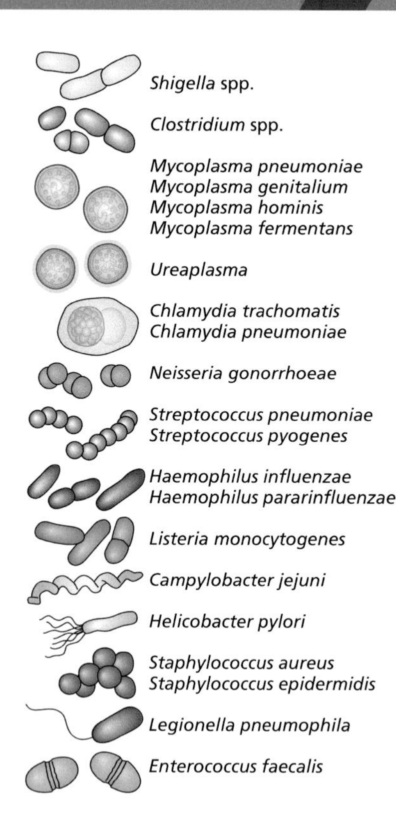

Figure 20.2 Bacterial species susceptible to Solithromycin (CEM-101). Solithromycin is a macrolide of the subclass fluoroketolide, which inhibits translation through its action on 23S rRNA.

specific in their activity have, unfortunately, not been supported for development. This restriction could perhaps be overcome in the future by advances in rapid diagnostics, enabling personalized infection identification and treatment. Second, when key genes have been identified as targets for drug development, small-molecule drugs proved ineffective, even in the short term. Small-molecule interference with enzymes resulted in rapid selection for resistance, which does not make them viable for clinical use either. This limitation may be able to be overcome in the future through design of molecular inhibitors that contact the target enzyme at more than one site, more like current antimicrobials.

Extensive understanding of bacterial pathogen genes and the antibiotics that have been used against them for decades has led to the development of some novel antibiotics. Unfortunately, three that were introduced in recent years were discontinued due to serious adverse events: Telithromycin; Temafloxacin; and Trovofloxacin. Also introduced in recent years was Fidaxomicin, available specifically for *Clostridioides difficile* colitis, despite issues with antibiotics for single species. Three antibiotics are available for community-acquired pneumonia and acute skin infections: Ceftaroline; Oritavancin; and Telavancin. A further two, Dalbavancin and Tedizolid, are also available for acute skin infections. Each of these faces market pressures from existing antibiotics, which are less expensive, however they provide alternatives when there is resistance to the existing repertoire of antibiotics. There are also several antibiotics in clinical trials. Solithromycin (CEM-101) is a fluoroketolide, a subclass of macrolides acting on the 23S rRNA subunit to inhibit translation. This fluoroketolide inhibits 50S ribosomal subunit formation and causes errors in translation. The structural modifications of Solithromycin increase its ribosomal binding and reduce its likelihood of development of resistance compared to macrolides. It has been demonstrated to act against many bacterial species (**Figure 20.2**). Zoliflodacin (AZD0914) is a spiropyrimidinetrione topoisomerase inhibitor that inhibits DNA biosynthesis and causes double-strand DNA breaks to accumulate in the bacterial cell. Although the target is similar to fluoroquinolone antibiotics, the mechanism of action of Zoliflodacin is different. It has been shown to act against several bacterial species (**Figure 20.3**). Gepotidacin (GSK2140944) also inhibits DNA replication like fluoroquinolones, but also via a different mechanism. It has demonstrated antimicrobial activity against some species as well (**Figure 20.4**).

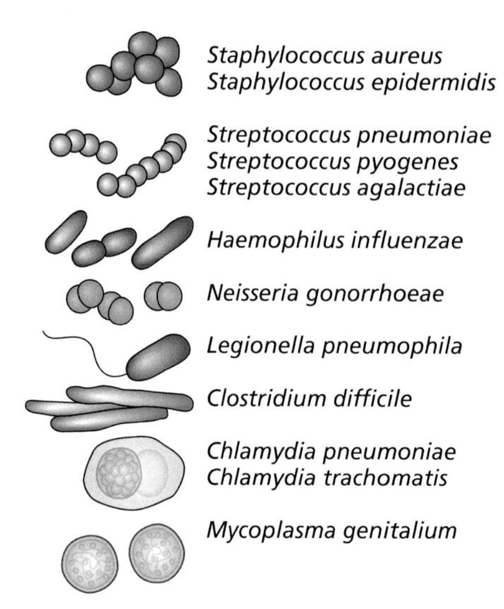

Figure 20.3 Bacterial species susceptible to Zoliflodacin (AZD0914). Zoliflodacin inhibits topoisomerase and thus inhibits the biosynthesis of DNA. It is a spiropyrimidinetrione antibiotic.

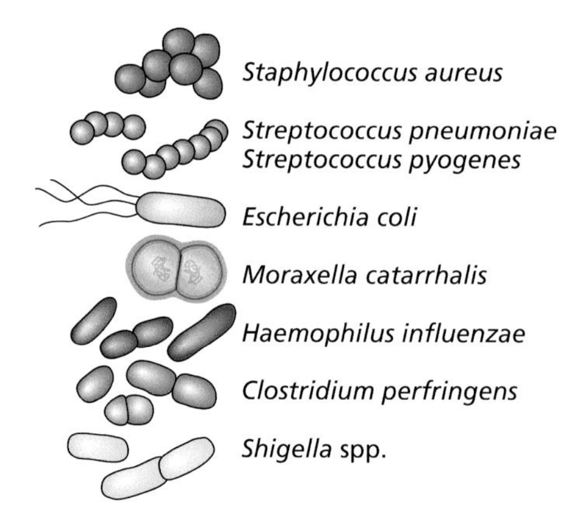

Figure 20.4 Bacterial species susceptible to Gepotidacin (GSK2140944). Gepotidacin inhibits DNA replication.

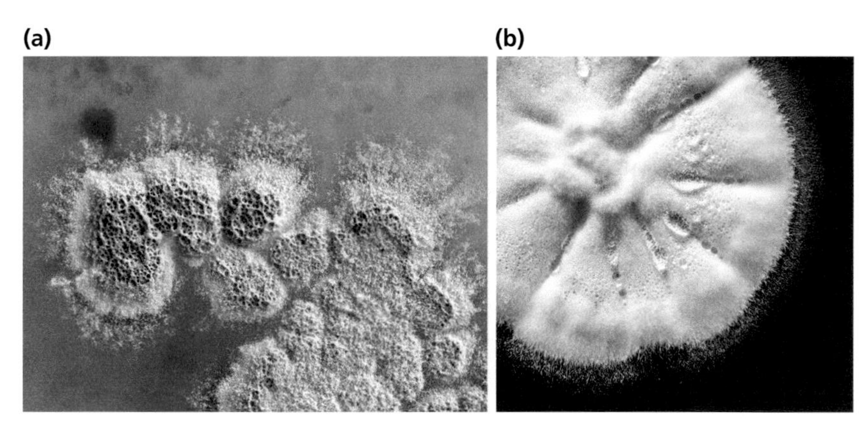

Figure 20.5 Sources of antibiotics: bacteria and fungi. Some bacteria, such as *Streptomyces coelicolor* (a), and fungi, such as *Penicillium chrysogenum* (b), produce antibiotics as defense mechanisms against other organisms in their environment. In panel (a), the antibiotic undecylprodigiosin is seen as the red areas diffusing into the medium. In panel (b), the spherical droplets of yellow on the surface are penicillin G. In some cases, antibiotics produced by organisms such as these have been exploited by humans, either in their natural form or as derivatives of the natural product. (Photographs from Dr. Jeremy Burgess and Daniela Beckmann, respectively, Science Photo Library.)

Genomics can aid in the search for new antibiotics

The majority of antibiotics originally came from or were derived from antimicrobial agents produced by bacteria and fungi (**Figure 20.5**). These agents originally evolved to rid the organisms of competitors in their niche and there are many more undiscovered antimicrobial compounds made by the microbial world. However, most of the bacterial species cannot yet be cultivated in the laboratory because the correct media and growth conditions are not known. Fortunately, metagenomics sequences can reveal the genes within unculturable bacterial and fungal species, including potential new antibiotics.

Several undertakings have been initiated to isolate bacteria from soil in an attempt to identify bacteria producing novel antimicrobials. These include major research projects as well as school projects. Some focus on simple microbiology techniques, while others also use genome sequencing methods. For example, whole-genome sequencing of *Pseudomonas* species from around plant roots has been used to identify novel antimicrobials made by these bacteria, which evolved against species in this niche but may be of use in combating human infections.

Researchers are also exploring unique environments for antibiotic producing bacteria, including those in the Antarctic and Atacama Desert in Chile. These sorts of investigations are also revealing that antibiotic resistance genes have been around far longer than our use of antibiotics. A β-lactamase gene within a metagenomics sequence was isolated from deep-sea sediment dated 10,000 years ago, reinforcing that these genes were present naturally as defenses against the antibiotics produced by other microbes. In addition to studies looking at the age of β-lactamase genes, others have used metagenomics to look at their diversity, discovering that it is much larger than previously believed, which may have implications for clinical control of infectious diseases with β-lactam-based antimicrobials.

Some old drugs are getting a new lease of life due to greater depth of understanding

β-Lactam antibiotics, such as penicillin, have been compromised in their usefulness by the acquisition of many bacterial species of **β-lactamases** conferring resistance. The β-lactam antibiotics are characterized by their ring

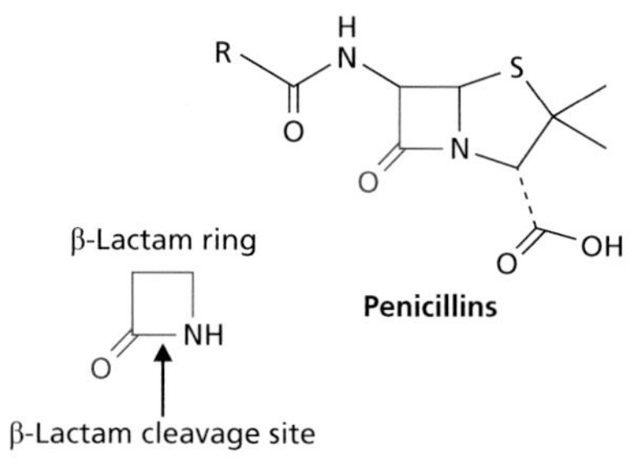

β-Lactam ring

β-Lactam cleavage site

Penicillins

Figure 20.6 The β-lactam ring and its cleavage by β-lactamase enzymes. Characteristic of β-lactam antibiotics such as penicillins is the presence of the β-lactam ring (red). This is the target of the β-lactamase enzyme for the inactivation of this class of antibiotics. β-lactamase enzymes cleave the bond in the ring between the nitrogen and the carbon that is bound to the oxygen.

structure, which is cleaved by the action of the β-lactamase enzyme (**Figure 20.6**). Understanding the horizontal transfer of the β-lactamase genes and analysis of the types of β-lactamases has led to developments of β-lactamase inhibitors that enable β-lactams to be used. The first of these was in the antibiotic Augmentin, released in 1984, which was a combination of Amoxicillin and clavulanic acid, a β-lactamase inhibitor. More recently, research has gone into development of new metallo-β-lactamase inhibitors designed to combat carbapenem-resistant Enterobacteriaceae infections. These Gram-negative bacteria express carbapenemase, inactivating last-line antibiotics used against them.

Two fairly new antibiotics, which are combinations of β-lactamase inhibitors and antibiotics, are being used in the treatment of complicated urinary tract infections and intra-abdominal infections. These are Ceftazidime-avibactam and Ceftolozane-tazobactam. Both antibiotic components of these drugs are third-generation cephalosporins, which would be susceptible to **extended-spectrum β-lactamases** (**ESBLs**), enzymes able to cleave the changed ring structure of **cephalosporins** (**Figure 20.7**). Cephalosporins are derivatives of β-lactam antibiotics that have changed the characteristic ring structure. Combining them with the inhibitors avibactam or tazobactam provides protection from ESBLs.

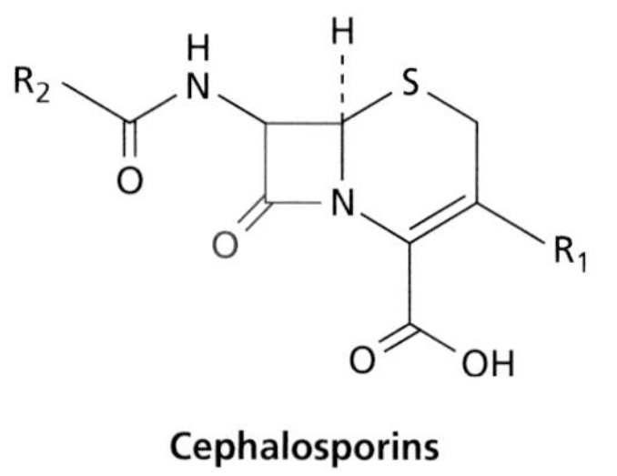

Cephalosporins

Figure 20.7 The structure of cephalosporin antibiotics. To overcome β-lactamase inactivation of β-lactam antibiotics, derivatives were created with an altered ring structure, the cephalosporins. Although they retain the β-lactam ring (red), the adjacent ring has an additional carbon. This provided resistance to standard β-lactamases. However, extended-spectrum β-lactamases are able to cleave the cephalosporin ring.

As discussed in Chapter 5, quorum sensing is often important for bacterial pathogenesis. Interfering with bacterial communication can therefore control infectious diseases. Autoinducer-2 is used by both Gram-negative and Gram-positive bacteria. It is known that existing drugs interfere with quorum sensing. For example, *P. aeruginosa* is not killed by azithromycin, but this antibiotic does interfere with *P. aeruginosa* quorum sensing. Quorum sensing is therefore an avenue for exploration, either through repurposing existing antibiotics or the development of new drugs to interfere with quorum sensing signals to disrupt virulence.

Bacterial genomics has led to the development of new vaccines

Genome sequencing of pathogens has made it possible to identify potential vaccine target antigens from sequence data, rather than immunological data. This process, a genome first approach to vaccine discovery, is called reverse vaccinology (see Chapter 11). The first licensed vaccine designed in this way targets *Neisseria meningitidis*. To overcome the inability to target a vaccine against the serogroup B meningococcal capsule, which mimics host antigens, the genome of *N. meningitidis* serogroup B strain MC58 was sequenced and analyzed to identify potential antigens for development of a vaccine. Following several years of research and clinical trials, a vaccine is now in use that originated with the genome sequencing project and the reverse vaccinology approach (**Figure 20.8**).

Investigations based on reverse vaccinology start with the genome sequence. The identified putative surface antigens from the sequence data are cloned and expressed as recombinant proteins. These proteins are then injected into animals, immunizing them to see if the proteins provide protection for the animals against bacterial infection. These experiments suggest that the genomics identified antigen is a promising candidate for vaccine development (see Figure 20.8).

In the future, reverse vaccinology may look at transcriptomics data to target antigens important for invasive disease, rather than colonization. This strategy would develop vaccines that restrict the bacteria to non-pathogenic lifestyles rather than those that harm the host.

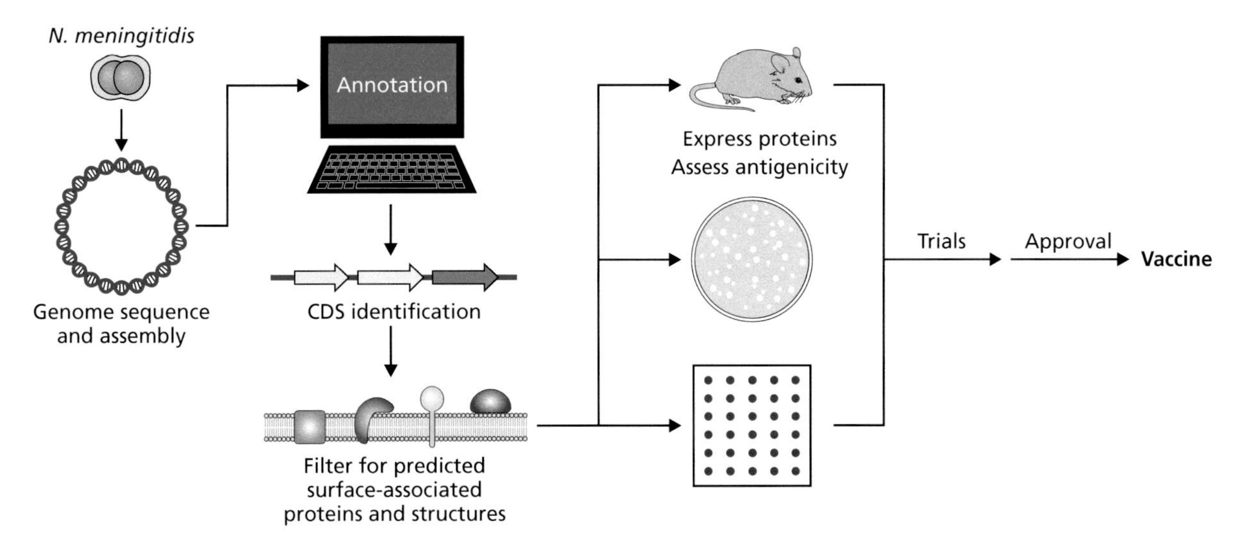

Figure 20.8 Reverse vaccinology development of a vaccine against *Neisseria meningitidis*. Starting with the bacteria, DNA is extracted for genome sequencing. The genome sequence data are assembled and annotated, with coding sequences (CDS) being identified. Since a vaccine is concerned with surface-associated proteins and structures, those are focused upon, being expressed and assessed for antigenicity and suitability as vaccine targets by *in vivo* and *in vitro* laboratory experiments. Candidates that are demonstrated to be effective undergo clinical trials and approvals before becoming a vaccine.

Reverse vaccinology is providing leads for several bacterial diseases

Reverse vaccinology is being applied to development of improved vaccines to combat leptospirosis, a **zoonotic disease** caused by *Leptospira* species. Zoonotic diseases are those that infect animals but can be transferred to humans as well. There are 15 infectious species of *Leptospira*, the diversity of which complicates the development of vaccines for animals and humans. Identification of antigens through comparative analysis of genomic data from several *Leptospira* species has been beneficial in providing leads for investigation.

Streptococcus agalactiae causes life-threatening infections in newborns, acquired at birth via transmission from the mother. Vaccination could reduce group B streptococcal disease. Eight genome sequences of *S. agalactiae* from different serotypes were compared to find the surface-associated and secreted proteins. Following expression of recombinant proteins and immunization of mice, four antigens were identified with good protective properties. One of the antigens is encoded as part of the core genome, whereas the other three are part of the accessory genome, being present in some *S. agalactiae*, but not all.

A. baumannii causes nosocomial infections, including meningitis and pneumonia. Antimicrobial resistance in this species is a clinical concern, therefore concentration is on vaccine development as a means of infection control. Fourteen genome sequences, including 11 multidrug-resistant strains, were comparatively analyzed to identify vaccine candidate antigens. This strategy found one surface antigen as a potential vaccine candidate.

Campylobacter jejuni and *Campylobacter coli* are leading causes of bacterial gastroenteritis. Reduction in poultry colonization would significantly reduce human disease and could potentially be achieved through avian vaccination. Antigens identified by reverse vaccinology were assessed against genomic data for more than 100 *Campylobacter* strains. Thus 14 candidate vaccine target antigens were selected for further investigation.

New drugs are being developed that will contain the virulence of bacteria

With the rise in antibiotic resistance, some researchers have turned to investigation of **antivirulence agents** as means of infectious disease control. These do not

(a)
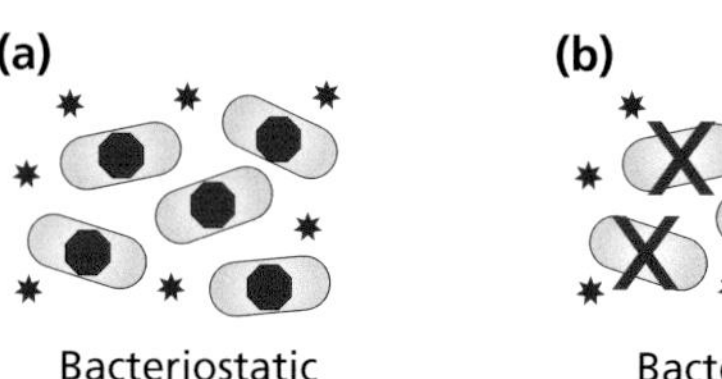
Bacteriostatic

(b)
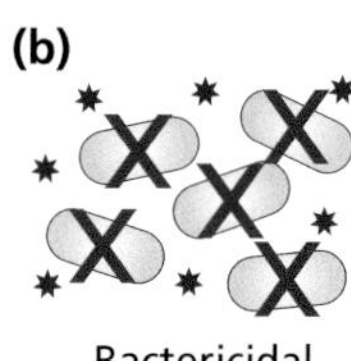
Bactericidal

(c)
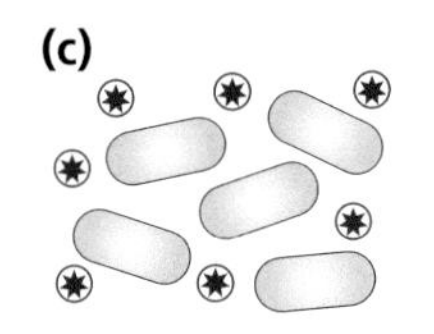
Antivirulence agent

(d)
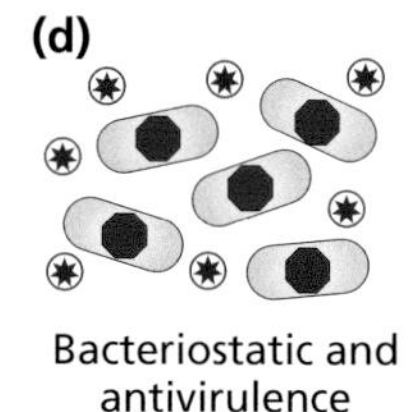
Bacteriostatic and antivirulence

(e)
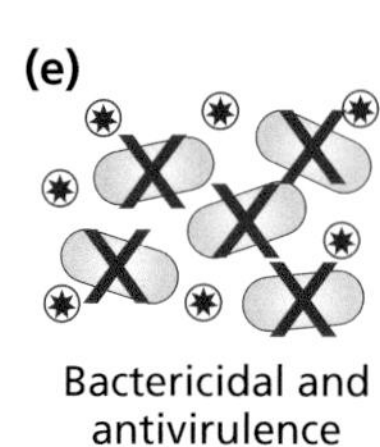
Bactericidal and antivirulence

Figure 20.8 Bacteriostatic, bactericidal, and antivirulence agents can work alone or in combination. Bacteriostatic agents stop the growth of bacteria (panel a) and bactericidal agents kill bacteria (panel b). Antivirulence agents do not act on bacteria directly, but are able to act against factors that cause disease, such as inactivating toxins (purple stars), as shown here (panel c). Antivirulence agents can be combined with bacteriostatic (panel d) or bactericidal (panel e) agents.

seek to stall bacterial growth like **bacteriostatic** antibiotics or eradicate bacteria like **bactericidal** antibiotics (**Figure 20.9**). Antivirulence agents act against the factors that cause disease and harm by the pathogens. They can work alone or in combination with antibiotics (see Figure 20.9). Often, they are designed against bacterial toxins, reducing tissue damage to the host yet also reducing selective pressures to develop resistance since the antivirulence agent is not impacting the survival of the bacteria. Targeting virulence may enable the host to clear the pathogen.

In a study published in 2018, researchers in Ohio reported their development of two small-molecule antivirulence agents against the transcription factor AgrA. The AgrA regulator is part of a two-component regulatory system (see Figure 13.17). It is blocked from binding to the promoter responsible for production of bacterial toxin. Homologues of AgrA are present in many toxin-producing Gram-positive pathogens, including methicillin-resistant *S. aureus* (MRSA), *Staphylococcus epidermidis*, *Streptococcus pyogenes*, and *Streptococcus pneumoniae*. Comparative assessment of genome sequence data made it possible to select the product of the *agrA* gene to develop the antivirulence agent. These small molecules against AgrA have been shown to work well in mouse models alone and in combination with antibiotics.

In addition to toxins, antivirulence agents can target bacterial adhesion. Since adhesion to host cells is important for colonization and infection, blocking this step can aid in preventing establishment of infection and promote clearance of the bacteria. For example, uropathogenic *E. coli* rely on adherence to stay in the urinary tract, where they cause kidney and bladder infections. In the bladder, the *E. coli* adhesion FimH binds to mannose on host cells; antivirulence agents that disrupt this are under development. When the *E. coli* cannot bind, they are washed out in the urine stream. Compounds targeting *E. coli* FmlH, the adhesion responsible for kidney cell adhesion, have also been pursued.

There have also been efforts to develop antivirulence agents against biofilm development. Infections that are of clinical importance as **nosocomial infections**, which are hospital-acquired infections, are referred to as **ESKAPE pathogens**, an acronym for the species *Enterococcus faecium*, *S. aureus*, *K. pneumoniae*, *A. baumannii*, *P. aeruginosa*, and *Enterobacter* species. Chronic biofilms by these species and other pathogens contribute to difficult to treat antimicrobial-resistant infections. Inhibition of biofilms is therefore important in combating infectious diseases, either by preventing biofilm formation or by dispersing established biofilms, or potentially both. Biofilm development proteins LecA and LecB from *P. aeruginosa* have been targeted for development of antivirulence agents. Unlike AgrA, the antivirulence agents against adhesion proteins FimH and FmlH and against biofilm proteins LecA and LecB are very much species specific, and the infecting organism would need to be identified. However, advances in rapid diagnostics may overcome these limitations (**Figure 20.10**).

Antivirulence target	Mechanism of action	Species
Toxin	Blocks toxin binding to host cells	Various
AgrA	Blocking two-component regulatory system	*S. aureus* *S. epidermidis* *S. pyogenes* *S. pneumoniae*
FimH	Disrupts bacteria binding to bladder cells	*E. coli*
FmlH	Disrupts bacteria binding to kidney cells	*E. coli*
LecAB	Disrupts biofilm development	*P. aeruginosa*

Figure 20.10 Summary of some of the antivirulence agents in development.

Monoclonal antibody therapy is useful for a variety of human diseases, including infectious diseases

In recent years, human monoclonal antibodies have been developed for the treatment of cancer, rheumatoid arthritis, Crohn's disease, asthma, and migraine. Monoclonal antibodies have also been developed against the toxins of important bacterial pathogens. Development of these makes use of the gene sequences and comparative analysis of the diversity of the toxin across the bacterial population in design of the antibody. Building on the success of other therapeutic antibodies, antitoxin antibodies have been developed against *C. difficile* toxin B, *Bacillus anthracis* toxin, and *S. aureus* α-hemolysin. Antibodies are also in development to block secretion of toxins by the bacterial Type 3 Secretion Systems (see Chapter 9). Molecules that bind to and inactivate Shiga toxins Stx1 and Stx2 of Enterohemorrhagic *E. coli* have been developed as well. Monoclonal antibodies against toxins and secreted proteins would act as antivirulence agents (**Figure 20.11**), similarly to the small molecules described above.

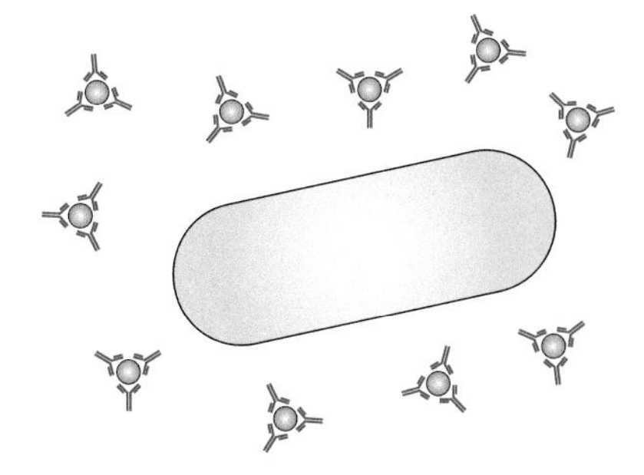

Figure 20.11 Monoclonal antibody therapy against infectious diseases. The toxins that have been secreted by this bacterial cell are inactivated by the monoclonal antibodies that have been specifically raised to bind to the toxin. Injection of the patient with the prepared monoclonal antibodies provides the antivirulence agent that protects the patient from the effects of the toxin.

Sequencing changes our understanding of the virulence factors that are important

S. pyogenes, group A *Streptococcus*, causes pharyngitis, frequently referred to as "strep throat," and superficial skin infections such as impetigo. It also causes toxic shock syndrome and necrotizing fasciitis. The leading cause of invasive group A streptococcal disease is serotype M1, *emm*1, although *emm*89 has also emerged despite its lack of a capsule. As for other bacterial capsules, the group A streptococcal capsule contributes to neutrophil phagocytosis resistance.

A study in Texas investigated the frequency of *S. pyogenes* disease by bacteria without capsules, looking at pharyngeal infections, skin and soft-tissue infections, and invasive disease. In this study, it was revealed that in 2016–2017, 50% of all isolates investigated did not have capsules. Where once it was believed the capsule was an essential virulence factor, careful surveillance and study of genetics uncovered that a significant proportion of disease is caused by capsule-negative group A streptococci. Almost 90% of these were from pharyngeal and skin and soft-tissue infections. These results suggest that the capsule is not necessary for this species to cause disease, but that it is unlikely that *S. pyogenes* without capsule will cause invasive disease, although not impossible.

Gene sequencing and genome sequencing improve the resolution of epidemiology of bacterial infectious diseases

Genomics applied to **epidemiology**, the study of the distribution of diseases, gives a high level of resolution of discrimination between infectious disease isolates. Local epidemiology investigations can be concerned with whether there is an outbreak within a hospital ward, for example, whereas global investigations

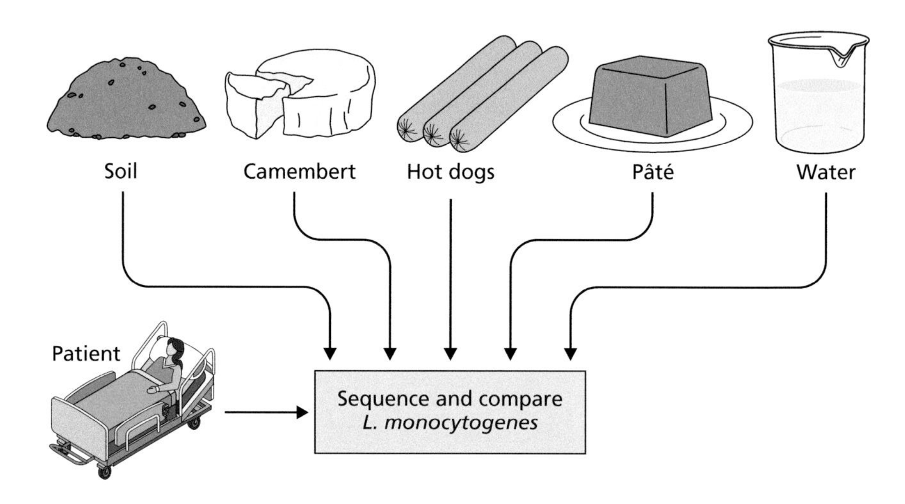

Figure 20.12 Identification of the *Listeria monocytogenes* contaminated source can stop disease spread. The bacteria can thrive in environmental and food sources such as soil, water, camembert, hot dogs, pâté, and others. Sequencing *L. monocytogenes* genome data from these sources and from a patient with the disease can identify the source and help contain spread to other individuals, perhaps through recall of contaminated foods.

can be concerned with the origins and distribution of antibacterial resistance markers, among other features.

Molecular epidemiology has been applied to the interesting question of the diseases caused by *S. pyogenes*, group A streptococci. In the winter months in temperate regions, where people spend time inside, *S. pyogenes* primarily causes pharyngitis. In summer months and in subtropical and tropical regions, *S. pyogenes* causes impetigo. Both seasonal occurrence and geography separate the diseases. Genetic analysis has been applied to see if this is reflected in the species population structure, to determine whether pharyngitis and impetigo strains are genetically distinct. Diversity in the sequence of the antiphagocytic M protein gene *emm* has been shown to correspond to these two disease outcomes.

Comparative genome analysis of the epidemiology of *C. difficile* has revealed several aspects of the nature of the disease. Transmission within wards and outbreaks have been confirmed based on genome sequence data. Dissemination of an epidemic strain of *C. difficile* across continents has been shown, as has the transmission from pigs to humans, all from assessment of the genomic data.

Genome sequence information from *Listeria monocytogenes* has been used to develop new typing methods with greater discriminatory power than previous tools. Bacterial isolate subtyping can identify if isolates from contaminated food, environmental samples, and human disease are related, aiding in infection control. Genomics has made a positive contribution in this way to the rapid identification of outbreaks of *L. monocytogenes* and tracing the contaminated food or environmental source to reduce the number of cases of disease (**Figure 20.12**).

Genome sequencing can improve infection control for surgical site infections

Patients who undergo surgery have a 0.6%–9.6% chance of surgical site infection. They are also at risk of development of nosocomial infections. As mentioned, outbreak detection is largely the same as it has been for about 100 years. **Surveillance** in the hospital, looking for infections, involves assessing a daily list of patients who test positive for infectious organisms. Catching overlaps in cases of patients with the same infections can be difficult. This can lead to an epidemiological investigation, however traditional typing takes weeks, involves sending samples away to a reference laboratory, and may not be sufficiently discriminatory.

Instead, bacterial sequencing within the hospital and epidemiological analysis can be used to identify outbreaks more quickly and with greater resolution. For example, the London School of Hygiene and Tropical Medicine has developed *Sequence First* outbreak detection. When a patient is identified with an infectious organism, it can be sequenced within a day. The sequence data and patient data together make a sample report that is available the morning of the following day. The report, identifying the organism, includes transmission likelihood information for the infectious disease and, coupled with the data from the rest of the hospital sequence information, gives early warning information for outbreaks. This can be expanded to provide data on outbreaks across hospitals, across nations, and across the globe.

Genome sequencing of methicillin-resistant *S. aureus* (MRSA) isolates from a special care baby unit has been conducted in the United Kingdom. This validated and expanded upon findings from conventional outbreak investigations that happened in the hospital. There were 12 infants with MRSA within 6 months in 2011 that could not be confirmed with conventional methods to be from an outbreak, however these were confirmed by genome sequencing, which found 26 related MRSA in total. These 26 MRSA included transmission within the special care baby unit, between mothers on the maternity ward, and in the community. It had been noted during conventional outbreak investigations that the special care baby unit outbreak persisted despite there being no obvious transmission chain between patients. This was revealed to be due to carriage by a member of hospital staff, but this was only evident from the sequence data.

Horizontal gene transfer between pathogens revealed by sequencing shows worrying trends in evolution

Bacteria of the genus *Neisseria* are naturally competent for transformation (see Chapter 3) and by preference take up DNA containing the *Neisseria* DNA uptake sequence (DUS). The DUS is present thousands of times in the genome (see Figure 17.9) and is present in pathogenic *N. meningitidis* and *Neisseria gonorrhoeae* and in non-pathogenic *Neisseria* species. Genetic material therefore horizontally transfers between species, although it has rarely been observed between the two pathogens. This is fortunate considering *N. meningitidis* can be rapidly fatal within 24 hours and *N. gonorrhoeae* has genes for multiple types of antibiotic resistance.

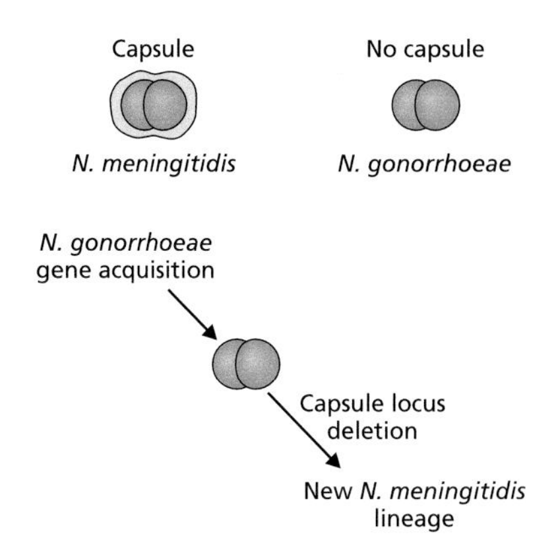

Figure 20.13 *Neisseria meningitidis* from the nasopharynx typically has a capsule. *N. gonorrhoeae* does not have a capsule. There is a lineage of *N. meningitidis*, typically isolated from the urogenital tract, that has acquired *N. gonorrhoeae* genes and deleted its capsule locus.

However, a lineage of *N. meningitidis* has emerged in the United States that has genes recognized by genome sequencing that originated in *N. gonorrhoeae*. *N. meningitidis* causes invasive, life-threatening meningitis and septicemia. It can also cause other infections more common to its relative *N. gonorrhoeae*, including urethritis. There have been some connections between cases of urethritis and meningococcal meningitis, possibly due to oral sex as a mode of transmission. In 2015, in Columbus, Ohio, it was noticed that a quarter of the *Neisseria* isolated from urethritis were *N. meningitidis* and not *N. gonorrhoeae* as might be expected from the isolation site. It has emerged that there is a lineage of *N. meningitidis* causing many urethritis cases and that this lineage is notable in the absence of the characteristic meningococcal capsule; the capsule locus has a multigene deletion (**Figure 20.13**). Capsule-deficient *N. meningitidis* tend to be isolated from infections at different body sites than those expressing a capsule, which colonize the nasopharynx and cause invasive meningococcal disease.

N. gonorrhoeae also does not have a capsule (see Figure 20.13). This similarity may contribute to the enhanced mucosal attachment of the *N. meningitidis* in the urethra. The urethritis *N. meningitidis* lineage genome sequence data has revealed that it has other features in common with *N. gonorrhoeae*. It is capable of nitrite dependent anaerobic growth, which is possible through the acquisition of *N. gonorrhoeae aniA* and *norB* genes. This anaerobic ability is also likely to contribute to enhanced urethral colonization.

Most concerning, one isolate of this *N. meningitidis* lineage has acquired the *N. gonorrhoeae* sequence of the *mtrR* promoter and coding sequence. This gene is responsible for the repression of expression of genes encoding an efflux pump that is associated in *N. gonorrhoeae* with resistance to several antibiotics, including azithromycin. In *N. meningitidis*, this efflux pump is constitutively expressed, meaning it is not controlled by MtrR repression. This means most

N. meningitidis are also not capable of high-level MtrCDE pump expression through removal of repression as in *N. gonorrhoeae*. The discovery of these *N. gonorrhoeae mtrR* sequences in *N. meningitidis* in the genomic data is a worrying trend, indicating that multidrug resistance traits could be transferred between the pathogens.

Genome sequencing is improving our understanding of infections that could impact transplant recovery

Infection following organ transplant is a major concern, particularly if the infection is a result of multidrug-resistant bacteria. There has been a rise in Gram-negative infections in live transplant patients, particularly carbapenem-resistant Enterobacteriaceae. It was found that as many as 67% of liver transplant patients are colonized with multidrug-resistant organisms with many of these developing infections. Through use of genome sequencing, it is better understood that the bacteria are dynamic in their nature. In some cases, patients are colonized prior to surgery, some are infected postoperatively from nosocomial infections, and some develop days after surgery from community-acquired infections. Resolution with genome sequencing can help inform treatments and within-hospital infection control measures.

Putting discoveries into practice

This chapter has presented and explored some novel ideas for alternatives to antibiotics and new ideas for controlling infectious diseases. Many of these are being put into practice and are in trials to be used in the near future. Included among these are bacteriophage therapies, which will be explored in the final chapter of this book. Research in the field of infectious disease control is advancing rapidly, motivated by the urgent need to address the growing antimicrobial resistance crisis, which not only impacts the healthcare of humans, but also the agricultural industry and veterinary medicine as well.

Key points

- New antimicrobials can be revealed through the study of bacterial genomes and understanding the role and function of essential genes and virulence genes.
- Metagenomics can aid in the search for new antibiotics through the identification of genes in bacterial species that cannot yet be cultured in laboratories.
- Pairing of β-lactam antibiotics and cephalosporin antibiotics with inhibitors of β-lactamases and extended-spectrum β-lactamases can overcome some types of antimicrobial resistance.
- Reverse vaccinology has used genomic sequence data as the basis for the development of new vaccines, where antigens are predicted from the genomic information.
- Novel approaches to control infectious diseases include development of antivirulence agents that prevent bacteria from causing disease.
- The resolution of genome sequence data provides details for epidemiology studies that can discriminate between lineages causing different diseases, identify transmission events, and trace the sources of infections so they can be contained.

- Surveillance of infections within hospitals is important for infection control; application of genome sequencing is improving rapid identification of outbreaks.
- Genome sequencing is changing our understanding of the importance of certain genes in disease, including revealing that the bacterial capsule is not necessary for species like *S. pyogenes* and *N. meningitidis* to cause infections.

Key terms

Define the following terms introduced in this chapter. Check your answers using the definitions in the Glossary. These terms are also available as Flashcards online.

β-Lactam	Broad-spectrum antibiotic	Extended-spectrum β-lactamases
β-Lactamases	Cephalosporins	Narrow-spectrum antibiotic
Antivirulence agent	Epidemiology	Nosocomial infections
Bactericidal	ESKAPE pathogens	Surveillance
Bacteriostatic	Etiological agent	Zoonotic disease

Questions and discussion topics

Self-study questions

Answer each question using 50–100 words or a table or labeled diagram. Advice on where to find answers to these questions is available online.

1. When were most antibiotics discovered? What are some of the reasons for antibiotic resistance to these drugs?
2. What is the difference between a narrow-spectrum antibiotic and a broad-spectrum antibiotic? Will there be a different outcome for the patient if both are equally able to kill the bacteria causing disease?
3. What are the sources of most antibiotics? Give some examples of natural antibiotic producers. Why do they make antibiotics?
4. Have antibiotic resistance genes been around longer than we have been using antibiotics? Explain your answer.
5. Give examples of how combining two drugs, an antibiotic and an inhibitor, is being used to tackle antibiotic resistance. How do these drugs work together?
6. Briefly explain the processes and principles involved in reverse vaccinology.
7. What is the difference between a drug that is an antivirulence agent and one that is an antibiotic? Could they work together? What is the advantage of targeting virulence?
8. Name two targets of antivirulence agents that have been under investigation for development. How does blocking these targets aid in controlling infectious diseases? Are these antivirulence agents broad spectrum?
9. Give some examples of how our understanding of bacterial pathogens has been changed through analysis of bacterial genome sequence data.
10. Describe how genome sequencing can be used in epidemiology.
11. What does infection surveillance involve in a hospital and how can sequencing be used to help?
12. What features of the genus *Neisseria* facilitate exchange of genes between species? How can genome sequencing clarify the identity of what may appear to be *N. gonorrhoeae* based on disease symptoms and isolation location, but is actually *N. meningitidis*?

Discussion topics

These topics are presented for discussion in study groups, as part of class discussions, or on your own. These questions go beyond what is directly covered in this part of the book. Use the research literature and other reading to explore these topics in more depth. Tips to help prepare for topic discussions are available online.

1 Explore how monoclonal antibodies are being used to tackle issues in human health, specifically those related to a specific infectious disease. Is the infection itself being targeted or is it a toxin or other virulence factor? What other treatments do patients undergoing the monoclonal antibody therapy require? How is their outcome improved through the additional use of the monoclonal antibodies?

2 As mentioned in this chapter, azithromycin does not kill *P. aeruginosa*, but it does interfere with its quorum sensing. How can azithromycin be exploited as an avenue to combat *P. aeruginosa* infections through quorum sensing interference? Investigate studies in the research literature that have looked at this specifically or at biofilm and quorum sensing disruption in general. Discuss the advantages of repurposing existing antibiotics, rather than trying to find new antibiotics.

3 Explore developments in antivirulence drugs or monoclonal antibody therapies. How effective have these been at demonstrating their potential to control infectious diseases? Have there been issues with resistance as there have been with antibiotics?

Online quiz questions

To further self-assess your understanding of the chapter material, please visit the following link, where you can participate in a range of interactive quiz questions:

www.routledge.com/cw/snyder

Further reading

Some old drugs are getting a new lease of life due to greater depth of understanding

Azimi S, Klementiev AD, Whiteley M, Diggle SP. Bacterial quorum sensing during infection. *Annu Rev Microbiol.* 2020; *74*: 201–219.

van Duin D, Bonomo RA. Ceftazidime/avibactam and ceftolozane/tazobactam: Second-generation β-lactam/β-lactamase inhibitor combinations. *Clin Infect Dis.* 2016; *63*(*2*): 234–241.

Bacterial genomics has led to the development of new vaccines

Serruto D, Bottomley MJ, Ram S, Giuliani MM, Rappuoli R. The new multicomponent vaccine against meningococcal serogroup B, 4CMenB: Immunological, functional and structural characterization of the antigens. *Vaccine.* 2012; *30*: B87–B97.

Reverse vaccinology is providing leads for several bacterial diseases

Chiang MH, Sung WC, Lien SP, Chen YZ, Lo AFY, Huang JH. Identification of novel vaccine candidates against *Acinetobacter baumannii* using reverse vaccinology. *Hum Vaccines Immunother.* 2015; *11*: 1065–1073.

Maione D, Margarit I, Rinaudo CD, Masignani V, Mora M, Scarselli M. Identification of a universal Group B *Streptococcus* vaccine by multiple genome screen. *Science.* 2005; *309*: 148–150.

Meunier M, Guyard-Nicodème M, Hirchaud E, Parra A, Chemaly M, Dory D. Identification of novel vaccine candidates against *Campylobacter* through reverse vaccinology. *J Immunol Res.* 2016; *2016*: 5715790.

New drugs are being developed that will contain the virulence of bacteria

Lau WYV, Taylor PK, Brinkman FSL, Lee AHY. Pathogen-associated gene discovery workflows for novel antivirulence therapeutic development. *EBioMedicine.* 2023; *88*: 104429.

Martínez OF, Duque HM, Franco OL. Peptidomimetics as potential anti-virulence drugs against resistant bacterial pathogens. *Front Microbiol.* 2022; *13*: 831037.

Parrino B, Schillaci D, Carnevale I, Giovannetti E, Diana P, Cirrincione G, Cascioferro S. Synthetic small molecules as anti-biofilm agents in the struggle against antibiotic resistance. *Eur J Med Chem.* 2019; *161*: 154–178.

Gene sequencing and genome sequencing improve the resolution of epidemiology of bacterial infectious diseases

Knight DR, Elliott B, Chang BJ, Perkins TT, Riley TV. Diversity and evolution in the genome of *Clostridium difficile*. *Clin Microbiol Rev.* 2015; *28*(*3*): 721–741.

Genome sequencing can improve infection control for surgical site infections

Harris SR, Cartwright EJ, Török ME, Holden MT, Brown NM, Ogilvy-Stuart AL. Whole-genome sequencing for analysis of an outbreak of methicillin-resistant *Staphylococcus aureus*: A descriptive study. *Lancet Infect Dis*. 2013; *13*(*2*): 130–136.

Köser CU, Holden MT, Ellington MJ, Cartwright EJ, Brown NM, Ogilvy-Stuart AL. Rapid whole-genome sequencing for investigation of a neonatal MRSA outbreak. *N Engl J Med*. 2012; *366*(*24*): 2267–2275.

Peacock SJ, Parkhill J, Brown NM. Changing the paradigm for hospital outbreak detection by leading with genomic surveillance of nosocomial pathogens. *Microbiology*. 2018; *164*(*10*): 1213–1219.

Chapter 21

Bacteriophages

Bacteriophages are abundant and ubiquitous, their nucleic acids are the most numerous on Earth, and they are prevalent across the globe. They are diverse in their size, shape, and structure, as well as in the organization of the genes in their genome. Bacteriophages are non-motile, relying on sheer numbers and chance to reach their targets on bacterial cells, which is necessary for their replication.

Bacteriophages have been studied for just over 100 years

Bacteriophages were first discovered in 1915, by Frederick Twort, who recognized that bacteriophages inhibit bacterial growth. It was 2 years later, in 1917, that Félix D'Hérelle isolated bacteriophages and used them to prevent *Salmonella* infection of chickens. There are believed to be 10^{31} bacteriophages on the planet; that is 10 nonillion, enough for 10 bacteriophages for each bacterial or archaeal cell on Earth. The Earth's oceans are the main reservoir of bacteriophages. In every milliliter of water in the ocean there are as many as one hundred million (10^8) bacteriophages. Another major reservoir are the guts of mammals, where each gram of fecal matter can also contain up to 100 million bacteriophages.

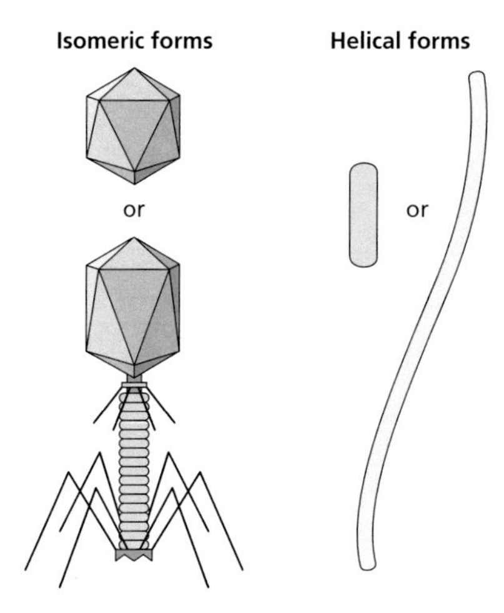

Figure 21.1 Basic shapes of bacteriophages. There are two basic morphologies of bacteriophages. The virion is either isomeric, in the shape of a polyhedral, such as an icosahedral, sometimes with a tail and tail fibers, or it is helical, in a spiral filamentous formation.

Bacteriophages are classified into the 19 recognized bacteriophage families by the type of genome they possess and the shape of the protein coat particle, called the **virion**. All bacteriophages have one type of nucleic acid, which can be double-stranded DNA, single-stranded DNA, or RNA. Most bacteriophages that have been identified and characterized contain double-stranded DNA. Only two bacteriophage families have RNA genomes, which is why we tend to discuss DNA genomes of bacteriophages. The few bacteriophages with RNA genomes can contain either single-stranded or double-stranded RNA. Of the remaining 17 families, 15 have double-stranded DNA genomes and two have single-stranded DNA genomes. The virion morphology of bacteriophages is either polyhedral, often icosahedral, forming an isomeric shape, or spiral, forming a helical shape that is filamentous (**Figure 21.1**). The icosahedral bacteriophage structures sometimes have tails, with or without tail fibers, that are involved in attachment to bacterial cells and delivery of the bacteriophage nucleic acids into the bacterial cell.

Five bacteriophage families have envelopes around their protein coats, meaning that they are surrounded by a phospholipid membrane derived from the bacterial cell. In this way, these bacteriophages are similar to enveloped viruses that infect eukaryotic cells. There are proteins embedded within bacteriophage encapsulating membranes that are encoded by the bacteriophage genome. The bacterial membrane proteins are excluded from the region of the bacterial membrane that will be used to encapsulate the bacteriophages (**Figure 21.2**). This process is similar to the formation of animal cell targeting viruses and the deposition of viral proteins in their encapsulating membranes.

In addition to morphology, bacteriophages are characterized by being lytic or lysogenic. Lytic bacteriophages are actively virulent and will immediately replicate within the bacterial cell. Lysogenic bacteriophages can sometimes integrate their nucleic acids into the bacterial chromosome. The bacteriophage genome is therefore replicated within the bacterial genome each time the bacterial cell divides. As discussed in Chapter 1, the bacteriophage genome integrated in the

DOI: 10.1201/9781003380436-28

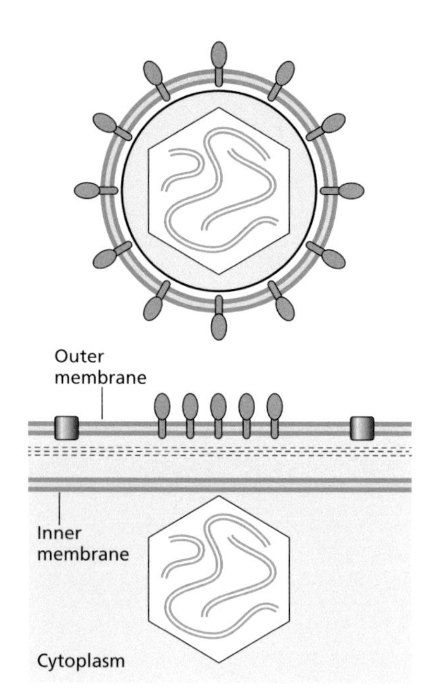

Figure 21.2 Bacteriophages with membrane envelopes. Some bacteriophages, such as φ6 shown here, are encapsulated by membrane envelopes, derived from the bacterial cell membrane. During formation (bottom panel), bacteriophage proteins (brown) are integrated into the bacterial membrane, excluding the bacterial surface proteins (red).

bacterial genome is called a prophage. Depending upon the nutritional status or stress experienced by the bacterial host cell, there may be a switch from the lysogenic to the lytic phase, which then produces bacteriophage particles again.

Bacteriophages cannot replicate without bacterial cells

Bacteriophage replication requires bacterial cells because bacteriophages lack their own processes for protein synthesis. Replication of the bacteriophage requires it to adhere to the bacterial surface, penetrate the bacterial cell, replicate its nucleic acids within the bacterial cell, form the new bacteriophage particles, and then release the bacteriophage particles from the cell (**Figure 21.3**). The attachment and penetration are specific for the bacterial strain or species, dependent on surface structures of the bacteria.

Adsorption, adhesion of the bacteriophage to the bacterial cell, is the first stage required for bacteriophage replication to occur. In the process of adsorption, bacteriophage proteins bind to receptors on the bacterial cell surface, such as surface proteins including nutrient transporters and outer membrane proteins. In Gram-negative bacteria, some bacteriophages recognize LPS, whereas other bacteriophages recognize the teichoic acids of the Gram-positive cell wall. For bacteriophages with a head and tail structure, adsorption to specific bacterial surface receptors involves the tail fibers.

Penetration is the process of the bacteriophage nucleic acids entering the bacterial cell. This may be through rupturing the cell wall and injecting the nucleic acids into the cytoplasm or finding another route through the membranes and cell wall, such as through a bacterial porin protein. Early in the lytic cycle, bacteriophage proteins digest the bacterial genome. This shuts off production of bacterial replication and transcription, allowing the bacteriophage free rein over the bacterial systems needed for its own replication. In the **eclipse phase**, the

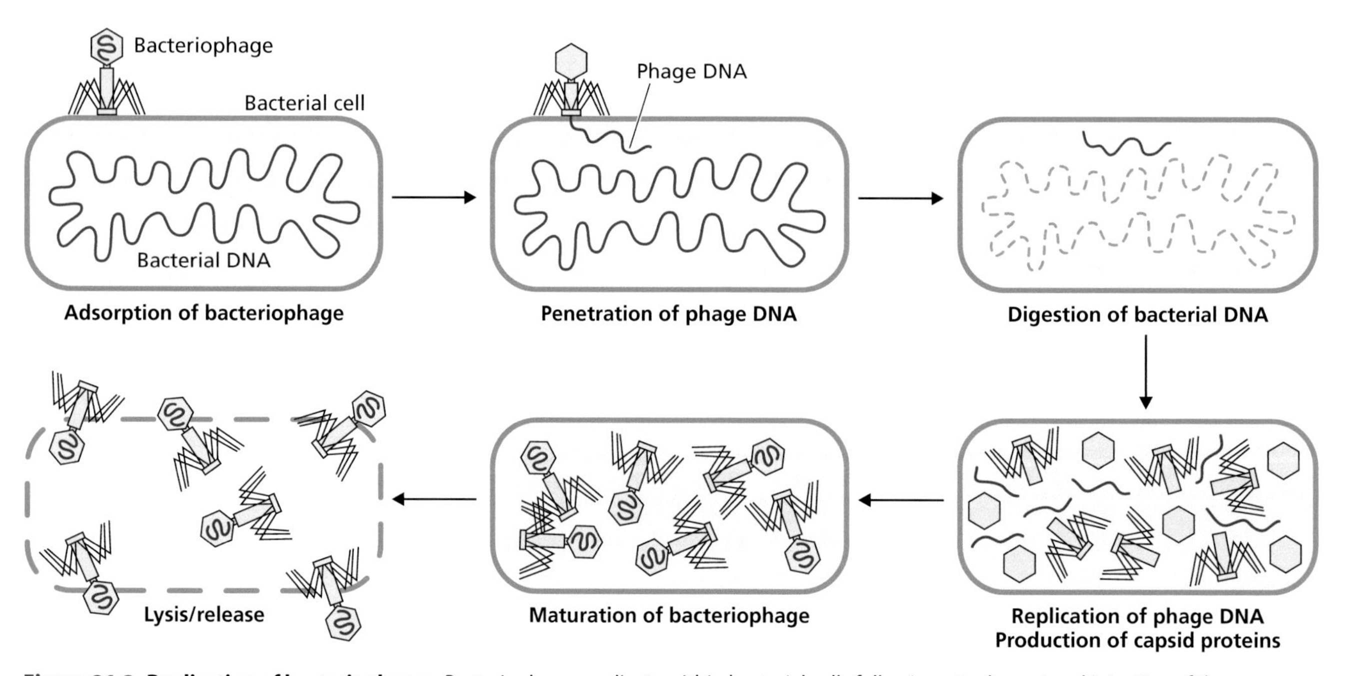

Figure 21.3 Replication of bacteriophages. Bacteriophages replicate within bacterial cells following attachment and injection of the bacteriophage genome into the cell. Once new bacteriophage particles are made, they are released from the bacterial cell via lysis.

new bacteriophages are made, including the nucleic acids and capsid, the protein coat of the bacteriophage particle. The mature bacteriophages are then formed, and the bacterial cell is lyzed to release them. Replication produces 100–200 new virions that are released upon lysis of the bacterial cell.

Some bacteriophages enter latency for a period before replication

Although lytic bacteriophages had been studied for quite some time, it wasn't until Esther Lederberg described the behavior of lambda (λ) bacteriophages in 1950 that the capacity for a lysogenic phase in some bacteriophages was appreciated. A lysogenic bacteriophage enters latency, meaning that it does not replicate immediately, but becomes quiescent within the bacterial cell (**Figure 21.4**). The bacteriophage genome is still replicated, but the entire bacteriophage is not. The nucleic acid enters the cell and rather than triggering the digestion of the bacterial DNA, the bacteriophage DNA joins it, integrating into the bacterial chromosome. No bacteriophage capsid proteins are made. The bacteriophage genome within the bacterial genome has become a prophage and is replicated as the chromosome is replicated, being copied into each daughter cell. Bacteria containing prophage are referred to as **lysogens**. This latent state can be interrupted, triggering the initiation of replication of the bacteriophage and a lytic cycle. This can occur spontaneously or be induced by UV radiation including sunlight, by antibiotics, by alkylating agents, and by cellular stress.

There are also intermediate types of bacteriophage infection that are neither lytic nor lysogenic, although these are perhaps less common. For example, chronic infections result in bacteriophages budding from the bacterial cell, rather than lysis (**Figure 21.5**). Pseudolysogenic infections fail to properly integrate into the chromosome and either segregate out from one daughter cell at each bacterial division or enter lytic replication of bacteriophages (see Figure 21.5). In pseudolysogeny, the bacteriophage fails to either replicate in a lytic cycle or integrate as a prophage in a lysogenic cycle. Instead, the bacteriophage nucleic acids remain extrachromosomal. Ultimately, conditions within the bacterial cell enable the lytic or lysogenic cycles to occur, or the bacteriophage nucleic acids are lost.

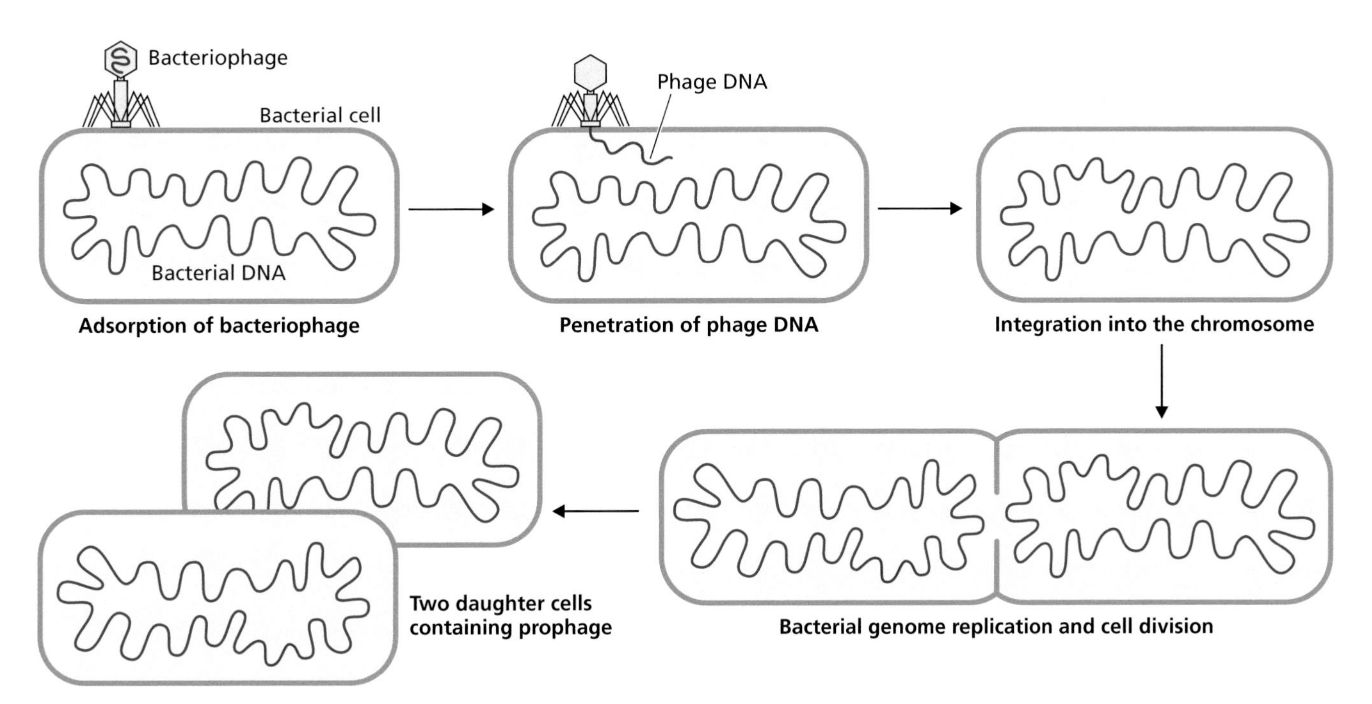

Figure 21.4 Lysogenic bacteriophages. Lysogenic bacteriophages integrate their nucleic acids into the bacterial chromosome.

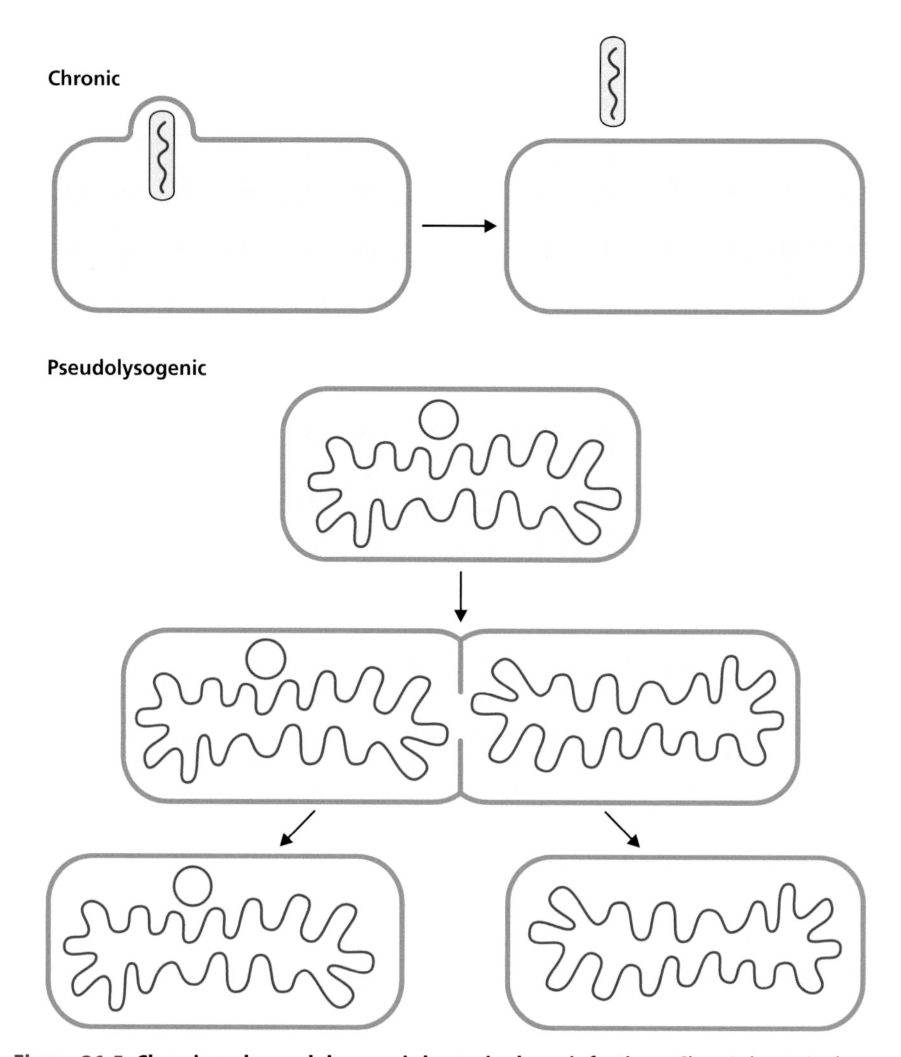

Figure 21.5 Chronic and pseudolysogenic bacteriophage infections. Chronic bacteriophage infections are similar to lytic infections, however the bacterial cells do not lyse. Instead, the bacteriophages bud from the bacterial membranes and leave the bacteria intact. Pseudolysogenic infections do not integrate into the chromosome, therefore the bacteriophage nucleic acids do not replicate with the bacterial DNA and do not segregate into daughter cells upon division, as they would in lysogens.

Integration and excision of bacteriophage DNA into the bacterial chromosome rely on both bacterial and bacteriophage factors. Bacteriophage DNA, such as λ described in more detail later, circularizes upon entry into the cell. The *attP* site on the circular bacteriophage DNA and *attB* on the circular bacterial DNA recombine, catalyzed by the bacteriophage protein Int, an integrase, and bacterial protein IHF, integration host factor (see Chapter 5). This recombination process forms *attL* and *attR* at either end of the integrated prophage (**Figure 21.6**). Upon induction of lysis, Int is again involved, along with Xis, a bacteriophage excision protein, which together remove the prophage and reform *attB* in the bacterial chromosome. This integration and excision system has been exploited in the development of cloning systems, such as Gateway, as discussed in Chapter 18.

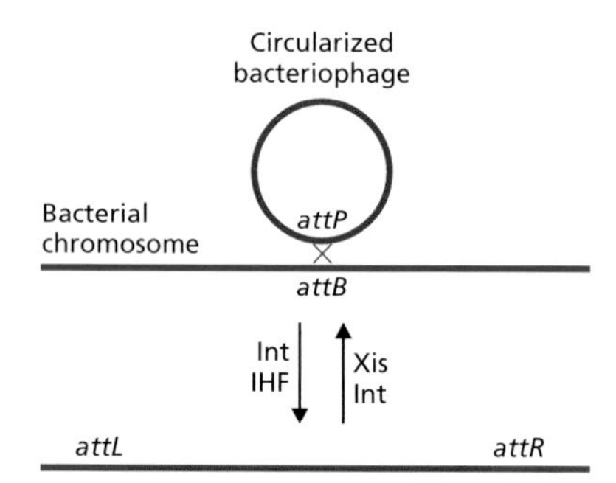

Figure 21.6 Integration and excision of the bacteriophage genome. To generate the prophage, the bacteriophage genome enters the bacterial cell cytoplasm as linear DNA, which then circularizes. The circular bacteriophage genome recombines with the bacterial genome at the *attP* and *attB* sites, respectively. Integration requires the bacteriophage Int and bacterial IHF proteins. Excision needs the additional Xis bacteriophage protein and reforms *attP* and *attB* from the *attL* and *attR* sequences that flank the prophage.

The MS2 bacteriophage has a very small genome and was the first genome sequenced

Bacteriophage MS2 is an RNA bacteriophage and has the distinction of being the first complete genome to be sequenced, in 1976. This sequencing used a two-dimensional fractionation method that was only possible with RNA and predates Sanger sequencing. The genome is only 3,569 bases, encoding four proteins, three of which overlap.

The bacteriophage MS2 is 27 nm across, made of one copy of the maturation protein, encoded by the gene *mat*, and 180 copies of the coat protein, encoded by the gene *cp*. These proteins make an icosahedral-shaped capsid (**Figure 21.7**). The receptor for MS2 bacteriophage is the conjugation pilus encoded by the F plasmid.

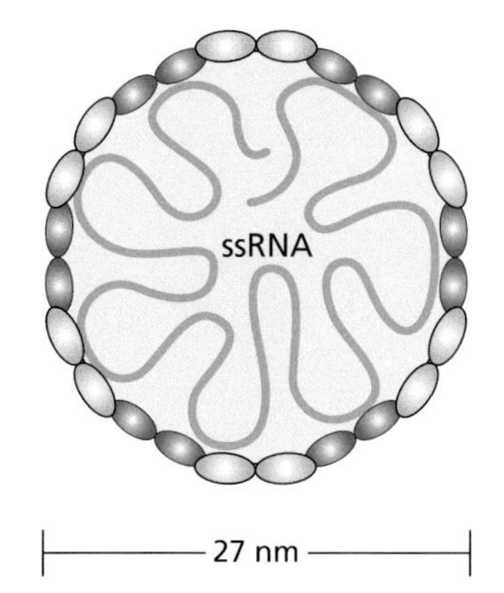

Figure 21.7 Bacteriophage MS2. The structure, morphology, and size of bacteriophage MS2. The capsid is made of the maturation protein and 180 copies of the coat protein. The genome is 3,569 bases of RNA, encoding four proteins.

Important discoveries about genetics have been made by studying bacteriophage λ

Bacteriophage λ was discovered in 1950 by Esther Lederberg and infects *Escherichia coli*. The virion structure has a capsid head, tail, and tail fibers (**Figure 21.8**). The head contains the double-stranded DNA, which is injected into the *E. coli* through the tail. There are approximately 1,000 proteins in one λ virion in total, with these being multiple copies of 12–14 different proteins.

The double-stranded DNA genome is 48,490 bases, plus 12 bases of single-stranded DNA at the 5′ ends of each strand. These "sticky ends" of the λ genome are the *cos* site, which is responsible for the circularization of the linear genome once it is within the bacterial cytoplasm. The circularized double-stranded DNA within the bacterial cell is therefore 48,502 bp.

Bacteriophage λ uses the bacterial maltose porin for adsorption and through this its genome gains entry into the bacterial cell (**Figure 21.9**). After passing through the maltose porin, LamB, into the periplasm, the bacteriophage DNA passes through the mannose permease complex in the inner membrane to reach the cytoplasm, where it circularizes. It is a model for the study of lysogeny and from the study of λ a great deal has been learned about insertion and excision from the chromosome, where the *attB* site on the bacterial genome in *E. coli* sits

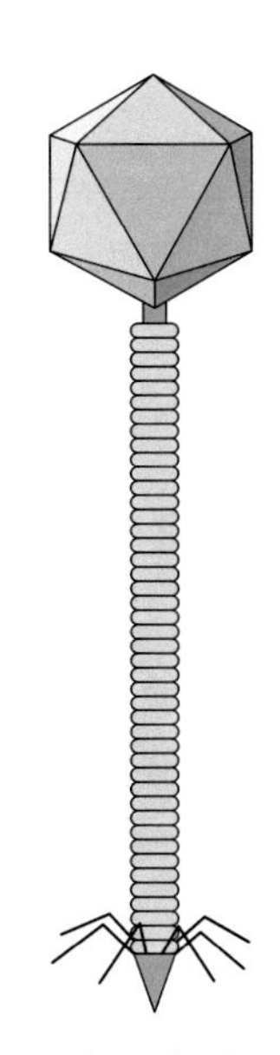

Figure 21.8 Bacteriophage λ. The structure, morphology, and size of bacteriophage λ. The capsid is made of many copies of proteins that form the head, tail, and tail fibers. The genome is 48,490 bases of double-stranded DNA, encoding 73 proteins.

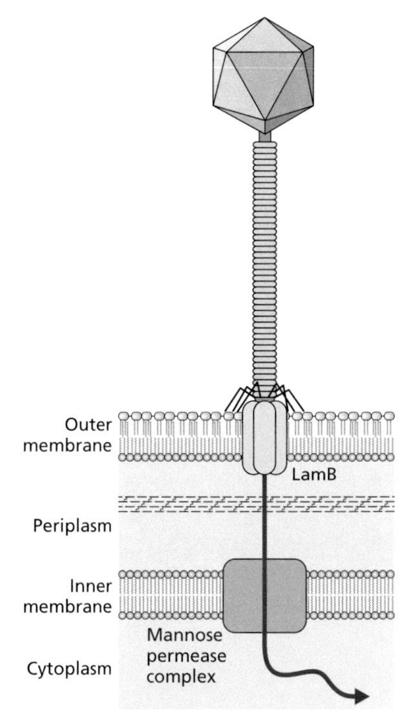

Figure 21.9 Bacteriophage λ genome entry via LamB. The receptor for bacteriophage λ on *E. coli* is the maltose porin protein LamB. Binding to this porin enables the genome to pass through the outer membrane. It then passes through the inner membrane via the mannose permease complex to reach the bacterial cytoplasm.

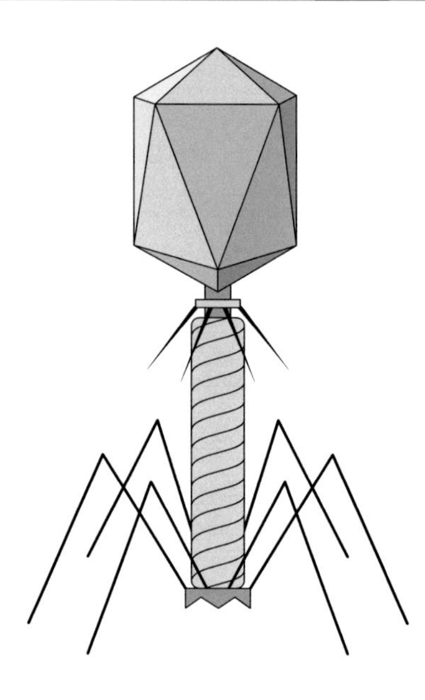

Figure 21.10 Bacteriophage T4. The structure and mechanism of nucleic acid injection into *E. coli* by bacteriophage T4. The tail of bacteriophage T4 has a contractile outer sheath and an inner core that is a rigid tube. When the tail fibers bind to the outer membrane, the sheath contracts and the tube pierces the outer membrane, peptidoglycan, and inner membrane, depositing the double-stranded DNA genome into the cytoplasm.

between the *gal* and *bio* operons. Additionally, study of the expression of the λ genes has been insightful in the understanding of genetic regulation in general.

The T4 bacteriophage has a characteristic morphology

The bacteriophage T4 infects *E. coli* and was discovered in 1944 with six other T phages, T1–T7. The T4 genome is 168 kb, present as double-stranded DNA, which encodes 289 proteins.

Structurally, the T4 bacteriophage also has an icosahedral head with a hollow tail and tail fibers, like λ, and has a lytic lifecycle when infecting *E. coli*. The tail passes into the bacterial cell to infect and deliver the DNA (**Figure 21.10**). The tail of the T4 bacteriophage pierces the bacterial cell to inject the bacteriophage genome into the cytoplasm. The tail is made of a sheath that is able to contract, surrounding a rigid tube. At the bottom of the tail is a base plate, to which are connected the tail fibers that secure the bacteriophages to the outer membrane. When the sheath contracts, the inner tube of the tail is driven through the *E. coli* outer membrane, peptidoglycan cell wall, and inner membrane.

This tail resembles the bacterial Type 6 Secretion System (see Chapter 9) in structure and function (**Figure 21.11**). The contractile nature of the T4 bacteriophage tail matches the contractile function of the Type 6 Secretion

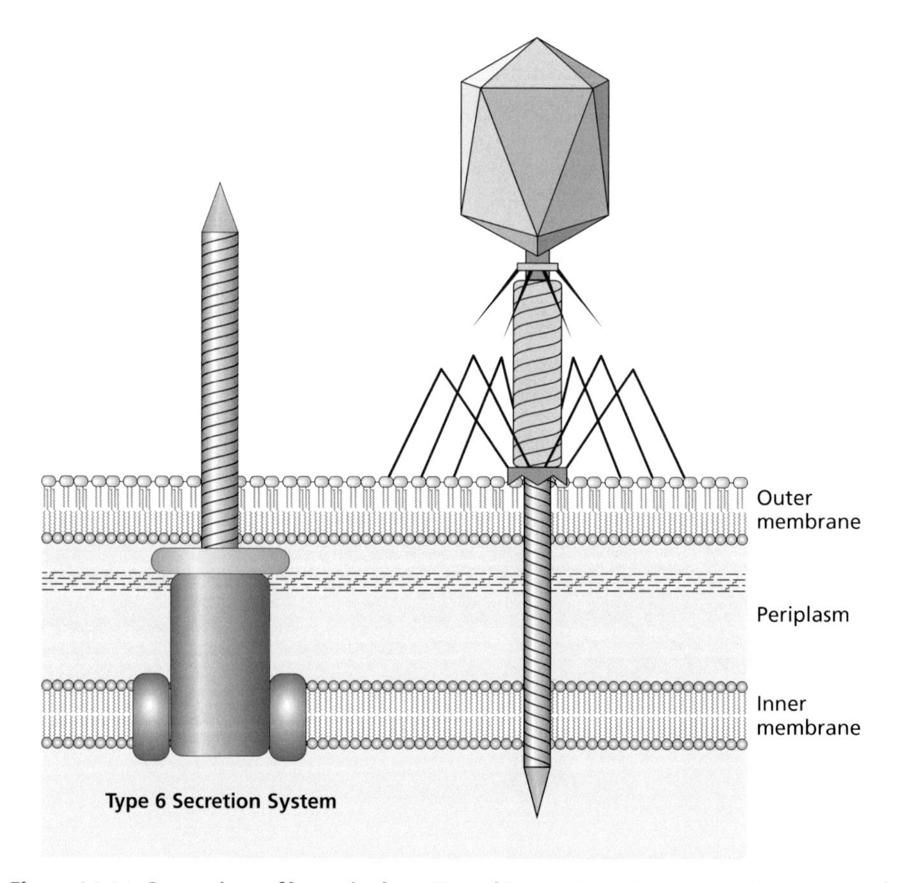

Figure 21.11 Comparison of bacteriophage T4 and Type 6 Secretion System. The structure of the T4 bacteriophage is similar to λ and to many other bacteriophages with a head and tail capsid structure. The tail of T4 is also similar to the bacterial Type 6 Secretion System, which evolved from bacteriophage structures.

System and the ability of this bacterial feature to stab competitor cells in its niche, including other bacterial cells, fungal cells, and eukaryotic cells.

The φX174 bacteriophage was the first DNA genome sequenced

The first complete DNA genome to be sequenced was from φX174, bacteriophages that infect *E. coli* and have a single-stranded DNA genome (**Figure 21.12**). This bacteriophage genome is 5,375 nucleotides and was the first sequence published with the Sanger method (see Chapter 11). The nature of the φX174 genome as circular DNA had previously been demonstrated by Walter Fiers and Robert Sinsheimer in 1962. The genome encodes 11 proteins, the genes for many of which overlap.

The adsorption target for bacteriophage φX174 is *E. coli* LPS (see Chapter 9). The capsid is an icosahedral shape without a tail. Bacteriophages with no tail rely on either bacterial structures such as pores or alteration of their own proteins to get their genomes into bacterial cells. For example, it was discovered in 2013 that the φX174 bacteriophage H proteins form together into a tube to deliver the bacteriophage DNA into the bacterial cell (**Figure 21.13**).

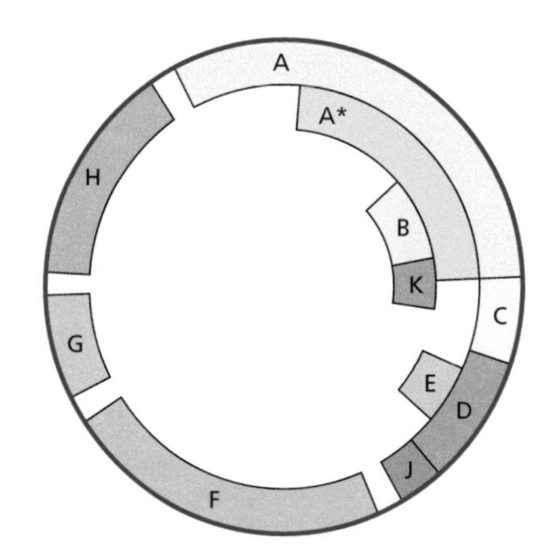

Figure 21.12 Bacteriophage φX174 genome. This is a representation of the genes encoded by the bacteriophage φX174 genome, which was the first DNA sequence to be determined using the Sanger sequencing method. This identified the coding regions of the bacteriophage.

Transduction is an important source of horizontal gene transfer for bacteria

Bacteriophage transduction is involved in bacterial evolution, as are transformation and conjugation. Prophage genes substantially increase the virulence potential of bacterial strains, contributing such genes as extracellular toxins, proteins involved in serum resistance, and in some cases antibiotic resistance genes. Not only do bacteriophages have the ability to integrate their own genetic material into the bacterial genome, but they can also mobilize other DNA. Some bacteriophages package the nucleic acids within the cell, whether it is bacteriophage in origin or bacterial.

Generalized transduction packages random pieces of DNA into bacteriophage capsids (**Figure 21.14**). This random DNA takes the place of bacteriophage genomic DNA and when infecting the next cell, it is the bacterial DNA that is transferred rather than the bacteriophage DNA. In **specialized transduction**, some bacterial DNA adjacent to the integrated prophage is excised with the prophage upon transition from lysogenic to lytic phase (see Figure 21.14).

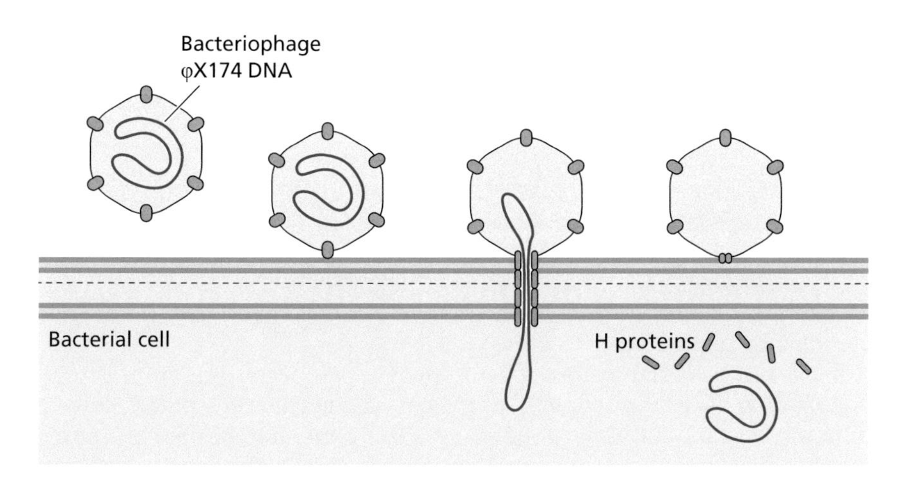

Figure 21.13 The temporary tail of bacteriophage φX174. To deliver its single-stranded circular DNA genome into the *E. coli* cytoplasm, bacteriophage φX174 targets LPS on the surface of the bacterial cell, then polymerizes copies of the H protein into a tube through the outer membrane, peptidoglycan, and inner membrane. The circular genome of the bacteriophage is then able to pass through to the bacterial cytoplasm, at which point the H proteins dissociate.

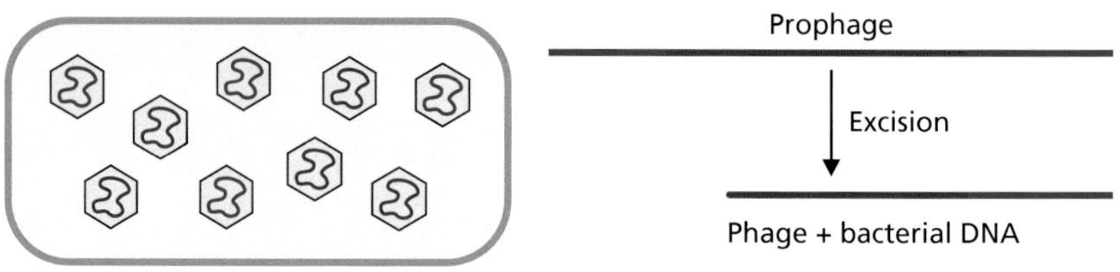

Figure 21.14 Generalized and specialized transduction. Bacterial DNA can end up packaged within bacteriophages. In generalized transduction (left), random DNA is packaged into the capsid, which can include the bacterial DNA (blue), as well as the prophage genome (red). In specialized transduction (right), some bacterial DNA (blue) may be excised with the prophage (red) and end up packaged into the bacteriophages. In both cases, bacterial DNA is carried by the bacteriophage virion to the next bacterial cell.

The bacterial DNA and bacteriophage DNA are taken as one piece into the bacteriophage capsid and released from the cell.

The prophages of group A streptococci are responsible for much of the genetic diversity of the species. Within the prophage genomes are genes that contribute to the pathogenicity of the various lineages of group A streptococci. Serotype M3 strains tend to cause deep tissue streptococcal infections, such as necrotizing fasciitis, as well as pharyngitis. Group A streptococcus serotype M3 strain MGAS315 has in its genome six different prophages containing six virulence factors. Prophage genes *speK, sla, sdn, speA, mf4*, and *ssa* encode streptococcal pyrogenic endotoxin K (SpeK), phospholipase A(2) (Sla), streptodornase (Sdn), streptococcal pyrogenic exotoxin A (SpeA), mitogenic factor 4 (Mf4), and streptococcal superantigen (Ssa). These are induced in expression upon co-culture with human epithelial pharyngeal cells, exposure to hydrogen peroxide, or damage of DNA by mitomycin C. The various conditions *in vitro* mimic environmental signals that would occur *in vivo*, resulting in differential expression of prophage genes in the various host niches.

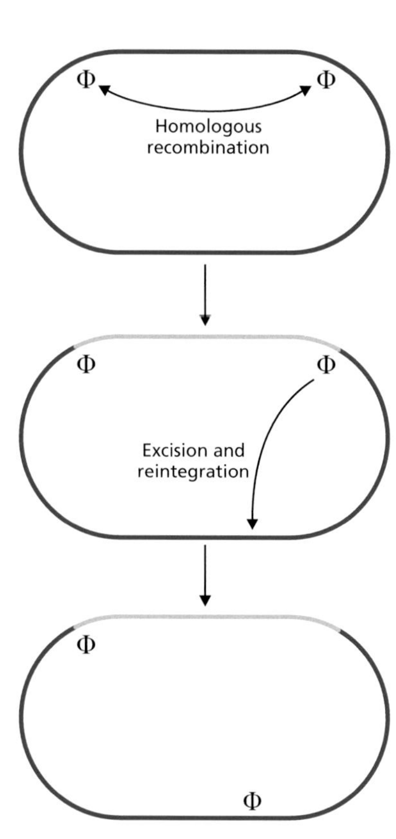

Figure 21.15 Bacteriophage-mediated chromosomal rearrangements. The presence of prophages within a bacterial chromosome or plasmid can result in rearrangements. This can occur due to homologous recombination between two prophage genomes, indicated here with Φ. This can also occur when the prophage excises and reintegrates elsewhere, sometimes taking some of the bacterial sequence with it in the process.

Bacteriophages also contribute to bacterial evolution through chromosomal rearrangements

Prophages are also involved in chromosomal rearrangements. Bacteriophage sequences can account for as much as 20% of a bacterial genome. Some of this sequence is intact bacteriophage genomic information capable of completing lytic replication, however there are other bacteriophage sequences within bacterial genomes that are remnants of prophages. These may have been left behind in the genome by incomplete excision, may have been rendered inactive as bacteriophages by gene degradation, or may have been horizontally transferred into the chromosome or plasmid. The prevalence of bacteriophage sequences among the bacterial nucleic acids increases the likelihood that these sequences will mediate rearrangements by homologous recombination.

In *Neisseria gonorrhoeae*, for example, there are multiple prophages integrated into the chromosome, which contains regions that are homologous to each other. Comparative analysis between completely assembled circular genome sequences of *N. gonorrhoeae* strains revealed large segments of reordered and rearranged DNA sequence, some of which has been mediated by the presence of integrated bacteriophage sequences.

Changes in bacterial genomes by prophages can occur due to homologous recombination between similar prophage sequences integrated into the chromosome or mobilization of prophages from one area of the chromosome to another via excision and reintegration (**Figure 21.15**). This latter process may result in the translocation of bacterial genetic information as well as prophage sequence, just as excision may sometimes result in specialized transduction of bacterial sequences to new cells (see Figure 21.14).

Bacteriophage and prophage genome evolution can provide interesting insights

As many more bacterial genomes are entering the public sequence databases, so too are many prophage genomes. This provides opportunities for the study of bacteriophages via bioinformatics. One interesting investigation has explored the prophages within *Helicobacter pylori*. This bacterial species colonizes approximately half of all humans, and the genotypes of this bacterial species can be linked to geographic origins of human migration across the globe. *H. pylori* and modern humans have evolved together with our ancestors carrying the bacteria with them in their stomachs as they migrated, passing the strains from mother to child. Like the bacterial genes, genes of the *H. pylori* resident prophages divide into populations with geographic distributions that correlate with human migration (see Chapter 17 and Figure 17.16).

Bacteria have evolved strategies to avoid bacteriophage infection

To avoid infection by bacteriophages, bacteria can take several approaches. One is to avoid infection entirely by changing the host cell receptor on the surface of the bacteria. If bacteriophage adsorption is prevented, then so is infection. This can be accomplished through mutation of the gene encoding the receptor itself or concealing the receptor with a physical barrier such as a capsule or other modification. Phase variation (see Chapter 12) can also play a role, in which the receptor is only expressed some of the time, reducing risk to the bacterial cell.

Release of **outer membrane vesicles (OMVs)** can also provide a survival strategy. OMVs are blebs of bacterial membrane that are shed from the surface of the living bacterial cells (**Figure 21.16**). These contain proteins and other features characteristic of the surface of the bacteria. Bacteriophages can be sequestered away from the bacterial cell by adsorption to the OMVs instead of the bacterial cell itself. These decoys reduce infection of bacterial cells and increase their chances of survival.

Another strategy is to block entry of the bacteriophage nucleic acids into the bacterial cell. This uses a superinfection exclusion system, which is actually encoded by bacteriophages. Superinfection exclusion systems stop bacteria that have been infected by bacteriophages from becoming infected by new competitor bacteriophages. The resident bacteriophage produces a membrane-associated protein that interferes with the injection of new bacteriophage nucleic acids into the cell.

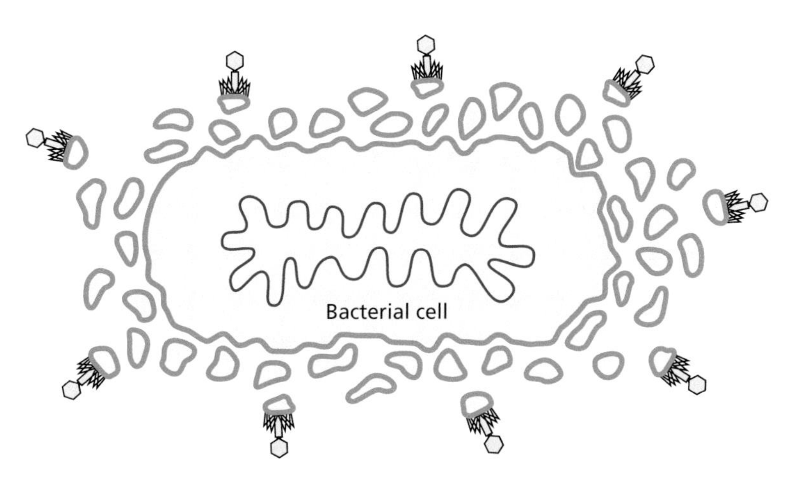

Figure 21.16 Outer membrane vesicles to survive bacteriophages. Blebs of the outer membrane have the same surface proteins and structures as the bacteria that they came from, therefore bacteriophage adsorption occurs on outer membrane vesicles (OMVs) as readily as it does on bacterial cells. This can help protect the bacterial cell, by diverting the bacteriophages to the OMVs.

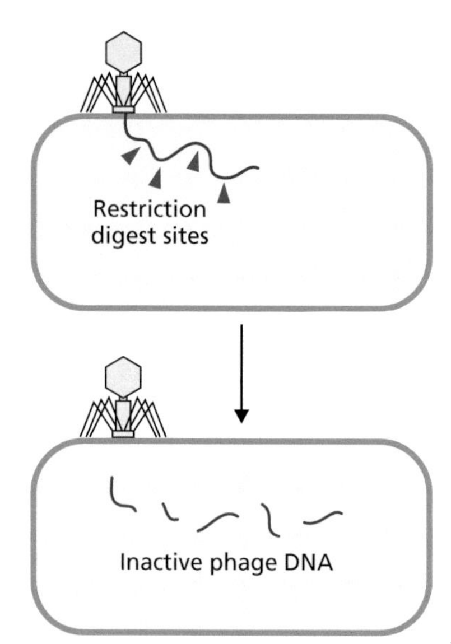

Figure 21.17 Restriction digestion of bacteriophage DNA. When the bacteriophage DNA enters the cell, it is subject to the restriction-modification system. If the bacteriophage genome contains any recognition sequences for restriction enzymes expressed by the bacteria, the bacteriophage DNA will be digested. It is therefore neither replicated nor integrated into the chromosome.

Even if bacteria become infected by bacteriophage nucleic acids, they can still fight back

Even if the nucleic acids of a bacteriophage make it into a bacterial cell, there are mechanisms whereby the bacteria can prevent the infection from destroying the bacteria. Restriction-modification systems are one means of protecting the bacterial cell from invading DNA, whatever the source. Therefore, when invading bacteriophage DNA enters the bacterial cytoplasm it is subject to restriction digestion by the resident bacterial enzymes. As described in Chapter 10, the bacterial DNA is protected from digestion by restriction-modification system methyltransferases, enzymes that modify specific recognition sites through methylation. The unmodified bacteriophage DNA is therefore not protected and is digested by the bacterial restriction enzymes before it can either replicate or integrate into the chromosome (**Figure 21.17**). Some bacteriophages are able to incorporate modified bases into their genomes as a mechanism of restriction-modification system resistance. To counter this bacteriophage genome modification, some bacterial restriction enzymes have recognition sites that include modified bases; although the sequence will be present in the bacterial genome, it will not be modified, therefore the enzyme only digests foreign modified DNA.

Bacteria have also evolved a type of immune system against foreign DNA, the CRISPR–Cas system that has been described previously (see Chapter 13). With this system, bacteria are able to build up a bank of foreign DNA sequences that can be recognized for elimination. This includes invading bacteriophage DNA. In addition to the CRISPR–Cas and restriction-modification systems that have been explored earlier in this book, there is another mechanism of resistance to bacteriophage infection that can act after entry of the nucleic acids into the cell.

As has been seen throughout this book, not all bacterial survival strategies involve survival of the individual bacterial cell, but rather take a population view. One means of surviving bacteriophage infection also takes this population level approach. An **abortive infection system** results in the death of the infected bacterial cell, however the surrounding population is spared (**Figure 21.18**). The abortive infection system is genetically encoded on a mobile element such as a plasmid or even a prophage. The gene products involved disrupt bacteriophage replication so that mature, infectious bacteriophages are not successfully made. The disruption can happen at various stages of the process, but the end result

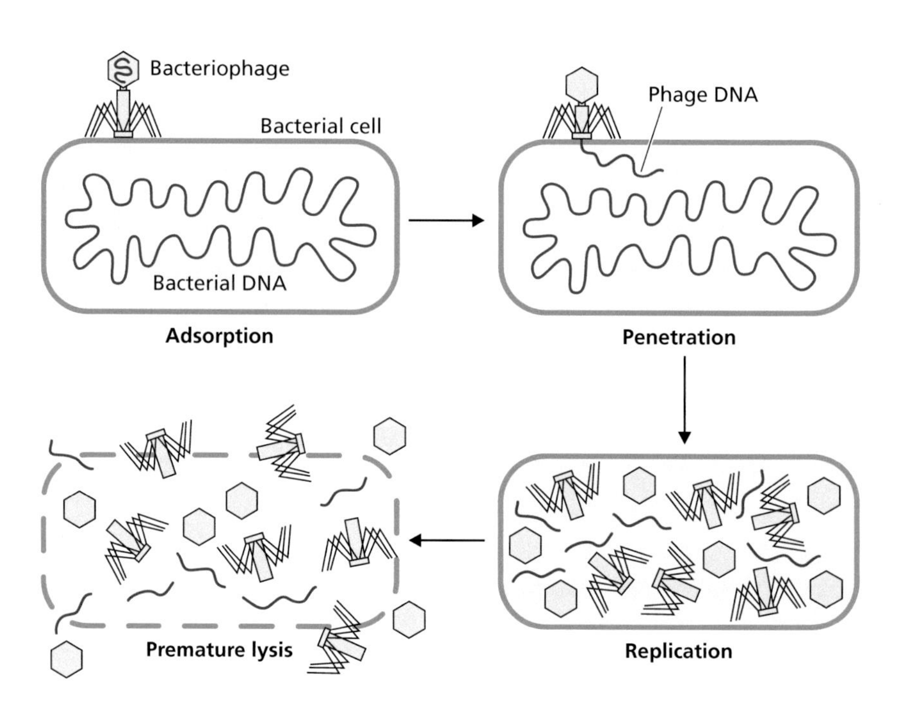

Figure 21.18 Abortive infection system. Bacterial population level resistance to bacteriophages means that the infected cell sacrifices itself, but the surrounding cells survive. The bacteriophage replication is interrupted, resulting in premature lysis of the bacterial cell. Because no intact bacteriophage particles are made, the neighboring bacterial cells in the population survive. Compare to Figure 21.3.

is that replication does not occur and infection does not spread in the bacterial population.

Bacteriophage resistance fights back and uses the bacteriophages for its own ends

The final mechanism of bacteriophage resistance to be discussed in this chapter is not so much resistance to bacteriophage infection, as a means of co-opting bacteriophages, allowing bacterial pathogenicity islands to hitchhike a ride to their next destination. This phenomenon has been well studied in *Staphylococcus aureus* where phage-inducible chromosomal islands, the *S. aureus* pathogenicity islands, are able to use incoming bacteriophages as a means of horizontally transferring to neighboring bacterial cells. These pathogenicity islands (see Chapter 6) have evolved mechanisms that enable the sequence of the island to be excised from the chromosome, replicated, and packaged into a bacteriophage capsid as a prophage would.

Upon infection with a bacteriophage, the *S. aureus* pathogenicity island encoded determinants limit the replication of the bacteriophage genome, in preference for its own excised sequence. The size of the capsid is also altered to the smaller capacity needed for its own DNA, further influencing the preference for mature particles with *S. aureus* pathogenicity island DNA, rather than bacteriophage DNA, packaged within.

When the capsids are assembled with the *S. aureus* pathogenicity island DNA within them, these particles are released with the lysis of the bacterial cell. This process stops the spread of the original infecting bacteriophages. Its DNA has been excluded in preference for that of the pathogenicity island. Instead of the bacteriophages spreading to new bacterial cells, the pathogenicity island will spread to the surrounding cells (**Figure 21.19**). These particles themselves do not induce a lytic phase when they infect. Instead, when the DNA enters the new bacterial cell, the pathogenicity island will be potentially integrated into the bacterial chromosome.

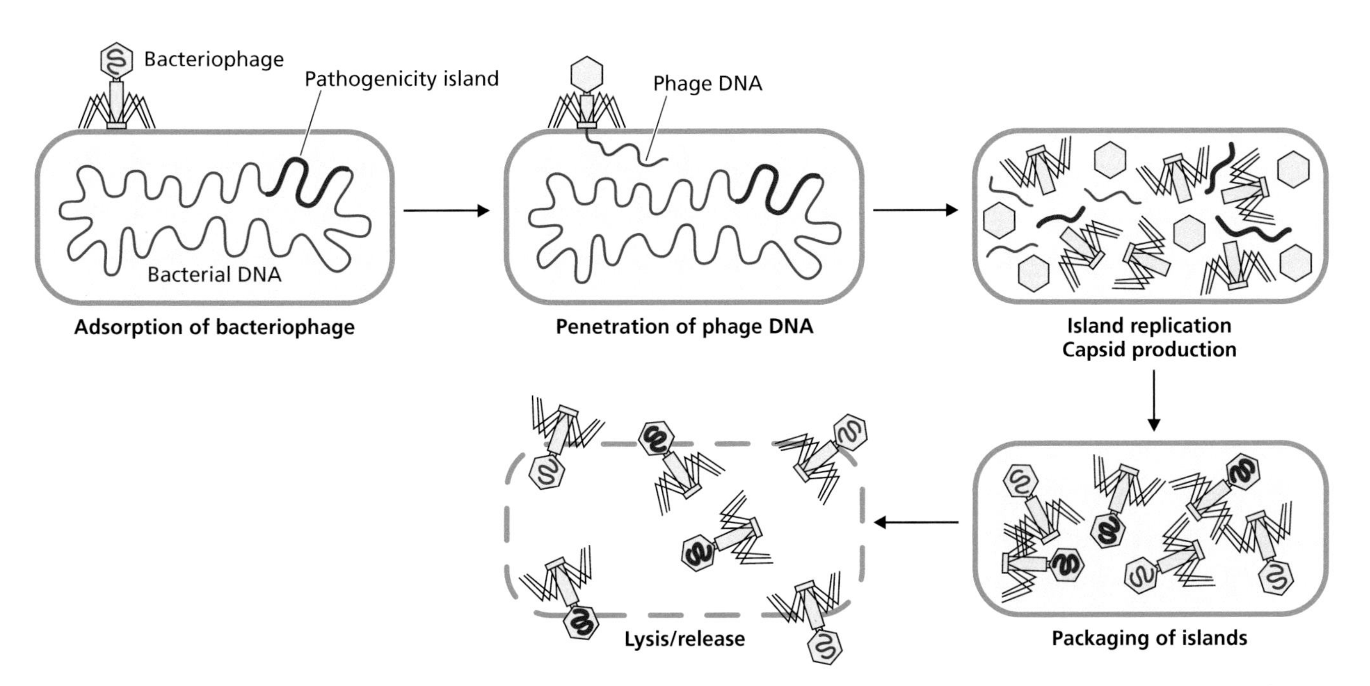

Figure 21.19 Phage-inducible chromosomal islands. Some pathogenicity islands use infection by bacteriophages as a means to spread to new bacterial cells. Rather than succumbing to infection, the bacterial cell shuts down the replication of the bacteriophage genome, excises the pathogenicity island, replicates it, and packages it into the bacteriophage capsid. The pathogenicity island is then released with the lysis of the bacterial cell, spreading to other bacteria.

Bacteriophage therapy is a potential alternative treatment for antimicrobial-resistant bacteria

Phage therapy uses lytic bacteriophages to kill their target bacteria, without harming the host organism infected by the bacteria. With the advent of broad-spectrum antibiotics in the early to mid-1900s, research into bacteriophage therapy was discontinued in many areas of the world. Notably, therapeutic use of bacteriophages is already integrated into healthcare in Eastern Europe and nations of the former Soviet Union, where it had been in use for decades before the resurgence in interest elsewhere in recent years, due to multidrug antibiotic resistance. The species and strain specificity of bacteriophages makes them both ideal for treatment, as they will not indiscriminantly eliminate other members of the microbiome like antibiotics, yet also more difficult to use. The specific etiological agent must be identified and its immunity to the bacteriophage treatment determined, which can be time-consuming.

There are, however, potential benefits of the use of bacteriophages to target bacterial infections. There have been cases of success with phage therapy, saving lives and showing promise in clinical trials. Lytic bacteriophages have been demonstrated to disrupt *in vitro* biofilms of *Acintobacter baumannii*. In 2017, a patient in the United States was successfully treated for *A. baumannii* pancreatic pseudocyst infection with bacteriophages administered intravenously. Phage therapy against *Pseudomonas aeruginosa* is promising, particularly to assist patients with cystic fibrosis suffering from chronic lung infections. Some of the bacteriophages considered for therapy target the bacterial efflux pumps, therefore bacteriophage resistance would render the *P. aeruginosa* susceptible to antibiotics.

Phage therapy has been suggested for use against inflammatory bowel disease (IBD). Included in the IBDs are Crohn's disease and ulcerative colitis. IBD is believed to develop due to a complex interaction involving human genetic characteristics and the gut microbiome. Although they do not infect host cells, bacteriophages in the gut can contribute to IBD through their interactions with human gut bacteria. An alteration of the diversity of bacteriophages in the gut impacts the population of gut flora. Therefore, it has been suggested that bacteriophage therapy can help adjust the gut bacteria for a positive impact on IBD.

Further research into bacteriophages and bacteriophage therapy is needed. The restriction of bacteriophages to specific bacteria may not be universal; there are many uncharacterized bacteriophages and broad-spectrum therapeutic phages that may yet be isolated. Ecological and metagenomics studies suggest that although many of the well-studied bacteriophages have narrow specificities for bacterial hosts, there are others with broad host ranges.

Mechanisms of defense against bacteriophages may result in resistance to phage therapy

There are concerns that bacteriophage resistance could develop against phage therapy. The naturally occurring bacteriophage defense systems in bacteria, such as the CRISPR-Cas9 and restriction endonucleases, target bacteriophages and could thwart their use therapeutically. Although a pathogen may be initially vulnerable to phage therapy, there is scope for development and acquisition of resistance with these systems and others.

To infect bacterial cells the bacteriophage must attach to the surface, generally via specific adsorption to bacterial cell surface receptors. Mutations in the receptor can stop bacteriophage binding. Binding can also be stopped via blebbing of the outer membrane (see Figure 21.16), whereby the receptor is lost from the outer membrane, by binding of a second protein made by either the bacteria or another bacteriophage that masks the recognition site on the surface protein, or covering the receptor protein with an extracellular blocking material such as the bacterial capsule. In some species, the receptor protein expression is phase variable, therefore those cells not expressing the receptor are selected for, so a portion of the bacterial population survives (**Figure 21.20**).

Resistance can also occur due to mechanisms that impact bacteriophage infection after the first step. As discussed previously, this can include the immunological-type systems like CRISPR-Cas9 (see Figure 13.18) and restriction digestion of bacteriophage genomes (see Figure 21.17), as well as premature lysis of infected cells before the bacteriophages can fully form (see Figure 21.18).

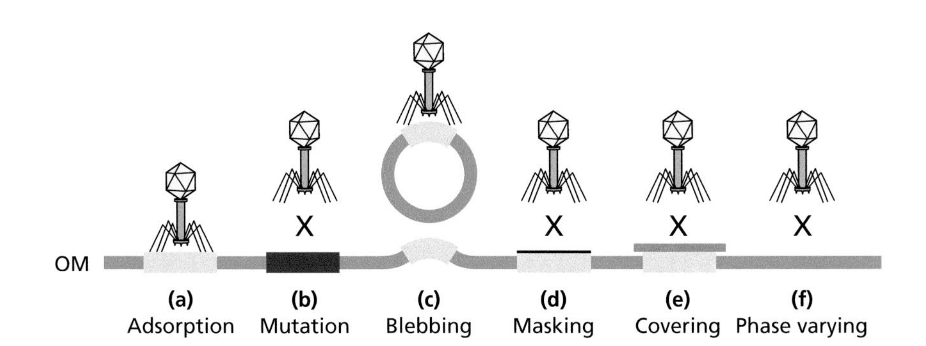

Figure 21.20 Potential mechanisms of resistance to bacteriophage therapy that interfere with adsorption. Infection of bacterial cells by bacteriophages starts with adsorption of the phage particle to the surface of the bacteria (a). Several mechanisms can prevent this from happening. Mutations can change the binding site for the bacteriophages, stopping them from infecting (b). Blebbing of the outer membrane will take outer membrane-bound proteins with the membrane bleb; this can remove the protein recognized for bacteriophage binding (c). The receptor protein can become masked by another protein, either produced by the bacteria or another bacteriophage (d). As with masking, covering the bacterial surface protein with an extracellular matrix-like capsule can prevent access by bacteriophages (e). In some species that engage in phase variation, it may occur that the bacteriophage receptor is on a phase variable protein, therefore for some cells there is no recognition site for the initial binding of the bacteriophages to initiate infection or phage therapy.

Key points

- Bacteriophages are classified by their shape and by the type of genome they carry.
- Bacteriophages replicate by one of two means, either a lytic cycle that immediately makes more phage particles or a lysogenic phase, where the bacteriophage genome is integrated into the bacterial genome and replicates there.
- The first genome ever sequenced was of bacteriophage MS2, an RNA genome of 3.5 kb.
- The bacteriophage λ uses the surface porin for maltose to pass its DNA through the outer membrane of *E. coli*.
- The bacteriophage T4 has a tail with a contractile sheath around a rigid core that it uses to pierce the outer membrane, peptidoglycan, and inner membrane of *E. coli* to inject its DNA into the bacterial cell.
- Bacterial DNA can be horizontally transferred between cells by bacteriophages via transduction, either generalized transduction, where random DNA is packaged in the capsid, or specialized transduction, where some bacterial DNA goes with the bacteriophage genome.
- Prophages may account for as much as 20% of a bacterial genome and may be responsible for chromosomal rearrangements.
- Bacterial cells have evolved several strategies to avoid or survive infection by bacteriophages.
- Bacteriophages have found some ways to circumvent bacterial survival strategies evolved against them.
- Some bacterial pathogenicity islands have evolved to use invading bacteriophages as a mechanism to spread to other bacterial cells via transduction.
- Bacteriophage therapy has been used for decades in some parts of the world and is being investigated for more widespread use in the face of multidrug-resistant infections.

Key terms

Define the following terms introduced in this chapter. Check your answers using the definitions in the Glossary. These terms are also available as Flashcards online.

Abortive infection system	Generalized transduction	Specialized transduction
Adsorption	Lysogens	Virion
Eclipse phase	Penetration	

Questions and discussion topics

Self-study questions

Answer each question using 50–100 words or a table or labeled diagram. Advice on where to find answers to these questions is available online.

1 How many different recognized bacteriophage families are there and how are they classified into these families?

2 What characteristic do five of the bacteriophage families share with some eukaryotic viruses? How is this characteristic achieved?

3 Draw the lytic phase of bacteriophages. Include the key important stages.

4 Draw the lysogenic phase that is possible for some bacteriophages. Include the key important stages.

5 In 1976, the first genome was sequenced. What was it? What were some of the key features of this genome?

6 Describe how the bacteriophage λ DNA gets inside the *E. coli* cell and what happens to it once it is there.

7 Describe how the tail of the T4 bacteriophage delivers DNA into *E. coli.*

8 Compare generalized transduction and specialized transduction. Why would bacteriophages from generalized transduction not be able to produce bacteriophage particles when infecting the next bacterial cell? Why might some of the bacteriophages in specialized transduction also be deficient in production of new bacteriophages?

9 How can bacteriophages contribute to chromosomal rearrangements?

10 Name at least two ways bacterial cells can avoid bacteriophage infections. Describe how one of them is able to prevent bacteriophages from infecting the bacterial cell.

11 Name at least two defense systems bacterial cells have that can stop their destruction by bacteriophages once penetration has occurred.

12 Briefly explain how bacteriophage therapy selectively kills specific bacterial cells, but does not harm the host organism microbiome or host organism cells.

Discussion topics

These topics are presented for discussion in study groups, as part of class discussions, or on your own. These questions go beyond what is directly covered in this part of the book. Use the research literature and other reading to explore these topics in more depth. Tips to help prepare for topic discussions are available online.

1 In 1917, Félix D'Hérelle used isolated bacteriophages to control *Salmonella* in chickens. Decades later, bacteriophages were harnessed again for the same purpose to control *Salmonella* on organic chicken farms. Investigate this or another biotechnology application of bacteriophages in agriculture or another industry. Discuss the success of this intervention, the type of bacteriophages used, and the potential for expansion in the future.

2 Some bacteriophages are able to switch between the lytic phase and the lysogenic phase. Using your drawings from questions 3 and 4, indicate at what point bacteriophages capable of lysogeny switch from the lysogenic phase in the drawing from 4 to the lytic phase in the drawing from 3. Likewise, at what stage does a bacteriophage change from lytic (3) to lysogenic (4)? Investigate and discuss what happens that triggers these switches.

3 Esther Lederberg discovered lambda (λ) bacteriophages and described lysogeny. She also made other contributions to microbiology and microbial techniques that contributed to bacterial genetics and genomics. The study of λ provided insight that was extrapolated across the field of genetics. Explore and discuss the contributions made by Esther Lederberg and at least one other scientist who made important contributions, but may not be well known, perhaps due to gender or race.

Online quiz questions

To further self-assess your understanding of the chapter material, please visit the following link, where you can participate in a range of interactive quiz questions:

www.routledge.com/cw/snyder

Further reading

Bacteriophages have been studied for just over 100 years

Ackermann H-W. Phage or phages. *Bacteriophage*. 2011; *1*(*1*): 52–53.

Nair A, Ghugare GS, Khairnar K. An appraisal of bacteriophage isolation techniques from environment. *Microb Ecol*. 2022; *83*(*3*): 519–535.

Some bacteriophages enter latency for a period before replication

Lederberg E. Lysogenicity in *Escherichia coli* strain K-12. *Microbial Genet Bul l*. 1950; *1*: 5–8.

The MS2 bacteriophage has a very small genome and was the first genome sequenced

Fiers W, Contreras R, Duerinck F, Haegeman G, Iserentant D, Merregaert J. Complete nucleotide sequence of bacteriophage MS2 RNA: Primary and secondary structure of the replicase gene. *Nature*. 1976; *260*: 500–507.

The φX174 bacteriophage was the first DNA genome sequenced

Sanger F, Air GM, Barrell BG, Brown NL, Coulson AR, Fiddes JC, Hutchison CA III, Slocombe PM, Smith M. Nucleotide sequence of bacteriophage φX174 DNA. *Nature*. 1977; *265*: 687–695.

Sun L, Young LN, Zhang X, Boudko SP, Fokine A, Zbornik E, Roznowski AP, Molineux IJ, Rossmann MG, Fane BA. Icosahedral bacteriophage ΦX174 forms a tail for DNA transport during infection. *Nature*. 2014; *505*: 432–435.

Transduction is an important source of horizontal gene transfer for bacteria

Banks DJ, Lei B, Musser JM. Prophage induction and expression of prophage-encoded virulence factors in group A *Streptococcus* serotype M3 strain MGAS315. *Infect Immun*. 2003; *71*(*12*): 7079–7086.

Bacteriophages also contribute to bacterial evolution through chromosomal rearrangements

Spencer-Smith R, Varkey EM, Fielder MD, Snyder LA. Sequence features contributing to chromosomal rearrangements in *Neisseria gonorrhoeae*. *PLoS One*. 2012; *7*(*9*): e46023.

Bacteriophage and prophage genome evolution can provide interesting insights

Falush D, Wirth T, Linz B, Pritchard JK, Stephens M, Kidd M. Traces of human migrations in *Helicobacter pylori* populations. *Science*. 2003; *299*(*5612*): 1582–1585.

Even if bacteria become infected by bacteriophage nucleic acids, they can still fight back

Lopatina A, Tal N, Sorek R. Abortive infection: Bacterial suicide as an antiviral immune strategy. *Ann Rev Virol*. 2020; *7*(*1*): 371–384.

Bacteriophage resistance that fights back and uses the bacteriophage for its own ends

Ram G, Chen J, Ross HF, Novick RP. Precisely modulated pathogenicity island interference with late phage gene transcription. *Proc Natl Acad Sci USA* 2014; *111*: 14536–14541.

Bacteriophage therapy is a potential alternative treatment for antimicrobial-resistant bacteria

Furfaro LL, Payne MS, Chang BJ. Bacteriophage therapy: Clinical trials and regulatory hurdles. *Front Cell Infect Microbiol*. 2018; *8*: 376.

Ling H, Lou X, Luo Q, He Z, Sun M, Sun J. Recent advances in bacteriophage-based therapeutics: Insight into the post-antibiotic era. *Acta Pharm Sin B*. 2022; *12*(*12*): 4348–4364.

Glossary

–10 promoter sequence A conserved sequence that sits approximately 10 bases 5′ of the transcription start site that is recognized by the sigma subunit of RNA polymerase

16S rRNA A segment of ribosomal RNA that is part of the 30S subunit of the ribosome

2D gel A technique used to resolve proteins by isoelectric point and molecular weight

23S rRNA A segment of ribosomal RNA that is part of the 50S subunit of the ribosome

30S subunit The smaller of two subunits that make up the ribosome in bacteria; the other is 50S and together they make 70S

–35 promoter sequence A conserved sequence present approximately 35 bases 5′ of the transcription start site that is recognized by the sigma subunit of RNA polymerase

5′-phosphoguanylyl-(3′-5′) guanosine *See* PGPG

5S rRNA A short segment of ribosomal RNA that is part of the 50S subunit of the ribosome

50S subunit The larger of two subunits that make up the ribosome in bacteria; the other is 30S and together they make 70S

70S ribosome The complete ribosome in bacteria, made of the proteins and rRNAs of the 50S subunit and the proteins and RNA of the 30S subunit

α-Amino acids Amino acids where the amine group is attached to the α-carbon

α-Complementation To indicate the successful cloning of an insert, this engineered system will generate blue colonies with no insert, and white ones with a cloned insert

α-Galactosidase An enzyme that breaks down lactose into its component sugars; glucose and galactose

α-Galactoside permease A transport protein that transports lactose into the bacterial cell

α Helix *See* Alpha helix

α-Subunit *See* Sigma subunit

β barrel *See* Beta barrel

β-Lactam A class of antibiotics containing a characteristic ring structure of three carbons, one with a double-bonded oxygen, and a nitrogen

β-Lactamases A class of enzymes that inactivate the class of β-lactam antibiotics by cleaving the β-lactam ring between the nitrogen and the carbon with adjacent double-bonded oxygen

β Sheet *See* Beta sheet

ABC family A class of efflux pump that uses ATP as the energy for transport across the membrane

ABC transporter A protein that is able to bind ATP and use the energy from it to transport a substrate across a membrane, present in Type 1 Secretion Systems

Abortive infection system Bacteriophage replication within a bacterial population is halted through incomplete replication within a bacterial cell, which sacrifices itself for the benefit of the population as a whole; no complete bacteriophage are generated; therefore, the infection is not spread

Accessory gene(s) A gene or genes that are found within a genome that are not necessarily present in all genomes of the same species

Accessory genome In a comparison of two or more bacterial genome sequences, those genes that are in only one or a subset of the genomes.

Acetyl CoA *See* Acetyl coenzyme A

Acetyl coenzyme A A protein that aids acetyltransferases by providing an acetyl group

Acetylation The addition of an acetyl group ($–CO–CH_3$) to a protein

Activator A protein that can enhance transcription of a gene

Acylation The addition of an acyl group to a protein

Adenine A nitrogenous base made of five carbons, five nitrogens, and five hydrogens, found in DNA and RNA

Adenosine diphosphate An adenine nitrogenous base that is bound to a sugar phosphate, which is bound to a second phosphate

Adenosine monophosphate An adenine nitrogenous base that is bound to a sugar phosphate

Adenosine triphosphate An adenine nitrogenous base that is bound to a sugar phosphate, which is bound to a second phosphate, which is bound to a third phosphate

ADP *See* Adenosine diphosphate

Adsorption Adhesion onto a surface, as in a bacteriophage attaching to a bacterial cell

Alloprotein An engineered protein not found in nature, which may include non-proteinogenic amino acids

Alpha helix A secondary amino acid structure formed by hydrogen bonding between amino acids such that a helix is made

Ameliorate In genetics, changes to nucleic acid sequence over time, generally as foreign DNA becomes similar to host DNA

Amine group A nitrogen and two hydrogens, found at one end of an amino acid

Amino acid The individual components that make up a protein, characterized by having a carboxyl group (COOH) and an amino group (NH_2)

Amino acid acceptor stem The fourth stem within a tRNA structure, to which an amino acid is attached

Amino acid identity *See* Identity (amino acid)

Aminoacyl site A place within the ribosome where tRNA is able to enter and interact with the mRNA

Aminoacyl-tRNA synthetase An enzyme that adds an amino acid to the 3′ adenine on a tRNA

AMP *See* Adenosine monophosphate

Annotation In genetics and genomics, the assigning of information to nucleic acid or amino acid sequence data, which may include predicted features and their functions

Antibody A Y-shaped protein made by eukaryotic organisms during immune responses against bacterial cells, targeting specific antigens

Anticodon The corresponding complementary bases to a codon, generally found within a tRNA

Anticodon loop The loop from the second of three stem loop structures in tRNA, starting from the 5′ end, which contains the anticodon

Anticodon stem The stem from the first of three stem loop structures in tRNA, starting from the 5′ end, which contains the anticodon

Antigen A molecule or molecular structure capable of being recognized by the immune system

Antigenic variation Changes to antigens, generally bacterial surface structures, to prevent recognition by the host immune system, allowing for immune evasion

Antiparallel Describes the relative orientation of a string of molecules, where there are two strands parallel to one another, yet orientated in opposite directions

Antisense RNA A small ncRNA that is encoded on the opposite strand as the gene that it regulates, making it perfectly complementary to the mRNA

Antisense strand *See* Negative strand

Anti-sigma factor A protein that binds to sigma factors, preventing it from binding to DNA

Anti-terminator A hairpin formation of mRNA that prevents the formation of a transcriptional terminator hairpin signal

Antivirulence agent A means to reduce the severity of an infectious disease by combating its virulence factors, such as secreted toxins

Aptamer RNA or DNA molecules that are able to form tertiary structures that are capable of binding to proteins or other targets

Archaea Organisms that are generally microscopic, yet are of a distinct evolutionary lineage from bacteria and eukaryotes

A-site *See* Aminoacyl site

Assembly In genomics, the process of taking fragmented sequence data and piecing it together into longer contiguous sequences, based on overlapping sequence homology

ATP *See* Adenosine triphosphate

Attenuation Reduction in production of complete transcripts

Attenuator A segment of mRNA sequence involved in attenuation of transcription

Autoinducer A molecule used in quorum sensing that is secreted by bacterial cells and sensed by bacterial cells to enable them to tally the presence of bacteria

Automated Sanger sequencing An enhancement of one of the first techniques for nucleotide sequencing, where use of fluorescently labeled nucleotides and computer-based detection automated much of the sequencing process

Autophosphorylate A protein that is able to add a phosphate, often from ATP, to itself

Autotransporter A Type 5 Secretion System protein that is able to transport itself out of the cell using a transporter domain that forms an outer membrane channel

Auxotroph A bacterial strain that has an auxotrophy

Auxotrophy Due to a genetic mutation, a bacterial cell requires a specific nutrient in the media to grow

Backbone In genetics, this refers to the sugar phosphate portion of a nucleic acid strand

Bacteria Organisms that are generally microscopic, single cells, of a distinct evolutionary lineage from eukaryotes and archaea

Bactericidal An antimicrobial that kills the bacterial cell

Bacteriocin Antimicrobial peptides produced by bacteria to eliminate competitor bacteria in a niche

Bacteriophage A small virus-like particle containing its own genetic material and capable of infecting bacteria

Bacteriostatic An antimicrobial that stops the growth of the bacterial cell

Basal body *See* Base complex

Base calling Identification of the individual nucleic acid sequence data from interpretation of the sequencing machine information; for example, this may be incorporation of bases in sequencing by synthesis and generation of light or release of hydrogen ions, or it may be differential through a nanopore

Base complex A multi-protein component of a Type 3 Secretion System that is embedded in the inner and outer membranes, spanning the periplasm

Basic local alignment search tool A computational method for finding sequences in a database similar to a query sequence

Beta barrel A tertiary amino acid structure formed by the stacking of beta sheets into a cylindrical structure

Beta sheet A secondary amino acid structure formed by hydrogen bonding between amino acids such that a sheet is made

Bioaugmentation Restoration of environmental sites contaminated with pollutants through the addition of microorganisms specifically chosen to address particular issues in that environment, with the goal of accelerating the bioremediation process

Bioenergy Generation of energy from biological sources, ideally renewable

Biofilm A community of microorganisms that together form a structure that is more resilient than the cells are individually

Bioinformaticians People who use computers to analyze biological data, bioinformatics

Bioinformatics Computational and software-based approaches to understanding biological data, such as DNA sequence data

Bioreactor A vessel for the long-term growth of microorganisms in a controlled environment

Bioremediation Restoration of environmental sites contaminated with pollutants through the introduction or assistance of microorganisms

Biostimulation Restoration of environmental sites contaminated with pollutants through the addition of nutrients and other factors that aid microorganisms naturally occurring in the environment and that are already conducting bioremediation, with the goal of accelerating these processes

Biotechnology The use of biological organisms for the improvement or betterment of human life

Biotreatment Removal of pollutants and contamination from waste water and other by-products of industry by microorganisms before they enter the environment

BLAST *See* Basic local alignment search tool

BLAST hits The results from a BLAST search; sequences in the database that match the query sequence

BLASTN *See also* BLAST; a specialized form of BLAST for comparing a nucleotide sequence with a nucleotide sequence database

BLASTP *See also* BLAST; a specialized form of BLAST for comparing an amino acid sequence with an amino acid sequence database

BLASTX *See also* BLAST; a specialized form of BLAST for comparing a nucleotide sequence with an amino acid sequence database by first translating the nucleotide sequence into its six-frame protein translation

Blunt ends Ends of a DNA double helix generated by a restriction enzyme, where the DNA is cut straight across so that there are no overhanging nucleotides

Branched chain amino acids Amino acids with branched side chains

Broad-spectrum antibiotic An antimicrobial agent that acts on a wide variety of bacterial species

cAMP *See* Cyclic AMP

cAMP receptor protein A protein, also known as CAP, that is able to bind to another molecule within the cell, cAMP, and is then able to bind to DNA, thus exerting repression of gene expression

CAP *See* Catabolite activator protein

Capsule An extracellular polysaccharide coating on the surface of some bacterial species

Carboxyl group A carbon, two oxygens, and a hydrogen, found at one end of an amino acid

Cas *See* CRISPR associated genes

Catabolite activator protein A protein that is able to bind to another molecule within the cell and able to bind to DNA, thus exerting catabolite repression

Catabolite repression The reduction or prevention of transcription due to the presence of molecules within the cell

CDS *See* Coding sequence

Cephalosporins A class of antibiotics that contain the characteristic β-lactam ring structure, as well as an altered adjacent ring structure that renders this class resistant to standard β-lactamases

Chaperone A protein that assists in the folding or stabilization of another protein

Chaperone-usher secretion A Type 5 Secretion System where three proteins work to make an outer membrane channel, properly fold a secreted protein, and export it

Cheater A bacterial phenotype that is able to continue in a hostile environment due to the presence of other organisms in the niche with a favorable phenotype that neutralize the hazards of the environment

ChIP-Seq The sequencing of chromatin immunoprecipitation to reveal the locations across the genome where proteins bind to the DNA

Chromatin Proteins that interact with DNA structurally, often causing it to become compact

Chromatin immunoprecipitation The identification of DNA bound by protein through use of antibodies directed against the protein to separate out the bound DNA

Chromatin immunoprecipitation sequencing *See* ChIP-Seq

Chromosome Discrete DNA that contains essential genetic material and is copied by replication; one or more chromosomes make up a genome

Cis Referring to chemical structures those being on the same side as one another; referring to genetics those being close

Clonal complexes Phylogenetic contexts for bacterial isolates as defined by sequence typing methods such as MLST

Clone In molecular genetics, this refers to bacterial cells that are in some way altered in the laboratory

Closed complex An assembly of RNA polymerase and DNA at a promoter region where hydrogen bonds between bases of opposite DNA strands remain intact

Clustered regularly interspersed short palindromic repeats A genetic system that enables bacteria to recognize foreign nucleic acid sequences such as bacteriophage

Coding region *See* Coding sequence

Coding sequence An open reading frame that has an appropriately located initation codon such that it could encode a gene

Codon A set of three nucleotides that corresponds to a sequence on a particular tRNA carrying a particular amino acid, thus the three letters encode a specific component of a protein

Codon table A listing of which three bases correspond to which amino acids (see Table 2.1)

Cold-shock proteins Proteins that protect the bacterial cell from the stresses experienced due to decrease in temperature

Colony PCR Use of a single bacterial colony, or DNA from one, as the template in PCR; *see also* PCR

Comparative genomics Analysis of genome sequence data when two or more genome sequences are evaluated against one another to determine similarities and differences between them

Comparative modeling *See* Homology modeling

Compatible cohesive ends Reference to the generation of restriction enzyme cut end overhangs that contain the same bases, even though they were generated by different enzymes

Complementation In molecular genetics, this adds back a gene that has been disrupted in a knockout mutant to check the phenotype is related to that gene and not another change in the genome

Concatemerization Generation of long, continuous DNA molecules that contain repeated or multiple copies of the same sequence

Conjugation The transfer of DNA from one bacterial cell to another by virtue of the F plasmid and conjugation pilus

Conjugation pili A specialist appendage of a bacterial cell that is encoded by the F plasmid and which is a conduit for transfer of DNA from donor cell to recipient cell

Consensus sequence A representative sequence arrived at based on the sequences that are generally present across a number of strains and/or species

Conserved domain A functional portion of a protein that is similar between proteins of similar functions, even if the whole of the protein does not share similarity

Conserved hypothetical protein A hypothetical protein that has homology with other hypothetical proteins in other bacterial species

Constitutive expression Expression of a transcript or protein at a relatively uniform level in the cell

Construct DNA that has been generated in the laboratory, generally through joining together different sequences

Contig A stretch of contiguous sequence data, often assembled from separate sequencing reactions that generated data with overlapping identical sequences

Convergent evolution Genetic sequence changes observed in separate bacterial genomes where the similarity in selective pressures results in similarity in gene changes

Core genes A gene or genes that are common to all genomes of a particular species

Core genome All of the genes that are common to all genomes of a particular species

Corepressor A small molecule that binds to a repressor protein and activates it so that it can bind to DNA

Coupled transcription–translation Where the process of transcription (making mRNA) and translation (making protein) happen concurrently for a given gene

CRISPR *See* Clustered regularly interspersed short palindromic repeats

CRISPR associated genes Genes associated with the CRISPR region of the genome, encoding proteins involved in the bacterial immune system

CRP *See* cAMP receptor protein

Cyclic AMP A form of AMP, adenosine monophosphate, that is present in the cell inversely to the presence of glucose

Cyclic di-GMP A secondary signal for the state of the bacterial cell, made from two GTPs

Cytosine A nitrogenous base made of four carbons, three nitrogens, one oxygen, and five hydrogens, found in DNA and RNA

Data depth This is a measure of how many times each individual nucleotide has been sequenced in all of the separate sequencing reactions

Deamination Through the addition of water, an amine group is lost, generally from cytosine resulting in uracil

Deformylase An enzyme that removes a formyl group (–COH) from a protein; the first step in removing fMet from the start of a peptide

Deoxyribonucleic acid (aka DNA) the genetic material of bacteria that stably encodes their traits, comprised of deoxyribose sugars, phosphates, and the nitrogenous bases adenine, cytosine, guanine, and thymine

Deoxyribonucleoside triphosphate (aka dNTP) a deoxyribose sugar, three phosphates, and a nitrogenous base

Deoxyribose A sugar containing five carbons, four oxygens, and 10 hydrogens that forms the backbone of DNA, together with phosphate

Diguanylate cyclases Enzymes that forms cyclic di-GMP from two GTPs

Dihydrouridine A modified nucleoside present in some tRNAs and rRNAs

Dimer Two of the same protein that complex with one another

Dinucleotide Two nucleotides joined by a phosphodiester bond between the sugar of one and the phosphate of another

Directional cloning Use of restriction enzymes so that an insert will only be ligated into a plasmid in one specific orientation

Disulfide bond A bond formed between two sulfides, often within proteins and formed between two cysteine amino acids

Divergent promoters When two nearby promoters are transcribed in opposite directions and away from each other, one on one strand and one on the other

D-loop The loop from the first of three stem loop structures in tRNA, starting from the 5′ end

DNA *See* Deoxyribonucleic acid

DNA gyrase A replication enzyme that unwinds the DNA double helix at the replication fork

DNA identity *See* Identity (DNA)

DNA polymerase An enzyme that is able to form a new strand of DNA based on a single strand of DNA as the template

dNTP *See* Deoxyribonucleoside triphosphate

Domain A functional part of a protein; *see also* Conserved domain

Dosage effect As replication copies DNA, some genes are present in multiple copies, which impacts upon their expression

Double digest Simultaneous use of two different restriction enzymes with different recognition sites

Downstream After a gene or other genetic feature, 3′ of the feature

D-stem The stem from the first of three stem loop structures in tRNA, starting from the 5′ end

E value Short for expect value, this is the number of alignments a BLAST query would generate in a search of the database by chance

Eclipse phase In bacteriophage replication, when bacteriophage particles and the genome are replicated and the bacterial systems are shut down

Effector A protein that is exported from the cell by a Type 3 Secretion System

Effector molecule A non-protein component, often a metabolite, that is able to influence transcription

Efflux pump system Multiprotein systems that export harmful substances out of the bacterial cell, including antibiotics

Elongation (transcription) In genetics, the phase of the process of transcription whereby RNA is made from a DNA template through the addition of nucleotide bases and the formation of phosphodiester bonds along the backbone

Elongation (translation) In genetics, the phase of the process of translation whereby a polypeptide chain is made based on an mRNA sequence

Endonucleolytic Action upon a nucleic acid within the strand

Enhancers Features within a sequence that increase transcription

Endosymbiont A bacterial cell that lives within a host cell

Epidemiology The study of the distribution of diseases, geographically and over time

E-site *See* Exit site

ESKAPE pathogens *Enterococcus faecium*, *Staphylococcus aureus*, *Klebsiella pneumoniae*, *Acinetobacter baumannii*, *Pseudomonas aeruginosa*, and *Enterobacter* species that are of clinical importance as pathogens

Essential gene A coding sequence required for life

Etiological agent That which is causing the disease, which can be a microorganism or toxin or other source

Eukaryote Also referred to as higher organisms, this form of life can be multicellular, with complex cell structures, of a distinct evolutionary lineage from bacteria and archaea

Evidence tag In annotation of sequence data, the scientific justification of the assigned annotation

Evolvability The capacity of the bacteria to evolve solutions to adapt

Evolve and resequence Growth of a bacterial culture for a period of time in a specific condition followed by sequencing of the genome again and comparison with the original data

Exit site A place within the ribosome where tRNA that has released the growing polypeptide chain goes before exiting the ribosome

Exonucleolytic Action upon a nucleic acid from the end of the strand

Exopolysaccharide Sugars secreted by bacteria into the surrounding environment that are a feature of biofilms

Expression platform The regulatory region of a riboswitch, which together with the aptomer region, can regulate expression of the mRNA

Expression profiling Assessment of the levels of transcription of genes

Extended −10 A conserved sequence sometimes present approximately 10 bases 5′ of the transcriptional start site that has high affinity for the sigma subunit of RNA polymerase; therefore no −35 region is needed

Extended spectrum β-lactamases A class of enzymes that have overcome the difference in the structure of cephalosporins that render them resistant to standard β-Lactamases

F plasmid A plasmid containing the genetic features needed to make a bacterial cell a conjugation donor

Factor for inversion stimulation A nucleoid protein that contributes to bending of the DNA

Facultative aerobes Bacteria that are able to switch their metabolism and continue to grow in the absence of oxygen

Facultative intracellular Bacteria that can live extracellularly or can live within cells

FASTA A genetics file format, which includes a first line with information about the sequence, followed by the nucleic acid sequence

Feedback regulation The expression of the gene encoding a regulator, which is regulated by the regulator itself

Fertility factor The genetic features needed to make a bacterial cell a conjugation donor

F-factor *See* Fertility factor

Fimbriae Thin, numerous bacterial surface structures made of protein

Fis *See* Factor for inversion stimulation

Flagella Long thick bacterial surface structures made of protein that are capable of rotating to cause cell motility

fMet *See N*-formylmethionine

Fold recognition *See* Protein threading

G+C The number of guanines plus the number of cytosines, an indicator calculation that can be performed on a sequence of DNA and give a figure characteristic of the DNA

Galactose A sugar generated from the breakdown of lactose by β-galactosidase

GC skew A calculation that can be performed on a DNA sequence, G-C/G+C, which can help identify the origin of replication

Gene Genetic feature encoding a protein that is characterized as having an initiation codon, a termination codon, and for which a protein product has been demonstrated

Gene conversion A mechanism used in antigenic variation where the sequence of an expressed gene locus is changed through recombination with sequences at silent loci

Generalized transduction Due to random packaging of nucleic acids into the bacteriophage head, some bacterial DNA may be packaged and therefore carried to new bacterial cells

Genetic map Determination of the relative location of genes on the chromosome, which can be created using timed conjugation experiments

Genome The genetic material, containing all of the traits typically found in an organism

Genotype The genetic traits possessed within an organism's genome

GlcNAc *See N*-acetyl glucosamine

Global alignment Matching the similarity of sequences so that similar nucleotides or amino acids align with one another over the whole of the sequence; an alignment will be generated even if there is no similarity

Global regulator A protein that can alter transcription of several genes

Glucose A sugar essential for energy in bacterial cells

Glycoprotein A protein with a sugar added to it

Glycosylation The addition of a sugar to a protein

Group I introns Regions that are capable of splicing themselves out of a nucleic acid

Guanine A nitrogenous base made of five carbons, five nitrogens, one oxygen, and five hydrogens, found in DNA and RNA

Heat inactivated Some enzymes can be rendered inactive, unable to perform their enzymatic activity, after exposure to certain temperatures

Heat-shock proteins Proteins that protect the bacterial cell from the stresses experienced due to increase in temperature

Heat-unstable protein A nucleoid protein that contributes to bending of the DNA

Heuristic A computer science term for finding an approximate solution quickly

Hfr *See* High frequency of recombination

HGT *See* Horizontal gene transfer

High frequency of recombination A designation for bacterial cells that have integrated the F plasmid into their genome

High-performance liquid chromatography A technique used to separate parts of a complex mixture for the purposes of identification and quantification

Histone-like nucleoid structuring protein A protein, which together with more of itself, binds to DNA and contributes to the compact nature of the nucleoid

H-NS *See* Histone-like nucleoid structuring protein

Holoenzyme The whole of an enzyme with all of its subunits

Homodimer Two identical proteins that come together to form a quaternary structure, often seen in DNA binding proteins

Homology A measure, usually in percentage, of the identity or similarity of two DNA or protein sequences

Homology modeling Generation of a three-dimensional model of a protein based on experimental evidence of the structure of a homologous protein and the amino acid sequence of the protein of interest

Horizontal gene transfer The process of movement of DNA from one bacterial cell to another that then results in the incorporation of the DNA into the new cell

HPLC *See* High-performance liquid chromatography

HU *See* Heat-unstable protein

Hydrogen bond A molecular connection between molecules formed by charge associated with hydrogen

Hydrolysis Through reaction with water, a chemical structure is split

Hydroxylation The addition of a hydroxyl group (OH) to a protein

Hypothetical protein A predicted protein sequence from a CDS that is proposed to encode a protein but for which there is no experimental evidence or sequence homology suggesting a function

Ident value Short for identity value, this is the identity score between the aligned BLAST query and target sequences

Identity (amino acid) A measure, usually expressed in percentage, of how much two protein sequences match

Identity (DNA) A measure, usually expressed in percentage, of how much two DNA sequences match

IHF *See* Integration host factor

In vitro A non-English term (note the italics) meaning in an artificial situation in a laboratory

In vivo A non-English term (note the italics) meaning in a natural situation within a living host or host cells

Indel A short addition, insertion, of one or a few bases or a short loss, deletion, of one or a few bases from DNA

Initial Darwinian Ancestor A hypothetical first organism on Earth, capable of passing on its genetic material

Initiation (transcription) In genetics, the beginning of the process of transcription, where transcription enzymes are recruited to promoters

Initiation (translation) In genetics, the beginning of the process of translation, where translation enzymes, mRNA, tRNA, and ribosome are recruited to begin making a protein

Initiation factor protein One of a set of proteins inolved in the initiation of translation

Inner membrane platform A protein component of a Type 2 Secretion System that is embedded in the inner membrane

Inosine A modified nucleoside present in some tRNAs

Integration host factor A nucleoid protein that contributes to bending of the DNA

Internalins Proteins that help bacteria to be internalized by host cells

Intracellular Bacteria that can live within other cells

Intrinsic transcriptional termination When the signal to stop transcription is a built in part of the sequence being transcribed

Introns Regions of a nucleic acid that are not present in the mature form of the DNA or RNA

Inversion When a segment of DNA has been flipped from one strand to another

IS elements Features within bacterial DNA that are capable of self-mobilization from one part of the DNA to another

Isoelectric point The pH at which a protein has no net electrical charge

Isogenic Two or more bacterial strains that have nearly identical genetic backgrounds, but which often differ by one or a discrete number of mutations

kb *See* Kilobase

Kilobase One thousand bases of nucleic acid

Knockout In molecular genetics, this is a mutation that gets rid of the functional sequence of a gene through deletion and/or disruption of the gene sequence

Lactose A disaccharide sugar made up of glucose and galactose

Lagging strand During replication, this is the strand of the DNA that must be copied in short segments that proceed away from the replication fork

LAMP *See* Loop-mediated isothermal amplification

Landmark database A small, highly curated database of sequences representing 27 genomes, including 11 bacterial species

Last Universal Common Ancestor (LUCA) The final organism in common between all types of life, before evolution into bacteria, eukaryotes, and archaea

LC/MS/MS *See* Liquid chromatography tandem mass spectrometry

Leader A segment of mRNA sequence involved in attenuation of transcription

Leading strand During replication, this is the strand of the DNA that can be copied directly, proceeding toward the replication fork

Library In molecular genetics, this is a collection of bacterial strains that are identical apart from a single discrete mutation

Ligase An enzyme that joins together the ends of a nucleic acid backbone; used in replication and molecular biology

Lineage Bacterial cells that originated from the same precursor cell and thus have a common ancestor

Lipid A The lipid portion of lipopolysaccharide and lipooligosaccharide, which has a hydrophobic region that is embedded in the outer membrane

Lipidation The addition of a lipid to a protein

Lipooligosaccharide Molecular structures that are on the surface of a few Gram-negative bacteria that form the outer membrane with phospholipids

Lipopolysaccharide Molecular structures that are on the surface of most Gram-negative bacteria that form the outer membrane with phospholipids

Lipoprotein A protein with a lipid component added to it

Liquid chromatography tandem mass spectrometry A technique combining separation by HPLC with mass analysis by MS; sensitivity and specificity is increased by using two mass spectrometers in tandem

Local alignment Matching the similarity of sequences so that similar nucleotides or amino acids align with one another in the segments of the sequence in which they are similar; if there is no similarity, no alignment is generated

Locus identification number Sequential numbering of features in a genome sequence with unique identifiers

Loop-mediated isothermal amplification A technique used to amplify DNA at a constant temperature using a series of nested primers that generate self-priming loops

LOS *See* Lipooligosaccharide

Low complexity regions Parts of a query sequence that will confuse a BLAST search due to the frequency of occurrence in databases, such as repetitive sequences

LPS *See* Lipopolysaccharide

LPS core The portion of lipopolysaccharide and lipooligosaccharide that is directly attached to the lipid A; in LPS the O-antigen is attached to the LPS core

LUCA *See* Last Universal Common Ancestor

Lysine acetyltransferase An enzyme that adds an acetyl group to a lysine residue within a protein

Lysine deacetylase An enzyme that removes an acetyl group from an acetylated lysine within a protein

Lysogenic phase Describes when a bacteriophage goes dormant within a bacteria, with its genome integrated into the bacterial chromosome as a prophage

Lysogens Bacterial cells containing one or more prophage genomes integrated into their chromosome(s)

Lysosome A eukaryotic organelle containing lysosomal enzymes that are capable of digesting bacterial cells

Lytic phase Describes when a bacteriophage replicates its genome and constructs new phage particles that break free and are capable of infecting other bacterial cells

MALDI-TOF-MS *See* Matrix-assisted laser desorption ionization-time-of-flight mass spectrometry

Mapping In the context of bioinformatics, alignment of sequence data to a reference sequence

Mass spectra The results produced by a mass spectrometer based on characteristics of the sample analyzed

Mass spectrometry Technology for the analysis of samples based on the mass of molecules within the sample

MATE efflux family A class of efflux pump that uses sodium import into the bacterial cell to fuel export out of the cell

Matrix-assisted laser desorption ionization-time-of-flight mass spectrometry A technique used to identify features of a sample, separating aspects of the sample to produce a characteristic mass spectrum

Max score Short for maximum score, this is a value assigned to a BLAST result based on aligned and matched sequences, taking account of those that mismatch and any gaps

Mb *See* Megabase

MCS *See* Multiple cloning site

Megabase One million bases of nucleic acid

Megaplasmid A very large plasmid

Membrane fusion protein A protein in the periplasm that bridges an inner membrane protein and an outer membrane protein, present in Type 1 Secretion Systems

Messenger RNA (aka mRNA) Ribonucleic acids that have been copied from DNA in a process called transcription that carry the genetic make-up of a gene or genes

Metagenomic sequencing The sequencing of DNA from a sample taken from a natural environment, without discrimination or selection for organisms

Methionine aminopeptidase An enzyme that removes the N-terminal methionine from a peptide chain after the action of deformylase

Methylation The addition of a methyl group ($-CH_3$), for example to a nucleic acid or protein

MFS family A class of efflux pump made of protein with several transmembrane helices

Microbiome The microorganisms present in a particular environmental niche, collectively

Microcolonies Small clusters of bacterial cells replicating in a localized area

MLD *See* mRNA-like domain

MLST *See* Multi-locus sequence typing

Molecular biology The investigation of nucleic acids and/or proteins, including their structure, interaction, and regulation

Molecular cloning The generation of recombinant DNA using molecular biology techniques

Molecular mass *See* Molecular weight

Molecular weight The mass of a protein or other molecule, based on the atomic weight of its parts

mRNA *See* Messenger RNA

mRNA-like domain Part of a tmRNA that is like mRNA and carries a short sequence including a C-terminal peptide tag and a termination codon

MS *See* Mass spectrometry

Multi-FASTA A genetics file format, which includes the information from more than one FASTA file, where there is more than one line with information about the sequence, followed by a description of the nucleic acid sequence

Multi-locus sequence typing A means of typing bacterial isolates that sequences a discrete set of genes and compares the data as a typing system

Multiple cloning site Engineered sites within plasmids that contain many, overlapping restriction enzyme recognition sites

Multiple sequence alignment Matching the similarity of three or more nucleotide or amino acid sequences along their length

MurNAc *See N*-acetyl muramic acid

Mutation A change in a DNA sequence

***N*-acetyl glucosamine** One of two repeating units that make up the peptidoglycan of the bacterial cell wall

***N*-acetyl muramic acid** One of two repeating units that make up the peptidoglycan of the bacterial cell wall

***N*-acyl homoserine lactones** Signal molecules produced by bacteria for quorum sensing

NAG *See* *N*-acetyl glucosamine

NAM *See* *N*-acetyl muramic acid

Nanopore An artificially created pore that is on the scale of nanometers or smaller

Narrow-spectrum antibiotic An antimicrobial agent that acts on a single species of bacteria or a very few species

ncRNA *See* Noncoding RNA

Needle A multi-protein component of aType 3 Secretion System that extends through the base complex out of the bacterial cell, through which effectors pass

Negative strand The strand of DNA complementary to that encoding a protein

Next-generation sequencing A collective term for advances in sequencing technology that allowed high-throughput generation of nucleotide sequence data rapidly, inexpensively, and easily

***N*-formylmethionine** At the start of production of any protein by the ribosome, this is the first amino acid

Nitrogenous base *See* Nucleoside base

Noncoding RNA A small RNA that is transcribed, but not translated, and may be involved in regulation of gene expression

Non-stop decay Degradation of an mRNA by RNases as a result of a stalled ribosome that has not encountered a signal to terminate translation

Non-synonymous A change in a DNA sequence that results in a change in amino acid when translated

Nosocomial infections Hospital-acquired infections

N-terminal acetyltransferase An enzyme that adds an acetyl group to the N-terminal end of a protein

Nuclease Enzymes that degrade nucleic acids

Nucleic acid Individual components that make up DNA or RNA, comprising a sugar, a phosphate, and a nitrogenous base

Nucleoid The condensed mass of chromosomal DNA within a bacterial cell that can be seen using a microscope

Nucleoside base An adenine, cytosine, guanine, thymine, or uracil that is not connected to a sugar phosphate group

Nucleotide base Adenine, cytosine, guanine, thymine, or uracil with a sugar phosphate

Nucleotidyl transferase An enzyme that adds the CCA motif to the 3′ end of a tRNA, forming the amino acid acceptor motif

O-antigen The portion of lipopolysaccharide that is farthest away from the surface of the cell and is the most variable in length and structure; not present in lipooligosaccharide

Obligate aerobes Bacteria that require oxygen to grow

Obligate intracellular Bacteria that can only live within other cells

Okazaki fragment The name given to the short segments of DNA made on the lagging strand in replication

Oligonucleotide A short string of nucleotides joined by phosphodiester bonds

OMVs *See* Outer membrane vesicles

Open complex An assembly of RNA polymerase and DNA at a promoter region where hydrogen bonds have been broken and the DNA strands have separated

Open reading frame A region between two termination codons

Operator A term for the region 5′ of a gene that can be bound by a regulatory protein that influences the level of transcription

Operon Two or more co-transcribed and co-regulated genes

Opsonization Immunological targeting of a bacteria for phagocytosis by covering it with antibodies or complement

ORF *See* Open reading frame

Origin of replication A location within a DNA chromosome where replication consistently begins

Orthologous regions Regions of two different genomes that share homology; not necessarily defined by the presence of coding sequences, *see also* Orthologue

Orthologue A homologous gene sequence between two bacterial species, where the two genes have a common ancestral origin, suggesting that the genes may have a common function

Outer membrane complex A protein component of a Type 2 Secretion System that is embedded in the outer membrane

Outer membrane vesicles In some bacterial species, portions of the outer bacterial membrane separate off from the bacterial cell as extracellular vesicles, also called blebs

Output domain The section of a response regulator protein in a two-component regulatory system that is activated through phosphorylation of the receiver domain, activating the regulatory function of the response regulator

Overexpression mutant A bacterial cell with a mutation that causes the gene to be expressed at higher levels than normal

Overlapping PCR Use of several rounds of PCR, with overlapping primer design, to generate a novel DNA sequence different from the original template, *see also* PCR

Oxidative folding The formation of disulfide bonds between cysteines in a protein

Pairwise alignment Matching the similarity of two nucleotide or amino acid sequences along their length

Palindromic A word, number, or sequence that can be read the same forward and backward, such as the word madam or the number 1234321

Pan-genome All of the genetic features found within all of the bacterial genome sequences of a particular species or set of organisms, including the core genes and all of the accessory genes and other features present

Paralogue A homologous gene sequence within a bacterial genome, where the two genes have a common origin, having arisen through a gene duplication event, but which may now have diverged in function

Pathogenicity island A region containing genes involved in virulence or pathogenicity, which may have been horizontally transferred

PCR *See* polymerase chain reaction

Penetration Insertion, in the context of bacteriophage, the bacteriophage genome entering the bacterial cell

Peptide bond The bond formed between two amino acids, with the carboxyl group of one amino acid joining to the amino group of another

Peptide chain A series of amino acids connected by peptide bonds

Peptidyl site A place within the ribosome where tRNA is transferred from the A-site; this is where the growing polypeptide chain remains attached to a tRNA

Persister A bacterial phenotype that is able to continue in a hostile environment due to its slow growth

Phagocytosis The ingestion of cells or particles by other cells, often immune cells engulfing bacterial cells

Phagosome A eukaryotic organelle that has engulfed a cell or particle

Phase variation Stochastic switching of gene expression, generally ON to OFF to ON, but also in levels of expression

Phasevarion A set of genes that are regulated by a single phase variable gene, usually a modification component of a restriction-modification system

Phenotype The genetic traits being expressed by an organism at a given time

Phosphodiester bond The bond formed between phosphates and sugars in the backbone of nucleic acids

Phosphodiesterases Enzymes that break down cyclic di-GMP

Phospholipids Molecular structures that make up the membranes with a hydrophilic head and hydrophobic tails

Phosphorelay An extended form of a two-component regulatory system that has four proteins along which are passed a phosphate before the final response regulator binds to DNA

Phosphorylation The addition of a phosphate group $(PO_3)^-$ to a protein

Photophores Specialized organs that host bacterial populations for the purposes of harnessing their light production

Physical map Determination of the base pair distances between restriction digest recognition sites in a chromosome

Pili Long thin bacterial surface structures made of protein and present in several copies per cell

Plasmid Small circular DNA that is separate from the chromosome(s), often containing non-essential traits

Plasmid copy number The number of plasmids present within a single bacterial cell

PMF *See* Proton motive force

Point mutation A single base change in a DNA sequence

Polishing Improving the accuracy of sequencing data by comparing the final data to the original base calling data

Polyadenylation The addition of adenines to the 3′ end of mRNA

Polycistronic units mRNA with two or more genes

Polylinker *See* Multiple cloning site

Polymerase chain reaction (aka PCR) A molecular technique that copies specific fragments of DNA in the laboratory

Positive strand The strand of DNA encoding a protein

Post-segregational killing The death of a bacterial cell as a consequence of loss of a plasmid that encoded a toxin-antitoxin system; the toxin persists in the cell, but the antitoxin does not and the gene that encoded it was lost with the plasmid

Post-translational modification Changes made to a peptide chain after it is generated by translation of the mRNA at the ribosome

PPi *See* Pyrophosphate

Pribnox box *See* –10 promoter sequence

Primary amino acid structure An unfolded string of amino acids in a linear sequence that are joined together through the process of translation

Primase A replication enzyme that forms a short RNA primer necessary for DNA polymerase to be able to replicate the DNA

Processivity The property of a particular DNA polymerase to be able to continue to extend the DNA, generating the second strand of the helix against the template strand

Profile sequence In PSI-BLAST, the starting set of protein sequences that are closely related to each other, these are used as the basis for searches

Promoter occlusion When the transcription from one promoter negates or significantly reduces the transcription from a secondary promoter oriented in the same direction

Promoter region A location within DNA that is recognized by RNA polymerase for the initiation of transcription

Prophage The genome, in whole or part, of bacteriophage that has been integrated into the bacterial chromosome

Protein A polypeptide chain of amino acids

Protein fold The localized three-dimensional structures formed by amino acid sequences

Protein threading Generation of a three-dimensional model of a protein based on experimental evidence of the structure of proteins with similar fold structures; useful when there are no homologous proteins with identified structures

Proteinogenic amino acid An amino acid that can be found in proteins

Proteolysis Degradation of a protein

Proteome All of the proteins present within a bacterial cell

Proteomics The study of the proteins present within a bacterial cell

Proton motive force The accumulation of protons on the surface of bacteria that forms an electrochemical gradient that can be used as an energy source for transport across the membranes

Pseudogene A genetic feature that has some, but not all characteristics of a coding sequence, suggesting that it may have been a gene that has degraded over time

Pseudoorganelle A genome-containing cellular structure found in *Thiomargarita magnifica* bacteria that is organelle-like, resembling the nuclei of cells of higher organisms

Pseudopilus A protein component of a Type 2 Secretion System that pushes folded proteins out of the bacterial cell through the outer membrane complex

Pseudouridine A modified nucleoside present in some tRNAs

P-site *See* Peptidyl site

PSK *See* Post-segregational killing

Psychrophilic Bacteria that are capable of growth and cell division at extremely low temperatures

Purine The nitrogenous bases adenine and guanine, characterized by a pyrimidine ring fused to an imidazole ring

Pyrimidine The nitrogenous bases cytosine, thymine, and uracil, characterized by a pyrimidine ring

Pyrophosphate Two phosphates released from a nucleotide triphosphate during the incorporation of a base into DNA or RNA by the corresponding polymerase

Quaternary amino acid structure A complex protein structure formed from two or more folded tertiary protein structures

Query cover Short for query coverage, this BLAST statistic addresses how much of the query along its length has been found in the database

Query sequence The DNA, RNA, or protein data used to interrogate a database; generally the sequence the researcher is interested in investigating

Quorum sensing The production and sensing of a chemical signal that allows the bacteria to determine the size of its population

R When discussing amino acids, the R stands for any side chain

Random mutagenesis The generation of mutants that not directed, but are randomly generated due to the methods used

Raw sequencing data Information collected from a sequencing technology machine prior to removal of poor quality data

RBS *See* Ribosome binding site

Read length This is a measure of how many nucleotides have been identified in each individual sequencing reaction

Reading frame Each set of three bases that makes up a codon in DNA; being read forward or reverse there are therefore six reading frames starting from sequential bases on the forward or reverse strand

Receiver domain The section of a response regulator protein in a two-component regulatory system that is phosphorylated by the histidine kinase

Reciprocal best hits Two sequences that are the best hits for one another in homology

Recombinant DNA Two DNA molecules from different sources that are combined into one DNA molecule in the laboratory

Recombinant protein A protein that is generated from recombinant DNA, such that the protein is made by a different species than the origin of the DNA encoding the protein

Recombination The exchange of DNA

Reductive genome evolution A process of eliminating genetic content that is not needed by the bacterial cell

Regions of difference Regions of DNA that are different between two or more sequences that are otherwise similar, often used when comparing bacterial genome sequences

Regulated expression Expression of a transcript of protein that is controlled so that there can be different levels of expression within the cell

Regulator A protein that can alter transcription of a gene

Regulatory network Connection between multiple genes in their expression due to common regulatory proteins and/or ncRNAs

Regulon The set of genetic features controlled by a common regulator

Release factor A protein that enters the A-site at the termination of translation and triggers disassociation of the translation machinery

Replication In genetics, the process of copying DNA, generally making a copy of a genome or chromosome

Replication fork During the process of replication, the two DNA strands are split apart; this is the split that moves along the chromosome as replication progresses

Repressor A protein that can stop or reduce transcription of a gene

Repressor protein A protein that can stop or reduce the transcription of a gene or genes, often through binding to DNA

Resequencing Conducting sequencing on a sample or bacterial species that has already been sequenced, often to determine changes that may have occurred over time

Response regulator A protein in a two-component regulatory system that is able to bind to DNA when it is phosphorylated

Restriction endonucleases *See* Restriction enzymes

Restriction enzymes Proteins that are able to cleave DNA at a specific site based on recognition of a specific sequence in the DNA

Restriction fragments Pieces of DNA generated by enzymatic digestion with a restriction enzyme

Restriction map *See* Physical map

Restriction modification system A restriction enzyme with a specific recognition sequence and a methylase that is able to methylate that recognition sequence to protect it from the restriction enzyme

Reverse complement The sequence on the opposite strand of double stranded DNA, where the sequence is both in reverse (running the other way) and complemented (due to base pairing)

Reverse vaccinology The design and development of a vaccine based on genome sequencing data and predicted antigens

Rho-dependent transcriptional terminator When the signal to stop transcription relies on the binding and action of a separate protein called Rho

Ribonucleic acid (aka RNA) genetic material used by bacteria to express their traits, comprised of ribose sugars, phosphates, and the nitrogenous bases adenine, cytosine, guanine, and uracil

Ribonucleotide triphosphate A ribose sugar, three phosphates, and a nitrogenous base

Ribose A sugar containing five carbons, five oxygens, and 10 hydrogens that forms the backbone of RNA, together with phosphate

Ribosomal RNA (aka rRNA) A type of ribonucleic acid that forms a complex with ribosomal proteins to form a ribosome, wherein translation of mRNA occurs

Ribosome A complex of proteins and ribosomal RNA essential for translating mRNA to protein

Ribosome binding site A location on mRNA where ribosomes bind to initiate translation

Ribosome recycling In genetics, the final phase of translation where the released ribosome is able to be involved in another translation initiation

Riboswitch Encoded regulatory factor for transcription or translation that alters expression by virtue of its sequence

Ribozyme An RNA with enzymatic activity

RNA *See* Ribonucleic acid

RNA polymerase An enzyme that is able to form a new strand of RNA based on a single strand of DNA as the template

RNA thermometers mRNA sequences that alter in their secondary structure in response to temperature changes, which affects expression of the encoded protein(s)

RNA World Hypothesis A scientific idea that RNA was used as the genetic material on Earth before there was DNA

RNase An enzyme that degrades RNA

RND family A class of efflux pump made of three proteins, an inner membrane protein, membrane fusion protein, and outer membrane protein

Rough A description that can be applied to Gram-negative bacteria with full-length LPS O-antigen structures

rRNA *See* Ribosomal RNA

Runaway transcription Generation of mRNA via transcription is more rapid than generation of proteins via translation, resulting in the RNA polymerase and ribosomes being distant on the mRNA

***Salmonella*-containing vacuole** A specialized host cell organelle created through the action of *Salmonella* Type 3 Secretion System effector proteins from the host cell vacuole specifically taking up a *Salmonella* cell

Saprophyte An organism that lives on decaying or dead organic matter

Scaffold Draft or incomplete sequence data that may contain unidentified nucleotides or errors that can be used to build more accurate sequence information by virtue of its relative locational information

Screening In molecular genetics, this is the process of identifying bacterial cells with a particular phenotype by testing each clone for a particular trait

SCV *See Salmonella*-containing vacuole

Sec pathway *See* Sec translocase

Sec translocase A protein system for transporting unfolded peptide chains across the cytoplasmic membrane

Secondary amino acid structure Basic structures formed by hydrogen bonding between the linear primary amino acids, including the α helix and β sheet

Secretion ATPase A protein component of a Type 2 Secretion System that provides energy from ATP in the cytoplasm

Secretion system Cellular machinery that exports proteins and other substrates out of the bacterial cell

Selection In molecular genetics, this is the process of identifying bacterial cells with a particular phenotype by generating media on which only the desired clones will grow

Selenocysteine Considered the 21st amino acid, it is encoded by a stop codon and other sequence features

Sense strand *See* Positive strand

Sensor kinase A membrane protein in a two-component regulatory system that autophosphorylates in response to stimuli

Sequence The order of nitrogenous bases in a DNA or RNA

Sequencing by synthesis Determination of the sequence of a nucleic acid using DNA polymerase to synthesize the second strand and monitor the incorporation of nucleotides

Sequencing platform A phrase that refers to the sequencing equipment, technology, and reagents collectively

Sessile cells Cells that are not motile, but rather are fixed in one place, generally specifically referring to one of two stages of the life cycle of *Caulobacter*

Shine–Dalgarno sequence A conserved sequence generally found 5′ of genes, which is recognized by the 16S rRNA within the ribosome

Shotgun sequencing Genome sequencing methodology that starts with random fragmentation of the genome before sequencing, with the goal of assembling the random sequence data using overlapping sequence data

Siderophore A molecule able to bind iron and scavenge it for the bacteria from the host environment

Sigma subunit One of six proteins that come together to form RNA polymerase; this subunit dictates promoter specificity

Signal recognition particle A series of amino acids at the start of a peptide chain that are recognized for transport across the membrane

Signal sequence A specific amino acid sequence in a peptide chain that is recognized for translocation of the protein across the membrane

Signature tagged mutagenesis (STM) A modification of transposon mutagenesis where a transposon library is made with each transposon carrying a unique signature tag that can be used for identification

Silent cassette A sequence that is not expressed, but resides in the genome as an alternative sequence for the expressed gene

Similarity (amino acid) A measure, usually expressed in percentage, of how much two proteins carry amino acids with attributes that match

Single-molecule sequencing The ability to generate sequencing data from a single molecule of DNA without amplification

Single-nucleotide polymorphism A change in one base of a DNA sequence

Small RNA A short piece of RNA that can serve a regulatory function within the cell

SmartBLAST A specialized BLAST that outputs the five best protein matches from a specially curated database of reference sequences, *see* landmark database

Smooth A description that can be applied to Gram-negative bacteria with short LPS O-antigen structures

SMR family A class of efflux pump involved in multidrug resistance that is made of several copies of a single protein

***S*-nitrosylated** When a thiol group (–SNO) has been added to a protein

SNP *See* Single-nucleotide polymorphism

Sonication Use of sound waves to disrupt a sample, including breaking open biological membranes to release proteins and DNA

Specialized transduction Due to imprecise excision of the prophage from the bacterial chromosome, some bacterial DNA may be packaged into the bacteriophage head and therefore carried to new bacterial cells

SRP *See* Signal recognition particle

Star activity Nonspecific, off-target cleavage of DNA by restriction enzymes due to variations in time and concentration

Start point of transcription The location on DNA where the generation of RNA based on the DNA sequence is started

Stem loop A secondary structure formed by DNA or RNA whereby complementary bases on the same strand hydrogen bond to one another, making the stem, and bending at non-complementary bases, making a loop

Stereoisomers Compounds containing the same elements, which are mirror images of each other structurally

Steric hindrance When the structures of two structures interfere with one another and therefore limit the form of the overall structure

Sticky ends Ends of a DNA double helix generated by a restriction enzyme, where the DNA cut is staggered so that there are overhanging nucleotides on either the 5′ or 3′ ends

STM *See* Signature tagged mutagnesis

Stochastic Referring to random occurrences that happen predictably

Stop codon One of three specific three-letter codes that signal the termination of translation

Strain A representative of a bacterial species that may slightly vary genetically from other members of the same species

Stringent response A regulated stress response of bacteria due to factors such as nutrient limitation, amino acid starvation, heat shock, iron depletion, or other stresses

Succinylation The addition of a succinyl group ($-CO-CH_2-CH_2-CO_2H$) to a protein

Supercoil The twisting of a DNA double helix upon itself, a process that compacts and condenses the DNA

Surveillance Tracking and monitoring of infections within a clinical setting, such as a hospital

Swarmer cells Cells that are motile, possessing flagella to enable motility, generally specifically referring to one of two stages of the life cycle of *Caulobacter*

Symbiosis A long-term interaction between two different bacterial species

Synonymous A change in a DNA sequence that results in the same amino acid when translated as it would have done before the change

Syntenic Describing sequences with synteny

Synteny A term used to describe regions that are similar between two or more sequences in terms of their organization and the features they contain, including genes and other genetic features, regardless of their sequence homology

T1SS–T10SS *See* Type 1–10 Secretion System

Target sequence The sequence from a database that is hit upon by a search, such as BLAST, as being similar to the query sequence

Tat pathway A system for the transport of folded proteins across the cytoplasmic membrane

TBLASTN *See also* BLAST; a specialized form of BLAST for comparing an amino acid sequence with a nucleotide sequence database by first translating the nucleotide sequence database into the six-frame protein translation amino acid sequences for each database entry

TBLASTX *See also* BLAST; a specialized form of BLAST for comparing a nucleotide sequence with a nucleotide sequence database, but which first generates six-frame protein translation amino acid sequences for both the query sequence and each database entry and then compares these protein translations

Temperature-sensitive mutant A bacterial cell with a mutation that only manifests when there is a change in temperature

Termination (transcription) In genetics, the end of the process of transcription, where transcription enzymes disassociate in response to a signal to stop transcription

Termination (translation) In genetics, the end of the process of translation, where the assembled components disassociate and translation stops

Termination codon *See* Stop codon

Termination of replication The region of the chromosome where a round of replication ends

Terminus In bacterial genetics, the location on the chromosome where replication ends

Tertiary amino acid structure Complex protein structures formed when secondary amino acid structures come together through hydrogen bonding, side chain interactions, ionic interactions, and disulfide bonds

Third base wobble The recognition that the third base of a codon can change but the encoded amino acid can stay the same

Thymine A nitrogenous base made of five carbons, two nitrogens, two oxygens, and six hydrogens, found in DNA and RNA

TLD *See* tRNA-like domain

T-loop The loop from the third of three-stem loop structures in tRNA, starting from the 5′ end

tmRNA *See* Transfer messenger RNA

Tn-seq The sequencing of a transposon library to reveal the regions of the genome required for various growth conditions

Tolerance Due to the physiological state of a population of bacteria, often in a biofilm, a proportion of the bacterial cells are able to survive antimicrobial killing

Tolerant *See* Tolerance

Tot score Short for total score, this takes account of a BLAST query having similarity to a subject in multiple regions

Trans Referring to chemical structures those being opposite one another; referring to genetics those being distant

Transcript The product of transcription, RNA made by RNA polymerase based in a DNA template

Transcription In genetics, the process whereby DNA is copied into RNA

Transcription interference When the transcription from one promoter influences the transcription from another promoter

Transcriptional regulation The process of altering transcription to increase or decrease the levels of RNA produced

Transcriptome All of the RNA transcripts within a cell

Transduction The transfer of DNA from one bacterial cell to another via a bacteriophage

Transfer messenger RNA (aka tmRNA) A specialized RNA within the bacterial cell that has activity like both a tRNA and mRNA and which releases the translation machinery stalled in non-stop decay and tags the peptide for degradation

Transfer RNA (aka tRNA) A type of ribonucleic acid that forms a specific structure capable of binding to an amino acid and of base pairing with the corresponding codon on a mRNA

Transformation The uptake of DNA from the environment by a bacterial cell, which may then incorporate the DNA into its own

Translation In genetics, the process whereby an mRNA sequence is used to assemble an amino acid sequence and form a protein

Translation initiation region The region on an mRNA that is recognized by a ribosome to begin translation

Translocation The movement of DNA from one location to another

Translocon A component of a Type 3 Secretion System that is at the tip of the needle and regulates the release of effectors

Transposon mutagenesis The use of transposons to disrupt the sequence of a bacterial chromosome, generating mutants

Transposons Features within bacterial DNA that are capable of self-mobilization from one part of the DNA to another and which carry additional genes

Transposon-sequencing *See* Tn-Seq

***Trans*-translation** The process of completing translation where there is no termination codon on the mRNA, using tmRNA that tags the erroneous protein for degradation and terminates translation

tRNA *See* Transfer RNA

tRNA-like domain Part of a tmRNA that is like tRNA and able to carry an amino acid to the ribosome

Tryptophan One of the amino acids within the bacterial cell that is incorporated into proteins

T-stem The stem from the third of three-stem loop structures in tRNA, starting from the 5′ end

twin-Arg pathway *See* Tat pathway

Twitching motility Movement of bacterial cells along surfaces via the retraction and extension of pili

Two-component regulatory system A system for transcriptional regulation that uses a protein to sense stimuli and a second protein to regulate transcription based in input from the sensor

Two partner secretion A Type 5 Secretion System where two proteins work to make an outer membrane channel and export the partner protein

Type 1 Secretion System A three-protein system with a ABC transporter, membrane fusion protein, and outer membrane protein that is capable of exporting substrates from Gram-negative bacteria

Type 2 Secretion System A four-protein system with an inner membrane platform, secretion ATPase, pseudopilus, and outer membrane complex that is capable of exporting substrates from Gram-negative bacteria

Type 3 Secretion System A three-part system with a base complex, needle, and translocon made of several proteins that is capable of exporting substrates from Gram-negative bacteria

Type 4 Secretion System A multi-protein system with homology to bacterial pili that is capable of exporting substrates from Gram-negative bacteria

Type 5 Secretion System Proteins that are capable of transporting themselves out of the Gram-negative bacterial cell

Type 6 Secretion System A multi-protein system with homology to bacteriophage that is capable of exporting substrates from Gram-negative bacteria

Type 7 Secretion System A system that aids in exporting substrates through the thick peptidoglycan layer of Gram-positive bacteria

Type 8 Secretion System A system of proteins in Gram-negative bacterial membranes that transports factors involved in Curli biosynthesis, biofilm formation, and cell aggregation

Type 9 Secretion System A multiprotein system that secretes effectors out of the cell, together with the Sec pathway, to generate gliding motility in environmental bacteria or an aspect of virulence in pathogens

Type 10 Secretion System A bacterial secretion system that has similarity to bacteriophage lysis proteins, with a holin in the inner membrane, a peptidoglycan cleavage enzyme in the periplasm, and a spanin that spans from the inner membrane to the outer membrane

Untranslated region The segment of mRNA 5′ of the initiation codon that is not translated into protein, yet is part of the mRNA

UP element A conserved sequence 5′ of a transcriptional start site that enhances the binding of RNA polymerase

Upstream Before a gene or other genetic feature, 5′ of the feature

Uracil A nitrogenous base made of four carbons, two nitrogens, two oxygens, and four hydrogens, found in RNA

UTR *See* Untranslated region

Vacuole A small membrane-bound space within a eukaryotic cell

Variable loop A loop that is present in some tRNAs, located between the anticodon stem and the T-stem

Virion The protein coat of a bacteriophage particle

Xeno nucleic acid An artificially created nucleic acid with a structure not found in nature, generally with sugars other than ribose or deoxyribose

XNA *See* Xeno nucleic acid

Zoonotic disease An infectious disease that primarily infect animals, but can be transferred to humans as well

Zwitterion A molecule that has a negative charge at one end, a positive charge at the other end, and a net zero charge

Glossary of Bacterial Species

Bacteria	What they do	Gram stain
Acinetobacter baumannii	An opportunistic human pathogen, associated with hospital-acquired infections. Of concern in immunocompromised individuals, particularly due to multi-drug resistance.	Gram-negative
Actinobacillus pleuropneumoniae	A pathogen in pigs causing a frequently lethal necrotizing hemorrhagic pneumonia.	Gram-negative
Agrobacterium tumefaciens	A plant pathogen that is capable of infecting a wide range of plants, causing tumor-like growth referred to as crown-gall diseases.	Gram-negative
Aliivibrio fischeri	An aquatic bacterium capable of bioluminscence that can be found living symbiotically with marine animals such as the bobtail squid. Formerly *Vibrio fischeri*.	Gram-negative
Anaeromyxobacter dehalogenans	A non-pathogenic bacterial species, present in soil and important for understanding soil dynamics.	Gram-negative
Anaplasma marginale	A pathogen of cattle and other ruminant animals, which is transmitted by ticks, causing progressive hemolytic anemia and death.	Gram-negative
Bacillus anthracis	A pathogen that causes three distinct types of human infection (cutaneous, inhalation, and gastrointestinal) and is capable of forming spores.	Gram-positive
Bacillus subtilis	A non-pathogen, spore-forming bacteria that has been isolated from a variety of environments, including the guts of animals, oceans, and soil.	Gram-positive
Bacteroides fragilis	Bacteria that are normally resident in the human gut as commensals, but can cause infection following surgery, trauma, or disease.	Gram-negative
Bacteroides ovatus	Bacteria that are normally resident in the human gut as commensals, but that appear to be involved in antibody responses in inflammatory bowel disease.	Gram-negative
Bordetella pertussis	A pathogen of humans that causes whooping cough, which can have fatal complications, including bronchopneumonia and acute encephalopathy.	Gram-negative
Borrelia burgdorferi	Transmitted by ticks, these bacteria cause Lyme disease in humans as a pathogen, but are not pathogenic in the small mammal reservoirs for these bacteria.	Gram-negative
Buchnera aphidicola	A bacterial endosymbiont of aphids, these Gram-negative bacteria do not have LPS or LOS.	Gram-negative
Burkholderia cepacia	Bacteria that frequently colonize fluids used in hospital wards, they pose health threats to hematology and oncology patients.	Gram-negative
Burkholderia dolosa	Bacteria that can be fatal in patients with cystic fibrosis.	Gram-negative
Burkholderia multivorans	Bacteria that can cause chronic disease and infection, particularly in patients with cystic fibrosis or other immunocompromised states.	Gram-negative
Burkholderia pseudomallei	The bacterial cause of the disease melioidosis, which can be potentially fatal.	Gram-negative
Campylobacter coli	A pathogenic bacterial species causing food poisoning in humans, including diarrhea.	Gram-negative
Campylobacter jejuni	One of the major causes of bacterial food poisoning from poultry, which not only causes significant disease, but can be fatal.	Gram-negative
Candidatus Nasuia deltocephalinicola	A symbiotic bacterium that lives within the leafhopper insect *Macrosteles quadripunctulatus* and has a very small genome of 112 kb and 16.6% G+C.	Gram-negative
Candidatus Zinderia insecticola	A symbiotic bacterium that lives within the spittlebug *Clastoptera arizonana*, it has a very small genome of 208 kb with 13.5% G+C.	Gram-negative
Caulobacter	A genus of bacteria that live in aquatic environments and are model organisms for the study of the cell cycle and synchronized cultures.	Gram-negative
Chlamydia pneumoniae	An obligate intracellular pathogen that is a common cause of bacterial pneumonia.	Gram-negative
Chlamydia trachomatis	An obligate intracellular pathogen that is the cause of the sexually transmitted disease chlamydia, amongst other diseases.	Gram-negative
Clavibacter michiganensis	A bacterial plant pathogen of economic importance due to the impact of disease on food crops such as potatoes and tomatoes.	Gram-positive

Bacteria	What they do	Gram stain
Clostridioides difficile	A pathogen of humans that is a major cause of hospital-acquired diarrhea; capable of forming spores. Formerly *Clostridium difficile*.	Gram-positive
Clostridium perfringens	A common cause of food poisoning, which causes abdominal cramps and diarrhea; capable of forming spores.	Gram-positive
Clostridium spp.	A genus of bacteria that are spore forming and typically anaerobic, some of which can opportunistically cause human disease. *Clostridioides difficile* was previously classified within this genus.	Gram-positive
Clostridium stercorarium	A bacterial species of interest for exploitation for biotechnology due to its ability to degrade polysaccharides in plant biomass, resulting in ethanol and acetate.	Gram-positive
Corynebacterium diphtheriae	A pathogen of humans that causes diphtheria, expressing diphtheria toxin.	Gram-positive
Coxiella burnetti	A pathogen that causes Q fever in humans and coxiellosis in animals.	Gram-negative
Dechloromonas aromatic	Bacteria of biotechnology interest because of their ability to oxidize benzene, a difficult pollutant to address.	Gram-negative
Enterobacter spp.	A genus of bacteria predominantly found in the intestines of animals that are able to thrive in aerobic and anaerobic environments; most have flagella.	Gram-negative
Enterobacteriaceae	A family of bacteria that includes species capable of causing disease in humans and animals, found in intestines as well as other environments.	Gram-negative
Enterococcus faecalis	A bacterial pathogen that can cause hospital-acquired urinary tract, wound, and other infections.	Gram-positive
Enterococcus faecium	A bacterial species that is a leading cause of nosocomial infections, especially in critically ill or immunocompromised patients.	Gram-positive
Escherichia coli	A diverse species of bacteria that is studied as a model in bacterial genetics and genomics, used in biotechnology, which is sometimes pathogenic.	Gram-negative
Flavobacterium johnsoniae	A non-pathogenic environmental species found in water and soil.	Gram-negative
Francisella novicida	A non-pathogenic species of interest due to its genetic similarity to the pathogen *Francisella tularensis*.	Gram-negative
Geobacter sulfurreducens	A bacterial species of interest for exploitation for biotechnology in generating fuel cells, bioremediation, and biofuel production.	Gram-negative
Haemophilus ducreyi	A bacterial pathogen that causes the sexually transmitted infection chancroid.	Gram-negative
Haemophilus influenzae	A species of bacteria that can cause a variety of human diseases, including meningitis, septicemia, and pneumonia.	Gram-negative
Haemophilus parainfluenzae	An opportunistic human pathogen that can cause some similar, yet less severe infections as *Haemophilus influenzae*.	Gram-negative
Helicobacter canadensis	A zoonotic bacterial species that is an emerging pathogen in humans, causing diarrhea.	Gram-negative
Helicobacter pylori	A bacterial pathogen of humans that is able to colonize and survive for decades in the stomach and upper digestive tract; cause of ulcers and other diseases.	Gram-negative
Klebsiella pneumoniae	A pathogen of humans that can harmlessly colonize the gut, but that causes hospital-acquired and community-acquired pneumonia.	Gram-negative
Lactobacillus	A genus of bacteria that produce lactic acid in their metabolism, they are found associated with animals, plant material, and fermented or spoiled foods.	Gram-positive
Lactobacillus plantarum	A bacterial species of interest to the food industry as a probiotic, it survives passage through the intestinal tract and produces lactic acid.	Gram-positive
Legionella pneumophila	A pathogen of humans that causes legionellosis, an acute pneumonia.	Gram-negative
Leptospira spp.	A genus of bacteria that cause leptospirosis, zoonotic disease. Can cause a wide range of symptoms and be underdiagnosed.	Gram-negative
Listeria monocytogenes	A pathogen associated with deli meats and cheeses, of concern to pregnant women, people over 65, and immunocompromised individuals.	Gram-positive
Mannheimia haemolytica	An opportunistic pathogen causing pneumonia in cattle and calves.	Gram-negative
Moraxella bovis	A pathogen causing keratoconjunctivitis (eye infection) in cattle with symptoms that range from mild to severe.	Gram-negative

Bacteria	What they do	Gram stain
Moraxella catarrhalis	A pathogen of humans causing otitis media (middle ear infection) and exacerbations of COPD (chronic obstructive pulmonary disease).	Gram-negative
Moraxella lancunata	A pathogen of humans that can cause eye infections and in rare cases some serious infections, usually associated with other stress or immunocompromised condition.	Gram-negative
Mycobacterium bovis	A pathogen found in cattle and related animals that can also cause tuberculosis in humans.	Acid fast[a]
Mycobacterium leprae	A pathogen that damages peripheral nerves and can cause severe deformities.	Acid fast
Mycobacterium tuberculosis	A pathogen that causes tuberculosis as well as other human diseases, which can result in death.	Acid fast
Mycoplasma fermentans	An opportunistic pathogen of immunocompromised humans, often seen as a co-infection associated with other diseases.	Gram-negative
Mycoplasma genitalium	A pathogen causing sexually transmitted infections in humans, which may have symptoms similar to gonorrhea or chlamydia, or no symptoms.	Gram-negative
Mycoplasma hominis	A pathogen associated with other bacteria in pelvic inflammatory disease and bacterial vaginosis, and also causing infections related to pregnancy.	Gram-negative
Mycoplasma gallicepticum	A pathogen causing respiratory infections in chickens and turkeys, as well as other birds.	Gram-negative
Mycoplasma pneumoniae	A pathogen that can cause mild pneumonia, sometimes referred to as "walking pneumonia," as well as more severe forms.	Gram-negative
Neisseria gonorrhoeae	A pathogen in humans that causes the sexually transmitted infection gonorrhea and infects other mucosal sites.	Gram-negative
Neisseria lactamica	A non-pathogenic commensal that colonizes the human nasopharynx; closely related to pathogens *N. gonorrhoeae* and *N. meningitidis*. A non-pathogenic commensal that colonizes the human nasopharynx; closely related to pathogens *N. gonorrhoeae* and *N. meningitidis*.	Gram-negative
Neisseria meningitidis	A pathogen in humans that colonizes the nasopharynx and can cause life-threatening invasive meningococcal meningitis and septicemia.	Gram-negative
Paracoccus denitrificans	A bacterial species from soil that can be grown aerobically or anaerobically on organic or inorganic energy sources. It may be the ancestor of eukaryotic mitochondria.	Gram-negative
Pseudoalteromonas haloplanktis	A bacterial species isolated from a sample of Antarctic seawater, which is able to live in cold temperatures.	Gram-negative
Pseudomonas aeruginosa	An opportunistic human pathogen, particularly of concern in immunocompromised individuals, it is commonly found in a variety of environments.	Gram-negative
Pseudomonas extremaustralis	A bacterial species from the Antarctic that is able to grow at low temperatures and is resistant to stress and high levels of polyhydroxybutyrate.	Gram-negative
Rhodobacter sphaeroides	A rod-shaped, metabolically diverse bacterial species that has potential in the development of biofuels, bioplastics, and bioremediation.	Gram-negative
Rickettsia prowazekii	Intracellular bacteria that cause the human disease typhus, transmitted by the human body louse.	Gram-negative
Rickettsia rickettsii	Intracellular bacteria that cause the human disease Rocky Mountain Spotted Fever, transmitted by ticks.	Gram-negative
Rickettsia spp.	A genus of intracellular bacteria that are responsible for spotted and typhus fevers.	Gram-negative
Salmonella enterica serovar *Enteritidis*	A pathogen causing illness following consumption of infected food, particularly poultry and eggs.	Gram-negative
Salmonella enterica typhimurium	A pathogen infecting humans and animals following ingestion of contaminated food or water.	Gram-negative
Serratia marcescens	An opportunistic pathogen that can cause a range of infections in humans and often produces a red pigment.	Gram-negative
Shewanella oneidensis	A bacterial species of interest in bioremediation due to its ability to reduce metals, including those in nuclear waste, causing uranium to precipitate out of water.	Gram-negative
Shigella spp.	A genus including bacterial pathogens that cause shigellosis, severe diarrhea that may contain blood or mucus.	Gram-negative

Bacteria	What they do	Gram stain
Sodalis glossinidius	An endosymbiont of the tsetse fly that has a mutualistic relationship with its host organism.	Gram-negative
Sorangium cellulosum	A species of bacteria with large genomes that produce secondary metabolites of interest in drug discovery and development.	Gram-negative
Staphylococcus aureus	A pathogen of humans that can live harmlessly on the skin, yet cause serious, life-threatening infections within the body.	Gram-positive
Staphylococcus epidermidis	An opportunistic pathogen that is commonly associated often with catheters and other indwelling devices.	Gram-positive
Streptococcus agalactiae	A pathogen, also known as GBS or group B Strep, that causes serious infections in newborn infants.	Gram-positive
Streptococcus pneumoniae	A pathogen of humans that causes pneumonia when the bacteria enter the lungs via inhalation.	Gram-positive
Streptococcus pyogenes	A pathogen, also known as GAS or group A Strep, that causes a range of infections in humans of varying severity.	Gram-positive
Streptococcus uberis	A pathogen of dairy cows that causes mastitis (infection of the udders).	Gram-positive
Streptomyces coelicolor	A spore-forming, soil-dwelling bacterial species that produces a variety of antibiotics.	Gram-positive
Thermus aquaticus	A bacterial species isolated from hot springs in Yellowstone National Park, capable of growth at high temperature, of significance when its thermostable DNA polymerase was used for PCR.	Gram-negative
Thiomargarita magnifica	The largest bacterial species identified, it is visible to the naked eye and has copies of its genome within organelle-like structures.	Gram-negative
Thiomargarita namibiensis	A bacterial species that was isolated off the Namibian coast and found to be much larger than most.	Gram-negative
Tropheryma whipplei	A pathogen of humans that causes Whipple's diseases as well as other infections.	Gram-positive
Ureaplasma	A genus of bacteria able to hydrolyze urea and can cause disease in humans; do not have a cell wall.	Gram-negative[b]
Vibrio cholerae	A pathogen of humans that lives in the aquatic environment, but that causes diarrhea and vomiting upon ingestion by humans.	Gram-negative
Vibrio harveyi	An aquatic bacterium capable of bioluminscence that is a pathogen of marine organisms.	Gram-negative
Vibrio parahaemolyticus	A pathogen of humans causing diseases associated particularly with consumption of raw seafood.	Gram-negative
Wolbachia	A genus of bacteria that infect insects, having complex interactions with their hosts.	Gram-negative
Yersinia pestis	A pathogen of humans and animals that also infects fleas, which transmit the bacteria; cause of plague.	Gram-negative
Yersinia pseudotuberculosis	A zoonotic bacterial species that is a human pathogen causing disease following ingestion of food or water contaminated with animal feces.	Gram-negative

[a] The cell surface of this genus means that it does not absorb Gram stain and cannot be classified by Gram staining.
[b] Stain as Gram-negative due to the absence of a cell wall.

Index

H

Q

R

S

X

Y

Z

Quick reference metric conversions

1000 µL = 1 mL	1000 mL = 1 L	1000 µg = 1 mg	1000 mg = 1 g
100 µL= 0.1 mL	100 mL = 0.1 L	100 µg = 0.1 mg	100 mg = 0.1 g
10 µL = 0.01 mL	10 mL = 0.01 L	10 µg = 0.01 mg	10 mg = 0.01 g
1 µL = 0.001 mL	1 mL = 0.001 L	1 µg = 0.001 mg	1 mg = 0.001 g

Useful concentration equations

A different equation for working out concentrations than $C_1V_1 = C_2V_2$

$$\frac{\text{Concentration you have} \times \text{Volume you add}}{\text{The final volume}} = \text{Concentration you want}$$

Abbreviated as:

$$\frac{\text{Have (Add)}}{\text{Volume}} = \text{Want}$$

Molarity

$$\frac{\text{Concentration (g/L)}}{\text{Molecular weight}} = \text{moles/ liter (M)}$$

Additionally:

Molecular weight in grams × volume in liters × molar concentration in M = amount to weigh and add to the volume to get the molar concentration.

When using this equation, be sure to use g, L, and M. For example, if you need 500 mL, put 0.5 L into the equation; likewise convert for other units like mg and mM.